PROBABILITY THEORY WITH APPLICATIONS

Second Edition

Mathematics and Its Applications

Volume 582

PROBABILITY THEORY WITH APPLICATIONS

Second Edition

By

M.M. RAO
University of California, Riverside, California

R.J. SWIFT
California State Polytechnic University, Pomona, California

e-ISBN: 0-387-27731-5
ISBN: 978-1-4419-3909-8 eISBN: 978-0-387-27731-8

Printed on acid-free paper.

AMS Subject Classifications: 60Axx, 60Exx, 60Fxx, 60Gxx, 62Bxx, 62Exx, 62Gxx, 62Mxx, 93Cxx

Softcover reprint of the hardcover 2nd edition 2006

9 8 7 6 5 4 3 2 1

springeronline.com

To the memory of my brother-in-law,
Raghavayya V. Kavuri
M.M.R.

To the memory of my parents,
Randall and Julia Swift
R.J.S.

Contents

Preface to Second Edition

The following is a revised and somewhat enlarged account of Probability Theory with Applications, whose basic aim as expressed in the preface to the first edition (appended here) is maintained. In this revision, the material and presentation is better highlighted with several (small and large) alterations made to each chapter. We believe that these additions make a better text for graduate students and also a reference work for a later study. We now discuss in some detail the subject of this text, as modified here. It is hoped that this will provide an appreciation for the view–point of this edition, as well as the earlier one, published over two decades ago.

In the present setting, the work is organized into three parts, the first being on the foundations of the subject, consists of Chapters 1–3. The second part concentrates on the analytical aspects of probability in relatively large chapters 4–5. The final part in Chapters 6–8 treats some serious and deep applications of the subject. The point of view presented here has the following focus. Parts I and II can be essentially studied independently with only cursory cross-references. Each part could easily be used for a quarter or semester long beginning graduate course in Probability Theory. The prerequisite is a graduate course in Real Analysis, although it is possible to study the two subjects concurrently. Each of these parts of this text also has applications and ideas some of which are discussed as problems that illustrate as well as extend the basic subject. The final part of the text can be used for a follow-up course on the preceding material or for a seminar thereafter. Numerous suggestions for further study and even several research problems are pointed out. We now detail some of these points for a better view of the treatment which is devoted to the mathematical content, avoiding nonmathematical views and concepts.

To accommodate the new material and not to substantially increase the size of the volume, we had to omit most of the original Chapter 6 and part of Chapter 7. Thus this new version has eight chapters, but it is still well focused and the division into parts makes the work more useful. We now turn to explaining the new format.

The first part, on foundations, treats the two fundamental ideas of probability, independence and conditioning. In Chapter 1 we recall the necessary results from Real Analysis which we recommend for a perusal. It is also important that readers take a careful look at the fundamental law of probability and the basic uniform continuity of characteristic functions.

Chapter 2 undertakes a serious study of (statistical) independence, which is a distinguishing feature of Probability Theory. Independence is treated in considerable detail in this chapter, both the basic strong and weak laws, as well as the convergence of series of random variables. The applications considered here illustrate such results as the Glivenko-Cantelli Theorem for empiric and density estimation, random walks, and queueing theory. There are also exercises (with hints) of special interest and we recommend that all readers pay particular attention to Problems 5 and 6, and also 7, 15 and 21 which explain the *very special nature of the subject and the concept of independence itself.*

The somewhat long third chapter is devoted to the second fundamental idea, namely conditioning. As far as we know, no other graduate text in probability has treated the subject of conditional probability in such detail and specificity. To mention some noteworthy points of our presentation, we have included: (i) the unsuspected, but spectacular, failure of the Vitali convergence theorem for conditional probabilities. This is a consequence of an interesting theorem of Blackwell and Dubins. We include a discussion and imposition of a restriction for a positive conclusion to prevail, (ii) the basic problem (still unresolved) of calculating conditional expected values (probabilities) when the conditioning is relative to random variables taking uncountably many values, particularly when the random variables arise from continuous distributions. In this setting, multiple answers (all natural) for the same question are exhibited via a Gaussian family. The calculations we give follow some work by Kac and Slepian, leading to paradoxes. These difficulties arise from the necessary calculation of the Radon-Nikodým derivative which is fundamental here, and for which no algorithmic procedure exists in the literature. A search through E. Bishop's text on the foundations of constructivism (in the way of L.E.J. Brower) shows that we do not yet have a solution or a resolution for the problems discussed. Thus our results are on existence and hence use "idealistic methods", which present, to future researchers in Bishop's words, "a challenge to find a constructive version and to give a constructive proof." Until this is fulfilled, we have to live with subjectively chosen solutions, for applications of our work in practice.

It is in this context, we detail in chapter 3, the Jessen-Kolmogorov-Bochner-Tulcea theorems on existence of arbitrary families of random variables on (suitable) spaces. We also include here the basic martingale limit theorems with applications to U-statistics, likelihood ratios, Markov processes and quasi-martingales. Several exercises, (about 50) add complements to the theory. These exercises include the concept of sufficiency, a martingale proof of the Radon-Nikodým theorem, aspects of Markov kernels, ergodic-martingale relations and many others. Thus here and throughout the text one finds that the exercises contain a large amount of additional information on the subject of probability. Many of these exercises can be omitted in a first reading but we strongly urge our readers to at least glance through them all and then return later for a serious study. Here and elsewhere in the book, we follow the lead of Feller's classics.

The classical as well as modern aspects of the so called analytical theory of probability is the subject of the detailed treatment of Part II. This part consists of the two chapters 4 and 5, with the latter being the longest in the text. These chapters can be studied with the basic outline of chapter 1 and just the notion of independence translated to analysis. The main aim of Chapter 4 is to use distribution theory (or image probabilities using random variables) on Euclidean spaces. This fully utilizes the topological structure of their ranges. Thus the basic results are on characteristic functions including the Lévy-Bochner-Cramér theorems and their multidimensional versions. The chapter concludes with a proof of the equivalence of convergences–pointwise a.e., in probability and in distribution–for sums of independent random variables. Regarding some characterizations, we particularly recommend Problems 4, 16, 26, and 33 in this chapter.

The second longest chapter of the text, is chapter 5 and is the heart of analytical theory. This chapter contains the customary central limit theory with Berry-Essen error estimation. It also contains a substantial introduction to infinite divisibility, including the Lévy-Khintchine representation, stable laws, and the Donsker invariance principle with applications to Kolmogorov-Smirnov type theorems. The basic law of the iterated logarithm, with H. Teicher's (somewhat) simplified proof, is presented. This chapter also contains interesting applications in several exercises. Noteworthy are Bochner's generalization of stable types (without positive definiteness) in Exercises 26-27 and Wendel's "elementary" treatment of Spitzer's identity in Exercise 33. We recommend that these exercises be completed by filling in the details of the proofs outlined there. We have included the m-dependent central limit theorem and an illustration to exemplify the applicability and limitations of the classical invariance principle in statistical theory. Several additional aspects of infinite divisibility and stability are also discussed in the exercises. These problems are recommended for study so that certain interesting ideas arising in applications of the subject can be learned by such an effort. These are also useful for the last part of the book.

The preceding parts I & II, prepare the reader to take a serious look at Part III, which is devoted to the next stage of our subject. This part is devoted to what we consider as very important in modern applications, both new and significant, in the subject. Chapters 6 and 7 are relatively short, but are concerned with the limit theory of nonindependent random sequences which demand new techniques. Chapter 6 introduces and uses stopping time techniques. We establish Wald's identities, which play key roles in sequential analysis, and the Doob optional stopping and sampling theorems, which are essential for key developments in martingale theory. Chapter 7 contains central limit theorems for a random number of random variables and the Birkhoff ergodic theorem. The latter shows a natural setting for strict stationarity of families of random variables and sets the stage for the last chapter of the text.

Chapter 8 presents a glimpse of the panorama of stochastic processes with some analysis. There is a significant increase and expansion of the last chapter of the first edition. It can be studied to get a sense of the expanding vistas of the subject which appear to have great prospects and potential for further research. The following items are considered to exhibit just a few of the many new and deep applications.

The chapter begins with a short existence proof of Brownian motion directly through (random) Fourier series and then establishes the continuity, nondifferentiability of its sample paths, stationarity of its increments, as well as the iterated logarithm law for it. These ideas lead to a study of (general) additive processes with independent, stable and strictly stationary increments. The Poisson process plays a key role very similar to Brownian motion, and points to a study of random measures with independent values on disjoint sets. We indicate some modern developments following the work of Kahane-Marcus-Pisier, generalizing the classical Paley-Zygmund analysis of random Fourier series. This opens up many possibilities for a study of sample continuity of the resulting (random) functions as sums, with just $0 < \alpha \leq 2$ moments. These ideas lead to an analysis of strongly stationary classes (properly) contained in strictly stationary families. The case $\alpha = 2$ is special since Hilbert space geometry is available for it. Thus the (popular) weakly stationary case is considered with its related (but more general) classes of weakly, strictly and strongly harmonizable processes. These are outlined along with their integral representations, giving a picture of the present state of stochastic analysis. Again we include several complements as exercises with hints in the way pioneered by Feller, and *strongly recommend to our readers to at least glance through them to have a better view of the possibilities and applications that are opened up here.* In this part therefore, problems 6 and 7 of Chapter 6; and problems 2, 6, and 10 of Chapter 7, and problems 8, 12, 15, and 16 in chapter 8 are interesting as they reveal the unfolding areas shown by this work.

This book gives our view of how Probability Theory could be presented and studied. It has evolved as a collaboration resulting from decades of research experience and lectures prepared by the first author and the experiences of the second author who, as a student, studied and learned the subject from the first edition and then subsequently used it as a research reference. His notes and clarifications are implemented in this edition to improve the value of the text. This project has been a satisfying effort resulting in a newer text that is offered to the public.

In the preparation of the present edition we were aided by some colleagues, friends and students. We express our sincere gratitude to Mary Jane Hill for her assistance and diligence with aspects of typesetting and other technical points of the manuscript. Our colleague Michael L. Green offered valuable comments, and Kunthel By, who read drafts of the early chapters with a student's perspective, provided clarifications. We would like to thank our wives Durgamba Rao and Kelly Swift, for their love, support, and understanding.

We sincerely thank all these people, and hope that the new edition will serve well as a graduate text as well as a reference volume for many aspiring and working mathematical scientists. It is our hope that we have succeeded, at least to some extent, to convey the beauty and magnificence of probability theory and its manifold applications to our audience.

Riverside, CA
Pomona, CA

M.M. Rao,
R.J. Swift

Preface to First Edition

The material in this book is designed for a standard graduate course on probability theory, including some important applications. It was prepared from the sets of lecture notes for a course that I have taught several times over the past 20 years. The present version reflects the reactions of my audiences as well as some of the textbooks that I used. Here I have tried to focus on those aspects of the subject that appeared to me to add interest both pedagogically and methodologically. In this regard, I mention the following features of the book: it emphasizes the special character of the subject and its problems while eliminating the mystery surrounding it as much as possible; it gradually expands the content, thus showing the blossoming of the subject; it indicates the need for abstract theory even in applications and shows the inadequacy of existing results for certain apparently simple real-world problems (See Chapter 6); it attempts to deal with the existence problems for various classes of random families that figure in the main results of the subject; it contains a more complete (and I hope more detailed) treatment of conditional expectations and of conditional probabilities than any existing textbook known to me; it shows a deep internal relation among the Lévy continuity theorem, Bochner's theorem on positive definite functions, and the Kolmogorov-Bochner existence theorem; it makes a somewhat more detailed treatment of the invariance principles and of limit laws for a random number of (ordered) random variables together with applications in both areas; and it provides an unhurried treatment that pays particular attention to motivation at every stage of development.

Since this is a textbook, essentially all proofs are given in complete detail (even at the risk of repetition), and some key results are given multiple proofs when each argument has something to contribute. On the other hand, generalization for its own sake is avoided, and as a rule, abstract-Banach-space-valued random variables have not been included (if they have been, the demands on the reader's preparation would have had to be much higher).

Regarding the prerequisites, a knowledge of the Lebesgue integral would be ideal, and at least a concurrent study of real analysis is recommended. The necessary results are reviewed in Chapter 1, and some results that are generally not covered in such a course, but are essential for our work, are given with proofs. In the rest of the book, the treatment is detailed and complete, in accordance with the basic purpose of the text. Thus it can be used for self-study by mature scientists having no prior knowledge of probability.

The main part of the book consists of Chapters 2-5. Even though I regard the order presented here to be the most natural, one can start, after a review of the relevant part of Chapter 1, with Chapter 2, 3 or 4, and with a little discussion of independence, Chapter 5 can be studied. The last four chapters concern applications and problems arising from the preceding work and partly generalizing it. The material there indicates some of the many directions along which the theory is progressing.

There are several exercises at the end of each chapter. Some of these are routine, but others demand more serious effort. For many of the latter type, hints are provided, and there are a few that complement the text (e.g., Spitzer's identity and aspects of stability in Chapter 5); for them, essentially complete details are given. I present some of these not only as good illustrations but also for reference purposes.

I have included in the list of references only those books and articles that influenced my treatment; but other works can be obtained from these sources. Detailed credits and priorities of discovery have not been scrupulously assigned, although historical accounts are given in the interest of motivation. For cross-referencing purposes, all the items in the book are serially numbered. Thus 3.4.9 is the ninth item of Section 4 of Chapter 3. In a given section (chapter) the corresponding section (and chapter) number is omitted.

The material presented here is based on the subject as I learned it from Professor M. D. Donsker's beautiful lectures many years ago. I feel it is appropriate here to express my gratitude to him for that opportunity. This book has benefited from my experience with generations of participants in my classes and has been read by Derek K. Chang from a student's point of view; his questions have resolved several ambiguities in the text. The manuscript was prepared with partial support from an Office of Naval Research contract and a University of California, Riverside, research grant. The difficult task of converting my handwritten copy into the finished typed product was ably done by Joyce Kepler, Joanne McIntosh, and Anna McDermott, with the care and interest of Florence Kelly. Both D. Chang and J. Sroka have aided me in proofreading and preparation of the Index. To all these people and organizations I wish to express my appreciation for this help and support.

(M.M. Rao)

List of Symbols

a.a.	almost all
a.e.	almost everywhere
ch.f.(s)	characteristic function(s)
d.f.(s)	distribution function(s)
iff	if and only if
i.i.d.	independent identically distributed
r.v.(s)	random variable(s)
m.g.f.	moment generating function
$A \Delta B$	symmetric difference of A and B
$\emptyset$	empty set
(a, b)	open interval
(Ω, Σ, P)	a probability space
$P(f^{-1}(A))$	$= P[f \in A](= (P \circ f^{-1}(A))$
χ_A	indicator of A
$\wedge$	minimum symbol
$\vee$	maximum symbol
$\mathbb{R}$	reals
$\mathbb{C}$	complex numbers
$\mathbb{N}$	natural numbers (=positive integers)
$\sigma(X_1, \ldots, X_n)$	sigma algebra generated by the r.v.s X_i, $i = 1, \ldots, n$
$\sigma^2(X)$	variance of the r.v. X
$\rho(X, Y)$	correlation of X and Y
$\mathcal{L}^0$	the set of scalar r.v.s on (Ω, Σ, P)
$\mathcal{L}^p$	the set of pth power integrable r.v.s. on (Ω, Σ, P)
L^p	the Lebesgue space of equivalence classes of r.v.s from $\mathcal{L}^p$
$\mathcal{P}$	usually a partition of a set
$L^p(\mathbb{R})$	the Lebesgue space on $\mathbb{R}$ with Lebesgue measure
$\nu \ll \mu$	ν is absolutely continuous relative to μ (measures)
$\nu \perp \mu$	ν is singular relative to μ

$\|\|X\|\|_p$	$= [E(\|X\|^p)]^{1/p}$ $= [\int_\Omega \|X\|^p \, dP]^{1/p}$ $=p$-norm of X
$[\alpha]$	integral part of the real number $\alpha \geq 0$
$\cong$	topological equivalence
$a_n \sim b_n$	means $a_n/b_n \to 1$ as $n \to \infty$
∂A	boundary of the set A
Log φ	distinguished logarithm of φ
sgn	signum function
$f_1 * f_2$	convolution of f_1 and f_2 in $L^1(\mathbb{R})$
$\binom{n}{k}$	the kth binomial coefficient in $(1+x)^n, n \in \mathbb{N}$, $0 \leq k \leq n$

Part I Foundations

The mathematical basis of probability, namely real analysis, is sketched with essential details of key results, including the fundamental law of probability, and a characterization of uniform integrability in Chapter 1 which is used frequently through out the book. Most of the important results on independence, the laws of large numbers, convergence of series as well as some key applications on random walks and queueing are treated in Chapter 2, which also contains some important complements as problems. Then a quite detailed treatment of conditional probabilities with applications to Markovian families, martingales, and the Kolmogorov-Bochner-Tulcea existence theorems on processes are included in Chapter 3. Also important additional results are in a long problems section. The basic foundations of modern probability are detailed in this part.

Chapter 1

Background Material and Preliminaries

In this chapter, after briefly discussing the beginnings of probability theory we shall review some standard background material. Basic concepts are introduced and immediate consequences are noted. Then the fundamental law of probability and some of its implications are recorded.

1.1 What Is Probability?

Before considering what probability is or what it does, a brief historical discussion of it will be illuminating. In a general sense, one can think of a probability as a long-term average, or (in a combinatorial sense) as the proportion of the number of favorable outcomes to the number of possible and equally likely ones (all being finite in number in a real world). If the last condition is not valid, one may give certain weights to outcomes based on one's beliefs about the situation. Other concepts can be similarly formulated. Such ideas are still seriously discussed in different schools of thought on probability.

Basically, the concept originates from the recognition of the uncertainty of outcome of an action or experiment; the assignment of a numerical value arises in determining the degree of uncertainty. The need for measuring this degree has been recognized for a very long time. In the Indian Jaina philosophy the uncertainty was explicitly stated as early as the fifth century B.C., and it was classified into seven categories under the name *syādvāda* system. Applications of this idea also seem to have been prevalent. There are references in medieval Hindu texts to the practice of giving alms to religious mendicants without ascertaining whether they were deserving or not. It was noted on observation that "only ten out of a hundred were undeserving," so the public (or the donors) were advised to continue the practice. This is a clear forerunner of what is now known as the frequency interpretation of probability.

References related to gambling may be found throughout recorded history. The great Indian epic, the *Mahābhārata*, deals importantly with gambling. Explicit numerical assignment, as in the previous example, was not always recorded, but its implicit recognition is discernible in the story. The Jaina case was discussed with source material by Mahalanobis (1954), and an interesting application of the *syādvāda* system was illustrated by Haldane (1957).

On the other hand, it has become customary among a section of historians of this subject to regard probability as having its roots in calculations based on the assumption of equal likelihood of the outcomes of throws of dice. This is usually believed to start with the correspondence of Fermat and Pascal in the 1650s or (occasionally) with Cardano in about 1550 and Galileo a little later. The Fermat-Pascal correspondence has been nicely dramatized by Rényi [see his book (1970) for references] to make it more appealing and to give the impression of a true beginning of probabilistic ideas.

Various reasons have been advanced as to why the concept of probability could not have started before. Apparently an unwritten edict for this is that the origins of the subject should be coupled approximately with the Industrial Revolution in Europe. Note also that the calculations made in this period with regard to probability, assume equal likelihood. However, all outcomes are not always equally likely. Thus the true starting point must come much later—perhaps with E. Borel, A. Liapounov, and others at the end of the nineteenth century, or even only with Kolmogorov's work of 1933, since the presently accepted broad based theory [1] started only then! Another brief personal viewpoint is expressed in the elementary text by Neuts (1973). We cannot go into the merits of all these historical formulations of the subject here. A good scholarly discussion of such a (historical) basis has been given in Maistrov's book (1974). One has to keep in mind that a considerable amount of subjectivity appears in all these treatments (which may be inevitable).

Thus the preceding sketch leads us to conclude that the concepts of uncertainty and prediction, and hence probabilistic ideas, started a long time ago. Perhaps they can be placed 2500 years ago or more. They may have originated at several places in the world. The methods of the subject have naturally been refined as time went on. Whether there has been cross- fertilization of ideas due to trade and commerce among various parts of the world in the early development is not clear, although it cannot be ruled out. But the sixteenth-seventeenth century "beginning" based on gambling and problems of dice cannot be taken as the sole definitive starting point of probability. With these generalities, let us turn to the present-day concept of probability that is the foundation for our treatment of the subject.

[1] As late as the early 1920s, R. von Mises summed up the situation, no doubt in despair, by saying, "Today, probability theory is not a mathematical science."

As is clear from the preceding discourse, probability is a numerical measure of the uncertainty of outcomes of an action or experiment. The actual assignment of these values must be based on experience and should generally be verifiable when the experiment is (if possible) repeated under essentially the same conditions. From the modern point of view, therefore, we consider all possible outcomes of an experiment and represent them by (distinct) points of a nonempty set. Since the collection of all such possibilities can be infinitely large, various interesting combinations of them, useful to the experiments, have to be considered. It is here that the modern viewpoint distinguishes itself by introducing an algebraic structure into the combinations of outcomes, which are called events. Thus one considers an algebra of events as the primary datum. This is evidently a computational convenience, though a decisive one, and it must and does include everything of conceivable use for an experiment. Then each event is assigned a numerical measure corresponding to the "amount" of uncertainty in such a way that this assignment has natural additivity and consistency properties. Once this setup is accepted, an axiomatic formulation in the style of twentieth-century mathematics in general becomes desirable as well as inevitable. This may also be regarded as building a mathematical model to describe the experiment at hand. A precise and satisfactory formulation of the latter has been given by Kolmogorov (1933), and the resulting analytical structure is almost universally accepted. In its manifold applications, some alterations have been proposed by de Finetti, Rényi, Savage, and others. However, as shown by the first author (Rao 1981) in a monograph on the modern foundations of the subject, the analytical structure of Kolmogorov actually takes care of these alterations when his work is interpreted from an abstract point of view. This is especially relevant in the case of conditional probabilities, which we discuss in detail in Chapter 3. Thus we take the Kolmogorov setup as the basis of this book and develop the theory while keeping in contact with the phenomenological origins of the subject as much as possible. Also, we illustrate each concept as well as the general theory with concrete (but not necessarily numerical) examples. This should show the importance and definite utility of our subject.

The preceding account implies that the methods of real analysis play a key role in this treatment. Indeed they do, and the reader should ideally be already familiar with them, although concurrent study in real analysis should suffice. Dealing with special cases that are immediately applicable to probability is not necessary. In fact, experience indicates that it can distort the general comprehension of both subjects. To avoid misunderstanding, the key results are recalled below for reference, mostly without proofs.

With this preamble, let us start with the axiomatic formulation of Kolmogorov. Let Ω be a nonempty point set representing all possible outcomes of an experiment, and let Σ be an algebra of subsets of Ω. The members of Σ, called events, are the collections of outcomes that are of interest to the experimenter. Thus Σ is nonempty and is closed under finite unions and complements, hence also under differences. Let $P : \Sigma \to \mathbb{R}^+$ be a mapping, called a probability, defined for all elements of Σ so that the following rules are satisfied.

(1) For each $A \in \Sigma, 0 \leq P(A)$ and $P(\Omega) = 1$.
(2) $A, B \in \Sigma$, $A \cap B = \emptyset$, implies $P(A \cup B) = P(A) + P(B)$.

From these two rules, we deduce immediately that (i) (taking $B = \emptyset$) $P(\emptyset) = 0$ and (ii) $A \supset B$, $A, B \in \Sigma$, implies $P(A - B) = P(A) - P(B)$. In particular, $P(A^c) = 1 - P(A)$ for any $A \in \Sigma$, where $A^c = \Omega - A$.

Such a P is called a "finitely additive probability." At this stage, one strengthens (2) by introducing a continuity condition, namely, countable additivity, as follows:

(2′) If $A_1, A_2, \ldots$ are disjoint events of Ω such that $A = \bigcup_{k=1}^{\infty} A_k$ is also an event of Ω, then $P(A) = \sum_{k=1}^{\infty} P(A_k)$.

Clearly (2′) implies (2), but trivial examples show that (2), is strictly weaker than (2′). The justification for (2′) is primarily operational in that a very satisfactory theory emerges that has ties at the deepest levels to many branches of mathematics. There are other cogent reasons too. For instance, a good knowledge of the theory with this "countably additive probability" enables one to develop a finitely additive theory. Indeed, every finitely additive probability function can be made to correspond uniquely to a countably additive one on a "nice" space, according to an isomorphism theorem that depends on the Stone space representation of Boolean algebras. For this and other reasons, we are primarily concerned with the countably additive case, and so *henceforth a probability function always stands for one that satisfies rules or axioms* (1) *and* (2′). The other concept will be qualified "finitely additive," if it is used at all.

If $P : \Sigma \to \mathbb{R}^+$ is a probability in the above sense and Σ is an algebra, it is a familiar result from real analysis that P can be uniquely extended to the σ-algebra (i.e., algebra closed under countable unions) generated by Σ (i.e., the smallest σ-algebra containing Σ). Hence we may and do assume for convenience that Σ is a σ-algebra, and the triple (Ω, Σ, P) is then called a *probability space.* Thus a probability space, in Kolmogorov's model, is a finite measure space whose measure function is normalized so that the whole space has measure one. Consequently several results from real analysis can be employed profitably in our study. However, this does not imply that probability theory is just a special case of the standard measure theory, since, as we shall see, it has its own special features that are absent in the general theory. Foremost of these, is the *concept of probabilistic (or statistical) independence.* With this firmly in hand, several modifications of the concept have evolved, so that the theory has been enriched and branched out in various directions. These developments, some of which are considered in Chapter 3, attest to the individuality and vitality of probability theory.

A concrete example illustrating the above discussion is the following:

Example 1. Let $\Omega_i = \{0, 1\}$ be a two-point space for each $i = 1, 2, \ldots$. This space corresponds to the ith toss of a coin, where 0 represents its tail and 1 its head and is known as a *Bernoulli trial*. Let $\Sigma_i = \{\emptyset, \{0\}, \{1\}, \Omega_i\}$ and $P_i(\{0\}) = q$ and $P_i(\{1\}) = p$, $0 < p = 1 - q < 1$. Then $(\Omega_i, \Sigma_i, P_i)$, $i = 1, 2, \ldots$, are identical copies of the same probability space. If $(\Omega, \Sigma, P)[= \bigotimes_{i \geq 1}(\Omega_i, \Sigma_i, P_i)]$ is the product measure space, then $\Omega = \{x : x = (x_1 x_2, \ldots), x_i = 0, 1 \text{ for all } i\}$, and Σ is the σ-algebra generated by the semiring $\mathcal{C} = \{I_n \subset \Omega : I_n$ consists of those $x \in \Omega$ whose first n components have a prescribed pattern$\}$. For instance, I_2 can be the set of all x in Ω whose first two components are 1. If $I_n (\in \mathcal{C})$ has the first n components consisting of k 1's and $n - k$ 0's, then $P(I_n) = p^k q^{n-k}$, and $P(\Omega) = 1$. [Recall that a semiring is a nonempty class $\mathcal{C}$ which is closed under intersections and if $A, B \in \mathcal{C}, A \subset B$, then there are sets $A_i \in \mathcal{C}$ such that $A = A_1 \subset \cdots \subset A_n = B$ with $A_{i+1} - A_i \in \mathcal{C}$.]

The reader should verify that $\mathcal{C}$ is a semiring and that P satisfies conditions (1) and (2′), so that it is a probability on Σ with the above-stated properties. We use this example for some other illustrations.

1.2 Random Variables and Measurability Results

As the definition implies, a probability space is generally based on an abstract point set Ω without any algebraic or topological properties. It is therefore useful to consider various mappings of Ω into topological spaces with finer structure in order to make available several mathematical results for such spaces. We thus consider the simplest and most familiar space, the real line $\mathbb{R}$. To reflect the structure of Σ, we start with the σ-algebra $\mathcal{B}$ of $\mathbb{R}$, generated by all open intervals. It is the Borel σ-algebra. Let us now introduce a fundamental concept:

Definition 1 A *random variable* f on Ω is a finite real-valued measurable function. Thus $f : \Omega \to \mathbb{R}$ is a random variable if $f^{-1}(\mathcal{B}) \subset \Sigma$, where $\mathcal{B}$ is the Borel σ-algebra of $\mathbb{R}$; or $\Leftrightarrow f^{-1}(A) = \{\omega : f(\omega) \in A\} \in \Sigma$, for $A = (-\infty, x), x \in \mathbb{R}$. (Also written $f^{-1}(-\infty, x), [f < x]$ for $f^{-1}(A)$.)

Thus a random variable is a function and each outcome $\omega \in \Omega$ is assigned a real number $f(\omega) \in \mathbb{R}$. This expresses the heuristic notion of "randomness" as a mathematical concept. A fundamental nature of this formulation will be seen later (cf. Problem 5 (c) of Chapter 2). The point of this concept is that it is of real interest when related to a probability function P. Its relation is obtained in terms of image probabilities, also called distribution functions in our case. The latter concept is given in the following:

Definition 2 If $f : \Omega \to \mathbb{R}$ is a random variable, then its *distribution function* is a mapping $F_f : \mathbb{R} \to \mathbb{R}^+$ given by

$$F_f(x) = P(f^{-1}(-\infty, x))(= P\{\omega : f(\omega) < x\} = P[f < x]), x \in \mathbb{R}.$$

Evidently P and f uniquely determine F_f. The converse implication is slightly involved. It follows from definitions that F_f is a nonnegative nondecreasing left continuous [i.e., $F_f(x - 0) = F_f(x)$] bounded mapping of $\mathbb{R}$ into $[0, 1]$ such that $\lim_{x \to -\infty} F(x) = F_f(-\infty) = 0$, $F_f(+\infty) = \lim_{x \to +\infty} F(x) = 1$. Now any function F with these properties arises from *some* probability space; let $\Omega = \mathbb{R}$, $\Sigma = \mathcal{B}$, $f =$ identity, and $P(A) = \int_A dF, A \in \mathcal{B}$. The general case of several variables is considered later. First let us present some elementary properties of random variables.

In the definition of a random variable, the probability measure played no part. Using the measure function, we can make the structure of the class of all random variables richer than without it. Recall that (Ω, Σ, P) is *complete* if for any *null set* $A \in \Sigma$ [i.e., $P(A) = 0$] every subset B of A is also in Σ, so that $P(B)$ is defined and is zero. It is known and easy to see that every probability space (indeed a measure space) can always be completed if it is not already complete. The need for completion arises from simple examples. In fact, let $f_1, f_2, \ldots$ be a sequence of random variables that forms a Cauchy sequence in measure, so that for $\varepsilon > 0$, we have $\lim_{m,n \to \infty} P[|f_n - f_m| > \varepsilon] = 0$. Then there may not be a unique random variable f such that

$$\lim_{n \to \infty} P[|f_n - f| > \varepsilon] = 0.$$

However, if (Ω, Σ, P) is complete, then there always exists such an f, and if f' is another limit function, then $P\{\omega : f(\omega) \neq f'(\omega)\} = 0$; i.e., the limit is unique outside a set of zero probability. Thus if, $\mathcal{L}^0$ is the class of random variables on (Ω, Σ, P), a complete probability space, then $\mathcal{L}^0$ is an algebra and contains the limits of sequences of random variables that are Cauchy in measure. (See Problem 3 on the structure of $\mathcal{L}^0$.) The following measurability result on functions of random variables is useful in this study. It is due to Doob and, in the form we state it, to Dynkin. As usual, $\mathcal{B}$ is the Borel σ-algebra of $\mathbb{R}$.

Proposition 3 *Let (Ω, Σ) and $(S, \mathcal{A})$ be measurable spaces and $f : \Omega \to S$ be measurable, i.e., $f^{-1}(\mathcal{A}) \subset \Sigma$. Then a function $g : \Omega \to \mathbb{R}$ is measurable relative to the σ-algebra $f^{-1}(\mathcal{A})$ [i.e., $g^{-1}(\mathcal{B}) \subset f^{-1}(\mathcal{A})$] iff (= if and only if) there is a measurable function $h : S \to \mathbb{R}$ such that $g = h \circ f$.* (This result is sometimes refered to, for convience, as the "Doob-Dynkin lemma.")

Proof One direction is immediate. For $g = h \circ f : \Omega \to \mathbb{R}$ is measurable implies $g^{-1}(\mathcal{B}) = (h \circ f)^{-1}(\mathcal{B}) = f^{-1}(h^{-1}(\mathcal{B})) \subset f^{-1}(\mathcal{A})$, since $h^{-1}(\mathcal{B}) \subset \mathcal{A}$

For the converse, let g be $f^{-1}(\mathcal{A})$-measurable. Clearly $f^{-1}(\mathcal{A})$ is a σ-algebra contained in Σ. It suffices to prove the result for g simple, i.e., $g = \sum_{i=1}^n a_i \chi_{A_i}, A_i \in f^{-1}(\mathcal{A})$. Indeed, if this is proved, then the general case is obtained as follows. Since g is measurable for the σ-algebra $f^{-1}(\mathcal{A})$, by the structure theorem of measurable functions there exist simple functions g_n, measurable for $f^{-1}(\mathcal{A})$, such that $g_n(\omega) \to g(\omega)$ as $n \to \infty$ for each $\omega \in \Omega$. Using the special case, there is an $\mathcal{A}$-measurable $h_n : S \to \mathbb{R}$, $g_n = h_n \circ f$, for each $n \geq 1$. Let $S_0 = \{s \in S : h_n(s) \to \tilde{h}(s), n \to \infty\}$. Then $S_0 \in \mathcal{A}$, and $g(\Omega) \subset S$. Let $h(s) = \tilde{h}(s)$ if $s \in S_0, = 0$ if $s \in S - S_0$. Then h is $\mathcal{A}$-measurable and $g(\omega) = h(f(\omega)), \omega \in \Omega$. Consequently, we need to prove the special case.

Thus let g be simple: $g = \sum_{i=1}^n a_i \chi_{A_i}$, and $A_i = f^{-1}(B_i) \in f^{-1}(\mathcal{A})$, for a $B_i \in \mathcal{A}$. Define $h = \sum_{i=1}^n a_i \chi_{B_i}$. Then $h : S \to \mathbb{R}$ is $\mathcal{A}$-measurable and simple. [Here the B_j need not be disjoint even if the A_i are. To have symmetry in the definitions, we may replace B_i by C_i, where $C_1 = B_1$ and $C_i = B_i - \cup_{j=1}^{i-1} B_j$ for $i > 1$. So $C_i \in \mathcal{A}$, disjoint, $f^{-1}(C_i) = A_i$, and $h = \sum_{i=1}^n a_i \chi_{C_i}$ is the same function.] Thus

$$h(f(\omega)) = \sum_{i=1}^n a_i \chi_{B_i}(f(\omega)) = \sum_{i=1}^n a_i \chi_{f^{-1}(B_i)}(\omega)$$

$$= \sum_{i=i}^n a_i \chi_{A_i}(\omega) = g(\omega), \quad \omega \in \Omega,$$

and $h \circ f = g$. This completes the proof.

A number of specializations are possible from the above result. If $S = \mathbb{R}^n$ and $\mathcal{A}$ is the Borel σ-algebra of $\mathbb{R}^n$, then by this result there is an $h : \mathbb{R}^n \to \mathbb{R}$, (Borel) measurable, which satisfies the requirements. This yields the following:

Corollary 4 *Let (Ω, Σ) and $(\mathbb{R}^n, \mathcal{A})$ be measurable spaces, and $f : \Omega \to \mathbb{R}^n$ be measurable. Then $g : \Omega \to \mathbb{R}$ is $f^{-1}(\mathcal{A})$-measurable iff there is a Borel measurable function $h : \mathbb{R}^n \to \mathbb{R}$ such that $g = h(f_1, f_2, \ldots, f_n) = h \circ f$ where $f = (f_1, \ldots, f_n)$.*

If $\mathcal{A}$ is replaced by the larger σ-algebra of all (completion of $\mathcal{A}$) Lebesgue measurable subsets of R^n, then h will be a Lebesgue measurable function. The above result will be of special interest in studying, among other things, the structure of conditional probabilities. Some of these questions will be considered in Chapter 3. The mapping f in the above corollary is also called a *multidimensional random variable* and f of the theorem, an *abstract random variable.* We state this concept for reference.

Definition 5 Let (Ω, Σ) be a measurable space and S be a separable metric space with its Borel σ-algebra. (E.g., $S = \mathbb{R}^n$ or $\mathbb{C}^n$ or $\mathbb{R}_+^n$). Then a mapping $f : \Omega \to S$ is called a *generalized* (or *abstract*) *random variable* (and *random vector* if $S = \mathbb{R}^n$ or $\mathbb{C}^n$) whenever $f^{-1}(B) \in \Sigma$ for each open (or closed) set $B \subset S$, and it is a random variable if $S = \mathbb{R}$. [See Problem 2b for an alternative definition if $S = \mathbb{R}^n$.]

As a special case, we get $f : \Omega \to \mathbb{C}$, where $f = f_1 + if_2$, $f_j : \Omega \to \mathbb{R}, j = 1, 2$, is a complex random variable if its real and imaginary parts f_1, f_2 are (real) random variables. To illustrate the above ideas, consider the following:

Example 6 Let (Ω, Σ, P) be the space as in the last example, and $f_n : \Omega \to \mathbb{R}$ be given as $f_n(\omega) = n$ if the first 1 appears on the nth component (the preceding are zeroes), $= 0$ otherwise. Since $\Sigma = 2^\Omega = \mathcal{P}(\Omega)$, it is clear that f_n is a random variable and in fact each function on Ω is measurable for Σ. This example will be further discussed in illustrating other concepts.

Resuming the theme, it is necessary to discuss the validity of the results on σ-algebras generated by certain simple classes of sets and functions. In this connection the monotone class theorem and its substitute, as introduced by E. B. Dynkin, called the (π, λ)-classes will be of some interest. Let us state the concept and the result precisely.

Definition 7 A nonempty collection $\mathcal{C}$ of subsets of a nonempty set Ω is called (i) a *monotone class* if $\{A_n, n \geq 1\} \subset \mathcal{C}$, A_n monotone $\Rightarrow \lim_n A_n \in \mathcal{C}$, (ii) a π-(or *product*) *class* if $A, B \in \mathcal{C} \Rightarrow A \cap B \in \mathcal{C}$, (iii) a λ- (or *latticial*) *class* if (a) $A, B \in \mathcal{C}, A \cap B = \emptyset \Rightarrow A \cup B \in \mathcal{C}$, (b) $A, B \in \mathcal{C}, A \supset B \Rightarrow A - B \in \mathcal{C}, \Omega \in \mathcal{C}$, and (c) $A_n \in \mathcal{C}$, $A_n \subset A_{n+1} \Rightarrow \cup A_n \in \mathcal{C}$; (iv) the smallest class of sets $\mathcal{C}$ containing a given collection $\mathcal{A}$ having a certain property (e.g., a monotone class, or a σ-algebra) is said to be *generated* by $\mathcal{A}$.

The following two results relate a given collection and its desirable generated class. They will be needed later on. Note that a λ-class which is a π-class is a σ-algebra. We detail some nonobvious (mathematical) facts.

Proposition 8 (a) *If $\mathcal{A}$ is an algebra, then the monotone class generated by $\mathcal{A}$ is the same as the σ-algebra generated by $\mathcal{A}$.*

(b) *If $\mathcal{A}$ is a λ-class and $\mathcal{B}$ is a π-class, $\mathcal{A} \supset \mathcal{B}$, then $\mathcal{A}$ also contains the σ-algebra generated by $\mathcal{B}$.*

Proof The argument is similar for both parts. Since the proof of (a) is in most textbooks, here we prove (b).

The proof of (b) is not straightforward, but is based on the following idea. Consider the collection $\mathcal{A}_1 = \{A \subset \Omega : A \cap B \in \mathcal{A}_0$ for all $B \in \mathcal{B}\}$. Here we take $\mathcal{A}_0 \supset \mathcal{B}$, and $\mathcal{A}_0$ is the smallest λ-class, which is the intersection of all such collections containing $\mathcal{B}$. The class $\mathcal{A}_1$ is not empty. In fact $\mathcal{B} \subset \mathcal{A}_1$. We observe that $\mathcal{A}_1$ is a λ-class. Clearly $\Omega \in \mathcal{A}_1$, $A_i \in \mathcal{A}_1$, $A_1 \cap A_2 = \emptyset \Rightarrow A_i \cap B, i = 1, 2$, are disjoint for all $B \in \mathcal{B}$, and $A_i \cap B \in \mathcal{A}_0$. Since $\mathcal{A}_0$ is a λ-class, $(A_1 \cup A_2) \cap B = (A_1 \cap B) \cup (A_2 \cap B) \in \mathcal{A}_0$, so that $A_1 \cup A_2 \in \mathcal{A}_1$. Similarly $A_1 \supset A_2 \Rightarrow A_1 \cap B - A_2 \cap B = (A_1 - A_2) \cap B \in \mathcal{A}_0$ and $A_1 - A_2 \in \mathcal{A}_1$.

The monotonicity is similarly verified. Thus $\mathcal{A}_1$ is a λ-class. Since $\mathcal{A}_0$ is the smallest λ-class, $\mathcal{A}_1 \supset \mathcal{A}_0 \supset \mathcal{B}$. Hence $A \in \mathcal{A}_0 \subset \mathcal{A}_1, B \in \mathcal{B} \Rightarrow A \cap B \in \mathcal{A}_0$.

Next consider $\mathcal{A}_2 = \{A \subset \Omega : A \cap B \in \mathcal{A}_0, \text{ all } B \in \mathcal{A}_0\}$. By the preceding work, $\mathcal{A}_2 \supset \mathcal{B}$ and, by an entirely similar argument, we can conclude that $\mathcal{A}_2$ is also a λ-class. Hence $\mathcal{A}_2 \supset \mathcal{A}_0 \supset \mathcal{B}$. This means with $A, B \in \mathcal{A}_0$, $A \cap B \in \mathcal{A}_0 \subset \mathcal{A}_2$, and hence $\mathcal{A}_0$ is a π-class. But by Definition 7, a collection which is both a π- and a λ-class is a σ-algebra. Thus $\mathcal{A}_0$ is a σ-algebra $\supset \mathcal{B}$. Then $\sigma(\mathcal{B}) \subset \mathcal{A}_0$, where $\sigma(\mathcal{B})$ is the generated σ-algebra by $\mathcal{B}$. Since $\mathcal{A}_0 \subset \mathcal{A}$, the proposition is proved.

The next result, containing two assertions, is of interest in theoretical applications.

Proposition 9 *Let $B(\Omega)$ be the space of real bounded functions on Ω and $\mathcal{H} \subset B(\Omega)$ be a linear set containing constants and satisfying* (i) *$f_n \in \mathcal{H}, f_n \to f$ uniformly $\Rightarrow f \in \mathcal{H}$, or (i′) $f \in \mathcal{H} \Rightarrow f^{\pm} \in \mathcal{H}$, where $f^+ = \max(f, 0)$ and $f^- = f^+ - f$, and* (ii) *$0 \leq f_n \in \mathcal{H}, f_n \uparrow f, f \in B(\Omega) \Rightarrow f \in \mathcal{H}$. If $\mathcal{C} \subset \mathcal{H}$ is any set which is closed under multiplication and $\Sigma = \sigma(\mathcal{C})$ is the smallest σ-algebra relative to which every element of $\mathcal{C}$ is measurable, then every $f(\in B(\Omega))$ which is Σ-measurable belongs to $\mathcal{H}$. The same conclusion holds if $\mathcal{C} \subset \mathcal{H}$ is not necessarily closed under multiplication, but $\mathcal{H}$ satisfies* (i′) *[instead of* (i)*] $\mathcal{C}$ is a linear set closed under infima, and $f \in \mathcal{C} \Rightarrow f \wedge 1 \in \mathcal{C}$.*

Proof The basic idea is similar to that of the above result. Let $\mathcal{A}_0$ be an algebra, generated by $\mathcal{C}$ and 1, which is closed under uniform convergence and is contained in $\mathcal{H}$. Clearly $\mathcal{A}_0$ exists. Let $\mathcal{A}_1$ be the largest such algebra. The existence of $\mathcal{A}_1$ is a consequence of the fact that the class of all such $\mathcal{A}_0$ is closed under unions and hence is partially ordered by inclusion. The existence of the desired class $\mathcal{A}_1$ follows from the maximal principle of Hausdorff.

If $f \in \mathcal{A}_1$, then there is a $k > 0$, such that if $|f| \leq k$ and if $p(\cdot)$ is any polynomial on $[-k, k]$, then $p(f) \in \mathcal{A}_1$. Also by the classical Weierstrass approximation theorem the function $h : [-k, k] \to \mathbb{R}$, $h(x) = |x|$, is the uniform limit of polynomials p_n on $[-k, k]$. Hence $p_n(f) \to |f|$ uniformly so that (by the uniform closure of $\mathcal{A}_1$) $|f| \in \mathcal{A}_1$ and $\mathcal{A}_1$ is a vector lattice.

Observe that $\mathcal{A}_1$ automatically satisfies (ii), since if $0 \leq g_n \in \mathcal{A}_1, g_n \uparrow g \in B(\Omega)$, then $g \in \mathcal{H}$, and if $\mathcal{A}_2$ is generated by $\mathcal{A}_1$ and g (as $\mathcal{A}_0$), then by the maximality of $\mathcal{A}_1, \mathcal{A}_2 = \mathcal{A}_1$. Thus $\mathcal{A}_1$ satisfies (i) and (ii) and is a vector lattice. The second part essentially has this conclusion as its hypothesis. Let us verify this. By (i′), if $f \in \mathcal{H}$, then $f^{\pm} \in \mathcal{H}$, so that $f^+ + f^- = |f| \in \mathcal{H}$. Hence if $f, g \in \mathcal{H}$, then $f \vee g = \frac{1}{2}(|f - g| + f + g) \in \mathcal{H}$, since $f - g \in \mathcal{H}$ (because $\mathcal{H}$ is a vector space). Thus $\mathcal{H}$ is a vector lattice. Consequently we consider vector lattices containing $\mathcal{C}$ and 1 which are subsets of $\mathcal{H}$. Next one chooses a maximal lattice (as above). If this is $\mathcal{A}_2'$, then it has the same properties as $\mathcal{A}_2$. Thus it suffices to consider $\mathcal{A}_2$ and prove that each f in $B(\Omega)$ which is Σ-measurable is in $\mathcal{A}_2$ $(\subset \mathcal{H})$.

Let $\mathcal{S} = \{A \subset \Omega : \chi_A \in \mathcal{A}_2\}$. Since $\mathcal{A}_2$ is an algebra, $\mathcal{S}$ is a π-class. Also $\mathcal{S}$ is closed under disjoint unions and monotone limits. Thus it is a λ-class as well, and by the preceding proposition it is a σ-algebra. If $0 \leq g[\in B(\Omega)]$ is $\mathcal{S}$-measurable, then there exist $0 \leq g_n \uparrow g$; g_n is an $\mathcal{S}$-measurable simple function. But then $g_n \in \mathcal{A}_2$ and so $g \in \mathcal{A}_2$ also. Since $\mathcal{A}_2$ is a lattice, this result extends to all $g \in B(\Omega)$ which are $\mathcal{S}$-measurable. To complete the proof, it is only necessary to verify $\Sigma = \sigma(\mathcal{C}) \subset \mathcal{S}$. Let $0 \leq f \in \mathcal{C}$ and $B = [f \geq 1] \in \Sigma$. We claim that $B \in \mathcal{S}$. In fact, let $g = f \wedge 1$. Then $g \in \mathcal{A}_2$ and $0 \leq g \leq 1$. Now $[g = 1] = B$ and $[g < 1] = B^c$. Thus $g^n \in \mathcal{A}_2$ and $g^n \downarrow 0$ on B^c, or $1 - g^n \uparrow 1$ on B^c. Since $1 - g^n \in \mathcal{A}_2$, and it is closed under bounded monotone limits, we have $1 - g^n \uparrow \chi_{B^c} \in \mathcal{A}_2 \Rightarrow B^c \in \mathcal{S}$, so that $B \in \mathcal{S}$. If $0 \leq f \in \mathcal{C}, B_a = [f \geq a] = [f/a \geq 1]$ for $a > 0$, then $f/a \in \mathcal{A}_2$ and by the above proof $B_a \in \mathcal{S}$ for each a. But such sets as B_a clearly generate Σ, so that $\Sigma \subset \mathcal{S}$. This completes the result in the algebra case.

In the lattice case $A, B \in \mathcal{S} \Rightarrow \chi_A \chi_B = \min(\chi_A, \chi_B) \in \mathcal{A}_2'$, so that $A \cap B \in \mathcal{S}$. Thus $\mathcal{S}$ is a π-class again. That it is a λ-class is proved as before, so that $\mathcal{S}$ is a σ-algebra. The rest of the argument holds verbatim. Since with each $f \in \mathcal{C}$, one has $f \wedge 1 \in \mathcal{C}$ we do not need to go to $\mathcal{A}_2'$, and the proof is simplified. This establishes the result in both cases.

1.3 Expectations and the Lebesgue Theory

If $X : \Omega \to \mathbb{R}$ is a random variable (r.v.) on $\mathcal{L}^1$, then X is said to have an *expected value* iff it is integrable in Lebesgue's sense, relative to P. This means $|X|$ is also integrable. It is suggestively denoted

$$E(X) = E_P(X) = \int_\Omega X dP, \tag{1}$$

the integral on the right being the (absolute) Lebesgue integral. Thus $E(X)$ exists, by definition, iff $E(|X|)$ exists. Let $\mathcal{L}^1$ be the class of all Lebesgue integrable functions on (Ω, Σ, P). Then $E : \mathcal{L}^1 \to \mathbb{R}$ is a positive linear mapping since the integral has that property. Thus for $X, Y \in \mathcal{L}^1$ we have

$$E(aX + bY) = aE(X) + bE(Y), \quad a, b \in \mathbb{R}, \tag{2}$$

and $E(1) = 1$ since $P(\Omega) = 1$, $E(X) \geq 0$ if $X \geq 0$ a.e. The operator E is also called the (*mathematical*) *expectation* on $\mathcal{L}^1$. It is clear that the standard results of Lebesgue integration are thus basic for the following work. In the next section we relate this theory to the distribution function of X.

To fix the notation and terminology, let us recall the key theorems of Lebesgue's theory, the details of which the reader can find in any standard text on real analysis [see, e.g., Royden (1968, 1988), Sion (1968), or Rao (1987, 2004)].

The basic Lebesgue theorems that are often used in the sequel, are the following:

Theorem 1 (Monotone Convergence) *Let* $0 \leq X_1 \leq X_2 \leq \ldots$ *be a sequence of random variables on* (Ω, Σ, P). *Then* $X = \lim_n X_n$ *is a measurable (extended) real valued function (or a "defective" random variable) and*

$$\lim_{n \to \infty} E(X_n) = E(X)$$

holds, where the right side can be infinite.

A result of equal importance is the following:

Theorem 2 (Dominated Convergence) *Let* $\{X_n, n \geq 1\}$ *be a sequence of random variables on* (Ω, Σ, P) *such that (i)* $\lim_{n \to \infty} X_n = X$ *exists at all points of* Ω *except for a set* $N \subset \Omega$, $P(N) = 0$, *(written* $X_n \to X$ *a.e.), and (ii)* $|X_n| \leq Y$, *an r.v., with* $E(Y) < \infty$. *Then* X *is an r.v. and* $\lim_n E(X_n) = E(X)$ *holds, all quantities being finite.*

The next statement is a consequence of Theorem 1.

Theorem 3 (Fatou's Lemma) *Let* $\{X_n, n \geq 1\}$ *be any sequence of nonnegative random variables on* (Ω, Σ, P). *Then we have* $E(\liminf_n X_n) \leq \liminf_n E(X_n)$.

In fact, if $Y_k = \inf\{X_n, n \geq k\}$, then Theorem 1 applies to $\{Y_k, k \geq 1\}$. Note that these theorems are valid if P is replaced by a nonfinite measure.

Many of the deeper results in analysis are usually based on inequalities. We present here some of the classical inequalities that occur frequently in our subject. First recall that a mapping $\phi : \mathbb{R} \to \mathbb{R}$ is called *convex* if for any $\alpha, \beta \geq 0, \alpha + \beta = 1$, one has

$$\phi(\alpha x + \beta y) \leq \alpha\phi(x) + \beta\phi(y), \quad x, y \in \mathbb{R}. \tag{3}$$

From this definition, it follows that if $\{\phi_n, n \geq 1\}$ is a sequence of convex functions $\alpha_i \in \mathbb{R}^+$, then $\sum_{n=1}^{m} \alpha_n \phi_n$ is also convex on $\mathbb{R}$, and if $\phi_n \to \phi$, then ϕ is convex. Further, from elementary calculus, we know that each twice-differentiable function ϕ is convex iff its second derivative ϕ'' is nonnegative. It can be shown that a measurable convex function on an open interval is necessarily continuous there. These facts will be used without comment. Hereafter "convex function" always stands for a *measurable convex function* on $\mathbb{R}$.

Let $\phi(x) = -\log x$, for $x > 0$. Then $\phi''(x) > 0$, so that it is convex. Hence (3) becomes

$$\log(\alpha x + \beta y) \geq \alpha \log x + \beta \log y = \log x^\alpha y^\beta, \quad x > 0, y > 0.$$

Since log is an increasing function, this yields for $\alpha \geq 0, \beta \geq 0, x > 0, y > 0$

$$x^\alpha y^\beta \leq \alpha x + \beta y, \quad \alpha + \beta = 1. \tag{4}$$

For any pair of random variables X, Y on (Ω, Σ, P), and $p \geq 1, q = p/(p-1)$, we define $||X||_p = [E(|X|^p)]^{1/p}, 1 \leq p < \infty$, and $||X||_\infty$ (= essential supremum of $|X|$) $= \inf\{k > 0 : P[|X| > k] = 0\}$. Then $|| \cdot ||_p$, $1 \leq p \leq \infty$, is a positively homogeneous invariant metric, called the *p-norm*; i.e., if $d(X, Y) = ||X - Y||_p$, then $d(\cdot, \cdot)$is a metric, $d(X + Z, Y + Z) = d(X, Y)$ and $d(aX, 0) = |a|d(X, 0), a \in \mathbb{R}$. We have

Theorem 4 *Let X, Y be random variables on (Ω, Σ, P). Then*
(i) **Hölder's Inequality**

$$||X||_p < \infty, \ ||Y||_q < \infty, \ (1/p) + (1/q) = 1, \ p \geq 1 \Rightarrow E(|XY|) \leq ||X_p||Y||_q; \tag{5}$$

(ii) **Minkowski's Inequality**

$$||X||_p < \infty, \ ||Y||_p < \infty, \ p \geq 1 \Rightarrow ||X + Y||_p \leq ||X||_p + ||Y||_p. \tag{6}$$

Proof (i) If $||X||_p = 0$, or $||Y||_q = 0$, then $X = 0$ a.e., or $Y = 0$ a.e., so that (5) is true and trivial. Now suppose $||X||_p > 0$, and $||Y||_q > 0$. If $p = 1$, then $q = \infty$, and we have $||Y||_\infty =$ ess sup $|Y|$, by definition (= k, say),

$$|E(XY)| = \left|\int_\Omega XY dP\right| \leq \int_\Omega |X||Y| dP \leq k \int_\Omega |X| dP$$

$$= kE(|X|) = ||Y||_\infty ||X||_1.$$

Thus (5) is true in this case. Let then $p > 1$, so that $q = p/(p-1) > 1$. In (4) set $\alpha = 1/p, \beta = 1/q, x = (|X|/||X||_p)^p(\omega)$, and $y = (|Y|/||Y||_q)^q(\omega)$. Then it becomes

$$\frac{|X| \cdot |Y|}{||X||_p ||Y||_q}(\omega) \leq \frac{1}{p} \frac{|X|^p}{||X||_p^p}(\omega) + \frac{1}{q} \frac{|Y|^q}{||Y||_q^q}(\omega), \quad \omega \in \Omega. \tag{7}$$

Applying the (positive) operator E to both sides of (7) we get

$$\frac{E(|XY|)}{||X||_p ||Y||_p} \leq \frac{1}{p} \frac{E(|X|^p)}{||X||_p^p} + \frac{1}{q} \frac{E(|Y|^q)}{||Y||_q^q} = \frac{1}{p} + \frac{1}{q} = 1.$$

This proves (5) in this case also, and hence it is true as stated.

(ii) Since $|X + Y|^p \leq 2^p \max(|X|^p, |Y|^p) \leq 2^p[|X|^p + |Y|^p]$, the linearity of E implies $E(|X + Y|^p) < \infty$, so that (6) is meaningful. If $p = 1$, the result follows from $|X + Y| \leq |X| + |Y|$. If $p = \infty$, $|X| \leq ||X||_\infty, |Y| \leq ||Y||_\infty$, a.e. Hence $|X + Y| \leq ||X||_\infty + ||Y||_\infty$, a.e., so that (6) again holds in this case.

Now let $1 < p < \infty$. If $||X+Y||_p = 0$, then (6) is trivial and true. Thus let $||X+Y||_p > 0$ also.

Consider

$$||X+Y||_p^p = \int_\Omega |X+Y|^{p-1}|X+Y|dP$$

$$\leq \int_\Omega |X+Y|^{p-1}|X|dP + \int_\Omega |X+Y|^{p-1}|Y|dP. \tag{8}$$

Since $(p-1) > 0$, let $q = p/(p-1)$. Then $(p-1)q = p$, and

$$E(|X+Y|^{(p-1)q}) < \infty.$$

Hence applying (5) to the two terms of (8) separately we get

$$||X+Y||_p^p \leq || \ |X+Y|^{p-1}||_q[||X||_p + ||Y||_p]$$

$$= ||X+Y||_p^{p/q}[||X||_p + ||Y||_p],$$

or

$$||X+Y||_p \leq ||X||_p + ||Y||_p.$$

This completes the proof.

Some specializations of the above result, which holds for any measure space, in the context of probability spaces are needed. Taking $Y = 1$ a.e. in (5) we get for $p \geq 1$

$$[E(|X|)]^p \leq E(|X|^p). \tag{9}$$

Hence writing $\phi(x) \equiv |x|^p$, (9) says that $\phi(E(|X|)) \leq E(\phi(X))$. We prove below that this is true for any continuous convex function ϕ, provided the respective expectations exist. The significance of (9) is the following. If X is an r.v., $s > 0$, and $E(|X|^s) < \infty$, then X is said to have the sth *moment* finite. Thus if X has pth moment, $p > 1$, then its expectation exists. More is true, namely, all of its lower-order moments exist, as seen from

Corollary 5 *Let X be an r .v., on a probability space, with sth moment finite. If $0 < r \leq s$, then $(E(|X|^r))^{1/r} \leq (E(|X|^s))^{1/s}$. More generally, for any $0 \leq r_i, i = 1,2,3$, if $\beta_{r_i} = E(|X|^{r_i})$, we have the* **Liapounov inequality**:

$$\log \beta_{r_1+r_2} \leq \frac{r_3}{r_2+r_3}\log\beta_{r_1} + \frac{r_2}{r_2+r_3}\log\beta_{r_1+r_2+r_3}. \tag{10}$$

Proof Since for any $0 \leq r \leq s$, $|x|^r \leq 1 \cdot \chi_{[|x|<1]} + |x|^s \cdot \chi_{[|x|\geq 1]} \leq 1 + |x|^s$, we get

$$E(|X|^r) \leq 1 + E(|X|^s) < \infty,$$

so that all lower-order moments exist. The inequality holds if we show that $\beta_r^{1/r}$ is a nondecreasing function of $r > 0$. But this follows from (9) if we let $p = s/r \geq 1$ and replace X by $|X|^r$ there. Thus

$$[E(|X|^r)]^{s/r} \leq E(|X|^{r(s/r)}) = E(|X|^s),$$

which is the desired result on taking the sth root.

For the Liapounov inequality (10), note that $\beta_0 = 1$, and on the open interval $(0, s)$, β_r is twice differentiable if $\beta_s < \infty$ [use the dominated convergence (Theorem) for differentiation relative to r under the integral sign], and

$$\beta_r'' = \frac{d^2\beta_r}{dr^2} = \int_\Omega (\log |X|)^2 |X|^r dP.$$

Let $y_r = \log \beta_r$. If $X \not\equiv 0$ a.e., then this is well defined and

$$\frac{d^2 y_r}{dr^2} = \frac{\beta_r \beta_r'' - (\beta_r')^2}{(\beta_r)^2} \geq 0,$$

because

$$\begin{aligned}|\beta_r'|^2 &= \int_\Omega (\log |X| \cdot |X|^{r/2})(|X|^{r/2}) dP \\ &\leq \int_\Omega (\log |X|)^2 |X|^r dP \cdot \int_\Omega |X|^r dP = \beta_r'' \beta_r,\end{aligned}$$

by the Hölder inequality with exponent 2. Thus y_r is also convex in r. Taking $\alpha = r_3/(r_2 + r_3), \beta = r_2/(r_2 + r_3)$ and $x = r_1, y' = r_1 + r_2 + r_3$ in (4) with $\phi(r) = y_r$, one gets $\alpha x + \beta y' = r_1 + r_2$, so that

$$y_{r_1+r_2} = y_{\alpha x + \beta y'} \leq \alpha y_x + \beta y_{y'} = \frac{r_3}{r_2 + r_3} y_{r_1} + \frac{r_2}{r_2 + r_3} y_{r_1+r_2+r_3},$$

which is (10). Note that the convexity of y_r can also be proved with a direct application of the Hölder inequality. This completes the proof.

The special case of (5) with $p = 2$ is the classical *Cauchy-Buniakowski-Schwarz* (or CBS) *inequality*. Due to its great applicational potential, we state it as

Corollary 6 (CBS Inequality) *If X, Y have two moments finite, then XY is integrable and*

$$|E(XY)| \leq E(|XY|) \leq [E(|X|^2) E(|Y|^2)]^{1/2}. \tag{11}$$

Proof Because of its interest we present an independent proof. Since X, Y have two moments, it is evident that $tX + Y$ has two moments for any $t \in \mathbb{R}$, and we have

$$0 \leq E((t|X| + |Y|)^2) = t^2 E(|X|^2) + 2tE(|XY|) + E(|Y|^2) < \infty.$$

This is a quadratic equation in t which is never negative. Hence it has no distinct real roots. Thus its discriminant must be nonpositive. Consequently,

$$4(E(|XY|))^2 - 4E(|X|^2)E(|Y|^2) \leq 0.$$

This is (11), and the proof is complete.

Remark The conditions for equality in (5), (6), (10), and (11) can be obtained immediately, and will be left to the reader. We invoke them later when necessary.

One can now present the promised generalization of (9) as

Proposition 7 (Jensen's Inequality) *If $\phi : \mathbb{R} \to \mathbb{R}$ is convex and X is an r.v. on (Ω, Σ, P) such that $E(X)$ and $E(\phi(X))$ exist, then*

$$\phi(E(X)) \leq E(\phi(X)). \tag{12}$$

Proof Let x_0, x_1 be two points on the line and x be an intermediate point so that $x = \alpha x_1 + \beta x_0$, where $0 \leq \alpha \leq 1$, $\alpha + \beta = 1$. Then by (3)

$$\phi(x) \leq \alpha\phi(x_1) + \beta\phi(x_0).$$

For definiteness, let $x_0 \leq x \leq x_1$ so that with $\alpha = (x - x_0)/(x_1 - x_0), \beta = (x_1 - x)/(x_1 - x_0)$, we get x. Hence the above inequality becomes

$$\phi(x) = \alpha\phi(x) + \beta\phi(x) \leq \frac{x - x_0}{x_1 - x_0}\phi(x_1) + \frac{x_1 - x}{x_1 - x_0}\phi(x_0),$$

so that

$$(x - x_0)(\phi(x) - \phi(x_1)) \leq (x_1 - x)(\phi(x_0) - \phi(x)).$$

By setting $y = x_1, y_0 = x$, this becomes

$$\phi(y) - \phi(y_0) \geq (y - y_0)\frac{\phi(y_0) - \phi(x_0)}{y_0 - x_0} = g(y_0)(y - y_0)(say). \tag{13}$$

In this inequality, written as $\phi(y) \geq \phi(y_0) + g(y_0)(y - y_0)$, the right side is called the *support line* of ϕ at $y = y_0$. Let $X(\omega) = y$, and $y_0 = E(X)$ in (13). Then $\phi(X)$ is an r.v., and taking expectations, we get

$$E(\phi(X)) - \phi(E(X)) \geq g(E(X))E(X - E(X)) = 0.$$

This is (12), and the result holds. [Note: $t_1 < t_2 \Rightarrow g(t_1) \leq g(t_2)$.[2]]

[2] This is not entirely trivial. Use (3) in different forms carefully. [See, e.g., G.H. Hardy, J.E. Littlewood, and G. Polyá (1934, p. 93).]

In establishing (10) we first showed that $\beta_r^{1/r} = [E(|X|^r)]^{1/r}$ is an increasing function of r. This has the following consequence:

Proposition 8 *For any random variable X, $\lim_{r\to\infty}, (E[|X|^r])^{1/r} = ||X||_\infty$.*

Proof If $||X||_\infty = 0, X = 0$ a.e., the result is true. So let $0 < k = ||X||_\infty \leq \infty$. Then, by definition, $P[|X| > k] = 0$. Hence

$$E(|X|^r) = \int_\Omega |X|^r dP \leq k^r \cdot 1,$$

so that for any $0 < t < k$,

$$k \geq [E(|X|^r)]^{1/r} \geq \left[\int_{[|X|\geq t]} |X|^r dP\right]^{1/r} \geq t \cdot (P[|X| \geq t])^{1/r}. \tag{14}$$

Letting $r \to \infty$ in (14), we get $k \geq \lim_{r\to\infty}(E(|X|^r))^{1/r} \geq t$. Since $t < k$ is arbitrary, the result follows on letting $t \uparrow k$.

Let X, Y be two random variables with finite second moments. Then we can define (a) the *variance* of X as

$$Var(X) = \sigma^2(X) = E(X - E(X))^2 [= E(X^2) - (E(X))^2] \geq 0, \tag{15}$$

which always exists since $\sigma^2(X) \leq E(X^2) < \infty$; and (b) the *covariance* of X, Y as

$$cov(X, Y) = E((X - E(X)(Y - E(Y))) = E(XY) - E(X)E(Y). \tag{16}$$

This also exists since by the CBS inequality,

$$|cov(X, Y)|^2 \leq E(X - E(X))^2 \cdot E(Y - E(Y))^2 = \sigma^2(X)\sigma^2(Y) < \infty. \tag{17}$$

The normalized covariance, called the *correlation*, between X and Y, denoted $\rho(X, Y)$, is then

$$\rho(X, Y) = \frac{cov(X, Y)}{\sigma(X)\sigma(Y)}, \tag{18}$$

where $\sigma(X), \sigma(Y)$ are the positive square roots of the corresponding variances. Thus $|\rho(X, Y)| \leq 1$ by (17). The quantity $0 \leq \sigma(X)$ is called the *standard deviation* of X. Note that if $E(X) = 0$, then $\beta_2 = \sigma^2(X)$, and generally $\beta_2 \geq \sigma^2(X)$, by (15).

Another simple but very useful inequality is given by

Proposition 9 (i) (**Markov's Inequality**) *If $\xi : \mathbb{R} \to \mathbb{R}^+$ is a Borel function and X is an r.v. on (Ω, Σ, P), then for any $\lambda > 0$,*

$$P[\xi(X) \geq \lambda] \leq \frac{1}{\lambda} E(\xi(X)). \tag{19}$$

(ii) (**Čebyšev's Inequality**) *If X has a finite variance, then*

$$P[|X - E(X)| \geq \lambda] \leq \frac{\sigma^2(X)}{\lambda^2} \tag{20}$$

Proof For (i) we have

$$E(\xi(X)) = \int_{\Omega} \xi(X) dP \geq \int_{[\xi(X) \geq \lambda]} \xi(X) dP \geq \lambda P[\xi(X) \geq \lambda].$$

(ii). In (19), replace X by $X - E(X)$, $\xi(x)$ by x^2 and λ by λ^2. Then ξ being one-to-one on $\mathbb{R}^+$, $[|X - E(X)|^2 > \lambda^2] = [|X - E(X)| > \lambda]$,and the result follows from that inequality.

Another interesting consequence is

Corollary 10 *If $X_1, \ldots, X_n$ are n random variables each with two moments finite, then we have*

$$\sigma^2(\sum_{i=1}^{n} X_i) = \sum_{i=1}^{n} \sigma^2(X_i) + 2 \sum_{1 \leq i < j \leq n} cov(X_i, X_j),$$

and if they are uncorrelated [i.e., $\rho(X_i, X_j) = 0$ for $i \neq j$] then

$$\sigma^2(\sum_{i=1}^{n} X_i) = \sum_{i=1}^{n} \sigma^2(X_i).$$

This follows immediately from definitions. The second line says that for uncorrelated random variables, the variance of the sum is the sum of the variances. We later strengthen this concept into what is called "independence" and deduce several results of great importance in the subject.

For future use, we include two fundamental results on multiple integration and differentiation of set functions.

Theorem 11 (i) (**Fubini-Stone**) *Let $(\Omega_i, \Sigma_i, \mu_i) i = 1, 2$, be a pair of measure spaces and (Ω, Σ, μ) be their product. If $f : \Omega \to \mathbb{R}$ is a measurable and μ-integrable function, then*

$$\int_{\Omega_1} f(\omega_1, \cdot) \mu_1(d\omega_1) \text{ is } \mu_2 - \text{measurable},$$

$$\int_{\Omega_2} f(\cdot, \omega_2) \mu_2(d\omega_2) \text{ is } \mu_1 - \text{measurable},$$

and, moreover,

$$\int_{\Omega} f(\omega_1,\omega_2)\mu(d\omega_1,d\omega_2) = \int_{\Omega_1}\int_{\Omega_2} f(\omega_1,\omega_2)\mu_2(d\omega_2)\mu_1(d\omega_1)$$
$$= \int_{\Omega_2}\int_{\Omega_1} f(\omega_1,\omega_2)\mu_1(d\omega_1)\mu_2(d\omega_2). \qquad (21)$$

(ii) (**Tonelli**) *If in the above μ_1, μ_2 are σ-finite and $f : \Omega \to \overline{\mathbb{R}}^+$ is measurable, or μ_i are arbitrary measures but there exists a sequence of μ-integrable simple functions $f_n : \Omega \to \mathbb{R}^+$ such that $f_n \uparrow f$ a.e. (μ), then again (21) holds even though both sides may now be infinite.*

The detailed arguments for this result are found in most standard texts [cf., e.g., Zaanen (1967), Rao (1987, 2004)]. The other key result is the following:

Theorem 12 (i) (**Lebesgue Decomposition**) *Let μ and ν be two finite or σ-finite measures on (Ω, Σ), a measurable space. Then ν can be uniquely expressed as $\nu = \nu_1 + \nu_2$, where ν_1 vanishes on μ-null sets and there is a set $A \in \Sigma$ such that $\mu(A) = 0$ and $\nu_2(A^c) = 0$. Thus ν_2 is different from zero only on a μ-null set. (Here ν_2 is called* **singular** *or* **orthogonal** *to μ and denoted $\mu \perp \nu_2$. Note also that $\nu_1 \perp \nu_2$ is written.)*

(ii) (**Radon-Nikodým Theorem**) *If μ is a σ-finite measure on (Ω, Σ) and $\nu : \Sigma \to \overline{\mathbb{R}}$ is σ-additive, and vanishes on μ-null sets (denoted $\nu \ll \mu$), then there exists a μ-unique function (or density) $f : \Omega \to \overline{\mathbb{R}}$ such that*

$$\nu(A) = \int_A f d\mu, \quad A \in \Sigma.$$

This important result is also proved in the above-stated references.

1.4 Image Measure and the Fundamental Theorem of Probability

As noted in the beginning of Section 2, the basic probability spaces often involve abstract sets without any topology. However, when a random variable (or vector) is defined on such (Ω, Σ, P), we can associate a distribution function on the range space which usually has a nice topological structure, as in Definition 2.2. Evidently the same probability space can generate numerous image measures by using different measurable mappings, or random variables. There is a fundamental relation between the expectation of a function of a random variable on the original space and the integral on its image space.

The latter is often more convenient in evaluation of these expectations than to work on the original abstract spaces.

A comprehensive result on these ideas is contained in

Theorem 1 (i) (**Image Measures**). *Let* (Ω, Σ, μ) *be a measure space with* $(S, \mathcal{A})$ *as a measurable space, and* $f : \Omega \to S$ *be measurable [i.e.,* $f^{-1}(\mathcal{A}) \subset \Sigma$*]. If* $\nu = \mu \circ f^{-1} : \mathcal{A} \to \overline{\mathbb{R}}^+$ *is the image measure, then for each* $g : S \to \mathbb{R}$ *measurable, we have*

$$\int_{\Omega} g \circ f(\omega)\mu(d\omega) = \int_{S} g(s)\nu(ds), \tag{1}$$

in the sense that if either side exists, so does the other and equality holds.

(ii) (**Fundamental Law of Probability**). *If* (Ω, Σ, μ) *is a probability space and* $X : \Omega \to \mathbb{R}$ *is a random variable with distribution function* F_X*, and* $g : \mathbb{R} \to \mathbb{R}$ *is a Borel function,* $Y = g(X)$*, then*

$$E(g(X)) = \int_{\mathbb{R}} g(x)dF_X(x)[= \int_{\mathbb{R}} ydF_Y(y)], \tag{2}$$

in the sense that if either side exists, so does the other with equality holding.

(iii) *In particular, for any* $p > 0$,

$$E(|X|^p) = \int_{-\infty}^{\infty} |x|^p dF_X(x) = p\int_0^{\infty} x^{p-1}[1 + F_X(-x) - F_X(x)]dx. \tag{3}$$

Proof (i) This very general statement is easily deduced from the definition of the image measure. Indeed, if $g(s) = \chi_A(s), A \in \mathcal{A}$, then the left side of (1) becomes

$$\int_{\Omega}(\chi_A \circ f)(\omega)\mu(d\omega) = \int_{\Omega} \chi_{f^{-1}(A)}\mu(d\omega) = \int_S \chi_A \nu(ds) = \int_S g(s)\nu(ds).$$

Thus (1) is true, and by the linearity of the integral and the (σ-additivity of ν) the same result holds if $g = \sum_{i=1}^n a_i \chi_{A_i}$, a simple function with $a_i \geq 0$. If $g \geq 0$ is measurable, then there exist simple functions $0 \leq g_n \uparrow g$, so that (1) holds by the Lebesgue monotone convergence theorem. Since any measurable $g = g^+ - g^-$ with $g^{\pm} \geq 0$ and measurable, the last statement implies the truth of (1) in general for which g^+ or g^- is integrable.

(ii) Taking $S = \mathbb{R}$, (μ is probability) we get $\nu(-\infty, x) = F_X(x)$, the distribution function of X. Thus (1) is simply (2). If $Y = g(X) : \Omega \to \mathbb{R}$, then clearly Y is a random variable. Replace X by Y, g by identity, and S by $\mathbb{R}$ in (1):

$$E(Y) = \int_{\mathbb{R}} y\, dF_Y(y),$$

which establishes all parts of (2).

(iii) This is just a useful application of (ii), stated in a convenient form. In fact, the first part of (3) being (2), for the last equation consider, with $Y = |X|$, and writing P for μ:

$$F_{|X|}(y) = P[|X| < y] = P[-y < X < y] = F_X(y) - F_X(-y+0).$$

Hence (2) becomes

$$\begin{aligned}
E(|X|^p) &= \int_0^\infty y^p dF_{|X|}(y) = \int_0^\infty y^p dF_X(y) - \int_0^\infty y^p dF_X(-y+0) \\
&= \lim_{M\to\infty} \{[y^p(F_X(y) - F_X(-y+0))]_{y=0}^M \\
&\quad - p\int_0^M y^{p-1}(F_X(y) - F_X(-y))\,dy\Big\} \\
&\qquad \text{(by integrating by parts and making a change of variable)} \\
&= \lim_{M\to\infty} p\int_0^M y^{p-1}\{(F_X(M) - F_X(-M)) - (F_X(y) - F_X(-y))\}dy \\
&= p\int_0^\infty y^{p-1}(1 + F_X(-y) - F_X(y))\,dy \quad \text{(by Theorem 1).} \qquad (4)
\end{aligned}$$

This is (3), and the proof is complete. In the last equality, $F_X(-\infty) = 0$ and $F_X(+\infty) = 1$ are substituted.

In the above theorem, it is clear that g can be complex valued since the stated result applies to $g = g_1 + ig_2$, where g_1, g_2 are real measurable functions. We use this fact to illustrate the following important concept on Fourier transforms of real random variables. Indeed if $X : \Omega \to \mathbb{R}$ is any random variable, $g : \mathbb{R} \to \mathbb{C}$ is a Borel function, then $g \circ X : \Omega \to \mathbb{C}$ is a complex random variable. If $g_t(x) = \cos tx + i \sin tx = e^{itx}$, then $g_t : \mathbb{R} \to \mathbb{C}$ is a bounded continuous function and $g_t(X)$ is a bounded complex random variable for all $t \in \mathbb{R}$.

Thus the following definition is meaningful:

$$\phi_X(t) = E(g_t(X)) = E(\cos tX) + iE(\sin tX), t \in \mathbb{R}. \qquad (5)$$

The mapping $\phi_X : \mathbb{R} \to \mathbb{C}$, defined for each random variable X, is called the *characteristic function* of X. It exists without any moment restrictions on X, and $\phi_X(0) = 1$, $|\phi_X(t)| \leq 1$. As an application of the above theorem we have

Proposition 2 *The characteristic function ϕ_X of a random variable X is uniformly continuous on $\mathbb{R}$.*

Proof By Theorem 1 ii, we have the identity

$$\phi_X(t) = E(e^{itX}) = \int_{\mathbb{R}} e^{itx} dF_X(x).$$

Hence given $\varepsilon > 0$, choose $L_\varepsilon > 0$ such that $F_X(L_\varepsilon) - F_X(-L_\varepsilon) \geq 1 - (\varepsilon/4)$. If $t_1 < t_2$, consider, with the elementary properties of Stieltjes integrals,

$$\begin{aligned}
|\phi_X(t_1) - \phi_X(t_2)| &= \left| \int_{\mathbb{R}} (e^{it_1x} - e^{it_2x}) dF_X(x) \right| \\
&\leq \int_{[|x| \leq L_\varepsilon]} \left| ix \int_{t_1}^{t_2} e^{itx}\, dt \right| dF_X(x) \\
&\quad + \left| \int_{[|x| > L_\varepsilon]} (e^{it_1x} - e^{it_2x}) dF_X(x) \right| \\
&\leq \int_{[|x| \leq L_\varepsilon]} |x| \int_{t_1}^{t_2} |e^{itx}| dt dF_X(x) + 2 \int_{[|x| > L_\varepsilon]} dF_X(x) \\
&\leq L_\varepsilon(t_2 - t_1) \int_{[|x| \leq L_\varepsilon]} dF_X(x) + 2[1 - F_X(L_\varepsilon) + F_X(-L_\varepsilon) \\
&< L_\varepsilon(t_2 - t_1) \cdot 1 + 2 \cdot \frac{\varepsilon}{4}. \qquad (6)
\end{aligned}$$

If $\delta_\varepsilon = \varepsilon/(2L_\varepsilon)$ and $|t_2 - t_1| < \delta_\varepsilon$, then (6) implies

$$|\phi_X(t_1) - \phi_X(t_2)| < \frac{\varepsilon}{2} + \frac{\varepsilon}{2} = \varepsilon.$$

This completes the proof.

This result shows that many properties of random variables on abstract probability spaces can be studied through their image laws and their characteristic functions with nice continuity properties. We make a deeper study of this aspect of the subject in Chapter 4. First, it is necessary to introduce several concepts of probability theory and establish its individuality as a separate discipline with its own innate beauty and elegance. This we do in part in the next two chapters, and the full story emerges as the subject develops, with its manifold applications, reaching most areas of scientific significance.

Before closing this chapter we present a few results on uniform integrability of sets of random variables. This concept is of importance in applications where an integrable dominating function is not available to start with. Let us state the concept.

Definition 3 An arbitrary collection $\{X_t, t \in T\}$ of r.v.s on a probability space (Ω, Σ, P) is said to be *uniformly integrable* if (i) $E(|X_t|) \leq k_0 < \infty$, $t \in T$, and (ii) $\lim_{P(A) \to 0} \int_A |X_t| dP = 0$ uniformly in $t \in T$.

The earliest occasion on which the reader may have encountered this concept is perhaps in studying real analysis, in the form of the Vitali theorem, which for *finite* measure spaces is a generalization of the dominated convergence criterion (Theorem 2.2). Let us recall this result.

Theorem 4 **(Vitali)** *Let $X_1, X_2, \ldots$ be a sequence of random variables on a probability space (Ω, Σ, P) such that $X_n \to X$ a.e. (or only in measure). If $\{X_n, n \geq 1\}$ is a uniformly integrable set, then we have*

$$\lim_{n\to\infty} \int_\Omega X_n dP = \int_\Omega X dP.$$

Actually the conclusion holds if only $E(|X_n|) < \infty, n \geq 1$, and (ii) of Definition 3 is satisfied for $\{X_n, n \geq 1\}$.

Note that if $|X_n| \leq Y$ and Y is integrable, then $\{X_n, n \geq 1\}$ is trivially uniformly integrable. The point of the above result is that there may be no such dominating function Y. Thus it is useful to have a characterization of this important concept, which is given by the next result. It contains the classical all-important *de La Vallée Poussin criterion* obtained in about 1915. It was brought to light for probabilistic applications by Meyer (1966).

Theorem 5 *Let $\mathcal{K} = \{X_t, t \in T\}$ be a set of integrable random variables on a probability space. Then the following conditions are equivalent* [(i)$\Leftrightarrow$ (iii) is due to de la Vallee Poussin]:

(i) *$\mathcal{K}$ is uniformly integrable.*

(ii)

$$\lim_{a\to\infty} \int_{[|X_t|>a]} |X_t| dP = 0 \text{ uniformly in } t \in T. \tag{7}$$

(iii) *There exists a convex function $\phi : \mathbb{R} \to \mathbb{R}^+$, $\phi(0) = 0$, $\phi(-x) = \phi(x)$, and $\phi(x)/x \nearrow \infty$ as $x \nearrow \infty$, such that* $\sup_{t\in T} E(\phi(X_t)) < \infty$.

Proof (i) $\Rightarrow$ (ii) By Proposition 3.9 (Markov's inequality) we have

$$P[|X_t| \geq a] \leq \frac{1}{a} E(|X_t|) \leq \frac{k_0}{a} \to 0 \text{ as } a \to \infty \tag{8}$$

uniformly in $t \in T$. Thus by the second condition of Definition 3, given $\varepsilon > 0$, there is a $\delta_\varepsilon > 0$ such that for any $A \in \Sigma, P(A) < \delta_\varepsilon \Rightarrow \int_A |X_t| dP < \varepsilon$ uniformly in $t \in T$. Let $A_t = [|X_t| \geq a]$ and choose $a > k_0/\delta_\varepsilon$ so that $P(A_t) < \delta_\varepsilon$ by (8), and hence $\int_{A_t} |X_t| dP < \varepsilon$, whatever $t \in T$ is. This is (7), and (ii) holds.

(ii) $\Rightarrow$ (iii) Here we need to construct explicitly a convex function ϕ of the desired kind. Let $0 \leq a_n < a_{n+1} \nearrow \infty$ be a sequence of numbers such that by (7) we have

$$\sup_t \int_{[|X_t|>a_n]} |X_t|\, dP < 2^{-n-1}, \quad n \geq 1. \tag{9}$$

The sequence $\{a_n, n \geq 1\}$ is determined by the set $\mathcal{K}$ but not the individual X_t. Let $N(n) =$ the number of a_k in $[n, n+1)$, $= 0$ if there is no a_k in this set, and put $\xi(n) = \sum_{k=0}^{n} N(k)$, with $N(0) = 0$. Then $\xi(n) \nearrow \infty$. Define

$$\phi(x) = \int_0^{|x|} \xi(t)\, dt, \; x \in \mathbb{R},$$

where $\xi(t)$ is a constant on $[k, k+1)$ and increases only by jumps. Clearly $\phi(\cdot)$ is convex, $\phi(-x) = \phi(x)$, $\phi(0) = 0$, $\phi(x)/x \geq \xi(k)((x-k)/x) \uparrow \infty$, for $k < x$ and $x, k \nearrow \infty$. We claim that this function satisfies the requirements of (iii).

Indeed, let us calculate $E(\phi(X_t))$. We have

$$\begin{aligned} E(\phi(X_t)) &= \sum_{k=1}^{\infty} \int_{[k-1\leq|X_t|<k]} \phi(X_t)\, dP \leq \sum_{k=1}^{\infty} \phi(k) P[k-1 \leq |X_t| < k] \\ &\leq \sum_{k=0}^{\infty} (\phi(k+1) - \phi(k)) P[|X_t| \geq k] \;\; [\text{ since } \phi(0) = 0] \\ &\leq \sum_{k=0}^{\infty} \xi(k) P[|X_t| \geq k]. \end{aligned} \tag{10}$$

However,

$$\begin{aligned} \int_{[|X_t|\geq a_n]} |X_t|\, dP &= \sum_{k=a_n}^{\infty} \int_{[k\leq|X_t|<k+1]} |X_t|\, dP \geq \sum_{k=a_n}^{\infty} kP[k \leq |X_t| < k+1] \\ &\geq \sum_{k=a_n}^{\infty} P[|X_t| > k]. \end{aligned}$$

Summing over n, we get with (9)

$$1 \geq \sum_{n=1}^{\infty} \int_{[|X_t|\geq a_n]} |X_t| dP \geq \sum_{n=1}^{\infty} \sum_{k=a_n}^{\infty} P[|X_t| > k] = \sum_{k=1}^{\infty} \xi(k) P[|X_t| > k]. \tag{11}$$

Thus (10) and (11) imply $\sup_t E(\phi(X_t)) \leq 1$, and (iii) follows.

(iii) $\Rightarrow$ (i) is a consequence of the Hölder inequality for Orlicz spaces since $\phi(\cdot)$ can be assumed here to be the so-called Young function. The proof is similar to the case in which $\phi(x) = |x|^p, p > 1$. By the support line property, the boundedness of $E(\phi(X_t)) \leq k < \infty$ implies that of $E(|X_t|) \leq k_1 < \infty$.

The second condition follows from $[q = p/(p-1)]$

$$\int_A |X_t| dP = \int_\Omega \chi_A |X_t| dP \leq ||X_t||_p \cdot ||\chi_A||_q \leq k^{1/p} \cdot (P(A))^{1/q} \to 0.$$

as $P(A) \to 0$. The general Young function has the same argument. However, without using the Orlicz space theory, we follow a little longer but an alternative and more elementary route, by proving (iii) $\Rightarrow$ (ii) $\Rightarrow$ (i) now.

Thus let (iii) be true. Then set $\tilde{k} = \sup_t E(\phi(X_t)) < \infty$. Given $\varepsilon > 0$, let $0 < b_\varepsilon = \tilde{k}/\varepsilon$ and choose $a = a_\varepsilon$ such that $|x| \geq a_\varepsilon \Rightarrow \phi(x) \geq |x| b_\varepsilon$, which is possible since $\phi(x)/x \nearrow \infty$ as $x \nearrow \infty$. Thus $\omega \in [|X_t| \geq a_\varepsilon] \Rightarrow b_\varepsilon |X_t|(\omega) \leq \phi(X_t(\omega))$, and

$$\int_{[|X_t|\geq a_\varepsilon]} |X_t|\, dP \leq \frac{1}{b_\varepsilon} \int_{[|X_t|\geq a_\varepsilon]} \phi(|X_t|)\, dP \leq \frac{\tilde{k}}{b_\varepsilon} = \varepsilon, \ t \in T.$$

This clearly implies (ii).

Finally, (ii) $\Rightarrow$ (i). It is evident that (7) implies that if $\varepsilon = 1$, then there is $a_1 > 0$ such that

$$\int_\Omega |X_t|\, dP = \int_{[|X_t|\leq a_1]} |X_t|\, dP + \int_{[|X_t|>a_1]} |X_t|\, dP \leq a_1 \cdot 1 + 1, \ t \in T.$$

So there is a $\bar{k}(\geq 1 + a_1) < \infty$ such that $\sup_t E(|X_t|) \leq \bar{k} < \infty$. To verify the second condition of Definition 3, we have for $A \in \Sigma$

$$\int_A |X_t|\, dP = \int_{A\cap[|X_t|>a]} |X_t|\, dP + \int_{A\cap[|X_t\leq a]} |X_t|\, dP \quad (a > 0)$$

$$\leq \int_{[|X_t|>a]} |X_t|\, dP + aP(A). \tag{12}$$

Given $\varepsilon > 0$, choose $a = a_\varepsilon > 0$, so that by (ii) the first integral is $< \varepsilon$ uniformly in t. For this a_ε, (12) becomes

$$\lim_{P(A)\to 0} \int_A |X_t|\, dP < \varepsilon \text{ for all } t \in T. \tag{13}$$

Since $\varepsilon > 0$ is arbitrary this integral is zero, and (i) holds. This completes the demonstration.

The following is an interesting supplement to the above, called *Scheffé's lemma*, it is proved for probability distributions on the line. We present it in a slightly more general form.

Proposition 6 (Scheffé) *Let $X, X_n \geq 0$ be integrable random variables on a probability space (Ω, Σ, P) and $X_n \to X$ a.e. (or in measure). Then $E(X_n) \to E(X)$ as $n \to \infty$ iff $\{X_n, n \geq 1\}$ is uniformly integrable, which is equivalent to saying that* $\lim_{n\to\infty} E(|X_n - X|) = 0$.

Proof If $\{X_n, n \geq 1\}$ is uniformly integrable, then $E(X_n) \to E(X)$ by the Vitali theorem (Theorem 4) even without positivity. Since $\{|X_n - X|, n \geq 1\}$ is again uniformly integrable and $|X_n - X| \to 0$ a.e. (or in measure), the last statement follows from the above theorem. Thus it is the converse which is of interest, and it needs the additional hypothesis.

Thus let $X, X_n \geq 0$ and be integrable. Then the equation

$$X_n + X = \min(X_n, X) + \max(X_n, X) \tag{14}$$

is employed in the argument. Since $\min(X_n, X) \leq X$, and $\min(X_n, X) \to X$ a.e., the dominated convergence theorem implies $E(\min(X_n, X)) \to E(X)$ as $n \to \infty$. Hence taking expectations on both sides of (14) and letting $n \to \infty$, we get $E(\max(X_n, X)) \to E(X)$ as well. On the other hand,

$$|X_n - X| = \max(X_n, X) - \min(X_n, X). \tag{15}$$

Applying the operator E to both sides of (15), and using the preceding facts on the limits to the right-side expressions, we get $E(|X_n - X|) \to 0$. This implies for each $\varepsilon > 0$ that there is an n_ε such that for all $n > n_\varepsilon$ and all $A \in \Sigma$,

$$\left| \int_A X_n \, dP - \int_A X \, dP \right| \leq \int_\Omega |X_n - X| \, dP < \varepsilon.$$

It follows that, because each finite set of integrable random variables is always uniformly integrable,

$$\lim_{P(A) \to 0} \int_A X_n \, dP \leq \lim_{P(A) \to 0} \int_A X \, dP + \varepsilon = \varepsilon \tag{16}$$

uniformly in n. Thus, because $\varepsilon > 0$ is arbitrary, $\{X_n, n \geq 1\}$ is uniformly integrable, as asserted.

In Scheffè's original version, it was assumed that $dP = f d\mu$, where μ is a σ- finite measure. Thus f is called the *density* of P relative to μ. If $g_n = f \cdot X_n \geq 0$, then $\int_\Omega g_n d\mu = \int_\Omega X_n \cdot f d\mu = \int_\Omega X_n dP$ is taken as unity, so that g_n itself is a probability density relative to μ. In this form $\{g_n, n \geq 1\}$ is assumed to satisfy $1 = \int_\Omega g_n d\mu \to \int_\Omega g d\mu = 1$ and $g_n \to g$ a.e. (or in measure). It is clear that the preceding result is another form of this result, and both are essentially the same statements. These results can be further generalized. (See, e.g., Problems 7–9.)

One denotes by $\mathcal{L}^p(\Omega, \Sigma, P)$, or $\mathcal{L}^p$, the class of all pth-power integrable random variables on (Ω, Σ, P). By the Hölder and Minkowski inequalities, it follows that $\mathcal{L}^p$ is a vector space, $p \geq 1$, over the scalars. Thus $f \in \mathcal{L}^p$ iff $||f||_p = [E(|f|^p)]^{1/p} < \infty$, and $|| \cdot ||_p$ is the p-norm, i.e., $||f||_p = 0$ iff $f = 0$ a.e., $||af + g||_p \leq |a| ||f||_p + ||g||_p$, $a \in \mathbb{R}$ (or $a \in \mathbb{C}$). When $||f - g||_p = 0$, so that $f = g$ a.e., one identifies the equivalence classes ($f \sim g$ iff $f = g$ a.e.). Then the quotient $L^p = \mathcal{L}^p / \sim$ is a normed linear space. Moreover, if $\{f_n, n \geq 1\} \subset \mathcal{L}^p$, $||f_m - f_n||_p \to 0$, as $n, m \to \infty$ then it is not hard to see that there is a P-unique $f \in \mathcal{L}^p$ such that $\|f - f_n\|_p \to 0$, so that $\mathcal{L}^p$ is complete. The space of equivalence classes $(L^p, \|\cdot\|_p)_{p\ge 1}$ is thus a complete normed linear (or Banach) space, called the *Lebesgue space*, for $1 \le p \le \infty$. It is customary to call the elements of L^p functions when a member of its equivalence class is meant. We also follow this custom.

Exercises

1. Let Ω be a nonempty set and $A \subset \Omega$. Then χ_A, called the *indicator* (or "characteristic," in older terminology) *function*, which is 1 on A, 0 on $\Omega - A = A^c$, is useful in some calculations on set operations. We illustrate its uses by this problem.

(a) If $A_i \subset \Omega, i = 1, 2$, and $A_1 \Delta A_2$ is the symmetric difference, show that $\chi_{A_1 \Delta A_2} = |\chi_{A_1} - \chi_{A_2}|$.

(b) If $A_n \subset \Omega, n = 1, 2, \ldots$, is a sequence, $A = \limsup_n A_n$ ($=$ the set of points that belong to infinitely many $A_n, = \cap_{k=1}^{\infty} \cup_{n\ge k} A_n$) and $B = \liminf_n A_n$ ($=$ the set of points that belong to all but finitely many $A_n, = \cup_{k=1}^{\infty} \cap_{n\ge k} A_n$), show that $\chi_A = \limsup_n \chi_{A_n}, \chi_B = \liminf_n A_n$, and $A = B$ (this common set is called the limit and denoted $\lim_n A_n$) iff $\chi_A = \lim_n \chi_{A_n}$.

(c) (E. Bishop) If $A_n \subset \Omega, n = 1, 2, \ldots$, define $C_1 = A_1, C_2 = C_1 \Delta A_2, \ldots$, $C_n = C_{n-1} \Delta A_n$. Show that $\lim_n C_n = C$ exists [in the sense of (b) above] iff $\lim_n A_n = \emptyset$. [Hint: Use the indicator functions and the results of (a) and (b). Verify that $|\chi_{C_{n+1}} - \chi_{C_n}| = \chi_{A_{n+1}}$.]

(d) If (Ω, Σ, P) is a probability space, and $\{A_n, n \ge 1\} \subset \Sigma$, suppose that $\lim_n A_n$ exists in the sense of (b). Then show that $\lim_n P(A_n)$ exists and equals $P(\lim_n A_n)$.

2. (a) Let (Ω, Σ, P) be a probability space and $\{A_i, 1 \le i \le n\} \subset \Sigma$, $n \ge 2$. Prove (**Poincare's formula**) that

$$P\left(\bigcup_{i=1}^{n} A_i\right) = \sum_{i=1}^{n} P(A_i) - \sum_{1\le i<j\le n} P(A_i \cap A_j) + \sum_{1\le i<j<k\le n} P(A_i \cap A_j \cap A_k)$$
$$- \ldots + (-1)^{n-1} P\left(\bigcap_{i=1}^{n} A_i\right) \ge \sum_{i=1}^{n} P(A_i) - \sum_{1\le i<j\le n} P(A_i \cap A_j).$$

Thus the first two terms usually underestimate the probability of $\cup_{i=1}^{n} A_i$.

(b) Let $(\Omega_i, \mathcal{A}_i), i = 0, 1, \ldots, n$, be measurable spaces, $f : \Omega_0 \to \times_{i=1}^{n} \Omega_i$ be a mapping. Establish that f is measurable iff each component of $f = (f_1, \ldots, f_n)$ is. [*Hint*: Verify $f^{-1}(\sigma(\mathcal{C})) = \sigma(f^{-1}(\mathcal{C}))$ for a collection $\mathcal{C}$ of sets.]

3. (a) Let $\{X_n, n \geq 1\}$ be a sequence of random variables on a probability space (Ω, Σ, P). Show that $X_n \to X$, a random variable, in probability iff

$$E\left(\frac{|X_n - X|}{1 + |X_n - X|}\right) \to 0 \text{ as } n \to \infty.$$

(b) If X, Y are any pair of random variables, and $\mathcal{L}^0$ is the set of all random variables, define

$$d(X, Y) = E\left(\frac{|X - Y|}{1 + |X - Y|}\right)$$

and verify that $d(\cdot, \cdot)$ is a metric on $\mathcal{L}^0$ and that $\mathcal{L}^0$ is an algebra of random variables.

(c) If $X \sim Y$ denotes $X = Y$ a.e., and $L^0 = \mathcal{L}^0/\sim$, show that $(L^0, d(\cdot, \cdot))$ is a complete linear metric space, in the sense that it is a vector space and each Cauchy sequence for $d(\cdot, \cdot)$ converges in L^0.

(d) Prove that $(L^p, \|\cdot\|_p)$ introduced in the last paragraph of Section 1.4 is complete.

4. Consider the probability space of Example 2.6. If $f : \Omega \to \mathbb{R}$ is the random variable defined there, verify that $E(f) = 1/p$ and $\sigma^2(f) = (1-p)/p^2$. In particular, if $p = 1/2$, then $E(f) = 2$, $\sigma^2(f) = 2$, so that the expected number of tosses of a fair coin to get the first head is 2 but the variance is also 2, which is "large."

5. (a) Let X be an r.v. on (Ω, Σ, P). Prove that $E(X)$ exists iff

$$\sum_{k=1}^{\infty} P[|X| \geq ak] < \infty$$

for some $a > 0$, and hence also for all $a > 0$.

(b) If $E(X)$ exists, show that it can be evaluated as

$$E(X) = \int_0^{\infty} P[X > x]dx - \int_0^{\infty} P[X < -x]dx.$$

[See Theorem 4.1iii.]

6. (a) Let X be a bounded random variable on (Ω, Σ, P). Then for any $\varepsilon > 0$ and any $r > 0$, verify that $E(|X|^r) \leq \varepsilon^r + a^r P[|X| \geq \varepsilon]$, where a is the bound on $|X|$. In particular, if $a = 1$, we have $E(X^2) - \varepsilon^2 \leq P[|X| \geq \varepsilon]$.

(b) Obtain an improved version of the one-sided Čebyšev's inequality as follows:

$$P[X > E(X) + \varepsilon] \leq \frac{Var(X)}{\varepsilon^2 + Var(X)}.$$

[*Hint*: Let $Y = X - E(X)$ and $\sigma^2 = \text{Var } X$. Set

$$f(x) = \left(\frac{\varepsilon x + \sigma^2}{\varepsilon^2 + \sigma^2}\right)^2.$$

Then if $B = [Y > \varepsilon]$, verify that $E(f(Y)) \geq P(B)$ and $E(f(Y)) = \frac{\sigma^2}{\varepsilon^2+\sigma^2}$.]

7. Let $\{X_n, n \geq 1\}$ be a sequence of r.v.s on (Ω, Σ, P) such that $X_n \to X$ a.e., where X is an r.v. If $0 < p < \infty$ and $E(|X_n|^p) < \infty, n \geq 1$, then $\{|X_n|^p, n \geq 1\}$ is uniformly integrable iff $E(|X_n - X|^p) \to 0$ as $n \to \infty$. The same argument applies to a more general situation as follows. Suppose $\phi : \mathbb{R}^+ \to \mathbb{R}^+$ is a symmetric function, $\phi(0) = 0$, and either ϕ is continuous concave increasing function on $\mathbb{R}^+$ or is a convex function satisfying $\phi(2x) \leq c\phi(x), x \geq 0$, for some $0 < c < \infty$. If $E(\phi(X_n)) \leq k < \infty$ and $E(\phi(X_n)) \to E(\phi(X))$,then $E(\phi(X_n - X)) \to 0$ as $n \to \infty$ and $\{\phi(X_n), n \geq 1\}$ is uniformly integrable. [*Hint*: Observe that there is a constant $0 < \tilde{c} < \infty$ such that in both the above convex and concave cases, $\phi(x+y) \leq \tilde{c}[\phi(x)+\phi(y)], x, y \in \mathbb{R}$. Hence $\tilde{c}[\phi(|X_n|)+\phi(|X|)] - \phi(|X_n - X|) \geq 0$ a.e.]

8. (Doob) Let (Ω, Σ, P) be a probability space, $\Sigma \supset \mathcal{F}_n \supset \mathcal{F}_{n+1}$ be σ-subalgebras, and $X_n : \Omega \to \mathbb{R}$ be $\mathcal{F}_n$-measurable (hence also measurable for Σ). Suppose that $\nu_n(A) = \int_A X_n dP, A \in \mathcal{F}_n$ satisfies for each $n \geq 1, \nu_n(A) \geq \nu_{n+1}(A), A \in \mathcal{F}_n$. Such sequences exist, as we shall see in Chapter 3. (A trivial example satisfying the above conditions is the following: $\mathcal{F}_{n+1} = \mathcal{F}_n = \Sigma$ all n, and $X_n \geq X_{n+1}$ a.e. for all $n \geq 1$.) Show that $\{X_n, n \geq 1\}$ is uniformly integrable iff (*) $\lim_n \nu_n(\Omega) > -\infty$. In the special example of a decreasing sequence for which (*) holds, deduce that there is a random variable X such that $E(|X_n - X|) \to 0$ as $n \to \infty$. [Hint: If $A_n^\lambda = [|X_n| > \lambda]$, verify that $P(A_n^\lambda) \to 0$ as $\lambda \uparrow \infty$ uniformly in n, after noting that $\int_{A_n} |X_n| dP \leq \nu_1(\Omega) + 2\nu_m(\Omega)$ for all $1 \leq n \leq m$. Finally, verify that

$$\int_{A_n} |X_n| dP \leq \nu(\Omega) + \int_{A_n^\lambda} |X_m| dP - \lim_k \nu_k(\Omega), \text{ for } m \leq n.]$$

9. [This is an advanced problem.] (a) Let (Ω, Σ, μ) be a measure space, $X_n : \Omega \to \mathbb{R}$, $n \geq 1$, be random variables such that (i) $X_n \to X$ a.e., $n \to \infty$, and (ii) $X_n = Y_n + Z_n, n \geq 1$, where the random variables Y_n, Z_n satisfy $(\alpha) Z_n \to Z$ a.e. and $\int_\Omega Z_n d\mu \to \int_\Omega Z d\mu \in \mathbb{R}, n \to \infty$, $(\beta) \lim_{n\to\infty} \int_A Y_n d\mu$ exists, $A \in \Sigma$, and (iii)

$$\lim_{m\to\infty} \lim_{n\to\infty} \int_{A_m} Y_n d\mu = 0$$

for any $A_m \downarrow \emptyset, A_m \in \Sigma$. Then $\lim_{n\to\infty} \int_\Omega X_n d\mu = \int_\Omega X d\mu$. If $\mu(\Omega) < \infty$, (iii) may be omitted here. [*Hints*: If $\lambda : A \mapsto \lim_{n\to\infty} \int_A Y_n d\mu$, then $\lambda : \Sigma \to \mathbb{R}$ is additive and vanishes on μ-null sets. (β) and (iii) $\Rightarrow \lambda$ is also σ-additive, so that $\lambda(A) = \int_A Y' d\mu$ for a μ-unique r.v. Y', since the Y_n, being integrable, vanish outside a fixed σ-finite set, and μ may thus be assumed σ-finite. It may be noted that (iii) is a consequence of (β) if $\mu(\Omega) < \infty$. Next, (β) also implies

$$\int_A Y_n d\mu \to \int_A Y' d\mu, A \in \Sigma,$$

so that it is "weakly convergent" to Y'. Let $F \in \Sigma, \mu(F) < \infty$. Then by the Vitali-Hahn-Saks theorem (cf. Dunford-Schwartz, III.7.2),

$$\lim_{\mu(E)\to 0} \int_E Y_n \cdot \chi_F d\mu = 0$$

uniformly in n. Also $Y_n\chi_F = (X_n - Z_n)\chi_F \to (X - Z)\chi_F$ a.e. Let $Y = X - Z$. These two imply $\int_\Omega |Y_n - Y|\chi_F d\mu \to 0$. Deduce that $Y = Y'$ a.e., and then $Y_n\chi_F \to Y\chi_F = Y'\chi_F$ in measure on each $F \in \Sigma$, with $\mu(F) < \infty$. Hence by another theorem in Dunford-Schwartz (III.8.12), $\int_\Omega |Y_n - Y| d\mu \to 0$. Thus using ($\alpha$), this implies the result. The difficulty is that the hypothesis is weaker than the dominated or Vitali convergence theorems, and the $X_n, n \geq 1$, are not uniformly integrable. The result can be extended if the X_n are vector valued.]

(b) The following example shows how the hypotheses of the above part can be specialized. Let X_n, g_n, h_n be random variables such that (i)$X_n \to X$ a.e., $g_n \to g$ a.e., and $h_n \to h$ a.e. as $n \to \infty$, (ii) $g_n \leq X_n \leq h_n$, $n \geq 1$, and (iii) $\int_\Omega g_n d\mu \to \int_\Omega g d\mu \in \mathbb{R}$, $\int_\Omega h_n d\mu \to \int_\Omega h d\mu \in \mathbb{R}, n \to \infty$. Then $\lim_{n\to\infty} \int_\Omega X_n d\mu = \int_\Omega X d\mu \in \mathbb{R}$. [Let $Y_n = X_n - g_n$, $Z_n = g_n$. Then (i) and (ii α) of (a) hold.

Now $0 \leq Y_n \leq h_n - g_n$ and $\int_\Omega (h_n - g_n) d\mu \to \int_\Omega (h - g) d\mu$ by hypothesis. Since $h_n - g_n \geq 0$ and we may assume that these are finite after some n, let us take n = 1 for convenience. As shown in Proposition 4.6, this implies the uniform integrability of $\{h_n - g_n, n \geq 1\}$, and (ii β) and (iii) will hold, since $\int_\Omega |(h_n - g_n) - (h - g)| d\mu \to 0$ is then true. Note that no order relation of the range is involved in (a), while this is crucial in the present formulation.] Observe that if $g_n \leq 0 \leq h_n$, we may take $g_n = -h_n$, replacing h_n by $\max(h_n, -g_n)$ if necessary, so that $|X_n| \leq h_n$ and $\int_\Omega h_n d\mu \to \int_\Omega h d\mu$ implies the h_n sequence, and hence the X_n sequence, is uniformly integrable as in Proposition 4.6. The result of (b) (proved differently) is due to John W. Pratt. The problem is presented here to show how uniform integrability can appear in different forms. The latter are neither more natural nor elegant than the ones usually given.

10. This is a slight extension of the Fubini-Stone theorem. Let (Ω_i, Σ_i), $i = 1, 2$, be two measurable spaces and $\Omega = \Omega_1 \times \Omega_2$, $\Sigma = \Sigma_1 \otimes \Sigma_2$ their products. Let $P(\cdot, \cdot) : \Omega_1 \times \Sigma_2 \to \mathbb{R}^+$ be such that $P(\omega_1, \cdot) : \Sigma_2 \to \mathbb{R}^+$ is a probability, $\omega_1 \in \Omega_1$ and $P(\cdot, A) : \Omega_1 \to \mathbb{R}^+$ be a Σ_1-measurable function for each $A \in \Sigma_2$. Prove that the mapping $Q : (A, B) \mapsto \int_A P(\omega_1, B)\mu(d\omega_1)$ for any probability $\mu : \Sigma_1 \to \mathbb{R}^+$ uniquely defines a probability measure on

(Ω, Σ), sometimes called a *mixture* relative to μ, and if $X : \Omega \to \mathbb{R}^+$ is any random variable, then the mapping $\omega_1 \mapsto \int_{\omega_2} X(\omega_1, \omega_2)P(\omega_1, d\omega_2)$ is $Q(\cdot, \Omega_2)$-measurable and we have the equation

$$\int_\Omega f(\omega)Q(d\omega) = \int_{\Omega_1}\int_{\Omega_2} f(\omega_1, \omega_2)P(\omega_1, d\omega_2)\mu(d\omega_1).$$

[If $P(\omega_1.\cdot)$ is independent of ω_1 then this reduces to Theorem 3.11(ii) and the proof is a modification of that result.]

11. (Skorokhod) For a pair of mixtures as in the preceding problem, the Radon-Nikodým theorem can be extended; this is of interest in probabilistic and other applications. Let $(\Omega_i, \Sigma_i), i = 1, 2$, be two measurable spaces and $P_i : \Omega_1 \times \Sigma_2 \to \mathbb{R}^+, \mu_i : \Sigma_1 \to \mathbb{R}^+$, and $Q_i : (A, B) \mapsto \int_A P_i(\omega_1, B)\mu_i(d\omega_1), i = 1, 2$, be defined as in the above problem satisfying the same conditions there. Then $Q_1 \ll Q_2$ on (Ω, Σ), the product measurable space iff $\mu_1 \ll \mu_2$ and $P_1(\omega_1, \cdot) \ll P_2(w_1, \cdot)$ for a.a.(ω_1). When the hypothesis holds (i.e., $Q_1 \ll Q_2$), deduce that

$$\frac{dQ_1}{dQ_2}(\omega_1, \omega_2) = \frac{dP(\omega_1, \cdot)}{dP_2(\omega_1, \cdot)}(\omega_2)\frac{d\mu_1}{d\mu_2}(\omega_1) \quad \text{a.e. } [\mu_1].$$

[*Hints*: If $Q_1 \ll Q_2$, then observe that, by considering the *marginal measures* $Q_i(\cdot, \Omega_2)$, we also have $\mu_1 \ll \mu_2$. Next note that for a.a.(ω_1),

$$\Big\{A \in \Sigma_2 : P_1(\omega_1, A) = \int_A \left[\frac{dQ_1}{dQ_2}(\omega_1, \omega_2) / \int_{\Omega_2} \frac{dQ_1}{dQ_2}(\omega_1, \omega_2)P_2(\omega_1, d\omega_2)\right] P_2(\omega_1, d\omega_2)\Big\}$$

is a monotone class and an algebra. Deduce that $P_1(\omega_1, \cdot) \ll P_2(\omega_1, \cdot)$, a.a,$(\omega_1)$. The converse is simpler, and then the above formula follows. Only a careful application of the "chain rule" is needed. Here the proof can be simplified and the application of monotone class theorem avoided if Σ_2 is assumed countably generated as was originally done.]

Chapter 2

Independence and Strong Convergence

This chapter is devoted to the fundamental concept of independence and to several results based on it, including the Kolmogorov strong laws and his three series theorem. Some applications to empiric distributions, densities, queueing sequences and random walk are also given. A number of important results, included in the problems section, indicate the profound impact of the concept of independence on the subject. All these facts provide deep motivation for further study and development of probability theory.

2.1 Independence

If A and B are two events of a probability space (Ω, Σ, P), it is natural to say that A is independent of B whenever the occurrence or nonoccurrence of A has no influence on the occurrence or nonoccurrence of B. Consequently the uncertainty about joint occurrence of both A and B must be higher than either of the individual events. This means that the probability of a joint occurrence of A and B should be "much smaller" than either of the individual probabilities. This intuitive feeling can be formalized mathematically by the equation

$$P(A \cap B) = P(A)P(B)$$

for a pair of events A, B. How should intuition translate for three events A, B, C if every pair among them is independent? The following ancient example, due to S. Bernstein, shows that, for a satisfactory mathematical abstraction, more care is necessary. Thus if $\Omega = \{\omega_1, \omega_2, \omega_3, \omega_4\}$, $\Sigma = \mathcal{P}(\Omega)$, the power set, let each point carry the same weight, so that

$$P(\{\omega_i\}) = \frac{1}{4}, i = 1, ..., 4.$$

Let $A = \{\omega_1, \omega_2\}$, $B = \{\omega_1, \omega_3\}$, and $C = \{\omega_4, \omega_1\}$. Then clearly $P(A \cap B) = P(A)P(B) = \frac{1}{4}$, $P(B \cap C) = P(B)P(C) = \frac{1}{4}$, and $P(C \cap A) = P(C)P(A) = \frac{1}{4}$.

But $P(A \cap B \cap C) = \frac{1}{4}$, and $P(A)P(B)P(C) = \frac{1}{8}$. Thus A, B, C are not independent. Also $A, (B \cap C)$ are not independent, and similarly $B, (C \cap A)$ and $C, (A \cap B)$ are not independent.

These considerations lead us to introduce the precise concept of mutual independence of a collection of events by not pairwise but by systems of equations so that the above anomaly cannot occur.

Definition 1 Let (Ω, Σ, P) be a probability space and $\{A_i, i \in I\} \subset \mathcal{P}(\Omega)$ be a family of events. They are said to be *pairwise independent* if for each distinct i, j in I we have $P(A_i \cap A_j) = P(A_i)P(A_j)$. If $A_{i_1}, \ldots, A_{i_n}$ are n (distinct) events, $n \geq 2$, then they are *mutually independent* if

$$P\left(\bigcap_{k=1}^{m} A_{i_k}\right) = \prod_{k=1}^{m} P(A_{i_k}) \tag{1}$$

holds simultaneously for each $m = 2, 3, \ldots, n$. The whole class $\{A_i, i \in I\}$ is said to be mutually independent if each finite subcollection is mutually independent in the above sense, i.e., equations (1) hold for each $n \geq 2$. Similarly if $\{\mathcal{A}_i, i \in I\}$ is a collection of families of events from Σ then they are mutually independent if for each n, $A_{i_k} \in \mathcal{A}_{i_k}$ we have the set of equations (1) holding for A_{i_k}, $k = 1, \ldots, m, 1 < m \leq n$. Thus if $A_i \in \mathcal{A}_i$ then $\{A_i, i \in I\}$ is a mutually independent family. [Following custom, we usually omit the word "mutually".]

It is clear that the (mutual) independence concept is given by a system of equations (1) which can be arbitrarily large depending on the richness of Σ. Indeed for each n events, (1) is a set of $2^n - n - 1$ equations, whereas the pairwise case needs only $\binom{n}{2}$ equations. Similarly "m-wise" independence has $\binom{n}{m}$ equations, and it does not imply other independences if $2 \leq m < n$ is a fixed number m. It is the strength of the (mutual) concept that allows all $n \geq 2$. This is the mathematical abstraction of the intuitive feeling of independence that experience has shown to be the best possible one. It seems to give a satisfactory approximation to the heuristic idea of independence in the physical world. In addition, this mathematical formulation has been found successful in applications to such areas as number theory, and Fourier analysis. **The notion of independence is fundamental to probability theory and distinguishes it from measure theory.** The concept translates itself to random variables in the following form.

Definition 2 Let (Ω, Σ, P) be a probability space and $\{X_i, i \in I\}$ be abstract random variables on Ω into a measurable space $(S, \mathcal{A})$. Then they are said to be *mutually independent* if the class $\{\mathcal{B}, i \in I\}$ of σ-algebras in Σ is mutually independent in the sense of Definition 1, where $\mathcal{B}_i = X_i^{-1}(\mathcal{A})$, the σ-algebra generated by $X_i, i \in I$. Pairwise independence is defined similarly.

Taking $S = \mathbb{R}(or\mathbb{R}^n)$ and $\mathcal{A}$ as its Borel σ-algebra, one gets the corresponding concept for real (or vector) random families.

It is perhaps appropriate at this place to observe that many such (independent) families of events or random variables on an (Ω, Σ, P) need not exist if (Ω, Σ) is not rich enough. Since $\emptyset$ and Ω are clearly independent of each event $A \in \Omega$, the set of equations (1) is non vacuous. Consider the trivial example $\Omega = \{0,1\}$, $\Sigma = \mathcal{P}(\Omega) = \{\emptyset, \{0\}, \{1\}, \Omega\}$, $P(\{0\}) = p = 1 - P(\{1\})$, $0 < p < 1$. Then, omitting the $\emptyset$, Ω, there are no other independent events, and if $X_i : \Omega \to \mathbb{R}$, $i = 1, 2$, defined as $X_1(0) = 1 = X_2(1)$ and $X_1(1) = 2 = X_2(0)$, then X_1, X_2 are distinct random variables, but they are not independent. Any other random variables defined on Ω can be obtained as functions of these two, and it is easily seen that there are no nonconstant independent random variables on this Ω. Thus (Ω, Σ, P) is not rich enough to support nontrivial (i.e., nonconstant) independent random variables. We show later that a probability space can be enlarged to have more sets, so that one can always assume the existence of enough independent families of events or random variables. We now consider some of the profound consequences of this mathematical formalization of the natural concept of mutual independence. It may be noted that the latter is also termed *statistical* (*stochastic* or *probabilistic*) *independence* to contrast it with other concepts such as *linear independence* and *functional independence*. [The functions X_1, X_2 in the above illustration are linearly independent but *not* mutually (or statistically) independent! See also Problem 1.]

To understand the implications of equations (1), we consider different forms (or consequences) of Definitions 1 and 2. First note that if $\{A_i, i \in I\} \subset \Sigma$ is a class of mutually independent events, then it is evident that $\{\sigma(A_i), i \in I\}$ is an independent class. However, the same cannot be said if the singleton A_i is replaced by a bigger family $\mathcal{G}_i = \{A^i_j, j \in J_i\} \subset \Sigma$, where each J_i has at least two elements, $i \in I$, as simple examples show. Thus $\{\sigma(\mathcal{G}_i), i \in I\}$ need not be independent. On the other hand, we can make the following statements.

Theorem 3 (a) *Let $\{\mathcal{A}, \mathcal{B}_i, i \in I\}$ be classes of events from (Ω, Σ, P) such that they are all mutually independent in the sense of Definition 1. If each $\mathcal{B}_i, i \in I$, is a π-class, then for any subset J of I, the generated σ-algebra $\sigma(\mathcal{B}_i, i \in J)$ and $\mathcal{A}$ are independent of each other.*

(b) *Definition 2 with $S = \mathbb{R}$ reduces to the statement that for each finite subset $i_1, \ldots, i_n$ of I and random variables $X_{i_1}, \ldots, X_{i_n}$, the collection of events $\{[X_{i_1} < x_1, \ldots, X_{i_n} < x_n], x_j \in \mathbb{R}, j = 1, ..., n, n \geq 1\}$ forms an independent class.*

Proof (a) Let $\mathcal{B} = \sigma(\mathcal{B}_i, i \in J), J \subset I$. If $A \in \mathcal{A}, B_j \in \mathcal{B}_j$,

$$j \in \{j_1, ..., j_n\} \subset J,$$

then

$$\{A, B_{j_1}, ..., B_{j_n}\}$$

are independent by hypothesis, i.e., (1) holds. We need to show that

$$P(A \cap B) = P(A)P(B), \quad A \in \mathcal{A}, B \in \mathcal{B}. \tag{2}$$

If B is of the form $B_1 \cap \ldots \cap B_m$, where $B_i \in \mathcal{B}_i$, $i \in J$, then (2) holds by (1). Let $\mathcal{D}$ be the collection of all sets B which are finite intersections of sets each belonging to a $\mathcal{B}_j, j \in J$. Since each $\mathcal{B}_j$ is a π-class, it follows that $\mathcal{D}$ is also a π-class, and by the preceding observation, (2) holds for $\mathcal{A}$ and $\mathcal{D}$, so that they are independent. Also it is clear that $\mathcal{B}_j \subset \mathcal{D}, j \in J$. Thus $\sigma(\mathcal{B}_j, j \in J) \subset \sigma(\mathcal{D})$. We establish (2) for $\mathcal{A}$ and $\sigma(\mathcal{D})$ to complete the proof of this part, and it involves another idea often used in the subject in similar arguments.

Define a class $\mathcal{G}$ as follows:

$$\mathcal{G} = \{B \in \sigma(\mathcal{D}) : P(A \cap B) = P(A)P(B), A \in \mathcal{A}\}. \tag{3}$$

Evidently $\mathcal{D} \subset \mathcal{G}$. Also $\Omega \in \mathcal{G}$, and if $B_1, B_2 \in \mathcal{G}$ with $B_1 \cap B_2 = \emptyset$, then

$$\begin{aligned} P((B_1 \cup B_2) \cap A) &= P(B_1 \cap A) + P(B_2 \cap A) \quad \text{(since the } B_i \cap A \text{ are disjoint)} \\ &= P(B_1)P(A) + P(B_2)P(A) \quad [\text{ by definition of (3)}] \\ &= P(B_1 \cup B_2)P(A). \end{aligned}$$

Hence $B_1 \cup B_2 \in \mathcal{G}$. Similarly if $B_1 \supset B_2$, $B_i \in \mathcal{G}$, then

$$\begin{aligned} P((B_1 - B_2) \cap A) &= P(B_1 \cap A) - P(B_2 \cap A) \text{ (since } B_1 \cap A \supset B_2 \cap A) \\ &= (P(B_1) - P(B_2))P(A) \\ &= P(B_1 - B_2)P(A). \end{aligned}$$

Thus $B_1 - B_2 \in \mathcal{G}$. Finally, if $B_n \in \mathcal{G}, B_n \subset B_{n+1}$, we can show, from the fact that P is σ-additive, that $\lim_n B_n = \cup_{n \geq 1} B_n \in \mathcal{G}$. Hence $\mathcal{G}$ is a λ-class. Since $\mathcal{G} \supset \mathcal{D}$, by Proposition 1.2.8b, $\mathcal{G} \supset \sigma(\mathcal{D})$. But (3) implies $\mathcal{G}$ and $\mathcal{A}$ are independent. Thus $\mathcal{A}$ and $\sigma(\mathcal{D})$ are independent also, as asserted. Note that since $J \subset I$ is an arbitrary subset, we need the full hypothesis that $\{\mathcal{A}, \mathcal{B}_i, i \in I\}$ is a mutually independent collection, and *not* a mere two-by-two independence.

(b) It is clear that Definition 2 implies the statement here. Conversely, let $\mathcal{B}_1$ be the collection of sets $\{[X_{i_1} < x], x \in \mathbb{R}\}$, and

$$\mathcal{B}_2 = \left\{ \bigcup_{j=2}^{n} [X_{i_j} < x_j], x_j \in \mathbb{R} \right\}$$

It is evident that $\mathcal{B}_1$ and $\mathcal{B}_2$ are π-classes. Indeed,

$$[X_{i_1} < x] \cap [X_{i_1} < y] = [X_{i_1} < \min(x, y)] \in \mathcal{B}_1,$$

and similarly for $\mathcal{B}_2$. Hence by (a), $\mathcal{B}_1$ and $\sigma(\mathcal{B}_2)$ are independent. Since $\mathcal{B}_1$ is a π-class, we also get, by (a) again, that $\sigma(\mathcal{B}_1)$ and $\sigma(\mathcal{B}_2)$ are independent. But $\sigma(\mathcal{B}_1) = X_{i_1}^{-1}(\mathcal{R})[= \sigma(X_{i_1})]$, and $\sigma(\mathcal{B}_2) = \sigma(\cup_{j=2}^{n} X_{i_j}^{-1}(\mathcal{R}))[= \sigma(X_{i_2}, \ldots, X_{i_n})]$, where $\mathcal{R}$ is the Borel σ-algebra of $\mathbb{R}$.

Hence if $A_1 \in \sigma(X_{i_1}), A_j \in X_{i_j}^{-1}(\mathcal{R})(= \sigma(X_{i_j})) \subset \sigma(\mathcal{B}_2)$, then A_1 and $\{A_2, ..., A_n\}$ are independent. Thus

$$P(A_1 \cap \ldots \cap A_n) = P(A_1) \cdot P(A_2 \cap \ldots \cap A_n). \tag{4}$$

Next consider X_{i_2} and $(X_{i_3}, \ldots, X_{i_n})$. The above argument can be applied to get

$$P(A_2 \cap \ldots \cap A_n) = P(A_2) \cdot P(A_3 \cap \ldots \cap A_n).$$

Continuing this finitely many times and substituting in (4), we get (1). Hence Definition 2 holds. This completes the proof.

The above result says that we can obtain (1) for random variables if we assume the apparently weaker condition in part (b) of the above theorem. This is particularly useful in computations. Let us record some consequences.

Corollary 4 *Let $\{\mathcal{B}_i, i \in I\}$ be an arbitrary collection of mutually independent π-classes in (Ω, Σ, P), and $J_i \subset I, J_1 \cap J_2 = \emptyset$. If*

$$\mathcal{G}_i = \sigma(\mathcal{B}_j, j \in J_i), i = 1, 2,$$

then $\mathcal{G}_1$ and $\mathcal{G}_2$ are independent. The same is true if $\mathcal{G}_i' = \pi(\mathcal{B}_j, j \in J_i), i = 1, 2$, are the generated π-classes.

If X, Y are independent random variables, f, g are any pair of real Borel functions on $\mathbb{R}$, then $f \circ X$, $g \circ Y$ are also independent random variables. This is because $(f \circ X)^{-1}(\mathcal{R}) = X^{-1}(f^{-1}(\mathcal{R})) \subset X^{-1}(\mathcal{R})$, and similarly $(g \circ Y)^{-1}(\mathcal{R}) \subset Y^{-1}(\mathcal{R})$; and $X^{-1}(\mathcal{R}), Y^{-1}(\mathcal{R})$ are independent σ-subalgebras of Σ. The same argument leads to the following:

Corollary 5 *If $X_1, \ldots, X_n$ are mutually independent random variables on (Ω, Σ, P) and $f : \mathbb{R}^k \to \mathbb{R}$, $g : \mathbb{R}^{n-k} \to \mathbb{R}$ are any Borel functions, then the random variables $f(X_1, \ldots, X_k)$, $g(X_{k+1}, \ldots, X_n)$ are independent; and $\sigma(X_1, \ldots, X_k)$, $\sigma(X_{k+1}, \ldots, X_n)$ are independent σ-algebras, for any $k \geq 1$.*

Another consequence relates to distribution functions and expectations when the latter exist.

Corollary 6 *If $X_1 \dots X_n$ are independent random variables on (Ω, Σ, P), then their joint distribution is the product of their individual distributions:*

$$F_{X_1,\dots,X_n}(x_1,\dots,x_n) = P[X_1 < x_1,\dots,X_n < x_n]$$
$$= \prod_{i=1}^{n} P[X_i < x_i]$$
$$= \prod_{i=1}^{n} F_{X_i}(x_i), x_i \in \mathbb{R}. \tag{5}$$

If, moreover, each of the random variables is integrable, then their product is integrable and we have

$$E\left(\prod_{i=1}^{n} X_i\right) = \prod_{i=1}^{n} E(X_i). \tag{6}$$

Proof By Theorem 3b, (1) and (5) is each equivalent to independence, and so the image functions $F_{X_1,\dots,X_n}$ and $\prod_{i=1}^{n} F_{X_i}$ are identical. In Definition 2.2 the distribution function of a single random variable is given. The same holds for a (finite) random vector, and $F_{X_1,\dots,X_n}$ is termed a *joint distribution function* of $X_1,\dots,X_n$. The result on image measures (Theorem 1.4.1) connects the integrals on the Ω-space with those on $\mathbb{R}^n$, the range space of $(X_1,\dots,X_n)$.

We now prove (6). Taking $f(x) = |x|$, $f : \mathbb{R} \to \mathbb{R}^+$ being a Borel function, by Corollary 5, $|X_1|,\dots,|X_n|$ are also mutually independent. Then by (5) and Tonelli's theorem,

$$E\left(\prod_{i=1}^{n} |X_i|\right) = \int_{\Omega} |X_1| \dots |X_n| dP$$
$$= \int_{\mathbb{R}_+^n} x_1 \dots x_n dG_{|X_1|,\dots,|X_n|}(x_1,\dots,x_n),$$

[by Theorem 1.4.1i with G as the image law]

$$= \int_{\mathbb{R}_+^n} x_1 \dots x_n dG_{|X_1|}(x_1) \dots, dG_{|X_n|}(x_n), \text{ [by (5)]},$$
$$= \prod_{i=1}^{n} \int_{\mathbb{R}^+} x_i dG_{|X_i|(x_i)}, \text{ (by Tonelli's theorem)}$$
$$= \prod_{i=1}^{n} E(|X_i|), \text{ [by Theorem 1.4.1i]}. \tag{7}$$

Since the right side is finite by hypothesis, so is the left side. Now that $\prod_{i=1}^{n} = |X_i|$ is integrable we can use the same computation above for X_i and $F_{X_1,\dots,X_n}(= \prod_{i=1}^{n} F_{X_i})$, and this time use Fubini's theorem in place of Tonelli's. Then we get (6) in place of (7). This proves the result.

Note. It must be remembered that a direct application of Fubini's theorem is *not* possible in the above argument since the integrability of $|\prod_{i=1}^{n} X_i|$ has to be established *first* for this result (cf. Theorem 1.3.11). In this task we need Tonelli's theorem for nonnegative random variables, and thus the proof cannot be shortened. Alternatively, one can prove (6) first for simple random variables with Theorem 3b, and then use the Lebesgue monotone (or dominated) convergence theorem, essentially repeating part of the proof for Tonelli's theorem.

We shall now establish one of the most surprising consequences of the independence concept, the zero-one law. If $X_1, X_2, \dots$ is a sequence of random variables, then $\bigcap_{n=1}^{\infty} \sigma(X_i, i \geq n)$ is called the *tail σ-algebra* of $\{X_n, n \geq 1\}$.

Theorem 7 (Kolmogorov's Zero-One Law) *Any event belonging to the tail σ-algebra of a sequence of independent random variables on (Ω, Σ, P) has probability either zero or one.*

Proof Denote by $\mathcal{T} = \bigcap_{n=1}^{\infty} \sigma(X_k, k \geq n)$, the tail σ-algebra of the sequence. Then by Theorem 3a, $\sigma(X_n)$ and $\sigma(X_k, k \geq n+1)$ are independent σ-algebras for each $n \geq 1$. But $\mathcal{T} \subset \sigma(X_k, k \geq n+1)$, so that $\sigma(X_n)$ and $\mathcal{T}$ are independent for each n. By Theorem 3a again $\mathcal{T}$ is independent of $\sigma(\sigma(X_n), n \geq 1) = \sigma(X_n, n \geq 1)$. However, $\mathcal{T} \subset \sigma(X_n, n \geq 1)$ also, so that $\mathcal{T}$ is independent of itself! Hence $A \in \mathcal{T}$ implies

$$P(A) = P(A \cap A) = P(A)P(A) = P(A)^2;$$

thus we must have $P(A) = 0$ or 1, completing the proof.

An immediate consequence is that any function measurable relative to $\mathcal{T}$ of the theorem must be a constant with probability one. Thus $\limsup_n X_n$, $\liminf_n X_n$ (and $\lim_n X_n$ itself, if this exists) of independent random variables are constants with probability one. Similarly if

$$A_n = \{\omega : |\sum_{k \geq n} X_k(\omega)| < \infty\},$$

then $\sum_{n=1}^{\infty} X_n(\omega)$ converges iff $\omega \in A_n$ for each n, i.e., iff $\omega \in A = \bigcap_{n=1}^{\infty} A_n$.
Since clearly $A_n \in \sigma(X_k, k \geq n), A \in \mathcal{T}$, so that $P(A) = 0$ or 1. Thus for independent X_n the series $\sum_{n=1}^{\infty} X_n$ converges with probability 0 or 1. The following form of the above theorem is given in Tucker (1967).

Corollary 8 *Let I be an arbitrary infinite index set, and $\{X_i, i \in I\}$ be a family of independent random variables on (Ω, Σ, P). If $\mathcal{F}$ is the directed (by inclusion) set of all finite subsets of I, the (generalized) tail σ-algebra is defined as*

$$\mathcal{T}_0 = \bigcap \{\sigma(X_i, i \notin J) : J \in \mathcal{F}\}. \tag{8}$$

Then P takes only 0 and 1 values on $\mathcal{T}_0$.

Proof The argument is similar to that of the theorem. Note that $\mathcal{T}_0$ and $\mathcal{B}_J = \sigma(X_i, i \in J)$ are independent for each $J \in \mathcal{F}$, as in the above proof. So by Theorem 3a, $\mathcal{T}_0$ and $\mathcal{B} = \sigma(\mathcal{B}_J, J \in \mathcal{F})$ are independent. But clearly $\mathcal{B} = \sigma(X_i, i \in I)$, so that $\mathcal{T}_0 \subset \mathcal{B}$. Hence the result follows as before.

Let us now show that independent random variables can be assumed to exist on a probability space by a process of *enlargement of the space by adjunction.* The procedure is as follows: Let (Ω, Σ, P) be a probability space. If this is not rich enough, let $(\Omega_i, \Sigma_i, P_i), i = 1, ..., n$, be n copies of the given space. Let $(\tilde{\Omega}, \tilde{\Sigma}, \tilde{P}) = (\times_{i=1}^n \Omega_i, \bigotimes_{i=1}^n \Sigma_i, \bigotimes_{i=1}^n P_i)$ be their Cartesian product. If $X_1, ..., X_n$ are random variables on (Ω, Σ, P), define a "new" set of functions $\tilde{X}_1, ..., \tilde{X}_n$ on (Ω, Σ, P) by the equations

$$\tilde{X}_i(\omega) = X_i(\omega_i), \quad \omega = (\omega_1, ..., \omega_n) \in \tilde{\Omega}, \quad i = 1, ..., n.$$

Then for each $a \in \mathbb{R}$,

$$\begin{aligned} \{\omega : \tilde{X}_i(\omega) < a\} &= \{\omega : X_i(\omega_i) < a\} \\ &= \Omega_1 \times \ldots \times \Omega_{i-1} \times [X_i < a] \times \Omega_{i+1} \times \ldots \times \Omega_n, \end{aligned}$$

which is a measurable rectangle and hence is in $\tilde{\Sigma}$. Thus $\tilde{X}_i$ is a random variable. Also, since $P_i = P$, we deduce that

$$\tilde{P}[\tilde{X}_1 < a_1, \tilde{X}_2 < a_2, ..., \tilde{X}_n < a_n] = \prod_{i=1}^n P[X_i < a_i], \tag{9}$$

by Fubini's theorem and the fact that $P_i(\Omega_i) = 1$. Consequently the $\tilde{X}_i$ are independent (cf. Theorem 3b) and each $\tilde{X}_i$ has the same distribution as X_i. Thus by enlargement of (Ω, Σ, P) to $(\tilde{\Omega}, \tilde{\Sigma}, \tilde{P})$, we have n independent random variables. This procedure can be employed for the existence of any finite collection of independent random variables without altering the probability structure (see also Problem 5 (a)). The results of Section 3.4 establishing the Kolmogorov- Bochner theorem will show that this enlargement can be used for *any* collection of random variables (countable or not). Consequently, we can and do develop the theory without any question of the richness of the underlying σ-algebra or of the existence of families of independent random variables.

The following elementary but powerful results, known as the *Borel-Cantelli lemmas*, are true even for the weaker pairwise independent events. Recall that $\limsup_n A_n = \{\omega : \omega \in A_n$ for infinitely many $n\}$. This set is abbreviated as $\{A_n, \text{ i.o.}\}$ [$= \{A_n$ occurs infinitely often$\}$].

Theorem 9 (i) (First Borel-Cantelli Lemma). *Let $\{A_n, n \geq 1\}$ be a sequence of events in (Ω, Σ, P) such that $\sum_{n=1}^{\infty} P(A_n) < \infty$. Then*

$$P(\limsup_n A_n) = P(A_n, \text{ i.o. }) = 0.$$

(ii) (Second Borel-Cantelli Lemma). *Let $\{A_n, n \geq 1\}$ be a sequence of pairwise independent events in (Ω, Σ, P) such that $\sum_{n=1}^{\infty} P(A_n) = \infty$. Then $P(A_n, i.o.) = 1$.*

(iii) *In particular, if $\{A_n, n \geq 1\}$ is a sequence of (pairwise or mutually) independent events, then $P(A_n, i.o.) = 0$ or 1 according to whether $\sum_{n=1}^{\infty} P(A_n)$ is $< \infty$ or $= \infty$.*

Proof (i) This simple result is used more often than the other more involved parts, since the events need not be (even pairwise) independent. By definition, $A = \limsup_n A_n = \bigcap_{n \geq 1} \bigcup_{k \geq n} A_k \subseteq \bigcup_{k \geq n} A_k$ for all $n \geq 1$. Hence by the σ-subadditivity of P, we have

$$P(A) \leq P(\bigcup_{k \geq n} A_k) \leq \sum_{k=n}^{\infty} P(A_k), n \geq 1.$$

Letting $n \to \infty$, and using the convergence of the series $\sum_{k=1}^{\infty} P(A_k)$, the result follows.

(ii) (After Chung, 1974) Let $\{A_n, n \geq 1\}$ be *pairwise* independent. By Problem 1 of Chapter 1, we have

$$A = [A_n, \text{ i.o.}] \text{ iff } \chi_A = \limsup_n \chi_{A_n}.$$

Hence

$$P(A) = 1 \text{ iff } P[\limsup_n \chi_{A_n} = 1] = 1.$$

Thus

$$P([\chi_{A_n} = 1, \text{ i.o. }]) = 1 \text{ iff } P\left(\left[\sum_{i=1}^{\infty} \chi_{A_n} = \infty\right]\right) = 1. \tag{10}$$

Now we use the hypothesis that the series diverges:

$$\sum_{n=1}^{\infty} P(A_n) = \sum_{n=1}^{\infty} E(\chi_{A_n}) = \lim_{n \to \infty} E(S_n) = +\infty, \tag{11}$$

where $S_n = \sum_{k=1}^{n} \chi_{A_k}$ and the monotonicity of S_n is used above. With (11) and the pairwise independence of A_n, we shall show that

$$P([\lim_{n\to\infty} S_n = \infty]) = 1,$$

which in view of (10) proves the assertion.

Now given $N > 0$, we have by Čebyšev's inequality, with $\varepsilon = N\sqrt{VarS_n}$,

$$P[|S_n - E(S_n)| > \varepsilon] \leq \frac{VarS_n}{\varepsilon^2} = \frac{1}{N^2}.$$

Equivalently,

$$P[E(S_n) - N\sqrt{VarS_n} \leq S_n \leq E(S_n) + N\sqrt{VarS_n}] \geq 1 - \frac{1}{N^2}. \tag{12}$$

To simplify this we need to evaluate $Var\ S_n$. Let $p_n = P(A_n)$. Then

$$E(S_n) = \sum_{k=1}^{n} E(\chi_{A_k}) = \sum_{k=1}^{n} p_k. \tag{13}$$

If $I_n = \chi_{A_n} - p_n$, then the I_n are orthogonal random variables. In fact, using the inner product notation,

$$\begin{aligned}(I_n, I_m) &= \int_\Omega (\chi_{A_n} - p_n)(\chi_{A_m} - p_m)dP \\ &= P(A_n \cap A_m) - p_n p_m = 0, \quad \text{if } n \neq m \text{ (by pairwise independence)} \\ &= p_n(1 - p_n), \quad \text{if } n = m. \end{aligned} \tag{14}$$

Thus

$$\begin{aligned} VarS_n &= E(S_n - E(S_n))^2 = E\left(\sum_{k=1}^{n} I_k\right)^2 \\ &= \sum_{k=1}^{n} E(I_k^2) = \sum_{k=1}^{n} p_k(1 - p_k), \quad [\text{by } (14)\ ], \\ &\leq \sum_{k=1}^{n} p_k = E(S_n), \quad [\text{by } (13)\ ]. \end{aligned} \tag{15}$$

Since by (11) $E(S_n) \nearrow \infty$, (15) yields $\sqrt{VarS_n / E(S_n)} \leq (E(S_n))^{-1/2} \to 0$. Thus given $N > 1$, and $0 < \alpha_1 = \alpha/N$ for $0 < \alpha < 1$, there exists $n_0 = n_0(\alpha, N)$ such that $n \geq n_0 \Rightarrow \sqrt{VarS_n/E(S_n)} \leq \alpha_1 < 1$. Since $\alpha_1 = \alpha/N$, we get

$$N\sqrt{VarS_n} \leq \alpha E(S_n), \quad n \geq n_0. \tag{16}$$

Consequently (12) implies, with $1 > \beta = 1 - \alpha > 0$ and the monotonicity of S_n, (i.e., $S_n \uparrow$)

$$P[\beta E(S_n) < \lim_n S_n] \geq P[\beta E(S_n) < S_n] \geq 1 - \frac{1}{N^2}, \quad n \geq n_0. \qquad (17)$$

Let $n \to \infty$, and then $N \to \infty$ (so that $\beta \to 1$); (17) gives $P[\lim_{n\to\infty} S_n = \infty] = 1$. This establishes the result because of (10).

(iii) This is an immediate consequence of (ii), and again gives a zero-one phenomenon! However, in the case of mutual independence, the proof is simpler than that of (ii), and we give the easy argument here, for variety. Let $A_n^c = B_n$. Then $B_n, n \geq 1$, are independent, since $\{\sigma(A_n), n \geq 1\}$ forms an independent class. Let $P(A_n) = \alpha_n$. To show that

$$P[A_n, \text{ i.o. }] = P\left[\bigcap_{k\geq 1}\bigcup_{n\geq k} A_n\right] = 1,$$

it suffices to verify that, for each $n \geq 1, P(\bigcup_{k>n} A_k) = 1$, or equivalently

$$P\left[\bigcap_{k>n} B_k\right] = 0, \quad n \geq 1.$$

Now for any $n \geq 1$,

$$\begin{aligned} P\left[\bigcap_{k>n} B_k\right] &= \lim_{m\to\infty} P\left[\bigcap_{k=n+1}^{m} B_k\right] \\ &= \lim_{m\to\infty} \prod_{k=n+1}^{m} (1 - \alpha_k) \quad \text{(by independence of } B_k \text{)} \\ &\leq \prod_{k=n+1}^{\infty} e^{-\alpha_k} \quad \left(\text{since } x \geq \int_0^x e^{-t}dt = 1 - e^{-x}\right) \\ &= \exp\left(-\sum_{k=n+1}^{\infty} \alpha_k\right) = 0 \quad \left(\text{since } \sum_{k=1}^{\infty} \alpha_k = \infty \text{ by hypothesis}\right). \end{aligned}$$

This completes the proof of the theorem.

Note 10 The estimates in the proof of (ii) yield a stronger statement than we have asserted. *One can actually show that*

$$P[\lim_{n\to\infty} \frac{S_n}{E(S_n)} = 1] = 1.$$

In fact, (12) implies for each n and N,

$$P\left[\frac{S_n}{E(S_n)} \leq 1 + \frac{N\sqrt{VarS_n}}{E(S_n)}\right] \geq 1 - \frac{1}{N^2}.$$

Since $VarS_n/(E(S_n))^2 \to 0$ as $n \to \infty$, for each fixed N, this gives

$$p[\limsup_n \frac{S_n}{E(S_n)} \leq 1] \geq 1 - \frac{1}{N^2}, \tag{18}$$

and letting $N \to \infty$, we get

$$\limsup_n \frac{S_n}{E(S_n)} \leq 1 \quad \text{a.e.}$$

On the other hand by (17), $P[\beta \leq S_n/E(S_n)] \geq 1 - 1/N^2$, $n \geq n_0$. Hence for each fixed N, this yields

$$P\left[\beta \leq \liminf_n \frac{S_n}{E(S_n)}\right] \geq 1 - \frac{1}{N^2}. \tag{19}$$

Now let $N \to \infty$ and note that $\beta \to 1$; then by the monotonicity of events in brackets of (19) we get $1 \leq \liminf_n [S_n/E(S_n)]$ a.e. These two statements imply the assertion.

Before leaving this section, we present, under a stronger hypothesis than that of Theorem 7, a zero-one law due to Hewitt and Savage (1955), which is useful in applications. We include a short proof as in Feller (1966).

Definition 11 If $X_1, \ldots, X_n$ are random variables on (Ω, Σ, P), then they are *symmetric* (or *symmetrically dependent*) if for each permutation $i_1, \ldots, i_n$ of $(1, 2, \ldots, n)$, the vectors $(X_{i_1}, \ldots, X_{i_n})$ and $(X_1, \ldots, X_n)$ have the same joint distribution. A sequence $\{X_n, n \geq 1\}$ is *symmetric* if $\{X_k, 1 \leq k \leq n\}$ is symmetric for each $n \geq 1$.

We want to consider some functions of $X = \{X_n, n \geq 1\}$. Now $X : \Omega \to \mathbb{R}^\infty = \times_{i=1}^\infty \mathbb{R}_i$, where $\mathbb{R}_i \equiv \mathbb{R}$ is an infinite vector. If $\mathcal{B}^\infty = \otimes_{i=1}^\infty \mathcal{B}_i$ is the (usual) product σ-algebra, then

$$\Sigma_0 = \sigma(X_n, n \geq 1) = X^{-1}(\mathcal{B}^\infty) = \sigma\left(\bigcup_{n=1}^\infty \sigma(X_1, \ldots, X_n)\right).$$

Let $g : \Omega \to \mathbb{R}$ be Σ_0-measurable. Then by Proposition 1.2.3 there is a Borel function $h : \mathbb{R}^\infty \to \mathbb{R}$ (i.e., h is $\mathcal{B}^\infty$-measurable) such that $g = h \circ X = h(X_1, X_2, \ldots)$. Thus if $\{X_n, n \geq 1\}$ is a symmetric sequence, then each Σ_0-measurable g is symmetric [1], so that

[1] In detail, this means if $g : \mathbb{R}^\infty \to \mathbb{R}$, then $g(X_1, \ldots, X_n, X_{n+1}, \ldots) = g(X_{i_1}, \ldots, X_{i_n}, X_{n+1}, \ldots)$ for each permutation $(i_1, \ldots, i_n)$ of $(1, 2, \ldots, n)$, each $n > 1$.

$$g = h(X_1, X_2, \ldots) = h(X_{i_1} \ldots, X_{i_n}, X_{i_{n+1}}, \ldots)$$

for each finite permutation. Let $A \in \Sigma_0$. Then A is a *symmetric event* if χ_A is a symmetric function in the above sense. The following result is true:

Theorem 12 (Hewitt-Savage Zero-One Law). *If $X_1, X_2, \ldots$ are independent with a common distribution, then every symmetric set in*

$$\Sigma_0 = \sigma(X_n, n \geq 1)$$

has probability zero or one.

Proof Recall that if $\rho : \Sigma_0 \times \Sigma_0 \to \mathbb{R}^+$ defined by $\rho(A, B) = P(A \triangle B)$ with $\triangle$ as symmetric difference, then (Σ, ρ) is a (semi) metric space on which the operations $\cup, \cap$, and $\triangle$ are continuous. Also, $\bigcup_{n=1}^{\infty} \sigma(X_1, \ldots, X_n) \subset \Sigma_0$ is a dense subspace, in this metric.

Hence if $A \in \Sigma_0$, there exists $A_n \in \sigma(X_1, \ldots, X_n)$ such that $\rho(A, A_n) \to 0$, and by the definition of $\sigma(X_1, \ldots, X_n)$ there is a Borel set $B_n \subset \mathbb{R}^n$ such that $A_n = [(X_1, \ldots, X_n) \in B_n]$. Since

$$\overline{X} = (X_{i_1}, \ldots, X_{i_n}, X_{n+1} \ldots) \text{ and } X = (X_1, \ldots, X_n, X_{n+1} \ldots)$$

have the same (finite dimensional) distributions, because the X_n are identically distributed, and we have for any $B \in \mathcal{B}^{\infty}$, $P(\overline{X} \in B) = P(X \in B)$. In particular, if the permutation is such that $\tilde{A}_n = [(X_{2n}, X_{2n-1}, \ldots, X_{n+1}) \in B_n]$, then A_n and $\tilde{A}_n$ are independent and $\rho(A, \tilde{A}_n) \to 0$ as $n \to \infty$ again.

Indeed, let τ be the 1-1 measurable permutation mapping $\tau A_n = \tilde{A}_n$ and $\tau A = A$ since A is symmetric. So

$$\rho(A, \tilde{A}_n) = \rho(\tau A, \tau A_n) = \rho(A, A_n) \to 0.$$

Hence also $A_n \cap \tilde{A}_n \to A \cap A = A$, by the continuity of $\cap$ in the ρ-metric. But

$$P(A_n \cap \tilde{A}_n) = P(A_n) P(\tilde{A}_n),$$

by independence.

Letting $n \to \infty$, and noting that the metric function is also continuous in the resulting topology, it follows that $A_n \to A$ in $\rho \Rightarrow P(A_n) \to P(A)$. Hence

$$\lim_{n \to \infty} P(A_n \cap \tilde{A}_n) = P(A \cap A) = \lim_{n \to \infty} P(A_n) \cdot P(\tilde{A}_n) = P(A)^2.$$

Thus $P(A) = P(A)^2$ so that $P(A) = 0$ or 1, as asserted.

Remarks (1) It is not difficult to verify that if $S_n = \sum_{k=1}^{n} X_k$, X_k as in the theorem, then for any Borel set B, the event $[S_n \in B \text{ i.o.}]$ is not necessarily a tail event but is a symmetric one. Thus this is covered by the above theorem, but not by the Kolmogorov zero-one law.

(2) Note 10, as well as part (ii) of Theorem 9, indicate how several weakenings of the independence condition can be formulated. A number of different extensions of Borel-Cantelli lemmas have appeared in the literature, and they are useful for special problems. The point here is that the concept of independence, as given in Definitions 1 and 2, leads to some very striking results, which then motivate the introduction of different types of dependences for a sustained study. In this chapter we present only the basic results founded on the independence hypothesis; later on we discuss how some natural extensions suggest themselves.

2.2 Convergence Concepts, Series, and Inequalities

There are four convergence concepts often used in probability theory. They are pointwise a.e., in mean, in probability, and in distribution. Some of these have already appeared in Chapter 1. We state them again and give some interrelations here. It turns out that for sums of independent (integrable) random variables, these are all equivalent, but this is a relatively deep result. A partial solution is given in Problem 16. Several inequalities are needed for the proof of the general case. We start with the basic Kolmogorov inequality and a few of its variants. As consequences, some important "strong limit laws" will be established. Applications are given in Section 2.4.

Definition 1 Let $\{X, X_n, n \geq 1\}$ be a family of random variables on a probability space (Ω, Σ, P).

(a) $X_n \to X$ *pointwise a.e.* if there is a set $N \in \Sigma$, $P(N) = 0$ and $X_n(\omega) \to X(\omega)$, as $n \to \infty$, for each $\omega \in \Omega - N$.

(b) The sequence is said to converge to X *in probability* if for each $\varepsilon > 0$, we have $\lim_{n\to\infty} P[|X_n - X| \geq \varepsilon] = 0$, symbolically written as $X_n \xrightarrow{P} X$ (or as $p\lim_n X_n = X$).

(c) The sequence is said to converge *in distribution* to X, often written $X_n \xrightarrow{D} X$ if $F_{X_n}(x) \to F_x(x)$ at all points $x \in \mathbb{R}$ for which x is a continuity point of F_X, where F_{X_n}, F_X are distribution functions of X_n and X (cf. Definition 2.2).

(d) Finally, if $\{X, X_n, n \geq 1\}$ have p-moments, $0 < p < \infty$, then the sequence is said to tend to X in *pth order mean*, written $X_n \xrightarrow{\mathcal{L}^p} X$, if $E(|X_n - X|^p) \to 0$. If $p = 1$, we simply say that $X_n \to X$ in mean.

The first two as well as the last convergences have already appeared, and these are defined and profitably employed in general analysis on arbitrary measure spaces. However, on finite measure spaces there are some additional relations which are of particular interest in our study. The third concept, on the other hand, is somewhat special to probability theory since distribution functions are image probability measures on $\mathbb{R}$. This plays a pivotal role in probability theory, and so we study the concept in greater detail.

Some amplification of the conditions for "in distribution" is in order. If $X = a$ a.e., then $F_X(x) = 0$ for $x < a$, $= 1$ for $x \geq a$. Thus we are asking that for $X_n \xrightarrow{D} a$, $F_{X_n}(x) \to F_x(x)$ for $x < a$ and for $x > a$ but not at $x = a$, the discontinuity point of F_X. Why? The restriction on the set is that it should be only a "continuity set" for the limit function F_X. This condition is arrived at after noting the "natural-looking" conditions proved themselves useless. For instance, if $X_n = a_n$ a.e., and $a_n \to a$ as numbers, then $F_{X_n}(x) \to F_X(x)$ for all $x \in \mathbb{R} - \{a\}$, but $F_{X_n}(a) \not\to F_X(a)$, since $\{F_{X_n}(a), n \geq 1\}$ is an oscillating sequence if there are infinitely many n on both sides of a. Similarly, if $\{X_n = a_n, n \geq 1\}$ diverges, it is possible that $\{F_{X_n}(x), n \geq 1\}$ may converge for each $x \in \mathbb{R}$ to a function taking values in the open interval $(0, 1)$. Other unwanted exclusions may appear. Thus the stipulated condition is weak enough to ignore such uninteresting behavior. But it is not too weak, since we do want the convergence on a suitable dense set of $\mathbb{R}$. (Note that the set of discontinuity points of a monotone function is at most countable, so that the continuity set of F_X is $\mathbb{R} - \{$ that countable set $\}$.) Actually, the condition comes from the so-called simple convergence on $C_{00}(\mathbb{R})$, the space of continuous functions with compact supports, which translates to the condition we gave for the distribution functions on $\mathbb{R}$ according to a theorem in abstract analysis. For this reason N. Bourbaki actually calls it the *vague convergence*, and others call it the *weak-star convergence*. We shall use the terminology introduced in the definition and the later work shows how these last two terms can also be justifiably used.

The first three convergences are related as follows:

Proposition 2 *Let X_n and X be random variables on (Ω, Σ, P). Then $X_n \to X$ a.e. $\Rightarrow X_n \xrightarrow{P} X \Rightarrow X_n \xrightarrow{D} X$. If, moreover, $X = a$ a.e., where $a \in \mathbb{R}$, then $X_n \xrightarrow{D} X \Rightarrow X_n \xrightarrow{P} X$ also. In general these implications are not reversible.* (Here, as usual, the limits are taken as $n \to \infty$.)

Proof The first implication is a standard result for any finite measure. In fact, if $X_n \to X$ a.e., then there is a set $N \in \Sigma$, $P(N) = 0$, and on $\Omega - N$, $X_n(\omega) \to X(\omega)$. Thus $\limsup_n X_n(\omega) = X(\omega)$, $\omega \in \Omega - N$, and for each $\varepsilon > 0$,

$$\left\{\omega : \limsup_n |X_n - X|(\omega) > \varepsilon\right\} \subset \bigcap_{k \geq 1} \bigcup_{n \geq k} \{\omega : |X_n(\omega) - X(\omega)| > \varepsilon\} \subset N.$$

Hence the set has measure zero. Since P is a finite measure, this implies

$$P\left(\lim_{k\to\infty}\bigcup_{n\geq k}\{\omega : |X_n(\omega) - X(\omega)| > \varepsilon\}\right) = \lim_{k\to\infty} P\left(\bigcup_{n\geq k}[|X_n - X| > \varepsilon]\right)$$

$$\leq P(N) = 0. \tag{1}$$

Consequently,

$$P([|X_n - X| > \varepsilon]) \leq P\left(\bigcup_{j\geq n}[|X_j - X| > \varepsilon]\right) \to 0 \text{ as } n \to \infty. \tag{2}$$

Thus $X_n \xrightarrow{P} X$, and the first assertion is proved.

For the next implication, let F_X, F_{X_n} be the distribution functions of X and X_n, and let a, b be continuity points of F_X with $a < b$. Then

$$[X < a] = [X < a, X_n < b] \cup [X < a, X_n \geq b]$$

$$\subset [X_n < b] \cup [X < a, X_n \geq b],$$

so that computing probabilities of these sets gives

$$F_X(a) \leq F_{X_n}(b) + P[X < a, X_n \geq b]. \tag{3}$$

Also, since $X_n \xrightarrow{P} X$, with $\varepsilon = b - a > 0$, one has from the inclusion

$$[X < a, X_n \geq b] \subset [|X_n - X| \geq b - a],$$

$$\lim_n P[X < a, X_n \geq b] = 0. \tag{4}$$

Thus (3) becomes

$$F_X(a) \leq \liminf_n F_{X_n}(b). \tag{5}$$

Next, by an identical computation, but with c, d $(c < d)$ in place of a, b and X_n, X in place of X, X_n in (3), one gets

$$F_{X_n}(c) \leq F_X(d) + P[X_n < c, X \geq d]. \tag{6}$$

The last term tends to zero as $n \to \infty$, as in (4). Consequently (6) becomes

$$\limsup_n F_{X_n}(c) \leq F_X(d). \tag{7}$$

From (5) and (7) we get for $a < b \leq c < d$,

$$F_X(a) \leq \liminf_n F_{X_n}(b) \leq \limsup_n F_{X_n}(b)$$

$$\leq \limsup_n F_{X_n}(c) \leq F_X(d). \tag{8}$$

Letting $a \uparrow b = c$ and $d \downarrow c$, where $b = c$ is a continuity point of F_X, (8) gives $\lim_n F_{X_n}(b) = F_X(b)$, so that $X_n \xrightarrow{D} X$, since such points of continuity of F_X are everywhere dense in $\mathbb{R}$.

If now $X = \alpha$ a.e., then for each $\varepsilon > 0$,

$$P([|X_n - \alpha| \geq \varepsilon]) = P([X_n \geq \alpha + \varepsilon]) + P([X_n \leq \alpha - \varepsilon])$$

$$= 1 - F_{X_n}(\alpha + \varepsilon) + F_{X_n}(\alpha - \varepsilon) \to 0 \text{ as } n \to \infty,$$

since

$$F_{X_n}(x) \to F_X(x) = \begin{cases} 0, & x < \alpha \\ 1, & x \geq \alpha, \end{cases}$$

and $\alpha \pm \varepsilon$ are points of continuity of F_X for each $\varepsilon > 0$. Thus $X \xrightarrow{P} \alpha$. This completes the proof except for the last comment, which is illustrated by the following simple pair of standard counter-examples.

Let X_n, X be defined on (Ω, Σ, P) as two-valued random variables such that $P([X_n = a]) = \frac{1}{2} = P([X_n = b]), a < b$, for all n. Next let $P([X = b]) = \frac{1}{2} = P([X = a])$. Then for each n, $\omega \in \Omega$, for which $X_n(\omega) = a$ (or b), we set $X(\omega) = b$ (or a), respectively. Thus $\{\omega : |X_n - X|(\omega) \geq \varepsilon\} = \Omega$ if $0 < \varepsilon < b - a$, and $X_n \not\to X$ in probability. But $F_{X_n} = F_X$, so that $X_n \xrightarrow{D} X$ trivially. This shows that the last implication cannot be reversed in general. Next, consider the first one. Let $\Omega = [0, 1]$, $\Sigma =$ Borel σ-algebra of Ω, and $P =$ Lebesgue measure. For each $n > 1$, express n in a binary expansion, $n = 2^r + k$, $0 \leq k \leq 2^r$, $r \geq 0$. Define $f_n = \chi_{A_n}$, where $A_n = [k/2^r, (k+1)/2^r]$. It is clear that f_n is measurable, and for $0 < \varepsilon < 1$,

$$P[|f_n - 0| > \varepsilon] \leq \frac{1}{2^r} < \frac{2}{n} \to 0.$$

But $f_n(\omega) \not\to 0$ for any $\omega \in \Omega$. This establishes all assertions. (If we are allowed to change probability spaces, keeping the same image measures of the random variables, these problems become less significant. Cf. Problem 5 (b).)

In spite of the last part, we shall be able to prove the equivalence to a subclass of random variables, namely, if the X_n form a sequence of partial sums of *independent* random variables. For this result we need to develop probability theory much further, and thus it is postponed until Chapter 4. (For a partial result, see Problem 16.) Here we proceed with the implications that do not refer to "convergence in distribution."

The following result is of interest in many calculations.

Proposition 3 **(F. Riesz)**. *Let* $\{X, X_n, n \geq 1\}$ *be random variables on* (Ω, Σ, P) *such that* $X_n \xrightarrow{P} X$. *Then there exists a subsequence* $\{X_{n_k}, k \geq 1\}$ *with* $X_{n_k} \to X$ *a.e. as* $k \to \infty$.

Proof Since for each $\varepsilon > 0$, $P[|X_n - X| \geq \varepsilon] \to 0$, let n_1 be chosen such that $n \geq n_1 \Rightarrow P[|X_n - X| \geq 1] < \frac{1}{2}$,and if $n_1 < n_2 < \ldots < n_k$ are selected, let $n_{k+1} > n_k$ be chosen such that

$$P[|X_n - X| \geq 1/2^k] < 1/2^{k+1}, \quad n \geq n_{k+1}. \tag{9}$$

If $A_k = [|X_{n_k} - X| \geq 1/2^{k-1}]$, $B_k = \bigcup_{n\geq k} A_n$, then for $\omega \in B_k^c$, $|X_{n_r} - X|(\omega) < 1/2^{r-1}$ for all $r \geq k$. Hence if $B = \lim_n B_n = \bigcap_{n=1}^{\infty} \bigcup_{k\geq n} A_k$, then for $\omega \in B^c$, $X_{n_r}(\omega) \to X(\omega)$ as $r \to \infty$. But we also have $B \subset B_n$ for all n, so that

$$P(B) \leq P(B_n) \leq \sum_{k\geq n} P(A_k) < \sum_{k\geq n} 2^{-k} = 2^{-n+1} \to 0 \text{ as } n \to \infty.$$

Thus $\{X_{n_r}, r \geq 1\}$ is the desired subsequence, completing the proof.

Remark We have not used the finiteness of P in the above proof, and the result holds on nonfinite measure spaces as well. (Also there can be infinitely many such a.e. convergent subsequences.) But the next result is strictly for (finite or) probability measures only.

Recall that a sequence $\{X_n, n \geq 1\}$ on (Ω, Σ, P) *converges P-uniformly* to X if for each $\varepsilon > 0$, there is a set $A_\varepsilon \in \Sigma$ such that $P(A_\varepsilon) < \varepsilon$ and on $\Omega - A_\varepsilon$, $X_n \to X$ uniformly. We then have

Theorem 4 (**Egorov**). *Let $\{X, X_n, n \geq 1\}$ be a sequence of random variables on (Ω, Σ, P). Then $X_n \to X$ a.e. iff the sequence converges to X P-uniformly.*

Proof One direction is simple. In fact, if $X_n \to X$ P-uniformly, then for $\varepsilon = 1/n_0$ there is an $A_{n_0} \in \Sigma$ with $P(A_{n_0}) < 1/n_0$ and $X_n(\omega) \to X(\omega)$ uniformly on $\Omega - A_{n_0}$. If $A = \bigcap_{n\geq 1} A_n$, then $P(A) = 0$, and if $\omega \in \Omega - A$, then $X_n(\omega) \to X(\omega)$, i.e., the sequence converges a.e. The other direction is non-trivial.

Thus let $X_n \to X$ a.e. Then there is an $N \in \Sigma$, $P(N) = 0$, and $X_n(\omega) \to X(\omega)$ for each $\omega \in \Omega - N$. If $k \geq 1$, $m \geq 1$ are integers and we define

$$A_{k,m} = \{\omega \in \Omega - N : |X_n(\omega) - X(\omega)| < \frac{1}{m} \text{ for all } n \geq k\},$$

then the facts that $X_n \to X$ on $\Omega - N$ and $A_{k,m} \subset A_{k+1,m}$ imply that $\Omega - N = \bigcup_{k=1}^{\infty} A_{k,m}$ for all $m \geq 1$. Consequently for each $\varepsilon > 0$, and each $m \geq 1$, we can find a large enough $k_0 = k_0(\varepsilon, m)$ such that $A_{k_0,m}$ has large measure, i.e., $P(\Omega - A_{k_0,m}) < \varepsilon/2^m$. If $A_\varepsilon = \bigcup_{m=1}^{\infty} A^c_{k_0(\varepsilon,m),m}$ then

$$P(A_\varepsilon) \leq \sum_{m=1}^{\infty} P(A^c_{k_0(\varepsilon,m),m}) < \sum_{m=1}^{\infty} \frac{\varepsilon}{2^m} = \varepsilon.$$

On the other hand, $n \geq k_0(\varepsilon, m) \Rightarrow |X_n(\omega) - X(\omega)| < 1/m$ for $\omega \in A_{k_0,m}$. Thus

$$n \geq k_0(\varepsilon, m) \Rightarrow \sup_{\omega \in A_\varepsilon^c} |X_n(\omega) - X(\omega)| \leq \sup_{\omega \in A_{k_0,m}} |X_n(\omega) - X(\omega)| \leq \frac{1}{m}$$

for every $m \geq 1$, so that $X_n \to X$ uniformly on A_ε^c. This completes the proof.

Also, the following is a simple consequence of Markov's inequality.

Remark *Let* $\{X, X_n, n \geq 1\} \subset \mathcal{L}^p(\Omega, \Sigma, P)$ *such that* $X_n \xrightarrow{\mathcal{L}^p} X$, $p > 0$. *Then* $X_n \xrightarrow{P} X$.

Proof Given $\varepsilon > 0$, we have

$$\begin{aligned} P[|X_n - X| > \varepsilon] &= P[|X_n - X|^p > \varepsilon^p] \\ &\leq \frac{1}{\varepsilon^p} E(|X_n - X|^p) \to 0 \text{ as } n \to \infty, \end{aligned}$$

by the pth mean convergence hypothesis. Note that there is generally no relation between mean convergence and pointwise a.e., since for the latter the random variables need not be in any $\mathcal{L}^p$, $p > 0$.

We now specialize the convergence theory if the sequences are partial sums of *independent* random variables, and present important consequences. Some further, less sharp, assertions in the general case are possible. Some of these are included as problems at the end of the chapter.

At the root of the pointwise convergence theory, there is usually a "maximal inequality," for a set of random variables. Here is a generalized version of Čebyšev's inequality. The latter was proved for only one r.v. We thus start with the fundamental result:

Theorem 5 (Kolmogorov's Inequality). *Let* $X_1, X_2, \ldots$ *be a sequence of independent random variables on* (Ω, Σ, P) *with means* $\mu_k = E(X_k)$ *and variances* $\sigma_k^2 = Var X_k$. *If* $S_n = \sum_{k=1}^n X_k$ *and* $\varepsilon > 0$, *then*

$$P[\max_{1 \leq k \leq n} |S_k - E(S_k)| \geq \varepsilon] \leq \frac{1}{\varepsilon^2} \sum_{k=1}^n \sigma_k^2. \tag{10}$$

Proof If $n = 1$, then (10) is Čebyšev's inequality, but the present result is deeper than the former. The proof shows how the result may be generalized to certain nonindependent cases, particularly to martingale sequences, to be studied in the next chapter.

Let $A = \{\omega : \max_{1 \leq k \leq n} |S_k(\omega) - E(S_k)| \geq \varepsilon\}$. We express A as a disjoint union of n events; such a decomposition appears in our subject on several occasions. It became one of the standard tools. [It is often called a process of *disjunctification* of a compound event such as A.] Thus let

$$A_1 = \{\omega : |S_1(\omega) - E(S_1)| \geq \varepsilon\}$$

and for $1 < k \leq n$,

$$A_k = \{\omega : |S_i(\omega) - E(S_i)| < \varepsilon, 1 \leq i \leq k-1, |S_k(\omega) - E(S_k) \geq \varepsilon\}.$$

In words, A_k is the set of ω such that $|S_k(\omega) - E(S_k)|$ exceeds ε for the first time. It is clear that the A_k are disjoint, $A_k \in \Sigma$, and $A = \bigcup_{k=1}^{n} A_k$. Let $Y_i = X_i - \mu_i$ and $\tilde{S}_n = \sum_{k=1}^{n} Y_k$, so that $E(\tilde{S}_n) = 0$, $Var\tilde{S}_n = VarS_n$. Now consider

$$\begin{aligned}
\int_{A_k} \tilde{S}_n^2 \, dP &= \int_{A_k} [\tilde{S}_k^2 + (\tilde{S}_n^2 - \tilde{S}_k^2)] \, dP \\
&= \int_{A_k} \tilde{S}_k^2 \, dP + 2\int_{A_k} \tilde{S}_k(Y_{k+1} + \ldots + Y_n) \, dP \\
&\quad + \int_{A_k} (Y_{k+1} + \ldots + Y_n)^2 \, dP, \text{ since } S_n = S_k + \sum_{i=k+1}^{n} Y_i, \\
&\geq \varepsilon^2 \int_{A_k} dP + 2\int_{\Omega} (\chi_{A_k}\tilde{S}_k)(Y_{k+1} + \ldots + Y_n) \, dP \qquad (11) \\
&= \varepsilon^2 P(A_k) + 2E\left(\chi_{A_k}\tilde{S}_k\right) E\left(\sum_{i=k+1}^{n} Y_i\right) \\
&\qquad \text{(since } \chi_{A_k}\tilde{S}_k \text{ and } Y_i, i \geq k+1, \text{ are independent)} \\
&= \varepsilon^2 P(A_k) \quad [\text{since} E(Y_i) = 0\ ].
\end{aligned}$$

Adding on $1 \leq k \leq n$, we get

$$Var(S_n) = Var(\tilde{S}_n) = \int_{\Omega} \tilde{S}_n^2 \, dP \geq \varepsilon^2 \sum_{k=1}^{n} P(A_k) = \varepsilon^2 P(A).$$

Since $VarS_n = \sum_{i=1}^{n} VarX_i$, by independence of the X_i, this gives (10), and completes the proof.

Remark The only place in the above proof where we use the independence hypothesis is to go from (11) to the next line to conclude that

$$E(\chi_{A_k}\tilde{S}_k(Y_{k+1} + \ldots + Y_n)) = 0.$$

Any other hypothesis that guarantees the nonnegativity of this term gives the corresponding maximal inequality. There are several classes of nonindependent random variables including (positive sub-) $\mathcal{L}^2$-martingale sequences giving such a result. This will be seen in the next chapter.

All the strong convergence theorems that follow in this section are due to Kolmogorov.

Theorem 6 *Let $X_1, X_2, \ldots$ be a sequence of independent random variables on (Ω, Σ, P) with means $\mu_1, \mu_2, \ldots,$ and variances $\sigma_1^2, \sigma_2^2, \ldots$. Let*

$$S_n = \sum_{k=1}^{n} (X_k - \mu_k)$$

and $\sigma^2 = \sum_{n=1}^{\infty} \sigma_n^2$. Suppose that $\sigma^2 < \infty$ and $\sum_{k=1}^{\infty} \mu_k$ converges. Then $\sum_{k=1}^{\infty} X_k$ converges a.e. and in the mean of order 2 to an r.v. X. Moreover, $E(X) = \sum_{k=1}^{\infty} \mu_k$, $Var X = \sigma^2$, and for any $\varepsilon > 0$,

$$P\left[\sup_{n\geq 1} |S_n| \geq \varepsilon\right] \leq \frac{\sigma^2}{\varepsilon^2} \tag{12}$$

Proof It should be shown that $\lim_n S_n$ exists a.e. If this is proved, since $\sum_{k=1}^{\infty} \mu_k$ converges, we get

$$\lim_n \sum_{k=1}^{n} X_k = \lim_{n\to\infty} S_n + \lim_{n\to\infty} \sum_{k=1}^{n} \mu_k = X \quad \text{exists a.e.}$$

But the sequence $\{S_n(\omega), n \geq 1\}$ of scalars converges iff it satisfies the Cauchy criterion, i.e., iff $\inf_m \sup_k |S_{m+k}(\omega) - S_m(\omega)| = 0$ a.e. Thus let $\varepsilon > 0$ be given, and by Theorem 5,

$$P\left[\max_{m\leq n\leq m+k} |S_n - S_m| \geq \varepsilon\right] \leq \frac{1}{\varepsilon^2} \sum_{i=m}^{m+k} \sigma_i^2 \leq \frac{1}{\varepsilon^2} \sum_{i\geq m} \sigma_i^2. \tag{13}$$

Hence letting $k \to \infty$ in (13) and noting that the events

$$\left[\max_{m\leq n\leq m+k} |S_n - S_m| \geq \varepsilon\right]$$

form an increasing sequence, we get

$$P\left[\sup_{n\geq 1} |S_{m+n} - S_m| \geq \varepsilon\right] \leq \frac{1}{\varepsilon^2} \sum_{k\geq m} \sigma_k^2. \tag{14}$$

It follows that

$$P\left[\inf_m \sup_{n\geq 1} |S_{m+n} - S_m| \geq \varepsilon\right] \leq \frac{1}{\varepsilon^2} \sum_{k\geq m} \sigma_k^2. \tag{15}$$

Letting $\varepsilon \nearrow \infty$, since $\sum_{k=1}^{\infty} \sigma_k^2 < \infty$, the right side of (15) goes to zero, so that $\limsup_{n,m} |S_n - S_m| < \infty$ a.e. But $|S_n| \leq |S_n - S_m| + |S_m|$, so

$$\begin{aligned}\limsup_{n\geq m}|S_n| &\leq \limsup_{n\geq m}[|S_n-S_m|+|S_m|]\\ &\leq |S_m|+\limsup_{n\geq m}|S_n-S_m|\\ &\leq |S_m|+\sup_{n\geq m}|S_n-S_m|<\infty \quad \text{a.e.}\end{aligned}$$

Thus $\limsup_n S_n$, $\liminf_n S_n$ must be finite a.e. Also

$$\left[\limsup_n S_n-\liminf_n S_n\geq 2\varepsilon\right]\subset\left[\sup_{n\geq m}|S_n-S_m|\geq\varepsilon\right], m\geq 1.$$

Hence by (14)

$$P\left[\limsup_n S_n-\liminf_n S_n\geq 2\varepsilon\right]\leq\frac{1}{\varepsilon^2}\sum_{k=m}^{\infty}\sigma_k^2\to 0 \tag{16}$$

as $m\to\infty$ for each $\varepsilon>0$. It follows that $\limsup_n S_n=\liminf_n S_n$ a.e., and the limit exists as asserted.

If we let $m=0$ in (14) and $X_0=0$, then (14) implies (12). It remains to establish mean convergence. In fact, consider for $m<n$, with $\tilde{X}_n=X_n-\mu_n$,

$$E((S_n-S_m)^2)=E((\tilde{X}_{m+1}+\ldots+\tilde{X}_n)^2)=\sum_{k=m+1}^{n}\sigma_k^2\to 0 \text{ as } m,n\to\infty. \tag{17}$$

Thus $S_n\to S$ in $\mathcal{L}^2(P)$, and hence also in $\mathcal{L}^1(P)$, since $\|f\|_1\leq\|f\|_2$ for any $f\in\mathcal{L}^2$. It follows that $E(S^2)=\lim_n E(S_n^2)=\lim_n\sum_{k=1}^n\sigma_k^2=\sigma^2$, and $E(S)=\lim_n E(S_n)=0$. But $X=S+\sum_{n=1}^\infty\mu_n$, so that $E(X)=\sum_{n=1}^\infty\mu_n$. This completes the proof.

Remarks (1) If we are given that $\lim_n\sum_{k=1}^n X_k$ exists in $\mathcal{L}^2$ and $\sum_{n=1}^\infty\mu_n$ converges, then $S_n=\sum_{k=1}^n(X_k-\mu_k)\to S$ in $\mathcal{L}^2$ also, so that $\sum_{k=1}^n\sigma_k^2=E(S_n^2)\to E(S^2)=\sigma^2$. Thus $\sum_{k=1}\sigma_k^2<\infty$. Hence by the theorem $\sum_{k=1}^\infty X_k$ also exists a.e.

(2) If the hypothesis of independence is simply dropped in the above theorem, the result is certainly false. In fact let $X_n=X/n$, where $E(X)=0$, $0<Var X=\sigma^2<\infty$, so that $\sum_{k=1}^\infty\mu_k=0$ and

$$\sum_{k\geq 1}\sigma_k^2=\sigma^2\sum_{n\geq 1}\frac{1}{n^2}<\infty.$$

But $\sum_{n=1}^\infty X_n=X\sum_{n=1}^\infty 1/n$, diverges a.e., on the set where $|X|>0$, a.e.

A partial converse of the above theorem is as follows.

Theorem 7 *Let $\{X_n,n\geq 1\}$ be a uniformly bounded sequence of independent random variables on (Ω,Σ,P) with means zero and variances*

$\{\sigma_n^2, n \geq 1\}$. If $\sum_{n=1}^{\infty} X_n$ converges on a set of positive measure, then $\sum_{n=1}^{\infty} \sigma_n^2 < \infty$, and hence the series actually converges a.e. on the whole space Ω.

Proof Let $X_0 = 0$ and $S_n = \sum_{i=1}^{n} X_i$. If A is the set of positive measure on which $S_n \to S$ a.e., then by Theorem 4 (of Egorov), there is a measurable subset $\tilde{A} \subset A$ of arbitrarily small measure such that if $B_0 = A - \tilde{A} \subset A$, we have $P(B_0) > 0$ and $S_n \to S$ on B_0 uniformly. Since S is an r.v., we can find a set $B \subset B_0$ of positive measure (arbitrarily close to that of B_0), and a positive number d such that $|S_n| \leq d < \infty$ on B. Thus if $\overline{A} = \bigcap_{n=0}^{\infty}[|S_n| \leq d]$, then $\overline{A} \in \Sigma$, $\overline{A} \supset B$, and $P(\overline{A}) \geq P(B) > 0$.

Let $A_n = \bigcap_{k=0}^{n}[|S_k| \leq d]$, so that $A_n \downarrow A$. If $C_n = A_n - A_{n+1}$, and $C_0 = \bigcup_{n=1}^{\infty} C_n$, which is a disjoint union, let $a_n = \int_{A_n} S_n^2 dP$. Clearly $a_n \leq d^2 P(A_n) \leq d^2$, so that $\{a_n, n \geq l\}$ is a bounded sequence. Consider

$$\begin{aligned} a_n - a_{n-1} &= \int_{A_n} S_n^2 dP - \int_{A_{n-1}} S_{n-1}^2 dP \\ &= \int_{A_{n-1}} (S_{n-1} + X_n)^2 dP \\ &- \int_{C_{n-1}} S_n^2 dP - \int_{A_{n-1}} S_{n-1}^2 dP \quad (\text{since} A_n = A_{n-1} - C_{n-1}) \\ &= \int_{A_{n-1}} X_n^2 dP + 2\int_{A_{n-1}} S_{n-1} X_n dP - \int_{C_{n-1}} S_n^2 dP. \end{aligned} \tag{18}$$

However,

$$\int_{A_{n-1}} X_n^2 dP = E(\chi_{A_{n-1}} X_n^2) = \sigma_n^2 P(A_{n-1}) \quad \text{by independence of } \chi_{A_{n-1}} \text{ and } X_n,$$

and

$$\int_{A_{n-1}} X_n S_{n-1} dP = E(X_n) E(\chi_{A_{n-1}} S_{n-1}) = 0,$$

since $E(X_n) = 0$. Thus by noting that $P(A_{n-1}) \geq P(A_n)$, (18) becomes, with these simplifications and the hypothesis that $|X_n| \leq c < \infty$ a.e.,

$$\begin{aligned} a_n - a_{n-1} &\geq \sigma_n^2 P(A_n) - \int_{C_{n-1}} (|S_{n-1}| + |X_n|)^2 dP \\ &\geq \sigma_n^2 P(A_n) - (c+d)^2 P(C_{n-1}) \\ &\quad (\text{since } |S_n| \leq |S_{n-1}| + |X_n| \leq d + c\,), \\ &\geq \sigma_n^2 P(\overline{A}) - (c+d)^2 P(C_{n-1}). \end{aligned}$$

Summing over $n = 1, 2, \ldots, m$, we get ($a_0 = 0$)

$$a_m \geq P(\overline{A}) \sum_{n=1}^{m} \sigma_n^2 - (c+d)^2 P(\bigcup_{n=1}^{m} C_{n-1}).$$

Hence recalling that $a_m \leq d^2$, one has

$$d^2 \geq P(\overline{A}) \sum_{n=1}^{m} \sigma_n^2 - (c+d)^2, \quad m \geq 1. \tag{19}$$

Since $P(\overline{A}) > 0$, (19) implies that $\sum_{n=1}^{\infty} \sigma_n^2 < \infty$. This yields the last statement and, in view of Theorem 6, completes the proof.

As an immediate consequence, we have

Corollary 8 *If $\{X_n, n \geq 1\}$ is a uniformly bounded sequence of independent random variables on (Ω, Σ, P) with $E(X_n) = 0, n \geq 1$, then $\sum_{n=1}^{\infty} X_n$ converges with probability 0 or 1.*

We are now in a position to establish a very general result on this topic.

Theorem 9 (Three Series Theorem). *Let $\{X_n, n \geq 1\}$ be a sequence of independent random variables on (Ω, Σ, P). Then $\sum_{n=1}^{\infty} X_n$ converges a.e. iff the following three series converge. For some (and then every) $0 < c < \infty$,*
(i) $\sum_{n=1}^{\infty} P([|X_n| > c])$,
(ii) $\sum_{n=1}^{\infty} E(X_n^c)$,
(iii) $\sum_{n=1}^{\infty} \sigma^2(X_n^c)$,
where X_n^c is the truncation of X_n at c, so that $X_n^c = X_n$ if $|X_n| \leq c$, and $= 0$ otherwise.

Proof Sufficiency is immediate. In fact, suppose the three series converge. By (i), and the first Borel-Cantelli lemma, $P[\limsup_n |X_n| > c] = 0$, so that for large enough n, $X_n = X_n^c$ a.e. Next, the convergence of (ii) and (iii) imply, by Theorem 6, $\sum_{n=1}^{\infty} X_n^c$ converges a.e. Since $X_n = X_n^c$ for large n, $\sum_{m=1}^{\infty} X_n$ itself converges a.e. Note that $c > 0$ is arbitrarily fixed.

Conversely, suppose $\sum_{i=1}^{\infty} X_i$ converges a.e. Then $\lim_n X_n = 0$ a.e. Hence if $A_{n,c} = [X_n \neq X_n^c] = [|X_n| > c]$ for any fixed $c > 0$, then the $A_{n,c}$ are independent and $P[\limsup_n A_{n,c}] = 0$. Thus by the second Borel-Cantelli lemma (cf. Theorem 1.9iii), $\sum_{n=1}^{\infty} P(A_{n,c}) < \infty$, which proves (i). Also, $\sum_{n=1}^{\infty} X_n^c$ converges a.e., since for large enough n, X_n^c and X_n are equal a.e. But now the X_n^c are uniformly bounded. We would like to reduce the result to Theorem 7. However, $E(X_n^c)$ is not necessarily zero. Thus we need a new idea for this reduction. One considers a sequence of independent random variables $\tilde{X}_n^c$ which are also independent of, but with the same distributions as, the

X_n^c-sequence. Now, the given probability space may not support two such sequences. In that case, we enlarge it by adjunction as explained after Corollary 8 in the last section. The details are as follows.

Let $(\tilde{\Omega}, \tilde{\Sigma}, \tilde{P}) = (\Omega, \Sigma, P) \otimes (\Omega, \Sigma, P)$, and let X_n^1, X_n^2; be defined on $\tilde{\Omega}$ by the equations

$$X_n^1(\omega) = X_n^c(\omega_1), X_n^2(\omega) = X_n^c(\omega_2), \quad \text{where } \omega = (\omega_1, \omega_2) \in \tilde{\Omega}. \tag{20}$$

It is trivial to verify that $\{X_n^1, n \geq 1\}$, $\{X_n^2, n \geq 1\}$ are two mutually independent sequences of random variables on $(\tilde{\Omega}, \tilde{\Sigma}, \tilde{P})$, $|X_n^i| \leq c$, $i = 1, 2$, and have the same distributions. Thus if $Z_n = X_n^1 - X_n^2$, $n \geq 1$, then $E(Z_n) = 0$, $Var Z_n = Var X_n^1 + Var X_n^2 = 2\sigma_n^2(X_n^c)$, and $\{Z_n, n \geq 1\}$ is a uniformly bounded (by 2c) independent sequence to which Theorem 7 applies. Hence, by that result, $\sum_{n=1}^{\infty} Var Z_n < \infty$, so that $\sum_{n=1}^{\infty} \sigma_n^2(X_n^c) < \infty$, which is (iii).

Next, if $Y_n = X_n^c - E(X_n^c)$, then $E(Y_n) = 0$, $Var Y_n = Var X_n^c$, so that $\sum_{n=1}^{\infty} \sigma^2(Y_n) < \infty$. Hence by Theorem 6, $\sum_{n=1}^{\infty} Y_n$ converges a.e. Thus we have $\sum_{n=1}^{\infty} E(X_n^c) = \sum_{n=1}^{\infty} X_n^c - \sum_{n=1}^{\infty} Y_n$, and both the series on the right converge a.e. Thus the left side, which is a series of constants, simply converges and (ii) holds. Observe that if the result is true for one $0 < c < \infty$, then by this part the three series must converge for every $0 < c < \infty$. This completes the proof.

Remarks (1) If any one of the three series of the above theorem diverges, then $\sum_{n \geq 1} X_n$ diverges a.e. This means the set $[\sum_{n=1}^{\infty} X_n$ converges$]$ has probability zero, so that the zero-one criterion obtains. The proof of this statement is a simple consequence of the preceding results (since the convergence is determined by $\sum_{k \geq n} X_k$ for large n), but not of Theorem 1.12.

(2) Observe that the convergence statements on series in all these theorems relate to *unconditional convergence*. It is *not absolute* convergence, as simple examples show. For instance, if $a_n > 0, \sum_{n=1}^{\infty} a_n = \infty$, but

$$\sum_{n=1}^{\infty} a_n^2 < \infty,$$

then the independent random variables $X_n = \pm a_n$ with equal probability on (Ω, Σ, P) satisfy the hypothesis of Corollary 8 and so $\sum_{n=1}^{\infty} X_n$ converges a.e. But it is clear that $\sum_{n=1}^{\infty} |X_n| = \sum_{n=1}^{\infty} |a_n| = \infty$ a.e. The point is that $X_n \in L^2(\Omega, \Sigma, P)$ and the series $\sum_{n=1}^{\infty} X_n$ converges unconditionally in $L^2(P)$, but not absolutely there if the space is infinite dimensional. In fact, it is a general result of the Banach space theory that the above two convergences are unequal in general.

(3) One can present easy sufficient conditions for absolute convergence of a series of random variables on (Ω, Σ, P). Indeed, $\sum_{n=1}^{\infty} X_n$ converges absolutely a.e. if $\sum_{n=1}^{\infty} E(|X_n|) < \infty$. This is true since $E(\sum_{n=1}^{\infty} |X_n|) = \sum_{n=1}^{\infty} E(|X_n|) < \infty$ by the Lebesgue dominated convergence theorem, and since $Y = \sum_{n=1}^{\infty} |X_n|$ is a (positive) r.v. with finite expectation, $P[Y > \lambda] \leq$

$E(Y)/\lambda \to 0$ as $\lambda \to \infty$, so that $0 \leq Y < \infty$ a.e. Here X_n need not be independent. But the integrability condition is very stringent. Such results are "nonprobabilistic" in nature, and are not of interest in our subject.

A natural question now is to know the properties of the limit r.v. $X = \sum_{n=1}^{\infty} X_n$ in Theorem 9 when the series converges. For example: if each X_n has a countable range, which is a simple case, what can one say about the distribution of X? What can one say about $Y = \sum_{n=1}^{\infty} a_n X_n$, where $\sum_{n=1}^{\infty} a_n^2 < \infty$, $E(X_n) = 0$, $E(X_n^2) = 1$, and X_n are independent?

Not much is known about these queries. Some special cases are studied, and a sample result is discussed in the problems section. For a deeper analysis of special types of random series, one may refer to Kahane (1985). We now turn to the next important aspect of averages of independent random variables, which has opened up interesting avenues for probability theory.

2.3 Laws of Large Numbers

Very early in Section 1.1 we indicated that probability is a "long-term average." This means that the averages of "successes" in a sequence of independent trials "converge" to a number. As the preceding section shows, there are three frequently used types of convergences, namely, the pointwise a.e., the stochastic (or "in probability") convergence, and the distributional convergence, each one being strictly weaker than the preceding one. The example following the proof of Proposition 2.2 shows that $X_n \xrightarrow{D} X$ does not imply that the $X_n(\omega)$ need to approach $X(\omega)$ for any $\omega \in \Omega$. So in general it is better to consider the a.e. and "in probability" types for statements relating to outcomes of ω. Results asserting a.e. convergence always imply the "in probability" statements, so that the former are called *strong laws* and the latter, *weak laws.* If the random variables take only two values $\{0, 1\}$, say, then the desired convergence in probability of the averages was first rigorously established by James Bernoulli in about the year 1713, and the a.e. convergence result for the same sequence was obtained by E. Borel only in 1909.

Attempts to prove the same statements for general random variables, with range space $\mathbb{R}$, and the success thus achieved constitute a general story of the subject at hand. In fact, P. L. Čebyšev seems to have devised his inequality for extending the Bernoulli theorem, and established the following result in 1882.

Proposition 1 (Čebyšev). *Let $X_1, X_2, \ldots$ be a sequence of independent random variables on (Ω, Σ, P) with means $\mu_1, \mu_2, \ldots$ and variances $\sigma_1^2, \sigma_2^2, \ldots$, such that if $S_n = \sum_{i=1}^{n} X_i$, one has $\sigma^2(S_n)/n^2 \to 0$ as $n \to \infty$. Then the sequence obeys the weak law of large numbers (WLLN), which means, given $\varepsilon > 0$, we have*

$$\lim_{n\to\infty} P\left[\left|\frac{S_n - E(S_n)}{n}\right| \geq \varepsilon\right] = 0. \tag{1}$$

Proof By Čebyšev's inequality (1) follows at once.

Note that if all the X_n have the same distribution, then they have equal moments, i.e., $\sigma_1^2 = \sigma_2^2 = ... = \sigma^2$, so that $\sigma^2(S_n) = \sum_{i=1}^n \sigma_i^2 = n\sigma^2$, and $\sigma^2(S_n)/n^2 = \sigma^2/n \to 0$ is automatically satisfied. The result has been improved in 1928 by A. Khintchine, by assuming just one moment. For the proof, he used a truncation argument, originally introduced in 1913 by A. A. Markov. Here we present this proof as it became a powerful tool. Later we see that the result can be proved, using the characteristic function technique, in a very elementary manner, and even with a slightly *weaker* hypothesis than the existence of the first moment [i.e., only with the existence of a derivative at the origin for its Fourier transform; that does *not* imply $E(X)$ exists].

Theorem 2 **(Khintchine)** *Let $X_1, X_2, \ldots$ be independent random variables on (Ω, Σ, P) with a common distribution [i.e., $P[X_n < x] = F(x), x \in \mathbb{R}$, for $n \geq 1$] and with one moment finite. Then the sequence obeys the WLLN.*

Proof We use the preceding result in the proof for the truncated functions and then complete the argument with a detailed analysis. Let $\varepsilon > 0, \delta > 0$ be given. Define

$$U_k^n = X_k \chi_{[|X_k| \leq n\delta]}, \quad V_k^n = X_k \chi_{[|X_k| > n\delta]}, \tag{2}$$

so that $X_k = U_k + V_k$. Let F be the common distribution function of the X_k. Since $E(|X_k|) < \infty$, we have $M = E(|X_k|) = \int_{\mathbb{R}} |x| dF(x) < \infty$, by the fundamental (image) law of probability. If $\mu = E(X_k) = \int_{\mathbb{R}} x dF(x)$ and $\mu_n' = E(U_k^n)$, then

$$\mu_n' = E(X_k \chi_{[|X_k| \leq n\delta]}) = \int_{[|x| \leq n\delta]} x dF(x)$$

and by the dominated convergence theorem, we have

$$\int_{[|x| \leq n\delta]} x dF(x) = \int_{\mathbb{R}} \chi_{[|X_k| \leq n\delta]} x dF(x) \to \int_{\mathbb{R}} x dF(x) = \mu \tag{3}$$

Thus there is N_1 such that $n \geq N_1 \Rightarrow |\mu_n' - \mu| < \varepsilon/2$. Note that μ_n' depends only on n, and hence not on k, because of the common distribution of the X_k. Similarly

$$Var U_k^n = E(U_k^n)^2 - (\mu_n')^2 \leq \int_{|x| \leq n\delta} x^2 dF(x) \leq n\delta \int_{\mathbb{R}} |x| dF(x) = n\delta M.$$

By hypothesis $U_1^n, U_2^n, ...$ are independent (bounded) random variables with means μ_n' and variances bounded by $n\delta M$. Let $T_m^n = U_1^n + \ldots + U_m^n$ and

$W_m^n = V_1^n + \ldots + V_m^n$. Then by the preceding proposition, or rather the Čebyšev's inequality,

$$P\left[\left|\frac{T_n^n - n\mu_n'}{n}\right| \geq \frac{\varepsilon}{2}\right] \leq \frac{4n^2\delta M}{n^2\varepsilon^2} = 4\delta M\varepsilon^{-2}. \tag{4}$$

On the other hand, adding and subtracting $n\mu$ and using the triangle inequality gives

$$\left|\frac{T_n^n - n\mu_n'}{n}\right| \geq \left|\frac{T_n^n - n\mu}{n}\right| - |\mu - \mu_n'|.$$

Thus if $n \geq N_1$, we have, with the choice of N_1 after (2), on the set

$$\left|\frac{T_n^n - n\mu}{n}\right| \geq \varepsilon,$$

the following:

$$\left|\frac{T_n^n - n\mu_n'}{n}\right| \geq \varepsilon - \frac{\varepsilon}{2} = \frac{\varepsilon}{2}.$$

Hence for $n \geq N_1$ this yields

$$P\left[\left|\frac{T_n^n - n\mu}{n}\right| \geq \varepsilon\right] \leq P\left[\left|\frac{T_n^n - n\mu_n'}{n}\right| \geq \frac{\varepsilon}{2}\right] \leq \frac{4\delta M}{\varepsilon^2}, \text{ by (4)}. \tag{5}$$

But by definition $S_n = T_n^n + W_n^n, n \geq 1$, so that

$$\omega \in \left[\left|\frac{S_n - n\mu}{n}\right| \geq \varepsilon\right] = \left[\left|\frac{T_n^n - n\mu + W_n^n}{n}\right| \geq \varepsilon\right]$$

$$\Rightarrow \omega \in \left[\left|\frac{T_n^n - n\mu}{n}\right| \geq \frac{\varepsilon}{2}\right] \cup [W_n^n \neq 0].$$

Thus

$$P\left[\left|\frac{S_n - n\mu}{n}\right| \geq \varepsilon\right] \leq P\left[\left|\frac{T_n^n - n\mu}{n}\right| \geq \frac{\varepsilon}{2}\right] + P[W_n^n \neq 0]. \tag{6}$$

However,

$$\begin{aligned} P[V_k^n \neq 0] &= P[|V_k^n| > 0] = P[|X_k| > n\delta] \\ &= \frac{1}{n\delta}\int_{[|X_k|>n\delta]} n\delta dP \leq \frac{1}{n\delta}\int_{[|X_k|>n\delta]} |X_k| dP \\ &= \frac{1}{n\delta}\int_{[|x|>n\delta]} |x| dP. \end{aligned} \tag{7}$$

Choose N_2 such that $n \geq N_2 \Rightarrow \int_{[|x|>n\delta]} |x| dF(x) < \delta^2$, which is possible since $M = E(|X_k|) < \infty$. Thus for $n \geq N_2$, $P[V_k^n \neq 0] \leq \delta^2/(n\delta) = \delta/n$ by (7). Consequently,

$$P[W_n^n \neq 0] \leq P\left(\bigcup_{k=1}^{n}[V_k^n \neq 0]\right) \leq \sum_{k=1}^{n} P[V_k^n \neq 0] < \delta. \tag{8}$$

If $N = \max(N_1, N_2)$ and $n \geq N$, then (5) and (8) give for (6)

$$P\left[\left|\frac{S_n - n\mu}{n}\right| \geq \varepsilon\right] \leq \frac{4\delta M}{\varepsilon^2} + \delta = \left[\frac{4M}{\varepsilon^2} + 1\right]\delta. \tag{9}$$

Letting $n \to \infty$ and then $\delta \to 0$ in (9), we get the desired conclusion.

It is important to notice that the independence hypothesis is used only in (4) in the above proof in deducing that the variance of T_n^n = the sum of variances of U_k^n. But this will follow if the U_k^n are uncorrelated for each n. In other words, we used only that

$$\begin{aligned} E(U_k^n U_j^m) &= \int_{[|x|\geq n\delta]} \int_{[|y|\leq m\delta]} xy dF_{X_k, X_j}(x, y) \\ &= \int_{[|x|\geq n\delta]} x dF_{X_k}(x) \int_{[|y|\leq m\delta]} y dF_{X_j}(y) = \mu_n' \mu_m'. \end{aligned}$$

Now this holds if X_i, X_j are independent when $i \neq j$. Thus the above proof actually yields the following stronger result, stated for reference.

Corollary 3 *Let $X_1, X_2, \ldots$ be a pairwise independent sequence of random variables on (Ω, Σ, P) with a common distribution having one moment finite. Then the sequence obeys the WLLN.*

In our development of the subject, the next result serves as a link between the preceding considerations and the "strong laws." It was obtained by A. Rajchman in the early 1930's. The hypothesis is weaker than pairwise independence, but demands the existence of a uniform bound on variances, and then yields a stronger conclusion. The proof uses a different technique, of interest in the subject.

Theorem 4 *Let $\{X_n, n \geq 1\}$ be a sequence of uncorrelated random variables on (Ω, Σ, P) such that $\sigma^2(X_n) \leq M < \infty$, $n \geq 1$. Then $[S_n - E(S_n)]/n \to 0$ in L^2-mean, as well as a.e. [The pointwise convergence statement is the definition of the strong law of large numbers (SLLN) of a sequence.]*

Proof The first statement is immediate, since

$$E\left(\left[\frac{S_n - E(S_n)}{n}\right]^2\right) = \frac{1}{n^2}\sum_{k=1}^{n}\sigma^2(X_k) \leq \frac{nM}{n^2} \to 0,$$

by the uncorrelatedness hypothesis of the X_n and the uniform boundedness of $\sigma^2(X_k)$. This, of course, implies by Proposition 1 that the *WLLN* holds for the sequence. The point is that the a.e. convergence also holds.

Consider now, by Čebyšev's inequality, for any $\varepsilon > 0$,

$$P[|S_{n^2} - E(S_{n^2})| > n^2\varepsilon] \leq \frac{Var S_{n^2}}{n^4\varepsilon^2} \leq \frac{M}{n^2\varepsilon^2},$$

so that

$$\sum_{n\geq 1} P[|S_{n^2} - E(S_{n^2})| > n^2\varepsilon] \leq \frac{M}{\varepsilon^2}\sum_{n\geq 1}\frac{1}{n^2} < \infty. \tag{10}$$

Hence by the first Borel-Cantelli lemma, letting $Y_k = X_k - E(X_k)$ and $\tilde{S}_n = \sum_{k=1}^n Y_k$ (so that the Y_k are orthogonal), one has $P([|\tilde{S}_{n^2}| > n^2\varepsilon], i.o.) = 0$, which means $\tilde{S}_{n^2}/n^2 \to 0$ a.e. This is just an illustration of Proposition 2.3.

With the boundedness hypothesis we show that the result holds for the full sequence $\tilde{S}_n/n$ and not merely for a subsequence, noted above.

For each $n \geq 1$, consider $n^2 \leq k < (n+1)^2$ and $\tilde{S}_k/k$. Then

$$\frac{|\tilde{S}_k|}{k} \leq \frac{|\tilde{S}_k - \tilde{S}_{n^2}| + |\tilde{S}_{n^2}|}{k} \leq \frac{|\tilde{S}_{n^2}|}{n^2} + \max_{n^2\leq k<(n+1)^2}\frac{|\tilde{S}_k - \tilde{S}_{n^2}|}{n^2} \tag{11}$$

and let $T_n = \max_{n^2\leq k<(n+1)^2}|\tilde{S}_k - \tilde{S}_{n^2}|$. Since as $k \to \infty$ the first term on the right $\to 0$ a.e., (shown above) it suffices to establish $T_n/n^2 \to 0$ a.e. To use the orthogonality property of the Y_k, consider T_n^2. We have

$$T_n^2 = \max_{n^2\leq k\leq(n+1)^2}|\tilde{S}_k - \tilde{S}_{n^2}|^2 \leq \sum_{k=n^2}^{(n+1)^2}|\tilde{S}_k - \tilde{S}_{n^2}|^2 = \sum_{k=n^2}^{(n+1)^2}\left(\sum_{i=n^2+1}^{k} Y_i\right)^2$$

and so for $n \geq 2$, since $\sigma^2(Y_i) = \sigma^2(X_i) \leq M$,

$$\begin{aligned} E(T_n^2) &\leq \sum_{k=n^2}^{(n+1)^2}\sum_{i=n^2+1}^{k}\sigma^2(Y_i) \\ &\leq \sum_{k=n^2}^{(n+1)^2}\sum_{i=n^2+1}^{k} M = \sum_{k=n^2}^{(n+1)^2}(k-(n^2+1))M \\ &= M(2n^2+n) \leq 3n^2M < (2n)^2M \end{aligned} \tag{12}$$

This crude estimate is sufficient to show, as before, that

$$P[|T_n| > n^2\varepsilon] \leq P[T_n^2 \geq n^4\varepsilon^2] \leq E(T_n^2)/n^4\varepsilon^2 \leq 4M/n^2\varepsilon^2, \tag{13}$$

by Markov's inequality and (12). Thus $\sum_{n>2} P[|T_n| > n^2\varepsilon] < \infty$ and the Borel-Cantelli lemma again yields $P[|T_n/n^2| > \varepsilon, i.o.] = 0$. Hence $T_n/n^2 \to 0$ a.e. and by (11) $\tilde{S}_k/k \to 0$ a.e., proving the result.

We now strengthen the probabilistic hypothesis from uncorrelatedness to mutual independence and weaken the moment condition. The resulting statement is significantly harder to establish. It will be obtained in two stages, and both are of independent interest. They have been proved in 1928 by A. Kolmogorov, and are sharp. We begin with an elementary but powerful result from classical summability theory.

Proposition 5 (Kronecker's Lemma). *Let $a_1, a_2, \ldots$ be a sequence of numbers such that $\sum_{n\geq 1}(a_n/n)$ converges. Then*

$$\frac{1}{n}\sum_{k=1}^{n} a_k \to 0 \text{ as } n \to \infty.$$

Proof Let $s_0 = 0$, $s_n = \sum_{k=1}^{n}(a_k/k)$, and $R_n = \sum_{k=1}^{n} a_k$. Then $s_n \to s$ by hypothesis. Also, $a_k = k(s_k - s_{k-1})$, so that

$$R_{n+1} = \sum_{k=1}^{n+1} ks_k - \sum_{k=1}^{n+1} ks_{k-1} = -\sum_{k=1}^{n} s_k + (n+1)s_{n+1}.$$

Hence

$$\frac{1}{n+1}R_{n+1} = s_{n+1} - \frac{n}{n+1}\frac{1}{n}\sum_{k=1}^{n} s_k \to s - 1 \cdot s = 0 \text{ as } n \to \infty$$

because $s_n \to s \Rightarrow$ for any $\varepsilon > 0$, there is $n_0[= n_o(\varepsilon)]$ such that $n > n_0 \Rightarrow |s_n - s| < \varepsilon$, and hence

$$\left|\frac{1}{n}\sum_{k=1}^{n} s_k - s\right| \leq \frac{|s_1 + \ldots + s_{n_0}|}{n} + \frac{|(s_{n_0+1} - s) + \ldots (s_n - s)|}{n} + \frac{n_0|s|}{n}$$

$$\leq \frac{|s_1 + \ldots + s_{n_0}|}{n} + \frac{(n - n_0)}{n}\varepsilon + \frac{n_0|s|}{n} \to \varepsilon \tag{14}$$

as $n \to \infty$. [This is called $(c, 1)$-convergence or *Cesàro summability* of s_n.] Since $\varepsilon > 0$ is arbitrary, the result follows.

Theorem 6 (First form of SLLN). *If $X_1, X_2, \ldots$ is a sequence of independent random variables on (Ω, Σ, P) with means zero and variances $\sigma_1^2, \sigma_2^2, \ldots$, satisfying $\sum_{n=1}^{\infty}(\sigma_n^2/n^2) < \infty$, then the sequence obeys the SLLN, i.e.,*

$$\frac{X_1 + X_2 + \ldots + X_n}{n} \to 0$$

a.e. as $n \to \infty$.

Proof Let $Y_n = X_n/n$. Then the $Y_n, n \geq 1$, are independent with means zero, and $\sum_{n=1}^{\infty}\sigma^2(Y_n) = \sum_{n=1}^{\infty}(\sigma_n^2/n^2) < \infty$. Thus by Theorem 2.6, $\sum_{n=1}^{\infty} Y_n$

converges a.e. Hence $\sum_{n=1}^{\infty}(X_n/n)$ converges a.e. By Kronecker's lemma, $(1/n)\sum_{k=1}^{n} X_k \to 0$ a.e., proving the theorem.

This result is very general in that there are sequences of independent random variables $\{X_n, n \geq 1\}$ with means zero and finite variances $\sigma_1^2, \sigma_2^2, \ldots$ satisfying $\sum_{n=1}^{\infty}(\sigma_n^2/n^2) = \infty$ for which the SLLN does not hold. Here is a simple example. Let $X_1, X_2, \ldots$ be independent two-valued random variables, defined as

$$P([X_n = n]) = P([X_n = -n]) = \frac{1}{2}.$$

Hence $E(X_n) = 0, \sigma^2(X_n) = n^2$, so that $\sum_{n=1}^{\infty}[\sigma^2(X_n)/n^2] = +\infty$. If the sequence obeys the SLLN, then $(\sum_{k=1}^{n} X_k)/n \to 0$ a.e. This implies

$$\frac{X_n}{n} = \frac{1}{n}\sum_{k=1}^{n} X_k - \frac{n-1}{n} \cdot \frac{1}{n-1}\sum_{k=1}^{n-1} X_k \to 0 \text{ a.e.},$$

hence $P[|X_n| \geq n, i.o.] = 0$. By independence, this and the second Borel-Cantelli lemma yield $\sum_{n=1}^{\infty} P[|X_n| \geq n] < \infty$. However, by definition $P[|X_n| \geq n] = 1$, and this contradicts the preceding statement. Thus $(1/n)\sum_{k=1}^{n} X_k \not\to 0$ a.e., and SLLN is not obeyed.

On the other hand, the above theorem is still true if we make minor relaxations on the means. For instance, if $\{X_n, n \geq 1\}$ is independent with means $\{\mu_n, n \geq 1\}$ and variances $\{\sigma_n^2; , n \geq 1\}$ such that (i) $\sum_{n=1}^{\infty} \sigma_n^2 < \infty$ and (ii) if either $\mu_n \to 0$ or just $(1/n)\sum_{k=1}^{n} \mu_k \to 0$ as $n \to \infty$, then $(1/n)\sum_{k=1}^{n} X_k \to 0$ a.e. Indeed, if $Y_n = X_n - \mu_n$, then $\{Y_n, n \geq 1\}$ satisfies the conditions of the above result. Thus $(1/n)\sum_{k=1}^{n} Y_k = (1/n)\sum_{k=1}^{n} X_k - (1/n)\sum_{k=1}^{n} \mu_k \to 0$ a.e. If $\mu_k \to \mu$, then $(1/n)\sum_{k=1}^{n} \mu_k \to \mu$ by (14). Here $\mu = 0$. The same holds if we only demanded $(1/n)\sum_{k=1}^{n} \mu_k \to 0$. In either case, then, $(1/n)\sum_{k=1}^{n} X_k \to 0$ a.e. However, it should be remarked that there exist independent symmetric two-valued X_n, $n \geq 1$, with $\sum_{n \geq 1}[\sigma^2(X_n)/n^2] = \infty$ obeying the SLLN. Examples can be given to this effect, if we have more information on the growth of the partial sums $\{S_n, n \geq 1\}$, through, for instance, the laws of the iterated logarithm. An important result on the latter subject will be established in Chapter 5.

The following is the celebrated SLLN of Kolmogorov.

Theorem 7 (Main SLLN). *Let $\{X_n, n \geq 1\}$ be independent random variables on (Ω, Σ, P) with a common distribution and $S_n = \sum_{k=1}^{n} X_k$. Then $S_n/n \to \alpha_0$, a constant, a.e. iff $E(|X_1|) < +\infty$, in which case $\alpha_0 = E(X_1)$. On the other hand, if $E(|X_1|) = +\infty$, then $\limsup_n(|S_n|/n) = +\infty$ a.e.*

Proof To prove the sufficiency of the first part, suppose $E(|X_1|) < \infty$. We use the truncation method of Theorem 2. For simplicity, let $E(X_1) = 0$, since otherwise we consider the sequence $Y_k = X_k - E(X_1)$. For each n, define

$$U_n = X_n\chi_{[|X_n|\le n]}, \quad V_n = X_n\chi_{[|X_n|>n]}.$$

Thus $X_n = U_n + V_n$ and $\{U_n, n \ge 1\}$, $\{V_n, n \ge 1\}$ are independent sequences. First we claim that $\limsup_n |V_n| = 0$ a.e., implying $(1/n)\sum_{k=1}^{n} V_k \to 0$ a.e. That is to say, $P([V_n \ne 0], i.o.) = 0$. By independence, and the Borel-Cantelli lemma, this is equivalent to showing $\sum_{n=1}^{\infty} P[|V_n| > 0] < \infty$.

Let us verify the convergence of this series:

$$\begin{aligned}
\sum_{n=1}^{\infty} P[V_n \ne 0] &= \sum_{n=1}^{\infty} P[|X_n| > n] \\
&= \sum_{n=1}^{\infty} P[|X_1| > n] \text{ (since the } X_i \text{ have the same distribution)} \\
&= \sum_{n=1}^{\infty}\sum_{k=n}^{\infty} P[k < |X_1| \le k+1] \\
&= \sum_{n=1}^{\infty} n a_{n,n+1} \text{ (where } a_{k,k+1} = P[k < |X_1| \le k+1]) \\
&= \sum_{n=1}^{\infty} \int_{[n<|X_1|\le n+1]} n dP \le \int_{\Omega} |X_1| dP = E(|X_1|) < \infty \qquad (15)
\end{aligned}$$

Next consider the bounded sequence $\{U_n, n \ge 1\}$ of independent random variables. If $\mu_n = E(U_n)$, then

$$\mu_n = \int_{[|X_1|\le n]} X_1 dP \to \int_{\Omega} X_1 dP = E(X_1) = 0$$

by the dominated convergence theorem. Hence $(1/n)\sum_{k=1}^{n} \mu_k \to 0$. Thus by the remark preceding the statement of the theorem, if $\sum_{n=1}^{\infty}[\sigma^2(U_n)/n^2] < \infty$, then $(1/n)\sum_{n=1}^{\infty} U_n \to 0$ a.e., and the result follows.

We verify the desired convergence by a computation similar to that used in (15). Thus

$$\begin{aligned}
\sigma^2(U_n) &= E(U_n^2) - \mu_n^2 \le E(U_n^2) \\
&= \int_{[|X_1|\le n]} X_1^2 dP \text{ (by the common distribution of the } X_n) \\
&\le \sum_{k=0}^{n-1} (k+1)^2 P[k < |X_1| \le k+1].
\end{aligned}$$

Hence

$$\sum_{n=1}^{\infty}\frac{\sigma^2(U^2)}{n^2} \le \sum_{n=1}^{\infty}\sum_{k=0}^{n-1}\frac{(k+1)^2}{n^2}a_{k,k+1} \ [\text{ using the notation of (15)}]$$

$$= \sum_{k=1}^{\infty} k^2 a_{k-1,k}\sum_{n=k}^{\infty}\frac{1}{n^2}$$

$$\le \sum_{k=1}^{\infty} k^2\cdot\frac{2}{k}a_{k-1,k} \ \left[\text{since } \sum_{n=k}^{\infty}\frac{1}{n^2} \le \frac{1}{k} + \int_{k+1}^{\infty}\frac{dx}{(x-1)^2} \le \frac{2}{k}\right]$$

$$= 2\sum_{k=1}^{\infty} k a_{k-1,k} \le 2E(|X_1|) + 2 < \infty \ [\text{ by(15)}].$$

Thus

$$\frac{1}{n}\sum_{k=1}^{n} X_k = \frac{1}{n}\sum_{k=1}^{n} U_k + \frac{1}{n}\sum_{k=1}^{n} V_k \to 0 \text{ a.e.}$$

as $n \to \infty$.

Conversely, suppose that $S_n/n \to \alpha_0$, a constant, a.e. We observe that

$$\frac{X_n}{n} = \frac{S_n}{n} - \frac{n-1}{n}\cdot\frac{1}{n-1}S_{n-1} \to 0 \text{ a.e.},$$

so that $\limsup_n(|X_n|/n) = 0$ a.e. Again by the Borel-Cantelli lemma, this is equivalent to saying that $\sum_{n=1}^{\infty} P[|X_n| > n] < \infty$. But

$$\sum_{n=1}^{\infty} P[|X_n| > n] = \sum_{n=1}^{\infty} P[|X_1| > n]$$

$$= \sum_{n=1}^{\infty} n a_{n,n+1} \ [\text{ as shown for (15)}]$$

$$= \sum_{n=1}^{\infty}(n+1)P[n < |X_1| \le n+1]$$

$$- \sum_{n=1}^{\infty} P[n < |X_1| \le n+1]$$

$$\ge E(|X_1|) - 2. \tag{16}$$

Hence $E(|X_1|) < \infty$. Then by the sufficiency $(1/n)S_n \to E(X_1)$ a.e., so that $\alpha_0 = E(X_1)$, as asserted.

For the last part, suppose that $E(|X_1|) = +\infty$, so that $E(|X_1|/\alpha) = +\infty$ for any $\alpha > 0$. Then the computation for (16) implies

$$\sum_{n=1}^{\infty} P[|X_n| > \alpha n] = +\infty,$$

since the X_n have the same distribution. Consequently, by the second Borel-Cantelli lemma, we have

$$P([|X_n| > \alpha n], i.o.) = 1. \tag{17}$$

But $|S_n - S_{n-1}| = |X_n| > \alpha n$ implies either $|S_n| > \alpha n/2$ or $|S_{n-1}| > \alpha n/2$. Thus (17) and this give

$$P\left(\left[|S_n| > \frac{\alpha n}{2}\right], i.o.\right) = 1. \tag{18}$$

Hence for each $\alpha > 0$ we can find an $A_\alpha \in \Sigma$, $P(A_\alpha) = 0$, such that

$$\limsup_n \frac{|S_n|}{n} > \frac{\alpha}{2} \quad \text{on } \Omega - A_\alpha.$$

Letting α run through the rationals and setting $A = \bigcup_{\alpha \in \text{ rationals}} A_\alpha$, we get $P(A) = 0$, and on $\Omega - A$, $\limsup_n(|S_n|/n) > k$ for every $k > 0$. Hence $\limsup_n(|S_n|/n) = +\infty$ a.e. This completes the proof of the theorem.

The above result contains slightly more information. In fact, we have the following:

Corollary 8 *Let $\{X_n, n \geq 1\}$ be as in the theorem with $E(|X_1|) < \infty$. Then $|S_n|/n \to |E(X_1)|$ in $L^1(P)$-mean (in addition to the a.e. convergence).*

Proof Since the X_n are i.i.d., so are the $|X_n|, n \geq 1$, and they are clearly independent. Moreover, by i.i.d. $P[-x < X_n < x] = P[-x < X_1 < x]$. Indeed

$$\begin{aligned} P[|X_n| < x] &= \int_\Omega \chi_{[|X_n|<x]} dP \\ &= \int_{\mathbb{R}} \chi_{[|\lambda|<x]} dF(\lambda) (\text{ since } X_n \text{ has } F \text{ as its d.f. for all } n \geq 1) \\ &= P[|X_1| < x] \text{ (by the image law)}. \end{aligned}$$

By the SLLN, $S_n/n \to E(X_1)$ a.e., so that $|S_n|/n \to |E(X_1)|$ a.e. Given $\varepsilon > 0$, choose $x_0 > 0$ such that

$$E(|X_n|\chi_{[|X_n|>x_0]}) = E(|X_1|\chi_{[|X_1|>x_0]}) < \varepsilon.$$

If $S'_n = \sum_{k=1}^n X_k \chi_{[|X_k| \leq x_0]}$ and $S''_n = S_n - S'_n$, then $\{S'_n/n, n \geq 1\}$ is uniformly bounded, so that it is uniformly integrable. But

$$\frac{1}{n} E(|S''_n|) \leq E(|X_1|\chi_{[|X_1|>x_0]}) < \varepsilon$$

uniformly in n. Thus $\{(1/n)S_n'', n \geq 1\}$ is also uniformly integrable. Consequently $\{(1/n)|S_n|, n \geq 1\}$ is a uniformly integrable set. Hence the result follows by Vitali's theorem and the limits must agree, as asserted.

Remark See also Problem 10 for similar (but restricted to finite measure or probability spaces) convergence statements of real analysis, without mention of independence.

These results and their methods of proofs have been extended in various directions. The idea of investigating the averages (both the WLLN and SLLN) has served an important role in creating the modern ergodic theory. Here the random variables X_n are derived from one fixed function $X_1 : \Omega \to \mathbb{R}$ in terms of a measurable mapping $T : \Omega \to \Omega$ $[T^{-1}(\Sigma) \subset \Sigma]$ which preserves measure, meaning $P = P \circ T^{-1}$, or $P(A) = P(T^{-1}(A)), A \in \Sigma$. Then

$$X_{n+1}(\omega) = (X_1 \circ T^n)(\omega), \omega \in \Omega, n \geq 1,$$

where $T^2 = T \circ T$ and $T^n = T \circ T^{n-1}, n \geq 1$. Since $X_1 : \Omega \to \mathbb{R}$ and $T : \Omega \to \Omega$ are both measurable, so that $(X_1 \circ T)^{-1}(\mathcal{B}) = T^{-1}(X^{-1}(\mathcal{B})) \subset T^{-1}(\Sigma) \subset \Sigma$, where $\mathcal{B}$ is the Borel σ-algebra of $\mathbb{R}$, X_2 is an r.v., and similarly X_n is an r.v. For such a sequence, which is no longer independent, the prototypes of the laws of large numbers have been proved. These are called *ergodic* theorems. The correspondents of weak laws are called mean ergodic theorems and those of the strong laws are termed individual ergodic theorems. This theory has branched out into a separate discipline, leaning more toward measure theoretic functional analysis than probability, but still retaining important connections with the latter. For a brief account, see Section 3 of Chapter 7.

Another result suggested by the above theorem is to investigate the growth of sums of independent random variables. How fast does S_n cross some prescribed bound? The laws of the iterated logarithm are of this type, for which more tools are needed. We consider some of them in Chapters 5 and later. We now turn to some applications.

2.4 Applications to Empiric Distributions, Densities, Queueing, and Random Walk

(A) Empiric Distributions

One of the important and popular applications of the SLLN is to show that the empiric distribution converges a.e. and uniformly to the distribution of the random variable. To make this statement precise, consider a sequence of random variables $X_1, X_2, \ldots$ on (Ω, Σ, P) such that $P[X_n < x] = F(x)$,

$x \in \mathbb{R}, n \geq 1$; i.e., they are identically distributed. If we observe "the segment" $X_1, \ldots, X_n$, then the *empiric distribution* is defined as the "natural" proportion for each outcome $\omega \in \Omega$:

$$F_n(x, \omega) = \frac{1}{n}\{\text{ number of } X_i(\omega) < x\}. \tag{1}$$

Equivalently, let us define

$$Y_j(\omega) = \chi_{[X_j < x]}(\omega), j = 1, ..., n.$$

Then

$$F_n(x, \cdot) = \frac{1}{n}(Y_1 + \ldots + Y_n). \tag{2}$$

We have the following important result, obtained in about 1933.

Theorem 1 **(Glivenko-Cantelli)**. *Let $X_1, X_2, \ldots$ be independent and identically distributed (i.i.d.) random variables on (Ω, Σ, P). Let F be their common distribution function, and if the first n-random variables are "observed" (termed a random sample of size n), let F_n be the empiric distribution determined by (1) [or (2)] for this segment. Then*

$$P\left[\lim_{n\to\infty} \sup_{-\infty<x<\infty} |F_n(x, \cdot) - F(x)| = 0\right] = 1. \tag{3}$$

Proof Since the X_i are identically distributed with a common distribution, the same is clearly true of the Y_i given by (2). Indeed,

$$P[Y_i = 1] = P[X_i < x] = F(x) = 1 - P[Y_i = 0] = 1 - P[X_i \geq x]$$

for all $i \geq 1$. Hence by the (special case of) SLLN, we get

$$F_n(x) = \frac{1}{n}\sum_{i=1}^{n} Y_i \to E(Y_1) = F(x) \text{ a.e.} \tag{4}$$

We need to prove the stronger assertion on a.e. uniform convergence in x for (4), which is (3). This is more involved and is presented in three steps.

1. Let $0 \leq k \leq r$ be integers and $x_{k,r}$ be a real number such that

$$F(x_{k,r}) \leq k/r \leq F(x_{k,r} + 0), \quad k = 1, 2, \ldots, r, \tag{5}$$

and set

$$x_{0,r} = -\infty, x_{r,r} = +\infty$$

for definiteness [and use $F(-\infty) = \lim_{x\to-\infty} F(x) = 0, F(+\infty) = \lim_{x\to+\infty} F(x) = 1$]. Also define

$$E_{k,r} = \left\{\omega : \lim_{n\to\infty} F_n(x_{k,r}, \omega) = F(x_{k,r})\right\},$$

and

$$H_{k,r} = \left\{\omega : \lim_{n\to\infty} F_n(x_{k,r}+0,\omega) = F(x_{k,r}+0)\right\}.$$

Then by (4), $P(E_{k,r}) = 1 = P(H_{k,r}), 1 \le k \le r$. Let

$$E_r = \bigcap_{k=1}^{r} E_{k,r} \cap \bigcap_{k=1}^{r} H_{k,r}.$$

2. We have $P(E_r) = 1$ and if $\tilde{E} = \bigcap_{r=1}^{\infty} E_r$, then $P(\tilde{E}) = 1$. In fact, if $A, B \in \Sigma$, $P(A) = 1 = P(B)$, then clearly

$$1 = P(A \cup B) = P(A) + P(B) - P(A \cap B) = 2 - P(A \cap B).$$

Hence $P(A \cap B) = 1$. By induction, with $A = E_{k,r}, B = H_{k,r}, k = 1, \ldots, r$, it follows that $P(E_r) = 1, r \ge 1$. Since $\tilde{E} = \bigcap_{n=1}^{\infty} B_n = \lim_{n\to\infty} B_n$, where $B_n = \bigcap_{r=1}^{n} E_r$, it also follows that $P(E) = \lim_n P(B_n) = 1$.

Let us express $\tilde{E}$ in a different form. First note that E_r is given by

$$E_r = \{\omega : \lim_{n\to\infty} \max_{\substack{1\le k\le r\\ 1\le j\le r}} [\ |F_n(x_{j,r},\omega) - F(x_{j,r})|,$$
$$|F_n(x_{k,r}+0,\omega) - F(x_{k,r}+0)|] = 0\}.$$

If we let

$$S = \left\{\omega : \lim_{n\to\infty} \sup_{x\in\mathbb{R}} |F_n(x,\omega) - F(x)| = 0\right\},$$

then $S \in \Sigma$, because if

$$\tilde{S} = \left\{\omega : \lim_{n\to\infty} \sup_{\substack{-\infty<r_i<\infty\\ r_i-rational}} |F_n(r_i,\omega) - F(r_i) = 0\right\},$$

clearly $S \subset \tilde{S}$, and by the density of rationals in $\mathbb{R}$, $\tilde{S} \subset S$ also follows. Since $\tilde{S} \in \Sigma$, so is $S \in \Sigma$. We need to establish the following result.

3. $\tilde{E} \subset S$, so that $1 = P(\tilde{E}) \le P(S) \le 1$. For, if we let $x \in (x_{k,r}, x_{k+1,r})$, then by the monotonicity of F_n and F, we get

$$F_n(x_{k,r}+0)(\omega) \le F_n(x)(\omega) \le F_n(x_{k+1,r})(\omega) \ a.a.(\omega),$$

$$F(x_{k,r}+0) \le F(x) \le F(x_{k+1,r}). \tag{6}$$

Let k, r be chosen in (5) such that $k \le r$ and $x_{k,r}$ satisfies

$$0 \le F(x_{k+1,r}) - F(x_{k,r}+0) \le \frac{k+1}{r} - \frac{k}{r} = \frac{1}{r}. \tag{7}$$

This is clearly possible since $F(x_{k+1,r}) \leq (k+1)/r$ and $F(x_{k,r}+0) \geq k/r$. Hence (6) may be written

$$F_n(x) - F(x) \leq F_n(x_{k+1,r}) - F(x_{k,r}+0)$$
$$\leq F_n(x_{k+1,r}) - F(x_{k+1,r}) + \frac{1}{r} \ a.e.,$$

and in a similar way

$$F_n(x) - F(x) \geq F_n(x_{k,r}+0) - F(x_{k+1,r})$$
$$\geq F_n(x_{k,r}+0) - F(x_{k,r}+0) - \frac{1}{r} \ a.e.$$

Combining these two sets of inequalities we get for a.a. (ω)

$$\sup_{\substack{-\infty<x<\infty \\ x_{k,r}\leq x\leq x_{k+1,r} \\ 1\leq k\leq r}} \{|F_n(x) - F(x)|\}(\omega)$$
$$\leq \max_{\substack{0\leq j\leq r-1 \\ 0\leq k\leq r-1}} \{|F_n(x_{k+1,r}) - F(x_{k+1,r})|(\omega),$$
$$|F_n(x_{j+1,r}+0) - F(x_{j+1,r}+0)|(\omega)\} + \frac{1}{r}.$$

Since $r \geq 1$ is arbitrary, the left-side inequality holds if the right-side inequality does, for almost all ω. Hence $\omega \in \tilde{E} \Rightarrow \omega \in \tilde{S} = S$. Thus $\tilde{E} \subset S$, and the theorem is proved.

Remark: The empiric distribution has found substantial use in the statistical method known as the "Bootstrap". In the theory of statistics, bootstrapping is a method for estimating the sampling distribution of an estimator by "resampling" with replacement from the original sample.

In the proof of the theorem, one notes that the detailed analysis was needed above in extending the a.e. convergence of (4) for each x to uniform convergence in x over $\mathbb{R}$. This extension does not involve any real probabilistic ideas. It is essentially classical analysis. If we denote by $\mathcal{C}$ the class of all intervals $(-\infty, x)$, and denote by

$$\mu_n(A)(\omega) = \int_A F_n(dx, \omega)$$

and similarly

$$\mu(A) = \int_A F(dx),$$

then $\mu_n(A)$ is a sample "probability" of A [i.e., $\mu_n(\cdot)(\omega)$ is a probability for each $\omega \in \Omega$, and $\mu_n(A)(\cdot)$ is a measurable function for each Borel set A]; and μ is an ordinary probability (that is, determined by the common image measure). Then (3) says the following:

$$P[\lim_{n\to\infty} \sup_{A\in\mathcal{C}} |\mu_n(A) - \mu(A)| = 0] = 1. \tag{8}$$

This form admits an extension if $X_1, X_2, \ldots$ are random vectors. But here the correspondent for $\mathcal{C}$ must be chosen carefully, as the result will not be true for all collections because of the special sets demanded in Definition 2.1 (see the counterexample following it). For instance, the result will be true if $\mathcal{C}$ is the (corresponding) family of all half-spaces of $\mathbb{R}^n$. But the following is much more general and is due to R. Ranga Rao, *Ann. Math. Statist.* **33** (1962), 659-680.

Theorem 2 *Let $X_1, X_2, \ldots$ be a sequence of independent random vectors on (Ω, Σ, P) with values in $\mathbb{R}^m$, and for each Borel set $A \subset \mathbb{R}^m$, we have $\mu(A) = P[X_n \in A]$, $n \geq 1$, so that they have the common image measure μ (or distribution). Let $\mu_n(A)$ be the empiric distribution based on the sample (or initial segment) of size n (i.e. on, $X_1, \ldots, X_n$) so that*

$$\mu_n(A) = \frac{1}{n}\{\text{ number of } X_k \in A, 1 \leq k \leq n\}.$$

If $\mathcal{C}$ is the class of measurable convex sets from $\mathbb{R}^m$ whose boundaries have zero measure relative to the nonatomic part of μ, then

$$P\left[\lim_{n\to\infty} \sup_{A\in\mathcal{C}} |\mu_n(A) - \mu(A)| = 0\right] = 1. \tag{9}$$

We shall not present a proof of this result, since it needs several other auxiliary facts related to convergence in distribution, which have not been established thus far. However, this result, just as the preceding one, also starts its analysis from the basic SLLN for its probabilistic part.

(B) **Density Estimation**

Another application of this idea is to estimate the probability density by a method that is essentially due to Parzen (1962).

Suppose that $P[X < x] = F_X(x)$ is absolutely continuous relative to the Lebesgue measure on $\mathbb{R}$, with density $f(u) = (dF_X/dx)(u)$, and one wants to find an "empiric density" of $f(\cdot)$ in the manner of the Glivenko-Cantelli theorem. One might then consider the "empirical density"

$$f_n(x,h) = \frac{F_n(x+h) - F_n(x)}{h}$$

and find conditions for $f_n(x,h) \to f(x)$ a.e. as $n \to \infty$ and $h \to 0$. In contrast to the last problem, we have two limiting processes here which need additional work. Thus we replace h by h_n so that as $n \to \infty$, $h_n \to 0$. Since $F_n(x)$ itself is an ω-function, we still need extra conditions. Writing $\tilde{f}_n(x)$ for $f_n(x, h_n)$, this quotient is of the form

$$\tilde{f}_n(x)(\cdot) = \int_{\mathbb{R}} K\left(\frac{x-t}{h_n}\right) F_n(dt, \cdot) = \frac{1}{nh_n}\sum_{j=1}^{n} K\left(\frac{x - X_j(\cdot)}{h_n}\right), \tag{10}$$

for a suitable nonnegative function $K(\cdot)$, called a *kernel.* The approximations employed in Fourier integrals [cf. Bochner (1955)], Chapter I) give us some clues. Examples of kernels $K(t)$ are (i) e^{-t^2}, (ii) $e^{-t}\chi_{[t\geq 0]}$, (iii) $\chi_{[0,1]}$, and (iv) $1/(1+t^2)$. In this way we arrive at the following result of Parzen. [Actually he assumed a little more on K, namely, that $\hat{K}$, the Fourier transform of K, is also absolutely integrable, so that the examples (ii) and (iii) are not admitted. These are included in the following result. However, the ideas of proof are essentially his.]

Theorem 3 *Let $X_1, X_2, \ldots$ be independent identically distributed random variables on (Ω, Σ, P) whose common distribution admits a uniformly continuous density f relative to the Lebesgue measure on the line. Suppose that $K : \mathbb{R} \to \mathbb{R}^+$ is a bounded continuous function, except for a finite set of discontinuities, satisfying the conditions: (i) $\int_{\mathbb{R}} K(t)dt = 1$ and (ii) $|tK(t)| \to 0$ as $|t| \to \infty$. Define the "empiric density" $f_n : \mathbb{R} \times \Omega \to \mathbb{R}^+$ by*

$$f_n(x) = \frac{1}{nh_n}\sum_{j=1}^{n} K\left(\frac{x - X_j}{h_n}\right), \quad x \in \mathbb{R} \tag{11}$$

where h_n is a sequence of numbers such that $nh_n^2 \to \infty$, but $h_n \to 0$ as $n \to \infty$. Then

$$P\left[\lim_{n\to\infty} \sup_{-\infty<x<\infty} |f_n(x) - f(x)| = 0\right] = 1. \tag{12}$$

Proof The argument here is somewhat different from the previous one, and it will be presented again in steps for convenience. As usual, let E be the expectation operator.

1. Consider

$$g_n(x) = E(f_n(x)) = \frac{1}{nh_n}\sum_{j=1}^{n} E\left(K\left(\frac{x - X_j}{h_n}\right)\right).$$

We assert that $g_n(x) \to f(x)$ uniformly in $x \in \mathbb{R}$ as $n \to \infty$. For using the i.i.d. hypothesis,

$$g_n(x) = \frac{1}{n}\sum_{j=1}^{n} \frac{1}{h_n}\int_{\mathbb{R}} K\left(\frac{x-y}{h_n}\right) f(y)dy. \tag{13}$$

If

$$v_n(x) = \frac{1}{h_n}\int_{\mathbb{R}} K\left(\frac{x-y}{h_n}\right) f(y)dy,$$

it suffices to show that $v_n(x) \to f(x)$ uniformly. Then since g_n is a $(C,1)$-average like v_n, it follows that $g_n(x) \to f(x)$ in the same sense. Since f is assumed to be uniformly continuous and integrable (and a probability density), it is easily seen that f is also bounded. Thus consider

$$v_n(x) - f(x) = \int_{\mathbb{R}} \frac{1}{h_n} K(\frac{t}{h_n})[f(x-t) - f(x)]dt. \tag{14}$$

But, given $\varepsilon > 0$, there is a $\delta_\varepsilon > 0$ such that $|f(x-t) - f(x)| \leq \varepsilon$ for $|t| \leq \delta$ by the uniform continuity of f. Thus

$$\begin{aligned}
|v_n(x) - f(x)| &\leq \int_{[|t|\leq\delta]} \frac{1}{h_n} K\left(\frac{t}{h_n}\right) \max_{|t|\leq\delta} |f(x-t) - f(x)|dt \\
&+ \int_{[|t|\geq\delta]} K\left(\frac{t}{h_n}\right) \frac{f(x-t)}{|t|} \frac{|t|}{h_n} dt \\
&+ f(x) \int_{[|t|\geq\delta]} K\left(\frac{t}{h_n}\right) \frac{1}{h_n} dt \\
&\leq \max_{|t|\leq\delta} |f(x-t) - f(x)| \int_{\mathbb{R}} K(u)du \\
&+ \sup_{|u|\geq\delta/h_n} |uK(u)| \frac{1}{\delta} \int_{\mathbb{R}} f(t)dt \\
&+ f(x) \int_{[|u|\geq\delta/h_n]} K(u)du \\
&\leq \varepsilon + \frac{1}{\delta} \sup_{|u|\geq\delta/h_n} |uK(u)| + M \int_{[|u|\geq\delta/h_n]} K(u)du,
\end{aligned} \tag{15}$$

since f is bounded. Letting $n \to \infty$, so that $h_n \to 0$, by (i) and (ii) both the second and third terms go to zero. Since the right side is independent of x, it follows that $v_n(x) \to f(x)$ uniformly in x, as $n \to \infty$.

2. We use now a result from Fourier transform theory. It is the following. Let $\hat{K}(u) = \int_{\mathbb{R}} e^{iux} K(x)dx$; then one has the inversion, in the sense that for almost every $x \in \mathbb{R}$ (i.e., except for a set of Lebesgue measure zero)

$$\lim_{a\to\infty}\frac{1}{2\pi}\int_{-a}^{a}\left(1-\frac{|u|}{a}\right)e^{-iux}\hat{K}(u)du = K(x). \tag{16}$$

Results of this type for distribution functions, called "inversion formulas," will be established in Chapter 4. If K is assumed integrable, then the above integral can be replaced by $(1/2\pi)\int_{\mathbb{R}} e^{-iux}\hat{K}(u)du = K(x)$ a.e. so (16) is the $(C,1)$-summability result for integrals, an exact analog for series that we noted in the preceding step.

Let $\phi_n(u) = (1/n)\sum_{j=1}^{n} e^{iuX_j}$. Then $e^{iuX_j} = \cos uX_j + i\sin uX_j$ is a bounded complex random variable and, for different j, these are identically distributed. Thus applying the SLLN to the real and imaginary parts, we get

$$\lim_{n\to\infty}\phi_n(u) = E(\phi_1(u)) = E(e^{iuX_1}) \ a.e.[P]. \tag{17}$$

If $\omega \in \Omega$ is arbitrarily fixed, then $\phi_n(u)$ can be regarded as

$$\phi_n(u)(\omega) = \int_{\mathbb{R}} e^{iux}F_n(dx,\omega), \tag{18}$$

where F_n is the empiric distribution of the X_n. Now using the "inversion formula" (16) for $\hat{K}$, we can express f_n as follows:

$$\begin{aligned}
&\lim_{a\to\infty}\tfrac{1}{2\pi}\int_{-a}^{a}\left(1-\frac{|u|}{a}\right)\hat{K}(h_n u)e^{-iux}\phi_n(u)(\omega)du \\
&= \frac{1}{n}\sum_{j=1}^{n}\lim_{a\to\infty}\frac{1}{2\pi}\int_{-a}^{a}\left(1-\frac{|u|}{a}\right)e^{-iu(x-X_j(\omega))}\hat{K}(h_n u)du \\
&= \frac{1}{n}\sum_{j=1}^{n}\lim_{\tilde{a}\to\infty}\frac{1}{2\pi}\int_{-\tilde{a}}^{\tilde{a}}\left(1-\frac{|r|}{\tilde{a}}\right)\exp\left(-i\frac{r}{h_n}t_j\right)\hat{K}(r)\frac{dr}{h_n} \\
&\qquad \text{with } t_j = x - X_j(\omega) \text{ and } \tilde{a} = h_n a, \\
&= \frac{1}{nh_n}\sum_{j=1}^{n}K\left(\frac{t_j}{h_n}\right) \\
&\qquad [\text{ by the inversion formula, a.e. (Lebesgue measure)}] \\
&= f_n(x)(\omega) \ [\text{ by}(12)].
\end{aligned} \tag{19}$$

We need this formula to get uniform convergence of $f_n(x)$ to $f(x)$.

3. The preceding work can be used in our proof in the following manner. By Markov's inequality

$$P\left[\lim_{n\to\infty}\sup_{x\in\mathbb{R}}|f_n(x)-f(x)| \geq \varepsilon\right] \leq \lim_{n\to\infty}\frac{E\left[\sup_{x\in\mathbb{R}}|f_n(x)-f(x)|\right]}{\varepsilon}, \tag{20}$$

where the limit can be brought outside of the P-measure by Fatou's lemma. (Note that the sup inside the square brackets is bounded by hypothesis and is a measurable function, by the same argument as in step 2 of the proof of Theorem 1. The existence of limit in (20) will be proved.) We now show that the right side of (20) is zero, so that (12) results. But if $||\cdot||_u$ is the uniform (or supremum) norm over $\mathbb{R}$, then

$$||f_n(\cdot) - f(\cdot)||_u \leq ||f_n(\cdot) - g_n(\cdot)||_u + ||g_n(\cdot) - f(\cdot)||_u, \tag{21}$$

and $x \mapsto g_n(x) = E(f_n(x))$ is a constant function (independent of ω). By step 1, the last term goes to zero as $n \to \infty$, and hence its expectation will go to zero by the dominated convergence since the terms are bounded. Thus it suffices to show that the expectation of the first term also tends to zero uniformly in x. Consider

$$||f_n(\cdot) - g_n(\cdot)||_u = \sup_{x \in \mathbb{R}} \left| \frac{1}{2\pi} \int_{-a}^{a} \left(1 - \frac{|u|}{a}\right) \hat{K}(h_n u) \right.$$

$$\left. \times \; e^{-iux}[\phi_n(u) - E(\phi_n(u))]du \right|,$$

where we used (19) and the fact that $g_n(x) = E(f_n(x))$, which is again obtained from (19) with $E(\phi_n(u))$. With the same computation, first using the Fubini theorem and then the dominated convergence theorem to interchange integrals on $[-a, a] \times \Omega$, we can pass to the limit as $a \to \infty$ through a sequence under the expectation. Thus

$$||f_n(\cdot) - g_n(\cdot)||_u \leq \lim_{a \to \infty} \frac{1}{2\pi} \int_{-a}^{a} \left(1 - \frac{|u|}{a}\right) \hat{K}(h_n u)|\phi_n(u) - E(\phi_n(u))|du. \tag{22}$$

But by (17), $|\phi_n(u) - E(\phi_n(u))| = |\phi_n(u) - E(e^{iuX_1})| \to 0$ a.e., and since these quantities are bounded, this is also true boundedly. Thus by letting $n \to \infty$ in both sides of (22) and noting that the limits on a and n are on independent sets, it follows that the right side of (22) is zero a.e. By the uniform boundedness of the left-side norms in (22), we can take expectations, and the result is zero.

Thus $E(||f_n(\cdot) - f(\cdot)||_u) \to 0$ as $n \to \infty$, and the right side of (20) is zero. This completes the proof.

Remark Evidently, instead of (17), even WLLN is sufficient for (22). Also, using the CBS-inequality in (22) and taking expectations, one finds that $Var(\phi_n) \leq M_1/n$ and this yields the same conclusion without even using WLLN. (However, this last step is simply the proof of the WLLN, as given by Čebyšev.) It is clear that considerable analysis is needed in these results, after employing the probabilistic theorems in key places. Many of the applications use such procedures.

(C) Queueing

We next present a typical application to queueing theory. Such a result was originally considered by A. Kolmogorov in 1936 and is equivalent to a one-server queueing model. It admits extensions and raises many other problems. The formulation using the current terminology appears to be due to D. V. Lindley.

A general queueing system consists of three elements: (i) customers, (ii) service, and (iii) a queue. These are generic terms; they can refer to people at a service counter, or planes or ships arriving at a port facility, etc. The arrival of customers is assumed to be random, and the same is true of the service times as well as waiting times in a queue. Let a_k be the interarrival time between the kth and the $(k+1)$th customer, b_k the service time, and W_k the waiting time of the kth customer. When customer one arrives, we assume that there is no waiting, since there is nobody ahead of this person. Thus it is reasonable to assume $a_0 = W_0 = 0$. Now $b_k + W_k$ is the length of time that the $(k+1)$th customer has to wait in the queue before the turn comes at the service counter. We assume that the interarrival times a_k are independent nonnegative random variables with a common distribution, and similarly, the b_k are nonnegative i.i.d. and independent of the a_k. As noted before, we can assume that the basic probability space is rich enough to support such independent sequences, as otherwise we can enlarge it by adjunction to accomplish this. The waiting times are also positive random variables. If $a_{k+1} > b_k + W_k$,then the $(k+1)$th customer obviously does not need to wait on arrival, but if $a_{k+1} \leq b_k + W_k$ then the person has to wait $b_k + W_k - a_{k+1}$ units of time. Thus

$$W_{k+1} = \max(W_k + b_k - a_{k+1}, 0), \quad k \geq 0. \tag{23}$$

If we let $X_k = b_{k-1} - a_k$, then the X_k are i.i.d. random variables, and (23) becomes $W_0 = 0$ and $W_{k+1} = \max(W_k + X_{k+1}, 0), k \geq 0$. Note that whenever $W_k = 0$ for some k, the server is free and the situation is like the one at the beginning, so that we have a recurrent pattern. This recurrence is a key ingredient of the solution of the problem of finding the limiting behavior of the W_k-sequence. It is called the *single server queueing* problem.

Consider $S_0 = 0$, $S_n = \sum_{k=1}^{n} X_k$. Then the sequence $\{S_n, n \geq 0\}$ is also said to perform a *random walk* on $\mathbb{R}$, and if $S_k \in A$ for some $k \geq 0$ and Borel set A, one says that the walk S_n visits A at step k. In the queueing situation, we have the following statement about the process $\{W_n, n \geq 0\}$.

Theorem 4 *Let $X_k = b_{k-1} - a_k$, $k \geq 1$, and $\{S_n, n \geq 0\}$ be as above. Then for each $n \geq 0$, the quantities W_n and $M_n = \max\{S_j, 0 \leq j \leq n\}$ are identically distributed random variables. Moreover, if $F_n(x) = P[W_n < x]$, then*

$$\lim_{n \to \infty} F_n(x) = F(x)$$

exists for each x, but $F(x) = 0$ is possible. If $E(X_1)$ exists, then $F(x) \equiv 0$, $x \in \mathbb{R}$, whenever $E(X_1) \geq 0$, and $F(\cdot)$ defines an honest distribution function when $E(X_1) < 0$, i.e., $F(+\infty) = 1$.

The last statement says that if $E(b_k) \geq E(a_k), k \geq 1$, so that the expected service time is not smaller than that of the interarrival time, then the line of customers is certain to grow longer without bound (i.e., with probability 1).

Proof For the first part of the proof we follow Feller (1966), even though it can also be proved by using the method of convolutions and the fact that W_k and X_{k+1} are independent. The argument to be given is probabilistic and has independent interest.

Since $W_0 = 0 = S_0$, we may express W_n in an alternative form as $W_n = \max\{(S_n - S_k) : 0 \leq k \leq n\}$. In fact, this is trivial for $n = 0$; suppose it is verified for $n = m$. Then consider the case $n = m + 1$. Writing $S_{n+1} - S_k = S_n - S_k + X_{n+1}$, we have with $\vee$ for "max"

$$\begin{aligned}\max_{0\leq k\leq m+1}(S_{m+1} - S_k) &= \left(\max_{0\leq k\leq m+1}(S_m - S_k) + X_{m+1}\right) \vee 0 \\ &= (W_m + X_{m+1}) \vee 0 = W_{m+1}. \end{aligned} \tag{24}$$

Hence the statement is true for all $m \geq 0$. On the other hand, $X_1, \ldots, X_n$ are i.i.d. random variables. Thus the joint distribution of $X_1, X_2, \ldots, X_n$ is the same as that of $X_1', X_2', \ldots, X_n'$,where $X_1' = X_n$, $X_2' = X_{n-1}, \ldots, X_n' = X_1$. But the joint distribution of $S_0', S_1', \ldots, S_n'$, where $S_n' = \sum_{k=1}^n X_k' (S_0' = 0)$, and that of $S_0, S_1, \ldots, S_n$ must also be the same. This in turn means, on substituting the unprimed variables, that $S_0, S_1, \ldots, S_n$ and S_0', $S_1' = S_n - S_{n-1}$, $S_2' = S_n - S_{n-2}, \ldots, S_n' = S_n - S_0$ are identically distributed. Putting these two facts together, we get $\max_{0\leq k\leq n} S_k'$ and $\max_{0\leq k\leq n}(S_n - S_k) = W_n$ are identically distributed. But the $S_0', S_1', \ldots, S_n'$ and $S_0, S_1, \ldots, S_n$ were noted to have the same distribution, so that $\max_{0\leq k\leq n} S_k' \overset{D}{=} \max_{0\leq k\leq n} S_k$ or M_n and W_n have the same distribution. This is the first assertion in which we only used the i.i.d. property of the X_n but not the fact that $X_n = b_{n-1} - a_n$.

The above analysis implies

$$F_n(x) = P[W_n < x] = P[M_n < x] = P\left[\max_{0\leq k\leq n} S_k < x\right].$$

But

$$\left[\max_{0\leq k\leq n} S_k < x\right] \downarrow \left[\lim_{n\to\infty} \max_{0\leq k\leq n} S_k < x\right],$$

so that

$$F(x) = \lim_{n\to\infty} F_n(x) = P\left[\lim_{n\to\infty} \max_{0\leq k\leq n} S_k \leq x\right] = P\left[\sup_{k\geq 0} S_k \leq x\right], \tag{25}$$

exists, and $0 \leq F(x) \leq 1, x \in \mathbb{R}$. Clearly $F(x) = 0$ for $x < 0$. On the other hand, if $E(|X_1|) = \infty$, then by Theorem 2.7, $\limsup_n |S_n| = +\infty$ a.e. which implies (since $S_0 = 0$) that either $\limsup_n S_n = +\infty$ a.e., so that $\sup_n S_n = +\infty$ a.e., or this can happen with probability zero. Since $F(x) = 0$ for $x < 0$, we only need to consider $x \geq 0$. Thus if $\sup_n S_n = +\infty$ a.e., then $1 - F(x) = P[\sup_{k\geq 0} S_k > x] = 1$, so that $F(x) = 0, x \in \mathbb{R}$. If $\sup_n S_n < \infty$ a.e., then $\lim_{x\to\infty} F(x) = 1$ and F is a distribution function. Note that, since $\sup_{n\geq 0} S_n$ is a symmetric function of the random variables X_n, which are i.i.d., we can deduce that $\sup_n S_n = \infty$ has probability zero or one by Theorem 1.12 so that (25) can be obtained in this way also.

Suppose that $E(|X_1|) < \infty$. Then we consider the cases (i) $E(X_1) > 0$, (ii) $E(X_1) < 0$, and (iii) $E(X_1) = 0$ separately for calculating the probability of A_x, where

$$A_x = \left[\sup_{n\geq 0} S_n \leq x\right].$$

Case (i): $\mu = E(X_1) > 0$: By the SLLN, $S_n/n \to E(X_1)$ a.e., so that for sufficiently large n, $S_n > E(X_1) \cdot \frac{n}{2}$ a.e. Thus

$$A_x = \bigcap_{n\geq 0} [S_n \leq x] \subset \left\{\omega : S_n(\omega) > \frac{n}{2}\mu, n \geq n_\omega\right\}^c$$

for any $x \in \mathbb{R}^+$, and hence $P(A_x) = 0$, or $F(x) = 0, x \in \mathbb{R}^+$, in this case.

Case (ii): $\mu = E(X_1) < 0$: Again by the SLLN, $S_n/n \to E(X_1)$ a.e., and given $\varepsilon > 0$, and $\delta > 0$, one can choose $N_{\varepsilon\delta}$ such that $n \geq N_{\varepsilon\delta}$ implies

$$P[|(S_n/n) - E(X_1)| \leq \varepsilon, n \geq N_{\varepsilon\delta}] \geq 1 - \delta \tag{26}$$

This may be expressed in the following manner. Let $\varepsilon > 0$ be small enough so that $E(X_1) + \varepsilon < 0$. Then for $0 < \delta < \frac{1}{2}$, choose $N_{\varepsilon\delta}$ such that with (26),

$$\begin{aligned} P[S_n < 0, n \geq N_{\varepsilon\delta}] &\geq P[S_n \leq n(\mu+\varepsilon), n \geq N_{\varepsilon\delta}] \\ &\geq P[n(\mu-\varepsilon) \leq S_n \leq (\mu+\varepsilon)n, n \geq N_{\varepsilon\delta}] \\ &\geq 1 - \delta. \end{aligned} \tag{27}$$

For this $N_{\varepsilon\delta}$, consider the finite set $S_1, S_2, \ldots, S_{N_{\varepsilon\delta}-1}$. Since these are real random variables, we can find an $x_\delta \in \mathbb{R}^+$ such that $x \geq x_\delta$ implies

$$P[S_1 < x, \ldots, S_{N_{\varepsilon\delta}-1} < x] > 1 - \delta. \tag{28}$$

If now

$$\tilde{A}_x = \bigcap_{n\geq N_{\varepsilon\delta}} [S_n < x], \quad B_x = \bigcap_{k=1}^{N_{\varepsilon\delta}-1} [S_k < x], \quad A_x = \bigcap_{n=1}^{\infty} [S_n < x],$$

then $A_x = \tilde{A}_x \cap B_x$, for $x \geq 0$. Hence we have

$$
\begin{aligned}
F(x) = P(A_x) &= P(\tilde{A}_x \cap B_x) \\
&= P(\tilde{A}_x) + P(B_x) - P(\tilde{A}_x \cup B_x) \\
&\geq 2(1-\delta) - 1 = 1 - 2\delta \quad [\text{ by (27) and (28) }].
\end{aligned}
$$

Since $0 < \delta < \frac{1}{2}$ is arbitrary, we conclude that $\lim_{x\to\infty} F(x) = 1$, and hence F gives an honest distribution in this case.

Case (iii): $E(X_1) = 0$: Now $S_n = \sum_{i=1}^n X_i$, $n \geq 1$, is a symmetrically dependent sequence of random variables and $S_0 = 0$. Thus $\sup_{n\geq 0} S_n \geq 0$ a.e., and since we can assume that $X_1 \not\equiv 0$ a.e., all the S_n do not vanish identically a.e. Consider the r.v. $Y = \limsup_n S_n$. Then $Y[= Y(S_n, n \geq 1)]$ is symmetrically dependent on the S_n and is measurable for the tail σ-algebra. Hence, by Theorem 1.12 it is a constant $= k_0$ a.e. It will be seen later (cf. Theorem 8 below) that, since $S_n/n \to 0$, a.e. by the SLLN, S_n takes both positive and negative values infinitely often. Thus $k_0 \geq 0$. But then

$$
\begin{aligned}
0 \leq Y &= \limsup_{n\geq 1} S_n \\
&= \limsup_{n\geq 1}(X_1 + \ldots + X_n) = X_1 + \limsup_{n\geq 2}(X_2 + \ldots + X_n) \\
&\stackrel{D}{=} X_1 + Y.
\end{aligned} \tag{29}
$$

Since $Y = k_0$ a.e. and X_1 is a real nonzero r.v., (29) can hold only if $k_0 = +\infty$.

Now $[\limsup_{n\geq 1} S_n = +\infty] \subset [\sup_{n\geq 0} S_n = \infty]$, and so we are back in the situation treated in case (i), i.e., $F(x) = 0, x \in \mathbb{R}$. This completes the proof of the theorem.

The preceding result raises several related questions, some of which are the following. When $E(X_1) < 0$, we saw that the waiting times $W_n \to W$ in distribution where W is a (proper) r. v. Thus, in this case, if Q_n is the number of customers in the queue when the service of the nth customer is completed, then Q_n is an r.v. But then what is the distribution of Q_n, and does $Q_n \stackrel{D}{\to} Q$? Since Q_n is no more than k iff the completion of the nth customer service time is no more than the interarrival times of the last k customers, we get

$$
P[Q_n < k] = P[W_n + b_n \leq a_{n+1} + \ldots + a_{n+k}]. \tag{30}
$$

The random variables on the right are all independent, and thus this may be calculated explicitly in principle. Moreover, it can be shown, since $W_n \stackrel{D}{\to} W$ and the b_n and a_n are identically distributed, that $Q_n \stackrel{D}{\to} Q$ from this expression.

Other questions, such as the distribution of the times that $W_n = 0$, suggest themselves. Many of these results use some properties of convolutions of the image measures (i.e., distribution functions) on $\mathbb{R}$, and we shall omit consideration of these specializations here.

All of the above discussions concerned a single-server queueing problem. But what about the analogous problem with many servers? This is more involved. The study of these problems has branched out into a separate discipline because of its great usefulness in real applications. Here we consider only one other aspect of the above result.

(D) **Fluctuation Phenomena**

In Theorem 4 we saw that the behavior of the waiting time sequence is governed by $S_n = \sum_{k=1}^n X_k$, the sequence of partial sums of i.i.d. random variables. In Section 2 we considered the convergence of sums of general independent random variables, but the surprising behavior of i.i.d. sums was not analyzed more thoroughly. Such a sequence is called a *random walk*. Here we include an introduction to the subject that will elaborate on the proof of Theorem 4 and complete it. The results are due to Chung and Fuchs. We refer to Chung (1974). For a detailed analysis of the subject, and its relation to the group structure of the range space, see Spitzer (1964).

Thus if $X_n, n \geq 1$, are i.i.d., and $\{S_n = \sum_{k=1}^n X_k, n \geq 1\}$ is a random walk sequence, let $Y = \limsup_n X_n$. We showed in the proof of Theorem 4 [Case (iii)] that $Y = X_1 + Y$ and Y is a "permutation invariant" r.v. Then this equation implies $Y = k_0$ a.e. ($= \pm\infty$ possibly), by the Hewitt-Savage zero-one law. If $X_1 = 0$ a.e., then by the i.i.d. condition, all $X_n = 0$ a.e., so that $S_n = 0$ for all n (and $Y = 0$ a.e.). If $X_1 \neq 0$ a.e., then $k_0 = -\infty$ or $+\infty$ only. If $k_0 = -\infty$, then clearly $-\infty \leq \liminf_n S_n \leq \limsup_n S_n = -\infty$, so that $\lim_{n\to\infty} S_n = -\infty$ a.e.; or if $k_0 = +\infty$, then $\liminf_n S_n$ can be $+\infty$, in which case $S_n \to +\infty$ a.e., or $\liminf_n S_n = -\infty < \limsup_n S_n = +\infty$. Since $\limsup_n(S_n) = -\liminf_n(-S_n)$, no other possibilities can occur. In the case $-\infty = \liminf_n S_n < \limsup_n S_n = +\infty$ a.e. (the interesting case), we can look into the behavior of $\{S_n, n \geq 1\}$ and analyze its fluctuations.

A state $x \in \mathbb{R}$ is called a *recurrent point* of the range of the sequence if for each $\varepsilon > 0$, $P[|S_n - x| < \varepsilon, i.o.] = 1$, i.e., the random walk visits x infinitely often with probability one. Let R be the set of all recurrent points of $\mathbb{R}$. A point $y \in \mathbb{R}$ is termed a *possible value* of the sequence if for each $\varepsilon > 0$, there is a k such that $P[|S_k - y| < \varepsilon] > 0$. We remark that by Cases (i) and (ii) of the proof of Theorem 4, if $E(X_1) > 0$ or < 0, then $\lim_{n\to\infty} S_n = +\infty$ or $= -\infty$ respectively. Thus fluctuations show up only in the case $E(X_1) = 0$ when the expectation exists. However, $E(|X_1|) < \infty$ will not be assumed for the present discussion.

Theorem 5 *For the random walk $\{S_n, n \geq 1\}$, the set R of recurrent values (or points) has the following description: Either $R = \emptyset$ or $R \subset \mathbb{R}$ is a* *closed subgroup. In the case $R \neq \emptyset$, $R = \{0\}$ iff $X_1 = 0$ a.e., and if $X_1 \neq 0$ a.e., we have either $R = \mathbb{R}$ or else $R = \{nd : n = 0, \pm 1, \pm 2, \ldots\}$, the infinite cyclic group generated by a number $d > 0$.*

Proof Suppose $R \neq \emptyset$. If $x_n \in R$ and $x_n \to x \in \mathbb{R}$, then given $\varepsilon > 0$, there is n_ε such that $n \geq n_\varepsilon \Rightarrow |x_n - x_1| < \varepsilon$. Thus letting $S_n(\omega) = x_n$, we get $|S_n(\omega) - x| < \varepsilon$, $n \geq n_\varepsilon(\omega)$, for almost all ω, and hence if $I = (x - \varepsilon, x + \varepsilon)$, then $P[S_n \in I, i.o.] = 1$. Since $\varepsilon > 0$ is arbitrary, $x \in R$, and so R is closed.

To prove the group property, let $x \in R$ and $y \in \mathbb{R}$ be a possible value of the random walk. We claim that $x - y \in R$. Indeed for each $\varepsilon > 0$, choose m such that $P[|S_m - y| < \varepsilon] > 0$. Since x is recurrent, $P[|S_n - x| < \varepsilon, i.o.] = 1$. Or equivalently $P[|S_n - x| < \varepsilon$, finitely many n only $] = 0$. Let us consider, since $[|S_n - x| < \varepsilon$ for finitely many $n] = [|S_n - x| \geq \varepsilon$ for all but finitely many $n]$,

$$
\begin{aligned}
&P[|S_n - x| < \varepsilon, \text{ finitely often }] \\
&\geq P[|S_m - y| < \varepsilon, |S_{m+n} - S_m - (x - y)| \geq 2\varepsilon, \text{ all } n \geq k_0],\ k_0 \geq 1, \\
&= P[|S_m - y| < \varepsilon] P\left[\left|\sum_{k=m+1}^{m+n} X_k - (x - y)\right| < 2\varepsilon, \text{ finitely often}\right], \\
&\quad \text{(by the independence of } S_m \text{ and } S_{m+n} - S_m). \qquad (31)
\end{aligned}
$$

By hypothesis $P[|S_m - y| < \varepsilon] > 0$, and this shows that the second factor of (31) is zero. But by the i.i.d. hypothesis, S_n and $S_{m+n} - S_m$ have the same distribution. Hence $P[|S_n - (x - y)| < 2\varepsilon$, finitely many $n] = 0$, and $x - y \in R$. Since $y = x$ is a possible value, $0 \in R$ always, and $x - (x - y) = y \in R$. Similarly $0 - y \in R$ and so R is a group. As is well known, the only closed subgroups of $\mathbb{R}$ are those of the form stated in the theorem,[2] and $R = \{0\}$ if $X_1 \equiv 0$ a.e. In the case that $X_1 \neq 0$ a.e., there is a possible value $y \in \mathbb{R}$ of the random walk, and $y \in R$ by the above analysis. Thus $R = \{0\}$ iff $X_1 \equiv 0$ a.e. It is of interest also to note that unless the values of the r.v. X_1 are of the form $nd, n = 0, \pm 1, \pm 2, \ldots, R = \mathbb{R}$ itself. This completes the proof.

It is clear from the above result that 0 plays a key role in the recurrence phenomenon of the random walk. A characterization of this is available:

Theorem 6 *Let $\{X_n, n \geq 1\}$ be i.i.d. random variables on (Ω, Σ, P) and $\{S_n, n \geq 0\}$ be the corresponding random walk sequence. If for an $\varepsilon > 0$ we*

[2] Indeed,if $R \neq \emptyset$, because it is a closed subgroup of $\mathbb{R}$, let $d = \inf\{x \in R, x > 0\}$. Then $d \geq 0$ and there exist $d_n \in R, d_n \downarrow d$. If $d = 0$, we can verify that $\{kd_n, k = 0, \pm 1, \pm 2, \ldots; n \geq 1\}$ is dense in $\mathbb{R}$ and $\subset R \Rightarrow R = \mathbb{R}$. If $d > 0$,then $\{nd, n = 0, \pm 1, \ldots\} \subset R$ and is all of R. There are no other kinds of groups. Note that if $R \neq \emptyset$ every possible value is also a recurrent value of the random walk.

have

$$\sum_{n=1}^{\infty} P[|S_n| < \varepsilon] < \infty, \tag{32}$$

then 0 *is not a recurrent value of* $\{S_n, n \geq 0\}$. *If, on the other hand,for every* $\varepsilon > 0$ *it is true that the series in (32) diverges, then* 0 *is recurrent.* [*It follows from (36) below that if the series (32) diverges for one* $\varepsilon > 0$, *then the same is true for all* $\varepsilon > 0$.]

Proof If the series in (32) converges, then the first Borel-Cantelli lemma implies $P[|S_n| < \varepsilon$, finitely often $] = 1$ so that $0 \notin R$. The second part is harder, since the events $\{[|S_n| < \varepsilon], n \geq 1\}$ are *not* independent. Here one needs to show that $P[|S_n| < \varepsilon, i.o.] = 1$. We consider the complementary event and verify that it has probability zero, after using the structure of the S_n sequence.

Consider for any fixed $k \geq 1$ the event A_m^k defined as

$$A_m^k = [|S_m| < \varepsilon, |S_n| \geq \varepsilon, n \geq m + k]. \tag{33}$$

Then A_m^k is the event that the S_n will not visit $(-\varepsilon, \varepsilon)$ after the $(m+k-1)$th trial, but visits at the mth trial [from the $(m+1)$th to $(m+k-1)$th trials, it may or may not visit]. Hence $A_m^k, A_{m+k}^k, A_{m+2k}^k, \ldots$ are disjoint events for $m \geq 1$ and fixed $k \geq 1$. Thus

$$\begin{aligned} \sum_{m=1}^{\infty} P(A_m^k) &= k + \sum_{m=k+1}^{\infty} P(A_m^k) \leq k + \sum_{\ell=1}^{k} \sum_{j=1}^{\infty} P(A_{\ell+jk}^k) \\ &= \sum_{\ell=1}^{k} P\left(\bigcup_{j=1}^{\infty} A_{\ell+jk}^k \right) \leq 2k. \end{aligned} \tag{34}$$

But for each $k \geq 1$, $[|S_n| < \varepsilon]$ and $[|S_n - S_m| \leq 2\varepsilon, n \geq m+k]$ are independent, and $A_m^k \supset [|S_m| < \varepsilon] \cap [|S_n - S_m| \geq 2\varepsilon, n \geq m + k]$, $k \geq 1$, since $|S_n| \geq (|S_n - S_m| - |S_m|) \geq 2\varepsilon - \varepsilon = \varepsilon$, on the displayed set. Hence, with independence, (34) becomes

$$\begin{aligned} &\sum_{m=1}^{\infty} P[|S_m| < \varepsilon] P[|S_n - S_m| \geq 2\varepsilon, n \geq m + k] \\ &= \sum_{m=1}^{\infty} P[|S_m| < \varepsilon, |S_n - S_m| \geq 2\varepsilon, n \geq m + k] \leq \sum_{m=1}^{\infty} P(A_m^k) \leq 2k. \end{aligned}$$

But

$$\begin{aligned} P[|S_n - S_m| \geq 2\varepsilon, n \geq m + k] &= P[|X_{m+1} + \ldots + X_n| \geq 2\varepsilon, n \geq m + k] \\ &= P[|S_n| \geq 2\varepsilon, n \geq k] \quad \text{(by the i.i.d. condition).} \end{aligned}$$

Hence

$$\sum_{m=1}^{\infty} P[|S_m| < \varepsilon] \cdot P[|S_n| \geq 2\varepsilon, n \geq k] \leq 2k.$$

Since we may take the second factor on the left out of the summation, and since the sum is divergent by hypothesis, we must have $P[|S_n| \geq 2\varepsilon, n \geq k] = 0$ for each k. Hence taking the limit as $k \to \infty$, we get

$$P[|S_n| > 2\varepsilon, \quad \text{finitely often }] = P\left[\bigcup_{k\geq 1}\bigcap_{n\geq k}[|S_n| \geq 2\varepsilon]\right] = 0,$$

or $P[|S_n| < \varepsilon, i.o.] = 1$ for any $\varepsilon > 0$. This means $0 \in R$ and completes the proof of the theorem.

Suppose, in the above, the $X_n : \Omega \to \mathbb{R}^k$ are i.i.d. random vectors and $S_n = \sum_{i=1}^{n} X_i$. If $|X_i|$ is interpreted as the maximum absolute value of the k components of X_i, and S_n visits $(-\varepsilon, \varepsilon)$ means it visits the cube $(-\varepsilon, \varepsilon)^k \subset \mathbb{R}^k$ (i.e., $|S_n| < \varepsilon$), then the preceding proof holds verbatim for the k-dimensional random variables, and establishes the corresponding result for the k-dimensional random walk. We state the result for reference as follows:

Theorem 7 *Let $\{X_n, n \geq 1\}$ be i.i.d. k-vector random variables on (Ω, Σ, P) and $S_n = \sum_{i=1}^{n} X_i$, $S_0 = 0$, where $k \geq 1$. Then 0 is a recurrent value of the k-random walk $\{S_n, n \geq 0\}$ iff for any $\varepsilon > 0$,*

$$\sum_{n=1}^{\infty} P[|S_n| < \varepsilon] = +\infty. \tag{35}$$

Moreover, the set of all recurrent values R forms a closed subgroup of the additive group $\mathbb{R}^k$.

The proof of the last statement is the same as that for Theorem 5, which has a more precise description of R in case $k = 1$.

If $R = \emptyset$, then the random walk is called *transient*, and is termed *recurrent*, (or *persistent*) if $R \neq \emptyset$.

We can now present a sufficient condition for the recurrence of a random walk, and this completes the proof of case (iii) of Theorem 4.

Theorem 8 *Let $S_n = X_1 + \ldots + X_n$, $\{X_n, n \geq 1$, i.i.d.$\}$ be a (real) random walk sequence on (Ω, Σ, P) such that $S_n/n \xrightarrow{P} 0$. Then the walk is recurrent.*

Remark As noted prior to Theorem 3.2, this condition holds for certain symmetric random variables without the existence of the first moment. On the other hand, if $E(|X_1|) < \infty$, then it is always true by the WLLN (or SLLN).

We shall establish the result with the weaker hypothesis as stated. The proof uses the linear order structure of the range of S_n. Actually the result itself is not valid in higher dimensions (≥ 3). It is true in 2-dimensions, but needs a different method with characteristic functions (cf. Problem 21.)

Proof We first establish an auxiliary inequality, namely, for each $\varepsilon > 0$,

$$\sum_{m=0}^{r} P[|S_m| < k\varepsilon] \leq 2k \sum_{m=0}^{r} P[|S_m| < \varepsilon], \quad r, k \geq 1, \text{ integers.} \tag{36}$$

If this is granted, the result can be verified (using an argument essentially due to Chung and Ornstein (1962)) as follows: We want to show that (32) fails. Thus for any integer $b > 0$, let $r = kb$ in (36). Then

$$\sum_{m=0}^{kb} P[|S_m| < \varepsilon] \geq \frac{1}{2k} \sum_{m=0}^{kb} P[|S_m| < k\varepsilon] \geq \frac{1}{2k} \sum_{m=0}^{kb} P\left[|S_m| < \frac{m\varepsilon}{b}\right], \tag{37}$$

because $(m/b) \leq k$. By hypothesis $S_m/m \xrightarrow{P} 0$, so that $P[|S_m|/m < \varepsilon/b] \to 1$ as $m \to \infty$. By the $(C, 1)$-summability,

$$\lim_{k\to\infty} \frac{1}{kb} \sum_{m=0}^{kb} P\left[\frac{|S_m|}{m} < \frac{\varepsilon}{b}\right] = 1, \quad \text{for each } b > 0.$$

Hence (37) becomes on letting $k \to \infty$

$$\sum_{m=0}^{\infty} P[|S_m| < \varepsilon] \geq \frac{b}{2} \lim_{k\to\infty} \frac{1}{kb} \sum_{m=0}^{kb} P\left[\frac{|S_m|}{m} < \frac{\varepsilon}{b}\right] = \frac{b}{2}.$$

Since $b > 0$ is arbitrary, (32) fails for each $\varepsilon > 0$, and so $\{S_n, n \geq 1\}$ is recurrent.

It remains to establish (36). Consider, for each integer m, $[m\varepsilon \leq S_n < (m+1)\varepsilon]$ and write it as a disjoint union:

$$[m\varepsilon \leq S_n < (m+1)\varepsilon] = \bigcup_{k=0}^{n} [m\varepsilon \leq S_n < (m+1)\varepsilon] \cap A_k, \tag{38}$$

where $A_0 = [m\varepsilon \leq S_0 < (m+1)\varepsilon]$ and for $k \geq 1$, $A_k = [S_k \in [m\varepsilon, (m+1)\varepsilon), S_j \notin [m\varepsilon, (m+1)\varepsilon), 0 \leq j \leq k-1]$. Thus A_k is the ω-set for which S_k enters the interval $[m\varepsilon, (m+1)\varepsilon]$ for the first time. Then

$$\begin{aligned}
&\sum_{n=0}^{r} P[S_n \in [m\varepsilon, (m+1)\varepsilon)] \\
&\qquad = \sum_{n=0}^{r} \sum_{k=0}^{n} P[(S_n \in [m\varepsilon, (m+1)\varepsilon) \cap A_k] \quad [\text{ by (38)}]
\end{aligned}$$

$$\leq \sum_{n=0}^{r}\sum_{k=0}^{n} P[A_k \cap [(S_n - S_k) \in (-\varepsilon, \varepsilon)]]$$

[since on$A_k, m\varepsilon \leq S_n, S_k < (m+1)\varepsilon \Rightarrow |S_n - S_k| < \varepsilon$]

$$= \sum_{n=0}^{r}\sum_{k=0}^{n} P(A_k)P[|S_n - S_k| < \varepsilon]$$

(since A_k is determined by $X_1, \ldots, X_k$

and hence is independent of $S_n - S_k$ for $n \geq k$)

$$= \sum_{n=0}^{r}\sum_{k=0}^{n} P(A_k)P[|S_{n-k}| < \varepsilon] \quad \text{(by the i.i.d. property)}$$

$$= \sum_{n=0}^{r} P(A_n) \sum_{k=n}^{r} P[|S_{r-k}| < \varepsilon]$$

$$\leq \sum_{j=0}^{r} P[|S_j| < \varepsilon] \quad \text{(since the } A_k \text{ are disjoint).} \tag{39}$$

Summing for $m = -k$ to $k - 1$, we get

$$\sum_{n=0}^{r}\sum_{m=-k}^{k-1} P[S_n \in [m\varepsilon, (m+1)\varepsilon)] = \sum_{n=0}^{r} P[S_n \in [-k\varepsilon, k\varepsilon)] \leq 2k \sum_{j=0}^{r} P[|S_j| < \varepsilon].$$

This proves the inequality (36), and hence also the theorem.

It is now natural to investigate several other properties of recurrent random walks, such as the distribution of the first entrance time T_A of the process into a given Borel set $A \subset \mathbb{R}$, finding conditions on X in order that $E(T_A) < \infty$ or $= \infty$, and $P[T_A < \infty] = 1$. Conversely, the recurrence and transience of a random walk determines the structure of the range space $\mathbb{R}$ or $\mathbb{R}^n$ on a general locally compact group G. However, these questions need for their consideration certain analytic tools that we have not yet developed. In particular, a detailed study of characteristic functions and distribution functions is an essential first step, and this is undertaken in Chapter 4. It is then necessary to study further properties of sums of independent but not necessarily identically distributed random variables, continuing the work of Section 2. Here the most striking result, which we have not yet touched upon, is the law of the iterated logarithm. This is a strong limit theorem, based on the existence of two moments, but for its proof we also need the work on the central limit problem. Thus the results of this chapter are those obtainable only by means of the basic techniques. We need to continue expanding the subject. First a weakening of the concept of independence is needed. Then one proceeds to a study of the central limit problem and the (distributional or) weak limit laws.

Exercises

1. (a) Let (Ω, Σ, P) be a probability space with Ω having at least three points. If $X : \Omega \to \mathbb{R}$ is a random variable taking three or more distinct values, verify that $1, X, X^2$ are linearly independent (in the sense of linear algebra) but will be stochastically independent only if X is two valued and X^2 is a constant with probability 1, in which case $1, X, X^2$ are not linearly independent. Give an example satisfying the latter conditions. On the other hand, if X, Y are stochastically independent and not both are constant, then they are linearly independent, whenever $X \neq 0$ and $Y \neq 0$.

(b) Consider $\Omega = \{1, 2, 3, 4, 5\}$ with $P(\{i\}) = 1/5$ for $i = 1, 2, 3, 4, 5$. Is it possible to find events A, B of Ω so that A and B are independent? The answer to this simple and interesting problem is no. A probability space (Ω, Σ, P) is called a *"dependent probability space"* if there are no nontrivial independent events in Ω, (Ω, Σ, P) is called an *independent space* otherwise. R. Shiflett and H. Schultz (1979) introduced this concept where they studied both finite and countably infinite settings for Ω. Show that if Ω is finite so that $\Omega = \{1, 2, \ldots, n\}$ with $P(\{i\}) = 1/n$ for $i = 1, 2, \ldots, n$ then (Ω, Σ, P) is a dependent probability space if and only if n is prime. Additional results on finite dependent spaces with uniform probabilities can be found in the article by Shiflett and Schultz and in the work of Eisenberg and Ghosh (1987). Recently, W.F. Edwards (2004) investigated the case of the space (Ω, Σ, P) with $\Omega = \{1, 2, 3, \ldots\}$ and the measure P not uniform as follows. Show that if $\Omega = \{1, 2, 3, \ldots\}$ with $P(\{i\}) = p_i \geq p_{i+1} = P(\{i + 1\})$ for all i and if

$$p_i \leq \sum_{k=1}^{\infty} p_{i+k}$$

then (Ω, Σ, P) is an independent space. The hypothesis in this last statement is sufficient but not necessary which can be seen by showing if $\Omega = \{1, 2, 3, \ldots\}$ with $p_i = (1 - r)r^{i-1}$ for $0 < r < 1, i = 1, 2, \ldots$, then (Ω, Σ, P) is an independent space. These results give an idea of the interest that is associated with the question, "Are there necessary and sufficient conditions for a probability space (Ω, Σ, P) to be dependent?"

(c) One result without a restriction on the cardinality of Ω can be obtained by showing that (Ω, Σ, P) is an independent probability space if and only if there exists a partition of Ω into four nontrivial events A, B, C and D

for which $P(A)P(B) = P(C)P(D)$. [A related idea was considered by Chen, Rubin and Vitale (1997) who show that if the collection of pairwise independent events are identical for two measures, then the measures coincide. These are just some of the ideas associated with independent probability spaces. This type of inquiry can be continued with a serious investigation.]

2. Let $\phi : \mathbb{R}^+ \to \mathbb{R}^+$ be an increasing continuous convex or concave function such that $\phi(0) = 0$, with $\phi(-x) = \phi(x)$, and in the convex case $\phi(2x) \leq c\phi(x)$, $x \geq 0$, $0 < c < \infty$. If $X_i : \omega \to \mathbb{R}$, $i = 1, 2$, are two random variables on (Ω, Σ, P) such that $E(\phi(X_i)) < \infty, i = 1, 2$, then verify that $E(\phi(X_1 + X_2)) < \infty$ and that the converse holds if X_1, X_2 are (stochastically) independent. [*Hint:* For the converse, it suffices to consider $|X_2| > n_0 > 1$. Thus

$$E(\phi(|X_1 + X_2|)) \geq E(\phi(|X_1| - |X_2|) \geq E(\phi(|X_1| - n)\chi_{A_n}))$$
$$= E(\phi(|X_1| - n))P(A_n)$$

for an $A_n = [|X_2| \leq n], n_0 \geq n \geq 0$. Note that the converse becomes trivial if $X_i \geq 0$ instead of the independence condition.]

3. The preceding problem can be strengthened if the hypothesis there is strengthened. Thus let X_1, X_2 be independent and $E(X_1) = 0$. If now $\phi : \mathbb{R}^+ \to \mathbb{R}^+$ there is restricted to a continuous convex function and $E(\phi(X_1 + X_2)) < \infty$, then $E(\phi(X_2)) \leq E(\phi(X_1 + X_2))$. If $E(X_2) = 0$ is also assumed, then $E(\phi(X_i)) \leq E(\phi(X_1 + X_2)), i = 1, 2$. [*Hint:* Use Jensen's inequality, and the fundamental law of probability, (Theorem 1.4.1) in $\phi(x) = \phi(E(X_2 + x)) \leq E(\phi(x + X_2))$ and integrate relative to $dF_{X_1}(x)$, then use Fubini's theorem.]

4. (a) Let $I = [0, 1]$, $\mathcal{B}$ = Borel σ-algebra of I, and P = Lebesgue measure on I. Let $X_1, \ldots, X_n$ be i.i.d. random variables on (Ω, Σ, P) with their common distribution $F(x) = P[X_1 < x] = x, 0 \leq x \leq 1$ (and $= 0$ for $x < 0$, $= 1$ for $x \geq 1$). Define $Y_1 = \min(X_1, ..., X_n)$, and if Y_i is defined, let $Y_{i+1} = \min\{X_k > Y_i : 1 \leq k \leq n\}$. Then (verify that) $Y_1 < Y_2 < \ldots < Y_n$ are random variables, called *order statistics* from the d.f. F, and are not independent. If $F_{Y_1,\ldots,Y_n}$ is their joint distribution, show that

$$F_{Y_1,\ldots,Y_n}(x_1, \ldots, x_n) = \begin{cases} n! \int \cdots \int_{0 < x_1 < \ldots < x_n < 1} dt_1 \ldots dt_n \\ 0, & \text{otherwise} \end{cases}.$$

From this deduce that, for $0 \leq a < b \leq 1, i = 1, \ldots, n$,

$$P[a \leq Y_i < b] = \frac{n!}{(i-1)!(n-i)!} \int_a^b y^{i-1}(1-y)^{n-i} dy,$$

and that for $0 \leq a < b < c \leq 1, 1 \leq i < j \leq n$,

$$P[a \le Y_i < b < Y_j < c]$$
$$= \frac{n!}{(i-1)!(j-i-1)!(n-j)!} \times \int\int_{0\le a\le y_1 \le b \le y_2 \le c} y_1^{i-1}(y_2 - y_1)^{j-i-1}(1-y_2)^{n-j}dy_2dy_1.$$

[Note that for $0 \le y_1 < y_2 < \ldots < y_n \le 1$, for small enough $\varepsilon > 0$ such that $[y_i, y_i + \varepsilon]$ are disjoint for $1 \le i \le n$, we have

$$P[y_i \le Y_i \le y_i + \varepsilon_i, 1 \le i \le n] = \sum_{\substack{\text{all permutations} \\ (i_1,\ldots,i_n) \text{ of } (1,2,\ldots,n)}} P[y_j \le X_{i_j} \le y_j + \varepsilon_j, 1 \le j \le n],$$

where the X_i are i.i.d. for each permutation, and that there are $n!$ permutations.]

(b) Let $Z_1, \ldots, Z_n$ be i.i.d, random variables on (Ω, Σ, P) with their common distribution F on $\mathbb{R}$ continuous and strictly increasing. If $X_i = F(Z_i)$, $1 \le i \le n$, show that $X_1, \ldots, X_n$ are random variables satisfying the hypothesis of **(a)**. Deduce from the above that if Z_i is the ith-order statistic of $(Z_1, \ldots, Z_n)$, then

$$P[\tilde{Z}_i < x] = \frac{n!}{(i-1)!(n-i)!} \int_0^{F(x)} y^{i-1}(1-y)^{n-j}dy.$$

Similarly, obtain the corresponding formulas of **(a)** for the $\tilde{Z}_i$-sequence.

5. (a) Following Corollary 1.8 we have discussed the adjunction procedure. Let $X_1, X_2, \ldots,$ be any sequence of random variables on (Ω, Σ, P). Let $F_i(x) = P[X_i < x], i = 1, 2, \ldots$. Then using the same procedure, show that there is another probability space $(\tilde{\Omega}, \tilde{\Sigma}, \tilde{P})$ and a mutually independent sequence of random variables $Y_1, Y_2, \ldots$ on it such that $P[Y_n < x] = F_n(x), x \in \mathbb{R}, n \ge 1$. [*Hint:* Since F_n is a d.f., let $\mu_n(A) = \int_A dF_n(x), A \subset \mathbb{R}$ Borel, $X_n =$ identity on $\mathbb{R}$. Then $(\mathbb{R}, \mathcal{B}, \mu_n)$ is a probability space and $\tilde{X}_n$ is an r.v. with F_n as its d.f. Consider, with the Fubini-Jessen theorem, the product probability space $(\tilde{\Omega}, \tilde{\Sigma}, \tilde{P}) = \bigotimes_{n\ge1}(\mathbb{R}_n, \mathcal{B}_n, \mu_n)$, where $\mathbb{R}_n = \mathbb{R}$, $\mathcal{B}_n = \mathcal{B}$. If $\tilde{\omega} = (x_1, x_2, \ldots) \in \tilde{\Omega} = \mathbb{R}^{\mathbb{N}}$, let $Y_n(\tilde{\omega}) = n$th coordinate of $\tilde{\omega}$ $[= x_n = \tilde{X}_n(\tilde{\omega})]$. Note that the Y_n are independent random variables on $(\tilde{\Omega}, \tilde{\Sigma}, \tilde{P})$ and $\tilde{P}[\tilde{Y}_n < x] = \mu_n[\tilde{X}_n < x] = F_n(x), x \in \mathbb{R}, n \ge 1$.]

(b) (Skorokhod) With a somewhat different specialization, we can make, the following assertion: Let $X_1, X_2 \ldots$ be a sequence of random variables on (Ω, Σ, P) which converge in distribution to an r.v. X. Then there is another probability space (Ω', Σ', P') and random variables $Y_1, Y_2, \ldots$ on it such that $Y_n \to Y$ a.e. and $P[X_n < x] = P'[Y_n < x], x \in \mathbb{R}$,for $n \ge 1$. Thus X_n, Y_n

have the same distributions and the (stronger) pointwise convergence is true for the Y_n-sequence. (Compare this with Proposition 2.2.) [*Sketch of proof:* Let $F_n(x) = P[X_n < x], F(x) = P[X < x], x \in \mathbb{R}, n \geq 1$. If Y_n, Y are inverses to F_n, F, then $Y_n(x) = \inf\{y \in \mathbb{R} : F_n(y) > x\}$; and similarly for Y. Clearly Y_n, Y are Borel functions on $(0,1) \to \overline{\mathbb{R}}$. Since $Y_n(x) < y$ iff $F_n(y) > x$, we have, on letting $\Omega' = (0,1)$, $\Sigma' =$ Borel σ-algebra of Ω', with P' as the Lebesgue measure, $P'[Y_n < y] = P'[x : x < F_n(y)] = F_n(y)$; and similarly for $P'[Y < y] = F(y)$. Since $F_n(x) \to F(x)$ at all continuity points of F, let x be a continuity point of F. If the F_n are strictly increasing, then $Y_n = F_n^{-1}$ and the result is immediate. In the general case, follow the argument of Proposition 2.2, by showing that for $a < b \leq c < d$,

$$Y(a) \leq \liminf_n Y_n(b) \leq \limsup_n Y_n(b) \leq \limsup_n Y_n(c) \leq Y(d),$$

and then setting $b = c$, a continuity point of F; let $a \uparrow b$ and $d \downarrow c$, so that $Y_n(c) \to Y(c)$. Since the discontinuities of F are countable and form a set of P' measure zero, the assertion follows. *Warning:* In this setup the Y_n will not be independent if Y is nonconstant (or X is nonconstant).]

(c) The following well-known construction shows that the preceding part is an illustration of an important aspect of our subject. Let (Ω, Σ) be a measurable space and $B_i \in \Sigma$ be a family of sets indexed by $D \subset \mathbb{R}$ such that for $i, j \in D, i < j \Rightarrow B_i \subset B_j$. Then there exists a unique random variable $X : \Omega \to \mathbb{R}$ such that $\{\omega : X(\omega) \leq i\} \subset B_i$ and $\{\omega : X(\omega) > i\} \subset B_i^c$. [Verify this by defining $X(\omega) = \inf\{i \in D : \omega \in B_i\}$ and that X is measurable for Σ.] If $P : \Sigma \to \mathbb{R}^+$ is a probability and D is countable, $\{B_i, i \in D\}$ is increasing P a.e. (i.e., for $i < j$, $P(B_i - B_j) = 0$), then the variable X above satisfies $\{\omega : X(\omega) \leq i\} = B_i$, a.e. and $\{\omega : X(\omega) \geq i\} = B_i^c$, $i \in D$. (See e.g., Royden (1968, 1988), 11.2.10.) Suppose that there is a collection of such families $\{B_i^n, i \in D = \mathbb{R}, n \geq 1\} \subset \Sigma$. Let X_n be the corresponding random variable constructed for each n, and let $F_n(x) = P(B_x^n)$ where $-\infty < x < \infty$. Show that $F_n = P \circ X_n^{-1}$, determined by the collection, and that for $n_1, \ldots, n_m, x_i \in \mathbb{R}, m \geq 1$ one has

$$P(B_{x_1}^{n_1} \cap \cdots \cap B_{x_m}^{n_m}) = F_{n_1,\ldots,n_m}(x_1, \ldots, x_m)$$

defines an m-dimensional (joint) distribution of $(X_{n_1}, \ldots, X_{n_m})$ so constructed. [This construction of distributions will play a key role in establishing a general family of random variables, or processes, later (cf., Theorem 3.4.10).

(d) Here is a concrete generation of independent families of random variables already employed by N. Wiener (cf. Paley and Wiener (1934), p. 143), and emphasized by P. Lévy ((1953), Sec. 2.3). It also shows where the probabilistic concept enters the construction. Let $Y_1, \ldots, Y_n$ be functions on (0,1) each represented by its decimal expansion

$$Y_n = \sum_{\nu=1}^{\infty} \frac{a_{n,\nu}}{10^\nu},$$

$a_{n,\nu}$ taking values $0, 1, \ldots, 9$ each with probability $\frac{1}{10}$, independent of one another. (This is where probability enters!) Then each Y_n is uniformly distributed and they are mutually independent. (Clearly binary or ternary etc. expansions can be used in lieu of decimal expansion. Unfortunately, no recipe exists for choosing $a_{n,\nu}$ here. A similar frustration was (reportedly) expressed by A. Einstein regarding his inability to find a recipe for a particular Brownian particle to be in a prescribed region, but only a probability of the event can be given. [cf., Science, **30** (2005), pp. 865-890, special issue on Einstein's legacy].) If $\{F_n, n \geq 1\}$ is a sequence of distribution functions on $\mathbb{R}$, let F_n^{-1} be the generalized inverse of F, as defined (in part (b)) above. Let $X_n = F_n^{-1}(Y_n), n \geq 1$. Then $\{X_n, n \geq 1\}$ is a sequence (of mutually independent) random variables with distributions F_n. [It is even possible to take a single uniformly distributed random variable Y by reordering $a_{n,\nu}$ into a single sequence $\{b_k, k \geq 1\}$ so that $Y = \sum_{k=1}^{\infty} \frac{b_k}{10^k}$, by excluding the terminating decimal expansions which are countable and hence constitute a set of (Lebesgue) measure zero, and then $X_n = F_n^{-1}(Y), n \geq 1$.] It should be observed that in the representation of X_n as a mapping of $(Y_1, \ldots, Y_n)$ [or of Y] by I_n which is one-to-one, there are infinitely many representations, while a unique distribution obtains if it is nondecreasing, such as F_n^{-1}. This fact is of interest in applications such as those implied in part (b) above.

The following example is considered by Wiener (in the book cited above, p. 146). Let Y_1, Y_2 be independent uniformly distributed random variables on (0,1) and define $R = (-\log Y_1)^{\frac{1}{2}}$, and $\theta = 2\pi Y_2$ and let $X_1 = R\cos\theta, X_2 = R\sin\theta$. Then the Jacobian is easily computed, and one has $dy_1 dy_2 = \frac{1}{\pi} e^{-(x_1^2 + x_2^2)}$ $dx_1 dx_2$ so that X_1, X_2 are independent normal random variables generated by Y_1, Y_2. Extending this procedure establish the following n-dimensional version. Let $Y_1, \ldots, Y_n$ be independent uniformly distributed random variables on (0,1), $\theta_k = 2\pi Y_{k+1}$ and $X_1 = R\sin\theta_{n-1} \ldots \sin\theta_2 \sin\theta_1; X_2 = R\sin\theta_{n-1} \ldots \sin\theta_2 \cos\theta_1, \ldots, X_{n-1} = R\sin\theta_{n-1}\cos\theta_{n-2}$, and $X_n = R\cos\theta_{n-1}$ where $R = (-2\log Y_1)^{\frac{1}{2}}$. The Jacobian is much more difficult, [use induction], but is nonvanishing, giving a one-to-one mapping. (With $R = 1$, the transformation has Jacobian to be $(-1)^n (\sin\theta_1)^n (\sin\theta_2)^n \ldots \sin\theta_{n-1}\cos\theta_n$ so that it is 1-1 between the open unit n-ball and the open rectangle $0 < \theta_i < \pi, i = 1, \ldots n$.) This shows that the Φ_n sequence (different from the F_n) can be somewhat involved, but the procedure is quite general as noted by N. Wiener whose use in a construction of Brownian motion is now legendary, and was emphasized by P. Lévy later. [In the last chapter we again consider the Brownian motion construction with a more recent and (hopefully) simpler method.]

6. **(a)** (Jessen-Wintner) If $\{X_n, n \geq 1\}$ is a sequence of independent countably valued random values on (Ω, Σ, P) such that $S_n = \sum_{k=1}^{n} X_k \to S$ a.e., then the distribution of S on $\mathbb{R}$ is either (i) absolutely continuous or singular relative to the Lebesgue measure or (ii) $P[S = j] > 0$ for a countable set of points $j \in \mathbb{R}$, and no mixed types can occur. [*Hints*: Let $G \subset \mathbb{R}$ be the group generated by the ranges of the X_n, so that G is countable. Note that for any Borel set B, the vector sum $G + B = \{x + y : x \in G, y \in B\}$ is again Borel. If $\Omega_0 = \{\omega : S_n(\omega) \to S(\omega)\}$, then let $A = \{\omega : S(\omega) \in (G + B) \cap \Omega_0\}$, and verify that A is a tail event, so that $P(A) = 0$ or 1 by Theorem 1.7. Indeed, if $g_1 - g_2 \in G$, then $g_1 \in G + B$ for some Borel set B iff $g_2 \in G + B$. Now if $S_n = S - (S - S_n) \in G$, then $S - S_n \in G + B$, and conversely. But $S - S_n \in \Omega_0$. Hence $A = [S - S_n \in G + B] \cap \Omega_0$, so that A is a tail event, and $P(A) = 0$ or 1. This implies either S is countably valued or else, since $P(\Omega_0) = 1$, $P[S \in G+B] = 0$ for each countable B. In this case $P[S \in B] = 0$ for each countable B, so that S has a continuous distribution, with range non-countable. Consequently, either the distribution of S is singular relative to the Lebesgue measure, or it satisfies $P[S \in G + B] = 0$ for all Borel B of zero Lebesgue measure. Since G is countable, this again implies $P[S \in G+B] = 0$, so that $P[S \in B] = 0$ for all Lebesgue null sets. This means the distribution of S is absolutely continuous. To see what type is the distribution of S, we have to exclude the other two cases, and no recipe is provided in this result. In fact this is the last result of Jessen-Wintner's long paper (1935).]

(b) To decide on the types above, we need to resort to other tricks, and some will be noted here. Let $\{X_n, n \geq 1\}$ be i.i.d. random variables with

$$P[X_1 = +1] = \frac{1}{2} = P[X_1 = -1].$$

Let $S_n = \sum_{k=1}^{n} X_k/2^k$. Then $S_n \to S$ a.e. (by Theorem 2.6). Also $|S| \leq 1$ a.e. Prove that the S distribution is related to that of $U - V$, where U and V are independent random variables on the Lebesgue unit interval $[0, 1]$, with the uniform distribution F, i.e., $F(x) = 0$ if $x \leq 0, = x$ if $0 < x \leq 1$, and $F(x) = 1$ for $x > 1$, and hence has an absolutely continuous distribution. [*Hints*: Note that if F_U, F_V are the distributions of U, V, then F_{U+V} can be obtained by the image law (cf. Theorem 1.4.1) as a convolution:

$$\begin{aligned} F_{U+V}(x) = P[U + V < x] &= \int_{\Omega} \chi_{[U+V<x]} dP \\ &= \int_{\mathbb{R}} \int_{\mathbb{R}} \chi_{[\lambda_1+\lambda_2<x]} F_U(d\lambda_1) F_V(d\lambda_2) \\ &\quad \text{(since } F_{U,V} = F_U \cdot F_V \text{ by independence)} \\ &= \int_{\mathbb{R}} F_U(d\lambda_1) F_V(x - \lambda_1). \end{aligned}$$

Thus F_{U+V} is continuous if at least one of F_U, F_V is continuous. Next verify that if $x = \sum_{k=1}^{\infty} \varepsilon_k/2^k$, where $\varepsilon_k = 0, 1$ is the dyadic expansion of $0 < x < 1$, then (as in the construction of Problem 5 (d) above)

$$\mu\{x : \varepsilon_k(x) = 0\} = \frac{1}{2} = \mu\{x : \varepsilon_k(x) = 1\}$$

with μ as the Lebesgue measure. Deduce that U has the same distribution as the identity mapping $I : (0,1) \to (0,1)$ with Lebesgue measure.](Explicit calculation with ch.f. is easier and will be noted in Exercise 4.11.)

(c) By similar indirect arguments verify the following: (i) If $\{X_n, n \geq 1\}$ is as above, then $S_n = \sum_{k=1}^{n} X_k/3^k \to S$ a.e. and S has a singular distribution. (ii) (P. Lévy) If $Y_n, n = 1, 2, \ldots$, are independent with values in a countable set $C \subset \mathbb{R}$, and if there is a convergent set of numbers $c_n \in C$ such that

$$\sum_{n=1}^{\infty} P[Y_n \in C - c_n] < \infty,$$

then $S = \sum_{k=1}^{\infty} Y_k$ exists a.e., and S takes only countably many values with positive probability.

(d) The proofs of Theorems 2.6 and 2.7 used the Kronecker lemma and the $(c,1)$-summability. Thus the Kolomogorov SLLN (Theorem 2.7) can be considered as a probabilistic analog of the classical $(c,1)$-summability in the sense that a sequence $\{X_n, n \geq 1\}$ of i.i.d. r.v.s on (Ω, Σ, P) obeys the $(c,1)$-pointwise a.e. iff $E(X_1) = \mu \in \mathbb{R}$ exists. Since classical analysis shows that $(c,1)$-summability implies (c,p)-summability for $p \geq 1$, one can expect a similar result for i.i.d sequences. In fact the following precise version holds. Let $p, \mu \in \mathbb{R}$, $p \geq 1$. Verify the following equivalences for i.i.d. r.v.s:

(i) $\{X_n, n \geq 1\}$ obeys the SLLN,
(ii) $E(X_1) = \mu$,
(iii) $\{X_n, n \geq 1\}$ obeys $(c,1)$-summability a.e. with limit μ,
(iv) $\{X_n, n \geq 1\}$ obeys (c,p)-summability a.e. with limit μ,

$$\text{i.e., } \lim_{n\to\infty} \frac{1}{\binom{n+p}{n}} \sum_{k=0}^{n-1} \binom{k+p-1}{k} X_{n-k} = \mu \text{ a.e.,}$$

(v) $\{X_n, n \geq 1\}$ obeys Abel mean a.e. with value μ,

$$\text{i.e., } \lim_{0 \leq \lambda \uparrow 1} (1-\lambda) \sum_{i=1}^{\infty} \lambda^i X_i = \mu \text{ a.e..}$$

[*Hints*: The classical theories on summability imply that (i) $\Rightarrow$ (iii) $\Rightarrow$ (iv) $\Rightarrow$ (v) and Theorem 2.7 gives (i) $\Leftrightarrow$ (ii). So it suffices to show (v) $\Rightarrow$ (ii). For ordinary sequences of reals, Abel convergence does not imply even $(c,1)$-convergence. (Here the converse holds if the sequence is bounded in addition, as shown by J.E. Littlewood.) But the i.i.d. hypothesis implies the converse a.e. as follows. Using the method of Theorem 2.9, called symmetrization, let $X_n^s = X_n - X_n'$ where X_n and X_n' are i.i.d. (one may use enlargement of the basic probability space as in the proof of 2.9, where X_n^s is denoted as Z_n there), and (v) can be expressed if $1 - \lambda = 1/m, m \geq 1$ as

$$\lim_{m\to\infty} \frac{1}{m}\sum_{i=1}^{\infty}(1-\frac{1}{m})^i X_i^s = \mu - \mu = 0,$$

or alternately

$$\lim_{m\to\infty} \frac{1}{m}\sum_{i=1}^{\infty} e^{i\log(1-1/m)} X_i^s = \lim_{m\to\infty} \frac{1}{m}\sum_{j=1}^{\infty} e^{-j/m} X_j^s = 0 \text{ a.e.}$$

Let

$$Y_m = \frac{1}{m}\sum_{j=1}^{m} e^{-j/m} X_j^s, \quad Z_m = \frac{1}{m}\sum_{j=m+1}^{\infty} e^{-j/m} X_j^s.$$

Then $Y_m + Z_m \to 0$ a.e. as $m \to \infty$, and Y_m, Z_m are independent. Verify that for each $\varepsilon > 0$, $P[|Z_m| \geq \varepsilon] \to 0$ as $m \to \infty$. Then using Slutzky's Theorem and stochastic calculus (Problems 9(b) and 11(c) below) suitably conclude that $Y_m \to 0$. Next $\overline{Y}_m = Y_m - \frac{1}{em}X_m^2 \to 0$ and finally that $X_m^s/m \to 0$ also as $m \to \infty$. [This needs some more work!] Then by the Borel-Cantelli lemma, deduce that $E(|X_1|) < \infty$, as in the proof of Theorem 3.7. Hence SLLN holds. Thus the equivalence follows. The above sketch is a paraphrase of T. L. Lai (1974). Can we replace mutual independence here by pairwise independence, as in Corollary 3.3 if we only ask for WLLN?]

7. This problem illustrates the strengths and limitations of our a.e. convergence statements. Let (Ω, Σ, P) be the Lebesgue unit interval, so that $\Omega = (0,1)$ and $P =$ Lebesgue measure on the completed Borel σ-algebra Σ. If $\omega \in \Omega$, expand this in decimals: $\omega = 0.x_1x_2\ldots$ so that if $X_n(\omega) = x_n$, then $X_n : \Omega \to \{0,1,\ldots,9\}$ is a r.v. Verify that $\{X_n, n \geq 1\}$ is an i.i.d. sequence with the common distribution F, given by $F(y) = (k+1)/10$, for $k \leq y < k+1$, $k = 0,1,\ldots,9$; $= 0$ if $y < 0$; $= 1$ for $y > 9$. Let $\delta_k(\cdot)$ be the Dirac delta function, and consider $\delta_k(X_n)$. Then $P[\delta_k(X_n) = 1] = 1/10, P[\delta_k(X_n) = 0] = 9/10$, and $\delta_k(X_n), n \geq 1$, are i.i.d., for each $k = 0,1,\ldots,9$. If $k_1, k_2, \ldots, k_r$ are a fixed r-tuple of integers such that $0 \leq k_i \leq 9$, define (cf. Problem 5 (d) also)

$$\varepsilon_{n,r} = \delta_{k_1}(X_{rn})\delta_{k_2}(X_{rn+1})\ldots\delta_{k_r}(X_{rn+r-1}).$$

Show that the $\varepsilon_{n,r}, n \geq 1$, are bounded uncorrelated random variables for which we have $(1/m)\sum_{n=1}^{m}\varepsilon_{n,r} \to 1/10^r$ a.e. as $m \to \infty$ (apply Theorem 3.4), $r = 1,2,\ldots$. This means for a.a. $\omega \in \Omega$, the ordered set of numbers $(k_1,\ldots,k_r)$ appears in the decimal expansion of ω with the asymptotic relative frequency of $1/10^r$. Every number $\omega \in \Omega$ for which this holds is called a *normal number*. It follows that $\sum_{n=1}^{m}\varepsilon_{n,r} \to \infty$ as $m \to \infty$ for a.a.(ω) (as in the proof of Theorem 4.4); thus $\varepsilon_{n,r} = 1$ infinitely often, which means that the given set $(k_1,\ldots,k_r)$ in the same order occurs infinitely often in the expansion of each normal number, and that almost all $\omega \in \Omega$ are normal. [This fact was established by E. Borel in 1909.] However, there is no known recipe to find which numbers in Ω are normal. Since the transcendental $(\pi - e) \in (0,1)$, it is not known whether $\pi - e$ is normal; otherwise it would have settled the old question of H. Weyl: Is it true or false that in the decimal expansion of the irrational number π, the integers $0, 1, \ldots, 9$ occur somewhere in their natural order? This question was raised in the 1920's to counter the assertion of the logicians of Hilbert's school asserting that every statement is either "true" or "false," i.e., has only two truth values. As of now we do not know the definitive answer to Weyl's question, even though π has been expanded to over 10^5 decimal places and the above sequence still did not appear! [See D. Shanks and J. W. Wrench, Jr. (1962). *Math. Computation* 16, 76-89, for such an expansion of π. On the other hand, it is known that 0.1234567891011121314151617..., using all the natural numbers, is normal. Recently two Japanese computer scientists seem to have shown that the answer is 'yes' after expanding π for several billions of decimal places. See, e.g. J.M. Borwain (1998), *Math. Intelligencer*, **20**, 14-15.]

8. The WLLN of Theorem 3.2 does not hold if (even) the symmetric moment does not exist. To see this, we present the classical *St. Petersburg game*, called a "paradox," since people applied the WLLN without satisfying its hypothesis. Let X be an r.v. such that

$$P[X = 2^n] = \frac{1}{2^n}, \quad \text{for } n \geq 1,$$

on (Ω, Σ, P). Let $\{X_n, n \geq 1\}$ be i.i.d. random variables with the distribution of X. If $S_n = \sum_{k=1}^n X_k$, show that $S_n/n \not\to \alpha$, as $n \to \infty$, for any $\alpha \in \mathbb{R}$, either in probability or a.e. for any subsequence. (Use the last part of Theorem 3.7.) The game interpretation is that a player tosses a fair coin until the head shows up. If this happens on the nth toss, the player gets 2^n dollars. If any fixed entrance fee per game is charged, the player ultimately wins and the house is ruined. Thus the "fair" fee will have to be "infinite," and this is the paradox! Show however, by the truncation argument, that $S_n/(n \log_2 n) \xrightarrow{P} 2$ as $n \to \infty$, where $\log_2 n$ is the logarithm of n to base 2. If the denominator is replaced by $h(n)$ so that $(n \log_2 n)/h(n) \to 0$, then $S_n/h(n) \xrightarrow{P} 0$ and a.e. In fact show that for any sequence of random variables $\{Y_n, n \geq 1\}$ there exists an increasing sequence k_n such that $P[|Y_n| > k_n, i.o.] = 0$, so that $Y_n/k_n \to 0$ a.e. Thus $n \log_2 n$ is the correct "normalization" for the St. Petersburg game. (An interesting and elementary variation of the St. Petersburg game can be found in D.K. Neal, & R.J. Swift, (1999) *Missouri J. Math. Sciences*, **11**, No. 2, 93-102.)

9. (Mann-Wald). A calculus of "in probability" will be presented here. (Except for the sums, most of the other assertions do not hold on infinite measure spaces!) Let $\{X_n, n \geq 1\}$ and $\{Y_n, n \geq 1\}$ be two sequences of random variables on (Ω, Σ, P). Then we have the following, in which ***no*** assumption of independence appears:

(a) $X_n \xrightarrow{P} X$, $Y_n \xrightarrow{P} Y \Rightarrow X_n \pm Y_n \xrightarrow{P} X \pm Y$, and $X_n Y_n \xrightarrow{P} XY$.

(b) If $f : \mathbb{R}^2 \to \mathbb{R}$ is a Borel function such that the set of discontinuities of f is measurable and is of measure zero relative to the Stieltjes measure determined by the d.f. $F_{X,Y}$ of the limit vector (X, Y) of **(a)**, then $f(X_n, Y_n) \xrightarrow{D} (X, Y)$ under either of the conditions: (i) $X_n \xrightarrow{P} X$, $Y_n \xrightarrow{P} Y$ or (ii) $\alpha X_n + \beta Y_n \xrightarrow{P} \alpha X + \beta Y$ for all real α, β. If f is continuous, then strengthen this to the assertion that $f(X_n, Y_n) \xrightarrow{P} f(X, Y)$ if condition (i) holds. [Hint: For (ii), use Problem 5(b) and the fact that $(X_n, Y_n) \xrightarrow{D} (X, Y)$ iff $\alpha X_n + \beta Y_n \xrightarrow{D} \alpha X + \beta Y$ for all real α, β.]

10. Suppose that for a sequence $\{X_n, n \geq 1, X\}$ in $L^1(P)$ we have $X_n \xrightarrow{D} X$. Show it is true that $E(|X|) \leq \liminf_n E(|X_n|)$, and if, *further*, the set is uniformly integrable, then $E(X) = \lim_n E(X_n)$. [*Hint*: Use Problem 5 (b) and the image probability Theorem 1.4.1. This strengthening of Vitali's convergence theorem (and Fatou's lemma) is a nontrivial contribution of Probability Theory to Real Analysis!]

11. (a) If X is an r.v. on (Ω, Σ, P), then $\mu(X)$, called a *median* of the distribution of X, is any number which satisfies the inequalities

$$P[X \leq \mu(X)] \geq \frac{1}{2}, \quad P[X \geq \mu(X)] \geq \frac{1}{2}.$$

Note that a median of X always exists [let $\mu(X) = \inf\{\alpha \in \mathbb{R} : P[X \leq \alpha] \geq \frac{1}{2}\}$ and verify that $\mu(X)$ is a median and $\mu(aX + b) = a\mu(X) + b$, for $a, b \in \mathbb{R}$]. If $X_n \xrightarrow{D} a_0$, $a_0 \in \mathbb{R}$, show that $\mu(X_n) \to a_0$

(b) A sequence $\{X_n, n \geq 1\}$ of random variables is *bounded in probability* if for each $\varepsilon > 0$ there is an $n_0[= n_0(\varepsilon)]$ and a constant $M_0[= M_0(\varepsilon)] > 0$ such that $P[|X_n| \geq M_\varepsilon] \leq \varepsilon$ for all $n \geq n_0$. Show that if $X_n \xrightarrow{D} X$ and $Y_n \xrightarrow{P} 0$ are two sequences of random variables, then $X_n Y_n \xrightarrow{P} 0$, $X_n + Y_n \xrightarrow{D} X$, as $n \to \infty$ and $\{X_n, n \geq 1\}$ is bounded in probability. If $\{X_n, n \geq 1\}$ has the latter property and $Y_n \xrightarrow{P} 0$, then $X_n Y_n \xrightarrow{P} 0$ as $n \to \infty$.

(c) (Cramér-Slutsky) Let $X_n \xrightarrow{D} X$, $Y_n \xrightarrow{D} a$, where $a \in \mathbb{R}$ and $n \to \infty$. Then $X_n Y_n \xrightarrow{D} aX$, and if $a \neq 0$, $X_n / Y_n \xrightarrow{D} X/a$, so that the distributions of aX and X/a are $F(x/a)$ and $F(ax)$ for $a > 0$, $1 - F(x/a)$ and $1 - F(ax)$ for $a < 0$. Here again the sequences $\{X_n\}$ and $\{Y_n\}$ need not be independent.

(d) Let $\{X_n, n \geq 1\}$ and $\{Y_n, n \geq 1\}$ be two sequences of random variables on (Ω, Σ, P) and $\alpha_n \downarrow 0$, $\beta_n \downarrow 0$ be numbers, such that $(X_n - a)/\alpha_n \xrightarrow{D} X$ and $(Y_n - b)/\beta_n \xrightarrow{D} Y$, where $a, b \in \mathbb{R}$, $b \neq 0$. Show that $(X_n - a)/\alpha_n Y_n \xrightarrow{D} X/b$. All limits are taken as $n \to \infty$.

12. (Kolmogorov). Using the method of proof of Theorem 2.7, show that if $\{X_n, n \geq 1\}$ is an independent sequence of bounded random variables on

(Ω, Σ, P), common bound M and means zero, then for any $d > 0$ we have, with $S_n = \sum_{k=1}^n X_k$,

$$P\left[\max_{1\le k\le n} |S_k| \le d\right] \le \frac{(2M+d)^2}{Var(S_n)}.$$

Deduce that if $Var(S_n) \to \infty$, then for each $d > 0$, $P[|S_n| \le d] \to 0$ as $n \to \infty$.

13. (Ottaviani). Let $\{X_n, n \ge 1\}$ be independent random variables on (Ω, Σ, P) and $\varepsilon > 0$ be given. If $S_n = \sum_{k=1}^n X_k$, $P[|X_k + \ldots + X_n| \le \varepsilon] \ge \eta > 0$, $1 \le k \le n$, show that

$$P\left[\max_{1\le k\le n} |S_k| \ge \varepsilon\right] \le \frac{1}{\eta} P[|S_n| \ge \frac{\varepsilon}{2}].$$

[Note that if $A_1 = [|S_1| \ge \varepsilon]$, and for $k > 1$, $A_k = [|S_k| \ge \varepsilon, |S_j| < \varepsilon, 1 \le j \le k-1]$, then $[|S_n| \ge \varepsilon/2] \supset \bigcup_k (A_k \cap [|X_{k+1} + \ldots + X_n| \le \varepsilon/2])$. The decomposition of $[\max_{k\le n} |S_k| \ge \varepsilon]$ is analogous to that used for the proof of Theorem 2.5.]

14. We present two extensions of Kolmogorov's inequality for applications. (a) Let $X_1, \ldots, X_n$ be independent random variables on (Ω, Σ, P) with means zero and variances $\sigma_1^2, \ldots, \sigma_n^2$. Then the following improved one-sided inequality [similar to that of Čebyšev's; this improvement in 1960 is due to A. W. Marshall] holds: for $\varepsilon > 0$, and $S_k = \sum_{i=1}^k X_i$, one has

$$P\left[\max_{1\le k\le n} S_k \ge \varepsilon\right] \le \frac{\sum_{k=1}^n \sigma_k^2}{\varepsilon^2 + \sum_{k=1}^n \sigma_k^2} \le \frac{1}{\varepsilon^2} \sum_{k=1}^n \sigma_k^2.$$

[*Hint*: Consider $f : \mathbb{R}^n \to \mathbb{R}$ defined by $f(x_1, \ldots x_n) = [\sum_{i=1}^n (\varepsilon x_i + \sigma_i^2)/(\varepsilon^2 + \sum_{i=1}^n \sigma_i^2)]^2$, and evaluate $E(f(X_1, \ldots, X_n))$ with the same decomposition as in Theorem 2.5. If $n = 1$, this reduces to Problem 6 (a) of Chapter 1.]

(b) Let $\{X_n, n \ge 1\}$ be independent random variables on (Ω, Σ, P) as above, with zero means and $\{\sigma_n^2, n \ge 1\}$ as respective variances. If $\varepsilon > 0$, $S_n = \sum_{k=1}^n X_k$, and $a_1 \ge a_2 \ge \ldots \to 0$, show that with simple modifications of the proof of Theorem 2.5,

$$P\left[\sup_{n\ge 1} a_n^{\frac{1}{2}} |S_n| > \varepsilon\right] \le \frac{1}{\varepsilon^2} \sum_{n\ge 1} (a_n - a_{n+1}) \sum_{k=1}^n \sigma_k^2.$$

[This inequality was noted by J. Hájek and A. Rényi.]

(c) If in (b) we take $a_k = (n_0 + k - 1)^{-2}$ for any fixed but arbitrary $n_0 \ge 1$, deduce that

$$P\left[\sup_{n\ge n_0} \frac{|S_n|}{n} > \varepsilon\right] \le \frac{1}{\varepsilon^2} \left(\frac{1}{n_0^2} \sum_{i=1}^{n_0} \sigma_i^2 + \sum_{n\ge n_0+1} \frac{\sigma_n^2}{n^2} \right).$$

Hence, if $\sum_{n\geq 1}(\sigma_n^2/n^2) < \infty$, conclude that the sequence $\{X_n, n \geq 1\}$ obeys the SLLN. (Thus we need not use Kronecker's lemma.)

15. In some problems of classical analysis, the demonstration is facilitated by a suitable application of certain probabilistic ideas and results. This was long known in proving the Weierstrass approximation of a continuous function by Bernstein polynomials. Several other results were noted by K. L. Chung for analogous probabilistic proofs. The following is one such: an inversion formula for Laplace transforms. Let $X_1(\lambda), \ldots, X_n(\lambda)$ be i.i.d. random variables on (Ω, Σ, P), depending on a parameter $\lambda > 0$, whose common d.f. F is given by $F(x) = 0$ if $x < 0$; and $= \lambda \int_0^x e^{-\lambda t} dt$ if $x \geq 0$. If $S_n(\lambda) = \sum_{k=1}^n X_k(\lambda)$, using the hints given for Problem 6(b) show that the d.f. of $S_n(\lambda)$ is F_n, where $F_n(x) = 0$ for $x < 0$, and $= [\lambda^n/(n-1)!] \int_0^x t^{n-1} e^{-\lambda t} dt$ for $x \geq 0$. Deduce that $E(S_n(\lambda)) = n/\lambda$, $Var S_n(\lambda) = n/\lambda^2$, so that $S_n(n/x) \xrightarrow{P} x$ as $n \to \infty$. Using the fundamental law of probability, verify that for any bounded continuous mapping $f : \mathbb{R}^+ \to \mathbb{R}^+$, or f Borel satisfying $E(f(S_n))^2 < k_0 < \infty$ (cf., also Proposition 4.1.3 later) then $E(f(S_n)) \to E(f(x)) = f(x)$, by uniform integrability, (use Scheffè's lemma, Proposition 1.4.6), where

$$E(f(S_n(\lambda))) = \frac{(-\lambda)^{n-1}}{(n-1)!} \lambda \int_0^\infty \frac{d^{n-1}}{d\lambda^{n-1}} (e^{-\lambda t} f(t)) dt.$$

Hence prove, using Problem 6(b), that for any continuous $f \in L^2(\mathbb{R}^+)$ if $\hat{f}$ is the Laplace transform of f $[\hat{f}(u) = \int_0^\infty e^{-ut} f(t) dt, u > 0]$ one has the inversion

$$f(t) = \lim_{n\to\infty} \left(\frac{-n}{t}\right)^{n-1} \cdot \frac{1}{(n-1)!} \frac{n}{t} \cdot \left(\frac{d^{n-1}\hat{f}}{d\lambda^{n-1}}\right) \left(\frac{n}{\lambda}\right) |_{\lambda=t} ,$$

the limit existing uniformly on compact intervals of $\mathbb{R}^+$. [Actually f can be in any $L^p(\mathbb{R}^+)$, $1 < p < \infty$, not just $p = 2$. The distribution of X_1 above is called the *exponential*, and that of $S_n(\lambda)$, the *gamma* with parameters (n, λ). More d.f.s are discussed in Section 4.2 later.] The result above is the *classical Post-Widder formula.*

16. Let $\{X_n, n \geq 1\}$ be independent random variables on (Ω, Σ, P) and $S_n = \sum_{k=1}^n X_k$. Then $S_n \to S$ a.e. iff $S_n \xrightarrow{P} S$. This result is due to P. Lévy. (We shall prove later that $S_n \xrightarrow{P} S$ can be replaced here by $S_n \xrightarrow{D} S$, but more tools are needed for it.) [*Hints*: In view of Proposition 2.2, it suffices to prove the converse. Now $S_n - S \xrightarrow{P} 0 \Rightarrow \{S_n, n \geq 1\}$ is Cauchy in probability, so for $1 > \varepsilon > 0$, there is an $n_0 [= n_0(\varepsilon)]$ such that $m, n > n_0 \Rightarrow P[|S_n - S_m| > \varepsilon] < \varepsilon$. Thus $P[|S_k - S_m| \geq \varepsilon] > 1 - \varepsilon$ for all $m < k \leq n$. Hence by Problem 13 applied to the set $\{X_j, j \geq m \geq n_0\}$, we get

$$P[\max_{m<k\leq n} |S_k - S_m| \geq 2\varepsilon] \leq \frac{1}{1-\varepsilon} P[|S_n - S_m| \geq \varepsilon] < \frac{\varepsilon}{1-\varepsilon}.$$

This implies upon first letting $n \to \infty$, and then letting $m \to \infty$, since the $0 < \varepsilon < 1$ is arbitrary, that $\{S_k, k \geq 1\}$ is pointwise Cauchy and hence converges a.e.]

17.(P. Lévy Inequalities). Let $X_1, \ldots, X_n$ be independent random variables on (Ω, Σ, P) and $S_j = \sum_{k=1}^{j} X_k$. If $\sigma(X)$ denotes a median (cf. Problem 11) of X, show that for each $\varepsilon > 0$ the following inequalities obtain:

(a) $P[\max_{1 \leq j \leq n}(S_j - \mu(S_j - S_n)) \geq \varepsilon] \leq 2P[S_n \geq \varepsilon]$;

(b) $P[\max_{1 \leq j \leq n} |S_j - \mu(S_j - S_n)| \geq \varepsilon] \leq 2P[|S_n| \geq \varepsilon]$.

[*Hints*: Use the same decomposition for max as we did before. Thus, let $A_j = [S_j - S_n \leq \mu(S_j - S_n)]$, so that $P(A_j) \geq \frac{1}{2}, 1 \leq j \leq n$, and

$$B_j = [S_j - \mu(S_j - S_n) \geq \varepsilon, \text{ for the first time at } j].$$

Then $B_j \in \sigma(X_1, \ldots, X_j), A_j \in \sigma(X_{j+1}, \ldots, X_n)$, and they are independent; $\bigcup_{j=1}^{n} B_j = B = [\max(S_j - \mu(S_j - S_n)) \geq \varepsilon]$, a disjoint union. Thus $P[S_n \geq \varepsilon] \geq \sum_{j=1}^{n} P(B_j \cap A_j) \geq \frac{1}{2}P(B)$, giving (a). Since $\mu(-X) = -\mu(X)$, write $-X_j$ for $X_j, 1 \leq j \leq n$, in (a) and add it to (a) to obtain (b). Hence if the X_j are also symmetric, so that $\mu(X) = 0$, (a) and (b) take the following simpler form:

(a') $P[\max_{1 \leq j \leq n} S_j \geq \varepsilon] \leq 2P[S_n \geq \varepsilon]$;

(b') $P[\max_{1 \leq j \leq n} |S_j| \geq \varepsilon] \leq 2P[|S_n| \geq \varepsilon]$.]

18. Let $\{X_n, n \geq 1\}$ be independent random variables on (Ω, Σ, P) with zero means and variances $\{\sigma_n^2, n \geq 1\}$ such that $\sum_{n \geq 1} \sigma_n^2 / b_n^2 < \infty$ for some $0 < b_n \leq b_{n+1} \nearrow \infty$. Then $(1/b_n) \sum_{k=1}^{n} X_k \to 0$ a.e. [*Hint*: Follow the proof of Theorem 3.6 except that in using Kronecker's lemma (Proposition 3.5) replace the sequence $\{n\}_{n \geq 1}$ there by the $\{b_n\}_{n \geq 1}$-sequence here. The same argument holds again.]

19. Let $\{X_n, n \geq 1\}$ be i.i.d. and be symmetric, based on (Ω, Σ, P). If $S_n = \sum_{k=1}^{n} X_k$, show that for each $\varepsilon > 0$,

$$\sum_{n \geq 1} P[|S_n| \geq n\varepsilon] < \infty \Rightarrow \{X_n, n \geq 1\} \subset L^2(P) \Leftrightarrow \sum_{n \geq 1} \sum_{j=1}^{n} P[|X_j| > n] < \infty.$$

[*Hints*: By Problem 17b' and the i.i.d. hypothesis, we have, with $S_0 = 0 = X_0$,

$$2P[|S_n| \geq \varepsilon] \geq P\left[\max_{0 \leq j \leq n} |S_j| \geq \varepsilon\right] \geq P\left[\max_{0 \leq j \leq n} |X_j| \geq 2\varepsilon\right]$$

$$= 1 - \prod_{j=1}^{n} P[X_j < 2\varepsilon], \tag{40}$$

since $[\max_{j\leq n}|S_j|\geq\varepsilon]\supset[\max_{j\leq n}|X_j|\geq 2\varepsilon]$. Summing and using the hypothesis with n for ε, and $\alpha_n=P[X_1<2n]$ in (40), we get

$$
\begin{aligned}
\infty > \sum_{n=1}^{\infty}(1-\alpha_n^n) &= \sum_{n=1}^{\infty}(1-\alpha_n)(1+\alpha_n+\ldots+\alpha_n^{n-1}) \\
&= \sum_{n=1}^{\infty} P[|X_1|\geq 2n]\sum_{j=0}^{n-1}\prod_{i=0}^{j} P[|X_i|<2n] \\
&= \sum_{n=1}^{\infty} P[|X_1|\geq 2n]\sum_{j=0}^{n-1} P\left[\max_{i\leq j}|X_i|<2n\right] \\
&\geq \sum_{n=1}^{\infty} nP[|X_1|\geq 2n] \\
&\quad \times \frac{1}{n}\sum_{j=0}^{n-1}(1-2P[|S_j|\geq j])\ \ [\text{ by } (40)]. \qquad (41)
\end{aligned}
$$

The convergence of the given series implies $P[|S_n|\geq n\varepsilon]\to 0$ as $n\to\infty$, and then by the $(C,1)$-summability the second term in (41) $\to 1$. Hence $\sum_{n=1}^{\infty} nP[|X_1|\geq 2n]<\infty$. But this is the same as the last series (by i.i.d.). Rewriting $P[|X_1|\geq 2n]$ as $\sum_{k\geq 2n} P[k\leq|X_1|<k+1]$ and changing the order of summation one gets $X_1\in L^2(P)$, and by the i.i.d. hypothesis

$$\{X_n, n\geq 1\}\subset L^2(P).$$

The converse here is similar, so that the last equivalence follows. It should be remarked that actually all the implications are equivalences. The difficult part (the first one) needs additional computations, and we have not yet developed the necessary tools for its proof. This (harder) implication is due to Hsu and Robbins (1947), and we establish it later, in Chapter 4.] Show, however, what has been given is valid if the symmetry assumption is dropped in the hypothesis.

20. In the context of the preceding problem, we say [after Hsu and Robbins (1947)] that a sequence $\{Y_n, n\geq 1\}$ of random variables on (Ω,Σ,P) *converges completely* if for each $\varepsilon>0$, $(*)\sum_{n=1}^{\infty} P[|Y_n|>\varepsilon]<\infty$. Show that complete convergence implies convergence a.e. Also, verify that (*) implies that the a.e. limit of Y_n is necessarily zero. Establish by simple examples that the converse fails. [For example, consider the Lebesgue unit interval and $Y_n=n\chi_{[0,1/n]}$.] Show, however, that the converse implication does hold if there is a probability space (Ω',Σ',P'), a sequence $\{Z_n, n\geq 1\}$ of independent random variables on it such that $P[Y_n<x]=P'[Z_n<x], x\in\mathbb{R}, n\geq 1$,

and $Z_n \to 0$ a.e. Compare this strengthening with Problem 5. [*Hint*: Note that $\limsup_n Z_n = 0$ a.e., and apply the second Borel-Cantelli lemma.]

21. The following surprising behavior of the symmetric random walk sequence was discovered by G. Pólya in 1921. Consider a symmetric random walk of a particle in the space $\mathbb{R}^k$. If $k = 1$, the particle moves in unit steps to the left or right, from the origin, with equal probability. If $k = 2$, it moves in unit steps in one of the four directions parallel to the natural coordinate axes with equal probability, which is 1/4. In general, it moves in unit steps in the $2k$ directions parallel to the natural coordinate axes each step with probability $1/2k$. Show that the particle visits the origin infinitely often if $k = 1$ or 2, and only finitely often for $k = 3$. (The last is also true if $k > 3$.) [*Hints*: If $e_1, \ldots, e_k$ are the unit vectors in $\mathbb{R}^k$, so that $e_i = (0, \ldots, 1, 0, \ldots, 0)$ with 1 in the ith place, and $X_n : \Omega \to \mathbb{R}^k$ are i.i.d., then

$$P[X_n = e_i] = P[X_n = -e_i] = 1/2k, i = 1, \ldots, k.$$

Let $S_n = \sum_{j=1}^{n} X_j$. Then if $k = 1$, the result follows from Theorem 4.7, and if $k = 2$ or 3, we need to use Theorem 4.8 and verify the convergence or divergence of (35) there. If $p_n = P[|S_n| = 0]$, so that the particle visits 0 at step n with probability p_n, then the particle can visit 0 only if the positive and negative steps are equal. Thus $p_n = 0$ for odd n and $p_{2n} > 0$. However, by a counting argument ("multinomial distribution "), we see that

$$p_{2n} = \sum_{j=0}^{n} \frac{(2n)!}{[j!(n-j)!]^2} \left(\frac{1}{2 \cdot 2}\right)^{2n} = 4^{-2n} \binom{2n}{n} \sum_{j=0}^{n} \binom{n}{j}^2 = 4^{-2n} \binom{2n}{n}^2.$$

Using Stirling's approximation, $n! \sim \sqrt{2\pi n} n^n e^{-n}$, one sees that $p_n \sim 1/n$, and so $\sum_{n \geq 1} p_{2n} = \infty$, as desired. If $k = 3$, one gets by a similar computation

$$\begin{aligned} p_{2n} &= \sum_{\substack{0 \leq i,j \leq n \\ i+j \leq n}} \frac{(2n)!}{[i!j!(n-i-j)!]^2} \left(\frac{1}{2 \cdot 3}\right)^{2n} \\ &= \sum_{0 \leq i+j \leq n} \left(\frac{n!}{i!j!(n-i-j)!} \cdot \frac{1}{3^n}\right) \binom{2n}{n} \frac{1}{2^{2n}}. \end{aligned}$$

Again simplification by Stirling's formula shows that $p_{2n} \sim 1/n^{3/2}$, so that $\sum_{n \geq 1} p_{2n} < \infty$ (in fact, the series is approximately $= 0.53$), and S_n is not recurrent. By more sophisticated computations, Chung and Fuchs in their work on random walk showed that the same is true if the X_n are just i.i.d., with $E(X_1) = 0, 0 < E(|X_1|^2) < \infty$, and no component of X_1 is degenerate. This problem also shows an intimate relation between the structure of random walks and the group theoretical properties of its range (or state space), and deeper connections with convolution operators on these spaces or the group. For a recent contribution on the subject, and several references to the related literature on the problems, the reader is referred to Rao (2004a).]

Chapter 3

Conditioning and Some Dependence Classes

This chapter is aimed at an extended study of two important classes of non-independent random families, namely, martingales and Markov processes. These classes are based on the general concept of conditioning; thus conditional expectations and probabilities are treated in considerable detail. The existence of such classes is deduced from a fundamental theorem of Kolmogorov and its generalization by Bochner. This is given in several different forms, and these ideas are illustrated with important applications as well as problems.

3.1 Conditional Expectations

The formative years of probability theory were naturally dominated by the then precisely defined notion of independence. However, the need for a relaxation of the conditions governing this concept was noticed while studying, for instance, the behavior of sums of independent random variables. But a useful concept generalizing this phenomenon was formulated by A. A. Markov only in 1906; and later in 1935 the martingale dependence was introduced by P. Lévy. Also in the early 1930s the strict sense and second-order stationary dependences were presented respectively by G. D. Birkhoff and A. Khintchine based on the needs of ergodic theory and harmonic analysis of random functions. This chapter consists of a study of the first two classes and their basic properties. We shall consider them further in the last three chapters, where some other dependence notions are briefly discussed. Of course, there are numerous other definable dependence classes, because events that are not independent are by default dependent. Many of these will not be considered in any detail.

The very definitions of both Markovian and martingale concepts depend on the notions of conditional probabilities and expectations. The latter concepts are so fundamental for the modern developments in probability theory that some authors wish to start the subject with conditional concepts. However, it appears desirable to follow the natural growth of our theory, and formulate each one precisely as an essentially unique solution of a functional equation. This is analogous to our mathematical formulation of the independence concept in Chapter 2, which employed a system of equations. We first motivate the concept, because it is not a simple or a particularly intuitive idea.

Let (Ω, Σ, P) be a probability triple that describes an experiment or a physical phenomenon mathematically. If an event A has been observed, how does one assign probabilities to the other events of Ω after incorporating this knowledge about A? Thus one should consider every event of Ω along with A, so that $\Sigma(A) = \{A \cap B : B \in \Sigma\}$ is the new class for which we need to define $P_A : \Sigma(A) \to \mathbb{R}^+$ as a probability. Assuming $P(A) > 0$, so that $P_A(A) = 1$ is desired, we see that $P_A(C) = P(C)/P(A), C \in \Sigma(A)$. Thus $P_A(B) = P(A \cap B)/P(A) \in [0, 1]$ is the correct assignment to B. If A, B are independent, then $P_A(B) = P(B)$, and thus A has no influence on B, as one would like to have for an extension. It is clear that $\Sigma(A)$ is a (trace) σ-algebra contained in Σ and $P_A : \Sigma(A) \to [0,1]$ is a probability measure. So $(A, \Sigma(A), P_A)$ is a new triple. Since $P_A(A^c) = 0$, we see that $P_A : \Sigma \to [0,1]$ is also defined, and is a measure. One calls P_A an *elementary conditional probability* on Σ relative to the event A satisfying $P(A) > 0$. If $X : \Sigma \to \mathbb{R}$ is an integrable random variable, we can define its *elementary conditional expectation*, given A, naturally as

$$E_A(X) = \int_\Omega X(\omega) P_A(d\omega) = \frac{1}{P(A)} \int_A X(\omega) P(d\omega),$$

where P is the probability measure of the original triple. If $P(A^c) > 0$ is also true, then $E_{A^c}(X)$ can be similarly defined, so that the elementary conditional expectation of X relative to A and A^c generally determines a two-valued function. In extending this argument for a countable collection $\mathcal{P} = \{A_n, n \geq 1\}$ of events such that $P(A_n) > 0, n \geq 1, \cup_{n \geq 1} A_n = \Omega$, and $A_n \cap A_m = \emptyset, n \neq m$ (thus $\mathcal{P}$ is a partition of Ω), we get

$$E_{\mathcal{P}}(X) = \sum_{n=1}^{\infty} \frac{1}{P(A_n)} \left(\int_{A_n} X(\omega) P(d\omega) \right) \chi_{A_n} \tag{1}$$

Then $E_{\mathcal{P}}(X)$ is called the (elementary) *conditional expectation* of the integrable r.v. X on (Ω, Σ, P) relative to the partition $\mathcal{P}$. This is an adequate definition as long as one deals with such countable partitions.

The above formulation is not sufficient if the knowledge of a part of the experiment cannot be expressed in terms of a countable set of conditions. Also, if $P(A) = 0$, then the above definition fails. The latter is of common enough occurrence as, for instance, in the case that X has a continuous distribution (so that $P[Y = a] = 0, a \in \mathbb{R}$) and $A_a = [Y = a]$; and one needs to define $E_{A_a}(X)$. These are nontrivial problems, and a general theory should address these difficulties. A satisfactory solution, combining (1) and taking the above points into consideration, has been formulated by Kolmogorov (1933) as follows. Let us first present this abstract concept as a natural extension of (1).

If $\mathcal{B} = \sigma(\mathcal{P})$, the σ-algebra generated by $\mathcal{P}$, then $E_{\mathcal{P}}(X)$ of (1) is clearly $\mathcal{B}$- measurable. Integrating (1) on $A \in \mathcal{P} \subset \mathcal{B}$, we get (because $A = \cup_{k \in J} A_k$ for some $J \subset \mathbb{N}$)

$$\int_A E_{\mathcal{P}}(X)dP = \sum_{n=1}^{\infty} \frac{1}{P(A_n)} \left(\int_{A_n} XdP \right) P(A_n \cap A) = \int_A XdP. \tag{2}$$

If we write $E^{\mathcal{B}}(X)$ for $E_{\mathcal{P}}(X)$ and note that $\mathcal{P}$ generated $\mathcal{B}$, (2) implies that the mapping $\nu_X : A \mapsto \int_A XdP, A \in \mathcal{B}$, is $P_{\mathcal{B}}$-continuous ($\nu_X \ll P_{\mathcal{B}}$), $P_{\mathcal{B}}$ being the restriction of P to $\mathcal{B} \subset \Sigma$, and (2) becomes

$$\nu_X(A) = \int_A E^{\mathcal{B}}(X)dP_{\mathcal{B}}, \ A \in \mathcal{B}. \tag{3}$$

This relation is generalized as follows:

Definition 1 Let (Ω, Σ, P) be a probability space, $X : \Omega \to \mathbb{R}$ be an r.v. which is integrable (at least X^+ or X^- is integrable), and $\mathcal{B} \subset \Sigma$ be any σ-algebra. Then a $\mathcal{B}$-measurable function $E^{\mathcal{B}}(X)$ satisfying the set of equations

$$\int_B E^{\mathcal{B}}(X)dP_{\mathcal{B}} = \int_B XdP, \ B \in \mathcal{B}, \tag{4}$$

is termed the *conditional expectation of* X *relative to* $\mathcal{B}$, and $P^{\mathcal{B}} : A \mapsto E^{\mathcal{B}}(\chi_A), A \in \Sigma$, is called the *conditional probability function* relative to $\mathcal{B}$. Thus $P^{\mathcal{B}}(\cdot)$ satisfies the functional equation

$$\int_B P^{\mathcal{B}}(A)dP_{\mathcal{B}} = P(A \cap B), \ A \in \Sigma, B \in \mathcal{B}. \tag{5}$$

The existence of $E^{\mathcal{B}}(X)$, hence $P^{\mathcal{B}}(\cdot) : A \mapsto E^{\mathcal{B}}(\chi_A), A \in \Sigma$, results from the fundamental Radon-Nikodým theorem (Theorem 1.3.12), since

$$\nu_X : A \mapsto \int_A XdP, \ A \in \Sigma,$$

defines a signed measure and $\nu_X \ll P_{\mathcal{B}}$. Thus $d\nu_x/dP_{\mathcal{B}} = E^{\mathcal{B}}(X)$ a.e. $[P_{\mathcal{B}}]$ exists and is $P_{\mathcal{B}}$-unique by that theorem. Any member of the $P_{\mathcal{B}}$-equivalence class is called a *version* of the conditional expectation $E^{\mathcal{B}}(X)$, and it is customary to call $E^{\mathcal{B}}(X)$ an r.v. when a version is meant.

This general concept could not have been formulated before the availability (before 1930) of the abstract Radon-Nikodým theorem. [Alternatively, if the martingale convergence for directed indexes is granted, then (1) can be extended; cf. Problem 30.] Since (3) is a special case of (4), the elementary definition is included in the present one. We note that in (4), if X is not integrable [i.e., $E(X^+) = \infty$ but $E(X^-) < \infty$, or $E(X^+) < \infty$ but $E(X^-) = \infty$], then $d\nu_X/dP_{\mathcal{B}} = +\infty$ (or $= -\infty$) on a set of positive $P_{\mathcal{B}}$-measure, so that $E^{\mathcal{B}}(X) = E^{\mathcal{B}}(X^+) - E^{\mathcal{B}}(X^-)$ is still defined but need not be a proper r.v. as compared with X. Thus the general case is deeper and not quite intuitive, in contrast to the elementary formulation (1). Similarly, $P^{\mathcal{B}} : \Sigma \to L^{\infty}(P)$, the conditional probability, is not a (scalar) measure, but it is a vector space-valued (bounded) function. These concepts constitute an enormous generalization of the classical expectation and probability notions. Just as the definition of independence is given by a system of equations, so are these conditional notions given by (4) and (5). We now present some simple properties and also show some of their individual characteristics that are not possessed by the unconditional concepts. These will give us a clear idea of their structure.

The first consequences are contained in the following:

Proposition 2 *Let (Ω, Σ, P) be a probability space, $\mathcal{B} \subset \Sigma$, a σ-algebra. Then the conditional expectation operator $E^{\mathcal{B}}$: $L^1(P) \to L^1(P)$ has the following properties a.e. Let $\{X, Y, XY\} \subset L^1(P)$: (i) $E^{\mathcal{B}}(X) \geq 0$ if $X \geq 0$, $E^{\mathcal{B}}(1) = 1$, (ii) $E^{\mathcal{B}}(aX + bY) = aE^{\mathcal{B}}(X) + bE^{\mathcal{B}}(Y)$, (iii) $E^{\mathcal{B}}(XE^{\mathcal{B}}(Y)) = E^{\mathcal{B}}(X)E^{\mathcal{B}}(Y)$, (iv) $|E^{\mathcal{B}}(X)| \leq E^{\mathcal{B}}(|X|)$, so that $\|E^{\mathcal{B}}(X)\|_1 \leq \|X\|_1$, (v) if $\mathcal{B}_1 \subset \mathcal{B}_2 \subset \Sigma$ are σ-algebras, then $E^{\mathcal{B}_1}(E^{\mathcal{B}_2}(X)) = E^{\mathcal{B}_2}(E^{\mathcal{B}_1}(X)) = E^{\mathcal{B}_1}(X)$, whence the operator $E^{\mathcal{B}_1}$ is always a contractive projection on $L^1(P)$, (vi) if $\mathcal{B} = \{\emptyset, \Omega\}$, then $E^{\mathcal{B}}(X) = E(X)$, and if $\mathcal{B} = \Sigma$, then $E^{\mathcal{B}}(X) = X$; also for any σ-algebra $\mathcal{B}_1 \subset \Sigma, E(E^{\mathcal{B}_1}(X)) = E(X)$ for all $X \in L^1(P)$ identically.*

Proof Definition 1 implies both (i) and (ii). Taken together, it says that $E^{\mathcal{B}}$ is a positive linear operator on $L^1(P)$.

For (iii), if $X_1 = E^{\mathcal{B}}(Y) \in L^1(\Omega, \mathcal{B}, P_{\mathcal{B}})$, we have to show that

$$\int_B E^{\mathcal{B}}(XX_1)dP_{\mathcal{B}} = \int_B XX_1 dP = \int_B X_1 E^{\mathcal{B}}(X)dP_{\mathcal{B}}, \quad B \in \mathcal{B}. \tag{6}$$

If $X_1 = \chi_A, A \in \mathcal{B}$, then $A \cap B \in \mathcal{B}$, so that (6) is true by (4). Thus by linearity of the Lebesgue integral, (6) holds if $X_1 = \sum_{i=1}^{n} a_i \chi_{A_i}, A_i \in \mathcal{B}$. If X, Y are positive, so that $X_1 \geq 0$, then there exist $\mathcal{B}$-measurable simple functions Y_n such that $0 \leq Y_n \uparrow X_1$, and by the monotone convergence theorem the last two terms of (6) will be equal. They thus agree for all X, Y in $L^1(P)$. Since the first two terms are always equal, (6) holds as stated, and (iii) follows.

Since $-|X| \leq X \leq |X|$, by (i) we get $|E^{\mathcal{B}}(X)| \leq E^{\mathcal{B}}(|X|)$ and integration yields (iv), on using (4). Similarly (vi) is immediate.

For (v), since $\mathcal{B}_1 \subset \mathcal{B}_2$ implies $E^{\mathcal{B}_1}(X)$ is $\mathcal{B}_1$-, hence $\mathcal{B}_2$-measurable, it follows that $E^{\mathcal{B}_1}(E^{\mathcal{B}_2}(X)) = E^{\mathcal{B}_1}(X)$, a.e. On the other hand,

$$\int_A E^{\mathcal{B}_1}(E^{\mathcal{B}_2}(X))dP_{\mathcal{B}_1} = \int_A E^{\mathcal{B}_2}(X)dP_{\mathcal{B}_2} = \int_A XdP, \quad \text{for } A \in \mathcal{B}_1 \subset \mathcal{B}_2$$

$$= \int_A E^{\mathcal{B}_1}(X)dP_{\mathcal{B}_1}, \quad A \in \mathcal{B}_1. \tag{7}$$

Identifying the extreme integrands which are $\mathcal{B}_1$-measurable, we get

$E^{\mathcal{B}_1}(E^{\mathcal{B}_2}(X)) = E^{\mathcal{B}_1}(X)$ a.e. Thus (v) holds, and if $\mathcal{B}_1 = \mathcal{B}_2$, $E^{\mathcal{B}_1}(E^{\mathcal{B}_1}(X)) = E^{\mathcal{B}_1}(X)$, so that $E^{\mathcal{B}_1} \cdot E^{\mathcal{B}_1} = E^{\mathcal{B}_1}$. This completes the proof.

Remark Property (iii) is often called an *averaging*, and (v) the *commutativity* property of the conditional expectation operator. Item (iv) is termed the *contractivity* property. Also, $E^{\mathcal{B}}(|X|) = 0$ a.e. iff $X = 0$ a.e., since

$$\int_{\Omega} E^{\mathcal{B}}(|X|)dP_{\mathcal{B}} = \int_{\Omega} |X|dP.$$

Thus $E^{\mathcal{B}}$ can be called a *faithful* operator.

Several of the standard limiting operations are also true for $E^{\mathcal{B}}$, as shown by the following result.

Theorem 3 *Let $\{X_n, n \geq 1\}$ be random variables on (Ω, Σ, P) and $\mathcal{B} \subset \Sigma$ be a σ-algebra. Then we have*

(i) (Monotone Convergence) $0 \leq X_n \uparrow X \Rightarrow 0 \leq E^{\mathcal{B}}(X_n) \uparrow E^{\mathcal{B}}(X)$ a.e.

(ii) (Fatou's Lemma) $0 \leq X_n \Rightarrow E^{\mathcal{B}}(\liminf_n X_n) \leq \liminf_n E^{\mathcal{B}}(X_n)$ a.e.

(iii) (Dominated Convergence) $|X_n| \leq Y$, $E(Y) < \infty$, *and* $X_n \to X$ a.e. $\Rightarrow E^{\mathcal{B}}(X_n) \to E^{\mathcal{B}}(X)$ a.e. *and in* $L^1(P)$. *If here* $X_n \xrightarrow{P} X$, *then* $E^{\mathcal{B}}(X_n) \to E^{\mathcal{B}}(X)$ *in* $L^1(P)$ *(but not necessarily a.e.).*

(iv) (Special Vitali Convergence) $\{X_n, n \geq 1\}$ *is uniformly integrable,* $X_n \to X$ a.e. $\Rightarrow \|E^{\mathcal{B}}(X_n) - E^{\mathcal{B}}(X)\|_1 \to 0$, *so that* $\{E^{\mathcal{B}}(X_n), n \geq 1\}$ *is uniformly integrable and* $E^{\mathcal{B}}(X_n)$ *converges in probability to* $E^{\mathcal{B}}(X)$. *(This convergence is again not necessarily a.e., and* **the full Vitali convergence theorem is not valid for the conditional expectations.**)

Proof The argument is a modification of the classical (unconditional) case.

(i) Since $E^{\mathcal{B}}(X_n) \leq E^{\mathcal{B}}(X_{n+1})$, by the preceding proposition, and since $\lim_n E^{\mathcal{B}}(X_n)$ exists and is $\mathcal{B}$-measurable, we have for any $A \in \mathcal{B}$

$$\int_A \lim_n E^{\mathcal{B}}(X_n)dP_{\mathcal{B}}$$

$$= \lim_n \int_A E^{\mathcal{B}}(X_n)dP_{\mathcal{B}} \text{ (by the classical monotone convergence),}$$

$$= \lim_n \int_A X_n dP \text{ [by (4)],}$$

$$= \int_A \lim_n X_n dP \text{ (by the classical monotone convergence),}$$

$$= \int_A X dP = \int_A E^{\mathcal{B}}(X)dP_{\mathcal{B}}, \text{ (by definition).}$$

Since $A \in \mathcal{B}$ is arbitrary and the extreme integrands are $\mathcal{B}$-measurable, they can be identified. Thus $\lim_n E^{\mathcal{B}}(X_n) = E^{\mathcal{B}}(X)$ a.e.

(ii) This is similar. Indeed, let $Y_n = \inf\{X_k : k \geq n\}$. Then $0 \leq Y_n \uparrow Y = \liminf_n X_n$. Hence by (i) and the monotonicity of $E^{\mathcal{B}}$ we have, since $Y_n \leq X_n$,

$$\begin{aligned} E^{\mathcal{B}}(\liminf_n X_n) = E^{\mathcal{B}}(Y) = \lim_n E^{\mathcal{B}}(Y_n) &= \liminf_n E^{\mathcal{B}}(Y_n) \\ &\leq \liminf_n E^{\mathcal{B}}(X_n) \text{ a.e.} \end{aligned}$$

Similarly, if $X_n \leq Z$ a.e., $E(|Z|) < \infty$, then $E^{\mathcal{B}}(\limsup_n X_n) \geq \limsup_n E^{\mathcal{B}}(X_n)$ a.e.

(iii) Since $-Y \leq X_n \leq Y$ a.e., $n \geq 1$, we can apply (ii) to $X_n + Y \geq 0$ and $X_n \leq Y$. Hence with

$$\liminf_n (X_n + Y) = \liminf_n X_n + Y = X + Y = \limsup_n (X_n + Y),$$

one has

$$\begin{aligned} E^{\mathcal{B}}(X) + E^{\mathcal{B}}(Y) &= E^{\mathcal{B}}\left(\liminf_n (X_n + Y)\right) \\ &\leq \liminf_n (E^{\mathcal{B}}(X_n) + E^{\mathcal{B}}(Y)) \\ &\leq \limsup_n (E^{\mathcal{B}}(X_n) + E^{\mathcal{B}}(Y)) = \limsup_n E^{\mathcal{B}}(X_n + Y) \\ &\leq E^{\mathcal{B}}\left(\limsup_n (X_n + Y)\right), \text{ (since } X_n + Y \leq 2Y \text{a.e.)}, \\ &= E^{\mathcal{B}}(X + Y) = E^{\mathcal{B}}(X) + E^{\mathcal{B}}(Y) \text{ a.e.} \end{aligned}$$

Thus there is equality throughout. Cancelling the (finite) r.v. $E^{\mathcal{B}}(Y)$, we get $\lim_n E^{\mathcal{B}}(X_n) = E^{\mathcal{B}}(X)$ a.e. Finally, as $n \to \infty$,

$$\begin{aligned} E(|E^{\mathcal{B}}(X_n - X)|) &\leq E(E^{\mathcal{B}}(|X_n - X|)) \text{ (by Proposition 2iv)} \\ &\leq E(|X_n - X|) \text{ (since } E = E^{\mathcal{B}_0}, \mathcal{B}_0 = \{\emptyset, \Omega\}) \\ &\to 0 \text{ (by the classical dominated convergence).} \end{aligned}$$

This yields the last statement also. The negative statement is clear since, e.g., if $\mathcal{B} = \Sigma$, then $X_n \xrightarrow{P} X \not\Rightarrow$ a.e. convergence.

(iv) Again we have

$$E(|E^{\mathcal{B}}(X_n - X)|) \leq E(E^{\mathcal{B}}(|X_n - X|)) = E(|X_n - X|) \to 0$$

as $n \to \infty$, by the classical Vitali theorem, since the X_n are uniformly integrable and the measure space is finite. The last statement is obvious. We obtain the negative result as a consequence of (the deeper) Theorem 5 below, finishing the proof.

If $\mathcal{B} \subset \Sigma$ is given by $\mathcal{B} = \sigma(Y)$ for an r.v. $Y : \Omega \to \mathbb{R}$, then for $X \in L^1(P)$ we *also write* $E^{\sigma(Y)}(X)$ *as* $E(X|Y)$, following custom. It is then read simply as the conditional expectation of X given (or relative to) Y. The following two properties are again of frequent use in applications, and the first one shows that Definition 1 is a true generalization of the independence concept.

Proposition 4 *Let X be an integrable r.v. on (Ω, Σ, P). If $\mathcal{B} \subset \Sigma$ is a σ- algebra independent of X, then $E^{\mathcal{B}}(X) = E(X)$ a.e. If Y is any r.v. on Ω, then there is a Borel mapping $g : \mathbb{R} \to \mathbb{R}$ such that $E(X|Y) = g(Y)$ a.e., so that the conditional expectation of X given Y is equivalent to a Borel function of Y.*

Proof If X and $\mathcal{B}$ are independent, then for any $B \in \mathcal{B}$, X and χ_B are independent. Hence

$$\begin{aligned}\int_B E(X)dP_{\mathcal{B}} &= E(X)P(B) = E(X) \cdot E(\chi_B)\\ &= E(X\chi_B) \text{ (by independence)}\\ &= \int_B XdP = \int_B E^{\mathcal{B}}(X)dP_{\mathcal{B}},\ B \in \mathcal{B}.\end{aligned}$$

Since the extreme integrands are $\mathcal{B}$-measurable, $E(X) = E^{\mathcal{B}}(X)$ a.e.

For the last statement, $E(X|Y)$ is $\sigma(Y)$-measurable. Then by Proposition 1.2.3, there is a Borel $g : \mathbb{R} \to \mathbb{R}$ such that $E(X|Y) = g(Y)$ a.e. This completes the proof.

Let us now return to the parenthetical statement of Theorem 3iv. The preceding work indicates that a conditional expectation operator has just about all the properties of the ordinary expectation. The discussion presented thus far is not delicate enough to explain the subtleties of the conditional concept unless we prove the above negative statement. This will be the purpose of the next important result, due to Blackwell and Dubins (1963). In this connection the reader should compare Exercise 10 of Chapter 2, and the following result, which is essentially also a converse to the conditional Lebesgue dominated convergence theorem. It reveals a special characteristic of probability theory, to be explained later on.

Theorem 5 *Let $\{X, X_n, n \geq 1\}$ be a set of integrable random variables on (Ω, Σ, P) such that $X_n \to X$ a.e., and, if $U = \sup_{n\geq 1} |X_n|, E(U) = +\infty$. Then there exists [after an enlargement of (Ω, Σ, P) by adjunction if* *necessary], a sequence $\{X', X_n', n \geq 1\}$ of integrable random variables on it having the same finite dimensional distributions as the X_n-sequence, i.e.,*

$$P[X_{i_1} < x_1, \ldots, X_{i_k} < x_k] = P[X_{i_1}' < x_1, \ldots, X_{i_k}' < x_k], \tag{8}$$

$x_i \in \mathbb{R}, 1 \leq i \leq k < \infty$, such that for some σ-algebra $\mathcal{B}_0 \subset \Sigma$, $E^{\mathcal{B}_0}(X_n') \to E^{\mathcal{B}_0}(X')$ at almost no point as $n \to \infty$. In other words, even if the X_n are uniformly integrable, the X_n' satisfy:

$$P[\lim_{n\to\infty} E^{\mathcal{B}_0}(X_n') = E^{\mathcal{B}_0}(X')] = 0. \tag{9}$$

[Thus the interchange of limit and $E^{\mathcal{B}_0}$ is not generally valid.]

Proof We first make a number of reductions before completing the argument. Since $X_n \to X$ a.e. implies that $\varphi(X_n) \to \varphi(X)$ a.e. for each real continuous φ on $\mathbb{R}$, taking $\varphi(x) = \max(x, 0) = x^+$, we get $X_n^{\pm} \to X^{\pm}$ a.e. Thus for the proof, it is sufficient to assume that $X_n \geq 0$ in addition. Let us then suppose that the given sequence is nonnegative from now on, and present the demonstration in steps.

I. It is enough to establish the result with $X = 0$ a.e. For let $Y_n = (X_n - X)^+$ and $Z_n = -(X_n - X)^-$, so that $X_n - X = Y_n + Z_n$. Clearly $E(Y_n) < \infty$, $E(-Z_n) < \infty$, and $Y_n \to 0$ a.e., $Z_n \to 0$ a.e. Also,

$$\sup_{n\geq 1} Y_n \geq \sup_{n\geq 1}(X_n - X) = \sup_{n\geq 1} X_n - X = U - X,$$

so that $E(\sup_{n\geq 1} Y_n) \geq E(U) - E(X) = +\infty$, by hypothesis. On the other hand, by definition $|Z_n| = (X_n - X)^- = X - X_n$ if $X_n < X$, and $= 0$ if $X_n \geq X$, so that (recalling $X_n, X \geq 0$)

$$\sup_{n\geq 1} |Z_n| \leq X \text{ and } E(X) < \infty.$$

Since $Z_n \to 0$ a.e., by Theorem 3iii, for every σ-algebra $\mathcal{B} \subset \Sigma$, $E^{\mathcal{B}}(Z_n) \to 0$ a.e. Consequently,

$$P\left[\lim_{n\to\infty} E^{\mathcal{B}}(X_n - X) = 0\right] = P\left[\lim_{n\to\infty} E^{\mathcal{B}}(Y_n) + \lim_{n\to\infty} E^{\mathcal{B}}(Z_n) = 0\right]$$

$$= P\left[\lim_{n\to\infty} E^{\mathcal{B}}(Y_n) = 0\right]. \tag{10}$$

Hence, if (Ω, Σ, P) is rich enough so that there is a σ-algebra $\mathcal{B} \subset \Sigma$ such that the right side of (10) is zero, it will follow that $E^{\mathcal{B}}(X_n) \to E^{\mathcal{B}}(X)$ at almost no point. Thus by the adjunction procedure the probability space can be enlarged to have this needed richness, so the result will follow if a sequence $Y_n \geq 0$, $Y_n \to 0$ a.e., but $E^{\mathcal{B}}(Y_n) \to 0$ almost nowhere, is constructed.

II. We now assert that it is even sufficient to find a suitable two-valued Y_n'-sequence. More explicitly, we only need to construct a sequence $\{Y_n', n \geq 1\}$ such that $Y_n' \to Y' = 0$, $P[Y_n' = a_n] = p_n = 1 - P[Y_n' = 0]$, $0 < p_n < 1$, $a_n > 0$, and for each $\omega \in \Omega$, $Y_n'(\omega) \cdot Y_m'(\omega) = 0$ if $n \neq m$, $\sum_{n=1}^{\infty} p_n = 1$, with $U' = \sup_{n \geq 1} Y_n' = \sum_{n=1}^{\infty} Y_n'$, satisfying

$$E(U') = \sum_{n=1}^{\infty} E(Y_n') = \sum_{n=1}^{\infty} a_n p_n = +\infty.$$

For let A_k be the event that $Y_k \geq U - 1$ for the first time, with the Y_k, U of the last step. This means

$$A_1 = [Y_1 \geq U - 1]$$

and

$$A_k = [Y_k \geq U - 1, Y_i < U - 1, 1 \leq i \leq k - 1] \text{ for } k > 1.$$

Then the A_k are disjoint, $\cup_{k \geq 1} A_k = [U \geq 1]$. Let $A_0 = \Omega - \cup_{k \geq 1} A_k$, and consider $Y_k \chi_{A_k}$, $k \geq 1$. By the structure theorem for measurable functions, there exists a sequence $0 \leq f_{kn} \uparrow Y_k \chi_{A_k}$, pointwise, with each f_{kn} as a simple function. Hence there is $n_0 = n_0(k)$ such that if $h_k = f_{kn_0} \leq Y_k \chi_{A_k}$, then

$$E(h_k) \geq \int_{A_k} Y_k dP - \frac{1}{2^k}, k \geq 1. \tag{11}$$

Here $h_k = \sum_{i=1}^{m} b_i \chi_{B_i}$ may be assumed to be in canonical form, so that $B_i \cap B_{i+1} = \emptyset, B_i \subset A_k, b_i \geq 0$. Hence $\sup_{k \geq 1} h_k = \sum_{k=1}^{\infty} h_k$, since the A_k are also disjoint. Thus

$$\begin{aligned} E\left(\sup_{k \geq 1} h_k\right) &= \sum_{k=1}^{\infty} E(h_k) \text{ (by monotone convergence)} \\ &\geq \sum_{k=1}^{\infty} \int_{A_k} Y_k dP - 1 \text{ [by (11)]} \\ &\geq \sum_{k=1}^{\infty} \int_{A_k} U dP - 2 \geq \int_{\Omega} U dP - 3 = +\infty. \end{aligned} \tag{12}$$

But $h_k \leq Y_k$; thus for each σ-algebra $\mathcal{B} \subset \Sigma$,

$$\left[\lim_k E^{\mathcal{B}}(h_k) > 0\right] \subset \left[\lim_k Y_k > 0\right],$$

and hence $P[\lim_k E^{\mathcal{B}}(h_k) = 0] \geq P[\lim_k Y_k = 0] \geq 0$, and it suffices to show that $P[\lim_k E^{\mathcal{B}}(h_k) = 0] = 0$ when $h_k \to 0$, each h_k being a positive simple function satisfying (12). Since each h_k is in the canonical form, it is a sum of a finite number of two-valued functions and is nonzero on only one set (a B_i here), and for different k, h_k lives on an A_k (a disjoint sequence). Now rearrange these two-valued functions into a single sequence, omit the functions which are identically zero, and let B_0 be the set on which all these vanish. If we add χ_{B_0} to the above set of two-valued functions, we get a sequence, to be called $Y_1', Y_2', \ldots$, which satisfies the conditions given at the beginning of the step. So it is now sufficient to prove the theorem in this case for a suitable σ-algebra $\mathcal{B}_0 \subset \Sigma$.

III. We now construct the required $\{Y_k', k \geq 1\}$ and $\mathcal{B}_0$. Since $0 < p_1 < 1$, choose an integer $k \geq 1$ such that $2^{-k} < p_1 \leq 2^{-k+1}$. Let $a_i > 0$ be numbers satisfying $\sum_{i=1}^{\infty} a_i p_i = +\infty$. Let $N_n = \{i \geq 2 : 2^{n+k} \leq a_i < 2^{n+k+1}\}, n \geq 1$. Note that $N_n = \emptyset$ is possible and N_n depends on k. Consider

$$r_n = \sum_{i \in N_n} p_i, \quad t_n = r_n + 2^{-(n+k)}, \quad r = \sum_{n \geq 1} r_n. \tag{13}$$

Let $T(= T_k) = \cup_n N_n = \{i \geq 2 : a_i \geq 2^{k+1}\}$. Set $r = \sum_{i \in T} p_i \leq 1 - p_1 < 1 - 2^{-k}$, and

$$t = \sum_{n \geq 1} t_n = r + \sum_{n \geq 1} 2^{-n-k} = r + 2^{-k} < 1.$$

Let $\{W, Z_n \geq 0\}_{n=1}^{\infty}$ be a set of mutually independent random variables on (Ω, Σ, P) (assumed rich enough, by enlarging it if necessary) whose distributions are specified as

$$P[W = 0] = 1 - t, \quad P[W = n] = t_n, n \geq 1,$$

$$P[Z_0 = 1] = \frac{(p_1 - 2^{-k})}{(1 - t)},$$

$$P[Z_0 = i] = \begin{cases} \frac{p_i}{(1-t)} & \text{if } i \geq 2, i \notin T \\ 0, & \text{otherwise,} \end{cases} \tag{14}$$

and for $n \geq 1$,

$$P[Z_n = 1] = (t_n 2^{n+k})^{-1},$$

$$P[Z = i] = \begin{cases} \frac{p_i}{t_n} & \text{for } n \in N_n \\ 0, & \text{otherwise.} \end{cases} \tag{15}$$

Let $\mathcal{B}_0 = \sigma(Z_n, n \geq 0)$. We next define the desired two-valued random variables from $\{W, Z_n, n \geq 0\}$, and verify that they and this $\mathcal{B}_0$ satisfy our requirements.

Consider the composed r.v. $V = Z \circ W (= Z_W)$. We assert that $P[V = n] = p_n$. Indeed,

$$P[V=1]=\sum_{n=0}^{\infty}P[Z_n=1,W=n]$$

$$=\sum_{n=0}^{\infty}P[Z_n=1]P[W=n] \text{ (by independence)}$$

$$=P[Z_0=1]P[W=0]+\sum_{n=1}^{\infty}(t_n 2^{n+k})^{-1}t_n$$

$$=p_1-2^{-k}+2^{-k}=p_1 \text{ [by (14) and (15)].}$$

Next for $i \geq 2, i \notin T$,

$$P[V=i]=P[W=0,Z_0=i]=(1-t)p_i(1-t)^{-1}=p_i,$$

and for $i \in T$

$$P[V=i]=P[W=n,Z_n=i]=p_i. \tag{16}$$

This proves the assertion. Define $Y_n' = a_n\chi_{[V=n]}$. Then the Y_n' have the two values $0, a_n > 0$, and, $[V=n]$ being disjoint events, for each n only one $Y_n'(\omega)$ is different from zero. Moreover, $Y_n' \to 0$ a.e., and

$$P[Y_n'=a_n]=P[V=n]=p_n, \tag{17}$$

as required in step II. Thus $\{Y_n', n \geq 0\}$ and $\mathcal{B}_0$ are the sequence and a σ-algebra, and it is to be shown that this satisfies the other condition of step II to complete the proof.

IV. $E^{\mathcal{B}_0}(Y_n') \to 0$ at almost no point of Ω.

For since $E^{\mathcal{B}_0}(Y_n') = a_nE^{\mathcal{B}_0}(\chi_{[V=n]}) = a_nP^{\mathcal{B}_0}[V=n]$, it suffices to show that $P[E^{\mathcal{B}_0}(Y_n') \geq 1, \text{ i.o. }] = P[P^{\mathcal{B}_0}[V=n] \geq 1/a_n, \text{ i.o.}] = 1$. This will be established by means of the second Borel-Cantelli lemma.

Since $\sigma(Z_n) \subset \mathcal{B}_0$, by Propositions 2v and 4, we have for $i \in N_n$,

$$E^{\sigma(Z_n)}(P^{\mathcal{B}_0}[V=i])(k) = P[V=i|Z_n=k]$$

$$=\frac{P[V=i,Z_n=k]}{P[Z_n=k]}$$

$$=\begin{cases}0, & k \neq i\\ t_n, & k=i.\end{cases}$$

Thus $a_nP[V=i|Z_n=i]=a_nt_n, i \in N_n$. But $t_n \geq 2^{-n-k}$ and, for $i \in N_n$, $a_i \geq 2^{n+k}$, so that $a_nt_n \geq 1$. Consequently if $A_n=[Z_n \neq 1]$ and

$$B_n=[a_nP[V=i|Z_n=i] \geq 1 \text{ for some } i \in N_n],$$

then for $n \geq 1, A_n \subset B_n$. Though the B_n are not necessarily independent, the A_n are mutually independent, since the $Z_n, n \geq 0$, are. Thus $[A_n,$

i.o.] $\subset [B_n$, i.o.], and it suffices to show that $P[A_n$, i.o.] $= 1$ by verifying that $\sum_{n=1}^{\infty} P(A_n) = \infty$ (cf. Theorem 2.1.9ii).

By (15), $P(A_n) = \sum_{i \in N_n} p_i/t_n = r_n/t_n$. To show that $\sum_{n\geq 1}(r_n/t_n) = \infty$, note that $r_n = t_n - 2^{-(n+k)}$, by (13). Now, if $r_n \geq 2^{-(n+k)}$ for infinitely many n, then $r_n/t_n \geq 1/2$ for all those n, so that $\sum_{n\geq 1}(r_n/t_n) = \infty$. If $r_n < 2^{-(n+k)}$ for $n \geq n_0$ and some n then $t_n < 2^{-(n+k-1)}$, so

$$\frac{r_n}{t_n} \geq 2^{n+k-1} r_n = \frac{1}{4} \sum_{i \in N_n} 2^{n+k+1} p_i \geq \frac{1}{4} \sum_{i \in N_n} a_i p_i. \tag{18}$$

Consequently,

$$\sum_{n \geq 1} \frac{r_n}{t_n} \geq \frac{1}{4} \sum_{n \geq n_0} \sum_{i \in N_n} p_i a_i = \frac{1}{4} \sum_{i \in T_0} p_i a_i \tag{19}$$

where $T_0 = \{i \geq 2 : a_i \geq 2^{n_0+k}\}$. But $\sum_{i \notin T_0} p_i a_i \leq 2^{n_0+k} \sum_{i \geq 1} p_i \leq 2^{n_0+k}$, and by choice $\sum_{i\geq 1} p_i a_i = +\infty$. Hence $\sum_{i \in T_0} p_i a_i = \infty$, so that the series in (19) diverges. Thus in all cases $\sum_{n\geq 1}(r_n/t_n) = +\infty$. It follows that $P[A_n$, i.o.] $= 1$, which completes the proof of the theorem.

The implications of this result will be discussed further after studying some properties of conditional probabilities. But it shows that a uniformly integrable sequence $\{X_n, n \geq 1\}$ such that $X_n \to X$ a.e. does *not* imply $E^{\mathcal{B}}(X_n) \to E^{\mathcal{B}}(X)$ a.e. for each σ-algebra $\mathcal{B} \subset \Sigma$ when $E(\sup_n |X_n|) = +\infty$. Thus the Vitali convergence theorem of Lebesgue integrals does not extend, with pointwise a.e. convergence, to conditional expectations. It is possible, however, to present conditions *depending on the given σ-algebra $\mathcal{B} \subset \Sigma$*, in order that a certain extension (of Vitali's theorem) holds. Let us say that a sequence $\{X_n, n \geq 1\}$ of random variables is *conditionally uniformly integrable relative to a σ-algebra $\mathcal{B} \subset \Sigma$* (or c.u.i.) if $\lim_{k\to\infty} E^{\mathcal{B}}(|X_n| \chi_{[|X_n| \geq k]}) = 0$ a.e., uniformly in n. If $\mathcal{B} = \{\emptyset, \Omega\}$, then this is the classical uniform integrability on a probability space (cf. Section 1.4). It is always satisfied if $|X_n| \leq Y$ and $E(Y) < \infty$, by Theorem 3iii, but may be hard to verify in a given problem. Note that if the $X_n, n \geq 1$, are integrable, then the c.u.i. implies the classical concept (use the ensuing result and Theorem 5), but the converse is not true. We have the following assertion, complementing the above remarks:

Proposition 6 *Let $\{X, X_n, n \geq 1\}$ be a sequence of integrable random variables on (Ω, Σ, P) and $\mathcal{B} \subset \Sigma$ be a σ-algebra. If this sequence is conditionally uniformly integrable relative to $\mathcal{B}$ and $X_n \to X$ a.e., then $E^{\mathcal{B}}(X_n) \to E^{\mathcal{B}}(X)$ a.e. [and in $L^1(P)$-norm by Theorem 3 already].*

Proof Since the X_n-sequence is conditionally uniformly integrable relative to $\mathcal{B}$, it is clear that the sequences $\{X_n^{\pm}, n \geq 1\}$ have the same property where, as usual, $X_n^+ = \max(X_n, 0)$ and $X_n^- = X_n^+ - X_n \geq 0$. Thus the hypothesis implies

$$U_m = \sup_{n\geq 1} E^{\mathcal{B}}(X_n^- \chi_{[X_n^- > m]}) \to 0, \quad V_m = \sup_{n\geq 1} E^{\mathcal{B}}(X_n^+ \chi_{[X_n^+ > m]}) \to 0 \tag{20}$$

a.e., as $m \to \infty$. But

$$\begin{aligned} E^{\mathcal{B}}(X_n) &= E^{\mathcal{B}}(X_n \chi_{[X_n^- \leq m]}) - E^{\mathcal{B}}(X_n \chi_{[X_n^+ > m]}) \\ &\geq E^{\mathcal{B}}(X_n \chi_{[X_n^- \leq m]}) - U_m \\ &= -E^{\mathcal{B}}(X_n^- \chi_{[X_n^- \leq m]}) - U_m \ \text{(since } X_n^+ \cdot X_n^- = 0). \end{aligned} \tag{21}$$

Hence, using the fact that $\liminf_n(-a_n) = -\limsup_n(a_n)$ for any $\{a_n, n \geq 1\} \subset \mathbb{R}$, (21) gives for all $m > 0$,

$$\begin{aligned} \liminf_n E^{\mathcal{B}}(X_n) &\geq -\limsup_n E^{\mathcal{B}}(X_n^- \chi_{[X_n^- \leq m]}) - U_m \\ &\geq -E^{\mathcal{B}}(-\limsup_n X_n^- \chi_{[X_n^- \leq m]}) - U_m \\ &\qquad \text{(by Theorem 3ii, since the } X_n^- \chi_{[X_n^- \leq m]} \text{ are bounded)} \\ &= E^{\mathcal{B}}\left(\liminf_n (-X_n^- \chi_{[X_n^- \leq m]}\right) - U_m \\ &= E^{\mathcal{B}}\left(\liminf_n X_n^- \chi_{[X_n^- \leq m]}\right) - U_m \ \text{(since } X_n^+ \cdot X_n^- = 0) \\ &\geq E^{\mathcal{B}}\left(\liminf_n X_n\right) - U_m \ \text{a.e.} \end{aligned} \tag{22}$$

Since $m > 0$ is arbitrary and $U_m \to 0$ as $m \to \infty$, (22) implies

$$\liminf_n E^{\mathcal{B}}(X_n) \geq E^{\mathcal{B}}\left(\liminf_n X_n\right) \text{ a.e.} \tag{23}$$

Considering X_n^+ and $-X_n$ in the above, we deduce that

$$\begin{aligned} \limsup_n E^{\mathcal{B}}(X_n) &= -\liminf_n E^{\mathcal{B}}(-X_n) \\ &\leq E^{\mathcal{B}}\left(-\liminf_n(-X_n)\right) \text{ a.e. [by (23)]} \\ &= E^{\mathcal{B}}\left(\limsup_n X_n\right) \text{ a.e.} \end{aligned} \tag{24}$$

Since $X_n \to X$ a.e., $\limsup_n X_n = \liminf_n X_n = X$ a.e. so that (23) and (24) imply

$$E^{\mathcal{B}}(X) \leq \liminf_n E^{\mathcal{B}}(X_n) \leq \limsup_n E^{\mathcal{B}}(X_n) \leq E^{\mathcal{B}}(X) \text{ a.e.} \tag{25}$$

Hence $\lim_n E^{\mathcal{B}}(X_n) = E^{\mathcal{B}}(X)$ a.e., as asserted.

If $\{X_n, n \geq 1\}$ is conditionally uniformly integrable relative to each σ-algebra $\mathcal{B} \subset \Sigma$, then the sequence constructed in the proof of Theorem 5 is prevented. Hence by that theorem $E(\sup_n |X_n|) < \infty$ must then be true. Combining these two results, we can deduce the following comprehensive result.

Theorem 7 *Let $\{X, X_n, n \geq 1\}$ be a sequence of integrable random variables on a probability space (Ω, Σ, P) such that $X_n \to X$ a.e. Then the following statements are equivalent:*

(i) $E(\sup_{n\geq 1} |X_n|) < \infty$.

(ii) *For each σ-algebra $\mathcal{B} \subset \Sigma$, $E^{\mathcal{B}}(X_n) \to E^{\mathcal{B}}(X)$ a.e.*

(iii) *For each σ-algebra $\mathcal{B} \subset \Sigma$, $\{X_n, n \geq 1\}$ is conditionally uniformly integrable relative to $\mathcal{B}$.*

If any one of these equivalent conditions is satisfied, then the $L^1(P)$-convergence also holds.

Now we ask: Is it possible to state a full conditional Vitali convergence assertion? One direction is Proposition 6. The converse appears slightly different; this is given as Problem 5. It is easy to give a sufficient condition on the sequence $\{X_n, n \geq 1\}$ in order that the hypothesis of the above theorem be verified:

Corollary 8 *Let $\varphi : \mathbb{R}^+ \to \mathbb{R}^+$ be an increasing function such that $\varphi(x)/x \nearrow \infty$ as $x \nearrow \infty$. If $\{X_n, n \geq 1\}$ is a set of random variables on (Ω, Σ, P) such that $X_n \to X$ a.e. and $E(\varphi(|X_n|)) < \infty$, suppose for each σ-algebra $\mathcal{B} \subset \Sigma$ there is a constant $C_{\mathcal{B}} > 0$ such that $E^{\mathcal{B}}(\varphi(|X_n|)) \leq C_{\mathcal{B}} < \infty$ a.e. Then $E^{\mathcal{B}}(X_n) \to E^{\mathcal{B}}(X)$ a.e. If the set $\{C_{\mathcal{B}} : C_{\mathcal{B}} \subset \Sigma\}$ is bounded, then $E(\sup_n |X_n|) < \infty$.*

Proof Let $\xi(x) = (\varphi(x)/x)$ and $A_m = [|X_n| > m]$. Then

$$E^{\mathcal{B}}(|X_n|\chi_{A_m}) \leq \frac{E^{\mathcal{B}}(\varphi(|X_n|\chi_{A_m}))}{\xi(m)} \leq \frac{C_{\mathcal{B}}}{\xi(m)} \to 0$$

as $m \to \infty$ (uniformly in n). Thus $\{X_n, n \geq 1\}$ is conditionally uniformly integrable relative to $\mathcal{B}$. If $C_{\mathcal{B}} \leq C < \infty$, all $\mathcal{B} \subset \Sigma$, then c.u.i. holds for all $\mathcal{B}$, and Theorem 7iii applies.

Using the order-preserving property of the conditional expectation operator $E^{\mathcal{B}}$, we can extend the classical proofs, as in Theorem 3, and obtain conditional versions of the inequalities of Hölder, Minkowski, and Jensen as follows.

Theorem 9 *Let X, Y be random variables on (Ω, Σ, P), $p \geq 1$, and $\mathcal{B} \subset \Sigma$ a σ-algebra. Then we have*

$$\text{(i)} \quad E^{\mathcal{B}}(|XY|) \leq [E^{\mathcal{B}}(|X|^p)]^{1/P}[E^{\mathcal{B}}(|Y|^q)]^{1/q} \text{ a.e.}, \quad \frac{1}{p} + \frac{1}{q} = 1. \tag{26}$$

$$\text{(ii)} \quad E^{\mathcal{B}}(|X+Y|^p) \leq [(E^{\mathcal{B}}(|X|^p)^{1/p} + (E^{\mathcal{B}}(|Y|^p)^p]^{1/p} \text{ a.e.}, \tag{27}$$

(iii) *If $\varphi : \mathbb{R} \to \mathbb{R}$ is a convex function such that $E(\varphi(X))$ exists, and $Y = E^{\mathcal{B}}(X)$ a.e., or $Y \leq E^{\mathcal{B}}(X)$ a.e. and φ is also increasing, then*

$$\varphi(E^{\mathcal{B}}(X)) \leq E^{\mathcal{B}}(\varphi(X)) \text{ a.e.} \tag{28}$$

Proof Since $E^{\mathcal{B}}$ is faithful (cf. remark after Proposition 2), these inequalities follow from the unconditional results. Briefly, (26) is true if $X = 0$ a.e. Thus let $0 < E^{\mathcal{B}}(|X|^p) = N_X^p < \infty$ a.e., and $0 < E^{\mathcal{B}}(|Y|^q) = N_Y^q < \infty$ a.e. Then the numerical inequality of Eq. (4) of Section 1.3, with ($\alpha = 1/p, \beta = 1/q$, and $1 < p < \infty$ there $(N_X^1 = N_X, N_Y^1 = N_Y)$ implies

$$\frac{|XY|}{N_X N_Y}(\omega) \leq \frac{1}{p}\frac{|X|^p}{N_X^p}(\omega) + \frac{1}{q}\frac{|Y|^q}{N_Y^q} \text{ a.a. } (\omega). \tag{29}$$

Note that N_X^p, N_Y^q are $\mathcal{B}$-measurable, so that by Proposition 2,

$$\frac{E^{\mathcal{B}}|XY|}{N_X N_Y}(\omega) \leq \frac{1}{p}\frac{E^{\mathcal{B}}(|X|^p)}{N_X^p}(\omega) + \frac{1}{q}\frac{E^{\mathcal{B}}(|Y|^q)}{N_Y^q}(\omega) = \frac{1}{p} + \frac{1}{q} = 1 \text{ a.e.}.$$

This is clearly equivalent to (i), called the conditional Hölder inequality. Similarly (ii) and (iii) are established. Because of its importance we outline (28) again in this setting.

Recall that the support line property of a convex function at y_0 is written as [cf. Eq. (13) of Section 1.3]

$$\varphi(y) \geq \varphi(y_0) + g(y_0)(y - y_0), \quad y \in \mathbb{R}, \tag{30}$$

where $g : \mathbb{R} \to \mathbb{R}$ is in fact the right (or left) derivative of φ at y_0 and is nondecreasing. (It is strictly increasing if φ is strictly convex.) Take $y = X(\omega), y_0 = E^{\mathcal{B}}(X)(\omega)$, in (30). Then we get

$$\varphi(X)(\omega) \geq g(E^{\mathcal{B}}(X))(X - E^{\mathcal{B}}(X))(\omega) + \varphi(E^{\mathcal{B}}(X))(\omega) \text{ a.a. } (\omega). \tag{31}$$

Since the hypothesis implies $g(E^{\mathcal{B}}(X))$ is $\mathcal{B}$-measurable, we have, again with the averaging property of $E^{\mathcal{B}}$ applied to the functional inequality (31),

$$E^{\mathcal{B}}(\varphi(X)) \geq g(E^{\mathcal{B}}(X))(E^{\mathcal{B}}(X) - E^{\mathcal{B}}(X)) + \varphi(E^{\mathcal{B}}(X)) \text{ a.e.}, \tag{32}$$

which is (28). Under the alternative set of hypotheses of (iii), one has from (32), since φ is now increasing,

$$E^{\mathcal{B}}(\varphi(X)) \geq \varphi(E^{\mathcal{B}}(X)) \geq \varphi(Y) \quad \text{a.e.} \tag{32'}$$

Note that (32) and (32′) are valid for any bounded r. v. X, and then the general case follows by the conditional monotone convergence theorem, so that $E^{\mathcal{B}}(\varphi(X))$ exists. This completes the proof.

Again equality conditions obtain in the above inequalities by an analysis of the proofs, just as in the classical unconditional case. For instance, it can be verified that in the conditional Jensen inequality (32), if φ is strictly convex and $\mathcal{B}$ is complete, equality holds when and only when X is $\mathcal{B}$-measurable. We leave the proof to the reader.

The preceding demonstration may lead one to think that similar results are true not only for operators such as $E^{\mathcal{B}}$, but also for a larger class of operators on the L^p-spaces. This is indeed true, and we present a family of mappings in the problem section (cf. Problems 6 and 7ii) to indicate this phenomenon.

The above theorem yields further structural properties of $E^{\mathcal{B}}$, as noted in the following:

Corollary 10 *For each σ-algebra $\mathcal{B} \subset \Sigma$ the operator $E^{\mathcal{B}}$ is a linear contraction on $L^p(\Omega, \Sigma, P)$, $1 \leq p \leq \infty$, i.e., it is linear and*

$$\|E^{\mathcal{B}}(X)\|_p \leq \|X\|_p, \quad X \in L^p(\Omega, \Sigma, P) = L^p. \tag{33}$$

Proof Since P is a finite measure $L^p \subset L^1$, ≥ 1, so that $E^{\mathcal{B}}$ is defined on all L^p, and is clearly linear. To prove the contraction property which we have seen for $p = 1$ in Proposition 2, note that for $p = +\infty$ this is clear, since $|X| \leq \|X\|_\infty$ a.e., and

$$|E^{\mathcal{B}}(X)| \leq E^{\mathcal{B}}(|X|) \leq \|X\|_\infty \cdot E^{\mathcal{B}}(1) = \|X\|_\infty \Rightarrow \|E^{\mathcal{B}}(X)\|_\infty \leq \|X\|_\infty.$$

Thus let $1 < p < \infty$. Then by (28), since $|X|^p$ is $\mathcal{B}$-measurable

$$\int_\Omega |E^{\mathcal{B}}(X)|^p dP_{\mathcal{B}} \leq \int_\Omega E^{\mathcal{B}}(|X|^p) dP_{\mathcal{B}} = \int_\Omega |X|^p dP_{\mathcal{B}} = \|X\|_p^p.$$

This implies (33), and that $E^{\mathcal{B}}(X) \in L^p$ for each $X \in L^p$, as asserted.

Remark The result implies that $E^{\mathcal{B}}(X_n) \to E^{\mathcal{B}}(X)$ in $L^p(P)$-mean whenever $X_n \to X$ in $L^p(P)$-mean.

It is clear that $E^{\mathcal{B}}(L^p(\Omega, \Sigma P)) = L^p(\Omega, \mathcal{B}, P_{\mathcal{B}})$. Thus it is natural to ask whether there are any other contractive projections on $L^p(\Omega, \Sigma, P)$ with range $L^p(\Omega, \mathcal{B}, P_{\mathcal{B}})$, or, considering the averaging property of $E^{\mathcal{B}}$ (cf. Proposition 2iii), are there any other contractive projections on $L^p \to L^p$ with the averaging property and having constants as fixed points? These questions lead to the corresponding nontrivial characterization problems for conditional expectations and are functional analytic in nature. To indicate the flavor of the problems, we present just a sample result. [It is not used in later work.]

Theorem 11 *Let $\mathcal{B} \subset \Sigma$ be a σ-algebra and consider a contractive projection Q on $L^1(\Sigma)$ $[= L^1(\Omega, \Sigma, P)]$ with range $L^1(\mathcal{B})$. Then $Q = E^{\mathcal{B}}$. Conversely, $E^{\mathcal{B}}$ is always a contractive projection on $L^1(\Sigma)$ with range $L^1(\mathcal{B})$ for any σ-algebra $\mathcal{B} \subset \Sigma$.*

Proof The second part has already been established. We need to show the more involved result $Q = E^{\mathcal{B}}$. For this we first *assert* that Q is a positive operator with the property $E(QX) = E(X), X \in L^1(\Sigma)$, and reduce the general case to this special result. Indeed, by the density of bounded functions in $L^1(\Sigma)$, and the fact that $L^1(\Sigma)$ is a vector lattice [i.e., $X \in L^1(\Sigma) \Rightarrow X = X^+ - X^-, 0 \leq X^{\pm} \in L^1(\Sigma)$], it suffices to establish the result for $0 \leq X \leq 1$ a.e. Since Q is identity on its range (because $Q^2 = Q$), and $1 \in L^1(\mathcal{B})$, it follows that $Q1 = 1$. Then

$$\begin{aligned} E(1 - X) &= \|1 - X\|_1 \geq \|Q(1 - X)\|_1 \text{ (since } Q \text{ is a contraction)} \\ &= E(|1 - QX|) = 1 - E(QX). \end{aligned}$$

This implies

$$0 \leq E(X) \leq E(QX) \leq E(|QX|) = \|QX\|_1 \leq \|X\|_1 = E(X).$$

There is equality throughout and this shows that $QX \geq 0$ a.e., as well as $E(X) = E(QX)$, proving our assertion.

We are now ready to establish the general result, namely, $Q = E^{\mathcal{B}}$. Since for each $X \in L^1(\Sigma), QX \in L^1(\mathcal{B})$, and the same for $E^{\mathcal{B}}(X)$, we have, for all $A \in \mathcal{B}$,

$$\begin{aligned} E(\chi_A QX) &= E(Q(\chi_A QX)) \text{ (by the assertion)}, \\ E(\chi_A E^{\mathcal{B}}(X)) &= E(E^{\mathcal{B}}(\chi_A X)) = E(\chi_A X) = E(Q(\chi_A X)), \end{aligned} \tag{34}$$

If we show that the right sides of these equations in (34) are equal, then we get the equality of the left sides giving the middle equality below:

$$E(\chi_A QX) = \int_A QX dP_{\mathcal{B}} = \int_A E^{\mathcal{B}}(X) dP_{\mathcal{B}} = E(\chi_A E^{\mathcal{B}}(X)),$$

and hence the arbitrariness of A in $\mathcal{B}$, the $\mathcal{B}$-measurability of $E^{\mathcal{B}}(X)$ and QX imply $E^{\mathcal{B}}X = QX$. Let us prove the above equality.

First, one obtains the stronger statement that $Q(\chi_A X) = Q(\chi_A QX), A \in \mathcal{B}$ and for all X such that $0 \leq X \leq \chi_A$. Indeed, by the positivity of Q established at the beginning of this proof, $0 \leq QX \leq Q\chi_A = \chi_A$, since $\chi_A \in L^1(\mathcal{B})$. Thus Q vanishes outside A, and hence, if $g = Q(\chi_A X) - Q(\chi_A QX)$,

$$\chi_A + g = Q[\chi_A(\chi_A + X - QX)] \leq Q[\chi_A + X - QX] = \chi_A \text{ a.e.} \tag{35}$$

Hence $g \leq 0$ a.e. Similarly $\chi_A - g \leq \chi_A$ and $g \geq 0$, so that $g = 0$ a.e. for all $0 \leq X \leq \chi_A$. If $0 \leq X \leq 1$, then consider $X\chi_A, X\chi_{A^c}$. By the special result of (35) and the fact that $Q(X\chi_A)\chi_{A^c} = 0 = Q(X\chi_{A^c})\chi_A$, and the linearity of Q we get [even without (35)]

$$\int_A (QX)dP_{\mathcal{B}} = \int_A Q(X\chi_A)dP_{\mathcal{B}} + \int_A Q(X\chi_{A^c})dP_{\mathcal{B}} = \int_A Q(X\chi_A)dP_{\mathcal{B}}$$

$$= \int_\Omega Q(X\chi_A)dP_{\mathcal{B}} = \int_\Omega X\chi_A dP = \int_A E^{\mathcal{B}}(X)dP.$$

Hence $QX = E^{\mathcal{B}}(X)$ for $0 \leq X \leq 1$. Then by linearity, this holds for all bounded random variables X, and by density of such X in $L^1(\Sigma)$, and the continuity of the operators $Q, E^{\mathcal{B}}$, the result holds for all $X \in L^1(\Sigma)$. This completes the proof.

With further work, one can prove that the above theorem is true if $L^1(\Sigma)$ is replaced by $L^p(\Sigma), 1 \leq p < \infty$. It is also true if Q is a contractive projection on $L^p(\Sigma), 1 \leq p < \infty, p \neq 2$, but $Q1 = 1$ a.e. Then *its range can be shown to be of the form* $L^p(\mathcal{B})$. The result again holds for $p = 2$ if Q is assumed positive in addition. These and related characterizations are not essential for the present work and are not pursued. A detailed account of the latter may be found in the first author's book (1981).

3.2 Conditional Probabilities

In Definition 1.1 we introduced the general concept of the conditional probability function relative to a σ-algebra $\mathcal{B} \subset \Sigma$ as any function $P^{\mathcal{B}} : \Sigma \to L^\infty_+(\Omega, \mathcal{B}, P)$ satisfying the functional equation

$$\int_B P^{\mathcal{B}}(A)dP_{\mathcal{B}} = P(A \cap B), \quad B \in \mathcal{B}, A \in \Sigma. \tag{1}$$

Since $P^{\mathcal{B}}(A) = E^{\mathcal{B}}(\chi_A)$ and $P^{\mathcal{B}}(A)$ is a $\mathcal{B}$-measurable function, unique outside of a $P^{\mathcal{B}}$-null set depending on A, one says that any member of the equivalence class is a version of the conditional probability (of A given $\mathcal{B}$), and $P^{\mathcal{B}}$ is referred to as the conditional probability. Some immediate consequences of the definition are recorded for reference:

Proposition 1 *Let (Ω, Σ, P) be a probability space, and $\mathcal{B} \subset \Sigma$ be a σ-algebra. Then the following assertions are true:*

(i) $A \in \Sigma \Rightarrow 0 \leq P^{\mathcal{B}}(A) \leq 1$ a.e., $P^{\mathcal{B}}(\Omega) = 1$ a.e., and $P^{\mathcal{B}}(A) = 0$ a.e. if $P(A) = 0$.

(ii) $\{A_n, n \geq 1\} \subset \Sigma, A_n$ *disjoint* $\Rightarrow P^{\mathcal{B}}(\cup_{n\geq 1} A_n) = \sum_{n=1}^{\infty} P^{\mathcal{B}}(A_n)$ *a.e.* $= \sup_n \sum_{i=1}^{n} P^{\mathcal{B}}(A_i)$ *a.e..*

(iii) If $\{A_n, n \geq 1\}$ *are as in (ii) and* $A = \cup_{n=1}^{\infty} A_n$*, then for* $1 \leq p < \infty$,

$$\left\| P^{\mathcal{B}}(A) - \sum_{n=1}^{m} P^{\mathcal{B}}(A_n) \right\|_p \to 0 \text{ as } m \to \infty.$$

These assertions are immediate from Theorem 1.3. Consider (iii) as an illustration:

$$\begin{aligned}\left\| P^{\mathcal{B}}(A) - \sum_{n=1}^{m} P^{\mathcal{B}}(A_n) \right\|_p^p &= \int_{\Omega} \left[P^{\mathcal{B}}\left(\cup_{n>m} A_n\right)\right]^p dP_{\mathcal{B}} \text{ [using (ii)]} \\ &= \int_{\Omega} \left|E^{\mathcal{B}}\left(\chi_{\cup_{n>m} A_n}\right)\right|^p dP_{\mathcal{B}} \\ &\leq \int_{\Omega} E^{\mathcal{B}}\left(\chi_{\cup_{n>m} A_n}\right) dP_{\mathcal{B}} \\ &\qquad \text{(by Corollary 1.10, Jensen's inequality)} \\ &= P\left(\cup_{n>m} A_n\right) \to 0 \text{ as } m \to \infty.\end{aligned}$$

Taking $\mathcal{B} = \Sigma$ (or $A_n \in \mathcal{B}$ for all n), one sees that the assertion fails if $p = +\infty$, since $P^{\mathcal{B}}(A) = E^{\mathcal{B}}(\chi_A) = \chi_A$ hence $\|\chi_{\cup_{k\geq n} A_k}\|_{\infty} = 1 \not\to 0$ as $n \to \infty$.

This proposition states that $P^{\mathcal{B}}(\cdot)$ has formally the same properties as the ordinary measure P. However, each property has an exceptional P-null set which varies with the sequence. Thus if $\mathcal{B}$ is not generated by a (countable) partition of Σ, then the collection of these exceptional null sets can have a union of positive P-measure. This indicates that there may be difficulties in treating $P^{\mathcal{B}}(\cdot)(\omega) : \Sigma \to [0,1]$ as a standard probability measure for almost all $\omega \in \Omega$. Indeed, there are counterexamples showing that $P^{\mathcal{B}}(\cdot)(\omega)$ cannot always be regarded as an ordinary probability function. We analyze this fact and the significance of property (iii) of the above proposition since the structure of conditional probability functions is *essential* for the subject.

Proposition 2 *There exists a probability space* (Ω, Σ, P) *and a* σ*-algebra* $\mathcal{B}_0 \subset \Sigma$ *such that for the conditional probability* $P^{\mathcal{B}_0}$, $P^{\mathcal{B}_0}(\cdot)(\omega) : \Sigma \to [0,1]$ *is not a probability measure for almost all* $\omega \in \Omega$.

Proof It was already seen that $P^{\mathcal{B}}$ exists satisfying the functional equation (1) for any σ-algebra $\mathcal{B} \subset \Sigma$. By definition $E^{\mathcal{B}}(\chi_A) = P^{\mathcal{B}}(A)$ and hence, by the linearity of $E^{\mathcal{B}}$, we have for each simple function $f = \sum_{i=1}^{n} a_i \chi_{A_i}, A_i \in \Sigma$,

$$E^{\mathcal{B}}(f) = \sum_{i=1}^{n} a_i P^{\mathcal{B}}(A_i) = \int_{\Omega} f(\omega) P^{\mathcal{B}}(d\omega) \quad \text{a.e.,} \tag{2}$$

where the integral for simple functions is defined by the sum. It is easily seen that the integral in (2) is well defined for all simple functions, and does not depend on the representation of f. Thus for such f, there is a P-null set N_f such that for $\omega \in \Omega - N_f$

$$E^{\mathcal{B}}(f)(\omega) = \int_{\Omega} f(\omega')P^{\mathcal{B}}(d\omega')(\omega). \tag{3}$$

We observe that, if $P^{\mathcal{B}}(\cdot)(\omega)$ is a measure for each $\mathcal{B} \subset \Sigma$ and $\omega \in \Omega - N_0$ for some fixed P-null set N_0 $(\supset N_f)$, then (3) will be a Lebesgue integral, and thus it can be extended to all measurable $f : \Omega \to \mathbb{R}^+$, using Theorem 1.3i on the left and the monotone convergence on the right. If follows from this that (3) holds for all P-integrable f, since then the right side of (3) is the standard (Lebesgue) integral.

Now suppose that our (Ω, Σ, P) is the probability space given in Theorem 1.5 with $\mathcal{B} = \mathcal{B}_0 \subset \Sigma$ there. If $P^{\mathcal{B}}(\cdot)(\omega)$ can be regarded as a probability measure, then by the standard Vitali convergence theorem for each sequence $\{f_n, n \geq 1\}$ of uniformly integrable functions such that $f_n \to f$ a.e.(P), we must have for each $\omega \in \Omega - N_0$, $(P(N_0) = 0)$,

$$\begin{aligned} E^{\mathcal{B}_0}(f)(\omega) = \int_{\Omega} f(\omega')P^{\mathcal{B}_0}(d\omega')(\omega) &= \lim_{n\to\infty} \int_{\Omega} f_n(\omega')P^{\mathcal{B}_0}(d\omega')(\omega) \\ &= \lim_{n\to\infty} E^{\mathcal{B}_0}(f_n)(\omega). \end{aligned} \tag{4}$$

However, if the above sequence is not dominated by an integrable function [i.e., $\sup_{n\geq 1} |f_n| \notin L^1(P)$] as in Theorem 1.5, then by that theorem, (4) must be false for almost all ω. It is sufficient to take for $\{f_n, n \geq 1\}$ the two-valued sequence of that same theorem. Consequently our assumption that

$$P^{\mathcal{B}_0}(\cdot)(\omega) : \Sigma \to [0, 1]$$

is a probability for a.a.(ω) cannot hold, and in fact $P^{\mathcal{B}_0}(\cdot)(\omega)$ is a measure for almost no $\omega \in \Omega$ This completes the proof. [We give another counterexample in Problem 13, using the axiom of choice.]

Motivated by the above proposition we introduce

Definition 3 If (Ω, Σ, P) is a probability space and $\mathcal{B} \subset \Sigma$ is a σ-algebra, then a mapping $\tilde{P}(\cdot, \cdot) : \Sigma \times \Omega \to [0, 1]$ is called a *regular conditional probability* if (i) $\tilde{P}(\cdot, \omega) : \Sigma \to [0, 1]$ is a probability for each $\omega \in \Omega - N_0$, $P(N_0) = 0$, (ii) $\tilde{P}(A, \cdot) : \Omega \to [0, 1]$ is a $\mathcal{B}$-measurable function for each $A \in \Sigma$, and $\tilde{P}$ satisfies (1), so that

$$\int_B \tilde{P}(A, \omega)P(d\omega) = P(A \cap B), \quad A \in \Sigma, B \in \mathcal{B}. \tag{5}$$

Since by the Radon-Nikodým theorem the mapping P satisfying this functional equation is P-unique, it follows that $P(A, \cdot) = P^{\mathcal{B}}(A)$ a.e., so that $\tilde{P}(\cdot, \cdot)$, if it exists, must be a version of $P^{\mathcal{B}}$. The preceding proposition asserts that a regular conditional probability need not always exist, while Eq. (2) of Section 1 shows that such a mapping exists if $\mathcal{B}$ is not "too large," e.g., if it is generated by a countable partition. Note that, by (3), *if there exists a regular probability function* $P^{\mathcal{B}}$, then $E^{\mathcal{B}}(\cdot)$ *is simply an integral relative to this measure.* This means (1) of Section 1.3 will be true in this general case also, but it is not valid (by Proposition 2) for all conditional probability functions. This circumstance raises the following two important questions in the subject:

(A) Under what restrictions on the probability space (Ω, Σ, P), and/or the σ-algebra $\mathcal{B} \subset \Sigma$, does there exist a regular conditional probability function?

(B) To what extent can the theory of conditional probability be developed without regard to regularity?

It is rather important and useful to know the solutions to these problems, because they are, in a sense, special to probability theory and also distinguish its individual character. We consider (A) in detail, but first let us record a few remarks concerning (B). Since in general $P^{\mathcal{B}}$ cannot be forced to behave like a scalar measure (by Proposition 2), let us return to Proposition 1iii, which does not depend on restrictions of the null sets. Indeed, this says that $P^{\mathcal{B}} : \Sigma \to L^p(\Omega, \mathcal{B}, P), 1 \leq p < \infty$, is σ-additive in the p-norm, and if we only look upon the L^p-spaces as vector lattices [i.e., for $f, g \in L^p, f \leq g$ iff $f(\omega) \leq g(\omega)$ a.a. (ω)], then the parenthetical statement of Proposition 1ii is that

$$P^{\mathcal{B}} : \Sigma \to L^p(\Omega, \mathcal{B}, P), \ \ 1 \leq p \leq \infty.$$

is σ-additive in the order topology. If $1 \leq p < \infty$, the norm and order topologies coincide for studying $P^{\mathcal{B}}$, and if $p = \infty$, the order topology is weaker than that of the norm. If we therefore regard $P^{\mathcal{B}}$ as a vector-valued mapping from Σ into the positive part of $L^p(\Omega, \Sigma, P)$, $p \geq 1$, then one can develop the theory with the σ-additivity of $P^{\mathcal{B}}$ in the order or norm topologies. This aspect is called the *vector measure* theory, and using the latter point of view, the integral in (3) *can* be developed for all $f \in L^p(\Sigma)$. Evidently, the classical Vitali convergence result is false in this generality, but fortunately the dominated convergence statement survives. Consequently, using the vector integration theory, it is possible to present a satisfactory answer to problem (B). The classical intuitive explanations with conditional probabilities and expectations have to be given with necessary and explicit care and restraint. Such a general theory has been discussed by the first author in a monograph (Rao (1993),(2005)). It turns out, however, that in most problems of practical interest, solutions found for (A) are sufficient, and so we turn to it. Because of the intuitive and esthetic appeal, a considerable amount of the literature is devoted to problem (A), and we present an important part of it here.

To consider (A), we have already noted that there exist regular conditional probability functions if the conditioning σ-algebra $\mathcal{B}$ is generated by a (countable) partition of Σ. As a motivation for the general study, let us present a classical example where $\mathcal{B}$ is a richer σ-algebra than that determined by a partition. Thus let $\Omega = \mathbb{R}^2, \Sigma =$ Borel σ-algebra of $\mathbb{R}^2$, $\mathcal{B} =$ Borel σ-algebra of $\mathbb{R}$, and let $f : \mathbb{R}^2 \to \mathbb{R}^+$ be a measurable function such that

$$\int_{\mathbb{R}} \int_{\mathbb{R}} f(x,y)dxdy = 1.$$

Let $P : A \mapsto \int_A f(x,y)dxdy, A \in \Sigma$, be the (Borel) probability measure, and π_1, π_2 be x- and y-coordinate projections on $\mathbb{R}^2$. If $\mathcal{B}_2 = \pi_2^{-1}(\mathcal{B}) \subset \Sigma$, then $\mathcal{B}_2$ is richer than a partition-generated σ-subalgebra of Σ. We now exhibit a "natural" regular conditional probability function $Q : \Sigma \times \Omega \to \mathbb{R}^+$ relative to $\mathcal{B}_2$. Define $g : \mathbb{R}^2 \to \overline{\mathbb{R}}^+$ by the equation

$$g(x|y) = \begin{cases} \frac{f(x,y}{f_2(y)} & \text{if } f_2(y) \neq 0 \\ +\infty, & \text{if } f_2(y) = 0 \end{cases}$$

where $f_2(y) = \int_{\mathbb{R}} f(x,y)dx$. Theorem 1.3.11i (Fubini) guarantees that $f_2(\cdot)$, and hence $g(\cdot|\cdot)$, is $\mathcal{B}$- and Σ-measurable. Since Σ is generated by the algebra of measurable rectangles $\mathcal{B} \times \mathcal{B}$, consider Q defined for such an $A = A_1 \times A_2 \in \mathcal{B} \times \mathcal{B}, \omega = (x,y) \in A$, by

$$Q(A,\omega) = \int_{A_1} g(x|y)dx = \int_{\pi_1(A)} g(x|\pi_2(\omega))dx, \quad \pi_2(\omega) \in A_2. \tag{6}$$

It is clear that Q is well-defined, $Q(\cdot,\omega)$ is σ-additive on $\mathcal{B} \times \mathcal{B}$, and $Q(A,\cdot)$ is $\mathcal{B}_2$-measurable. If Q is shown to satisfy (5), then we can conclude that it is a regular conditional probability function. To see this consider, for any $B_2 \in \mathcal{B}_2 = \pi_2^{-1}(\mathcal{B})(B_2 = \mathbb{R} \times B$ for a unique $B \in \mathcal{B})$,

$$\begin{aligned}
\int_{B_2} Q(A,\omega)P(d\omega) &= \int_{B_2} \left(\int_{A_1} g(x|\pi_2(\omega))dx \right) P(d\omega) \\
&= \int_B \int_{\pi_1(A)} g(x|v)dx \int_{\mathbb{R}} f(u,v)dudv \\
&\qquad \text{(by Tonelli's theorem)} \\
&= \int_B \int_{\pi_1(A)} g(x|v)f_2(v)dxdv = \int_{\pi_1(A)\times B} f(x,v)dxdv \\
&= P(\pi_1(A) \times \pi_2(B_2)) = P(A \cap B_2).
\end{aligned} \tag{7}$$

Now both sides of (7) are σ-additive in $A(\in \mathcal{B} \times \mathcal{B})$ for each fixed $B_2 \in \mathcal{B}_2$. Since $\mathcal{B} \times \mathcal{B}$ generates $\mathcal{B} \otimes \mathcal{B} = \Sigma$, the Hahn extension theorem for σ-finite measures implies that (7) holds on Σ for each B_2 in $\mathcal{B}_2$. Hence $Q(\cdot,\cdot)$ is a regular conditional probability function as asserted. Note that $Q(A,\cdot)$ is thus a *version* of $P_2^{\mathcal{B}}(A), A \in \Sigma$.

It is clear that, if $\Omega = \mathbb{R}^n \times \mathbb{R}^m$ in the above, the same procedure holds for the Q defined by a more general $g(x_1,\ldots,x_n|y_1,\ldots,y_n)$ for any $1 \le m,n < \infty$. Here the structure of Borel algebras of $\mathbb{R}^n$ is used. Also, $\{\mathbb{R}^n\}_{n\ge 1}$ are the range spaces of random vectors. It is shown later that the above considerations can be extended to the image measures (i.e., distribution functions) on such spaces.

Guided by the above example, we can give the regularity concept of Definition 3 in the following slightly more convenient form.

Definition 4 Let (Ω, Σ, P) be a probability space and $\mathcal{B}$, $\mathcal{S}$ be two σ-subalgebras of Σ without any particular inclusion relationship between them. A mapping $\tilde{P} : \mathcal{S} \times \Omega \to \mathbb{R}^+$ is a regular conditional probability (in the extended sense) if (i) $\tilde{P}(\cdot,\omega) : \mathcal{S} \to \mathbb{R}^+$ is a probability, for each $\omega \in \Omega$, (ii) $\tilde{P}(A,\cdot) : \Omega \to \mathbb{R}^+$ is $\mathcal{B}$-measurable, for each $A \in \mathcal{S}$, and (iii)

$$\int_B \tilde{P}(A,\omega)P(d\omega) = P(A \cap B), A \in \mathcal{S}, B \in \mathcal{B}. \tag{8}$$

If $\mathcal{S} = \Sigma$, then this definition essentially reduces to the preceding one. For this reason, the phrase , "in the extended sense" will be dropped hereafter, since there is no conflict or ambiguity by this action. If $B = \Omega$ in (8), and if $\tilde{P}(\cdot,\cdot)$ satisfies (i) and (ii) of the above definition, then any P on $\mathcal{S}$ for which (8) holds is often called an *invariant probability* for the "kernel" or the "transition probability" $\tilde{P}(\cdot,\cdot)$. For this interpretation, $(\Omega, \mathcal{S})$ and $(\tilde{\Omega}, \mathcal{B})$ can be completely different. The generalization is of interest for an extended study of Markov processes. However, for the present, we use Definition 3. Its significance is made evident when the pair $(\mathcal{S}, \mathcal{B})$ is specialized and image conditional measures are brought into play (cf. Theorem 1.4.1).

Let $X : \Omega \to \mathbb{R}^n, n \ge 1$, be a random variable (or vector if $n > 1$), and let $\mathcal{R}$ be the Borel σ-algebra of $\mathbb{R}^n$ with $\mathcal{S} = X^{-1}(\mathcal{R}) \subset \Sigma$. Then $\tilde{P} : \mathcal{S} \times \Omega \to \mathbb{R}^+$ becomes $Q_X(D,\omega) = \tilde{P}(X^{-1}(D),\omega), D \in \mathcal{R}$, and (8) reduces to

$$\int_B Q_X(D,\omega)P(d\omega) = \int_B \tilde{P}(X^{-1}(D),\omega)P(d\omega) = P(X^{-1}(D) \cap B). \tag{9}$$

Such a $Q_X : \mathcal{R} \times \Omega \to \mathbb{R}^+$ is called a *regular conditional distribution* (= *image regular conditional probability*) when $\tilde{P}$ is regular in the sense of Definition 4. Since $Q(\cdot,\omega) = (P^{\mathcal{B}} \circ X^{-1}(\cdot))(\omega) : \mathcal{R} \to \mathbb{R}^+$ is a version of the image conditional probability function $P^{\mathcal{B}} \circ X^{-1} : \mathcal{R} \to L^1(\Omega, \mathcal{B}, P)$, it is important to know about the existence of such a Q. In other words, when does a function $Q_X : \mathcal{R} \times \Omega \to \mathbb{R}^+$ exist, satisfying the conditions: (i) $Q_X(D,\cdot)$ is $\mathcal{B}$-measurable $(\mathcal{B} \subset \Sigma)$, (ii) $Q_X(\cdot,\omega) : \mathcal{R} \to [0,1]$ is a probability, and (iii) $Q_X(D,\cdot) = P^{\mathcal{B}_0}X^{-1}(D)(\cdot)$ a.e.$[P]$?

If $X = (X_1, X_2)$ and $f(\cdot,\cdot)$ is the density function of X relative to the planar Lebesgue measure [i.e., if $F(x,y) = P[X_1 < x, X_2 < y]$, then

$F(x,y) = \int_{-\infty}^{x}\int_{\infty}^{y} f(u,v)dudv]$, and $(\Omega,\Sigma) = (\mathbb{R}^2, \mathcal{B}\otimes\mathcal{B})$, then, as we have shown, Q defined by (6) is indeed such a regular conditional distribution. Without assuming such a representation, we can show that a regular conditional distribution always exists for a random vector. The result is due to Doob (1953).

Theorem 5 *Let* (Ω,Σ,P) *be a probability space and* $X:\Omega\to\mathbb{R}^n$ *be a random vector. If* $\mathcal{B}\subset\Sigma$ *is a* σ*-algebra, then a regular conditional distribution* $Q_X:\mathcal{B}\times\Omega\to\mathbb{R}^+$ *for* X *given* $\mathcal{B}$ *exists.*

Proof The argument uses the fact that rationals are dense in $\mathbb{R}$ (or use any dense denumerable set in $\mathbb{R}^n$), and properties (i) and (ii) of Proposition 1 are available. Let $\{r_i, i\geq 1\}\subset\mathbb{R}$ be an enumeration of the dense set which we take as rationals, and consider for each $\omega\in\Omega$,

$$F_n(r_{i_1},\ldots,r_{i_n};\omega) = P^{\mathcal{B}}[X_k < r_{i_k}, k=1,\ldots,n](\omega), \tag{10}$$

where $\mathcal{B}$ is the given σ-subalgebra. By Proposition 1, there is a P-null set $N(r_{i_1},\ldots,r_{i_n})$ such that, if $\omega\notin N(r_{i_1},\ldots,r_{i_n})$ then $F_n(\cdots;\omega)$ given by (10) is nonnegative and nondecreasing, and if

$$N = \cup_{\{r_{i_k},1\leq k\leq n\}} N(r_{i_1},\ldots,r_{i_n})$$

so that $P(N) = 0$, then $F_n(\cdots;\omega)$ is also left continuous and in fact is an n-dimensional distribution function for $\omega\notin N$. If $x_k\in\mathbb{R}$ is arbitrary, define for $\omega\notin N$,

$$F_n(x_1,\ldots,x_n;\omega) = \lim_{r_{i_k}\uparrow x_k} F(r_{i_1},\ldots,r_{i_n};\omega), \tag{11}$$

and if $\omega\in N$, let F_n be equal to any fixed distribution function G_n. For definiteness, let us take $G_n(x_1,\ldots,x_n) = P[X_1 < x_1,\ldots,X_n < x_n]$. Then we see that $F_n(\cdots;\omega)$ is an n-dimensional distribution function for each $\omega\in\Omega$. Thus one can define (for each Borel $B\subset\mathbb{R}^n$, i.e., $B\in\mathcal{R}$)

$$Q_X(B,\omega) = \int\cdots\int_B dF_n(x_1,\ldots,x_n;\omega), \omega\in\Omega. \tag{12}$$

It is clear that $Q_X(\cdot,\cdot):\mathcal{R}\times\Omega\to\mathbb{R}^+$ is well defined, $Q_X(\cdot,\omega)$ is σ-additive on $\mathcal{B}$, and $Q_X(B,\cdot)$ is measurable relative to $\mathcal{B}$, where $\mathcal{R}$ is the Borel σ-algebra of $\mathbb{R}^n$. Thus $Q_X(\cdot,\cdot)$ will be a regular conditional distribution of X if it is shown that $Q_X(B,\omega) = P^{\mathcal{B}}(X^{-1}(B))(\omega)$ for a.a. $\omega\in\Omega$, i.e., that Q_X is a version of $P^{\mathcal{B}}\circ X^{-1}$ on $\mathcal{R}$.

For this purpose, consider the class $\mathcal{C}\subset\mathcal{R}$ defined by

$$\mathcal{C} = \{B\in\mathcal{R}: Q_X(B,\omega) = P^{\mathcal{B}}(X^{-1}(B))(\omega), a.a.\omega\in\Omega\}.$$

By definitions (10) and (12), if $\mathcal{S}$ is the semiring of right-open left-closed intervals (or rectangles), then $\mathcal{S}\subset\mathcal{B}$. Also $\mathcal{S}$ is a π-class, since it is closed under intersections. Moreover, by the monotone convergence theorem and (10)–(12) it follows immediately that $\mathcal{C}$ is a λ-class. Hence by Proposition 1.2.8, $\sigma(\mathcal{S}) = \mathcal{R} \subset \mathcal{C}$. Thus $Q_X(\cdot,\cdot)$ is a version of $P^{\mathcal{B}} \circ X^{-1}$, completing the proof.

It is gratifying to know that there is always a version of $P^{\mathcal{B}} \circ X^{-1}$ which is a regular conditional distribution on $(\mathbb{R}^n, \mathcal{R})$ for each random variable $X : \Omega \to \mathbb{R}^n$ and σ-algebra $\mathcal{B} \subset \Sigma$. Note that, in this description we are only concentrating on $\mathcal{S} = X^{-1}(\mathcal{R}) \subset \Sigma$ and $P^{\mathcal{B}} : \mathcal{S} \to L^1(\Omega, \mathcal{B}, P)$. But usually $\mathcal{S}$ is a much smaller σ-ring contained in Σ. Can we say, on $\mathcal{S}$, $P^{\mathcal{B}}$ itself has a version which is a regular conditional probability? In other words, can we transport the regularity property of $Q_X(\cdot,\cdot)$ to $P^{\mathcal{B}}$ on $\mathcal{S}$? In general, the answer is "NO." To see that there is a problem here, note that if $X(\Omega) = B_0 \subset \mathbb{R}^n$, where $B_0 \neq \mathbb{R}^n$, then by definition of inverse image, $X^{-1}(D) = \emptyset$ for all $D \subset \mathbb{R}^n - B_0$. Consequently the mapping $X^{-1} : \mathcal{R} \to \mathcal{S}$ is *not* one-to-one. If $B_0 = \mathbb{R}^n$ or if $B_0 \in \mathcal{R}$, then we may replace $\mathcal{R}$ by the trace σ-subalgebra $\mathcal{R}(B_0) = \{A \cap B_0 : A \in \mathcal{R}\}$, and then $X^{-1} : \mathcal{R}(B_0) \to \mathcal{S}$ is a σ-homomorphism, and the above pathology is immediately eliminated. Further, we can provide an affirmative answer easily. This observation is also due to Doob.

Proposition 6 *Let (Ω, Σ, P) be a probability space $\mathcal{B} \subset \Sigma$ a σ-algebra, and $X : \Omega \to \mathbb{R}^n$ a random vector. If $X(\Omega) = B_0 \in \mathcal{R}$ and $\mathcal{S} = X^{-1}(\mathcal{R}(B_0))$, where $\mathcal{R}$ is the Borel σ-algebra of $\mathbb{R}^n$, then there exists a version ν_X of $P^{\mathcal{B}} : \mathcal{S} \to L^1(\Omega, \mathcal{B}, P)$ $[\nu_X(A,\omega) = P^{\mathcal{B}}(A)(\omega)$ a.a. (ω), $A \in \mathcal{S}]$ that is a regular conditional probability. (Thus ν_X is defined only on $\mathcal{S} \subset \Sigma$.)*

Proof Since by hypothesis $B_0 \in \mathcal{R}$, and by definition $X^{-1}(B_0) = X^{-1}(\mathbb{R}^n)$, let $Q_X(\cdot,\cdot)$ be a regular conditional distribution of X given $\mathcal{B}$, guaranteed by the above theorem. It follows that $Q_X(B_0,\omega) = P^{\mathcal{B}} \circ X^{-1}(B_0)(\omega) = P^{\mathcal{B}}(X^{-1}(\mathbb{R}^n))(\omega) = 1$ a.a.(ω). Let $A \in \mathcal{S}$. Since $X : \Omega \to B_0$ is onto, X^{-1} is one-to-one on $\mathcal{R}(B_0) \to \mathcal{S}$. Hence there exists a $B_1 \in \mathcal{R}(B_0)$ with $A = X^{-1}(B_1)$. [In fact, if there is also a $B_2 \in \mathcal{R}(B_0)$ with $A = X^{-1}(B_2)$, then

$$B_1 = X(X^{-1}(B_1)) = X(A) = X(X^{-1}(B_2)) = B_2,$$

because the onto property of X is equivalent to $X[X^{-1}(D)] = D$ for all $D \subset B_0$.] Thus $\nu_X(A,\omega) = Q_X(B,\omega)$ with $A = X^{-1}(B)$, $\omega \in \Omega - N$, $P(N) = 0$, unambiguously defines ν_X for almost all $\omega \in \Omega$. Since X^{-1} preserves all set operations, it follows that $\nu_X(\cdot,\omega)$ is a probability for all $\omega \in \Omega - N$, and then $\nu_X(A,\omega) = Q_X(B,\omega) = P^{\mathcal{B}}(X^{-1}(B))(\omega) = P^{\mathcal{B}}(A)(\omega)$ a.a.(ω), $A \in \mathcal{S}$. Since $Q_X(B_0^c,\omega) = 0$ and $P^{\mathcal{B}}(X^{-1}(D)) = 0$ a.e. for all $D \in \mathcal{R}(B_0^c)$, we have $\nu_X(A,\omega) = P^{\mathcal{B}}(A)(\omega)$ a.a. (ω), $A \in \mathcal{S}$, so that ν_X is a version of $P^{\mathcal{B}}$ on $\mathcal{S}$. This completes the proof.

In the above result, suppose $\mathcal{B} = \sigma(Y)$, the σ-algebra generated by a random vector $Y : \Omega \to \mathbb{R}^m$. Then (cf. Proposition 1.4) there is a Borel function $g : \mathbb{R}^m \to \mathbb{R}$ such that for each $A = [X < a] \in \mathcal{S} = \sigma(X)$, we have

$$P^{\mathcal{B}}(A)(\omega) = g_A(Y)(\omega) = g_A(Y(\omega)) \quad \text{a.a. } \omega \in \Omega. \tag{13}$$

If $Y(\omega) = y(\in \mathbb{R}^m)$, then one expresses (13) *symbolically* as

$$P^{\mathcal{B}}([X < a])(\omega) = P\{\omega' : X(\omega') < a | Y(\omega) = y\}, \tag{14}$$

and the right-side quantity has *no other significance except as an abbreviation of the left side.* If $X : \Omega \to \mathbb{R}^m$ is a coordinate variable, in the sense that its range is $\mathbb{R}^n$—this happens in many applications—then the conditional probability in (14) is regular, by Proposition 6, and it is a constant on the set $B = \{\omega : Y(\omega) = y\}$. Thus when $P^{\mathcal{B}}$ is regular, one denotes the right side of (14), indifferently and unfortunately, as

$$F_{X|Y}(a|y) = P\{\omega' : X(\omega') < a | Y = y\} = F_X^{\omega}(a|B), \quad \omega \in B. \tag{15}$$

We call $F_{X|Y}(\cdot|\cdot)$ a conditional distribution function of X given Y [or $\mathcal{B} = \sigma(Y)$]. This terminology is "meaningful", since $F_{X|Y}(\cdot|y)$ is a distribution function, and $F_{X|Y}(x|\cdot)$ is a Borel function. So,

$$F_{X|Y}(a|y) = P^{\mathcal{B}}(X^{-1}(-\infty, a))(\omega),$$

and the right side always defines a conditional probability distribution by Theorem 5; and (15) says that $F_{X|Y}(\cdot|y)$ is a distribution function (can be chosen to be such for *each* y), and $F_{X|Y}(a|\cdot)$ is Borel measurable. Hence we may state the following (but we show later there is room for troubles in this calculation!):

Proposition 7 *Let (Ω, Σ, P) be a probability space and X, Y be a pair of random vectors on Ω into $\mathbb{R}^n$ and $\mathbb{R}^m$, respectively. If $F_{X,Y}$ is their joint distribution, F_Y that of Y and $F_{X|Y}(\cdot|\cdot)$ the conditional distribution, then*

$$F_{X,Y}(x, y) = \int_{-\infty}^{y} F_{X|Y}(x|t) F_Y(dt), \quad y \in \mathbb{R}^m, \tag{16}$$

and the integral is an m-fold symbol. Also, $F_{X|Y}(x|y) = F_X(x), x \in \mathbb{R}^n$, if X, Y are independent. Moreover,

$$\int_A \int_B h(x, y) F_{X,Y}(dx, dy) = \int_A \left[\int_B h(x, y) F_{X|Y}(dx|y) \right] F_Y(dy), \tag{17}$$

for all Borel sets $A \subset \mathbb{R}^n$, $B \subset \mathbb{R}^m$ and a bounded Borel function $h : \mathbb{R}^n \times \mathbb{R}^m \to \mathbb{R}$.

Proof For notational convenience, we take $m = n = 1$. Let $B_x = (-\infty, x)$, $B_y = (-\infty, y)$. Then by definitions of $F_{X,Y}, F_{X|Y}, F_Y$, we have, since $\mathcal{B}_Y = \sigma(Y)$ and $Y^{-1}(B_y) \in \mathcal{B}_y$,

$$\begin{aligned}
F_{X,Y}(x,y) &= P(X^{-1}(B_x)\cap Y^{-1}(B_y)) \\
&= \int_{Y^{-1}(B_y)} P^{\mathcal{B}_Y}(X^{-1}(B_x))(\omega)P_{\mathcal{B}_Y}(d\omega), \\
&= \int_{B_y} \tilde{P}(B_x|t)F_Y(dt) \quad \text{(by the image probability law)} \\
&= \int_{-\infty}^{y} F_{X|Y}(x|t)F_Y(dt) \quad [\text{by (15)}].
\end{aligned}$$

Thus (16) holds. If X, Y are independent, then $P^{\mathcal{B}}(A) = E^{\mathcal{B}}(\chi_A) = P(A)$ a.e., by Proposition 1.4. Hence (15) becomes $F_{X|Y} = F_X(x), x \in \mathbb{R}$.

To prove (17), since $F_{X|Y}(\cdot|y)$ is a distribution function, the mapping $\nu(\cdot|y) : A \mapsto \int_A F_{X|Y}(dx|y), A \in \mathcal{R}$ (Borel σ-algebra) is a probability measure. Consider for any bounded Borel function $h : \mathbb{R}\times\mathbb{R} \to \mathbb{R}$,

$$\mathcal{C} = \left\{ A \in \mathcal{R} : \int_B \int_A h(x,y)F_{X|Y}(dx,dy) \right.$$
$$\left. = \int_B \int_A h(x,y)\nu(dx|y)F_Y(dy), \text{ all } B \in \mathcal{R} \right\}.$$

It is clear that $\mathcal{C}$ contains all intervals of the form $[a,b)$ and their intersections, so that the semiring $\mathcal{S}$ of such intervals is in $\mathcal{C}$ and $\mathbb{R} \in \mathcal{C}$. It is a π-class, and by the monotone convergence theorem it follows that $\mathcal{C}$ is also a λ-class. Hence by Proposition 1.2.8, $\mathcal{C} \supset \sigma(\mathcal{S}) = \mathcal{R}$. Since $\mathcal{C} \subset \mathcal{R}$, we have $\mathcal{C} = \mathcal{R}$, and (17) is verified. Note that for the argument here and for (17), h can be any $F_{X|Y}$-integrable function and need not be bounded. This completes the proof.

Suppose that the distribution functions $F_{X,Y}, F_{X|Y}$ and F_Y are absolutely continuous with densities $f_{X,Y}(\cdot,\cdot), f_{X|Y}(\cdot|y)$, and $f_Y(\cdot)$. Then (17) implies

$$\begin{aligned}
&\int_{\mathbb{R}^n}\int_{\mathbb{R}^m} h(x,y)f_{X,Y}(x,y)dydx \\
&\qquad = \int_{\mathbb{R}^m}\left[\int_{\mathbb{R}^n} h(x,y)f_{X|Y}(x|y)dx\right] f_Y(y)dy \\
&\qquad = \int_{\mathbb{R}^n}\int_{\mathbb{R}^m} h(x,y)f_{X|Y}(x|y)f_Y(y)dydx, \\
&\qquad\qquad \text{(by Fubini's theorem).} \qquad (18)
\end{aligned}$$

Since this equation is true for all bounded Borel h, it follows from the Lebesgue theory that $f_{X,Y}(x,y) = f_{X|Y}(x|y)f_Y(y)$ for a.a.(x,y) (Lebesgue). Hence the example given for regular conditional distributions [cf. (6)] is recovered.

We restate (17) and (18) in a different form for the Lebesgue-Stieltjes measures, for reference, as follows:

Proposition 8 *Let (Ω, Σ, P) be a probability space and X, Y be random vectors on Ω into $\mathbb{R}^n$, $\mathbb{R}^m$. Suppose $P_{X,Y}, P_Y$ are the Lebesgue-Stieltjes measures on $\mathbb{R}^n \times \mathbb{R}^m$ and $\mathbb{R}^m$, respectively. If $Q(\cdot|y)$ is the regular conditional distribution of X given $Y = y$, then for any Borel sets $A \subset \mathbb{R}^n, B \subset \mathbb{R}^m$, we have*

$$\begin{aligned} P_{X,Y}(\pi_1^{-1}(A) \cap \pi_2^{-1}(B)) &= P_{X,Y}(A \times B) \\ &= \int_{\pi_2^{-1}(B)} Q(\pi_1^{-1}(A)|y) P_Y \circ \pi_2^{-1}(dy) \\ &= \int_B P_{X|Y}(A|y) P_Y(dy), \end{aligned} \tag{19}$$

where $\pi_1 : \mathbb{R}^{m+n} \to \mathbb{R}^n$, $\pi_2 : \mathbb{R}^{m+n} \to \mathbb{R}^m$ are coordinate projections. If, further, $P_{X,Y}$ is absolutely continuous relative to the Lebesgue measure with density $f_{X,Y} : \mathbb{R}^n \times \mathbb{R}^m \to \overline{\mathbb{R}}^+$, then P_Y, the marginal of $P_{X,Y}$ [i.e., $P_{X,Y}(\mathbb{R}^n \times \cdot) = P_Y(\cdot)$], also has a density f_Y [so that $f_Y(y) = \int_{\mathbb{R}^n} f_{X,Y}(x, y)dx$] and

$$Q(\cdot|y) = P_{X|Y}(\pi_1^{-1}(\cdot)|y)$$

is absolutely continuous relative to the Lebesgue measure. A version of its density is $f_{X|Y} : \mathbb{R}^n \to \overline{\mathbb{R}}^+$, and it satisfies

$$f_{X,Y}(x, y) = f_{X|Y}(x|y) f_Y(y) \quad a.a.(x, y) \in \mathbb{R}^n \times \mathbb{R}^m. \tag{20}$$

Moreover, if $h : \mathbb{R}^n \times \mathbb{R}^m \to \mathbb{R}$ is a bounded Borel function, then

$$E(h(X, Y)|Y)(\omega) = \int_{\mathbb{R}^n} h(x, Y(\omega)) f_{X|Y}(x|Y)(\omega)) dx \quad a.a. \omega \in \Omega, \tag{21}$$

and (cf Proposition 1.2vi)

$$E(h(X, Y)) = \int_{\mathbb{R}^m} \int_{\mathbb{R}^n} h(x, y) P_{X|Y}(dx|y) P_Y(dy). \tag{22}$$

All these statements have already been proved above. Because of (21), the conditional Hölder and Minkowski inequalities (cf. Theorem 1.9) can also be obtained using the corresponding classical procedures with Lebesgue-Stieltjes integrals, once the regular conditional probability theory is available, i.e., if Theorem 5 is given.

If X, Y are random variables, as above, having an absolutely continuous (joint) distribution function with density $f_{X,Y}$, then for any real a, b, and $\delta_1 > 0, \delta_2 > 0$, we have

$$P[a \le X < a+\delta_1, b \le Y < b+\delta_2] = \int_a^{a+\delta_1} \int_b^{b+\delta_2} f_{X,Y}(x,y)dydx \to 0, \quad (23)$$

as $\delta_1 \to 0$ or $\delta_2 \to 0$. Thus, e.g., $P[a \le X < a+\delta_1, Y = b] = 0$. However, by (14) and (20) one has

$$P_{X|Y}[a \le X < a+\delta_1 | Y = b] = \int_a^{a+\delta_1} f_{X|Y}(x|b)dx, \quad \delta_1 > 0. \quad (24)$$

On the other hand, using the naive approach [cf. (1) of Section 1]:

$$P[a \le X < a+\delta_1 | b \le Y < b+\delta_2] = \frac{P[a \le X < a+\delta_1, b \le Y < b+\delta_2]}{P[b \le Y < b+\delta_2]}$$

$$= \frac{(1/\delta_2)\int_a^{a+\delta_1}\int_b^{b+\delta_2} f_{X,Y}(x,y)dydx}{(1/\delta_2)\int_b^{b+\delta_2} f_Y(y)dy}.$$

Letting $\delta_2 \to 0$ and using the classical Lebesgue differentiation theorem one gets [the left side symbolizing the limit to be denoted as the left side of (24)] the following:

$$P[a \le X < a+\delta_1 | Y = b] = \lim_{\delta_2 \downarrow 0} P[a \le X < a+\delta_1 | b \le Y < b+\delta_2]$$

$$= \int_a^{a+\delta_1} \frac{f_{X,Y}(x,b)}{f_Y(b)} dx \quad \text{a.e. (Lebesgue)}$$

$$= \int_a^{a+\delta_1} f_{X|Y}(x|b)dx, \quad (25)$$

which is (24). It is worth noting an *important difference* between (24) and (25). We now explain the lack of *unique constructive procedures* here.

Initially, we defined $P(A|B)$ as $P(A \cap B)/P(B)$, which is unambiguous *only if* $P(B) > 0$, while for the abstract equation (1), $P^{\mathcal{B}}(A)$ is defined for a *family* $\mathcal{B}$ of events with $B \in \mathcal{B}$. We then specialized the latter to obtain (24) directly. However, we have *not presented a recipe* for calculating $P^{\mathcal{B}}(A)$, or the regular conditional probability $Q(A,\omega)$ [a version of $P^{\mathcal{B}}(A)$], when it exists. In fact, this is a nontrivial problem. As (25) demonstrates, in any given situation, $Q(A,\omega)$ or $P_{X|Y}(A|y)$ should be calculated using *additional information* that may be available. Also, (19) clearly shows that this is an abstract differentiation problem of $P_{X,Y}$ relative to P_Y. The problem is relatively simple if $P[Y = b] > 0$. But it is important and nontrivial in the general case, where Y is also a vector and $P[Y = b] = 0$. The theory of differentiation of integrals enters crucially here. For an overview of the latter work, the reader may consult a lucid survey by Bruckner (1971), and for further abstract results, Hayes and Pauc (1970). The point is that, especially in the vector case of the conditioning variable Y, the limit in (25) may either (i) not exist, or (ii) exist *but depend on the approximation sequence* $B_n \downarrow B = [Y = b]$. If the approximation sequence, called the differentiation basis, is not the correct one, and if the limit exists with some such "basis," the result can be another "conditional density." When an appropriate differentiation basis is used and the derivative exists, then it will be a version of the Radon-Nikodým integrand by the general theory of differentiation. These points merit the following elucidation.

We first indicate the correct meaning of the differentiation basis, state a positive result, and then present a set of examples showing how different limits appear for different approximation sequences. Thus if (Ω, Σ, μ) is a general σ-finite measure space, then a family $\mathcal{F} \subset \Sigma$ of sets of *positive finite* μ-measure is called a *differentiation basis* if the following two conditions are satisfied: (i) for each $\omega \in \Omega$, there is at least one generalized sequence (or net) $\{F_\alpha, \alpha \in I\} \subset \mathcal{F}$ such that $F_\alpha \rightsquigarrow \omega$ (read "F_α contracts to ω") in the Moore-Smith sense (i.e., there is α_0 such that $\alpha > \alpha_0 \Rightarrow \omega \in F_\alpha$, where $>$ is the ordering of the index I), and (ii) every cofinal[1] subsequence of a sequence of a contracting $\{F_\alpha, \alpha \in I\}$ also contracts to ω. Here $\omega \in F_\alpha$ is not necessary. If there is topology in Ω, $F_\alpha \rightsquigarrow \omega$ can be interpreted in other appropriate ways. The general existence result here is as follows:

Proposition 9 (Existence of a Basis) *If* (Ω, Σ, μ) *is a Carathéodory-generated measure space and* μ *is* σ*-finite, then there always exists a differentiation basis* $\mathcal{F} \subset \Sigma$ *with the following property* (called a **Vitali property**): *for each* $A \subset \Omega$ *and* $\varepsilon > 0$ *there exists a sequence* $\{F_n, n \geq 1\} \subset \mathcal{F}$ *such that* (i) *the* F_n *are disjoint,* (ii) $\mu^*(A - \cup_n F_n) = 0$, *and* (iii) $\mu(\cup_n F_n - \tilde{A}) < \varepsilon$, *where* $\tilde{A}$ *is a measurable cover of* A *and* μ^* *is the outer measure generated by* μ.

This result will not be proved. [It is quite involved, and indeed is related to the existence of a "lifting map" on (Ω, Σ, μ). A complete proof can be found in books on differentiation, and a recent version (and discussion) of it is given in the first author's monograph (Rao (1993), (2005), Section 3.4).] We do not need the result here except to draw the readers attention to the existence of a nontrivial problem with calculations of regular conditional (probabilities or) distributions.

A Set of Examples 10 Let (Ω, Σ, P) be a probability space and $X : \mathbb{R}^+ \times \Omega \to \mathbb{R}$ be a mapping such that $X(t, \cdot)$ is a random variable for each $t \in \mathbb{R}^+$ and $X(\cdot, \omega)$ is differentiable for almost all $\omega \in \Omega$. In particular, let $Y = X'(0, \cdot)$, the derivative of X at $t = 0$. It is an r.v. We stipulate that for each finite set $t_1 < t_2 < \ldots < t_n$ and $a_i \in \mathbb{R}$, the r.v. $Z_a = \sum_{i=1}^n a_i X(t_i, \cdot)$ has the distribution

[1] Recall that in a partially ordered set I, a subset J is called *cofinal* if for each $i \in I$ there is a $j \in J$ with $j \geq i$.

$$P[Z_a < z] = \frac{1}{\sqrt{2\pi\sigma_a^2}} \int_{-\infty}^{z} e^{\frac{-u^2}{2\sigma_a^2}} du, \quad \sigma_a^2 > 0. \tag{26}$$

The σ_a^2 is determined (for simplicity) by the condition that if $a_i = 1, a_j = 0, i \neq j$, in $a = (a_1, \ldots, a_i, \ldots, a_n)$, then [the above implies $E(X(t)) = 0$] $E(X^2(t)) = 1, t \geq 0$. Also, we assume that the function $\{X(t, \cdot), t \geq 0\}$ is "ergodic." For our purposes it suffices to say that this implies that Y and $X(t, \cdot)$ are mutually independent. (A definition of ergodic process is given after Theorem 7.3.3.) It follows from the Kolmogorov existence theorem, to be proved below in Section 4, that such families of random variables exist on suitable probability spaces (and $\{X(t, \cdot), t \geq 0\}$ will be called a Gaussian process). For now we can and will assume that there is such a space (Ω, Σ, P) and a mapping X with the above properties. A problem of considerable practical interest is to find the "conditional density of Y when the process (or family $X(t, \cdot), t \geq 0$) has started at a, i.e., $X(0, \cdot) = a$." From (26) it follows that $X(t, \cdot)$ has a continuous distribution for all t, so that

$$P[X(t, \cdot) = a] = 0$$

for $t \geq 0$. Hence we are in the predicament discussed following (25). Since Y is obtained by a linear operation on $X(t, \cdot)$, it is easily verified that $E(Y) = 0$. Let $E(Y^2) = \alpha^2 > 0 (\alpha^2 < \infty$ always), and

$$P[Y < y] = \frac{1}{\sqrt{2\pi\alpha^2}} \int_{-\infty}^{y} e^{-v^2/2\alpha^2} dv. \tag{27}$$

We shall now calculate the conditional density of Y given $X(0, \cdot) = a$ with different approximations of $A = [X(0, \cdot) = a]$ and show that the naive approach gives completely different answers depending on the approximations. This example is adapted from Kac and Slepian (1959).

(i) *Approximation* 1. Let $A_\delta = [a \leq X(0, \cdot) < a + \delta]$ for $\delta > 0$. Then $P(A_\delta) > 0$, and as $\delta \to 0$, $A_\delta \to A$ in the Moore-Smith sense. Hence, if $p(\cdot)$ is the density of Y,

$$\begin{aligned} P[Y < y | A_\delta] &= \frac{P[Y < y, A_\delta]}{P(A_\delta)} = P[Y < y] \quad \text{(by independence)} \\ &= \int_{-\infty}^{y} p(u) du, \end{aligned} \tag{28}$$

so that with (27) the conditional density is obtained as $\delta \to 0$ [since the right side of (28) does not depend on δ] :

$$p_{Y|X(0)}(u|a) = p(u) = \frac{1}{\sqrt{2\pi\alpha^2}} e^{\frac{-u^2}{2\alpha^2}}.$$

However, in this approximation, the fact that $X(0)$ is part of $\{X(t), t \geq 0\}$ and that $X(0) = \lim_{t \to 0} X(t)$ is not used; i.e., part of the information is ignored. So we remedy this in the next set of approximations.

(ii) *Approximation* 2(m). Let $\delta > 0$ and m be a real number that is the slope of a straight line through $(0, a)(y_t = a + mt, t \geq 0)$. Let $A_\delta^m = [X(t) :$ $X(t)$ passes through the line $y = a + mt$, of length δ, for some $t \geq 0]$. Thus $A_\delta^m = \{\omega : X(t, \omega) = a + mt$, for some $0 \leq t \leq \delta/(1+m^2)^{1/2}\}$, A_δ^m is an event.

Again $P(A_\delta^m) > 0$, and for each m, $A_\delta^m \to A$ as $\delta \to 0$. We now calculate the "conditional density" $p_m(\cdot|a)$ and show that for each m it is a *different* function. First let $Y > m$, and using the procedure of (28) and differentiating relative to y to obtain the density, we get (on noting that Y and $X(0)$ are independent and that

$$\left|\frac{X(t) - (a - mt)}{t} - Y\right| \to 0$$

a.e, as $\delta \to 0$; now the approximation of sets A_δ^m depends on the values of Y)

$$p_m(y|a) =$$

$$\lim_{\delta \to 0} \frac{\frac{1}{\delta}\int_{a-(y-m)\delta}^{a} p(y) \cdot f(x)dx}{\frac{1}{\delta}\int_m^\infty p(y)dy \int_{a-(y-m)\delta}^{a} f(x)dx + \frac{1}{\delta}\int_{-\infty}^{m} p(y)dy \int_a^{a-(y-m)\delta} f(x)dx}, \tag{29}$$

where $f(\cdot)$ is the density of $X(0)$ as in (26) and $p(\cdot)$ is the density of Y given by (27). Here since $Y > m$, the approximation obtains only when $a - (y-m)\delta \leq X(0) \leq a$. To simplify (29), let us find the limits of the numerator $(= N_\delta)$ and denominator $(= D_\delta)$ separately:

$$\begin{aligned}\lim_{\delta \to 0}(N_\delta) &= \lim_{\delta \to 0}(y - m)p(y)\frac{1}{(y-m)\delta}\int_{a-(y-m)\delta}^{a} f(x)dx \\ &= (y - m)p(y)(-f(a)) \quad \text{[by Lebesgue's differentiation} \\ &\qquad \text{theorem because } f(\cdot) \text{ is continuous]}.\end{aligned}$$

Similarly,

$$\begin{aligned}\lim_{\delta \to 0}(D_\delta) &= \lim_{\delta \to 0}\left[\int_m^\infty (y-m)p(y)dy[(y-m)\delta]^{-1}\int_{a-(y-m)\delta}^{a} f(x)dx\right. \\ &\quad \left. + \int_{-\infty}^{m}(y-m)p(y)dy[(y-m)\delta]^{-1}\int_a^{a-(y-m)\delta} f(x)dx\right] \\ &= \int_m^\infty (y-m)p(y)dy(-f(a)) + \int_{-\infty}^{m}(y-m)p(y)dy \cdot f(a) \\ &= -f(a)\left[\int_m^\infty (y-m)p(y)dy + \int_m^\infty (y+m)p(y)dy\right. \\ &\quad \left. - \int_{-m}^{m}(y-m)p(y)dy\right] \quad [\text{ since } p(y) = p(-y)]\end{aligned}$$

$$= -f(a)\left[2\int_m^\infty yp(y)dy - \int_{-m}^m yp(y)dy + m\int_{-m}^m p(y)dy\right]$$

$$= -f(a)\left[\frac{1}{\sqrt{2\pi\alpha^2}}\int_m^\infty y\cdot e^{-y^2/2\alpha^2}dy - 0\right.$$

$$\left. + m\int_{-m}^m p(y)dy\right] \quad [\text{ again using } p(y) = p(-y)]$$

$$= -f(a)\left[\sqrt{\frac{2\alpha^2}{\pi}}e^{-m^2/2\alpha^2} + m\int_{-m}^m p(y)dy\right].$$

Using similar calculations for $Y < m$, and combining it together with the above for (29), one finds

$$p_m(y|a) = \frac{|y-m|e^{-y^2/2\alpha^2}}{2\alpha^2 e^{-m^2/2\alpha^2} + m\int_{-m}^m e^{-y^2/2\alpha^2}dy}. \tag{30}$$

Giving m a different value each time, we get *uncountably many limits* for the conditional density of Y given $X(0) = a$ here. Similarly with other types of approximations, still other $p(\cdot|a)$ can be obtained. Which one should be taken as the correct value? It is seen that not all these $\{A_\delta^m, \delta > 0\}$ qualify to be the differentiation bases. (If $m \to 0$, then one may choose a net out of this to be such a basis and the result will be independent of m.) It is thus *necessary that one verify first whether the approximating family is indeed a differentiation basis, and then calculate the conditional distribution.* Without this, one may end up with "spurious" densities, and the conditional distribution theory does lead to ambiguities. Thus, the result *depends* on the basis. Unfortunately, there is no known method to verify this fact.

The preceding discussion, examples, and comments do not complete the sketch of the subject unless the following two natural questions are also considered in addition to (A) and (B) treated above:

(C) If we obtain a conditional probability distribution, using a (suitable) differentiation basis, does it satisfy the functional equation (9) or (16)?

(D) Can we find general conditions on the probability space (Ω, Σ, P), on which the r.v. X is defined, so that the family of regular conditional distributions $\{Q_X(\cdot,\omega), \omega \in \Omega\}$ of (9) is unique outside of a P-null set (with the differentiation procedure noted above)?

The affirmative answer to (C) has already been indicated in the preceding discussion. This is a consequence of the general differentiation theory that can be found, for instance, in the references of Bruckner (1971) or Hayes and Pauc (1970). There is also a positive solution to (D), but it lies somewhat deeper. This is related to the theory of "disintegration of measures." Here topology plays a key role. Thus if Ω is a locally compact space, Σ is its Borel σ-algebra, and P is a regular probability on Σ, and if $(\tilde{\Omega}, \tilde{\Sigma})$ is another such (locally compact) Borelian measurable space (for us $\tilde{\Omega} = \mathbb{R}^n, \tilde{\Sigma} = \mathcal{R}$) and $X : \Omega \to \tilde{\Omega}$ is *P-proper* [*i.e.*, X is measurable for $(\tilde{\Sigma}', \Sigma)$, where $\tilde{\Sigma}'$ is the P-completion of $\tilde{\Sigma}$, and $f(X)$ is P-integrable for each continuous $f : \Omega \to \mathbb{R}$ with compact support], then any pair $Q_X^i(\cdot, \tilde{\omega}), \tilde{\omega} \in \tilde{\Omega}, i = 1, 2,$ of regular conditional probabilities (9) (which always exist) can be obtained through the differentiation process. Moreover, they are equal outside a P-null set if for each $\tilde{\omega}, Q_X^1(\cdot, \tilde{\omega})$ and $Q_X^2(\cdot, \tilde{\omega})$ have their supports contained in $X^{-1}(\{\tilde{\omega}\})$ [satisfy (9)] and $\rho(Q_X^1(A, \cdot)) = \rho(Q_X^2(A, \cdot)), A \in \Sigma$. Here $\rho : L^\infty(\tilde{\Sigma}, P \circ X^{-1}) \to L^\infty(\tilde{\Sigma}, P \circ X^{-1})$ is a "lifting map" uniquely determined by (and determining) the differentiation basis. In other words, Proposition 9 yields this ρ. In this theory for certain "nice" bases, ρ has stronger properties, called a "strong lifting," meaning that if $\tilde{A} \in \tilde{\Sigma}$ is open, then $\rho(\chi_{\tilde{A}}) \geq \chi_{\tilde{A}}$ must hold, and then each $Q_X^i(\cdot, \tilde{\omega})$ automatically has its support in $X^{-1}(\{\tilde{\omega}\})$. This is a deeper result, and is proved in the book by Ionescu Tulcea (1969, Chapter IX, Section 5, Theorem 5), and the relevant result is also given in the first author's book (1979, Chapter III, Section 6, Theorem 2). We therefore omit further discussion here, except to note that this aspect of the theory is quite delicate. Note that even these "nice" versions depend on ρ, and hence on a differentiation basis. A lifting is seldom unique, depending upon the axiom of choice.

Remarks 11 Many special attempts have been made in the literature to make this work "easy" through axiomatic means or other assumptions. One of the extensive developments is the new axiomatic theory of probability, started by Rényi in the middle 1950s and detailed in his book (1970). The basic idea is to take the concept of conditional probability as a mapping $P : \Sigma \times \mathcal{B} \to [0,1]$ such that $P(\cdot, B)$ is a probability on the σ-algebra Σ for each $B \in \mathcal{B}$, a collection of measurable sets for which $\emptyset \notin \mathcal{B}$, and it satisfies two other conditions in the second variable. Then it is shown to be of the form $\mu(A \cap B)/\mu(B), A \in \Sigma, B \in \mathcal{B}$, for a ($\sigma$-finite) measure $\mu : \Sigma \to \overline{\mathbb{R}}^+$, and the theory develops quickly without the difficulties noted above. However, problems of the type discussed in (26)–(30) do not fit in this theory, and thus it is not general enough to substitute for Kolmogorov's model. The second approach was advanced by Tjur (1974). This is based on classical methods for $P(A|Y = y) = \lim_\alpha P(A \cap B_\alpha)/P(B_\alpha)$, where the net $\{B_\alpha, \alpha \in I\}$ converges to $[Y = y]$ in the Moore-Smith sense. By assuming that the probability space (or the range space of random variables) is nice enough, a general approach to calculating this limit using the methods of differential geometry is developed. However, here also the functions $P(\cdot|y)$ *depend on the limit procedure used*, and it is not possible to decide on the correct function from the family of such functions. While these attempts may have individual merits, they do not include some simple and natural applications. Consequently, we stay with the

Kolmogorov model. The abstract definitions by means of the functional equations [cf. (1) and (5)] appear to be the most general ones available. Therefore, the rest of the work in this book is based mainly on the Kolmogorov axioms of Chapter 1. *A possible way to reconcile these difficulties is to consider the constructive analysis, as reformulated by E. Bishop (1967). As of now there is no constructive method to calculate Radon-Nikodým derivatives (hence conditional expectations). So this is a question to be left for the future.* For a more detailed analysis and discussion, one may refer to a recent account in the first author's book (Rao, (2005)).

As a final item of this section, let us introduce an extension of the previous chapter's independence concept for conditional probability measures without invoking regularity. This leads to some important applications.

Definition 12 Let (Ω, Σ, P) be a probability space and $\{\mathcal{B}, \mathcal{B}_\alpha, \alpha \in I\}$ be a family of σ-subalgebras of Σ. The $\mathcal{B}_\alpha$, $\alpha \in I$, are said to be *conditionally independent given* $\mathcal{B}$ when I has cardinality at least 2, if for each (distinct) finite set $\alpha_1, \ldots, \alpha_n$ of I, and any $A_{\alpha_i} \in \mathcal{B}_{\alpha_i}, i = 1, \ldots, n$, we have the system of equations

$$P^{\mathcal{B}}\left(\bigcap_{i=1}^{n} A_{\alpha_i}\right) = \prod_{i=1}^{n} P^{\mathcal{B}}(A_{\alpha_i}) \quad \text{a.e. }, n \geq 2. \tag{31}$$

Similarly, a family $\{X_\alpha, \alpha \in I\}$ of random variables is (mutually) *conditionally independent given* $\mathcal{B}$, if the σ-algebras $\mathcal{B}_\alpha = \sigma(X_\alpha)$ have that property.

It is clear that this reduces to (unconditional or mutual) independence if $\mathcal{B} = \{\emptyset, \Omega\}$, since then $P^{\mathcal{B}}$ becomes P. Consequently it is reasonable to extend many of the results of the last chapter, and those based on them, with the conditional probability function. However, some new and unfamiliar forms of the "expected " results appear. We briefly illustrate this point here, and note some alternative (operational) forms of conditional independence. In what follows, $P^{\mathcal{B}}(\cdot)$ is also denoted as $P(\cdot|\mathcal{B})$ for notational convenience.

Proposition 13 *Let* $\mathcal{B}, \mathcal{B}_1, \mathcal{B}_2$ *be* σ*-subalgebras from* (Ω, Σ, P). *Then the following are equivalent:*

(i) $\mathcal{B}_1, \mathcal{B}_2$ *are conditionally independent given* $\mathcal{B}$.

(ii) $P(B_1|\sigma(\mathcal{B} \cup \mathcal{B}_2)) = P(B_1|\mathcal{B})$ *a.e.*, $B_1 \in \mathcal{B}_1$.

(iii) $P(B_2|\sigma(\mathcal{B} \cup \mathcal{B}_1)) = P(B_2|\mathcal{B})$ *a.e.*, $B_2 \in \mathcal{B}_2$.

(iv) *For all* $X : \Omega \to \mathbb{R}^+, \mathcal{B}_1$*-measurable*, $E(X|\sigma(\mathcal{B} \cup \mathcal{B}_2)) = E(X|\mathcal{B})$ *a.e.*

Proof (i) $\Rightarrow$ (ii) If $B \in \mathcal{B}$ and $B_2 \in \mathcal{B}_2$, then $B \cap B_2 \in \sigma(\mathcal{B} \cup \mathcal{B}_2)$ and is a generator of the latter σ-algebra. So it suffices to prove the equation of (ii) when integrated on such a generator, since both sides are measurable relative to $\sigma(\mathcal{B} \cup \mathcal{B}_2)$. Thus

$$\int_{B\cap B_2} P(B_1|\sigma(\mathcal{B}\cap\mathcal{B}_2))dP = \int_{B\cap B_2} E(\chi_{B_1}|\sigma(\mathcal{B}\cup\mathcal{B}_2))dP$$

$$= \int_{B\cap B_2} \chi_{B_1} dP = \int_B \chi_{B_1\cap B_2} dP$$

$$= \int_B E^{\mathcal{B}}(\chi_{B_1\cap B_2})dP_{\mathcal{B}} = \int_B P^{\mathcal{B}}(B_1\cap B_2)dP_{\mathcal{B}}$$

$$= \int_B P^{\mathcal{B}}(B_1)\cdot P^{\mathcal{B}}(B_2)dP_{\mathcal{B}} \ \text{[by (i)]}$$

$$= \int_B E^{\mathcal{B}}(\chi_{B_2}P^{\mathcal{B}}(B_1))dP_{\mathcal{B}}$$

(by the averaging property of $E^{\mathcal{B}}$)

$$= \int_B \chi_{B_2}P^{\mathcal{B}}(B_1)dP = \int_{B\cap B_2} P^{\mathcal{B}}(B_1)dP.$$

Hence the extreme integrands agree a.e.[P] on $\sigma(\mathcal{B}\cup\mathcal{B}_2)$ which is (ii), because all such sets generate the latter σ-algebra (or use Proposition 1.2.8).

(ii) $\Rightarrow$ (i) Since $\mathcal{B}\subset\sigma(\mathcal{B}\cup\mathcal{B}_2)$, and hence $E^{\mathcal{B}} = E^{\mathcal{B}}E^{\sigma(\mathcal{B}\cup\mathcal{B}_2)}$, (ii) yields for a.a.$(\omega)$,

$$P^{\mathcal{B}}(B_1\cap B_2) = E^{\mathcal{B}}(E^{\sigma(\mathcal{B}\cup\mathcal{B}_2)}(\chi_{B_1}\cdot\chi_{B_2}))$$

$$= E^{\mathcal{B}}(\chi_{B_2}\cdot E^{\sigma(\mathcal{B}\cup\mathcal{B}_2)}(\chi_{B_1}))\ [\text{ since } \mathcal{B}_2\subset\sigma(\mathcal{B}\cup\mathcal{B}_2)]$$

$$= E^{\mathcal{B}}(\chi_{B_2}P(B_1|\sigma(\mathcal{B}\cup\mathcal{B}_2))$$

$$= E^{\mathcal{B}}(\chi_{B_2}\cdot P(B_1|\mathcal{B}))\quad \text{(by hypothesis)}$$

$$= P^{\mathcal{B}}(B_2)\cdot P^{\mathcal{B}}(B_1)\quad \text{a.e.,}$$

since $P^{\mathcal{B}}(B_1)(= P(B_1|\mathcal{B}))$ is $\mathcal{B}$-measurable. Thus (i) holds.

By interchanging the subscripts 1 and 2 in the above proof, (i) $\Leftrightarrow$ (iii) follows. Finally (iv) $\Rightarrow$ (ii) trivially, and then (ii) $\Rightarrow$ (iv) for all step functions by the linearity of conditional expectation operators. Since each $\mathcal{B}_1$-measurable $X\geq 0$ is a limit of an increasing sequence of $\mathcal{B}_1$-simple functions, and the monotone convergence criterion holds for these operators, (iv) holds for all such X. This completes the proof.

Using the (π,λ)-class theorem (cf. Proposition 1.2.8) just as in the unconditional case, we deduce the following result from the above.

Corollary 14 *Let $X_1,\ldots,X_n$ be random variables on (Ω,Σ,P) and $\mathcal{B}\subset\Sigma$ be a σ-algebra. Then the set $\{X_1,\ldots,X_n\}$ is (mutually) conditionally* *independent iff for all* $1 < m \leq n$,

$$P^{\mathcal{B}}\left(\bigcap_{i=1}^{m}[X_i < x_i]\right) = \prod_{i=1}^{m} P^{\mathcal{B}}([X_i < x_i]) \text{ a.e., } x \in \mathbb{R}. \tag{32}$$

Another consequence of the concept is given by

Corollary 15 *Let* $\mathcal{B}_1, \mathcal{B}_2$ *be independent* σ*-algebras from* (Ω, Σ, P). *If* $\mathcal{B}$ *is a* σ*-subalgebra of* $\mathcal{B}_1$ *(or* $\mathcal{B}_2$*), then* $\mathcal{B}_1$ *and* $\mathcal{B}_2$ *are conditionally independent given* $\mathcal{B}$*, so that*

$$P^{\mathcal{B}}(A \cap B) = P^{\mathcal{B}}(A)P^{\mathcal{B}}(B)\ [= P^{\mathcal{B}}(A) \cdot P(B)] \quad \text{a.e.}, A \in \mathcal{B}_1, B \in \mathcal{B}_2, \mathcal{B} \subset \mathcal{B}_1. \tag{33}$$

Proof For any $C \in \mathcal{B}$ we have, with $\mathcal{B} \subset \mathcal{B}_1$,

$$\int_C P^{\mathcal{B}}(A \cap B)dP_{\mathcal{B}} = P(A \cap B \cap C) = P(A \cap C)P(B)$$

$$\text{(since } \mathcal{B}_1, \mathcal{B}_2 \text{ are independent and } A \cap C \in \mathcal{B}_1\text{)}$$

$$= P(B)E(\chi_{A\cap C}) = \int_C \chi_A \cdot P(B)dP$$

$$= \int_C E^{\mathcal{B}}(\chi_A) \cdot P(B)dP_{\mathcal{B}} = \int_C P^{\mathcal{B}}(B)P^{\mathcal{B}}(A)dP_{\mathcal{B}}$$

[because $P^{\mathcal{B}}(B)$ is $P(B)$ by the independence of B and $\mathcal{B}$].

Hence $P^{\mathcal{B}}(A \cap B) = P^{\mathcal{B}}(A)P^{\mathcal{B}}(B)$ a.e. Now the case that $\mathcal{B} \subset \mathcal{B}_2$ being similar, the result holds as stated.

Remark: The following consequence is worthy of special mention: Let X, Y, Z be random variables on (Ω, Σ, P) with $\mathcal{F}_1 = \sigma(X)$, $\mathcal{F}_2 = \sigma(Y)$ and $\mathcal{F}_3 = \sigma(Z)$. If X is independent of Y and Z, and Y is integrable, then $E(Y|X, Z) = E(Y|Z)$ a.e., i.e., $E^{\sigma(X,Z)}(Y) = E^{\sigma(Z)}(Y)$ a.e. so that X and Y are conditionally independent given Z. This follows from from the corollary by taking $\mathcal{B} = \mathcal{F}_3$, $\mathcal{B}_1 = \mathcal{F}_1$ and $\mathcal{B}_2 = \mathcal{F}_2$ [and similarly X and Z are conditionally independent given Y]. For this alternative (distributional) relation, it is not necessary to demand integrability of X, Y and Z. For the relation with conditional expectation, the integrability of Y or Z is needed.

The Kolmogorov zero-one law takes the following form:

Proposition 16 *If* $\{X_n, n \geq 1\}$ *is a sequence of random variables on* (Ω, Σ, P) *that is conditionally independent relative to a* σ*-algebra* $\mathcal{B} \subset \Sigma$*, and if* $\mathcal{T} = \bigcap_{k=1}^{\infty} \sigma(X_n, n \geq k)$ *is its tail* σ*-algebra, then* $\mathcal{B}$ *and* $\mathcal{T}$ *are equivalent,* *in the sense that for each* $A \in \mathcal{T}$ *there is a* $B \in \mathcal{B}$ *such that* $A = B$ *a.e.*[P].

Proof The argument is analogous to the earlier result. Indeed, by the preceding corollary we have, for each n, the σ-algebras $\sigma(X_1, \ldots, X_n)$ and $\sigma(X_k, k \geq n+1)$ are conditionally independent given $\mathcal{B}$. Hence $\sigma(X_k, \ldots, X_n)$ and $\mathcal{T}$ have the same property for each n. We deduce that $\sigma(X_k, k \geq 1)$ and $\mathcal{T}$ are conditionally independent given $\mathcal{B}$. Since $\mathcal{T} \subset \sigma(X_k, k \geq 1)$, it follows that $\mathcal{T}$ is conditionally independent of itself given $\mathcal{B}$. Thus $A \in \mathcal{T} \Rightarrow P^{\mathcal{B}}(A \cap A) = (P^{\mathcal{B}}(A))^2$, so that $P^{\mathcal{B}}(A) = 0$ or 1 a.e., and it is a $\mathcal{B}$-measurable indicator. So for some $B \in \mathcal{B}, P^{\mathcal{B}}(A) = \chi_B$ a.e. Since then both are Radon-Nikodým derivatives of P relative to $P^{\mathcal{B}}$, it follows that $A = B$ a.e. [P], completing the proof.

Since in the unconditional case $\mathcal{B} = \{\emptyset, \Omega\}$, this says that if $\{X_n, n \geq 1\}$ are mutually independent, $\mathcal{T}$ and $\{\emptyset, \Omega\}$ are equivalent, which is Kolmogorov's zero-one law. Another interesting observation is that if the $X_n, n \geq 1$, are independent with the same distribution and if $\mathcal{P}$ is the σ-algebra of permutable events relative to $\{X_n, n \geq 1\}$, by definition of permutability, we can conclude that $\{X_n, n \geq 1\}$ will still be conditionally independent given $\mathcal{P}$, since in Definition 11 only finitely many X_n will appear each time. Consequently by the above proposition $\mathcal{P}$ and $\mathcal{T}$ are equivalent. Since each event of $\mathcal{T}$ has probability zero or one, so must each event of $\mathcal{P}$. Thus each permutable event determined by the independent X_n with the same distribution has probability zero or one which is the Hewitt-Savage law (cf. Theorem 2.1.12).

3.3 Markov Dependence

Using the concepts developed in the preceding two sections, we can consider various classes of dependent random families. Thus in this and the next sections we introduce two fundamental classes of such dependences, namely, Markovian and martingale families. The first one, to be discussed here, was introduced by A. A. Markov in 1906 for the case that the range of each random variable is a finite set, so that no difficulties with conditional probabilities arise. The general case emerged later with the studies of Kolmogorov, Doeblin, P. Lévy, Doob, Feller, Hunt, and others. It is one of the most active areas of probability theory. The second area, martingales, of equal importance and activity, will be considered in Section 5, after some existence theory in Section 4.

The concept of Markovian dependence is an extension of that of independence given in Chapter 2, and so we introduce it here by following Definition 2.1.1, and then present equivalent versions for computational convenience.

Definition 1 Let (Ω, Σ, P) be a probability space and I be an ordered set. If $\{\mathcal{B}_\alpha, \alpha \in I\}$ is a net of σ-subalgebras of Σ, consider the "past" and "future" σ-algebras $\mathcal{G}_\alpha = \sigma(\mathcal{B}_\gamma, \gamma \leq \alpha)$ and $\mathcal{G}^\alpha = \sigma(\mathcal{B}_{\gamma'}, \gamma' \geq \alpha)$. Then the net is said to be *Markovian* if for each $\alpha \in I$ (i.e., "present"), the σ-algebras $\mathcal{G}_\alpha$ and $\mathcal{G}^\alpha$ are conditionally independent given $\mathcal{B}_\alpha$, so that ("$\leq$" is the ordering of I)

$$P^{\mathcal{B}_\alpha}(A \cap B) = P^{\mathcal{B}_\alpha}(A) \cdot P^{\mathcal{B}_\alpha}(B) \ \ a.e., A \in \mathcal{G}_\alpha, \ B \in \mathcal{G}^\alpha. \tag{1}$$

In particular, if $\{X_\alpha, \alpha \in I\}$ is a set of random variables on Ω, then it is called Markovian if the σ-algebras $\{\sigma(X_\alpha), \alpha \in I\}$ form a Markovian family in the above sense. If $I \subset \mathbb{R}$ $(\mathbb{R}^n)$, then $\{X_\alpha, \alpha \in I\}$ is called a *Markov process* (*Markov random field*).

Using the result of Proposition 2.12 and Corollary 2.13, it is possible to present several equivalent forms of the above definition. We do this below. Note that (1) can be stated *informally* as follows: The σ-algebras $\{\mathcal{B}_\alpha, \alpha \in I\}$, or the random variables $\{X_\alpha, \alpha \in I\}$, form a Markovian family if the past and future are conditionally independent given the present. Since $P^{\mathcal{B}_\alpha}$ is not necessarily a genuine probability measure (it is generally a vector measure), as discussed at length earlier in Section 2, the above statement is only informal. For a finer analysis (using the classical Lebesgue integration theory) we need to *assume* that the $P^{\mathcal{B}_\alpha}$ are regular. However, several preliminary considerations can be presented without such restrictions, and we proceed to discuss them now. Again set $P^{\mathcal{B}}(\cdot) = P(\cdot|\mathcal{B})$ whenever it is convenient.

Proposition 2 *Let $\{X_t, t \in T\}$ be a family of random variables on (Ω, Σ, P), with $T \subset \mathbb{R}$. Let $\mathcal{B}_1 = \sigma(X_t) \subset \Sigma, t \in T$. Then the following statements are equivalent:*

(i) *The family is Markovian.*

(ii) *For each $t_1 < t_2 < \ldots < t_{n+1}, t_j \in T, n \geq 1$, if $\tau_n = \{t_1, \ldots, t_n\}, \mathcal{G}_{\tau_n} = \sigma(X_{t_1}, \ldots, X_{t_n})$, then $P(A|\mathcal{G}_{t_n}) = P(A|\mathcal{B}_{t_n})$* a.e., and $A \in \mathcal{B}_{t_{n+1}}$.

(iii) *If $t_1 < t_2, \mathcal{G}_{t_1} = \sigma(X_t, t \leq t_1, t \in T), t_i \in T, i = 1, 2$, then $P(A|\mathcal{G}_{t_1}) = P(A|\mathcal{B}_{t_1})$* a.e., and $A \in \mathcal{B}_{t_2}$.

(iv) *If $\mathcal{G}^t = \sigma(X_s, s \geq t, s \in T), t \in T$, and $\mathcal{G}_t$ is as in* (iii), *then $P(A|\mathcal{G}_{t_1}) = P(A|\mathcal{B}_{t_1})$* a.e., and $A \in \mathcal{G}^{t_1}$.

(v) *If $\mathcal{G}_t, \mathcal{G}^t$ are as above and Z is any bounded $\mathcal{G}^t$-measurable function, then $E^{\mathcal{G}_{t_1}}(Z) = E^{\mathcal{B}_{t_1}}(Z)$* a.e.

Note Observe that since the definition of a Markov family is symmetrical in that, intuitively, the "past" and "future" are conditionally independent given the "present," the above proposition gives the following alternative interpretation, namely; the family is Markovian implies that, for predicting the future behavior of the process, given the whole past and present the behavior depends only on the present, and conversely. This is the natural interpretation of (iii) $\Leftrightarrow$ (i) $\Leftrightarrow$ (iv). These are obviously only one-sided statements. But

Definition 1, which is equivalent to these, on the other hand, shows that if $\{X_t, t \in T\}$ is Markovian, then $\{X_{-t}, t \in T\}$ is also Markovian if $T = \mathbb{R}$ or $T \subset \mathbb{R}$ is a symmetric set relative to the origin. Thus a Markovian family remains Markovian if the time direction (i.e., of T, the index set) is reversed. It is useful to have these alternative forms of the concept. Note also that if $T_1 \subset \mathbb{R}$ and $\alpha : T_1 \to T$ is an isotone mapping (so that it preserves order), then $\{X_{\alpha(i)}, i \in T_1\}$ is also Markovian if the family $\{X_t, t \in T\}$ is. In particular, each subfamily of a Markovian family is Markovian. Also, if $g_t : \mathbb{R} \to \mathbb{R}$ is a one-to-one and onto mapping such that $g_t^{-1}(\mathcal{R}) = \mathcal{R}$ the Borel σ-algebra of $\mathbb{R}$, then the process $\{Y_t = g_t(X_t), t \in \mathbb{R}\}$ is Markovian whenever $\{X_t, t \in \mathbb{R}\}$ is such, since $\sigma(Y_t) = Y_t^{-1}(\mathcal{R}) = X_t^{-1}(g_t^{-1}(\mathcal{R})) = \sigma(X_t)$. However, it is possible to introduce other definitions of reversibility. For instance, if the range space of $\{X_t, t \in T\}$ is a set of integers (or only positive integers), then one can ask whether the probability of taking values from i to j $(i \leq j)$ by the process is the same as that from j to i. This need not be true in general, and when true it is called a "symmetry" or "path reversibility." This special case will be designated as such.

Proof (i) $\Rightarrow$ (ii) Taking $I = T \subset \mathbb{R}$ in Definition 1, for each $t_n \in T$, $\mathcal{G}_{t_n}$ and $\mathcal{G}^{t_n}$ are conditionally independent, given $\mathcal{B}_{t_n}$. Since $\mathcal{B}_{t_{n+1}} \subset \mathcal{G}^{t_n}$, it follows that $\mathcal{G}_{t_n}$ and $\mathcal{B}_{t_{n+1}}$ are also conditionally independent, given $\mathcal{B}_{t_n}$. Then by Proposition 2.12iii

$$P(A|\mathcal{G}_{t_n}) = P(A|\mathcal{B}_{t_n}) \text{ a.e.,} \quad A \in \mathcal{B}_{t_{n+1}}. \tag{2}$$

Consider $\tau_n = \{t_1, \ldots t_n\}$. Clearly $\mathcal{G}_{\tau_n} \subset \mathcal{G}_{t_n}$ and $\mathcal{B}_{t_n} \subset \mathcal{G}_{\tau_n}$. Hence applying the operator $E^{\mathcal{G}_{\tau_n}}$ to both sides of (2) and noting that $E^{\mathcal{B}_{\tau_n}} \prec E^{\mathcal{G}_{\tau_n}}$, i.e., $E^{\mathcal{G}_{\tau_n}} E^{\mathcal{G}_{t_n}} = E^{\mathcal{G}_{\tau_n}}$, we get [since the right side of (2) is $\mathcal{B}_{t_n}$-measurable]

$$\begin{aligned} P(A|\mathcal{G}_{\tau_n}) &= E^{\mathcal{G}_{\tau_n}}(\chi_A) = E^{\mathcal{G}_{\tau_n}}(E^{\mathcal{G}_{t_n}}(\chi_A)) \\ &= E^{\mathcal{G}_{\tau_n}}(P(A|\mathcal{B}_{t_n})) = P(A|\mathcal{B}_{t_n}) \text{ a.e.,} \ A \in \mathcal{B}_{t_{n+1}} \quad \text{[by (2)]}. \end{aligned}$$

Hence (ii) follows.

(ii) $\Rightarrow$ (iii) First note that if $\mathcal{F}$ is the collection of all finite subsets of T_1 (i.e., $\alpha \in \mathcal{F}$ iff $\alpha = \{u_1, \ldots, u_n\} \subset T_1$ for some $1 \leq n < \infty$), where $T_1 = \{t \in T, t \leq t_1\}$, then $\mathcal{G}_{t_1} = \sigma(\cup_{\alpha \in \mathcal{F}} \mathcal{G}_\alpha)$, with $\mathcal{G}_\alpha = \sigma(X_t, t \in \alpha)$. Indeed, since each $\mathcal{G}_\alpha \subset \mathcal{G}_{t_1}$, $\alpha \in \mathcal{F}$, $\mathcal{G}_{t_1}$ contains the right-side σ-algebra. But $\mathcal{G}_{t_1} = \sigma(\cup_{t \in T_1} \mathcal{B}_t)$, and each $\mathcal{B}_t \subset \mathcal{G}_\alpha$ for some $\alpha \in \mathcal{F}$. Hence $\mathcal{G}_{t_1}$ is contained in the right side. Now (ii) implies, by Proposition 2.12iii again, that $\mathcal{G}_\alpha$ and $\mathcal{B}_{t_2}$ are conditionally independent given $\mathcal{B}_{t_1}$ for each $\alpha \in \mathcal{F}$. The argument proceeds as in the case of independent events (see the proof of Theorem 2.1.3). Thus to use Proposition 2.12ii, we introduce two families of sets $\mathcal{D}$ and $\mathcal{C}$ as follows. Let $\mathcal{D}$ be the class of all finite intersections of events each belonging to a $\mathcal{G}_\alpha, \alpha \in \mathcal{F}$. Then it results in $\mathcal{G}_\alpha \subset \mathcal{D}$, $\alpha \in \mathcal{F}$. But this clearly implies $\mathcal{G}_{t_1} \subset \sigma(\mathcal{D})$. It suffices to show, for (iii), that $\mathcal{B}_{t_2}$ and $\sigma(\mathcal{D})$ are conditionally independent given $\mathcal{B}_{t_1}$. For this consider the class

$$\mathcal{C} = \{B \in \Sigma : P(B \cap A|\mathcal{B}_{t_1}) = P(B|\mathcal{B}_{t_1}) \cdot P(A|\mathcal{B}_{t_1}) \text{ a.e., all } A \in \mathcal{B}_{t_2}\}.$$

Evidently $\mathcal{D} \subset \mathcal{C}$ (since $\mathcal{D}$ and $\mathcal{B}_{t_2}$ are conditionally independent). It is easy to verify that [$\mathcal{D}$ is a π-class and that] $\mathcal{C}$ is a λ-class, as in the independent case. Hence $\mathcal{C} \subset \sigma(\mathcal{D})$. Since $\mathcal{C}$ and $\mathcal{B}_{t_2}$ are conditionally independent given $\mathcal{B}_{t_1}$, the same is true of $\sigma(\mathcal{D})$, and hence of $\mathcal{G}_{t_1}$ and $\mathcal{B}_{t_2}$ given $\mathcal{B}_{t_1}$. This shows that (iii) is true. The remaining implications can now be deduced quickly.

(iii) $\Rightarrow$ (iv) Writing t for t_1, in (iii), we get that $\mathcal{G}_t$ and $\mathcal{B}_{t_2}$ are conditionally independent given $\mathcal{B}_t$ for any $t_2 \in T$, $t_2 > t$. From this, by Proposition 2.12ii, we deduce that $\mathcal{G}_t$ and $\sigma(X_{t_i}, i = 1, \ldots, n, t_i > t, t_i \in T)$ are conditionally independent given $\mathcal{B}_t$. Then, by the argument of the preceding paragraph, $\mathcal{G}_t$ and $\mathcal{G}^t$ are conditionally independent given $\mathcal{B}_t$, which is (iv).

(iv) $\Rightarrow$ (i) By Proposition 2.12iii, $\mathcal{G}_t$ and $\mathcal{G}^t$ are conditionally independent given $\mathcal{B}_t$. Thus (i) holds by Definition 1.

(iv) $\Rightarrow$ (v) This is true if $Z = \chi_A, A \in \mathcal{G}^t$. By linearity, the result holds if $Z = \sum_{i=1}^n a_i \chi_{A_i}$, $A_i \in \mathcal{G}^t$. By the conditional monotone convergence criterion, it is true for any $\mathcal{G}^t$-measurable $Z \geq 0$. This implies (v) for general bounded Z, since then $Z = Z^+ - Z^-$, $Z^\pm$ being $\mathcal{G}^t$-measurable. That (v) $\Rightarrow$ (iv) is trivial. This completes the proof of the proposition.

If $\{X_t, t \in T\}$ is a family of independent random variables on (Ω, Σ, P) and $T \subset \mathbb{R}$, then it is evident that this forms a Markovian class. An equally simple example is that any monotone class $\{\mathcal{B}_t, t \in T\}, T \subset \mathbb{R}$, of σ-algebras from (Ω, Σ, P) is Markovian. In fact, if $\mathcal{B}_t \subset \mathcal{B}_{t'}$, for $t < t'$ and $\mathcal{G}_t, \mathcal{G}^t$ are the σ-algebras generated by $\{\mathcal{B}_s, s \in T, s \leq t\}$ and $\{\mathcal{B}_\alpha, \alpha \in T, \alpha \geq t\}$, then $\mathcal{G}_t \subset \mathcal{B}_t \subset \mathcal{G}^t$, so that for any $A \in \mathcal{G}_t$, $B \in \mathcal{G}^t$, we have

$$\begin{aligned} P^{\mathcal{B}_t}(A \cap B) &= E^{\mathcal{B}_t}(\chi_A \chi_B) = \chi_A E^{\mathcal{B}_t}(\chi_B) \text{ a.e.} \\ &= E^{\mathcal{B}_t}(\chi_A) \cdot E^{\mathcal{B}_t}(\chi_B) = P^{\mathcal{B}_t}(A) \cdot P^{\mathcal{B}_t}(B) \text{ a.e.} \end{aligned}$$

The decreasing case is similar. A less simple result is the following:

Example 3 (a) Let $\{X_n, n \geq 1\}$ be a sequence of independent random variables on (Ω, Σ, P) and $Y_n = \sum_{k=1}^n X_k$. Then $\{Y_n, n \geq 1\}$ is Markovian.

To verify this, let $\mathcal{B}_n = \sigma(Y_n)$ and $\mathcal{A}_n = \sigma(X_n)$. Then $\mathcal{A}_1 = \mathcal{B}_1$ and $\mathcal{B}_n \subset \sigma(\mathcal{B}_{n-1} \cup \mathcal{A}_n) \subset \sigma(\mathcal{A}_1 \cup \mathcal{A}_2 \ldots \cup \mathcal{A}_n)$. If

$$\mathcal{G}_n = \sigma(\mathcal{B}_1 \cup \ldots \cup \mathcal{B}_n),\ \mathcal{G}^n = \sigma\left(\bigcup_{k \geq n} \mathcal{B}_k\right),$$

then $\mathcal{G}^n = \sigma(\mathcal{B}_n \cup \bigcup_{k \geq n+1} \mathcal{A}_k)$. Also, $\mathcal{A}_k$ is independent of $\mathcal{B}_n$ for $k \geq n+1$. Moreover, if $\mathcal{D}$ is the class of events of the form $C_1 \cap C_2$, where $C_1 \in \mathcal{B}_n$ and

$C_2 \in \sigma(\bigcup_{k\geq n+1} \mathcal{A}_k)$, then $\mathcal{D}$ is a π-class generating $\mathcal{G}^n$. Hence it suffices to verify the truth of (1) for all $A \in \mathcal{G}_n$ and $B = C_1 \cap C_2 \in \mathcal{D}$ by the above proposition [see the proof of (ii) $\Leftrightarrow$ (iii)]. Thus for these A, B we have

$$\begin{aligned}
P^{\mathcal{B}_n}(A \cap B) &= E^{\mathcal{B}_n}(\chi_A \cdot \chi_{C_1} \cdot \chi_{C_2}) \\
&= E^{\mathcal{B}_n}(\chi_A \cdot \chi_{C_1} \cdot E(\chi_{C_2})) \\
&\quad \text{(since } C_2 \text{ is independent of } \mathcal{B}_n \text{ and also of } \mathcal{G}_n, \\
&\quad \text{Corollary 2.15 applies)} \\
&= E^{\mathcal{B}_n}(\chi_A) \cdot \chi_{C_1} \cdot E(\chi_{C_2}) \text{ (since } C_1 \in \mathcal{B}_n) \\
&= E^{\mathcal{B}_n}(\chi_A) \cdot E^{\mathcal{B}_n}(\chi_{C_1}) \cdot E^{\mathcal{B}_n}(\chi_{C_2}) \text{ a.e.,} \\
&= E^{\mathcal{B}_n}(\chi_A) \cdot E^{\mathcal{B}_n}(\chi_{C_1} \cdot \chi_{C_2}) \\
&\quad \left[\text{since } C_1 \in \mathcal{G}_n, C_2 \in \sigma\left(\bigcup_{k\geq n+1} \mathcal{A}_k\right), \mathcal{B}_n \subset \mathcal{G}_n, \right. \\
&\quad \left. \text{and Corollary 2.15 applies again} \right] \\
&= E^{\mathcal{B}_n}(\chi_A) \cdot E^{\mathcal{B}_n}(\chi_B) = P^{\mathcal{B}_n}(A) \cdot P^{\mathcal{B}_n}(B) \text{ a.e.}
\end{aligned}$$

This proves the Markovian property of $\{Y_n, n \geq 1\}$.

(b) The following consequence of the above illustration is useful in some applications (cf., e.g., in Problem 33 of Chapter 5 later). Thus if $X_1, \ldots, X_n$ are independent random variables, and $Y_n = \sum_{k=1}^n X_k$, as in the illustration, so that $\{Y_n, \mathcal{F}_n = \sigma(X_1, \ldots, X_n), n \geq 1\}$ is a Markov process, we assert that, with $n = 2$ for simplicity and letting $\mu_i = P \circ X_i^{-1}(\cdot), i = 1, 2, \mu = P \circ (X_1 + X_2)^{-1}(\cdot)$, the following obtains:

$$P[X_1 + X_2 \in B | X_1](\omega) = \mu_2(B - X_2)(\omega), \text{ a.e. } (= \mu_2^{\omega}(B - X_1(\omega)), \qquad (*)$$

for each Borel set $B \subset \mathbb{R}$.

Indeed, if $A \subset \mathbb{R}$ is any Borel set, then

$$\begin{aligned}
P[X_1 \in A, X_1 + X_2 \in B] &= \int\int_{[X_1 \in A] \cap [X_1 + X_2 \in B]} dP \\
&= \int\int_{[x_1 \in A] \cap [x_1 + x_2 \in B]} dF_{X_1, X_2}(x_1, x_2), \\
&\quad \text{[by Theorem 1.4.1 (ii),]}
\end{aligned}$$

$$= \int_{[x_1 \in A]} \int_{[x_1+x_2 \in B]} dF_{X_1}(x_1) dF_{X_2}(x_2),$$

[by independence of X_1, X_2,]

$$= \int_{[x_1 \in A]} dF_{X_1}(x_1) \int_{B-\{x_1\}} d\mu_2(x_2)$$

$$= \int_A d\mu_1(x_1)\, \mu_2(B - x_1)$$

$$= \int_{X_1^{-1}(A)} P(B - X_1(\omega))\, dP_{\mathcal{B}}, \qquad (+)$$

From (1) of Section 2.1, the left side $= \int_{X_1^{-1}(A)} P[X_1 + X_2 \in B | X_1]\, dP_{\mathcal{B}}$ where $\mathcal{B} = \sigma(X_1)$. It follows that (*) holds since $X^{-1}(A)$ is a generator of $\mathcal{B}$. This can be extended for $n \geq 2$ to obtain the following, which we leave to the reader (cf. also Chung (1974), P. 308):

$$P(Y_n \in B|Y_{n-1}) = P(Y_n \in B|Y_1, \ldots, Y_{n-1})[= P(Y_n \in B|X_1, \ldots, X_{n-1})]$$
$$= P \circ X_n^{-1}(B - S_{n-1}), \text{ a.e.} \qquad (++)$$

Recall that if $\mathcal{B} = \sigma(X)$ and Y is integrable, then $E^{\mathcal{B}}(Y)$ is also written $E(Y|X)$. If $Y = \chi_A$, then the latter is often denoted $P(A|X)$. With this notation, the Markovian property given by Proposition 2 can be stated as follows. The class $\{X_t, t \in T\}$, $T \subset \mathbb{R}$, is a Markov process iff any one of the following equivalent conditions holds: For any $s < t$ in T and Borel set $A \subset \mathbb{R}$,

(i) $$P([X_t \in A]|X_r, r \leq s) = P([X_t \in A]|X_s) \text{ a.e.} \qquad (3)$$

(ii) For any $t_1 < t_2 < \ldots < t_{n+1}$ in T and Borel set $A \subset \mathbb{R}$,

$$P([X_{t_{n+1}} \in A]|X_{t_1}, \ldots, X_{t_n}) = P([X_{t_{n+1}}, \in A]|X_{t_n}), n \geq 1, \text{ a.e.}; \qquad (4)$$

(iii) For any $s_1 < s_2 < \ldots < s_n < t < t_1 < \ldots < t_m$ in T and Borel sets $A_i \subset \mathbb{R}, B_j \subset \mathbb{R}$,

$$P\left(\bigcap_{i=1}^{n}[X_{s_i} \in A_i] \cap \bigcap_{j=1}^{m}[X_{t_j} \in B_j]|X_t\right)$$

$$= P\left(\bigcap_{i=1}^{n}[X_{s_i} \in A_i]|X_t\right) P\left(\bigcap_{j=1}^{m}[X_{t_j} \in B_j]|X_t\right) \text{ a.e.} \qquad (5)$$

This interesting form leads us to derive a fundamental property of Markov processes. We recall from Theorem 2.5 that for any $\mathcal{B}$, a σ-subalgebra of Σ of

(Ω, Σ, P), and a random variable $X : \Omega \to \mathbb{R}$, a regular conditional distribution for X given $\mathcal{B}$ always exists on $\mathbb{R}$. It is a version of the image measure of $P^{\mathcal{B}}$, i.e., of $P^{\mathcal{B}} \circ X^{-1}(\cdot)$. Thus we have

Proposition 4 *If* $\{X_t, t \in T \subset \mathbb{R}\}$ *is a Markov process on a probability space* (Ω, Σ, P) *and* $r < s < t$ *from* T, *then the following relation, called the* **Chapman- Kolmogorov equation**, *holds:*

$$P([X_t < \lambda])]|X_r) = E(P([X_t < \lambda]|X_s)|X_r) \text{ a.e.}, \lambda \in \mathbb{R}. \tag{6}$$

If a version of the (image or) regular conditional distribution of

$$P([X_t \in A]|X_s)(\omega)$$

is denoted $Q_{s,t}(A, X_s(\omega))$, *then (6) can be expressed as*

$$Q_{r,t}(A, X_r(\omega)) = \int_{\mathbb{R}} Q_{s,t}(A, y) Q_{r,s}(dy, X_r(\omega)), \quad \text{Borel } A \subset \mathbb{R}, \tag{7}$$

for almost all $\omega \in \Omega$, *the exceptional null set depending on* r, s, t *and* A.

[Often $Q_{s,t}(A, X_s(\omega))$ is written as $p(X_s(\omega), s; A, t)$ in (7) and interpreted as the probability of the motion of a particle ω starting at time s from the state $X_s(\omega)$ and moving into a position or state in the set A at time $t > s$.]

Proof. Consider the process $\{X_u, r \leq u \leq s < t, X_t : r, u, s, t \text{ in } T\}$. Then by (3)

$$P([X_t < \lambda]|X_u, r \leq u \leq s) = P([X_t < \lambda]|X_s) \text{ a.e.} \tag{8}$$

Hence applying the operator $E(\cdot|X_r)$ to both sides of (8), and noting that $\sigma(X_r) \subset \sigma(X_u, r \leq u \leq s)$, one gets, by Proposition 1.2,

$$P([X_t < \lambda]|X_r) = E(P([X_t < \lambda]|X_s)|X_r) a.e.,$$

which is (6). Since $P([X_t < \lambda]|X_s)$ is $\sigma(X_s)$-measurable and Theorem 1.4.1 holds for $P^{\mathcal{B}}$ because of Theorem 1.3 (i), we have

$$E(Y|X_r)(\omega) = \int_{\mathbb{R}} y Q_{r,s}(dy, X_r(\omega)) \quad \text{a.a. } (\omega)$$

for a $\sigma(X_s)$-adapted bounded Y, so that (7) follows from this and (6), completing the proof.

For convenience of terminology, one calls the range space of a Markov process (or any family of random variables) the *state space* of the process. If the latter is at most countable, then a Markov process is called a *Markov chain.* If the range is a finite set, we say that the Markov process is a *finite Markov chain.*

The preceding proposition implies that for every Markov process its family of conditional probability functions $\{p(\cdot, t; \cdot, s), s < t \text{ in } T\}$ must satisfy the Chapman-Kolmogorov equation (6) or (7). It would be surprising if a Markov process can be characterized by this property, in the sense that the only conditional probability functions satisfying (6) or (7) are those given by a Markov process. Unfortunately this is *not* true, as was first pointed out by P. Lévy already in 1949. The following simple example, due to W. Feller, illustrates this point.

Counterexample 5 We present a non-Markovian process whose conditional probabilities satisfy (7). Let the state space be $\{1, 2, 3\}$. Consider the probability space $\Omega = \{\omega_i, 1 \leq i \leq 9\}$. Here the points ω_i of $\mathbb{R}^3$ have integer coordinates and are specified by $\omega_i = (1, 2, 3)$, $\omega_2 = (1, 3, 2)$, $\omega_3 = (3, 1, 2)$, $\omega_4 = (3, 2, 1)$, $\omega_5 = (2, 1, 3)$, $\omega_6 = (2, 3, 1)$, $\omega_7 = (1, 1, 1)$, $\omega_8 = (2, 2, 2)$, $\omega_9 = (3, 3, 3)$. Let Σ be the power set of Ω and $P(\{\omega_i\}) = \frac{1}{9}, 1 \leq i \leq 9$. On this probability space, consider a "process" X_1, X_2, X_3 where $X_i(\omega) =$ ith coordinate of ω. Thus, for instance, $X_1(\omega_4) = 3$, $X_3(\omega_6) = 1$, etc. Then $X_i : \Omega \to \{1, 2, 3\}$, and

$$P[X_i = n] = \frac{1}{3}, \quad i = 1, 2, 3, \quad n = 1, 2, 3,$$

$$P[X_i = m, X_j = n] = \begin{cases} \frac{1}{9} & \text{if } i \neq j, 1 \leq i, j \leq 3, 1 \leq m, n \leq 3 \\ \frac{1}{3} & \text{if } i = j, m = n \\ 0 & \text{if } i = j, m \neq n. \end{cases}$$

For any (i, j) and (m, n) we have

$$\begin{aligned} Q_{i,j}(m, n) &= P([X_j = m] | X_i = n) \\ &= \frac{P[X_j = m, X_i = n]}{P[X_i = n]} \\ &= \begin{cases} \frac{1}{3} & \text{if } i \neq j \\ 1 & \text{if } i = j, m = n \\ 0 & \text{otherwise.} \end{cases} \end{aligned}$$

Also, $\{X_1, X_2, X_3\}$ are pairwise independent. They are *not* Markovian. To see the latter,

$$P([X_3 = 2] | X_1 = 1, X_2 = 1) = 0 \neq P([X_3 = 2] | X_2 = 1),$$

so that the "future" depends not only on the "present" but also on the "past." But for $1 \leq i < j \leq 3$, the $Q_{i,j}(\cdot, \cdot)$ satisfy (7), since

$$Q_{1,2}(m, n) = \sum_{\ell=1}^{3} Q_{1,1}(m, \ell) Q_{1,2}(\ell, n) = 1 \cdot \frac{1}{3} + 0 = \frac{1}{3},$$

$$Q_{1,3}(m,n) = \sum_{\ell=1}^{3} Q_{1,2}(m,\ell)Q_{2,3}(\ell,n) = \frac{1}{3}\cdot\frac{1}{3} + \frac{1}{3}\cdot\frac{1}{3} + \frac{1}{3}\cdot\frac{1}{3} = \frac{1}{3},$$

and similarly other combinations are verified. Note that (7) holds *identically* (not a.e.).

The preceding example can be extended to an infinite family of random variables with the same (non-Markovian) properties. Let us assume that the basic probability space (Ω, Σ, P) is rich enough to support the following structure. [Actually, it is possible to enlarge the space to suit the needs. For this, if $(\tilde{\Omega}, \tilde{\Sigma}, \tilde{P})$ is as in the example, then we let $(\Omega, \Sigma, P) = \bigotimes_{i\geq 1}(\tilde{\Omega}_i, \tilde{\Sigma}_i, \tilde{P}_i)$, $\tilde{\Omega}_i = \tilde{\Omega}$, etc., as a product space, and the correctness of this procedure is a simple consequence of Theorem 3 of the next section.] Let $\{X_n, n \geq 1\}$ be random variables such that X_1, X_2, X_3 are the ones defined above, and let each block of the following three r.v.s have the same distributions as these. Thus for any $m \geq 1$,

$$P[X_{3m+1} = i_1, X_{3m+2} = i_2, X_{3(m+1)} = i_3]$$

$$= P[X_1 = i_1, X_2 = i_2, X_3 = i_3],$$

where $1 \leq i_1, i_2, i_3 \leq 3$ are integers. It then follows immediately that the family $\{X_n, n \geq 1\}$ is non-Markovian, but $p_{i_1,i_2} = P[X_{k+1} = i_2 | X_k = i_1] = 1/3, k \geq 1$, and (7) holds. Here one defines X_n at $\omega \in \Omega$, $\omega = (\tilde{\omega}_1, \tilde{\omega}_2, \ldots), \tilde{\omega}_i \in \Omega_i'$, by the equation

$$X_{3m+j}(\omega) = X_j(\tilde{\omega}), j = 1, 2, 3;\ \tilde{\omega} \in \tilde{\Omega}.$$

This completes the description of the example.

Note It is of interest to observe that the commutativity properties of conditional expectation operators (cf. Proposition 1.2), with $\tilde{Q}_1(A) = P[X_1 \in A], A \subset \mathbb{R}$ Borel, $Q_n(A, X_n) = Q_{n+1,n}(A, X_n)$, and the Markov property of $\{X_n, n \geq 1\}$ together imply, in the context of Proposition 2 above

$$E(f(X)) = \int_{\mathbb{R}} \tilde{Q}_1(dx_1) \int_{\mathbb{R}} Q_1(dx_2, x_1) \ldots \int_{\mathbb{R}} f(x) Q_n(dx_{n+1}, x_n), \quad (9)$$

for any bounded Borel $f : \mathbb{R} \to \mathbb{R}$ and any random variable X on $(\Omega, \mathcal{B}_{n+1}, P)$, where $\mathcal{B}_{n+1} = \sigma(X_1, \ldots, X_{n+1})$, of the family. Indeed, we have

$$E(f(X)|\mathcal{B}_{n+1}) = \int_{\mathbb{R}} f(x)(P^{\mathcal{B}_{n+1}} \circ X_{n+1}^{-1})(dx), \quad (10)$$

by regularity of $P^{\mathcal{B}_n} \circ X^{-1}$. But $E(f(X)|\mathcal{B}_n) = E(E(f(X)|\mathcal{B}_{n+1}|\mathcal{B}_n)$, since $\mathcal{B}_n \subset \mathcal{B}_{n+1}$. Hence with the Markovian property of $\{X_1, X_2, \ldots, X_{n+1}\}$, we get $P^{\mathcal{B}_n} \circ X_{n+1}^{-1} = P^{\sigma(X_n)} \circ X_{n+1}^{-1} = Q_n(\cdot, \cdot)$, and the last integral of (9) is just (10). Similarly (with the Markovian property again),

$$E(f(X)|\mathcal{B}_{n-1}) = \int_{\mathbb{R}} P^{\mathcal{B}_{n-1}} \circ X_n^{-1}(dy) \int_{\mathbb{R}} f(x) P^{\mathcal{B}_n} \circ X_{n+1}^{-1}(dx)$$

$$= \int_{\mathbb{R}} Q_{n-1}(dx_n, x_{n-1}) \int_{\mathbb{R}} f(x) Q_n(dx, x_n). \tag{11}$$

Since $E(f(X)) = EE^{\mathcal{B}_1} E^{\mathcal{B}_2} \dots E^{\mathcal{B}_n}(f(X))$, by iterating (11) n times we get (9). In particular, if $f(X) = f_1(X_1) \dots f_{n+1}(X_{n+1})$, then (9) reduces to

$$E\left(\prod_{i=1}^{n+1} f_i(X_i)\right) = \int_{\mathbb{R}} f_1(x_1)\tilde{Q}_1(dx_1) \int_{\mathbb{R}} f_2(x_2) Q_1(dx_2, x_1) \times$$

$$\dots \int_{\mathbb{R}} f_{n+1}(x_{n+1}) Q_n(dx_{n+1}, x_n), \tag{12}$$

for any real bounded Borel functions $f_1, \dots, f_{n+1}$ and the image measure $\tilde{Q}_1$ of P given by $\tilde{Q}_1(A) = P(X_1^{-1}(A)) = P[X_1 \in A]$. This $\tilde{Q}_1$ is called the *initial distribution* of the process $\{X_n, n \geq 1\}$, being the image measure of the first, or initial, random variable X_1. Taking $f_i = \chi_{A_i}$ $A_i \subset \mathbb{R}$, Borel, (12) yields an important expression for the joint distribution of $(X_1, \dots, X_{n+1})$:

$$P[X_1 \in A_1, \dots, X_{n+1} \in A_{n+1}] = \int_{A_1} \tilde{Q}_1(dx_1) \int_{A_2} Q_1(dx_2, x_1) \times$$

$$\dots \int_{A_{n+1}} Q_n(dx_{n+1}, x_n). \tag{13}$$

Setting $A_1 = \mathbb{R}, A_i = \mathbb{R}, i \geq 3$, this relation in conjunction with (7) gives the distribution of X_2 and all the marginal distributions of any subset of $\{X_k, k \geq 1\}$. For instance, $\tilde{Q}_3(\cdot)$ is given by [since $Q_n(\mathbb{R}, x_{n-1}) = 1$]

$$\tilde{Q}_3(A) = P[X_1 \in \mathbb{R}, X_2 \in \mathbb{R}, X_3 \in A, X_4 \in \mathbb{R}, \dots, X_{n+1} \in \mathbb{R}]$$

$$= \int_{\mathbb{R}} \tilde{Q}_1(dx_1) \int_{\mathbb{R}} Q_1(dx_2, x_1) \cdot Q_2(A, x_2)$$

$$= \int_{\mathbb{R}} \tilde{Q}_1(dx_1) \cdot Q_{1,2}(A, x_1) \quad \text{[by (7)]},$$

for all Borel sets $A \subset \mathbb{R}$. Similarly others are obtained. Here $Q_{1,2}(\cdot, x_1)$ is the same as in (7) with $r = 1, t = 2$ there. *Even though the conditional distributions* $\{Q_{r,s}(\cdot,\cdot), r < s\}$ *do not uniquely correspond to a Markov process, as shown in the counterexample above, the (absolute) finite dimensional or joint distributions (as n varies) given by* (13) *are uniquely defined and, as demonstrated in the next section, they determine a Markov process.* In this sense both (7) and (13) play crucial roles in the Markov process work.

In the preceding discussion $Q_{r,t}(A, x)$ is a version of $P([X_t \in A]|X_r = x)$, and is a regular conditional distribution of X_t given $X_r = x$. Since generally there exist several versions of the conditional probability, how should one choose a family $\{Q_{r,t}(\cdot, x), r < t, x \in \mathbb{R}\}$ in order that all these measures simultaneously satisfy (13)? There is no problem if the family is finite or even countable, since we can find a fixed null set and arrange things so that this is possible. Also, if the state space is at most countable (i.e., the Markov chain case), we can take the exceptional set as empty and answer the problem affirmatively. In the general case, no such method is available. To include all these cases, one **assumes** the desired property and develops the general theory. This is called the (Markov) transition probability (nonconstructive or idealistic procedure) family defined as follows [we write $Q(\cdot, t; x, r)$ in place of $Q_{r,t}(\cdot, x)$ for convenience].

A mapping $Q : \mathcal{B} \times \mathbb{R}^+ \times \mathbb{R} \times \mathbb{R}^+ \to [0,1]$, where $\mathcal{B}$ is the Borel σ-algebra of the state space $\mathbb{R}$, is a (Markov) **transition probability** if for each $0 \leq r \leq t, x \in \mathbb{R}$,

(i) $Q(\cdot, t; x, r) : \mathcal{B} \to [0,1]$ is a probability,

(ii) $Q(A, t; \cdot, r) : \mathbb{R} \to [0,1]$ is $\mathcal{B}$ -measurable for each $A \in \mathcal{B}, 0 \leq r \leq t$,

(iii) for each $0 \leq r < s < t$, one has

$$Q(A, t; x, r) = \int_{\mathbb{R}} Q(A, t; y, s) Q(dy, s; x, r), A \in \mathcal{B}, x \in \mathbb{R}, \tag{14}$$

identically in all the variables shown. [For this definition $\mathbb{R}^+$ can be replaced by a subinterval of $\mathbb{R}$. But for the following work we use $\mathbb{R}^+$.]

If μ is an initial probability on $\mathcal{B}$ then substituting μ and Q in (13) one can generate an n-dimensional probability measure on $\mathbb{R}^n$. We also take

$$Q(A, t; \cdot, t) = \chi_A$$

as a boundary condition. If a Markov process $\{X_t, t \geq 0\}$ on (Ω, Σ, P) is given, then $Q(A, t; x, s)$ is a version of $P^{\sigma(X_s)}(X_t^{-1}(A))(\omega)$, and so they are equal a.e. Thus (14) is just the Chapman-Kolmogorov equation, which is now *assumed to be true identically* (without any exceptional sets). It follows from the work of the next section that there exists a Markov process on a probability space (Ω, Σ, P) such that $\mu(A) = P[X_0 \in A]$, and $P([X_t \in A]|X_s)(\omega) = Q(A, t; X_s(\omega), s)$ for a.a.(ω). [The exceptional null sets depend on A, s, t, in general, and this is why we *assume* the conditions (i)-(iii) above identically.]

A consequence of the above (strengthened) conditions is that one can transform the Markov process theory into one of functional operations by means of (14):

Proposition 6 *Let $B(\mathbb{R}, \mathcal{B})$ $[= B(\mathbb{R}),$ say$]$ be the space of real bounded Borel functions on $\mathbb{R}$ with the uniform norm: $\|f\| = \sup\{|f(x)| : x \in \mathbb{R}\}$. If $\{Q(\cdot, \cdot; \cdot, \cdot)\}$ is a transition probability family, as above, and for each $0 \leq s \leq t, U_{s,t}$ is defined as*

$$(U_{s,t} f)(x) = \int_{\mathbb{R}} f(y) Q(dy, t; x, s), f \in B(\mathbb{R}), x \in \mathbb{R}, \tag{15}$$

then $U_{s,t} : B(\mathbb{R}) \to B(\mathbb{R})$ *is a positive contractive linear mapping, i.e.,* $U_{s,t}1 = 1$, $||U_{s,t}f|| \le ||f||$, *and* $U_{s,t}f \ge 0$ *for* $f \ge 0$. *Moreover,* $\{U_{s,t} : 0 \le s \le t\}$ *forms a generalized semigroup (or satisfies an* ***evolution equation****), in that*

$$U_{r,s}U_{s,t} = U_{r,t}, \;\; 0 < r < s < t, \;\; U_{r,r} = id. \tag{16}$$

Conversely, every such family of (evolution) operators on $B(\mathbb{R})$ *uniquely determines a (Markov) transition probability family on* $\mathcal{B} \times \mathbb{R}^+ \times \mathbb{R} \times \mathbb{R}^+$.

Proof Since f is bounded, the integral in (15) is well defined. If f is a simple function in $B(\mathbb{R})$, it is clear that $U_{s,t}f$ is a bounded Borel function. By the dominated convergence theorem, the result follows for all $f \in B(\mathbb{R})$. Thus $U_{s,t} : B(\mathbb{R}) \to B(\mathbb{R})$ is linear, maps positive elements into positive elements, and $||U_{s,t}f|| \le ||f||$, since $Q(\mathbb{R}, t; x, s) = 1$. Also, by (14)

$$\begin{aligned}(U_{r,t}f)(x) &= \int_{\mathbb{R}} f(y) \left[\int_{\mathbb{R}} Q(dy, t; u, s)Q(du, s; x, r)\right] \quad (r \le s \le t)\\ &= \int_{\mathbb{R}} Q(du, s; x, r) \int_{\mathbb{R}} f(y)Q(dy, t; u, s)\\ &\quad \text{(by a form of Fubini's theorem)}\\ &= \int_{\mathbb{R}} (U_{s,t}f)(u)Q(du, s; x, r) = (U_{r,s}(U_{s,t}f))(x).\end{aligned}$$

Hence $U_{r,t} = U_{r,s}U_{s,t}$ proving (16), since $U_{r,r} = id$ is obvious.

In the opposite direction, let U be as given, and for each $A \in \mathcal{B}, x \in \mathbb{R}, s \le t$, define $Q(A, t; x, s) = (U_{s,t}\chi_A)(x)$. Then $Q(\cdot, t; x, s)$ is additive by the linearity of $U_{s,t}$, $Q(A, t; \cdot, s)$ is $\mathcal{B}$-measurable, and $1 = U_{s,t}1 = Q(\mathbb{R}, t; x, s)$. To see it is σ-additive on $\mathcal{B}$, let f_n be any sequence of continuous functions with compact supports such that $f_n \downarrow 0$ pointwise. Then $A_n^\varepsilon = \{x : f_n(x) \ge \varepsilon\}$ is compact and $\bigcap_{n \ge 1} A_n^\varepsilon = \emptyset$. Hence there is an n_0 such that $A_{n_0}^\varepsilon = \emptyset$ (finite intersection property).

Thus $||f_n|| \le \varepsilon, n > n_0$. If we consider the linear functional $(U_{s,t}(\cdot))(x)$, then the norm condition implies that $||U_{s,t}f_n|| \le ||f_n|| \to 0$ as $n \to 0$. Since it is positive also, it defines an integral on $C_{00}[\subset B(\mathbb{R})]$, the space of continuous functions with compact supports. If $\tilde{Q}(\cdot, t; x, s)$ represents (Riesz representation theorem) this $(U_{s,t}(\cdot))(x)$, then $Q = \tilde{Q}$ on all compact sets in $\mathcal{B}$. But the latter generate $\mathcal{B}$. Thus the standard results in measure theory imply $Q = \tilde{Q}$ on all of $\mathcal{B}$, and hence Q is a transition family, since then (16) becomes (14). This completes the proof.

In a moment we present an important application of the above results implying that all these assumptions are automatically satisfied for that class. We must first observe that the relation (16) is a natural one in Markov process work. Our concept of Markovian family as given by Definition 1 (cf. Proposition 2 also) is based on the distributions (or probability measures) of the random variables. A weaker concept will be to ask for conditions on a few moments. This inquiry leads to an introduction of a wide-sense Markov family, of some interest in applications. Since the conditional expectation is a contractive projection on $L^1(\Sigma)$ (cf. Theorem 1.11), the corresponding wide-sense operation will use the Hilbert space geometry, and is given as follows.

Definition 7 (a) Let $\{X_t, t \in T\}$ be a family of square integrable(complex) random variables on a probability space (Ω, Σ, P) and $T \subset \mathbb{R}$. We define a *correlation characteristic* $\rho(s,t)$ as

$$\rho(s,t) = \begin{cases} \frac{E(X_s \overline{X}_t)}{E|X_s|^2} & \text{if } E(|X_s|^2) > 0,\ s < t \text{ in } T \\ 0 & \text{otherwise.} \end{cases}$$

[$\rho(s,t)$ will be the correlation if $E(|X_s|^2)$ is replaced by $[E(|X_s|^2)E(|\overline{X}_t|^2)]^{1/2}$ and all the X_t have zero means; $\overline{X}_t$ is the complex conjugate of X_t.]

(b) If $t_1 < \ldots < t_n$ are points of T, and $\mathfrak{M}_n^k = sp\{X_{t_k}, \ldots, X_{t_n}\}$ is the linear span, let $\hat{E}(\cdot|\mathfrak{M}_n^k)$ be the orthogonal projection of $L^2(\Sigma)$ onto $\mathfrak{M}_n^k$. Then the given family is called a *Markov process in the wide sense* if for each such collection of points $t_1 < \ldots < t_n$, $n \geq 1$, of T,

$$\hat{E}(X_{t_n}|\mathfrak{M}_{n-1}^1) = \hat{E}(X_{t_n}|\mathfrak{M}_{n-1}^{n-1}) \quad \text{a.e.}$$

This may be equivalently expressed as

$$\hat{E}(X_{t_n}|X_{t_1}, \ldots, X_{t_{n-1}}) = \hat{E}(X_{t_n}|X_{t_{n-1}}) \quad \text{a.e.} \tag{17}$$

Note that if $\mathfrak{M}_n^1$ is replaced by $L^2(\sigma(X_{t_1}, \ldots, X_{t_n})) \subset L^2(P)$, then (17) becomes $E(X_{t_n}|X_{t_1}, \ldots, X_{t_{n-1}}) = E(X_{t_n}|X_{t_{n-1}})$ a.e., and the concept reduces to the ordinary Markovian definition in $L^2(P)$, by Proposition 2; the latter should be called a *strict-sense Markovian concept* for a distinction. But this qualification is usually omitted. [Then $\hat{E}(\cdot|\mathfrak{M}_1^1)$ becomes $E(\cdot|X_{t_1})$.]

The following characterization of (17), due to Doob, reminds us of the "evolution" equation, formula (16) again.

Proposition 8 *Let $\{X_t, t \in T\} \subset L^2(\Omega, \Sigma, P)$ be a family of random variables, with $T \subset \mathbb{R}$. Then it is a Markov process in the wide sense iff its correlation characteristic satisfies the following functional equation, for $r < s < t$ in T:*

$$\rho(r,t) = \rho(r,s)\rho(s,t). \tag{18}$$

Proof For $s < t$, consider X_s and $Y_s = X_t - \overline{\rho}(s,t)X_s$. Then for any complex scalar a we have

$$\begin{aligned} E(aX_s\overline{Y}_s) &= aE(X_s\overline{X}_t) - a\rho(s,t)E(X_s\overline{X}_s) \\ &= a\rho(s,t)E(|X_s|^2) - a\rho(s,t)E(|X_s|^2) = 0, \end{aligned} \tag{19}$$

so that aX_s and Y_s are orthogonal vectors in the Hilbert space $\mathcal{H} = L^2(\Omega, \Sigma, P)$. Consider $\mathfrak{M}_s$ as the linear span of X_s, and $\mathfrak{M}_s^\perp$ as its orthogonal complement, so that $\mathcal{H} = \mathfrak{M}_s \bigoplus \mathfrak{M}_s^\perp$ and each vector in $\mathcal{H}$ can be uniquely expressed as a sum of two mutually orthogonal vectors, one from $\mathfrak{M}_s$ and another from $\mathfrak{M}_s^\perp$. In particular, if the (orthogonal) projection in $\mathcal{H}$ onto $\mathfrak{M}_s$ is denoted $Q_s[= \hat{E}(\cdot|X_s)]$, then setting $a = \overline{\rho}(s,t)$ in the conclusion of (19), we get (adding Y_s and aX_s there)

$$X_t = Q_s X_t + (I - Q_s)X_t = \overline{\rho}(s,t)X_s + (X_t - \overline{\rho}(s,t)X_s),$$

which implies $Q_s X_t = \overline{\rho}(s,t)X_s$ [and $(I - Q_s)X_t = X_t - \overline{\rho}(s,t)X_s$]. But by the wide-sense Markovian hypothesis, if $Q_{r,s}$ is the projection on $\mathfrak{M}_s^r = sp\{X_r, X_s\}$, the linear span, then

$$Q_{r,s}X_t = \hat{E}(X_t|X_r, X_s) = \hat{E}(X_t|X_s) = Q_s X_t = \overline{\rho}(s,t)X_s, \text{ a.e.} \tag{20}$$

However, $(I - Q_{r,s})X_t$ is orthogonal to $\mathfrak{M}_s^r$, so that it is orthogonal to both X_r and X_s. By (20), $(I - Q_{r,s})X_t = X_t - \overline{\rho}(s,t)X_s = Y_s$ is orthogonal to X_r (and X_s). Thus

$$\begin{aligned} 0 = E(X_r\overline{Y}_s) &= E(X_r\overline{X}_t) - \rho(s,t)E(X_r\overline{X}_s) \\ &= \rho(r,t)E(|X_r|^2) - \rho(s,t)\cdot\rho(r,s)E(|X_r|^2). \end{aligned} \tag{21}$$

Cancelling $0 < E(|X_r|^2)$, this gives (18).

Conversely, suppose (18) is true for any $r < s < t$. Then (21) is true; it says that X_r and Y_s are orthogonal to each other for any $r < s < t$ $(r, s, t \in T)$. Also by (19), X_s and Y_s are always orthogonal. Hence from the arbitrary nature of r, Y_s is orthogonal to $X_s, X_{r_1}, \ldots, X_{r_k}$ for $r_1 < r_2 < \ldots < r_k < s < t$ in T. This means $Y_s \in \mathfrak{M}_s^\perp \cap (\mathfrak{M}_s^{r_1})^\perp$ so that

$$\hat{E}(X_t|X_{r_1}, X_{r_2}, \ldots, X_{r_k}, X_s) = \hat{E}(X_t|X_s) = Q_s X_t (= \rho(s,t)X_s) \text{ a.e.}$$

Thus the process is wide-sense Markov, completing the proof.

An Application We now present an important and natural application of the preceding considerations showing that the various assumptions made above are fulfilled for a class of problems. Let $\{X_k, 1 \leq k \leq n\}$ be a sequence of i.i.d. random variables (also called a *random sample*), with a continuous strictly increasing distribution function F. Consider the *order statistics* $\{X_i^*, 1 \leq i \leq n\}$ based on the above family, meaning $X_1^* = \min(X_1, \ldots, X_n), \ldots, X_{n-1}^* =$ the second largest of $(X_1, \ldots, X_n)$, and $X_n^* = \max(X_1, \ldots, X_n)$. Since $X_1^* < X_2^* < \cdots < X_n^*$ with probability one, they are *not* independent. Also, the X_i need have no moments.

Regarding the "process" $\{X_i^*, 1 \leq i \leq n\}$, it has the remarkable Markov property already observed by Kolmogorov in 1933.

Theorem 9 *Let $X_1, \ldots, X_n$ be i.i.d. random variables on (Ω, Σ, P) with a continuous strictly increasing distribution function F. If $\{X_i^*, 1 \le i \le n\}$ is the process of order statistics formed from the X_i, let $Y_i = F(X_i^*)$ and $Z_i = -\log F(X_{n+1-i}^*), 1 \le i \le n$. Then the three sequences $\{X_i^*, 1 \le i \le n\}, \{Y_i, 1 \le i \le n\}$, and $\{Z_i, 1 \le i \le n\}$ form strict-(=ordinary-) sense Markov families with $(1, 2, \ldots, n)$ as the parameter (=index) set. Moreover, the Z_i-process is one of independent increments, and the Y_i-process is both a strict and a wide-sense Markov family. If $1 \le i_i < i_2 < \cdots < i_k \le n$ and $dF/dx = F'$ exists, then the joint distribution of $X_{i_1}^*, \ldots, X_{i_k}^*$ has a density $g_{i_1,\ldots,i_k}(\cdot, \ldots, \cdot)$ relative to the Lebesgue measure, and is given for $-\infty < \lambda_1 < \lambda_2 < \cdots < \lambda_k < \infty$, by*

$$
\begin{aligned}
&g_{i_1,\ldots,i_k}(\lambda_1, \ldots, \lambda_k) \\
&\quad = \frac{n!}{(i_1-1)!(i_2-i_1-1)!\cdots(n-i_k)!} \\
&\qquad \times (F(\lambda_1))^{i_1-1}(F(\lambda_2)-F(\lambda_1))^{i_2-i_1-1}(F(\lambda_3)-F(\lambda_2))^{i_3-i_2-1} \\
&\qquad \times \cdots \times (1-F(\lambda_k))^{n-i_k} \times F'(\lambda_1)F'(\lambda_2)\cdots F'(\lambda_k),
\end{aligned} \tag{22}
$$

and $g_{i_1,\ldots,i_k} = 0$ for all other values of λ_i.

Proof Let us first recall that if $X_t : \Omega \to B(t \in T$, interval) is a Markov process with $(B, \mathcal{B})$ as measurable space, and $\phi_t : B \to A$ is a one-to-one and onto mapping, where $(A, \mathcal{A})$ is a measurable space, with $\phi_t^{-1}(\mathcal{A}) = \mathcal{B}, t \in T$, then $Y_t = \phi_t(X_t) : \Omega \to A$ is also a Markov process. Indeed, $\mathcal{C}_t = Y_t^{-1}(\mathcal{A}) = X_t^{-1}(\phi_t^{-1}(\mathcal{A})) = X_t^{-1}(\mathcal{B}) = \sigma(X_t) \subset \Sigma$, and, by Definition 1, X_t is a Markov process, which means the set of σ-algebras $\{\sigma(X_t) = \mathcal{C}_t, t \in T\}$ is Markovian. Consequently the Y_t-process is Markovian. We identify them with X_i^*, Y_i, Z_i.

Taking $\phi_t = F : \mathbb{R} \to (0, 1)$, all $t = 1, 2, \ldots, n$, in the above so that ϕ_t is one-to-one, onto, and the Borel σ-algebras are preserved, it follows that $\{X_i^*, 1 \le i \le n\}$ is Markovian iff $\{Y_i, 1 \le i \le n\}$ is; and then the same is true of $\{Z_i, 1 \le i \le n\}$, since, if $\psi(x) = -\log x, \psi : (0, 1) \to \mathbb{R}^+$ is also one-to-one and onto, with similar properties. Because of this we shall establish directly the Markovian property of the Z_i-sequence. [Similarly one can prove it directly for the Y_i-sequence also; see Problem 22 for another argument.]

Note that if ξ_k denotes the kth largest of the numbers $X_1(\omega), \ldots, X_n(\omega)$, then $X_k^*(\omega) = \xi_k(X_1(\omega), \ldots, X_n(\omega))$. Consequently by hypothesis on F, it follows that $Y_i = F(X_i^*) = F(\xi_i(X_1, \ldots, X_n)) = \xi_i(F(X_1), \ldots, F(X_n))$. Hence Y_i is the ith-order statistic of $F(X_1), \ldots, F(X_n)$. Similarly, Z_k is the kth-order statistic given by $Z_k = -\log F(X_{n+1-k}^*) = \xi_k(\zeta_1, \ldots, \zeta_n)$, where

$$\zeta_i = -\log F(X_i).$$

This follows from the fact that $-\log y$ is decreasing as $y > 0$ increases, and hence the kth largest from below to Z_k corresponds to the kth smallest from above to the X_i- or $F(X_i)$-sequence. If $G(\cdot)$ is the distribution of the ζ_i (which are clearly independently and identically distributed), then

$$G(z) = P[\zeta_k < z] = \begin{cases} 0, & z \leq 0 \\ P\left[\log \frac{1}{F(X_i)} < z\right], & z > 0. \end{cases}$$

Hence for $z > 0$,

$$\begin{aligned} G(z) &= P\left[\frac{1}{F(X_i)} < e^z\right] = P[F(X_i) > e^{-z}] \\ &= 1 - P[X_i \leq F^{-1}(e^{-z})] = 1 - F(F^{-1}(e^{-z})) = 1 - e^{-z}, \end{aligned}$$

since F is strictly increasing and X_i has F as its distribution function. Thus $G(\cdot)$ has a density g given by

$$g(z) = G'(z) = \begin{cases} e^{-z}, & \text{if } z > 0 \\ 0, & \text{if } z \leq 0, \end{cases} \tag{23}$$

and $Z_1, \ldots, Z_n$ are order statistics of an r.v. ζ whose density is (23). Note that in the above, we have also obtained the d.f. of $\eta_k = F(X_k)$. In fact, $0 \leq \eta_k \leq 1$ a.e., and its d.f. is given by

$$H(x) = P[\eta_k < x] = P[X_k < F^{-1}(x)] = F(F^{-1}(x)) = \begin{cases} 0 & \text{if } x \leq 0 \\ x & \text{if } 0 < x < 1 \\ 1 & \text{if } x \geq 1. \end{cases} \tag{24}$$

These facts will be useful for the following computations.

Let us now derive the joint distribution of the order statistics. Since $-\infty < X_1^* < \cdots < X_n^* < \infty$ a.e. (equalities hold only with probability zero, since F is continuous), the range of these functions is the set $A = \{(x_1, \ldots, x_n) : -\infty < x_1 < x_2 < \cdots < x_n < \infty\} (\subset \mathbb{R}^n)$, which is Borel. Let $B \subset A$ be any Borel subset. Then

$$\begin{aligned} &P[(X_1^*, \ldots, X_n^*) \in B] \\ &\quad = \sum_\sigma P[(X_{\sigma_1}, X_{\sigma_2}, \ldots, X_{\sigma_n}) \in B] \\ &\qquad [\text{the sum is over all permutations } \sigma \text{ of } (1, \ldots, n)] \\ &\quad = \sum_\sigma \int \cdots \int_B dF(\lambda_1) \cdots dF(\lambda_n), \quad (\text{since the } X_i \text{ are i.i.d.}) \\ &\quad = n! \int \cdots \int_B dF(\lambda_1) \cdots dF(\lambda_n). \end{aligned} \tag{25}$$

If F' exists, then the above can be written as

$$= \int \cdots \int_B (n!) F'(\lambda_1) \cdots F'(\lambda_n) \, d\lambda_1 \cdots d\lambda_n.$$

But this is true for all Borel subsets B of A. Hence by the Radon-Nikodým theorem, $(X_1^*, \ldots, X_n^*)$ has density, say, f^*, given by

$$f^*(\lambda_1, \ldots, \lambda_n) = \begin{cases} n!\, F'(\lambda_1) \cdots F'(\lambda_n) & \text{if } -\infty < \lambda_1 < \cdots < \lambda_n < +\infty, \\ 0 & \text{otherwise.} \end{cases} \tag{26}$$

If $1 \leq i_1 < i_2 < \cdots < i_k \leq n$ is a subset of the set $(1, 2, \ldots, n)$, then the marginal density of $X_{i_1}^*, \ldots, X_{i_k}^*$, say, $g_{i_1,\ldots,i_k}(x_1, \ldots, x_k)$, is obtained from (26) by integrating f^*, for fixed $x_1 < x_2 < \cdots < x_k$, relative to all the λ over the following set

$$\{(\lambda_1, \ldots, \lambda_n) : -\infty < \lambda_1 < \cdots < \lambda_{i_1 - 1} < x_1 < \lambda_{i_1+1} < \cdots \\ < \lambda_{i_k - 1} < x_k < \lambda_{i_k+1} < \cdots < \lambda_n\}.$$

This gives (22). The details of this elementary integration are left to the reader. It remains to establish the special properties of the Y_i- and Z_i-sequences.

First consider the Z_i. Let $U_k = Z_k - Z_{k-1}$ (and $Z_0 = 0$ a.e.), $k = 1, \ldots, n$. Then $U_k \geq 0$, and $Z_1 = U_1, Z_2 = U_1 + U_2, \ldots, Z_n = \sum_{k=1}^n U_k$. The mapping from the range space of $(Z_1, \ldots, Z_n)$ to that of $(U_1, \ldots, U_n)$ is one-to-one, and we shall compute the joint distribution of $U_1, \ldots, U_n$ and show that they are mutually independent. Since, as seen in Example 3, each partial sum sequence of independent random variables is a Markov process, both properties announced for the Z_i-sequence are established at once. Now with (23), the joint distribution of $Z_1, \ldots, Z_n$ is given from (26) as

$$\tilde{f}^*(\lambda_1, \ldots, \lambda_n) = \begin{cases} n! \exp\left(-\sum_{i=1}^n \lambda_i\right), & \text{if } 0 < \lambda_1 < \cdots < \lambda_n < \infty, \\ 0 & \text{otherwise.} \end{cases}$$

The Jacobian J of the above mapping from $(Z_1, \ldots, Z_n) \to (U_1, \ldots, U_n)$ is given in their range spaces, with $z_k = \sum_{i=1}^k u_i, 1 \leq k \leq n$, which is triangular, and one finds

$$J = \partial(z_1, \ldots, z_n)/\partial(u_1, \ldots, u_n) = 1.$$

Thus for any n-dimensional Borel set B in the range space of the U_i, if A is its preimage in the range of $Z_1, \ldots, Z_n$, then

$$\begin{aligned} P[(U_i, \ldots, U_n) \in B] &= P[(Z_1, \ldots, Z_n) \in A] \\ &= \int \overset{\cdots}{\underset{A}{}} \int \tilde{f}^*(z_1, \ldots, z_n)\, dz_1 \cdots dz_n \\ &= \int \overset{\cdots}{\underset{B}{}} \int \tilde{f}^*\left(u_1, u_1 + u_2, \ldots, \sum_{i=1}^n u_i\right) |J|\, du_1 \cdots du_n \\ &= n! \int \overset{\cdots}{\underset{B}{}} \int \left(\prod_{j=0}^{n-1} e^{-(n-j)u_{j+1}}\right) du_1 \cdots du_n. \end{aligned} \tag{27}$$

Since B is an arbitrary Borel set in the positive orthant of $\mathbb{R}^n$, the range space of $(U_1, \ldots, U_n)$, it follows from (27), by means of the Radon-Nikodým theorem, that the U_k have a density function h given by

$$h(u_1, \ldots, u_n) = \begin{cases} n! \prod_{j=0}^{n-1} e^{-(n-j)u_{j+1}} & \text{if } u_1 > 0, \ldots, u_n > 0 \\ 0 & \text{otherwise.} \end{cases}$$

Since the density factors, it results that the set $\{U_1, \ldots, U_n\}$ is a mutually independent family (U_j with density = constant $\cdot e^{-(n-j)u_{j+1}}$ for $u_{j+1} \geq 0$). Thus the Z_i-sequence is Markovian, and is in fact of independent increments. It only remains to verify the wide sense Markov property for the real Y_i-sequence.

Let $r(i_1, i_2) = E(Y_{i_1} Y_{i_2})$. Now using (24) in (26) or (22) for $1 \leq i_1 < i_2 \leq n$, we get, on substitution, that the densities g_{i_1} and g_{i_1,i_2} of $Y_{i_1}, (Y_{i_1}, Y_{i_2})$ to be

$$g_{i_1}(x) = \begin{cases} \frac{n!}{(i_1-1)!(n-i_1)!} x^{i_1-1}(1-x)^{n-i_1}, & \text{if } \quad 0 < x < 1 \\ 0 & \text{elsewhere,} \end{cases}$$

and

$$g_{i_1,i_2}(x, y) = \begin{cases} \frac{n!}{(i_1-1)!(i_2-i_1-1)!(n-i_2)!} x^{i_1-1}(y-x)^{i_2-i_1-1}(1-y)^{n-i_2}, & \\ & \text{if } \quad 0 < x < y < 1 \\ 0 & \text{otherwise.} \end{cases}$$

Hence

$$\begin{aligned} r(i_1, i_2) &= \int\int_{0<x<y<1} xy g_{i_1,i_2}(x, y) \, dx \, dy \\ &= \frac{n!}{(i_1 - 1)!(i_2 - i_1 - 1)!(n - i_2)!} \int_0^1 y^{i_2-i_1}(1-y)^{n-i_2} \, dy \\ &\quad \times \int_0^y x^{i_1} \left(1 - \frac{x}{y}\right)^{i_2-i_1-1} dx \\ &= \frac{i_1(i_2 + 1)}{(n+1)(n+2)}. \end{aligned}$$

Similarly,

$$\begin{aligned} r(i_1, i_1) = E(Y_{i_1}^2) &= \frac{n!}{(i_1 - 1)!(n - i_1)!} \int_0^1 x^{i_1+1}(1-x)^{n-i_1} \, dx \\ &= \frac{i_1(i_1 + 1)}{(n+1)(n+2)}. \end{aligned}$$

Thus

$$\rho(i_1, i_2) = \frac{r(i_1, i_2)}{r(i_1, i_1)} = \frac{i_2 + 1}{i_1 + 1}.$$

Hence for $1 \leq i_1 < i_2 < i_3 \leq n$, we get $\rho(i_1, i_2)\rho(i_2, i_3) = \rho(i_1, i_3)$. By Proposition 8, this shows that the Y_i-sequence is also wide-sense Markov. The proof is complete.

It is easy to prove the Markovian property of the order statistics (for which the existence of F' need not be assumed) using the Y_i-sequence instead of the Z_i-sequence employed in the above proof. This alternative method of interest is given later as Problem 22. The point of the above result is to recognize this (Markov) property for order statistics of a (finite) random sample from *any* continuous (strictly increasing) distribution function. This illuminates the structure of the problem and admits further analysis. (cf. Theorem 5.4.8, 7.2.5.)

For a deeper study, the class of Markov processes must be restricted to satisfy some regularity conditions. There are numerous specialized works, both intensive and extensive, on the subject. Proposition 6, for instance, exhibits an intimate relation of these processes to the theory of semigroups of operators on various function spaces. We will not enter into these special relations in the present work. We now turn, instead, to showing that such processes exist under quite broad and reasonable conditions.

3.4 Existence of Various Random Families

In all the preceding discussion, we have assumed that there exists a family of random variables on a probability space (Ω, Σ, P). When does it really exist? This fundamental question will now be answered. Actually there are two such basic existence results, corresponding to the independent and dependent families. These were proved in the 1930s. They are due to B. Jessen and A. Kolmogorov, respectively. Both results were extended subsequently. We discuss them here and present the details of a relevant part of this work that implies the existence of all the random families studied in our book. Readers pressed for time may glance through this section and return to it later for a detailed study.

To introduce some terminology and to give a motivation, consider (Ω, Σ, P). An indexed collection $\{X_t, t \in T\}$ of random variables $X_t : \Omega \to \mathbb{R}$ is termed a *stochastic (or random) process or family*. If $T \subset \mathbb{R}$, let $t_1 < \ldots < t_n$ be n points from T, $n \geq 1$, and consider the n-dimensional distributions (or image measures) of $X_{t_1}, \ldots, X_{t_n}$ given by

$$F_{t_1,\ldots,t_n}(x_1, \ldots, x_n) = P[X_{t_1} < x_1, \ldots, X_{t_n} < x_n]. \tag{1}$$

As n varies, the set $\{F_{t_1,\ldots,t_n}, t_i \in T, n \geq 1\}$ clearly satisfies the following system of equations, called the *compatibility conditions*:

$$\lim_{x_n \to +\infty} F_{t_1,\ldots,t_n}(x_1,\ldots,x_n) = F_{t_1,\ldots,t_{n-1}}(x_1,\ldots,x_{n-1}),$$

$$\lim_{x_n \to -\infty} F_{t_1,\ldots,t_n}(x_1,\ldots,x_n) = 0, \tag{2}$$

and if $(i_1,\ldots,i_n)$ is a permutation of $(1,\ldots,n)$, then

$$F_{t_{i_1},\ldots,t_{i_n}}(x_{i_1},\ldots,x_{i_n}) = F_{t_1,\ldots,t_n}(x_1,\ldots,x_n). \tag{3}$$

The mysterious condition (3) simply states that the intersection of the sets inside $P(\cdot)$ in (1) is commutative and thus the determination of F does not depend on the ordering of the indices of T. Equations (2) and (3) can be put compactly into a single relation, namely: if for any set $\alpha_n = (t_1,\ldots,t_n)$ of T, $\mathbb{R}^n = \times_{t_i \in \alpha_n} \mathbb{R}_{t_i}, \mathbb{R}_{t_i} = \mathbb{R}, \mathcal{B}^n = \bigotimes_{t_i \in \alpha_n} \mathcal{B}_{t_i}, \mathcal{B}_{t_i} = \mathcal{B}$, the Borel σ-algebra of $\mathbb{R}$, and $\pi_{n,n+1} : \mathbb{R}^{n+1} \to \mathbb{R}^n$, the coordinate projection [i.e., $\pi_{n,n+1}(x_1,\ldots,x_{n+1}) = (x_1,\ldots,x_n)$], then $\mathbb{R}^{n+1} = \pi^{-1}_{n,n+1}(\mathbb{R}^n)$, $\mathcal{B}^{n+1} \supset \pi^{-1}_{n,n+1}(\mathcal{B}^n)$. Moreover, if we define

$$P_{\alpha_n}(A) = \int_A dF_{t_1,\ldots,t_n}(x_1,\ldots,x_n), \quad A \in \mathcal{B}^n,$$

then (2) and (3) can be combined into the single statement that

$$P_{\alpha_n}(A) = P_{\alpha_{n+1}}(\pi^{-1}_{n,n+1}(A)), \quad A \in \mathcal{B}^n, \tag{4}$$

where $P_{\alpha_n} = P \circ (X_{t_1},\ldots,X_{t_n})^{-1}$. Thus (4) is the compatibility relation.

The preceding discussion shows that a given random family always induces a set of mutually compatible finite-dimensional probability distributions (=image laws) on $\{\mathbb{R}^n, n \geq 1\}$. Kolmogorov (and, in the important case of independence, Jessen) showed that, conversely, such a compatible collection of finite-dimensional distributions determines a random family on some probability space whose image laws are the given (finite-dimensional) distributions. We precisely state and prove these statements in this section. First let us give two simple examples of a compatible family of probability distributions.

Let μ be any measure on $(\mathbb{R}, \mathcal{B})$ (e.g., μ is the Lebesgue or counting measure). If $f_n, n \geq 1$, are any nonnegative $\mathcal{B}$-measurable mappings such that $\int_{-\infty}^{\infty} f_n d\mu = 1, n \geq 1$, consider the family

$$P_{\alpha_n}(A) = \int_A \cdots \int f_1(t_1)\ldots f_n(t_n)\mu(dt_1)\ldots\mu(dt_n), \quad A \in \mathcal{B}^n,$$

where $\alpha_n = (1,2,\ldots,n)$. Then $\{P_{\alpha_n}, n \geq 1\}$ clearly satisfies (2) and (3), or (4). It is, of course, possible to construct more complicated sets $\{P_{\alpha_n}, n \geq 1\}$ with other specific measures such as the Gaussian (to be detailed later). The measures defined by the right side of (13) in the preceding section can be shown to be a compatible family also. A class of distributions $\{P_{\alpha_n}, n \geq 1\}$ which cannot be so factored is the following "multinomial" (discrete) distribution:

$$P_{\alpha_n}(A) = \sum_{(x_1,\ldots,x_n)\in A} f_{1,\ldots,n}(x_1,\ldots,x_n), \quad A \in \mathcal{B}^n,$$

where $\alpha_n = (1, 2, \ldots, n)$ and

$$f_{1,\ldots,n}(x_1,\ldots,x_n) = \begin{cases} \frac{n!}{x_1!x_2!\ldots x_n!} p_1^{x_1} p_2^{x_2} \ldots p_n^{x_n} \\ \quad \text{if } 0 < p_1 < 1, x_i \geq 0, \ \sum_{i=1}^n x_i = n, \ x_i \text{ integer}, \\ \quad \sum_{i=1}^n p_i = 1, \\ 0 \quad \text{otherwise.} \end{cases}$$

It can be verified that $\{P_{\alpha_n}, n \geq 1\}$ is a compatible family. Many others can be exhibited. See also the interesting Exercise 5(c) of Chapter 2.

We turn to the basic existence theorems noted earlier. One of the problems is the (formidable) notation itself. Thus a special effort is made to minimize this unavoidable difficulty. Let $(\Omega_i, \Sigma_i, P_i)_{i\in I}$ be a family of probability spaces. (Here $I = \mathbb{N}$ or $I = T \subset \mathbb{R}^+$ is possible. But I can be taken as any index set, since this has no effect on the notational or conceptual problem.) If $\Omega = \times_{i\in I}\Omega_i$ is the cartesian product, let $\pi_i : \Omega \to \Omega_i$ be the coordinate projection, so that $\omega = (\omega_i, i \in I) \in \Omega$ implies $\pi_i(\omega) = \omega_i \in \Omega_i, i \in I$. Similarly, if $\alpha_n = (i_1, \ldots, i_n)$ is a finite subset of I, $\Omega_{\alpha_n} = \times_{i\in\alpha_n}\Omega_i$ then $\pi_{\alpha_n} : \Omega \to \Omega_{\alpha_n}$ with $\pi_{\alpha_n}(\omega) = (\omega_{i_1}, \ldots, \omega_{i_n})$. Let $\Sigma_{\alpha_n} = \bigotimes_{i\in\alpha_n} \Sigma_i$ be the product σ-algebra of Ω_{α_n}. The subsets $\{\omega : \pi_i(\omega) \in A\} \subset \Omega$ for $A \in \Sigma_i$ and $\{\omega : \pi_{\alpha_n}(\omega) \in B_n\} \subset \Omega$, for $B_n \subset \Sigma_{\alpha_n}$ are called *cylinder sets* of Ω with bases A and B_n, respectively. If $\mathcal{F}$ is the collection of all finite subsets of I and α_n, α_m are two elements of $\mathcal{F}$, we say $\alpha_m < \alpha_n$ if $\alpha_m \subset \alpha_n$ and for any β_1, β_2 in $\mathcal{F}$ there is a $\gamma \in \mathcal{F}$ such that $\beta_1 < \gamma, \beta_2 < \gamma$.(Simply set $\gamma = \beta_1 \cup \beta_2 \in \mathcal{F}$). With this definition $(\mathcal{F}, <)$ becomes a *directed* set, so that for β_1, β_2 in $\mathcal{F}$, $\beta_1 < \beta_2$ is meaningful, and for any two γ_1, γ_2 there is a larger element γ_3 $(\gamma_i < \gamma_3, i = 1, 2)$.

If α_1, α_2 are in $\mathcal{F}$, $\alpha_1 < \alpha_2$, then we can define another coordinate projection between $\Omega_{\alpha_1}, \Omega_{\alpha_2}$ namely, $\pi_{\alpha_1\alpha_2} : \Omega_{\alpha_2} \to \Omega_{\alpha_1}$. [In (4), $\pi_{n,n+1}$ is such a $\pi_{\alpha_1\alpha_2}$.] For example, if $\alpha_1 = (i_1, i_2)$, $\alpha_2 = (i_1, i_2, i_3)$, and $\omega_{\alpha_2} = (\omega_{i_1}, \omega_{i_2}, \omega_{i_3})$ with $\omega_i \in \Omega_i$, then $\pi_{\alpha_1\alpha_2}(\omega_{\alpha_2}) = (\omega_{i_1}, \omega_{i_2}) = \omega_{\alpha_1} \in \Omega_{\alpha_1}$. It is now easy to verify that for any $\alpha_1 < \alpha_2 < \alpha_3$, we have $\pi_{\alpha_1\alpha_3} = \pi_{\alpha_1\alpha_2} \circ \pi_{\alpha_2\alpha_3}$, $\pi_{\alpha\alpha} =$ identity, and $\pi_{\alpha_1} = \pi_{\alpha_1\alpha_2} \circ \pi_{\alpha_2}$. Moreover, the relation between the product σ-algebras is that $\pi_{\alpha_1\alpha_2}^{-1}(\Sigma_{\alpha_1}) \subset \Sigma_{\alpha_2}$. Thus each $\pi_{\alpha_1\alpha_2}$ is $(\Sigma_{\alpha_2}, \Sigma_{\alpha_1})$-measurable. If for each $\alpha \in \mathcal{F}$, we define $P_\alpha = \bigotimes_{i\in\alpha} P_i$ by the Fubini theorem (cf. 1.3.11), then $\{P_\alpha, \alpha \in \mathcal{F}\}$ forms a compatible family, in the sense that for each $\alpha_1 < \alpha_2$ in $\mathcal{F}$ we have $P_{\alpha_2}(\pi_{\alpha_1\alpha_2}^{-1}(A)) = P_{\alpha_1}(A), A \in \Sigma_{\alpha_1}$, satisfying (4). This is a consequence of the image measure theorem (cf. 1.4.1). With this notation, the following result is obtained:

Proposition 1 *Let* $\{(\Omega_i, \Sigma_i, P_i), i \in I\}$ *be a family of probability spaces and* $\Omega = \times_{i \in I} \Omega_i$, $\Sigma_\alpha = \bigotimes_{i \in \alpha} \Sigma_i$, $P_\alpha = \bigotimes_{i \in \alpha} P_i$, $\alpha \in \mathcal{F}$, *where* $\mathcal{F}$ *is the directed set of all finite subsets of the nonempty index set* I. *If we let* $\Sigma_0 = \bigcup_{\alpha \in \mathcal{F}} \pi_\alpha^{-1}(\Sigma_\alpha)$ *be the class of all cylinder sets of* Ω, *then* Σ_0 *is an algebra of* Ω *and there exists a unique* **finitely additive** *function (a "probability") $P : \Sigma_0 \to [0, 1]$ such that $P \circ \pi_\alpha^{-1} = P_\alpha$ for each $\alpha \in \mathcal{F}$, with $\{P_\alpha, \alpha \in \mathcal{F}\}$ as a compatible family of probability measures on $\{\Sigma_\alpha, \alpha \in \mathcal{F}\}$.*

Proof Since inverse mappings preserve all set operations (i.e., unions, intersections, differences, and complements) for any collection [which need not be countable, e.g., $\pi_\alpha^{-1}(\bigcap_\beta A_\beta) = \bigcap_\beta \pi_\alpha^{-1}(A_\beta)$], it follows that $\pi_\alpha^{-1}(\Sigma_\alpha)$ is a σ-algebra on Ω, $\alpha \in \mathcal{F}$, and if $\alpha_1 < \alpha_2$, then $\pi_{\alpha_1}^{-1}(\Sigma_{\alpha_1}) \subset \pi_{\alpha_2}^{-1}(\Sigma_{\alpha_2})$. This immediately implies that Σ_0 is an algebra (and in general *not* a σ-algebra). We now define P on Σ_0. The compatibility of the P_α was already verified preceding the statement of the proposition, and it is essential for the definition of P.

Let $A \in \Sigma_0$. Then $A \in \pi_\alpha^{-1}(\Sigma_\alpha)$ and perhaps also $A \in \pi_\beta^{-1}(\Sigma_\beta)$. Thus there are $B_1 \in \Sigma_\alpha, B_2 \in \Sigma_\beta$ such that $A = \pi_\alpha^{-1}(B_1) = \pi_\beta^{-1}(B_2)$. Let $\gamma = \alpha \cup \beta \in \mathcal{F}$, so that $\gamma > \alpha, \gamma > \beta$. Then we know from the definition of the π_α above that $\pi_\alpha = \pi_{\alpha\gamma} \circ \pi_\gamma$ and $\pi_\beta = \pi_{\beta\gamma} \circ \pi_\gamma$. Consequently A can be represented in two ways:

$$A = \pi_\alpha^{-1}(B_1) = \pi_\gamma^{-1}(\pi_{\alpha\gamma}^{-1}(B_1))$$

or

$$A = \pi_\beta^{-1}(B_2) = \pi_\gamma^{-1}(\pi_{\beta\gamma}^{-1}(B_2)) \tag{5}$$

Comparing the last quantities of both lines of (5), we get

$$\pi_\gamma^{-1}(\pi_{\alpha\gamma}^{-1}(B_1) - \pi_{\beta\gamma}^{-1}(B_2)) = \emptyset.$$

But $\pi_\gamma : \Omega \to \Omega_\gamma$ is onto $[\pi_\gamma(\Omega) = \Omega_\gamma]$, and this implies $\pi_\gamma^{-1}(B) = \emptyset$ iff $B = \emptyset$. Hence $\pi_{\alpha\gamma}^{-1}(B_1) = \pi_{\beta\gamma}^{-1}(B_2)$. By the compatibility of the P_α noted above,

$$P_\alpha(B_1) = P_\gamma(\pi_{\alpha\gamma}^{-1}(B_1)) = P_\gamma(\pi_{\beta\gamma}(B_2)) = P_\beta(B_2). \tag{6}$$

Hence if we set $P(A) = P_\alpha(B_1) = P_\beta(B_2)$ then P is unambiguously defined, and $P : \Sigma_0 \to [0, 1]$. Moreover, $P_\alpha(B_1) = P(\pi_\alpha^{-1}(B_1)) = P(A)$, so that $P_\alpha = P \circ \pi_\alpha^{-1}$. If A_1, A_2 are disjoint sets in Σ_0, then $A_1 \in \pi_{\alpha_1}^{-1}(\Sigma_{\alpha_1}), A_2 \in \pi_{\alpha_2}^{-1}(\Sigma_{\alpha_2})$, so that if $\gamma = \alpha_1 \cup \alpha_2$, then A_1, A_2 are in $\pi_\gamma^{-1}(\Sigma_\gamma)$ $[\supset \pi_{\alpha_1}^{-1}(\Sigma_{\alpha_1}) \cup \pi_{\alpha_2}^{-1}(\Sigma_{\alpha_2})]$. Hence $A_i = \pi_\gamma^{-1}(D_i)$, $D_i \in \Sigma_\gamma, i = 1, 2$, disjoint, $P(\Omega) = 1$, and

$$P(A_1 \cup A_2) = P_\gamma(D_1 \cup D_2) = P_\gamma(D_1) + P_\gamma(D_2) = P(A_1) + P(A_2).$$

Thus P is a finitely additive "probability." The uniqueness of P is now evident. Note that we have not used the product measure property of P_α in the above argument, and the result is thus valid for *any* compatible family of $\{P_\alpha, \alpha \in \mathcal{F}\}$. The proof is complete.

The main problem now is to show (under suitable conditions) that P is σ-additive on Σ_0, so that it has a unique σ-additive extension $\overline{P}$ to $\sigma(\Sigma_0)$, $(= \Sigma$, say). The triple (Ω, Σ, P) is then the desired probability space, giving our first existence result. We now establish this property for the product case.

Proposition 2 *The function $P : \Sigma_0 \to [0,1]$ of the preceding proposition is σ-additive, when each P_α is a product probability.*

Proof Since P is already shown to be additive, σ-additivity is equivalent to showing that P is continuous at $\emptyset$ from above, i.e., if $A_n \downarrow \emptyset$, $A_n \in \Sigma_0$, then $P(A_n) \to 0$. This is further equivalent to showing that for any decreasing sequence $\{A_n, n \geq 1\} \subset \Sigma_0$, $P(A_n) \geq \delta > 0$ implies $\bigcap_{n=1}^{\infty} A_n \neq \emptyset$, since $P(A_n) \geq P(A_{n+1})$ is implied by additivity. This is verified as follows.

Since $\{A_n \downarrow\} \subset \Sigma_0 = \bigcup_{\alpha \in \mathcal{F}} \pi_\alpha^{-1}(\Sigma_\alpha)$, there exists a sequence $\{\alpha_n, n \geq 1\} \subset \mathcal{F}$ such that $A_n \in \pi_{\alpha_n}^{-1}(\Sigma_{\alpha_n}), n \geq 1$. But we have seen that for $\alpha < \beta$, $\pi_\beta^{-1}(\Sigma_\beta) \supset \pi_\alpha^{-1}(\Sigma_\alpha)$. Thus replacing the sequence $\{\alpha_n, n \geq 1\}$ by $\{\beta_n, n \geq 1\}$ with $\beta_1 = \alpha_1, \ldots, \beta_n = \bigcup_{k=1}^{n} \alpha_k \in \mathcal{F}$, if necessary, one can assume that $\alpha_1 < \alpha_2 < \ldots$. With this, each $A_n = \pi_{\alpha_n}^{-1}(B_n)$ for some $B_n \in \Sigma_{\alpha_n}$, and $P(A_n) = P_\alpha(B_n)$. Note that A_n is a cylinder: $A_n = B_n \times \times_{i \in I - \alpha_n} \Omega_i$. Since $P_{\alpha_n} = \bigotimes_{i \in \alpha_n} P_i$ is a product measure, by the Fubini theorem, one has, if $\alpha_n = (i_1, \ldots, i_n)$, that the function

$$h_n : \omega_{i_1} \mapsto \int_{\Omega_{i_2}} \cdots \int_{\Omega_{i_n}} \chi_{B_n}(\omega_{i_1}, \ldots, \omega_{i_n}) P_{i_2}(d\omega_{i_2}) \ldots P_{i_n}(d\omega_{i_n})$$

is P_{i_1}-measurable, and $P_{\alpha_n}(B_n) = \int_{\Omega_{i_n}} h_n(\omega_{i_1}) P_{i_1}(d\omega_{i_1}) = P(A_n) \geq \delta > 0$. Also, $0 \leq h_n \leq 1$ and the decreasing sequence $\{h_n\}$ has its integrals bounded below by $\delta > 0$, it follows by the Lebesgue bounded convergence theorem that there exists an $\omega_{i_1}^0 \in \Omega_1$ such that $h_n(\omega_{i_1}^0) \not\to 0$ as $n \to \infty$. Next apply the same argument to the function

$$g_n : \omega_{i_2} \mapsto \int_{\Omega_{i_3}} \cdots \int_{\Omega_{i_n}} \chi_{B_n}(\omega_{i_1}^0, \omega_{i_2}, \ldots, \omega_{i_n}) P_{i_3}(d\omega_{i_3}) \ldots P_{i_n}(d\omega_{i_n})$$

and deduce that there is an $\omega_{i_2}^0 \in \Omega_{i_2}$ such that $g_n(\omega_{i_2}^0) \not\to 0$. Repeating this procedure, we see that $\chi_{B_n}(\omega_{i_1}^0, \ldots, \omega_{i_k}^0, \omega_{i_{k+1}} \ldots, \omega_{i_n})$ cannot be zero for *all* points $(\omega_{i_{k+1}} \ldots, \omega_{i_n}) \in \Omega_{i_{k+1}} \times \ldots \times \Omega_{i_n}$. Thus there exists an $\omega^0 \in \Omega$ such that $\omega^0 = (\omega_{i_1}^0, \ldots, \omega_{i_k}^0, \omega_{i_{k+1}} \ldots) \in A_n$, for any $n > k$. If $\beta = \bigcup_{i=1}^{\infty}, \alpha_i$ we can choose ω^0 such that its countable set of components corresponding to the countable set β, and the rest arbitrarily in $\times_{i \in I - \beta} \Omega_i$. Then by the form of A_n (that they are cylinders), $\omega^0 \in A_n$ for all $n \geq 1$, so that $\omega^0 \in \bigcap_{n=1}^{\infty} A_n$. This shows that $P(A_n) \geq \delta > 0$ implies $\bigcap_n A_n \neq \emptyset$, and hence P is σ-additive on Σ_0. Thus $(\Omega, \Sigma, \overline{P})$ exists, where $\overline{P}$ is the unique extension of P onto

$\Sigma = \sigma(\Sigma_0)$, by the Hahn extension theorem. This completes the proof.

Remark The space $(\Omega, \Sigma, \overline{P})$ is called the *product* of $\{(\Omega_i, \Sigma_i, P_i), i \in I\}$, and usually it is denoted $(\Omega, \Sigma, \overline{P}) = \bigotimes_{i\in I}(\Omega_i, \Sigma_i, P_i)$.

The preceding two propositions can be combined to yield the following result about the existence of a class of random families. This is obtained by B. Jessen, noted above, in a different form.

Theorem 3 (Jessen). *Let $\{(\Omega_i, \Sigma_i, P_i), i \in I\}$ be any family of probability spaces. Then there exists a probability space (Ω, Σ, P), their product, and a family of random variables (i.e., measurable mappings) $\{X_i, i \in I\}$, where $X_i : \Omega \to \Omega_i$ is defined as $X_i(\omega) = \omega_i \in \Omega_i$ for each $\omega = (\omega_i, i \in I)$ (X_i is the coordinate mapping), such that* (a) *they are all mutually independent, and* (b) *for any measurable rectangle $A_{i_1} \times \ldots \times A_{i_n} \in \Sigma_{i_1} \times \ldots \Sigma_{i_n}$ one has*

$$P\{\omega : X_{i_1}(\omega) \in A_{i_1}, \ldots, X_{i_n}(\omega) \in A_{i_n}\} = \prod_{k=1}^{n} P_{i_k}(A_{i_k}), \quad n > 1. \tag{7}$$

Proof The first part is a restatement of Proposition 2, and since $X_i^{-1}(\Sigma_i) \subset \Sigma_0$, it follows that each X_i is measurable. That the X_i are independent and that (7) is valid now follow immediately. Indeed, each X_i is a coordinate function, and hence $X_i^{-1} = \pi_i^{-1}$ in a different notation, so that we have

$$P[X_{i_1} \in A_{i_1}, \ldots, X_{i_n} \in A_{i_n}]$$

$$= P\left[\bigcap_{k=1}^{n} \pi_{i_k}^{-1}(A_{i_k})\right]$$

$$= P[\pi_\alpha^{-1}(A_\alpha)]$$

[here α is $(i_1, \ldots, i_n)$ and $A_\alpha = A_{i_1} \times \ldots \times A_{i_n} \in \Sigma_\alpha$]

$$= P_\alpha(A_\alpha)$$

$$= \prod_{k=1}^{n} P_{i_k}(A_{i_k}) \text{ (since } P_\alpha \text{ is a product measure)} \tag{8}$$

$$= \prod_{k=1}^{n} P[\pi_{i_k}^{-1}(A_{i_k})]. \tag{9}$$

Thus (8) is (7), and (9) proves independence, since it is true on Σ_0 and hence by the (π, λ) criterion (cf. Theorem 2.1.3), on all of Σ. This completes the proof of the theorem.

If we take each (Ω_i, Σ_i) as the Borelian line $(\mathbb{R}, \mathcal{B})$ and $P_i : \mathcal{B} \to [0,1]$ as any probability (or a distribution) function, then the above result ensures the existence of arbitrary families of mutually independent random variables. Thus all the random families considered in Chapter 2 exist. Regarding the generality of the preceding result, it must be *emphasized that the spaces (Ω_i, Σ_i) are abstract and that no topological conditions entered in the discussion.* Now looking at the proof of the key Proposition 2 (and hence of Theorem 3), one feels that the full force of independence was not utilized. In fact, one can use the same argument for certain (Markovian-type) dependence families, including those of the preceding section. Such an extension is presented for the case when the index set I is $\mathbb{N}$, the natural numbers. This result is due to C. Ionescu Tulcea (1949). [It may be noted in passing that if I is uncountable, then (Ω, Σ, P) will not be separable even when each Ω_i is finite.]

As a motivation, let us first record the following simple result when I is countable.

Lemma 4 *Let $\{(\Omega_i, \Sigma_i, P_i), i \in \mathbb{N}\}$ be probability spaces and (Ω, Σ, P) be their product, as given by Theorem 3. If $A_i \in \Sigma_i$ and $A = \times_{i\in\mathbb{N}} A_i$, then $A \in \Sigma$ and*

$$P(A) = \lim_{n\to\infty} \prod_{i=1}^{n} P_i(A_i) \quad \left[= \prod_{i=1}^{\infty} P_i(A_i)\right]. \tag{10}$$

Proof Let $\alpha_n = (1, \ldots, n)$ and $\pi_i : \Omega \to \Omega_i$ be the coordinate projections, so that $\pi_{\alpha_n} : \Omega \to \Omega_{\alpha_n} = \times_{i=1}^{n} \Omega_i$. If $B_n = A_1 \times \ldots \times A_n$ and $\Omega^{(n)} = \times_{i>n} \Omega_i$, then $\pi_{\alpha_n}^{-1}(B_n) = B_n \times \Omega^{(n)} \in \Sigma$ and $\pi_{\alpha_n}^{-1}(B_n) \supset \pi_{\alpha_{n+1}}^{-1}(B_{n+1})$. Also,

$$A = \lim_{n\to\infty} \pi_{\alpha_n}^{-1}(B_n) = \bigcap_{n=1}^{\infty} \pi_{\alpha_n}^{-1}(B_n),$$

so that $A \in \Sigma$, since $\Sigma = \sigma(\Sigma_0)$ is the σ-algebra generated by the cylinder set algebra Σ_0 of Ω. Hence by Proposition 2,

$$P(A) = \lim_{n\to\infty} P(\pi_{\alpha_n}^{-1}(B_n)) = \lim_{n\to\infty} P_{\alpha_n}(B_n)$$

$$= \lim_{n\to\infty} \prod_{i=1}^{n} P_i(B_i) \quad \text{[by (7)]}.$$

This establishes (10), and hence the lemma.

It is of interest to note that the algebra of cylinder sets Σ_0, as defined in Proposition 1, can be described in the present case as follows. If $A_n \in \Sigma_n$, then $\pi_n^{-1}(A_n) \in \Sigma_0$. But $\pi_n^{-1}(A_n) = \Omega_1 \times \ldots \times \Omega_{n-1} \times A_n \times \Omega^{(n)} = \pi_{\alpha_n}^{-1}(D_n)$, where $D_n = \Omega_1 \times \ldots \times \Omega_{n-1} \times A_n$, and $\alpha_n = (1, \ldots, n)$. Thus if $\alpha_m < \alpha_n$ is again written for $\alpha_m \subset \alpha_n$ and $\Sigma_{\alpha_n} = \bigotimes_{i=1}^{n} \Sigma_i$, we have $\Sigma_0 = \bigcup_{n\geq 1} \tilde{\Sigma}_{\alpha_n}$, where $\tilde{\Sigma}_{\alpha_n} = \pi_{\alpha_n}^{-1}(\Sigma_{\alpha_n})$. This reduces the compatibility condition to saying, since $\tilde{\Sigma}_{\alpha_n} \subset \tilde{\Sigma}_{\alpha_{n+1}}$, that $\tilde{P}_{\alpha_n} = \tilde{P}_{\alpha_{n+1}} | \tilde{\Sigma}_{\alpha_n}$ where $\tilde{P}_{\alpha_n}(\pi_{\alpha_n}^{-1}(D_n)) = P_{\alpha_n}(D_n)$

for $D_n \in \Sigma_{\alpha_n}$; i.e., the image measures $\tilde{P}_{\alpha_n}$ of P_{α_n} on $\tilde{\Sigma}_{\alpha_n}$ are extensions of each other as we go from $\tilde{\Sigma}_{\alpha_n}$ to $\tilde{\Sigma}_{\alpha_{n+1}}$. Hence by Proposition 1, there exists a unique additive mapping on the algebra Σ_0 into $[0,1]$ if only we are given probabilities P_{α_n} on Σ_{α_n} such that

$$P_{\alpha_{n+1}}(A_1 \times \ldots \times A_n \times \Omega_{n+1}) = P_{\alpha_n}(A_1 \times \ldots \times A_n), \ A_i \in \Sigma_i, \qquad (11)$$

$1 \leq i \leq n$. The P_{α_n} need not be product measures. Indeed (11) is just the statement that $\tilde{P}_{\alpha_n} = \tilde{P}_{\alpha_{n+1}}|\tilde{\Sigma}_{\alpha_n}$. (Verify this.)

Suppose now that we are given an *initial probability* P_1 on Σ_1. Let us then say that P_{α_n} is a *productlike* measure on Σ_{α_n}, where $\alpha_n = (1, \ldots, n)$ as before, if there exist mappings $P_n(\cdot,\cdot) : \Sigma_n \times \Omega_{\alpha_{n-1}} \to \mathbb{R}^+$ such that

(i) $P_n(A_n;\cdot)$ is $\Sigma_{\alpha_{n-1}}$-measurable for each $A_n \in \Sigma_n$, and
(ii) $P_n(\cdot;\omega_{\alpha_{n-1}})$ is a probability for each $\omega_{\alpha_{n-1}} \in \Omega_{\alpha_{n-1}} = \times_{i=1}^{n-1}\Omega_i, n > 1$,
in terms of which

$$P_{\alpha_n}(A_1 \times \ldots \times A_n) = \int_{A_1} P_1(d\omega_1)\int_{A_2} P_2(d\omega_2;\omega_1)\ldots\int_{A_n} P_n(d\omega_n;\omega_1,\ldots,\omega_{n-1}), \qquad (12)$$

for each measurable rectangle $A_1 \times \ldots \times A_n$ of Σ_{α_n}.

The classical (unsymmetric) Fubini theorem implies that (12) is well defined and has a unique σ-additive extension to Σ_{α_n}, $n > 1$. Moreover, condition (ii) on $P_n(\cdot;\cdot)$ implies that the P_{α_n} satisfy (11). In other words, if $\tilde{P}_{\alpha_n}$ is the image measure of P_{α_n} (thus $\tilde{P}_{\alpha_n} \circ \pi_{\alpha_n}^{-1} = P_{\alpha_n}$), then $\{\tilde{P}_{\alpha_n}, \tilde{\Sigma}_{\alpha_n}, n > 1\}$ is a compatible family of probability measures. Here $\tilde{P}_{\alpha_1} = P_1$ is the initial probability. If each $P_n(\cdot;\omega_{\alpha_{n-1}})$ is independent of the second variable $\omega_{\alpha_{n-1}}$ then P_{α_n} is simply the product probability. But what is the basis of (12)? Are there any $P_n(\cdot;\cdot)$ other than the absolute probabilities to talk about in connection with the generalization? Indeed, such measures are regular conditional probabilities if each (Ω_i, Σ_i) is $(\mathbb{R}, \mathcal{B})$, the Borelian line, by Proposition 2.8. Thus such productlike measures can exist without being product measures. Also (12) is obtained from the commutative property of conditional expectations and the image probability theorem, since we have $\tilde{\Sigma}_{\alpha_1} \subset \tilde{\Sigma}_{\alpha_2} \cdots \subset \tilde{\Sigma}_{\alpha_n}$ and for any bounded $f : \Omega \to \mathbb{R}$, measurable for $\Sigma_{\alpha_{n+1}}$,

$$E(f) = E(E^{\tilde{\Sigma}_{\alpha_1}}(\cdots(E^{\tilde{\Sigma}_{\alpha_n}}(f))\cdots)). \qquad (13)$$

The right side of (13) is just the right side of (12) when *regular conditional probabilities exist.* We now present the desired extension using the above notation.

Theorem 5 (Tulcea) *Let $\{(\Omega_i, \Sigma_i), i \in \mathbb{N}\}$ be a family of measurable spaces and suppose that $P_{\alpha_n} : \Sigma_{\alpha_n} \to [0,1]$ is a productlike probability for each $n \geq 1$ with $P_{\alpha_1} = P_1$ as the initial probability. Then there exists a unique* *probability P on (Ω, Σ) such that $P_{\alpha_n} = P \circ \pi_{\alpha_n}^{-1}, n \geq 1$, where π_{α_n} is the coordinate projection of Ω into Ω_{α_n}, (Ω, Σ) being the (product) measurable space introduced above.*

Proof It was already noted that $\{P_{\alpha_n}, n \geq 1\}$ defined by (12) forms a compatible family. Hence there exists a finitely additive $P : \Sigma_0 \to [0, 1]$ such that $P \circ \pi_{\alpha_n}^{-1} = P_{\alpha_n}, n \geq 1$. To show that it is σ-additive, we consider an arbitrary sequence $\tilde{C}_n \supset \tilde{C}_{n+1}$, $\{\tilde{C}_n, n \geq 1\} \subset \Sigma_0, \bigcap_{n=1}^{\infty} \tilde{C}_n = \emptyset$, and verify that $P(\tilde{C}_n) \to 0$. If this is false, then there exists a $\delta > 0$ and $P(\tilde{C}_n) \geq \delta > 0$ for all n. Proceeding exactly as in Proposition 2, it may be assumed that there exist $n_1 < n_2 < \ldots$ such that $C_i \in \tilde{\Sigma}_{\alpha_{n_i}}, i \geq 1$, and then $P_{\alpha_{n_i}}(C_i) \geq \delta$, where C_i is the base of the cylinder $\tilde{C}_i$. Then

$$h_i : \omega_1 \mapsto \int_{\Omega_2} P_2(d\omega_2; \omega_1) \int_{\Omega_3} P_3(d\omega_3; \omega_1, \omega_2) \ldots \int_{\Omega_{n_i}} \chi_{C_i}(\omega_1, \ldots, \omega_{n_i})$$

$$\times\, P_{n_i}(d\omega_n; \omega_1, \ldots, \omega_{n_i-1})$$

is P_1-measurable, $1 \geq h_i \geq h_{i+1} > 0$, and $P_{\alpha_{n_i}}(C_i) = \int_{\Omega_i} h_i(\omega_1) P_1(d\omega_1) \geq \delta > 0$. Here all are Lebesgue integrals and, by the monotone convergence, there is an $\omega_1^0 \in \Omega_1$ such that $h_i(\omega_1^0) \not\to 0$ as $i \to \infty$. We then repeat this argument and deduce, as in the proof of Proposition 2, that $\bigcap_{n\geq 1}^{\infty} \tilde{C}_n \neq \emptyset$. Thus P must be σ-additive on Σ_0. The rest of the argument is the same in both cases, and this terminates the proof.

One of the interesting features of this result is that Jessen's theorem extends to certain nonindependence (but just productlike) cases. Again no topological conditions exactly intervene, but the *existence of regular conditional probabilities,* is assumed instead. We now show that this result implies the existence of Markov processes that were discussed in the last section.

Theorem 6 *Let $\{(\Omega_i, \Sigma_i), i \in \mathbb{N}\}$ be a sequence of measurable spaces and $\Omega = \times_{i=1}^{\infty} \Omega_i, \Sigma = \sigma(\Sigma_0)$, with $\Sigma_0 = \bigcup_{n\geq 1} \pi_{\alpha_n}^{-1}(\Sigma_{\alpha_n})$ as the algebra of cylinder sets. If $P_1 : \Sigma_1 \to [0, 1]$ is an initial probability, for each $n > 1$, let $P_n(\cdot;\cdot) : \Sigma_n \times \Omega_{n-1} \to [0, 1]$ be a (Markov) transition probability in the sense that*

(i) *$P_n(\cdot; \omega_{n-1})$ is a probability measure on Σ_n for each $\omega_{n-1} \in \Omega_{n-1}$,*
(ii) $P_n(A; \cdot)$ is *Σ_{n-1}-measurable for each $A \in \Sigma_n$.*

Then there is a unique probability $P : \Sigma \to [0, 1]$, a Markov process $\{X_n, n \geq 1\}$ on (Ω, Σ, P) such that $P[X_1 \in A] = P_1(A), A \in \Sigma_1$, and for each $A_{\alpha_n} \in \Sigma_n$ of the form $A_{\alpha_n} = A_1 \times \ldots \times A_n$, $\alpha_n = (1, 2, \ldots, n)$, we have

$$P(\pi_{\alpha_n}^{-1}(A_{\alpha_n})) = \int_{A_1} P_1(d\omega_1) \int_{A_2} P_2(d\omega_2; \omega_1) \ldots \int_{A_n} P_n(d\omega_n; \omega_{n-1}). \quad (14)$$

In fact, $X_n(\omega) = \omega_n \in \Omega_n, n \geq 1$, *defines the above Markov process on* (Ω, Σ, P) *with values in the spaces* $\{\Omega_n, n \geq 1\}$.

Proof If P_{α_n} denotes the measure defined by the right side of (14), then it is a productlike probability, since, comparing it with (12), we can take

$$P_k(\cdot; \omega_1, \ldots, \omega_{k-1})$$

as $P_k(\cdot; \omega_{k-1})$. Consequently, the existence of a unique probability P on (Ω, Σ) follows from Theorem 5. Since $X_n : \Omega \to \Omega_n$ are coordinate functions (indeed $X_n = \pi_n$) it is clear that $X_n^{-1}(\Sigma_n) \subset \Sigma$ and the X_n are measurable. Thus $\{X_n, n \geq 1\}$ is a random process. To see it is a Markov process, we need to show that

$$P([X_n \in A_n]|X_1, \ldots, X_{n-1}) = P([X_n \in A_n]|X_{n-1}) \text{ a.e., } n > 1, \tag{15}$$

for each $A_n \in \Sigma_n$. Let $C = [X_n \in A_n]$, and $B \in \sigma(X_1, \ldots X_{n-1}) = \mathcal{B}$. We can restrict B to the generators of $\mathcal{B}$ and verify (15) on any such B. Hence, writing $P(C|X_1, \ldots, X_{n-1})$ as $E^{\mathcal{B}}(\chi_C)$ and expressing $B = \bigcap_{i=1}^{n-1} \pi_i^{-1}(A_i), A_i \in \Sigma_i$, we have

$$\int_B E^{\mathcal{B}}(\chi_C) P_{\mathcal{B}}(d\omega) = \int_B \chi_C dP = P(B \cap C)$$

$$= \int_{\bigcap_{i=1}^n \pi_i^{-1}(A_i)} dP$$

[since $C = X_n^{-1}(A_n) = \pi_n^{-1}(A_n)$ and B is as above]

$$= \int_{A_1} P_1(d\omega_1) \int_{A_2} P_2(d\omega_2; \omega_1) \ldots \int_{A_n} P_n(d\omega_n; \omega_{n-1}) \text{ [by (14)]}$$

$$= \int_{A_1 \times \ldots \times A_{n-1}} P_{\alpha_{n+1}}(d\omega_1, \ldots, d\omega_{n-1}) P_n(A_n; \omega_{n-1})$$

[where $P_{\alpha_{n-1}}$ on Σ_{α_n} is given by (12)]

$$= \int_B P([X_n \in A_n]|X_{n-1})(\omega) P_{\mathcal{B}}(d\omega) \text{ (by the image probability law).}$$

Since the extreme integrands are $\mathcal{B}$-measurable and B is a generator, we can identify them P-uniquely. But this is (15) in a different notation, and completes the proof of the theorem.

This result implies that all random families considered in the last section exist. Regarding both Theorems 5 and 6, the reader may have noticed the special role played by the availability of a minimum value in the index set.

This is no accident. If the index set is as general as in Jessen's theorem, how should we proceed in establishing a corresponding result? The existence of a minimal element allowed us to simplify the compatibility condition, which is essential in proving the existence of a finitely additive function $P : \Sigma_0 \to [0,1]$ (cf. Proposition 1). After that, the result in Proposition 2 did not use this. In the general case, therefore, we need the (strengthened) compatibility of the P_α and a precise version is given below. It enables us to assert the existence of (Markov) processes $\{X_t, t \in T\}$, where $T \subset \mathbb{R}$ is any index set.

If $\{(\Omega_i, \Sigma_i), i \in I\}$ is a family of measurable spaces, where I is an index set, let $\mathcal{F}$ be the directed set (by inclusion) of all finite subsets of I and $\Omega_\alpha = \times_{i\in\alpha}\Omega_i$, $\Sigma_\alpha = \bigotimes_{i\in\alpha}\Sigma_i, \alpha \in \mathcal{F}$, be as in Proposition 1. For each $\alpha \in \mathcal{F}$, suppose a probability $P_\alpha : \Sigma_\alpha \to [0,1]$ is given. Then the system $\{P_\alpha, \alpha \in \mathcal{F}\}$ is termed *generalized productlike* if for each α, β in $\mathcal{F}$, $\alpha < \beta$, such that $\beta = (\alpha, i_1, \ldots, i_k) \subset I$ (is a finite set), we have

$$P_\beta(A_\beta) = \int_{A_\alpha} P_\alpha(d\omega_\alpha) \int_{A_{i_1}} P_{i_1}(d\omega_{i_1}; \omega_\alpha) \times \\ \cdots \int_{A_{i_k}} P_{i_k}(d\omega_{i_k}; \omega_\alpha, \omega_{i_1}, \ldots, \omega_{i_{k-1}}), \qquad (16)$$

where $A_\beta = A_\alpha \times A_{i_1} \times \ldots \times A_{i_k}$ with $A_\alpha \in \Sigma_\alpha$, $A_{i_j} \in \Sigma_{i_j}$, and for each i_j

(a) $P_{i_j}(A_{i_j}; \cdot)$ is $\Sigma_\alpha \bigotimes \Sigma_{i_1} \bigotimes \ldots \bigotimes \Sigma_{i_{j-1}}$-measurable
(b) $P_{i_j}(\cdot\,; \omega_\alpha, \omega_{i_1}, \ldots, \omega_{i_{j-1}})$ is a Probability on Σ_{i_j}.

This definition reduces to (12) if I has a minimal element and I is countable, but is stronger otherwise. Also, if $A_{i_1} = \Omega_{i_1}, \ldots, A_{i_k} = \Omega_{i_k}$, so that $A_\beta = \pi_{\alpha\beta}^{-1}(A_\alpha)$, where $\pi_{\alpha\beta} : \Omega_\beta \to \Omega_\alpha$ is the coordinate projection mapping, then (16) implies $P_\beta(A_\beta) = P_\alpha(A_\alpha)$, or equivalently, $P_\beta(\pi_{\alpha\beta}^{-1}(A_\alpha)) = P_\alpha(A_\alpha)$. In fact, for *any* $\alpha < \beta$, from $\mathcal{F}$, (16) implies that $P_\beta \circ \pi_{\alpha\beta}^{-1} = P_\alpha$. This is precisely the compatibility condition for the family $\{P_\alpha, \alpha \in \mathcal{F}\}$. It is then immediately obtained from the argument used in the proof of Proposition 1 that there is a unique additive set function P on the algebra of cylinder sets Σ_0 into $[0,1]$ such that $P \circ \pi_\alpha^{-1} = P_\alpha, \alpha \in \mathcal{F}$, where $\pi_\alpha : \Omega \to \Omega_\alpha$ is the coordinate projection. With this strengthening, Theorem 5 may be restated as follows:

Theorem 7 *Let $\{(\Omega_i, \Sigma_i), i \in I\}$ be measurable spaces, and $\{P_\alpha, \alpha \in \mathcal{F}\}$ be a system of generalized productlike probabilities on $\{\Sigma_\alpha, \alpha \in \mathcal{F}\}$ of the given family. Then there exists a unique (σ-additive) probability P on the σ-algebra Σ generated by the cylinder sets Σ_0 of the spaces $\{(\Omega_i, \Sigma_i), i \in I\}$ such that $P_\alpha = P \circ \pi_\alpha^{-1}, \alpha \in \mathcal{F}$.*

The proof is almost identical to that of Theorem 5, and is left to the reader.

Using this form, a continuous parameter version of Theorem 6 can be obtained quickly. However, the "one-step" transition probabilities employed in that result will not be meaningful here, since there is no such "step" in the continuous parameter case. A precise version of the result is as follows, and one needs the full force of the Chapman-Kolmogorov equation [cf. Proposition 3.4, particularly Eq. (7) there] which is now **assumed to hold everywhere.**

Theorem 8 *Let* $\{(\Omega_t, \Sigma_t), t \in T \subset \mathbb{R}\}$ *be a family of measurable spaces and* (Ω, Σ) *be their product.* (i) *If* T *has a minimal element* t_0*, let* P_0 *be the initial probability on* Σ_{t_0}*, or* (ii) *if there is no minimal element, let* P_t *be a probability on* Σ_t*,for each* $t \in T$*. Let* $r < s < t$ *be points in* T*, and let there be given Markov transition probabilities* $p_{r,s}(\cdot,\cdot) : \Sigma_s \times \Omega_r \to \mathbb{R}^+$*,* $p_{s,t} : \Sigma_t \times \Omega_s \to \mathbb{R}^+$*, and* $p_{r,t} : \Sigma_t \times \Omega_r \to \mathbb{R}^+$ *such that they satisfy the Chapman-Kolmogorov equation identically:*

$$p_{r,t}(A, x) = \int_{\Omega_s} p_{s,t}(A, y) p_{r,s}(dy, x), \quad r < s < t, x \in \Omega_r. \tag{17}$$

Then there exists (in either case) a unique probability P *on* Σ *and a Markov process* $\{X_t, t \in T\}$ *on* (Ω, Σ, P) *with values in* $\{\Omega_t, t \in T\}$ *such that* (i') $P[X_{t_0} \in A] = P_0(A), A \in \Sigma_{t_0}$ *or* (ii') $P[X_t \in B] = P_t(B), B \in \Sigma_t$*, respectively, and*

$$P \circ \pi_\alpha^{-1}(A_\alpha) = \int_{A_{t_1}} p_{t_1}(d\omega_{t_1}) \int_{A_{t_2}} p_{t_2}(d\omega_{t_2}; \omega_{t_1}) \cdots \int_{A_{t_n}} p_{t_n, t_{n-1}}(d\omega_{t_n}; \omega_{t_{n-1}}), \tag{18}$$

where $A_\alpha = A_{t_1} \times \ldots \times A_{t_n} \in \Sigma_\alpha$*,* $\alpha = (t_1, \ldots, t_n) \subset T$*. In fact,* $X_t(\omega) = \omega_t \in \Omega_t, t \in T, \omega = (\omega_t, t \in T) \in \Omega$*, defines the above Markov process.*

This result follows immediately from Theorem 7, and the fact that (17) implies the consistency of the system defined by (18). Note that if the minimal element exists, then we can always start for each $\alpha \in \mathcal{F}$ from the minimal $t_0 \in T$. The modifications are simple, and again are omitted (given as Problem 23).

Observe that we strengthen (17) so as to be valid for *all* $x \in \Omega_r$, not just a.a. $x \in \Omega_r$. This is useful in constructing Markov processes from *each* given starting point; i.e., in case (i'), P_{t_0} may be replaced by $P_{t_0}^{x_0} [= P_{t_0}(\cdot|x_0)$ with $P_{t_0}(A|x_0) = \chi_A(x_0)]$. With this result we now have the complete existence theory for the work of Section 3. However, the demand for regular conditional probabilities is not always fulfilled naturally unless we have topologies in the Ω_t-spaces with which to use the theory of Section 2. In those cases (with topologies) we can present a more general and natural proposition for applications. This is due to Kolmogorov (1933), and its extension to Bochner (1955). These results are sufficient for all the processes we deal with in this book, and in fact for essentially all stochastic theory. We thus turn to them. They again may be skipped at a first reading.

Let $\{\Omega_t, t \in T\}$ be a family of Hausdorff topological spaces and T a nonempty index set. Let $\Omega = \times_{t\in T}\Omega_t$ be the cartesian product space with the (Tychonov) product topology. If for each $t \in T$, Σ_t is a σ-algebra of Ω_t containing all of its compact sets, let $\Sigma_0 = \bigcup_{\alpha\in\mathcal{F}} \pi_\alpha^{-1}(\Sigma_\alpha)$ be the algebra of cylinder sets of Ω where $\pi_\alpha : \Omega \to \Omega_\alpha(= \times_{t\in\alpha}\Omega_t)$ is the coordinate projection for each α in $\mathcal{F}$, the directed set (by inclusion) of all **finite** subsets of T, as in the preceding discussion. The point of interest in this case is the family of all cylinder sets of Ω with compact bases from the Ω_α, as it plays a key role in the existence theorem. Thus let $\mathcal{C} = \{C \subset \Omega : C = \pi_\alpha^{-1}(K), K \subset \Omega_\alpha \text{ compact}\}$. Clearly $\mathcal{C} \subset \Sigma_0$. Even though the elements of $\mathcal{C}$ are not necessarily compact (since Ω need not be a compact space), the following technical property is available.

Proposition 9 *If* $\{C_n, n \geq 1\} \subset \mathcal{C}$ *such that* $\bigcap_{k=1}^n C_k \neq \emptyset$ *for each* $n \geq 1$, *then* $\bigcap_{n=1}^\infty C_n \neq \emptyset$.

Proof Since each C_n has a compact base, $C_n = K_{\alpha_n} \times \times_{t\in T-\alpha_n}\Omega_t$, where $K_{\alpha_n} \subset \Omega_{\alpha_n}$ is compact, so that $C_n = \pi_{\alpha_n}^{-1}(K_{\alpha_n})$. Let $T_1 = \bigcup_n \alpha_n$. Then T_1 is a countable subset of the index set T. Thus for each $t \in T_1$, let α_{n_t} be an element of T_1 which contains t, and $K_{\alpha_{n_t}}$ be the (compact) base of C_{n_t}. Since each $C_n \neq \emptyset$, $K_{\alpha_{n_t}} \neq \emptyset$ also. Let $\omega^0 \in \Omega$ be a point such that $\omega_t^0 \in K_{\alpha_{n_t}}, t \in T_1$. Since $\omega^0 = (\omega_t^0, t \in T)$, this is possible. But Ω is not generally compact. Thus we manufacture a compact subset $K \subset \Omega$ and select a suitable net in K, whose cluster point (there is at least one) in K will lie in each C_n. This will finish the argument.

Let $K = \times_{t\in T_1}K_{\alpha_{n_t}} \times \times_{t\in T-T_1}\{\omega_t^0\}$. Since each member is nonvoid and compact, K is a (nonvoid) compact subset of Ω in its product topology. Now let $\mathcal{D}$ be the collection of all (nonvoid) finite subfamilies of $\mathcal{C}_0 = \{C_n, n \geq 1\}$, directed by inclusion. Next for each $\mathcal{E} \in \mathcal{D}$, set $T_\mathcal{E} = \bigcup\{\alpha : \pi_\alpha^{-1}(K_\alpha) \in \mathcal{E}\}$. Then $T_\mathcal{E}$ is a finite subset of T_1. By the finite intersection property of $\mathcal{C}_0$ for each $\mathcal{E} \in \mathcal{D}$, we have $\bigcap_{G\in\mathcal{E}} G \cap \bigcap_{t\in T_\mathcal{E}} C_{n_t} \neq \emptyset$. Let $\omega^\mathcal{E}$ be any point in this intersection. Since $T_\mathcal{E} \subset T_1$, for each $t \in T_\mathcal{E}$ the tth coordinate $\omega_t^\mathcal{E}(= \pi_{\{t\}}(\omega^\mathcal{E}))$ is in $K_{\alpha_{n_t}}$. Let us select a point $\overline{\omega}^\mathcal{E}$ of Ω for each $\mathcal{E}$ by the rule

$$\overline{\omega}_t^\mathcal{E} = \begin{cases} \omega_t^\mathcal{E} & t \in T_\mathcal{E} \\ \omega_t^0 & t \in T - T_\mathcal{E}. \end{cases}$$

Then $\overline{\omega}^\mathcal{E} \in \cap_{G\in\mathcal{E}} G \cap K$, since $\omega_t^0 \in K_{\alpha_{n_t}}$ for $t \in T_1$ and $T_\mathcal{E} \subset T_1$. Hence $\{\overline{\omega}^\mathcal{E}, \mathcal{E} \in \mathcal{D}\} \subset K$ is a net and, since K is compact, so that it is closed (since the product topology of Ω is Hausdorff), the net has a cluster point $\omega^* \in K$. If C_n from $\mathcal{C}$ is any set, then there is an $\mathcal{E}$ in $\mathcal{D}$ such that $\{C_n\} \subset \mathcal{E}$, and $\overline{\omega}^\mathcal{E} \in \bigcap_{G\in\mathcal{E}} G \Rightarrow \overline{\omega}^\mathcal{E} \in C_n$. Since $\{\overline{\omega}^\mathcal{E}, \mathcal{E} \in \mathcal{D}\} \subset K$, it is also in C_n for all $\mathcal{E}$ sufficiently large or "refined," so that the net enters C_n for each n eventually. Hence $\omega^* \in C_n$ for all n, and $\bigcap_{n\geq 1} C_n \neq \emptyset$, as asserted.

With this topological property of cylinders, it is possible to present the following result, which is somewhat more general than the original formulation of Kolmogorov's (1933) existence theorem, but is a specialization of its 1955 extension of Bochner's theorem. Because of this circumstance, we call it the *Kolmogorov-Bochner* theorem, and it is an interpolation of both these results. The classical version will also be stated for comparison. After establishing the result we discuss its relation to Ionescu Tulcea's theorem and its significance for the subject.

Theorem 10 (Kolmogorov-Bochner) *Let $\{(\Omega_t, \Sigma_t), t \in T\}$ be a family of topological measurable spaces, where Ω_t is Hausdorff, Σ_t is a σ-algebra containing all compact subsets of Ω_t, and T is a nonempty index set. Let $\mathcal{F}$ be the class of all finite subsets of T, directed by inclusion, and*

$$\Omega = \times_{t\in T}\Omega_t, \quad \Sigma = \bigotimes_{t\in T}\Sigma_t \left[= \sigma\left(\bigcup_{\alpha\in\mathcal{F}} \pi_\alpha^{-1}(\Sigma)\right)\right].$$

[Here $\pi_\alpha : \Omega \to \Omega_\alpha = \times_{t\in\alpha}\Omega_t$ is the coordinate projection.] For each $\alpha \in \mathcal{F}$, let $P_\alpha : \Sigma \to [0,1]$ be a Radon probability [i.e., a probability that satisfies $P_\alpha(A) = \sup\{P_\alpha(K) : K \subset A, \text{ compact }\}, A \in \Sigma_\alpha = \bigotimes_{t\in\alpha}\Sigma_t$.] Suppose that $\{P_\alpha, \alpha \in \mathcal{F}\}$ is a compatible family, in the sense that for each $\alpha < \beta$, with α, β in $\mathcal{F}$, $P_\beta \circ \pi_{\alpha\beta}^{-1} = P_\alpha$, where $\pi_{\alpha\beta} : \Omega_\beta \to \Omega_\alpha$ is the coordinate projection. Then there exists a unique probability $P : \Sigma \to [0,1]$ such that $P_\alpha = P \circ \pi_\alpha^{-1}, \alpha \in \mathcal{F}$, and a family of Ω_t -valued random variables $\{X_t, t \in T\}$ such that

$$P[X_{t_1} \in A_{t_1}, \ldots, X_{t_n} \in A_{t_n}] = \int_{A_{t_1}} \cdots \int_{A_{t_n}} P_\alpha(d\omega_{t_1}, \ldots, d\omega_{t_n}), \tag{19}$$

where $A_{t_i} \in \Sigma_{t_i}, \alpha = (t_1, \ldots, t_n) \in \mathcal{F}$. Here P on Σ need not be a Radon probability, but only has the following weak approximation property:

$$P(A) = \sup\{P(C) : C \subset A, \ C \text{ a compact based cylinder }\}, \tag{20}$$

for each cylinder set A in Σ.

Proof As noted in the proof of Proposition 1, $P : \Sigma_0 \to [0,1]$ defined, for each $A \in \Sigma_0, A = \pi_\alpha^{-1}(A_\alpha)$ for some $\alpha \in \mathcal{F}$, by the equation $P(A) = P_\alpha(A_\alpha)$, is unambiguous, and is finitely additive there. Also, $P(\Omega) = 1$. [See Eqs. (5) and (6).] We need to establish that P is σ-additive on Σ_0, so that by the classical (Hahn) extension theorem it will have a unique Σ-additive extension to $\Sigma = \sigma(\Sigma_0)$.

First let us verify (20). If $A \in \Sigma_0$, so that $A = \pi_\alpha^{-1}(A_\alpha), \alpha \in \mathcal{F}$ and $A_\alpha \in \Sigma_\alpha$, then

$$\begin{aligned} P(A) &= P_\alpha(A_\alpha) \\ &= \sup\{P_\alpha(K) : K \subset A_\alpha, \text{ compact }\} \ (\text{ since } P_\alpha \text{ is Radon }) \end{aligned}$$

$$= \sup\{P(\pi_\alpha^{-1}(K)) : \pi^{-1}(K) \subset A, K \text{ compact}\}$$

$$\leq \sup\{P(C) : C \subset A, C \text{ a compact based cylinder}\}. \tag{21}$$

But for each $C \subset A, P(C) \leq P(A)$, so that the opposite inequality also holds in (21), and hence (20) is true, as stated. We now establish σ-additivity.

Let $\{A_n, n \geq 1\} \subset \Sigma_0$ and $A_n \downarrow \emptyset$. It is sufficient to show, by the additivity of P on Σ_0 that $P(A_n) \downarrow 0$, which will imply σ-additivity. Let $\varepsilon > 0$. Since $A_n \supset A_{n+1}$ and $A_n \in \pi_\alpha^{-1}(\Sigma_\alpha)$ for some $\alpha_n \in \mathcal{F}$, we may clearly assume by the directedness of $\mathcal{F}$ that $\alpha_n < \alpha_{n+1}, n \geq 1$. Then by (20), there exists a compact based cylinder $C_n \subset A_n$ such that

$$P(A_n) \leq P(C_n) + \frac{\varepsilon}{2^n}, \quad n \geq 1. \tag{22}$$

But the C_n need not be monotone. So let $B_n = \bigcap_{k=1}^n C_k \subset A_n$. Then $B_n \in \Sigma_0$. We assert that the $P(B_n)$ also approximate the $P(A_n)$. To see this, consider for $n = 2$,

$$\begin{aligned} P(C_1 \cup C_2) &+ P(C_1 \cap C_2) \\ &= P(C_1) + P(C_2) \text{ (by additivity of } P \text{ on } \Sigma_0) \\ &\geq P(A_1) + P(A_2) - \left(\frac{\varepsilon}{2} + \frac{\varepsilon}{2^2}\right) \text{ [by (22)]}. \end{aligned}$$

But $C_1 \cup C_2 \subset A_1 \cup A_2 = A_1$, since $A_n \downarrow$, and so $P(A_1) - P(C_1 \cup C_2) \geq 0$. Hence writing $B_2 = C_1 \cap C_2$, we get

$$P(B_2) \geq P(A_2) - \left(\frac{\varepsilon}{2} + \frac{\varepsilon}{2^2}\right).$$

By a similar computation with B_2 and C_3 one gets for $B_3 = B_2 \cap C_3$,

$$P(B_3) \geq P(A_3) - \left(\frac{\varepsilon}{2} + \frac{\varepsilon}{2^2} + \frac{\varepsilon}{2^3}\right),$$

and by induction

$$P(B_n) \geq P(A_n) - \sum_{k=1}^n \frac{\varepsilon}{2^k}. \tag{23}$$

But $\cap_{n=1}^\infty B_n = \cap_{n=1}^\infty C_k \subset \cap_{n=1}^\infty A_n = \emptyset$. Since the C_k are compactly based, by Proposition 9, there exists an $n_0 \geq 1$ such that $\cap_{n=1}^{n_0} C_k = \emptyset$. Thus $B_{n_0} = \emptyset$. Hence (23) implies for all $n \geq n_0$,

$$P(A_n) \leq \sum_{k=1}^n \frac{\varepsilon}{2^k} < \varepsilon,$$

and since $\varepsilon > 0$ is arbitrary, $\lim_n P(A_n) = 0$. Thus P is σ-additive.

Finally, let $X_t : \omega \mapsto \omega_t$. Since $\pi_t : \Omega \to \Omega_t$ is continuous, as a coordinate projection mapping (and $X_t = \pi_t$), it follows that $X^{-1}(\Sigma_t) \subset \Sigma$, and X_t is a random variable on Ω with values in Ω_t. If $\alpha = (\alpha_{t_1}, \ldots, \alpha_{t_n}), A_\alpha = A_{t_1} \times \ldots \times A_{t_n}$ then

$$P(\pi_\alpha^{-1}(A_\alpha)) = P_\alpha(A_\alpha) = \int_{A_{t_1}} \cdots \int_{A_{t_n}} P_\alpha(d\omega_{t_1}, \ldots, d\omega_{t_n}),$$

which is (19). Thus the proof is complete.

If each $\Omega_t = \mathbb{R}, \Sigma_t = \mathcal{B}$, the Borel σ-algebra of $\mathbb{R}$, and P_α is given by a distribution function $F_{t_1,\ldots,t_n}$ so that

$$P_\alpha(A_\alpha) = \int_{A_{t_1}} \cdots \int_{A_{t_n}} dF_{t_1,\ldots,t_n}(x_1, \ldots, x_n),$$

then the above result reduces to the classical theorem of Kolmogorov. We state it for a convenient reference.

Theorem 11 (Kolmogorov) *Let $T \subset \mathbb{R}, t_1 < \ldots < t_n$ be n points from T. For each such n-tuple, let there be given an n-dimensional distribution function $F_{t_1,\ldots,t_n}$ such that the family $\{F_{t_1,\ldots,t_n}, t_i \in T, n \geq 1\}$ is compatible, in the sense that Eqs. (2) and (3) hold. Let $\Omega = \mathbb{R}^T = \times_{t\in T}\mathbb{R}_t, \mathbb{R}_t = R$, and Σ be the smallest σ-algebra containing all the cylinder sets of the form $\{\omega \in \Omega : \omega_t < a\}$ for each $t \in T$ and $a \in \mathbb{R}$. Then there exists a unique probability $P : \Sigma \to [0,1]$ and a stochastic process (or a random family) $\{X_t, t \in T\}$ on (Ω, Σ, P) such that*

$$P[X_{t_1} < x_1, \ldots, X_{t_n} < x_n] = F_{t_1,\ldots,t_n}(x_1, \ldots, x_n), t_i \in T, x_i \in \mathbb{R}, n \geq 1. \tag{24}$$

The process is actually defined by the coordinate functions $X_t : \omega \mapsto \omega_t$ for each $\omega \in \Omega, t \in T$.

Discussion Since by our work in Section 2, regular conditional distributions exist under the hypothesis of *this* theorem, the result is equivalent to Theorem 7 when each $\Omega_i = \mathbb{R}$ and $\Sigma_i = \mathcal{B}$ there. In the present case, it is clearly better to stay with the (regular, that is, Radon or Lebesgue-Stieltjes) probability functions without going to the conditional distributions. In the context of Markov processes, it is appropriate to consider results such as Theorems 5-7 without invoking topological hypotheses. The general conditions for the existence of regular conditional distributions are known, thus far, only when each Ω_i is either a complete separable metric (also called Polish) space, or each (Ω_i, Σ_i) is a Borel space (cf. Problem 14). However, Theorem 10 is less restrictive than either of these conditions. Note also that the random families constructed in Theorems 6, 8, and 10 need not take values in a fixed set. They can vary from point to point in T. This freedom is useful in applications.

The original Bochner version extending Theorem 10 is in terms of abstract "projective systems." It and related generalizations, which need a more elaborate framework, have been given in the first author's (1981, 1995) monographs, and they will be omitted here. The present theorems suffice for our current needs. Now we turn to another important dependence class for analysis.

3.5 Martingale Sequences

In considering the first general dependence class, namely, Markovian families, we did not impose any integrability conditions on the random variables. There the concept involved first conditional probability measures, but for essentially all the analysis their regularity is also demanded. However, if the families are restricted to integrable random variables, then one can employ the concept of conditional expectations and no regularity hypotheses enter. With this point in mind we can introduce a large new dependence class, called martingales, and some of its relatives. The motivation for this concept comes from some gambling systems. Suppose in a certain game of chance, a gambler's successive fortunes are $X_1, X_2, \ldots$ at times $1, 2, \ldots$ Then it is reasonable (and fair) to hope that the "expected fortune" on the $(n+1)$th game, having known the fortunes of the first n games, is the present (or the nth) one. In terms of conditional expectations, this says that $E(X_{n+1}|X_1, \ldots, X_n) = X_n, n \geq 1$, a.e. The asymptotic behavior of the X_n-sequence is of interest in probabilistic analysis. Thus we state the concept precisely, and study the limit properties of the process.

Definition 1 (a) Let (Ω, Σ, P) be a probability space and $\mathcal{B}_n \subset \Sigma$ be an increasing sequence of Σ-algebras. If $X_n : \Omega \to \mathbb{R}$ is an integrable random variable on (Ω, Σ, P), and is $\mathcal{B}$-measurable for each n (also called $\mathcal{B}_n$-*adapted*), then the adapted sequence $\{X_n, \mathcal{B}_n, n \geq 1\}$ is a *martingale* whenever the following set of equations holds:

$$E^{\mathcal{B}_n}(X_{n+1}) = X_n \quad \text{a.e., } n \geq 1, \tag{1}$$

and $\{\mathcal{B}_n, n \geq 1\}$ is termed a (stochastic) *base* of the martingale. In case $\mathcal{B}_n = \sigma(X_1, \ldots X_n)$, it is the *natural base*, and then $\{X_n, n \geq 1\}$ itself is sometimes referred to as a martingale, omitting any mention of the (natural) base. Also (1) is expressed suggestively as

$$E(X_{n+1}|X_1, \ldots X_n) = X_n \quad \text{a.e., } n \geq 1. \tag{2}$$

(b) An adapted integrable sequence $\{X_n, \mathcal{B}_n, n \geq 1\}$ on (Ω, Σ, P) is called a *submartingale* (*supermartingale*) if

$$E^{\mathcal{B}_n}(X_{n+1}) \geq X_n \quad \text{a.e.} \quad (\leq X_n \quad \text{a.e. }), n \geq 1. \tag{3}$$

In the gambling interpretation, a submartingale is a *favorable* game and a supermartingale is an *unfavorable* game, to the player. A martingale is therefore a *fair* game. Thus a sequence $\{X_n, \mathcal{B}_n, n \geq 1\}$ which is both a sub- and supermartingale is a martingale. Note that $E(X_n)$ is a constant for martingales, and is nondecreasing for submartingales (nonincreasing for supermartingales) by (1) and (3). For instance, $E^{\mathcal{B}_n}(X_{n+1}) = X_n$ a.e. implies $E(X_{n+1}) = E(E^{\mathcal{B}_n}(X_{n+1})) = E(X_n), n \geq 1$.

An immediate consequence of the above definition is the following:

Proposition 2 *Let $\{X_n, \mathcal{B}_n, n \geq 1\} \subset L^1(P)$. Then it is a martingale iff X_n can be expressed as $X_n = \sum_{k=1}^n Y_k$, where $E^{\mathcal{B}_k}(Y_{k+1}) = 0$ a.e., $k \geq 1$. Moreover, for a martingale sequence $\{X_n, \mathcal{B}_n, n \geq 1\}$, if each X_n is in $L^2(P)$, then its increments $\{Y_{n+1} = X_{n+1} - X_n, n \geq 1, Y_1 = X_1\}$ form an orthogonal sequence. [The $Y_n, n \geq 1$ is also termed a martingale difference sequence.]*

Proof For a martingale $\{X_n, \mathcal{B}_n, n \geq 1\}$, if we set $Y_n = X_n - X_{n-1}, n > 1$, and $Y_1 = X_1$, then

$$0 = E^{\mathcal{B}_n}(X_{n+1} - X_n) = E^{\mathcal{B}_n}(Y_{n+1}) \quad \text{a.e., } n \geq 1.$$

Conversely, if the condition holds, then for each $n \geq 1$, since $X_{n+1} = \sum_{k=1}^{n+1} Y_k = Y_{n+1} + \sum_{k=1}^n Y_k = Y_{n+1} + X_n$, then

$$E^{\mathcal{B}_n}(X_{n+1}) = E^{\mathcal{B}_n}(Y_{n+1} + X_n) = X_n + E^{\mathcal{B}_n}(Y_{n+1}) = X_n \quad \text{a.e.}$$

Hence $\{X_n, \mathcal{B}_n, n \geq 1\}$ is a martingale.

If the martingale is square integrable, then for the increments sequence $\{Y_n, n \geq 1\}$ we have, with $m < n$,

$$\begin{aligned} E(Y_n Y_m) &= E(E^{\mathcal{B}_{n-1}}(Y_n Y_m)) \quad \text{(by Proposition 1.2)} \\ &= E(Y_m E^{\mathcal{B}_{n-1}}(Y_n)) = 0, \quad \text{since } E^{\mathcal{B}_{n-1}}(Y_n) = 0. \end{aligned}$$

Hence $\{Y_n, n \geq 1\} (\subset L^2(P))$ is orthogonal, as asserted.

A simple example of a martingale is the sequence $\{S_n, n \geq 1\}$ of partial sums $S_n = \sum_{k=1}^n X_k$ of independent integrable random variables X_n with zero means.

Large classes of martingales can be recognized by means of the next result:

Proposition 3 *Let X be any integrable random variable on (Ω, Σ, P) and $\mathcal{B}_n \subset \mathcal{B}_{n+1} \subset \Sigma$ be a sequence of σ-algebras. If $X_n = E^{\mathcal{B}_n}(X)$, then $\{X_n, \mathcal{B}_n, n \geq 1\}$ is a martingale and moreover, it is a uniformly integrable set in $L^1(P)$.*

Proof Consider

$$E^{\mathcal{B}_n}(X_{n+1}) = E^{\mathcal{B}_n}(E^{\mathcal{B}_{n+1}}(X))$$
$$= E^{\mathcal{B}_n}(X)$$

(by the commutative property of $E^{\mathcal{B}_n}$, since $\mathcal{B}_n \subset \mathcal{B}_{n+1}$)

$$= X_n \quad \text{a.e., } n \geq 1.$$

Hence it is a martingale.

For the second statement, note that

$$||X_n||_1 = ||E^{\mathcal{B}_n}(X)||_1 \leq ||X||_1, \quad n \geq 1, \tag{4}$$

and for each $a > 0$, with the Markov inequality,

$$P[|X_n| > a] \leq \left(\frac{1}{a}\right) E(|X_n|) \leq \frac{||X||_1}{a} \quad \text{[by (4)]}, \tag{5}$$

and the right side tends to zero uniformly in n as $a \to \infty$. Then

$$\int_{[|X_n|>a]} |X_n| dP = \int_{[|X_n|>a]} |E^{\mathcal{B}_n}(X)| dP$$
$$\leq \int_{\Omega} \chi_{[|X_n|>a]} E^{\mathcal{B}_n}(|X|) dP$$

(by the conditional Jensen inequality)

$$= \int_{\Omega} E^{\mathcal{B}_n}(\chi_{[|X_n|>a]}|X|) dP, \quad \text{(because } X_n \text{ is } \mathcal{B}_n - \text{adapted)},$$
$$= \int_{\Omega} \chi_{[|X_n|>a]} |X| dP \to 0, \text{ (uniformly in } n),$$

as $a \to \infty$ through a sequence, by the dominated convergence theorem. Hence by Theorem 1.4.5, $\{X_n, n \geq 1\}$ is uniformly integrable, as desired.

To gain some facility with operations on martingales and submartingales let us establish some simple results on transformations of these sequences.

Lemma 4 *Let $\{X_n, \mathcal{B}_n, n \geq 1\}$ be a (sub-) martingale and $\phi : \mathbb{R} \to \mathbb{R}$ be an (increasing) convex function. If $E(\phi(X_{n_0})) < \infty$, for some $n_0 > 1$, then $\{\phi(X_k), \mathcal{B}_k, 1 \leq k \leq n_0\}$ is a submartingale.*

Proof The assertions follow from the conditional Jensen inequality. Indeed, by hypothesis

$$E^{\mathcal{B}_n}(X_{n+1}) \geq X_n \quad \text{a.e.}$$

Hence in both cases of martingales or submartingales,

$$\phi(X_n) \le \phi(E^{\mathcal{B}_n}(X_{n+1})) \text{ (with equality in the martingale case)}$$
$$\le E^{\mathcal{B}_n}(\phi(X_{n+1})) \text{ (by Theorem 1.9iii)}, \quad (6)$$

provided $E(\phi(X_n)) < \infty$ for all $n < n_0$. Since ϕ is convex, there is a support line such that $ax + b \le \phi(x)$ for some real a, b and all x. Hence $E(\phi(X_n)) \ge aE(X_n) + b > -\infty$ for each $n \ge 1$. Thus if $E(\phi(X_n)) < \infty$, then $|E(\phi(X_n))| < \infty$ for $1 \le n \le n_0$, and (6) implies $\{\phi(X_n), \mathcal{B}_n, 1 \le n \le n_0\}$ is a submartingale, as asserted.

Taking $\phi(x) = x^+$, we get $\{X_n^+, \mathcal{B}_n, n \ge 1\}$ to be a positive submartingale for any (sub-) martingale $\{X_n, \mathcal{B}_n, n \ge 1\}$. In the martingale case $\{X_n^{\pm}, \mathcal{B}_n, n \ge 1\}$, and hence $\{|X_n|, \mathcal{B}_n, n \ge 1\}$ are submartingales. Taking $\phi(\cdot)$ as a nonincreasing concave function, we get for any (super-) martingale $\{X_n, \mathcal{B}_n, n \ge 1\}$ that $\{\phi(X_n), \mathcal{B}_n, n \ge 1\}$ is a supermartingale if $E(\phi(X_n))$ exists. In general, if $\{X_n, \mathcal{B}_n, n \ge 1\}$ is a submartingale, $\{-X_n, \mathcal{B}_n, n \ge 1\}$ is a supermartingale. Thus it suffices to consider sub- (or super-) martingales, and the other (super) case can then be deduced from the first.

Another property, useful in some computations, is contained in

Lemma 5 *Let $\{X_n, \mathcal{B}_n, n \ge 1\}$ be an $L^1(P)$-bounded martingale, so that $\sup_{n\ge1} E(|X_n|) < \infty$. Then $X_n = X_n^{(1)} - X_n^{(2)}, n \ge 1$, and $\{X_n^{(i)}, \mathcal{B}_n, n \ge 1\}, i = 1, 2$, are positive martingales.* [This is a *Jordan-type* martingale decomposition.]

Proof Let $\mu_n : A \mapsto \int_A X_n dP, A \in \mathcal{B}_n$. Then μ_n is a signed measure and $|\mu_n|(A) = \int_A |X_n| dP$, the variation measure of μ_n on $\mathcal{B}_n$. Thus $|\mu_n|(\cdot)$ is σ-additive. Since $\{|X_n|, \mathcal{B}_n, n \ge 1\}$ is a submartingale, as a consequence of the preceding lemma, $|\mu_n|(A) \le |\mu_{n+1}|(A), A \in \mathcal{B}_n$, by (3). Also, $\sup_n |\mu_n|(\Omega) < \infty$. Hence if $\nu_n(A) = \lim_{k\to\infty} |\mu_k|(A), A \in \mathcal{B}_n$, which exists by the monotonicity of $|\mu_n|$, then $\nu_n : \mathcal{B}_n \to \mathbb{R}^+$ is an additive bounded set function, since $|\mu_n|(\cdot)$ has those properties. Also the σ-additivity of $|\mu_n|(\cdot)$ in addition implies the same property of ν_n, since evidently $\mathcal{B}_n \subset \mathcal{B}_{n+1}$ and $\nu_n(A) = \nu_{n+1}(A)$ for all $A \in \mathcal{B}_n$. This is a standard application of results in measure theory [cf. Halmos (1950, p. 170)]. We include a short proof for completeness.

Thus for the σ-additivity, if $\varepsilon > 0$ and $n_0 \ge 1$ are given, there exists an n_ε such that $n_0 \le n_\varepsilon \Rightarrow \nu_{n_0}(\Omega) < |\mu_{n_\varepsilon}|(\Omega) + \varepsilon/2$, by definition of ν_{n_0} since $\Omega \in \mathcal{B}_{n_0}$. On the other hand, $|\mu_{n_\varepsilon}|(\cdot)$ is σ-additive, so that $\sum_{j=1}^{\infty} |\mu_{n_\varepsilon}|(A_j) = |\mu_{n_\varepsilon}|(\Omega)$ for any measurable partition $\{A_j, j \ge 1\}$ of Ω in $\mathcal{B}_{n_0} = \mathcal{B}_{n_\varepsilon}$. The convergence of this series implies the existence of a $j_0(\varepsilon)$, and since $|\mu_n|(A) \le |\mu_{n+1}|(A)$ and $\nu_n(A) = \lim_{m\to\infty} |\mu_n|(A)$ implies $\mu_n(A) < \nu_n(A)$, such that the following inequalities hold:

$$|\mu_{n_\varepsilon}|(\Omega) - \frac{\varepsilon}{2} \le \sum_{j=1}^{j_0(\varepsilon)} |\mu_{n_\varepsilon}|(A_j)$$

$$\le \sum_{j=1}^{j_0(\varepsilon)} \nu_{n_0}(A_j) \quad (\text{since } A_j \in \mathcal{B}_{n_0}). \tag{7}$$

But by choice of n_ε, we have with (7),

$$\nu_{n_0}(\Omega) - \varepsilon < |\mu_{n_\varepsilon}|(\Omega) - \frac{\varepsilon}{2} \le \sum_{j=1}^{j_0(\varepsilon)} \nu_{n_0}(A_j). \tag{8}$$

Since ε, n_0, are arbitrary, (8) plus additivity of $\nu_{n_0} \Rightarrow \nu_{n_0}(\Omega) = \sum_{j=1}^{\infty} \nu_{n_0}(A_j)$, and hence ν_n is σ-additive on $\mathcal{B}_n$ for each n . But each $|\mu_n|$ is P-continuous, and thus it follows that ν_n is also P-continuous. By the Radon-Nikodým theorem (cf. 1.3.12ii), we may define $X_n^{(1)} = d\nu_n/dP_{\mathcal{B}_n}$ on $\mathcal{B}_n$. Then the fact that $\nu_n(A) = \nu_{n+1}(A)$ for all $A \in \mathcal{B}_n$ implies

$$\int_A X_n^{(1)} dP_{\mathcal{B}_n} = \int_A X_{n+1}^{(1)} dP_{\mathcal{B}_{n+1}} = \int_A E^{\mathcal{B}_n}(X_{n+1}^{(1)}) dP_{\mathcal{B}_n} \quad A \in \mathcal{B}_n. \tag{9}$$

This means $\{X_n^{(1)}, \mathcal{B}_n, n \ge 1\}$ is a positive martingale. But we also have $\nu_n(A) \ge |\mu_n|(A) \ge \mu_n(A)$, so that

$$\int_A X_n^{(1)} dP_{\mathcal{B}} \ge \int_A X_n dP_{\mathcal{B}}, \quad A \in \mathcal{B}_n. \tag{10}$$

The integrands are $\mathcal{B}_n$-measurable. Hence $X_n^{(2)} = X_n^{(1)} - X_n \ge 0$ a.e., and

$$E^{\mathcal{B}_n}(X_{n+1}^{(2)}) = E^{\mathcal{B}_n}(X_{n+1}^{(1)}) - E^{\mathcal{B}_n}(X_{n+1}) = X_n^{(1)} - X_n = X_n^{(2)} \text{ a.e.}$$

Thus $\{X_n^{(2)}, \mathcal{B}_n, n \le 1\}$ is also a positive martingale, and this finishes the proof.

The next result extends Kolmogorov's inequality in two ways. The extensions are due to Doob, and Hájek and Rényi. These are frequently used in martingale analysis. [Here (11), (12), and (14) are due to Doob.]

Theorem 6 *Let* $\{X_n, \mathcal{B}_n, n \ge 1\}$ *be a submartingale on* (Ω, Σ, P). *Then* (i) $\lambda \in \mathbb{R}$ *implies*

$$\lambda P\left[\max_{k\le n} X_k \ge \lambda\right] \le \int_{[\max_{k\le n} X_k \ge \lambda]} X_n dP \le E(X_n^+) \tag{11}$$

and

$$\lambda P\left[\min_{k\le n} X_k \le \lambda\right] \ge \int_{[\min_{k\le n} X_k\le\lambda]} X_n dP - E(X_n - X_1) \ge E(X_1) - E(X_n^+). \tag{12}$$

(ii) *If, moreover,* $X_n \ge 0$ *a.e. for each* n, *we have for* $\lambda \ge 0$, *and* $a_n \downarrow 0$,

$$\lambda P\left[\sup_{k\ge1} a_k X_k > \lambda\right] \le \sum_{n=1}^{\infty}(a_n - a_{n+1})E(X_n), \tag{13}$$

and for $1 \le p < \infty, q = p/(p-1)$

$$E\left(\max_{1\le k\le n} X_k^p\right) \le \begin{cases} q^p E(X_n^p) & \text{if } p > 1 \\ \frac{e}{e-1}[1 + E(X_n \log^+ X_n)] & \text{if } p = 1. \end{cases} \tag{14}$$

where $\log^+ \alpha = \log\alpha$ *if* $\alpha > 1$ *and* $\log^+ \alpha = 0$ *if* $\alpha \le 1$, *e being the base of 'log'.*

Proof (i) As in the proof of Theorem 2.2.5, we decompose $M = [\max_{k\le n} X_k \ge \lambda]$ into suitable disjoint events and estimate their probabilities. Thus, let $M_1 = [X_1 \ge \lambda]$, and for $1 < k \le n$, set $M_k = [X_k \ge \lambda, X_i < \lambda, 1 \le i \le k-1]$. Then $M_i \in \mathcal{B}_i$, disjoint, and $M = \bigcup_{k=1}^n M_k$. Hence

$$\begin{aligned} \int_M X_n dP &= \sum_{k=1}^n \int_{M_k} X_n dP \\ &= \sum_{k=1}^n \int_{M_k} E^{\mathcal{B}_k}(X_n)\, dP \quad \text{since } M_k \in \mathcal{B}_k \\ &\ge \sum_{k=1}^n \int_{M_k} X_k dP \text{ (by the submartingale property)} \\ &\ge \lambda \sum_{k=1}^n \int_{M_k} dP = \lambda P\left(\bigcup_{k=1}^n M_k\right) = \lambda P(M). \end{aligned}$$

This gives (11). For (12), we consider $N = [\min_{k\le n} X_k \le \lambda]$, and set $N_1 = [X_1 \le \lambda]$. If $1 < k \le n$, let $N_k = [X_k \le \lambda, X_i > \lambda, 1 \le i \le k-1]$. Thus $N_k \in \mathcal{B}_k$, disjoint, and $N = \bigcup_{k=1}^n N_k$. Hence $N_k \subset (\bigcup_{i=1}^{k-1} N_i)^c$ and

$$\begin{aligned} E(X_1) &= \int_{N_1} X_1 dP + \int_{N_1^c} X_1 dP \\ &\le \lambda P(N_1) + \int_{N_1^c} X_2 dP \end{aligned}$$

(since $N_1^c \in \mathcal{B}$ and $\{X_1, X_2\}$ is a submartingale for $\{\mathcal{B}_1, \mathcal{B}_2\}$)

$$= \lambda P(N_1) + \int_{N_2} X_2 dP + \int_{N_1^c \cap N_2^c} X_2 dP \quad (\text{since } N_2 \subset N_1^c)$$

$$\leq [P(N_1) + P(N_2)] + \int_{N_2} X_2 dP + \int_{N_1^c \cap N_2^c} X_3 dP$$

[since $(N_1 \cup N_2)^c \in \mathcal{B}_2$ and $\{X_2, X_3\}$ is a submartingale for $\{\mathcal{B}_2, \mathcal{B}_3\}$]

$$\vdots$$

$$\leq \lambda \sum_{i=1}^{n} P(N_i) + \int_{(\bigcup_{i=1}^{n} N_i)^c} X_n dP$$

$$= \lambda P(N) + E(X_n) - \int_N X_n dP.$$

This gives (12) and (i) follows.

(ii) The argument is again similar to the above. Since $X_n \geq 0$, $a_n > 0$, let $M = [\sup_{k \geq 1} a_k X_k > \lambda]$. Set $M_1 = [a_1 X_1 > \lambda]$ and for $k > 1, M_k = [a_k X_k > \lambda, a_i X_i < \lambda, 1 \leq i \leq k-1]$. As before $M_k \in \mathcal{B}_k, M = \bigcup_{k \geq 1} M_k$, disjoint union. If the right side of (13) is infinite, then the inequality is true. Thus let it be finite. Set $S = \sum_{n=1}^{\infty} (a_n - a_{n+1}) X_n$. Consequently, $E(S) < \infty$, and, since $X_n \geq 0$, $a_n - a_{n+1} > 0$, we have

$$E(S) \geq \sum_{n=1}^{\infty} (a_n - a_{n+1}) \int_M X_n dP$$

$$= \sum_{n=1}^{\infty} (a_n - a_{n+1}) \sum_{k=1}^{\infty} \int_{M_k} X_n dP = \sum_{k=1}^{\infty} \sum_{n=1}^{\infty} (a_n - a_{n+1}) \int_{M_k} X_n dP$$

$$\geq \sum_{k=1}^{\infty} \sum_{n \geq k} (a_n - a_{n+1}) \int_{M_k} X_k dP$$

(since $M_k \in \mathcal{B}_k$ and the sequence is a positive submartingale)

$$= \sum_{k=1}^{\infty} a_k \int_{M_k} X_k dP \geq \lambda \sum_{k=1}^{\infty} P(M_k) = \lambda P(M).$$

Hence (13) obtains.

To establish (14), let $Y = \max_{1 \leq k \leq n} X_k$. Then $Y \in L^p(P)$. If $p > 1$, by Theorem 1.4.1iii, we have $Y \geq 0$ because $X_k \geq 0$, and

$$E(Y^p) = \int_{\mathbb{R}} |Y|^p dF_Y(y) = p \int_0^{\infty} y^{p-1} [1 + F_Y(-y) - F_Y(y)] dy$$

$$= p\int_0^\infty y^{p-1}P[Y \geq y]dy$$

$$\leq p\int_0^\infty y^{p-2}\int_{[Y\geq y]} X_n dP dy \quad \text{[by (11)]}$$

$$= p\int_\Omega X_n(\omega)\left(\int_0^{Y(\omega)} y^{p-2}dy\right)P(d\omega) = \frac{p}{p-1}\int_\Omega X_n Y^{p-1}dP$$

$$\leq q||X_n||_p||Y^{p-1}||_q \text{ (by Hölder's inequality)}$$

$$= q||X_n||_p(||Y||_p)^{p/q}. \tag{15}$$

If $||Y||_p = 0$, then the inequality is true; if $||Y||_p^{p/q} > 0$, dividing both sides by this number, (15) reduces to (14) in this case.

If $p = 1$, we let $Z = (Y-1)\chi_{[Y\geq 1]}$ and calculate, (since $(Y-1)\chi_{[Y<1]} < 0$),

$$E(Y-1) \leq E(Z) = \int_0^\infty P[Z \geq y]dy \text{ (by Theorem 1.4.1iii again)}$$

$$\leq \int_0^\infty P[Y \geq y+1]dy$$

$$\leq \int_0^\infty \frac{1}{y+1}\int_{[Y\geq y+1]} X_n dP dy \quad \text{[by (11)]}$$

$$= \int_\Omega X_n(\omega)\int_1^{Y(\omega)} \frac{dy}{y}P(d\omega)$$

$$= \int_\Omega (X_n \log^+ Y)(\omega)P(d\omega). \tag{16}$$

But $a\log b \leq a\log^+ a + a\log(b/a) \leq a\log^+ a + (b/e)$ for any $a > 0, b > 0$, since $a\log(b/a)$ has a maximum for $a = b/e$ for each fixed $b > 0$. Thus (16) becomes

$$E(Y) - 1 \leq E(X_n \log^+ X_n) + \left(\frac{1}{e}\right)E(Y).$$

Hence

$$E(Y)[\frac{(e-1)}{e}] \leq 1 + E(X_n \log^+ X_n),$$

which is (14) if $p = 1$, and this completes the proof of the theorem.

Remarks (1) Letting $n \to \infty$ in (11) and (12), we get the following inequalities, which are useful in some computations. For a submartingale $\{X_n, \mathcal{B}_n, n \geq 1\}$, and $\lambda \in \mathbb{R}$, we have [cf. the last parts of proof of (11) and (12)]

$$\lambda P[\sup_{n\geq 1} X_n > \lambda] \leq \liminf_n \int_{[\sup_{n\geq 1} X_n > \lambda]} X_n^+ dP \leq \lim_n \int_\Omega X_n^+ dP \tag{17}$$

and

$$\lambda P[\inf_{n\geq 1} X_n < \lambda] \geq \limsup_n \int_{[\inf_{n\geq 1} X_n < \lambda]} X_n^+ dP + E(X_1) - \lim_n \int_\Omega X_n^+ dP. \tag{18}$$

(2) The inequalities (11) and (13) reduce to Kolmogorov's inequality, as they should, under the following identification. Let $Y_1, \ldots, Y_n$ be independent random variables with $E(Y_i) = \mu_i$ and $VarY_i = \sigma_i^2$. If $S_n = \sum_{k=1}^n (Y_k - \mu_k)$ and $X_n = S_n^2$, then $\{X_k, k \geq 1\}$ is a positive submartingale, so that (11) gives for $\lambda \neq 0$,

$$P\left[\sup_{k\leq n} X_k > \lambda^2\right] \leq \frac{1}{\lambda^2} E(X_n) = \frac{1}{\lambda^2} E(S_n^2) = \frac{1}{\lambda^2} \sum_{i=1}^n \sigma_i^2.$$

Similarly in (13), let $a_1 = \ldots = a_n = 1$ and $a_k = 0, k > n, X_n = S_n^2$. Then (13) becomes the above inequality. On the other hand, if for any $n_0 > 1$ we let $a_k = (n_0 + k - 1)^{-2}, Z_k = S_{n_0+k-1}^2, \mathcal{B}_n = \sigma(S_1, \ldots, S_n)$, considering the positive submartingale $\{Z_k, \mathcal{B}_{n_0+k-1}, k \geq 1\}$, we get

$$P\left[\sup_{n\geq n_0} \left|\frac{S_n}{n}\right| > \lambda\right] \leq \frac{1}{\lambda^2}\left(\frac{1}{n_0^2}\sum_{i=1}^{n_0}\sigma_i^2 + \sum_{n>n_0}\frac{\sigma_n^2}{n^2}\right). \tag{19}$$

(3) If $\sum_{n\geq 1}(\sigma_n^2/n^2) < \infty$, then letting $n_0 \to \infty$ in (19) and noting

$$\frac{1}{n_0^2}\sum_{i=1}^{n_0}\sigma_i^2 \leq \left(\frac{1}{n_0^2}\sum_{i=1}^{k}\sigma_i^2\right) + \left(\sum_{i=k+1}^{n_0}\frac{\sigma_i^2}{i^2}\right) \to 0$$

as $n_0 \to \infty$ and then $k \to \infty$, so one gets $P[\lim_{n\to\infty}(S_n/n) = 0] = 1$. This is Theorem 2.3.6, and is one form of the SLLN. Thus the general study leads to other applications.

The fundamental questions in the analysis of martingale sequences concern their pointwise a.e. convergence behavior. Two forms of the martingale convergence theorem are available, due independently to Doob, and Andersen and Jessen. They are now presented, each with two proofs. Our treatment also shows their equivalence, and gives a better insight into the structure of these processes.

Theorem 7 (Andersen-Jessen) *Let (Ω, Σ, P) be a probability space and $\nu : \Sigma \to \mathbb{R}$ be σ-additive, so that it is also bounded. If ν is P-continuous, and $\mathcal{B}_n \subset \mathcal{B}_{n+1} \subset \Sigma$ are σ-algebras, let $\nu_n = \nu|\mathcal{B}_n, P_n = P|\mathcal{B}_n$ be the restrictions. Let $X_n = d\nu_n/dP_n$ be the Radon-Nikodým derivatives for each $n \geq 1$.*

Then $X_n \to X_\infty$ *a.e., and in* $L^1(P)$. *Moreover,* $X_\infty = d\nu_\infty/dP_\infty$ *a.e., where* $\mathcal{B}_\infty = \sigma(\bigcup_{n\geq 1}\mathcal{B}_n), \nu_\infty = \nu|\mathcal{B}_\infty$ *and* $P_\infty = P|\mathcal{B}_\infty$.

First Proof To begin with, we establish that $X_n \to X'$ a.e., and then that $X' = X_\infty$ a.e. Thus let $X_* = \liminf_n X_n$ and $X^* = \limsup_n X_n$. Then $X_* \leq X^*$ and are $\mathcal{B}$-measurable. For the first assertion it suffices to show that if $B = [X_* < X^*]$, then $P(B) = 0$. Equivalently, if $B_{r_1 r_2} = [X_* \leq r_1 < r_2 \leq X^*]$, so that $B = \bigcup\{B_{r_1 r_2} : r_1, r_2 \text{ rationals}\}$, then $P(B_{r_1 r_2}) = 0$, since the union is countable. [We used the fact that ν is P-continuous $\Rightarrow \nu_n$ is P_n-continuous and so, $X_n = d\nu_n/dP_n$.]

Let a, b be in $\mathbb{R}$ and consider $H_a = [X_* \leq a], K^b = [X^* \geq b]$. Then $H_a \in \mathcal{B}_\infty, K^b \in \mathcal{B}_\infty$. We assert that

$$\text{(i)}\quad \nu(H_a \cap A) \leq aP(H_a \cap A), A \in \mathcal{B}_\infty,$$

$$\text{(ii)}\quad \nu(K^b \cap A) \geq bP(K^b \cap A), A \in \mathcal{B}_\infty. \tag{20}$$

Indeed, let $a_n \searrow a$, and define $H_n = [\inf_{k\geq 1} X_{n+k} < a_n](\in \mathcal{B}_\infty)$. Using the by now standard decomposition of H_n, as in the proof of Theorem 6, we express H_n as a disjoint union. Thus let $H_{n1} = [X_{n+1} < a_n]$, and for $k > 1$, let H_{nk} be the event that X_{n+k} is not above a_n at $n + k$ for the first time so that

$$H_{nk} = [X_{n+k} < a_n, X_{n+j} \geq a_n, 1 \leq j \leq k-1] \in \mathcal{B}_{n+k}.$$

Then $H_n = \bigcup_{k\geq 1} H_{nk}$, disjoint union, and $H_n \supset H_{n+1}, H_a = \bigcap_{n\geq 1} H_n$, since $a_n \searrow a$. But $A \in \bigcup_{n\geq 1}\mathcal{B}_n \Rightarrow A \in \mathcal{B}_{n_0}$ for some n_0, and thus for $n \geq n_0$ we have $H_{nk} \cap A \in \mathcal{B}_{n+k}, k \geq 1$. Hence

$$\begin{aligned}
\nu(H_n \cap A) &= \sum_{k\geq 1} \nu(H_{nk} \cap A) \quad (\text{since } \nu \text{ is } \sigma-\text{additive}) \\
&= \sum_{k\geq 1} \nu_{n+k}(H_{nk} \cap A) \quad (\text{since } \nu|\mathcal{B}_{n+k} = \nu_{n+k}) \\
&= \sum_{k\geq 1} \int_{H_{nk}\cap A} X_{n+k}\, dP \leq a_n P(H_n \cap A).
\end{aligned} \tag{21}$$

Since $|\nu|(\Omega) < \infty$, on letting $n \to \infty$ (21) reduces to (i) of (20) if $A \in \bigcup_{n\geq 1}\mathcal{B}_n$. For the general case, let $\mu(A) = aP(H_a \cap A) - \nu(H_a \cap A)$. Then μ is a real σ-additive function on the algebra $\bigcup_{n\geq 1}\mathcal{B}_n$ (since $\mathcal{B}_n \subset \mathcal{B}_{n+1}$), and hence by the classical Hahn extension theorem it has a unique σ-additive extension onto the σ-algebra generated by this algebra, namely, $\mathcal{B}$. Hence (i) is true as stated. (ii) is similarly established, or it follows from (i) if we replace a, ν, X_n by $-b, -\nu$ and $-X_n$.

Now in (20) let $a = r_1, b = r_2$, where $r_1 < r_2$. Then $B_{r_1 r_2} = H_{r_1} \cap K^{r_2}$, and (i) and (ii) of (20) yield

$$r_1 P(B_{r_1 r_2}) \geq \nu(B_{r_1 r_2}) \geq r_2 P(B_{r_1 r_2}).$$

But this is possible only if $P(B_{r_1 r_2}) = 0$. Hence $P(B) = 0$ and $X_* = X^*$ a.e. Let X' be the common value, so that $X_n \to X'$ a.e.

To see that $X' = X_\infty$ a.e., note that $\{X_n, n \geq 1\}$ is uniformly integrable. Indeed, since $|\nu|(\Omega) < \infty$, $|\nu_n|(\Omega) \leq |\nu|(\Omega)$, so that $\sup_n \int_\Omega |X_n| dP < \infty$. Also,

$$\lim_{P(A)\to 0} \int_A X_n dP = \lim_{P(A)\to 0} \nu_n(A) = \lim_{P(A)\to 0} \nu(A) = 0$$

uniformly in n. Hence by Definition 1.4.3, the set is uniformly integrable, and by the Vitali convergence (Theorem 1.4.4),

$$\int_A X_\infty dP_\infty = \nu_\infty(A) = \lim_{n\to\infty} \nu_n(A) = \lim_{n\to\infty} \int_A X_n dP_\infty = \int_A X' dP_\infty, A \in \mathcal{B}_m.$$

Since $m \geq 1$ is arbitrary, this shows that $\nu_\infty(A) = \int_A X' dP, A \in \bigcup_{m\geq 1} \mathcal{B}_m$, and then as in the preceding argument the σ-additive function $\nu_\infty(\cdot) - \int_{(\cdot)} X' dP$, which vanishes on this algebra, also vanishes on $\mathcal{B}_\infty$. Thus

$$\int_A X_\infty \, dP_\infty = \int_A X' \, dP_\infty, \quad A \in \mathcal{B}_m. \tag{22}$$

Since X_∞, X' are $\mathcal{B}_\infty$-adapted, $X' = X_\infty$ a.e. Moreover, by the same Vitali theorem, $E(|X_n - X_\infty|) \to 0$ as $n \to \infty$. This proves the theorem completely.

Second Proof By hypothesis, for $m < n$, $A \in \mathcal{B}_m, \Rightarrow \nu_m(A) = \nu_n(A) = \nu_\infty(A)$. Hence

$$\int_A X_m \, dP_m = \int_A X_n \, dP_n = \int_A E^{\mathcal{B}_m}(X_n) \, dP_m, \quad A \in \mathcal{B}_m. \tag{23}$$

Since the extreme integrands are $\mathcal{B}_m$-measurable, $X_m = E^{\mathcal{B}_m}(X_n)$ a.e., and similarly $X_m = E^{\mathcal{B}_m}(X_\infty)$ a.e. Thus $\{X_n, \mathcal{B}_n, 1 \leq n \leq \infty\}$ is a martingale sequence. Since $\mathcal{B}_\infty = \sigma(\bigcup_{n\geq 1} \mathcal{B}_n)$, it follows that $\bigcup_{n\geq 1} L^1(\Omega, \mathcal{B}_n, P) \subset L^1(\Omega, \mathcal{B}_\infty, P)$, and clearly the former space is dense in the latter. Also, $X_n \in L^1(\Omega, \mathcal{B}_n, P), X_\infty \in L^1(\Omega, \mathcal{B}_\infty, P)$. By density, given $\varepsilon > 0$, there exists $Y_\varepsilon \in L^1(\Omega, \mathcal{B}_{n_0}, P)$ for some n_0 such that $E(|X_\infty - Y_\varepsilon|) < \varepsilon/2$. Since $E^{\mathcal{B}_n}(Y_\varepsilon) = Y_\varepsilon$ a.e. for all $n \geq n_0$, we have for $n \geq m \geq n_0$

$$\begin{aligned} |X_n - X_m| &\leq |E^{\mathcal{B}_n}(X_\infty) - Y_\varepsilon| + |E^{\mathcal{B}_m}(X_\infty) - Y_\varepsilon| \quad \textit{a.e.} \\ &\leq E^{\mathcal{B}_n}(|X_\infty - Y_\varepsilon| + E^{\mathcal{B}_m}(|X_\infty - Y_\varepsilon|) \text{ a.e.,} \\ &\quad \text{(by conditional Jensen's inequality)} \\ &\leq 2 \sup_n E^{\mathcal{B}_n}(|X_\infty - Y_\varepsilon|) \quad \textit{a.e.} \end{aligned} \tag{24}$$

But $\{E^{\mathcal{B}_n}(|X_\infty - Y_\varepsilon|), n \geq 1\}$ is a (positive) martingale. Hence by (17) for any $\lambda > 0$ we get from (24)

$$
\begin{aligned}
P\left[\limsup_{m,n\to\infty}|X_n - X_m| > \lambda\right] &\le P\left[\limsup_{n\to\infty} E^{\mathcal{B}_n}(|X_\infty - Y_\varepsilon|) > \lambda/2\right] \\
&= \lim_{k\to\infty} P\left[\sup_{n\ge k} E^{\mathcal{B}_n}(|X_\infty - Y_\varepsilon|) > \lambda/2\right] \\
&\le \lim_{n\to\infty} P\left[\sup_{1\le k\le n} E^{\mathcal{B}_k}(|X_\infty - Y_\varepsilon|) > \lambda/2\right] \\
&\le \lim_{n\to\infty} (2/\lambda) E(E^{\mathcal{B}_n}(|X_\infty - Y_\varepsilon|)) \quad \text{[by(17)]} \\
&= (2/\lambda) E(|X_\infty - Y_\varepsilon|) < \varepsilon/\lambda.
\end{aligned}
$$

Letting $\varepsilon \searrow 0$ and then $\lambda \to \infty$, we get $|X_n - X_m| \to 0$ a.e. as $n, m \to \infty$. Hence $X_n \to X'$ a.e., and by Fatou's lemma $E(|X'|) \le \liminf_n E(|X_n|) \le |\nu|(\Omega) < \infty$. Next we apply the same argument as for (22), and deduce that the set $\{X_n, n \ge 1\}$ is uniformly integrable, and hence $X' = X_\infty$ a.e. as well as $E(|X_n - X_\infty|) \to 0$. This finishes the proof.

The preceding argument shows, for any integrable r.v. Z such that $X_n = E^{\mathcal{B}_n}(Z), n \ge 1$ (cf. also Proposition 3), that $\{X_n, \mathcal{B}_n, n \ge 1\}$ is a uniformly integrable martingale. Conversely, given any uniformly integrable martingale $\{Y_n, \mathcal{B}_n, n \ge 1\}$, define

$$\nu_n : A \mapsto \int_A Y_n \, dP, \quad A \in \mathcal{B}_n.$$

Then (23) shows that $\nu_n = \nu_{n+1} | \mathcal{B}_n$. Hence we may define $\nu : \bigcup_{n\ge 1} \mathcal{B}_n \to \mathbb{R}$ by setting $\nu(A) = \nu_n(A)$ if $A \in \mathcal{B}_n$, and this gives ν unambiguously on the algebra $\bigcup_{n\ge 1} \mathcal{B}_n$, and it is additive there. The uniform integrability now additionally implies that ν is σ-additive on this algebra, and hence has a unique σ-additive extension to $\mathcal{B}_\infty$. Thus for each martingale $\{X_n, \mathcal{B}_n, n \ge 1\}$, we can associate a compatible system $\{\Omega, \mathcal{B}_n, \nu_n, P, n \ge 1\}$ which determines a signed measure space $(\Omega, \mathcal{B}_\infty, \nu)$. Here $\pi_m : \Omega \to \Omega_m = \Omega$ are identity mappings. This exhibits an inherent relation between martingale theory and the existence theory of Kolmogorov and Bochner (cf. Theorem 4.10). This seemingly simple connection actually is much deeper between these two theories. An aspect of this is exemplified in the second proof below. However, if the sequence $\{X_n, \mathcal{B}_n, n \ge 1\}$ is merely a martingale (but not uniformly integrable), then also $\nu : \bigcup_{n\ge 1} \mathcal{B}_n \to \mathbb{R}$ is uniquely defined, but is only a finitely additive function. Finally, note that in Theorem 7, the sequence $\{X_n, \mathcal{B}_n, 1 \le n \le \infty\}$ is a martingale, so that $X_n = E^{\mathcal{B}_n}(X_\infty), n \ge 1$. If there is such an r.v. $X \in L^1(P)$ with $X_n = E^{\mathcal{B}_n}(X)$, then the martingale $\{X_n, \mathcal{B}_n, E^{\mathcal{B}_\infty}(X), n \ge 1\}$ is said to be *closed* on the right.

We shall now present the general martingale convergence theorem, again with two proofs. The first one is direct, in the sense that it is based only on Theorem 6i after a preliminary simplification. The second one, based on an application of Theorem 4.10, is a reduction of the proof to that of the preceding theorem of Andersen and Jessen. It therefore shows that both these results are *equivalent*, although this connection lies somewhat deeper. There are several other proofs of both these theorems (and also of their equivalence assertion), but we shall present a relatively simple argument. However, all these different proofs have independent interest, since they lead to various extensions of the subject.

Theorem 8 (Doob) *Let $\{X_n, \mathcal{B}_n, n \geq 1\}$ be a martingale on (Ω, Σ, P) and* $\sup_n E(|X_n|) < \infty$. *Then $X_n \to X_\infty$ a.e. and $E(|X_\infty|) \leq \liminf_n E(|X_n|)$.*

First Proof Here the convergence assertion follows if we express each $X_n = X_n^{(1)} - X_n^{(2)}$ with $\{X_n^{(i)}, \mathcal{B}_n, n \geq 1\}$ (since $\sup_n E(|X_n|) < \infty$) as positive martingales for $i = 1, 2$, by Lemma 5, and prove that $X_n^{(i)} \to X_\infty^{(i)}$ a.e. Thus the result obtains if each positive martingale $\{Y_n, \mathcal{B}_n, n \geq 1\}$ is shown to converge a.e. Since $\{e^{-Y_n}, \mathcal{B}_n, n \geq 1\}$ is a positive uniformly bounded submartingale, and $Y_n \to Y_\infty$ a.e. iff $e^{-Y_n} \to e^{-Y_\infty}$ a.e., it is clearly enough to establish that each bounded positive submartingale converges a.e. Since $L^\infty(P) \subset L^2(P)$, this will follow if we demonstrate that each positive square integrable submartingale $\{Z_n, \mathcal{B}_n, n \geq 1\}$ satisfying $E(Z_n^2) \leq K < \infty$ converges a.e. Now by Lemma 4, $\{Z_n^2, \mathcal{B}_n, n \geq 1\}$ is a submartingale, and if $a_n = E(Z_n^2)$, then $a_n \uparrow a, a \leq K < \infty$, as $n \to \infty$, because

$$\begin{aligned} a_{n+1} &= E(Z_{n+1}^2) \geq E(E^{\mathcal{B}_n}(Z_n^2)) \\ &= E(Z_n^2) = a_n \geq 0, \text{ by submartingale property,} \end{aligned}$$

so the expectations of a submartingale form an increasing sequence. Thus for $n > m$,

$$0 \leq a_n - a_m = E(Z_n^2 - Z_m^2) = E(Z_n - Z_m)^2 + 2E(Z_m(Z_n - Z_m)), \quad (25)$$

and both terms on the right are nonnegative since by the submartingale hypothesis $E^{\mathcal{B}_m}(Z_m(Z_n - Z_m)) = Z_m(E^{\mathcal{B}_m}(Z_n) - Z_m) \geq 0$ a.e. Now let $n \to \infty$, and then $m \to \infty$; the left side of (25) tends to zero, and hence so does the right side. Thus $E(Z_n - Z_m)^2 \to 0$, implying that $Z_n \to Z_\infty$ in $L^2(P)$. Using this, we can deduce the pointwise convergence.

Let $m \geq 1$ be fixed and consider $\{Z_n - Z_m, \mathcal{B}_n, n > m\}$. This is clearly a submartingale. Hence by Theorem $6i$, given $\lambda > 0$, we have

$$\lambda P\left[\max_{m<k\leq n}(Z_k - Z_n) \geq \lambda\right] \leq E(|Z_n - Z_m|)$$

and

$$-\lambda P\left[\min_{m<k\leq n}(Z_k - Z_m) \leq -\lambda\right] \geq E(|Z_{m+1} - Z_m|) - E(|Z_n - Z_m|).$$

Let $n \to \infty$; this implies, on subtraction,

$$P\left[\sup_{k>m}(Z_k - Z_m) \geq \lambda\right] + P\left[\inf_{k>m}(Z_k - Z_m) \leq -\lambda\right]$$
$$\leq \frac{2}{\lambda}[E(|\, Z_{m+1} - Z_m|) + E(|\, Z_m + Z_\infty|)]. \tag{26}$$

If $m \to \infty$ in (26), since the L^2-convergence of the Z_n implies their L^1-convergence, one gets

$$\lim_{m\to\infty} P\left[\sup_{k\geq m}|Z_k - Z_m| \geq \lambda\right] = 0. \tag{27}$$

Hence $|Z_k - Z_m| \to 0$ a.e., and so $\{Z_k(\omega), k \geq 1\}$ is a scalar Cauchy sequence for a.a.(ω). It follows that $Z_n \to Z_\infty$ a.e. [and in $L^1(P)$]. Thus, recapitulating the argument, what we have shown implies that $X_n \to X_\infty$ a.e. for the original martingale.

The preceding result also implies $|\, X_n| \to |\, X_\infty|$ a.e., and then by the Fatou inequality one gets $E(|\, X_\infty|) \leq \liminf_n E(|\, X_n|)$. This proves the result completely.

Second Proof By Lemma 5, it is again sufficient to consider only positive martingales $\{X_n, \mathcal{B}_n, n \geq 1\}$ and show its a.e. convergence. Thus hereafter we can and do assume that $X_n \geq 0$ a.e.

The key idea now is to identify the given martingale with another martingale on a nice topological space and show that the latter converges a.e. by means of Theorem 7. We then transport the result to the original space. To implement this, let $S_n = \mathbb{R}^+, \mathcal{R}_n = \mathcal{R}$, the Borel σ-algebra of $\mathbb{R}^+$, and consider the product measurable space $(S, \mathcal{I}) = \times_{n\in\mathbb{N}}(S_n, \mathcal{R}_n)$, as in the last section, so that S is the cartesian product space with its product topology and $\mathcal{I}$ is the σ-algebra of cylinder sets of S. As usual, let $\pi_n : S \to S_{\alpha_n} = \times_{i\in\alpha_n} S_i$, where $\alpha_n = (1, 2, \ldots, n)$. We define a probability measure on $\mathcal{I}$ as follows. Let $f : \omega \mapsto (X_1(\omega), X_2(\omega), \ldots) \in S$ be the mapping from $\Omega \to S$. If $A = \times_{i=1}^{\infty} A_i = \bigcap_{n=1}^{\infty}(A_1 \times \cdots \times A_n \times S^{(n)}) \in \mathcal{I}$ (cf. Lemma 4.4), then

$$f^{-1}(A) = \bigcap_{i=1}^{\infty}[X_i^{-1}(A_i)] \in \Sigma.$$

It follows that f is measurable, and so we can define $\mu : \mathcal{I} \to [0, 1]$ as the image probability of P under f, i.e., $\mu(A) = P(f^{-1}(A)), A \in \Sigma$. Then $(S, \mathcal{I}, \mu)$ is a probability space, and if $\mathcal{F}_n = \pi_{\alpha_n}^{-1}(\mathcal{I}_{\alpha_n})$, where $\mathcal{I}_{\alpha_n} = \bigotimes_{i\in\alpha_n} \mathcal{R}_i$, we have $\mathcal{F}_n \subset \mathcal{F}_{n+1} \subset \mathcal{I}$. Also $\pi_n(f(\omega)) = X_n(\omega), \omega \in \Omega, n \geq 1$, and, of course, $\pi_n : S \to \mathbb{R}^+$ is a positive random variable. By the image law result (cf. Theorem 1.4.1), it follows that

$$\int_S \pi_n(s)\mu(ds) = \int_{f^{-1}(S)} X_n(\omega)P(d\omega), \qquad n \geq 1. \tag{28}$$

Hence $\{\pi_n, \mathcal{F}_n, n \geq 1\}$ is a positive adapted integrable sequence on $(S, \mathcal{I}, \mu)$, and moreover, the integrals are constants by the right side of (28)(since X_n is a martingale sequence). Actually $\{\pi_n, \mathcal{F}_n, n \geq 1\}$ is also a martingale on $(S, \mathcal{I}, \mu)$.

To verify this, let $A \in \mathcal{F}_n$. Then $f^{-1}(A) \in \mathcal{B}_n$, and

$$\begin{aligned}
&\int_A \pi_n(s)\, d\mu_{\mathcal{F}_n}(s) \\
&\quad = \int_{f^{-1}(A)} X_n(\omega)\, dP_{\mathcal{B}_n}(\omega) \quad \text{[as in (28)]} \\
&\quad = \int_{f^{-1}(A)} X_{n+1}(\omega)\, dP_{\mathcal{B}_{n+1}}(\omega) \\
&\quad\quad \text{(by the martingale property of the } X_n) \\
&\quad = \int_A \pi_{n+1}(f(\omega))\, dP \circ f^{-1}(\omega), \\
&\quad\quad \text{(by the image law 1.4.1, and } A \in \mathcal{F}_n \subset \mathcal{F}_{n+1} \subset \mathcal{I}) \\
&\quad = \int_A \pi_{n+1}(s)\, d\mu_{\mathcal{F}_{n+1}}(s) = \int_A E^{\mathcal{F}_n}(\pi_{n+1})\, d\mu_{\mathcal{F}_n}. \qquad (29)
\end{aligned}$$

Since $A \in \mathcal{F}_n$ is arbitrary, $\{\pi_n, \mathcal{F}_n, n \geq 1\}$ is a positive martingale.

Finally, let $\nu_n : \mathcal{F}_n \to \mathbb{R}^+$ be defined by $\nu_n(A) = \int_A \pi_n(s)\, d\mu(s)$. Then ν_n is σ-additive on $\mathcal{F}_n$ for each n. Also, this gives a unique additive set function ν on all the cylinder sets of Ω since, $\nu_n = \nu_{n+1}|\, \mathcal{F}_n$, by (29). On the other hand, let $G_{\alpha_n} = \nu_n \circ \pi_{\alpha_n}^{-1}$. Then G_{α_n} is a finite measure on $\mathcal{I}_{\alpha_n}$, and so is a Lebesgue-Stieltjes or regular measure (i.e., by the standard real analysis theory, each open set has G_{α_n}-measure finite, and it can be approximated from inside by compact sets, even intervals). If the constant value $E(X_n) = a$ is taken as $a = 1$, by dividing if necessary, then G_{α_n} is even an n-dimensional distribution function on S_{α_n}. Hence by Theorem 4.10, the compatible family $\{G_{\alpha_n}, n \geq 1\}$ uniquely determines a σ-additive function $\zeta : \mathcal{I} \to [0, a]$ such that $G_{\alpha_n} = \zeta \circ \pi_{\alpha_n}^{-1} = \nu \circ \pi_{\alpha_n}^{-1}, n \geq 1$. It follows that ν is a σ-additive and uniquely (by extension) defined on $\mathcal{I}, \nu_n = \nu|\mathcal{F}_n, \pi_n = d\nu_n/d\mu_{\mathcal{F}_n}$. Hence by Theorem 7, $\pi_n \to \pi_\infty$ a.e. (and $\pi_\infty = d\nu/d\mu$ also.) Thus there is a set $A \in \mathcal{I}$ with $\mu(A) = 0$, and $\pi_n(s) \to \pi_\infty(s)$, for all $s \in S - A$. Let $N = f^{-1}(A)$, so that $P(N) = \mu(A) = 0$, and—if $\omega \notin N$, then $f(\omega) \notin A$—we have

$$X_n(\omega) = \pi_n(f(\omega)) \to \pi_\infty(f(\omega)) = X_\infty(\omega), \quad \omega \in \Omega - N. \qquad (30)$$

The last statement now follows by Fatou's lemma, as before. This completes the second demonstration of the theorem.

Discussion Theorem 8 includes the result of Theorem 7. In fact, if $X_n = d\nu_n/dP_n$, then $\{X_n, \mathcal{B}_n, n \geq 1\}$ is a uniformly integrable martingale, $\sup_n E(|\, X_n|) < \infty$. Hence $X_n \to X_\infty$ a.e., by Theorem 8, and the uniform integrability implies $E(|X_n - X_\infty|) \to 0$. It is then easily inferred that $d\nu_\infty/dP_\infty = X_\infty$ a.e. However, in Theorem 8 one is given only that $\{X_n, \mathcal{B}_n, n \geq 1\}$ is an $L^1(P)$-bounded martingale. Thus if $\nu_n : A \mapsto \int_A X_n \, dP, A \in \mathcal{B}_n$, then $\nu_n = \nu_{n+1}|\mathcal{B}_n$ and $\nu : \bigcup_{n\geq 1} \mathcal{B}_n \to \mathbb{R}$ is well defined by $\nu|\mathcal{B}_n = \nu_n$, but, in general, it is only finitely additive. Examples can be given to show that ν is even only "purely finitely additive," so that Theorem 7 cannot be applied directly. But the second proof of Theorem 8 shows that with a product space representation [this idea is borrowed from Ionescu Tulcea (1969)] the preceding noted hurdle is not real, and the result can be reduced to that of Theorem 7. Here the crucial pointwise convergence, which interestingly enough used the classical Kolmogorov result (really Theorem 4.11 sufficed), followed from Theorem 7. In this way Theorem 8 is also a consequence of Theorem 7. The work needed in this latter reduction is evidently nontrivial. For this reason, it was believed for some time that Theorem 7 is strictly weaker than Theorem 8. These various demonstrations also show that the martingale convergence theorems are essentially related to the differentiation of set functions. In fact, if Theorem 8 (its first proof) is granted, then we can actually deduce the Radon-Nikodým theorem from it. This will become clear shortly (see Problem 28). In this sense, the extra sets of proofs given for the above two theorems are inherently useful and illuminating.

It is now time to present analogous convergence statements for submartingales. Even though these results can be proved independently, it is fortunate that the submartingale convergence can be deduced from the martingale convergence itself. For this deduction, one needs the following simple decomposition, which turns out to have, especially in the continuous index case, a profound impact on the subject. In the discrete case it is called the *Doob decomposition,* after its inventor and we shall present it. But its continuous parameter analog (called the *Doob-Meyer decomposition*) is much more involved, and is not considered here.

Theorem 9 *Let $\{X_n, \mathcal{B}_n, n \geq 0\} \subset L^1(\Omega, \Sigma, P), \mathcal{B}_n \subset \mathcal{B}_{n+1}$, be any adapted process (so that X_n is $\mathcal{B}_n$-measurable). Then there exists a martingale $\{Y_n, \mathcal{B}_n, n \geq 1\}$ and an integrable adapted process $\{A_n, \mathcal{B}_{n-1}, n \geq 1\} \subset L^1(\Omega, \Sigma, P), A_0 = 0$, such that*

$$X_n = Y_n + A_n, \qquad n \geq 0, \text{ a.e.}, \tag{31}$$

uniquely. Further, A_n is increasing iff the given process is a submartingale.

Proof The decomposition is obtained constructively. Set $A_0 = 0$, and define recursively for $n \geq 1$

$$A_n = E^{\mathcal{B}_{n-1}}(X_n) - X_{n-1} + A_{n-1} = E^{\mathcal{B}_{n-1}}(X_n - X_{n-1}) + A_{n-1}, \tag{32}$$

and let $Y_n = X_n - A_n$. Then $\{A_n, \mathcal{B}_{n-1}, n \geq 1\}$ is adapted, integrable, $Y_n \in L^1(P)$, and

$$\begin{aligned}E^{\mathcal{B}_n}(Y_{n+1}) &= E^{\mathcal{B}_n}[X_{n+1} - E^{\mathcal{B}_n}(X_{n+1}) + X_n - A_n] \quad [\text{by } (32)]\\ &= E^{\mathcal{B}_n}(X_n) - A_n = X_n - A_n = Y_n \quad \text{a.e.},\end{aligned}$$

since X_n is $\mathcal{B}_n$-adapted and A_n is $\mathcal{B}_{n-1} \subset \mathcal{B}_n$-adapted. Thus $\{Y_n, \mathcal{B}_n, n \geq 1\}$ is a martingale and the decomposition (31) holds.

For uniqueness, let $X_n = Y_n + A_n = Y_n' + A_n'$ be two such decompositions. Then $Y_n - Y_n' = A_n' - A_n = B_n$, say, defines a process such that $B_0 = 0$, and the left side gives a martingale, while $\{A_n' - A_n, \mathcal{B}_{n-1}, n \geq 1\}$ is adapted. Since $B_0 = 0, Y_0 = Y_0'$ a.e., and for $n \geq 1$, one has

$$\begin{aligned}E^{\mathcal{B}_n}(B_{n+1}) = E^{\mathcal{B}_n}(Y_{n+1} - Y_{n+1}') &= Y_n - Y_n' \quad \text{(by the martingale property)}\\ &= Y_{n+1} - Y_{n+1}'\\ &\quad \text{(because } B_{n+1} \text{ is } \mathcal{B}_n\text{-adapted).}\end{aligned}$$

Thus $B_n = B_{n+1}$, and so $0 = B_0 = B_1 = \cdots = B_n$ a.e. This shows $Y_n = Y_n'$ a.e., and then $A_n = A_n'$ a.e., $n \geq 1$. Hence the decomposition in (31) is unique.

If $\{X_n, \mathcal{B}_n, n \geq 1\}$ is a submartingale, then $E^{\mathcal{B}_n}(X_{n+1}) \geq X_n$ a.e., so that $A_1 \geq 0$, and by definition (32), $A_n \geq A_{n-1}$ a.e., $n \geq 1$. Conversely, if $A_n \geq A_{n-1} \geq 0$, then (31) implies

$$\begin{aligned}E^{\mathcal{B}_n}(X_{n+1}) &= E^{\mathcal{B}_n}(Y_{n+1}) + A_{n+1} = Y_n + A_{n+1} \quad \text{a.e.}\\ &\geq Y_n + A_n = X_n \quad \text{a.e.,} \quad n \geq 1.\end{aligned}$$

Hence $\{X_n, \mathcal{B}_n, n \geq 1\}$ is a submartingale, as asserted.

The submartingale convergence can now be established.

Theorem 10 *Let $\{X_n, \mathcal{B}_n, n \geq 1\}$ be a submartingale with* $\sup_n E(|X_n|) < \infty$. *Then $X_n \to X_\infty$ a.e., and $E(|X_\infty|) \leq \liminf_n E(|X_n|)$.*

Proof By the above theorem, $X_n = Y_n + A_n$, where $A_n \geq A_{n-1} \geq 0$ a.e., and $\{Y_n, \mathcal{B}_n, n \geq 1\}$ is a martingale. Hence

$$\begin{aligned}0 \leq E(A_n) = E(X_n) - E(Y_n) &= E(X_n) - E(Y_1)\\ &\leq \sup_n E(X_n^+) - E(Y_1) < \infty.\end{aligned} \tag{33}$$

Since $A_n \nearrow A_\infty$ a.e., by the monotone convergence theorem, the inequality (33) implies that $E(A_\infty) < \infty$. Hence

$$E(|Y_n|) \leq E(|X_n|) + E(A_n) \leq \sup_n E(|X_n|) + E(A_\infty) < \infty, \tag{34}$$

by hypothesis. Thus $\sup_n E(|Y_n|) < \infty$, and Theorem 8 implies $Y_n \to Y_\infty$ a.e. Consequently $X_n = Y_n + A_n \to Y_\infty + A_\infty = X_\infty$, say, a.e. The last inequality between expectations follows again by Fatou's lemma, completing the proof.

Remark In (33) only the weaker hypothesis that $E(X_n^+) \leq K_0 < \infty$ is used, but in (34) we needed $E(|X_n|) \leq K_1 < \infty$. However, these two are equivalent conditions. In fact, $|X_n| = 2X_n^+ - X_n$ and $E(X_n) \geq E(X_1)$ for submartingales. Thus

$$E(|X_n|) \leq 2E(X_n^+) - E(X_1) \leq 2\sup_n E(X_n^+) - E(X_1), \tag{35}$$

and hence if $E(X_n^+) \leq K_0 < \infty$, then $E(|X_n|) \leq K_1 < \infty$ [since $X_1 \in L^1(P)$]. On the other hand, $X_n^+ \leq |X_n|$, so that the opposite implication is always true.

We now present a result on characterizations of convergences.

Theorem 11 *Let $\{X_n, \mathcal{B}_n, n \geq 1\}$ be a submartingale on (Ω, Σ, P), and $\mathcal{B}_\infty = \sigma(\bigcup_{n\geq 1} \mathcal{B}_n)$. Then the following statements are equivalent:*

(i) *The sequence is uniformly integrable.*
(ii) *The sequence is Cauchy in $L^1(P)$ on (Ω, Σ, P).*
(iii) $\limsup_n E(|X_n|) = K < \infty, X_n \leq E^{\mathcal{B}_n}(X_\infty), n \geq 1$, *a.e., where $X_n \to X_\infty$ a.e., and $K = E(|X_\infty|), \{X_n, \mathcal{B}_n, 1 \leq n \leq \infty\}$ is a submartingale.*
(iv) *There exists a symmetric convex function $\phi : \mathbb{R}^+ \to \mathbb{R}^+, \phi(0) = 0, \phi(x)/x \nearrow \infty$ as $x \nearrow \infty$, and* $\sup_n E(\phi(X_n)) < \infty$.

Proof (i)⇔(ii) By uniform integrability $\sup_n E(|X_n|) < \infty$, so that by Theorem 10, $X_n \to X_\infty$ a.e. Then $|X_n - X_\infty| \to 0$ a.e., and $\{|X_n - X_\infty|, n \geq 1\}$ is uniformly integrable. Thus by the classical Vitali convergence (cf. Theorem 1.4.4), $E(|X_n - X_\infty|) \to 0$ as $n \to \infty$ and (ii) holds. That (ii)⇒(i) is a standard fact in Lebesgue integration, independent of martingale theory.

(ii)⇒(iii) Since the Cauchy convergence implies $\{E(|X_n|), n \geq 1\}$ is convergent (hence bounded), it follows as before (by Theorem 10) that $X_n \to X_\infty$ a.e., and also $E(|X_n|) \to E(|X_\infty|)$ as $n \to \infty$. To prove the submartingale property, consider for $A \in \mathcal{B}_m$, $(m < n)$

$$\begin{aligned}\int_A X_m \, dP_{\mathcal{B}_m} &\leq \int_A X_n \, dP_{\mathcal{B}_n} \\ &= \int_A X_n \, dP \to \int_A X_\infty \, dP \\ &\quad \text{(by the Vitali convergence,} \\ &\quad \text{since the } X_n \text{ are uniformly integrable)} \\ &= \int_A E^{\mathcal{B}_m}(X_\infty) \, dP_{\mathcal{B}_m}. \end{aligned} \tag{36}$$

Since the extreme integrands are $\mathcal{B}_m$-measurable and $A \in \mathcal{B}_m$ is arbitrary, it follows that $X_m \leq E^{\mathcal{B}_m}(X_\infty)$ a.e. Thus $\{X_n, \mathcal{B}_n, 1 \leq n \leq \infty\}$ is a submartingale, proving (iii).

(iii)⇒(i) Since $X_n \leq E^{\mathcal{B}_n}(X_\infty) \Rightarrow X_n^+ \leq E^{\mathcal{B}_n}(X_\infty^+)$ a.e., and $E(X_n^+) \leq E(X_\infty^+) < \infty$, by (35), $\sup_n E(|X_n|) < \infty$. Hence by Theorem 10, $X_n \to X_\infty$ a.e., so that we also have $|X_n| \to |X_\infty|$ a.e., and the relation $E(|X_n|) \to E(|X_\infty|)$ implies by Proposition 1.4.6 (Scheffé's lemma) that $\{X_n, n \geq 1\}$ is uniformly integrable. Thus (i) holds.

(i)⇔(iv) This was already proved as part of Theorem 1.4.5, and does not depend on martingale methods or results. The proof is complete.

For martingales the result takes the following form:

Theorem 12 *Let $\{X_n, \mathcal{B}_n, n \geq 1\}$ be a martingale on (Ω, Σ, P) and $\mathcal{B}_\infty = \sigma(\bigcup_{n\geq 1} \mathcal{B}_n)$. Then the following are equivalent statements:*

(i) *The sequence is uniformly integrable.*

(ii) *The sequence is Cauchy in $L^1(P)$.*

(iii) *The sequence is $L^1(P)$-bounded, so that $X_n \to X_\infty$ a.e., and $\{X_n, \mathcal{B}_n, 1 \leq n \leq \infty\}$ is a martingale.*

(iv) *The sequence satisfies $K = \sup_n E(|X_n|) < \infty$ (i.e., $L^1(P)$-bounded), so that $X_n \to X_\infty$ a.e., and $E(|X_\infty|) = K$.*

(v) *There exists a symmetric convex function $\phi : \mathbb{R}^+ \to \mathbb{R}^+, \phi(0) = 0, \phi(x)/x \nearrow \infty$ as $x \nearrow \infty$, and $\sup_n E(\phi(X_n)) < \infty$.*

Proof (i)⇒(ii)⇒(iii) and (i)⇒(iv) have the same proof as in the preceding result with equality in (36). That (iii)⇒(i) follows from the fact that $X_n = E^{\mathcal{B}_n}(X_\infty)$ a.e., by the present hypothesis, $|X_n| \leq E^{\mathcal{B}_n}(|X_\infty|)$ a.e., by the conditional Jensen inequality, and that $\{E^{\mathcal{B}_n}(|X_\infty|), n \geq 1\}$ is a uniformly integrable set. [This property was noted before, and is an immediate consequence of Theorem 1.4.5. Indeed, if $Y_n = E^{\mathcal{B}_n}(|X_\infty|)$,

$$\int_{[Y_n > a]} Y_n \, dP_{\mathcal{B}_n} = \int_{[Y_n > a]} |X_\infty| \, dP \to 0 \quad \text{as } a \to \infty,$$

uniformly in n, because $X_\infty \in L^1(P)$ and

$$P[Y_n > a] \leq (1/a)E(Y_n) \leq (1/a)E(|X_\infty|) \to 0$$

as $a \to \infty$, uniformly in n.] Since $\{|X_n|, \mathcal{B}_n, n \geq 1\}$ is a submartingale, and $E(|X_n|) \leq E(|X_{n+1}|)$, (iv) implies that $\lim_n E(|X_n|) = K = E(|X_\infty|)$, where $X_n \to X_\infty$ a.e. by Theorem 8 (thus $|X_n| \to |X_\infty|$ a.e.). The preceding equality and Proposition 1.4.6 together imply (i), as in the last proof.

On the other hand (i)⇔(v) is contained in Theorem 1.4.5, and does not involve martingale theory. This establishes all equivalences.

Remark A difference between the statements of Theorems 11 and 12 is in part (iii) of the first result, which became parts (iii) and (iv) in the martingale case. This distinction is basic. In fact, $\{X_n, \mathcal{B}_n, 1 \leq n \leq \infty\}$ *can be* a submartingale *without* being uniformly integrable, whereas this complication cannot occur for martingales, as proved above. Here is a simple counter-example supporting the former statement.

Let (Ω, Σ, P) be the Lebesgue unit interval, and $\mathcal{B}_n \subset \Sigma$ be the σ-algebra determined by the partition $(0,1/n]$, $(1/n, 1/(n-1)], \ldots, (1/2, 1)$. If $f_n = -n\chi_{(0,1/n]}$, then $\{f_n, \mathcal{B}_n, n \geq 1\}$ is a negative adapted integrable sequence such that $E(f_n) = -1, f_n \to 0$ a.e., and for any $A \in \mathcal{B}_n$ we have

$$\int_A f_{n+1}\, dP = \begin{cases} 0 \text{ if } & A \cap (0,1/n] = \emptyset \\ -1 \text{ if } & A \supset (0,1/n] \end{cases}$$

$$= \int_A f_n\, dP.$$

Hence $E^{\mathcal{B}_n}(f_{n+1}) = f_n$ a.e., and if $f_\infty \equiv 1$, then $E^{\mathcal{B}_n}(f_\infty) = 1 > 0 \geq f_n$ a.e. Thus $\{f_n, \mathcal{B}_n, 1 \leq n \leq \infty\}$ is a submartingale. But Theorem 11iii is not true for this sequence and $\{f_n, n \geq 1\}$ is not uniformly integrable. Note that $\{f_n, \mathcal{B}_n, n \geq 1\}$ is a convergent martingale, while $\{f_n, \mathcal{B}_n, 1 \leq n \leq \infty\}$ is *not* a martingale.

The following consequence of the above result has some interest.

Corollary 13 *Let $\{X_n, n \geq 1\}$ be a sequence of r.v.s on (Ω, Σ, P) such that (i) $X_n \to X_\infty$ a.e. and (ii) $|X_n| \leq Y, E(Y) < \infty$. If $\{\mathcal{B}_n, n \geq 1\}$ is any increasing sequence of σ-subalgebras of Σ, and $Z_n = E^{\mathcal{B}_n}(X_n)$, then $Z_n \to Z_\infty$ a.e., and in $L^1(P)$, where $Z_\infty = E^{\mathcal{B}_\infty}(X_\infty), \mathcal{B}_\infty = \sigma(\bigcup_{n\geq 1} \mathcal{B}_n)$.*

Proof By (ii), Z_n, Z_∞ are integrable. Also, by Theorem 12iii, $\{E^{\mathcal{B}_n}(X_\infty), 1 \leq n \leq \infty\}$ is a uniformly integrable martingale, so that $E^{\mathcal{B}_n}(X_\infty) \to X_\infty$, a.e., and in $L^1(P)$. Let $U_m = \sup_{n\geq m} |X_n - X_\infty| \leq 2Y, m \geq 1$. Then by hypothesis $U_m \to U_\infty = \lim_{m\to\infty} \sup_{n\geq m} |X_n - X_\infty| = 0$ a.e., and dominatedly. In particular, $E(U_m) \leq 2E(Y) < \infty$, and $U_m \downarrow 0$ a.e. On the other hand, if $n \geq m$,

$$\begin{aligned} |Z_n - Z_\infty| &\leq E^{\mathcal{B}_n}(|X_n - X_\infty|) + |(E^{\mathcal{B}_n} - E^{\mathcal{B}_\infty})(X_\infty)| \\ &\leq E^{\mathcal{B}_n}(U_m) + |E^{\mathcal{B}_n}(X_\infty) - E^{\mathcal{B}_\infty}(X_\infty)|. \end{aligned} \quad (37)$$

The last term of (37) tends to zero a.e. as well as in $L^1(P)$, and the first term also goes to zero by the conditional dominated convergence criterion. Hence $Z_n \to Z_\infty$ a.e. Thus $E(E^{\mathcal{B}_n}(U_n)) = E(U_n) \to 0$ by the Lebesgue dominated convergence, so that $E(|Z_n - Z_\infty|) \to 0$ also, as $n \to \infty$. This proves all the statements.

It is possible to present convergence results for *decreasing indexed* (or *reverse*) martingales and submartingales. These are slightly easier than Theorems 7 and 8 on probability spaces. It must be noted that there are analogous results if the probability space (Ω, Σ, P) is replaced by a nonfinite (σ-finite) space (Ω, Σ, μ), and then the comparison of difficulties will be reversed. However, we do not treat this case here. (See Problems 39, 40 and 42.)

Theorem 14 *Let $\mathcal{B}_n \supset \mathcal{B}_{n+1}, n \geq 1$, be a sequence of σ-subalgebras from (Ω, Σ, P), and $\{X_n, \mathcal{B}_n, n \geq 1\}$ be a decreasing indexed martingale in that $X_n \in L^1(P)$ and $E^{\mathcal{B}_{n+1}}(X_n) = X_{n+1}$* a.e., $n \geq 1$ *[equivalently, if $\mathcal{B}_{-n} \supset \mathcal{B}_{-n-1}, X_{-n} \in L^1(P)$, and $X_{-n} = E^{\mathcal{B}_{-n}}(X_{-n-1})$ a.e.]. If $\mathcal{B}_\infty = \lim_n \mathcal{B}_n (= \bigcap_{n=1}^\infty \mathcal{B}_n)$, then $X_n \to X_\infty$ a.e. and in $L^1(P)$-norm, so that $\{X_n, \mathcal{B}_n, 1 \leq n \leq \infty\}$ is a martingale.*

Proof We follow the argument of the first proof of Theorem 8, for convenience. If $\nu_k : A \mapsto \int_A X_k \, dP, A \in \mathcal{B}_k$, then ν_k is a signed measure and the martingale property implies $\nu_1 | \mathcal{B}_k = \nu_k, k \geq 1$. Since a signed measure is bounded, and by the Jordan decomposition $\nu_1 = \nu_1^+ - \nu_1^-$, we let $\nu_k = \tilde{\nu}_k^+ - \tilde{\nu}_k^-$, where $\tilde{\nu}_i^\pm : \mathcal{B}_i \to \mathbb{R}^+$ is a (finite) measure such that $\nu_1^\pm | \mathcal{B}_i = \tilde{\nu}_i^\pm, i \geq 1$. Evidently, $\tilde{\nu}_k^\pm$ is $P_k (= P|\mathcal{B}_k)$-continuous. By the Radon-Nikodým theorem, there exist $X_k^{(1)} = d\tilde{\nu}_k^+ / dP_k, X_k^{(2)} = d\tilde{\nu}_k^- / dP_k$ such that $X_k = X_k^{(1)} - X_k^{(2)}$ (because $\nu_k = \tilde{\nu}_k^+ - \tilde{\nu}_k^-$), and $\{X_k^{(i)}, \mathcal{B}_k, k \geq 1\}, i = 1, 2$, are positive decreasing martingales. Hence to prove $X_k \to X_\infty$ a.e., it suffices to prove that each positive decreasing martingale converges a.e. [Note that the proof of Jordan-type decomposition for decreasing indexed martingales is simpler than that for the increasing case (cf. Lemma 5), since in the latter there need be no σ-additive ν on $\sigma(\bigcup_{n \geq 1} \mathcal{B}_n)$ such that $\nu | \mathcal{B}_k = \nu_k$. Even though the Jordan decomposition is valid for finitely additive set functions, their restrictions $\tilde{\nu}_k^\pm$ (of $\nu^\pm$ to $\mathcal{B}_k$) are not suitable, and to obtain a useful splitting, one needs the computations given in Lemma 5.]

The proof is now essentially the same as in Theorem 8. Thus briefly, if $\{X_n, \mathcal{B}_n, n \geq 1\}$ is a positive decreasing martingale, then $\{e^{-X_n}, \mathcal{B}_n, n \geq 1\}$ is a positive bounded decreasing submartingale by Lemma 4, and $X_n \to X_\infty$ a.e. iff $e^{-X_n} \to e^{-X_\infty}$ a.e. If $\{Z_n, \mathcal{B}_n, n \geq 1\}$ is a positive decreasing $L^2(P)$-bounded submartingale, then $E(Z_n^2) = a_n \downarrow a \geq 0$ as $n \to \infty$. Next (25) implies, on considering $0 \leq a_m - a_n \to 0$, letting first $n \to \infty$ and later $m \to \infty$, that $E(Z_n - Z_m)^2 \to 0$, so that $Z_n \to Z_\infty$ in $L^2(P)$. If $Y_n = Z_n - Z_\infty$, then $\{Y_n, \mathcal{B}_n, n \geq 1\}$ is a submartingale such that $Y_n \to 0$ in $L^2(P)$. With this the work of (26), (27) holds, since the maximal inequalities in the decreasing case take the form for any $\lambda > 0$,

$$\lambda P \left[\max_{m \leq k \leq n} Y_k \geq \lambda \right] \leq E(Y_m^+),$$

$$-\lambda P \left[\min_{m \leq k \leq n} Y_k \leq -\lambda \right] \geq E(Y_n) - E(Y_m^+).$$

[Compare with (11) and (12).] Hence

$$P\left[\sup_{n\geq k\geq m}|Y_k|\geq\lambda\right]\leq[2E(|Y_m|)-E(Y_n)]/\lambda\to 0.$$

as $n\to\infty$ and then $m\to\infty$. It follows from this that $Y_n\to Y_\infty$ a.e. and that $Y_\infty=0$ a.e., [since $Y_n\to 0$ in $L^2(P)$]. Hence $Z_n\to Z_\infty$ a.e. and in $L^2(P)$. This proves that $X_n\to X_\infty$, a.e.

The uniform integrability follows from the fact that

$$\begin{aligned}&\int_{[|X_n|>\lambda]}|X_n|\,dP\\&\quad\leq\int_{[|X_n|>\lambda]}X_n^{(1)}\,dP+\int_{[|X_n|>\lambda]}X_n^{(2)}\,dP,\qquad X_n^{(1)}-X_n^{(2)}=X_n\\&\quad\quad\text{(as defined in the first paragraph)}\\&\quad=\nu_1^+[|X_n|>\lambda]+\nu_1^-[|X_n|>\lambda]\quad(\text{since }[|X_n|>a]\in\mathcal{B}_n)\\&\quad=|\nu_1|[|X_n|>\lambda]\quad(|\nu_1|\text{ being the variation measure of }v_1).\end{aligned}\tag{38}$$

But $P[|X_n|>\lambda]\leq E(|X_1|)/\lambda\to 0$ as $\lambda\to\infty$, uniformly in n, and ν_1 is P-continuous. Thus $\{X_n,\mathcal{B}_n,1\leq n\leq\infty\}$ is a martingale and the $L^1(P)$-convergence is an immediate consequence of the Vitali convergence (cf. Theorem 1.4.4). The proof is finished.

Just as in the increasing case, the decreasing indexed submartingale convergence can also be deduced from the corresponding martingale result.

Theorem 15 *Let $\{X_n,\mathcal{B}_n,n\geq 1\}$ be a submartingale on (Ω,Σ,P), where $\mathcal{B}_n\supset\mathcal{B}_{n+1}$ are σ-subalgebras of Σ. Then $X_n\to X_\infty$ a.e. Moreover, the sequence $\{X_n,\mathcal{B}_n,1\leq n\leq\infty\}$ is a submartingale iff $E(X_n)\geq K>\infty$, or equivalently the submartingale is uniformly integrable.* (The last condition is automatic for martingales, but not for submartingales.)

Proof For the convergence statement, one uses a form of Theorem 9, then reduces to the result of Theorem 14 as follows. Define $a_1=0$, and recursively for $n>1$,

$$a_{n+1}=E^{\mathcal{B}_{n+1}}(X_n)-X_{n+1},\quad A_n=\sum_{k\geq n}a_{k+1}\quad\text{a.e.}\tag{39}$$

Then $A_n\geq 0,\mathcal{B}_{n+1}$-adapted, and A_n decreases as n increases. First suppose that $E(X_n)\geq K>-\infty$. Since $\infty>E(X_1)\geq E(X_2)\geq\cdots\geq K$, it follows that $\lim_n E(X_n)\geq K$, and hence by (39) and the Lebesgue dominated convergence theorem,

$$E(A_1)=\sum_{k\geq 1}E(a_{k+1})=\lim_{n\to\infty}\sum_{k=1}^{n}E[E^{\mathcal{B}_{k+1}}(X_k)-X_{k+1}]$$

$$= \lim_{n\to\infty} \sum_{k=1}^{n} [E(X_k) - E(X_{k+1})]$$
$$= E(X_1) - \lim_{n\to\infty} E(X_n) < \infty.$$

Hence $X_n' = X_n - A_n$ is well defined, and $\{X_n', \mathcal{B}_n, n \geq 1\}$ is readily seen to be a martingale. By Theorem 14, $X_n' \to X_\infty'$ a.e., since $E((X_n')^+) \leq E(X_1^+) + E(A_1) < \infty$. But $0 \leq A_n \downarrow$, integrable, so that $A_n \to A_\infty$ a.e., and $X_n = X_n' + A_n \to X_\infty' + A_\infty = X_\infty$ a.e. Because $0 \leq A_n \leq A_1$ and $E(A_1) < \infty$, both $\{X_n', n \geq 1\}$ and $\{A_n, n \geq 1\}$ are uniformly integrable. Hence so is $\{X_n, n \geq 1\}$. We conclude that $\{X_n, \mathcal{B}_n, 1 \leq n \leq \infty\}$ is a submartingale because

$$E^{\mathcal{B}_\infty}(X_n) = E^{\mathcal{B}_\infty}(X_n' + A_n) \geq E^{\mathcal{B}_\infty}(X_n' + A_\infty) \quad (\text{since } A_n \geq A_\infty)$$
$$= X_\infty' + A_\infty = X_\infty \quad \text{a.e.}$$

On the other hand, if $\{X_n, \mathcal{B}_n, 1 \leq n \leq \infty\}$ is a submartingale, then $E(X_\infty) > -\infty$, and so $\lim_n E(X_n) \geq E(X_\infty) > -\infty$. This establishes the last half. For the first part, let $\alpha > -\infty$, and define $Y_n^\alpha = \max(X_n, \alpha)$. Then $\{Y_n^\alpha, \mathcal{B}_n, n \geq 1\}$ is a decreasing submartingale (by Lemma 4) such that $\lim_{n\to\infty} E(Y_n^\alpha) \geq \alpha > -\infty$. Thus by the preceding, $Y_n^\alpha \to Y_\infty^\alpha$ a.e., and hence there is a set $N_\alpha, P(N_\alpha) = 0$, such that $Y_n^\alpha(\omega) \to Y_\infty^\alpha(\omega), \omega \in \Omega - N_\alpha$. Because $Y_n^\alpha(\omega) = X_n(\omega)$ if $\inf_n X_n(\omega) \geq \alpha$, it follows that $\lim_n X_n(\omega)$ exists for almost all ω in the set $\{\omega : \inf_n X_n(\omega) \geq \alpha\}$. Note also that if $\alpha_n > \alpha_{n+1} \downarrow -\infty$, then $N_{\alpha_1} \subset N_{\alpha_2}$ and hence, if $\Omega_0 = \bigcap_{n=1}^{\infty} N_{\alpha_n}^c$, then for each $\omega \in \Omega_0, X_n(\omega) \to X_\infty(\omega)$. Since $E(X_n) \leq E(X_1) \leq E(X_1^+) < \infty$, and $\{X_n^+, \mathcal{B}_n, n \geq 1\}$ is a submartingale, $X_n^+ \to X_\infty^+ \geq 0$ a.e. and $E(X_\infty^+) < \infty$. Thus we have $-\infty \leq X_\infty < \infty$ a.e. With this the theorem is completely proved.

The following is an immediate consequence of Corollary 13 and Theorems 14 and 15:

Corollary 16 *Let $\{Y_n, -\infty < n < \infty\}$ be a sequence of r.v.s on (Ω, Σ, P) such that for all $n, |Y_n| \leq |Z|$, where $Z \in L^1(P)$. If $\mathcal{B}_n \subset \mathcal{B}_{n+1} \subset \Sigma$ are σ-algebras, and $\lim_{n\to\infty} Y_n = Y_\infty, \lim_{n\to-\infty} Y_n = Y_{-\infty}$ exist a.e., then the following are true with $\mathcal{B}_\infty = \sigma(\bigcup_n \mathcal{B}_n), \mathcal{B}_{-\infty} = \bigcap_n \mathcal{B}_n$:*

(i) $\lim_{n\to\infty} E^{\mathcal{B}_n}(Y_n) = E^{\mathcal{B}_\infty}(Y_\infty), \lim_{n\to-\infty} E^{\mathcal{B}_n}(Y_n) = E^{\mathcal{B}_{-\infty}}(Y_{-\infty})$ a.e.
(ii) $\lim_{n\to\infty} E^{\mathcal{B}_n}(Z) = E^{\mathcal{B}_\infty}(Z), \lim_{n\to-\infty} E^{\mathcal{B}_n}(Z) = E^{\mathcal{B}_{-\infty}}(Z)$ a.e.

These limits also hold in $L^1(P)$-mean.

Two Applications We present here two useful applications of the preceding results. Some others appear in the problems section. There are numerous refinements of the foregoing theory and the subject is one of the most active areas in probability. A few extensions will be treated later, for instance, in Chapter 7.

In the theory of *nonparametric statistics,* a class of random sequences, called U-statistics (originally introduced by W. Hoeffding in the late 1940's), together with their limiting behavior plays a key role. The latter can be studied effectively with martingale theory. Let us define these and deduce their convergence properties.

Let $\{X_n, n \geq 1\}$ be an i.i.d. sequence of random variables, or more generally an *exchangeable* or *permutable* or*symmetrically dependent* sequence, in the sense that for each $n \geq 1, \{X_{i_1}, \ldots, X_{i_n}\}$ and $\{X_1, \ldots, X_n\}$ have the same joint distributions for any permutation $(i_1, \ldots, i_n)$ of $(1,2,\ldots, n)$ (cf. Definition 2.1.11). If $f : \mathbb{R}^k \to \mathbb{R}$ is a Borel function such that $f(x_{i_1}, \ldots, x_{i_k}) = f(x_1, \ldots, x_k)$, so that it is symmetric and if either f is bounded or

$$f(X_{i_1}, \ldots, X_{i_k}) \in L^1(P),$$

then the *U-statistics* are the random variables $\{U^f_{k,n}, n \geq k\}$ defined by

$$U^f_{k,n} = \sum_{1 \leq i_1 < \cdots < i_k \leq n} f(X_{i_1}, \ldots, X_{i_k}) \Big/ \binom{n}{k} \ . \tag{40}$$

Thus if $k = 1, f(x) = x$, then $U^f_{1,n} = \overline{X}_n$, the sample mean, and for $k = 2, f(x_1, x_2) = \frac{1}{2}(x_1 - x_2)^2$, one gets $U^f_{2,n}$ = the sample variance. Other specializations yield sample central moments, etc. (For example, for the third moment,

$$\begin{aligned} f(x_1, x_2, x_3) = \tfrac{1}{6}\ &[(x_1 - x_2)^2(x_1 + x_2 + 2x_3) + (x_1 - x_3)^2(x_1 + x_3 + 2x_2) \\ &+ (x_2 - x_3)^2(x_2 + x_3 + 2x_1)], \end{aligned}$$

so that $E(U^f_{3,n}) = E(X^3)$ for the i.i.d. case, and complicated higher symmetric functions are possible candidates for other parameter estimations. The matter is not pursued further.)

This sequence has the following general property:

Proposition 17 *Let $\{U^f_{k,n}, n \geq k\}$ be a sequence of U-statistics of a symmetrically dependent family $\{X_n, n \geq 1\}$ on (Ω, Σ, P) and a symmetric Borel $f : \mathbb{R}^k \to \mathbb{R}$ such that $E(|f(X_{i_1}, \ldots, X_{i_k})|) < \infty$. If $\mathcal{F}_{k,n} = \mathcal{F}_n = \sigma(U^f_{k,m}, m \geq n)$, then $\{U^f_{k,n}, \mathcal{F}_n, n \geq k\}$ forms a decreasing martingale and $U^f_{k,n} \to U^f_{k,\infty} = E^{\mathcal{F}_\infty}(f(X_1, \ldots, X_k))$ a.e., and in $L^1(P)$. If the X_n are i.i.d., then $U^f_{k,\infty} = E(f(X_1, \ldots, X_k))$ a.e., and hence is a constant.*

Proof First note that, for symmetrically dependent r.v.s, by definition for each subset $i_1 < i_2 < \cdots < i_k$ of $(1,2,\ldots, n)$, $1 \leq k \leq n < \infty$, the joint distributions of $(X_{i_1}, \ldots, X_{i_k})$ and $(X_1, \ldots, X_k)$ are identical. Hence if

$f: \mathbb{R}^k \to \mathbb{R}$ is a symmetric Borel function, then the r.v.s $f(X_{i_1}, \dots, X_{i_k})$ and $f(X_1, \dots, X_k)$ are identically distributed for any k-subset of $(1, \dots, n)$. If the latter r.v.s are integrable, then they have the same conditional expectations. This follows from a simple computation. Indeed, let X, Y be identically distributed and $C \subset \mathbb{R}$ be a Borel set. Thus $A = X^{-1}(C), B = Y^{-1}(C)$ are in Σ, and $P(A) = P(B)$. If $D \in \mathcal{F}_n$, we have

$$\begin{aligned} \int_D E^{\mathcal{F}_n}(\chi_A)\, dP_{\mathcal{F}_n} &= P(A \cap D) = P(B \cap D) \quad \text{(by the equidistributedness)} \\ &= \int_D E^{\mathcal{F}_n}(\chi_B)\, dP_{\mathcal{F}_n}. \end{aligned}$$

Since the extreme integrands are $\mathcal{F}_n$-measurable, $E^{\mathcal{F}_n}(\chi_A) = E^{\mathcal{F}_n}(\chi_B)$ a.e. Considering linear combinations, we deduce that the same equality holds for all simple functions, and then by approximation $E^{\mathcal{F}_n}(Y) = E^{\mathcal{F}_n}(Z)$. Hence in our case this implies

$$E^{\mathcal{F}_{n+1}}(f(X_{i_1}, \dots, X_{i_k})) = E^{\mathcal{F}_{n+1}}(f(X_1, \dots, X_k)) \quad \text{a.e.} \tag{41}$$

Since $\mathcal{F}_{n+1} = \sigma(U^f_{k,m}, m \geq n+1)$, the first r.v. in the sequence which is measurable for $\mathcal{F}_{n+1}$ is $U^f_{k,n+1}$. But (41) yields, on choosing the $\binom{n}{k}$ possible k-subsets of the integers of $(1, \dots, n)$ and $\binom{n+1}{k}$ of $(1, \dots, n+1)$ and averaging,

$$E^{\mathcal{F}_{n+1}}(U^f_{k,n}) = E^{\mathcal{F}_{n+1}}(U^f_{k,n+1}) = U^f_{k,n+1} \quad \text{a.e.,} \quad n \geq k. \tag{42}$$

It follows that $\{U^f_{k,n}, \mathcal{F}_n, n \geq 1\}$ is a decreasing martingale and $E(|U^f_{k,n}|) \leq E(|f(X_1, \dots, X_k)|) < \infty$ by hypothesis. By Theorem 14, this martingale is uniformly integrable, so that

$$U^f_{k,n} \to U^f_{k,\infty} = E^{\mathcal{F}_\infty}(U^f_{k,k}) \quad \text{a.e.,} \quad \text{and in } L^1(P), \text{as } n \to \infty.$$

In case the r.v.s are i.i.d. also, then $\mathcal{F}_\infty$, being the tail σ-algebra, is degenerate, so that $U^f_{k,\infty}$ is constant by Theorem 2.1.12. The $L^1(P)$-convergence implies that $E(U^f_{k,n})$ tends to $E(f(X_1, \dots, X_k))$, as asserted.

Taking $k = 1, f(x) = x$, the above result implies the SLLN for symmetrically dependent r.v.s, since then $U^f_{1,n} = (1/n)\sum_{i=1}^n X_i$. We state it for reference as follows:

Corollary 18 *Let $\{X_i, i \geq 1\}$ be symmetrically dependent integrable r.v.s on (Ω, Σ, P). Then, if $\mathcal{F}_n = \sigma(X_k, k \geq n)$ and $\mathcal{F}_\infty = \bigcap_n \mathcal{F}_n$, we have*

$$\frac{1}{n}\sum_{i=1}^n X_i \to E^{\mathcal{F}_\infty}(X_1) \quad \text{a.e.} \quad \text{and in } L^1(P), \text{ as } n \to \infty. \tag{43}$$

Remark If the X_n are i.i.d., then the above result becomes, since X_1 is independent of $\mathcal{F}_\infty$, so that $E^{\mathcal{F}_\infty}(X_1) = E(X_1)$, the sufficiency part of the SLLN given in Theorem 2.3.7. The necessity part, however, is *not* implied by the martingale theory and one has to refer to the original work presented there.

The decreasing martingales always appear if one formulates "ergodic sequences" as martingales. The latter are discussed in Section 7.3. Those results are generalizations of the SLLN to other dependent (including ergodic) sequences. Here we consider *another application* to likelihood ratios, which plays a useful role in the theory of statistical inference. One uses instead the increasing indexed martingales for these problems.

Let $X_1, X_2, \ldots$ be a sequence of random variables on a measurable space (Ω, Σ). Suppose one is given two probability measures P and Q on Σ, and it is desired to determine the correct underlying measure, P or Q, based on a single realization of the sequence. The problem is made precise as follows. Suppose $\mathcal{B}_n = \sigma(X_1, \ldots, X_n) \subset \Sigma$ and $P_n = P|\mathcal{B}_n, Q_n = Q|\mathcal{B}_n$. Let $Y_n = dQ_n^c/dP_n$, the Radon-Nikodým density of the P_n-continuous part of Q_n relative to P_n. It is called the *likelihood ratio* for certain historical reasons. If n is increased without bound, then a variational principle (in this context it is called a "Neyman-Pearson lemma") says that the event $E_{n,k} = [Y_n \geq k]$ has P-probability "small" and Q-probability "large" for a suitable k. Hence if the realization $\omega \in E_{n,k}$ for large n, then we take Q to be the correct probability and reject P and the opposite action otherwise. The discussion on the appropriateness and goodness of this rule belongs to the theory of statistical inference, and it is not possible to go into that topic. (For a mathematical treatment and consequences, see eg., Rao (2000), Chapter II.) Here we establish the convergence properties of the likelihood ratio sequence $\{Y_n, n \geq 1\}$ defined above.

Theorem 19 *Let $X_1, X_2, \ldots$ be a sequence of random variables on a measurable space (Ω, Σ) and suppose that it is governed by one of two probabilities P or Q on Σ. If $\mathcal{B}_n = \sigma(X_1, \ldots, X_n)$, and $Y_n = dQ_n^c/dP_n$ is the likelihood ratio, as defined above, then $\{Y_n, \mathcal{B}_n, n \geq 1\}$ is a nonnegative supermartingale on $(\Omega, \Sigma, P), Y_n \to Y_\infty$ a.e. $[P]$, and $Y_\infty = dQ_\infty^c/dP_\infty$, where $Q_\infty = Q|\mathcal{B}_\infty, P_\infty = P|\mathcal{B}_\infty$ with $\mathcal{B}_\infty = \sigma(X_n, n \geq 1)$.*

Proof By hypothesis, if $Q_n = Q_n^c + Q_n^s$ is the Lebesgue decomposition of Q_n on $\mathcal{B}_n$ relative to P_n into the absolutely continuous part Q_n^c and the singular part Q_n^s, so that $Q_n^s(B_0) > 0$ only on B_0 with $P_n(B_0) = 0$, we have for any $B \in \mathcal{B}_n$,

$$\begin{aligned}\int_B Y_n \, dP_n &= Q_n^c(B) = Q_n(B) - Q_n^s(B) \\ &= Q_{n+1}(B) - Q_n^s(B) \quad (\text{since } Q_{n+1}|\mathcal{B}_n = Q_n)\end{aligned}$$

$$= Q_{n+1}^c(B) + Q_{n+1}^s(B) - Q_n^s(B)$$
$$= \int_B Y_{n+1}\, dP_{n+1} + [Q_{n+1}^s(B) - Q_n^s(B)]. \tag{44}$$

However, if $B_0 \subset B$ is maximal with $Q_n^s(B_0) > 0$ [B_0 exists], and similarly $\tilde{B}_0 \subset B$ with $Q_{n+1}^s(\tilde{B}_0) > 0$ [so that $P_{n+1}(\tilde{B}_0) = 0, P_n(B_0) = 0$, and since $P_{n+1}|\mathcal{B}_n = P_n, P_{n+1}(B_0) = 0$ also], it follows that $\tilde{B}_0 \supset B_0$. Hence $Q_{n+1}^s(B) - Q_n^s(B) = Q_{n+1}(N_{n+1} \cap B) - Q_n(N_n \cap B)$, where N_n is the singular set of Q_n, so that $N_n \subset N_{n+1}$, and thus it is nonnegative. Consequently, (44) becomes

$$\int_B Y_n\, dP_n \geq \int_B Y_{n+1}\, dP_{n+1} = \int_B E^{\mathcal{B}_n}(Y_{n+1})\, dP_n, \quad B \in \mathcal{B}_n. \tag{45}$$

Since the extreme integrands are $\mathcal{B}_n$-measurable, and B in $\mathcal{B}_n$ is arbitrary, (45) implies that $E^{\mathcal{B}_n}(Y_{n+1}) \leq Y_n$ a.e. and $\{Y_n, \mathcal{B}_n, n \geq 1\}$ is a supermartingale, as asserted. Since $\int_\Omega Y_n\, dP_n = E(Y_n) \leq Q(\Omega) = 1$, Theorem 10 (rather, its counterpart for supermartingales) implies that $Y_n \to Y_\infty$ a.e. It remains to identify the limit. This can be done using the method of proof of Theorem 8. For variety, we present here an alternative argument reducing the result to Theorem 8 itself.

Let $\mu = P + Q : \Sigma \to \mathbb{R}^+$, which is a finite measure dominating both P and Q. If $g = dQ/d\mu, f_0 = dP/d\mu$, and if $Q = Q^c + Q^s$ is the Lebesgue decomposition relative to P, then let $g_1 = dQ^c/d\mu, g_2 = dQ^s/d\mu$, so that $g = g_1 + g_2$ a.e.$[\mu]$. Note that $[g_1 > 0] \subset [f_0 > 0]$ and $P([g_2 > 0] \cap [f_0 > 0]) = 0$. Let $N_0 = [(g_2/f_0) > 0] = [(g_2/f_0) = +\infty]$, so that $P(N_0) = 0$. If $f = dQ^c/dP$, then by the chain rule for Radon-Nikodým derivatives one has

$$f = \frac{dQ^c}{d\mu} \Big/ \frac{dP}{d\mu} = \frac{g_1}{f_0} \quad \text{a.e.}[\mu]. \tag{46}$$

Consequently for each $\alpha > 0$

$$[g \geq \alpha f_0] = [(g_1 + g_2)/f_0 \geq \alpha] = [g_1/f_0 \geq \alpha] \cup N_0. \tag{47}$$

It follows from (47) that $[(g/f_0) \geq \alpha] = [(g_1/f_0) \geq \alpha]$ a.e. $[P]$. This also shows that these sets do not depend on the auxiliary dominating measure μ, and any other such μ' can be used. Since $f = g_1/f_0$, by (46) one has $0 \leq f(\omega) \leq (g/f_0)(\omega) < \infty$ for a.a.(ω). Replacing Σ by $\mathcal{B}_n$ and $\mathcal{B}_\infty$, we deduce that $f_n = (g_n/f_{0,n})$ and $f_\infty = (g_\infty/f_{0,\infty})$ a.e.$[P]$. But $\{g_n, \mathcal{B}_n, 1 \leq n \leq \infty\}$ and $\{f_{0,n}, \mathcal{B}_n, 1 \leq n \leq \infty\}$ are martingales on (Ω, Σ, μ) by Theorem 8, so that $g_n \to g_\infty$ a.e. $[\mu]$ and in $L^1(\mu)$; similarly $f_{0,n} \to f_{0,\infty}$ a.e. $[\mu]$ and in $L^1(\mu)$, with $f_{0,\infty} = 0$ a.e. only if the measure $P_\infty = 0$. Since this is not the case $[Q(\Omega) = 1 = P(\Omega)]$, we get $f_n = g_n/f_{0,n} \to f_\infty = g_\infty/f_{0,\infty} = (dQ_\infty/d\mu_\infty)/(dP_\infty/d\mu_\infty)$ a.e., and the last ratio is dQ_∞^c/dP_∞ a.e.$[\mu]$ (hence $[P]$), by (46). This proves the theorem completely.

The preceding result is usually given in terms of the image measures, which are distribution functions. This can be translated as follows. Let F_n, G_n be the n-dimensional d.f.s of P_n and Q_n, i.e.,

$$F_n(x_1, \ldots, x_n) = P[X_1 < x_1, \ldots, X_n < x_n]$$

and

$$G_n(x_1, \ldots, x_n) = Q[X_1 < x_1, \ldots, X_n < x_n].$$

Suppose that $F_n : \mathbb{R}^n \to [0,1], G_n : \mathbb{R}^n \to [0,1]$ are absolutely continuous relative to the Lebesgue measure of $\mathbb{R}^n$, with densities f_n and g_n. Then they are Borel functions and $f_n(X_1(\cdot), \ldots, X_n(\cdot)) : \Omega \to \mathbb{R}^+$ is a random variable. By (46) and (47) and the ensuing discussion, it follows that

$$Y_n(\omega) = \frac{g_n(x_1, \ldots, x_n)}{f_n(x_1, \ldots, x_n)} = \frac{g_n(X_1(\omega), \ldots, X_n(\omega))}{f_n(X_1(\omega), \ldots, X_n(\omega))} \quad \text{a.a.}(\omega). \tag{48}$$

Here $x_i = X_i(\omega)$ is the value of X_i for the realization ω. Thus the ratios (g_n/f_n) form a supermartingale and their values can be calculated in any problem. It should be noted, however, that each $g_n/f_n : \mathbb{R}^n \to \mathbb{R}^+$ is defined and as n changes their domains vary and are not nested. Thus $\{g_n/f_n, n \geq 1\}$ cannot be meaningfully described as a supermartingale on the spaces $\{\mathbb{R}^n, n \geq 1\}$, though *informally* it may be and is so described in the literature. The rigorous definition (and meaning) is that given by (48).

An interesting consequence of the above theorem is the following result, which we present as a final item of this chapter.

Theorem 20 *Let (Ω, Σ, P) be a probability space and $\nu : \Sigma \to \mathbb{R}$ be a σ-additive function (a signed measure). If $\mathcal{B}_n \subset \mathcal{B}_{n+1} \subset \Sigma$ are σ-algebras, $\nu_n = \nu|\mathcal{B}_n, P_n = P|\mathcal{B}_n$, and $X_n = d\nu_n^c/dP_n$, where ν_n^c is the P_n-continuous part of ν_n relative to P_n, then $X_n \to X_\infty$ a.e. $[P_\infty], X_\infty = d\nu_\infty^c/dP_\infty$, where $\mathcal{B}_\infty = \sigma(\bigcup_{n \geq 1} \mathcal{B}_n)$. Moreover, the adapted sequence $\{X_n, \mathcal{B}_n, n \geq 1\}$ is a* ***quasi − martingale****, in the sense that*

$$E\left(\sum_{n=1}^{\infty} |E^{\mathcal{B}_n}(X_{n+1}) - X_n|\right) < \infty. \tag{49}$$

Remark The term "quasi-martingale" was used for a more general process in the middle 1960s by D. L. Fisk and it was also termed an *F-process* and a (*)-*process.* It is an adapted process satisfying (49), called a *star condition.* The term has not yet become standard. We use it here since in the discrete case they all coincide. Clearly every martingale and every $L^1(P)$-bounded sub- or supermartingale (see below) is a quasi-martingale. This class presents a nice generalization of these important processes. The main convergence result above is originally due to E. S. Andersen and B. Jessen, who gave a direct proof as in Theorem 8.

Proof Let $\nu = \nu^+ - \nu^-$ be the Jordan decomposition of ν on Σ. Then on (Ω, Σ), the finite measures $\nu^\pm$ and P satisfy the hypothesis of Theorem 7. (It is irrelevant that ν^+, ν^- are not probabilities. That they are finite measures is sufficient.) If $\tilde{\nu}_n^\pm = \nu^\pm|\mathcal{B}_n$ and $Y_n = d(\tilde{\nu}_n^+)^c/dP_n, Z_n = d(\tilde{\nu}_n^-)^c/dP_n,$

then $X_n = Y_n - Z_n$ and $\{Y_n, \mathcal{B}_n, n \geq 1\}$ and $\{Z_n, \mathcal{B}_n, n \geq 1\}$ are positive supermartingales and hence converge a.e. Thus $X_n \to (Y_\infty - Z_\infty) = X_\infty$ a.e. and

$$X_\infty = \frac{d(\nu_\infty^+)^c}{dP_\infty} - \frac{d(\nu_\infty^-)^c}{dP_\infty} = \frac{d\nu_\infty^c}{dP_\infty} \quad \text{a.e.}$$

It is now asserted that the X_n-sequence satisfies (49). In fact,

$$\begin{aligned}
&E\left(\sum_{n=1}^{\infty} |E^{\mathcal{B}_n}(X_{n+1} - X_n)|\right) \\
&\leq E\left(\sum_{n=1}^{\infty} |E^{\mathcal{B}_n}(Y_{n+1} - Y_n)|\right) + E\left(\sum_{n=1}^{\infty} |E^{\mathcal{B}_n}(Z_{n+1} - Z_n)|\right) \\
&= \sum_{n=1}^{\infty} [E(Y_n - E^{\mathcal{B}_n}(Y_{n+1})) + E(Z_n - E^{\mathcal{B}_n}(Z_{n+1}))], \\
&\qquad (\text{since } E^{\mathcal{B}_n}(Z_{n+1}) \geq Z_n \text{ and similarly for } Y_n), \\
&= E(Y_1) - \lim_{n\to\infty} E(Y_n) + E(Z_1) - \lim_{n\to\infty} E(Z_n) \\
&\leq E(Y_1 + Z_1) - E(Y_\infty + Z_\infty) \quad (\text{by Fatou's lemma}) \\
&= E(|X_1|) - E(|X_\infty|) < \infty.
\end{aligned}$$

This establishes the result.

It is an interesting (and not too difficult a) fact that every $L^1(P)$-bounded adapted sequence satisfying (49) (i.e., every quasi-martingale) can be expressed as a difference of a pair of nonnegative supermartingales. Thus the preceding result admits a further extension of the martingale theory. We omit a specialized study of the subject here.

This chapter, more than anything else, shows vividly the individual characteristics of probability theory and its potential growth as well as its interactions with many other areas of analysis and applications. Several other interesting consequences of, and complements to, this theory are included in the following collection of exercises.

Exercises

1. Let X, Y be two square integrable random variables on (Ω, Σ, P). If, further, $E(X|Y) = Y$ a.e. and $E(Y|X) = X$ a.e., show that $X = Y$ a.e.. The same result is true even if X, Y are only integrable, but the proof is slightly more involved. (For the latter use Theorem 1.4.5iii and Theorem 1.9iii.)

2. Let $\{X_n, n \geq 1; Y\}$ be a set of r.v.s in $L^2(\Omega, \Sigma, P)$, and $\mathcal{B} = \sigma(X_n, n \geq 1)$. Consider the subspace $L^2(\Omega, \mathcal{B}, P)$ of $L^2(\Omega, \Sigma, P)$. Each X_n is in the subspace, but Y need not be. Show that there is a unique $Y_0 \in L^2(\Omega, \mathcal{B}, P)$ such that $E(|Y - Y_0|^2) = \inf\{E(|Y - X|^2) : X \in L^2(\Omega, \mathcal{B}, P)\}$, and that $Y_0 = E^{\mathcal{B}}(Y)$. This Y_0 is called the *best (nonlinear) predictor* of Y based on $\{X_n, n \geq 1\}$ and it is a function of the latter (by Proposition 1.4).

3. Let $\phi : \mathbb{R}^+ \to \mathbb{R}$ be a continuous symmetric convex function, and Y be an r.v. on (Ω, Σ, P) such that $E(\phi(2Y))$ exists. (Here 2 can be replaced by any $\alpha > 1$.) If $a = E(Y)$, and $X = \{X_n, n \geq 1\}$ is a random vector, let $Z = E(Y|X)[= E^{\mathcal{B}}(Y), \mathcal{B} = \sigma(X_n, n \geq 1)]$. Show that

$$E(\phi(Y - a)) \geq E(\phi(Z - a)),$$

with strict inequality if ϕ is strictly convex and Y is not $\mathcal{B}$-measurable. In particular, if $\phi(x) = x^2$, then we have Var $Y \geq$ Var Z, so that in estimating the mean a by Y or Z, the latter is preferable to the former, which does not use the information about the X_n and thus has a larger variance. This is of interest in statistical applications, and the inequality when $\phi(x) = x^2$ has been established in the middle 1940s by D. Blackwell and C. R. Rao independently.

4. Let X_1, X_2 be a pair of r.v.s on (Ω, Σ, P) and $\mu_i = P \circ X_i^{-1}, i = 1, 2$, and $\nu = P \circ (X_1, X_2)^{-1}$ be the image measures on $\mathbb{R}$ and $\mathbb{R}^2$. If $\mu = \mu_1 \otimes \mu_2$ is the product measure, ν is μ-continuous (is this automatic?), and $h(x_1, x_2) = d\nu/d\mu(x_1, x_2)$ a.e. $[\mu]$, show that for any bounded Borel function $g : \mathbb{R}^2 \to \mathbb{R}$, the mapping $f : x_2 \mapsto E(g(X_1, x_2)h(X_1, x_2))$ is μ_2-measurable and integrable and, moreover, $f(X_2) = E(g(X_1, X_2)| X_2)$ a.e. $[P]$.

5. Let $\{X_n, n \geq 1\}$ be a uniformly integrable sequence of r.v.s on (Ω, Σ, P) and $\mathcal{B} \subset \Sigma$ be a σ-algebra. If $E^{\mathcal{B}}(X_n) \to E^{\mathcal{B}}(X)$ a.e., then show that $\{X_n, n \geq 1\}$ is conditionally uniformly integrable relative to $\mathcal{B}$ in the sense defined prior to Proposition 1.6. [*Hint*: If $\nu_n^A(C) = \int_C E^{\mathcal{B}}(|X_n|\chi_A)\, dP$, then $\lim_{P(A)\to 0} \nu_n^A(C) = 0$ uniformly in n and C. Note that there is a "derivation basis" $\mathcal{F} \rightsquigarrow \{\omega\}$, (cf. Proposition 2.9) such that $(D_{\mathcal{F}}\nu_n^A)(\omega) \to E^{\mathcal{B}}(|X_n|\chi_A)(\omega)$ a.a.(ω). These two assertions imply that

$$\lim_{k\to\infty} E^{\mathcal{B}}(|X_n|\chi_{[|X_n|>k]}) = 0, \text{ uniformly in } n.]$$

6. Let (Ω, Σ, μ) be a $(\sigma-)$ finite space and $L^p(\mu)$ be the usual real Lebesgue space on it, $1 \le p < \infty$. An operator $T : L^p(\mu) \to L^p(\mu)$ is termed *positive* if $Tf \ge 0$ a.e. for each $f \ge 0$ a.e. Establish the following statements:

(a) If T is a positive linear operator on $L^p(\mu) \to L^p(\mu)$, then (i) $|T(f)| \le T(|f|)$ a.e., (ii) $f_n \ge 0, f_n \le g \in L^p(\mu) \Rightarrow \sup_n T(f_n) \le T(\sup_n f_n)$ a.e., and (iii) a positive linear operator $T : L^p(\mu) \to L^p(\mu)$ is always continuous (=bounded).

(b) Let $T : L^p(\mu) \to L^p(\mu)$ be as above, and $|f_n| \le g \in L^p(\mu)$. If $f_n \to f$ a.e., then $T(f_n) \to T(f)$ a.e., and in $L^p(\mu)$. (In other words, the assertions of Theorem 1.3 for conditional expectation operators extend to a more general class. Here T need not be faithful or a projection.)

7. (i) If $\{X_n, n \ge 1\} \subset L^p(P)$ on a probability space (Ω, Σ, P), then X_n is said to converge weakly to $X \in L^p(P)$ if for each $Y \in L^q(P)$, where $p^{-1} + q^{-1} = 1, 1 \le p < \infty, E(X_nY) \to E(XY)$ as $n \to \infty$. Show that for any σ-algebra $\mathcal{B} \subset \Sigma, E^{\mathcal{B}}(X_n) \to E^{\mathcal{B}}(X)$ weakly if $X_n \to X$ weakly. [*Hint*: Verify, by Proposition 1.2, that $E(YE^{\mathcal{B}}(X)) = E(XE^{\mathcal{B}}(Y))$.] This property of $E^{\mathcal{B}}$ is called "self-adjointness."

(ii) Let $\{X_n, n \ge 1\}$ be as in (i) and $T : L^p(P) \to L^p(P)$ be a continuous linear operator. It is known that there is a continuous linear $T^* : L^q(P) \to L^q(P)$ such that $E(YT(X)) = E(XT^*(Y)), X \in L^p(P), Y \in L^q(P)$, and if $1 < p < \infty$, then $(T^*)^* = T$. Show that if $X_n \to X$ weakly in $L^p(P), 1 < p < \infty$, then $TX_n \to TX$ in the same sense. (This reduces to (i) if $T = E^{\mathcal{B}}$.)

8. Let (Ω, Σ) be a measurable space, and $\lambda_i : \Sigma \to [0,1], i = 1, 2$, be two probability measures such that $\lambda_2 \ll \lambda_1$. If $\mathcal{B} \subset \Sigma$ is a σ-algebra, X is a nonnegative r.v. on Ω and $E_i^{\mathcal{B}} : L^1(\lambda_i) \to L^1(\lambda_i), i = 1, 2$, are the corresponding conditional expectation operators, show that $E_2^{\mathcal{B}}(X) = E_1^{\mathcal{B}}(X_g/E_1^{\mathcal{B}}(g))$a.e.$[\lambda_1]$, where $g = d\lambda_2/d\lambda_1$ a.e.$[\lambda_1]$. Interpret the result with $X = \chi_A$ if the conditional probabilities $P_i^{\mathcal{B}}(\cdot), i = 1, 2$, are regular. Deduce that $E_2(X|g) = E_1(X|g)$ a.e.$[\lambda_1]$, for all bounded r.v.s X, so that the (distinct) conditional expectations $E_i(\cdot|g), i = 1, 2$, agree on $L^1(\lambda_1)$.

9. Let (Ω, Σ) be a measurable space and $\mathcal{P} = \{P_\theta, \theta \in I\}$ be a family of probability measures on Σ. A σ-algebra $\mathcal{B} \subset \Sigma$ is said to be *sufficient* (also termed a "sufficient subfield") for the family $\mathcal{P}$ if the conditional expectation operator $E_\theta^{\mathcal{B}}$ on $L^1(P_\theta)$ is the same for all $\theta \in I$, so that $E_\theta^{\mathcal{B}}(X)$ is invariant as θ varies over I, for any bounded r.v. X on Ω. Clearly $\mathcal{B} = \Sigma$ is sufficient for $\mathcal{P}$, since then $E_\theta^{\mathcal{B}}$ =identity. Also, the last part of Problem 8 above implies that if $P_{\theta_1} \ll P_{\theta_2}$ and $g = dP_{\theta_1}/dP_{\theta_2}$, then $\mathcal{B} = \sigma(g)$ is sufficient for the pair $(P_{\theta_1}, P_{\theta_2})$. This extends as follows:

(a) Suppose that $\{P_\theta, \theta \in I\}$ is a dominated family, in the sense that there is a finite (or σ-finite) measure $\lambda : \Sigma \to \mathbb{R}^+$ such that $P_\theta \ll \lambda$ for all $\theta \in I$. [Note that if λ is σ-finite, then there exist $A_n \in \Sigma$, disjoint, $\Omega = \bigcup_{n\geq 1} A_n, 0 < \lambda(A_n) < \infty$, by definition, so that

$$\mu(\cdot) = \sum_{n\geq 1} \frac{1}{2^n} \frac{\lambda(A_n \cap \cdot)}{1 + \lambda(A_n)}$$

is a finite measure and $\lambda \ll \mu \ll \lambda$; thus in the domination questions we may assume that $\lambda(\Omega) < \infty$, and we do so here.] Let $g_\theta(\omega) = (dP_\theta/d\lambda)(\omega)$ a.e.$[\lambda]$. Then a σ-algebra $\mathcal{B} \subset \Sigma$ is sufficient for $\{P_\theta, \theta \in I\}$ iff each $g_\theta(\cdot)$ is $\mathcal{B}$-measurable, $\theta \in I$. [*Hint*: If $E_\theta^{\mathcal{B}}(X) = E^{\mathcal{B}}(X)$, for all $\theta \in I$, then the domination hypothesis implies that $E^{\mathcal{B}}(X) = E_\lambda^{\mathcal{B}}(X)$ a.e.$[\lambda]$, and hence, if $\tilde{g}_\theta(\omega) = (d\tilde{P}_\theta/d\lambda_{\mathcal{B}})(\omega)$, where $\tilde{P}_\theta = P_\theta|\,\mathcal{B}$, then taking $X = \chi_A, A \in \Sigma$, we have

$$\begin{aligned} P_\theta(A \cap B) &= \int_B E_\theta^{\mathcal{B}}(\chi_A)\, d\tilde{P}_\theta = \int_B E_\lambda^{\mathcal{B}}(\chi_A)(\omega)\tilde{g}_\theta(\omega)\, d\lambda_{\mathcal{B}}(\omega) \\ &= \int_B E^{\mathcal{B}}(\chi_A \tilde{g}_\theta(\omega))\, d\lambda_{\mathcal{B}}(\omega) = \int_{B\cap A} \tilde{g}_\theta(\omega)\, d\lambda(\omega), \quad B \in \mathcal{B}. \end{aligned}$$

Hence $dP_\theta/d\lambda = \tilde{g}_\theta$ a.e. It is $\mathcal{B}$-measurable, and $\tilde{g}_\theta = g_\theta$ a.e.$[\lambda]$. The converse, that g_θ is $\mathcal{B}$-measurable $\Rightarrow E_\theta^{\mathcal{B}}(X) = E^{\mathcal{B}}(X)$, is similar. If $\mathcal{B}$ is the smallest σ-algebra relative to which all the $\{g_\theta, \theta \in I\}$ are measurable, it is called the *minimal sufficient σ-algebra*].

(b) If $X_i : \Omega \to \mathbb{R}, i = 1, \ldots, n$, are r.v.s and $Q_n(A, \theta) = P_\theta(X^{-1}(A))$ is the (image) probability distribution on $\mathbb{R}^n, X = (X_1, \ldots, X_n)$, suppose that $\mu_n(= \lambda \circ X^{-1})$ dominates $Q_n(\cdot, \theta), \theta \in I$. Let $T : \mathbb{R}^n \to \mathbb{R}^m$ be a Borel function. Then $T(X) : \Omega \to \mathbb{R}^m$ is called a *statistic*. It is termed a *sufficient statistic* for $Q_n(\cdot, \theta)$ if $\mathcal{B} = \sigma(T(X))$ is sufficient for $\{P_\theta, \theta \in I\}$, in the sense of (a). Show, from (a) that $T(X)$ is sufficient for $\{Q_n(\cdot, \theta), \theta \in I\}$ iff

$$(dQ_n(\cdot, \theta)/d\mu_n(\cdot))(x) = c(x) \cdot p_n(T(x), \theta), \quad \text{a.e.}[\mu_n], \quad \theta \in I,$$

where $c(\cdot)$ is a Borel function independent of θ and $p_n(\cdot, \theta)$ depends on x only through the function T. [This is called a *factorization criterion*. The concept of sufficiency originated with R. A. Fisher in the early 1920s and developed later in increasing generality by J. Neyman, P. R. Halmos and L. J. Savage, and R. R. Bahadur.]

(c) If $\mathcal{P} = \{P_\theta, \theta \in I\}$ is not dominated, results similar to the above are not always valid. However, we can assert the following:

(i) If $\{\mathcal{B}_n, n \geq 1\} \subset \Sigma$ is a monotone sequence of σ-algebras each $\mathcal{B}_n$ being sufficient for $\mathcal{P}$, then $\mathcal{B} = \lim_n \mathcal{B}_n$ is also sufficient for $\mathcal{P}$. (We need to use Theorem 5.9 or 5.14 appropriately.)

(ii) If $\{\mathcal{B}_n, n \geq 1\}$ is an arbitrary sequence of sufficient σ-algebras from Σ for $\mathcal{P}$, and each $\mathcal{B}_n$ contains all the P_θ-null sets for each $\theta \in I$, then

$\bigcap_{n\geq 1} \mathcal{B}_n = \mathcal{B}$ is also sufficient for $\mathcal{P}$.

In the other direction, if $\mathcal{B}_1$ is generated by a countable collection of sets, and $\mathcal{B}_2$ is sufficient, then $\sigma(\mathcal{B}_1 \cup \mathcal{B}_2)$ is sufficient for $\mathcal{P}$. [This result uses not only Theorem 5.9, and (i) above for the sequence $\{\bigcap_{k=1}^n \mathcal{B}_k, n \geq 1\}$, but another result on the a.e. convergence of $\{S_n f, n \geq 1\}, f \in L^2(P_\theta)$, where $S_n = E^{\mathcal{B}_n} E^{\mathcal{B}_{n-1}} \cdots E^{\mathcal{B}_1}$. The result of this part is in D. L. Burkholder *Ann. Math. Statist.*, *32*(1961), 1191-1200, and should be consulted on details. A thorough discussion of sufficiency is given in Chapter 6 of *Conditional Measures and Applications*, (1993, 2nd. ed. 2005) by the first author.]

10. This problem contains two illustrations of sufficiency discussed in the preceding exercise:

(a) Let $\{(\Omega, \Sigma, P_\theta), \theta \in I\}$ be a family of probability spaces, $(X_1, \ldots, X_n)$ be a vector of symmetrically dependent r.v.s such that the distribution (or image probability) function on $\mathbb{R}^n$ has a density $f_n(\cdot; \theta)$ relative to the Lebesgue measure. [The hypothesis implies $f_n(x_{i_1}, \ldots, x_{i_n}; \theta) = f_n(x_1, \ldots, x_n; \theta)$ for each permutation $(i_1, \ldots, i_n)$ of $(1, \ldots, n)$.] Let $X_{(1)} < \cdots < X_{(n)}$ be the order statistics of the given vector, $T(X) = T(X_1, \ldots, X_n) = (X_{(1)}, \ldots, X_{(n)})$. Show that $\sigma(T(X))$ is generated by $\{T(X)^{-1}(A) : A \subset \mathbb{R}^n, \chi_A$ is symmetric Borel$\}$, and that $T(X)$ is sufficient for X, by verifying that, for all $\theta \in I$,

$$E^{\sigma(T(X))} : g(X) \mapsto \frac{1}{n!} \sum g(X_{i_1}, \ldots, X_{i_n}),$$

where the summation ranges over all $n!$ permutations of $(1, \ldots, n)$ and $g : \mathbb{R}^n \to \mathbb{R}$ is any bounded Borel function.

(b) Let $\{(\Omega, \Sigma, P_\theta), \theta \in I\}$ be as above, and $X : \Omega \to \mathbb{R}$ be an r.v. such that

$$P_\theta[X < x] = \int_{-\infty}^{x} A(\theta) \exp\left\{\sum_{j=1}^{k} B_j(\theta) T_j(t)\right\} C(t) \mu(dt), \quad \theta \in I,$$

for suitable $A(\theta) \geq 0, C(t) > 0$, $B_j(\theta) \in \mathbb{R}$, and $\mu(\cdot)$ is σ-finite on $\mathbb{R}$.

Such a collection is called an *exponential family* of distributions. Show by use of Problem 9b that for any $X_1, \ldots, X_n$ i.i.d. having the above distribution, $T(X) = (\sum_{i=1}^n T_j(X_i), j = 1, \ldots, k)$ constitutes a sufficient statistic for θ, based on $X = (X_1, \ldots, X_n), n \geq 1$.

11. If X_1, X_2 on (Ω, Σ, P) are jointly (or bivariate) normally distributed, so that

$$P[X_1 < x_1, X_2 < x_2] = \int_{-\infty}^{x_1} \int_{-\infty}^{x_2} [4\pi^2 \sigma_1^2 \sigma_2^2 (1 - \rho^2)]^{-1/2}$$

$$\times \exp\left\{-\frac{1}{2(1-\rho^2)}\left(\frac{t_1^2}{\sigma_1^2}-2\rho\frac{t_1t_2}{\sigma_1\sigma_2}+\frac{t_2^2}{\sigma_2^2}\right)\right\} dt_1dt_2,$$

with $0 < \sigma_i^2 < \infty, |\rho| < 1$, let $Z = X_1+X_2$. Show that $E^{\sigma(Z)}(X_1) = aZ$ a.e.$[P]$ is a representation of the conditional operator, where a is some real number. Verify that $E(Z)^2 \leq \sigma_1^2/a_1^2$ with $a_1 = (\sigma_1^2+\rho\sigma_1\sigma_2)(\sigma_1^2+2\rho\sigma_1\sigma_2+\sigma_2^2)^{-1}$. [First compute the d.f. of (X_1, Z).] Is $a = a_1$?

12. In Problem 9c (converse), we assumed a σ-algebra $\mathcal{B}_1$ to be countably generated. This is still more general than that generated by a partition. Thus, if $\mathcal{B}$ is countably generated, show that there exist at most continuum many atoms A_t in $\mathcal{B}$ (i.e., $A \subset A_t, A \in \mathcal{B} \Rightarrow A = \emptyset$ or $A = A_t$) such that $\bigcup_t A_t = \Omega$, and each $B \in \mathcal{B}$ is a union of these atoms.

13. We now state a form of the classical example showing the nonexistence of a regular conditional probability. Let (Ω, Σ, P) be the Lebesgue unit interval, and $A \subset \Omega$ be a nonmeasurable set of Lebesgue outer measure one and inner measure zero. Let $\tilde{\Sigma}$ be the σ-algebra generated by Σ and A. Verify that $\tilde{A} \in \tilde{\Sigma}$ iff $\tilde{A} = (A \cap B) \cup (A^c \cap D), B, D \in \Sigma$. Let $\tilde{P}(\tilde{A}) = \alpha P(B) + (1-\alpha)P(D), 0 < \alpha < 1$. Show that $\tilde{P}$ is well defined on $\tilde{\Sigma}$, is a measure, $\tilde{P}|\Sigma = P$, and that $\tilde{P}(\cdot|\Sigma)$ on $\tilde{\Sigma}$, defined by the functional equation (2.1) is not regular, i.e., that $\tilde{P}(\cdot|\Sigma)(\omega) : \tilde{\Sigma} \to [0,1]$ is a measure for a.a. $\omega \in \Omega$ is not true [even though $\tilde{P}(A|\Sigma)(\cdot)$ is measurable for Σ]. However, $\tilde{P}(\cdot|\Sigma)$ is a vector-valued σ-additive set function satisfying Proposition 1.1ii and 1.1iii, but the Vitali convergence theorem is false for it. [In contrast to Proposition 2.2, this result depends on the axiom of choice through the existence of a Lebesgue nonmeasurable set A.]

14. In the opposite direction of the above result, here is a positive assertion. If $(S, \mathcal{I})$ is a measurable space, it is called a *Borel space* if there exists a Borel subset $A \subset \mathbb{R}$ and a one-to-one bimeasurable, onto, mapping $f : S \to A$ [so f is $(\mathcal{I}, \mathcal{R}(A))$-measurable and the inverse $f^{-1} : A \to S$ is $(\mathcal{R}(A), \mathcal{I})$-measurable, where $\mathcal{R}(A)$ is the trace σ-algebra of the Borel σ-algebra $\mathcal{R}$ of $\mathbb{R}$.] It is known that every separable complete metric space with its Borel σ-algebra is a Borel space. [Cf. Parthasarathy (1967), Section V.2, on Borel spaces.] If $\mathcal{B} \subset \Sigma$ is any σ-algebra of a probability space $(\Omega, \Sigma, P), (S, \mathcal{I})$ is a Borel space, and $X : \Omega \to S$ is an r.v. [i.e., $X^{-1}(\mathcal{I}) \subset \Sigma$], then a conditional distribution $P^{\mathcal{B}} \circ X^{-1} = \tilde{P}(\cdot,\cdot) : \mathcal{I} \times \Omega \to [0, 1]$ exists, and if $X(\Omega) \in \mathcal{I}$, then $P^{\mathcal{B}}$ itself is regular on $\Sigma_1 = X^{-1}(\mathcal{I})$. [*Hint*: Consider $Y = f \circ X : \Omega \to A$, and by Theorem 2.5, $P^{\mathcal{B}} \circ Y^{-1}$ gives a version which is a regular conditional distribution, $Q(\cdot,\cdot)$, and let $\tilde{P}(\cdot,\cdot) = Q(f(\cdot),\cdot)$. Since $f = (f^{-1})^{-1}$, it preserves countable operations. The second part is as in Proposition 2.6.]

15. We present here a useful complement to the preceding problem. Let (Ω, Σ, P) be a probability space, $\mathcal{B} \subset \Sigma$ a countably generated σ-algebra.

Show that there is a regular conditional probability $Q(\cdot,\cdot):\Sigma\times\Omega\to[0,1]$ such that (i) $Q(\cdot,\omega)$ is σ-additive for all $\omega\in\Omega-N, N\in\mathcal{B}, P(N)=0$, (ii)$Q(A,\cdot)$ is $\mathcal{B}$-measurable, and (iii) for each $\omega\in\Omega-N$, if $A(\omega)=\bigcap\{A:\omega\in A\in\mathcal{B}\}$, then $A(\omega)\in\mathcal{B}$ and $Q(A(\omega),\omega)=1$. [If $\mathcal{P}$ is the countable collection with $\sigma(P)=\mathcal{B}$, then $Q(A,\cdot)=\chi_A$ a.e., $A\in\mathcal{P}$, and then the same holds if A is replaced by the atom $A(\omega)$ of $\mathcal{B}$ which is $\mathcal{B}$-measurable. The result is somewhat similar to the partition case.]

16. This problem is a kind of converse to the preceding one. Let (Ω,Σ,P) be a probability space and $\mathcal{B}\subset\Sigma$ be a σ-algebra containing a countably generated σ-algebra $\mathcal{B}_0\subset\mathcal{B}$. Let $P^{\mathcal{B}}$ be the conditional probability function. If $P^{\mathcal{B}}$ is regular and if on each $\mathcal{B}_0$-atom $P^{\mathcal{B}}$ is indecomposable (so it is a $P^{\mathcal{B}}$-atom), then verify that $P^{\mathcal{B}}$ can be decomposed as follows: there is an $N\in\mathcal{B}, P(N)=0$, and $\Omega-N=\bigcup_{t\in I}A_t, A_t$ is a $\mathcal{B}_0$-atom, the cardinality of I being at most of the continuum, and

$$P^{\mathcal{B}}(A)=\sum_{t\in I}\chi_{A_t}P(A\cap A_t)/P(A_t)\quad\text{a.e.},\quad A\in\Sigma.$$

[This nontrivial result is in D. Blackwell, *Ann. Math. 43,* (1942), 560-567.]

17. (a) Let $\{A_n,n\geq 1\}$ be a partition of Ω in $(\Omega,\Sigma,P), P(A_n)>0$. If $B\in\Sigma, P(B)>0$, then show that the following, called *Bayes formula,* is valid:

$$P(A_n|\,B)=\left([P(A_n)P(B|\,A_n)]\Big/\sum_{n\geq 1}P(A_n)P(B|\,A_n)\right).$$

(b) If $(X,Y):\Omega\to\mathbb{R}^2$ is a random vector on (Ω,Σ,P) whose distribution has density (relative to Lebesgue measure) $f_{X,Y}$ and if f_X,f_Y are the (marginal) densities of X and Y, suppose $f_X(x)>0, f_Y(y)>0$ a.e. (Lebesgue). Then verify the continuous "analog" of the above formula if $f_{X|Y}, f_{Y|X}$ are supposed to satisfy:

$$f_{X|Y}(x|\,y)=f_{Y|X}(y|\,x)f_X(x)/f_Y(y)\quad\text{a.e. (Lebesgue).}$$

These two formulas find some applications in statistical inference.

18. Consider the family $\{X(t),0\leq t\leq 1\}$ of 2.10 (the counterexample). To find the conditional density of the a.e. derivative Y of X at 0 given $X(0)=a$, let $B_\delta=[(X(t)-a)^2+t^2\leq\delta^2$, for some $0<t\leq\delta]$. Verify that B_δ is measurable and the conditional density of Y (the derivative of X at $t=0$) given $X(0)=a$, by means of approximations using the B_δ, is

$$\lim_{\delta\to 0}P([Y<y]|\,B_\delta)=\int_{-\infty}^{y}\frac{\sqrt{1+u^2}e^{-u^2/2\alpha^2}}{\int_{-\infty}^{\infty}\sqrt{1+v^2}e^{-v^2/2\alpha^2}dv}du,\quad -\infty<y<\infty,$$

where $\alpha^2 = E(Y^2) > 0$. Thus we have yet another form of the density depending on the derivation basis which is different from all the others!

19. Here is an application which will be analyzed further. In certain problems of population genetics, gene combinations under random influences can be described as follows. From generation to generation genes combine between sexes I and II, which may be identified as a game between two players with a fixed fortune. In this terminology the game can be stated as: A random portion X_n of the genes (or fortune) is retained by I at the nth generation (or stage) and Y_n is the random proportion given by II to I at the nth generation. Thus if Z_n is the fortune of I at stage n, and $\tilde{Z}_n$ that of II, then we assume that $Z_0 + \tilde{Z}_0 = 1$ (by normalization), so that we get $Z_n = X_n Z_{n-1} + Y_n(1 - Z_{n-1})$ and $\tilde{Z}_n = 1 - Z_n$. Suppose that $(X_n, Y_n) : \Omega \to \mathbb{R}^2$ are independent bivariate random vectors and $P[0 \leq X_n \leq 1, 0 \leq Y_n \leq 1] = 1$. Also, let (X_n, Y_n) be independent of Z_{n-1}. Then show that $\{Z_n, n \geq 1\}$ is a Markov process. (Compare with Example 3.3.)

20. (Continuation) Suppose that the $(X_n, Y_n), n = 1, 2, \ldots$, are i.i.d. as (X, Y) and $P[|X - Y| = 1] < 1$. If $\mu_k(n) = E(Z_n^k)$, show that $\lim_{n\to\infty} \mu_k(n) = \alpha_k$ exists for each $k = 1, 2, \ldots$ (use induction), and that

$$\alpha_{k+1}(1 - E[(X - Y)^{k+1}]) = \sum_{i=0}^{k} \binom{k+1}{i} E[(X - Y)^i Y^{k+1-i}] \alpha_k.$$

Thus the limiting moments do not depend on the initial distribution of the r.v. Z_0. What happens if $P[|X - Y| = 1] = 1$? [It results from the Helly-Bray theorem of next chapter that $Z_n \xrightarrow{D} Z$ and $\alpha_k = E(Z^k)$. The last equation follows from Problem 10 of Chapter 2 also, if the convergence result is given, since $\{Z_n, n \geq 1\}$, being a bounded set, is uniformly integrable. This result (and the next one) appears in DeGroot and the first author (1963).]

21. (Continuation) Let $Z_n \xrightarrow{D} Z$ as in the preceding problem, keeping the same assumptions and notation. For some applications, indicated in Problem 19 above, it is of interest to find the limit Z-distribution. If F_n is the distribution of Z_n and F that of Z, using the relation $Z_n = X_n Z_{n-1} + Y_n(1 - Z_{n-1})$ and knowing that $F_n \to F$, show that for each n (the intervening conditional probabilities are regular and)

$$F_n(x) = \int_0^1 P([X_n Z_{n-1} + Y_n(1 - Z_{n-1}) < x] | Z_{n-1} = t) F_{n-1}(dt),$$

and, since (X_n, Y_n) has the distribution of (X, Y) deduce that

$$F(x) = \int_0^1 P[Xt + Y(1-t) < x] F(dt) \tag{+}$$

is the integral equation determining F, which depends only on the distribution of (X, Y) and not on the initial distribution of Z_0. If (X, Y) has a density $g(\cdot, \cdot)$, so that $F' = f_Z$ also exists a.e., show that f_Z is the unique solution of the integral equation

$$f_Z(x) = \int_0^1 h(x;t) f_Z(t)\, dt, \tag{$*$}$$

where

$$h(x;t) = \int_{a(x,t)}^{b(x,t)} (1-t)^{-1} g\left(u, \frac{x-tu}{1-t}\right) du,$$

with $a(x,t) = \max[0, (x+t-1)/t]$ and $b(x,t) = \min[(x/t), 1]$. It is of interest to solve the integral equation (*) for general classes of $g(\cdot,\cdot)$, but this does not seem easy. The reader is urged to work on this interesting open problem.

As an example, verify that if X, Y are independent and have densities g_X and g_Y, given (for $p > 0, q > 0$) by the *beta density*

$$g_X(x) = \frac{\Gamma(p+q)}{\Gamma(p)\Gamma(q)} x^{p-1}(1-x)^{q-1},$$
$$g_Y(y) = \frac{\Gamma(q+p)}{\Gamma(q)\Gamma(p)} y^{q-1}(1-y)^{p-1},$$

$0 < x < 1, 0 < y < 1$, then f_Z of $(*)$ is also a beta density, given as

$$f_Z(u) = \frac{\Gamma(2(p+q))}{\Gamma(p+q)^2} u^{p+q-1}(1-u)^{p+q-1}, \quad 0 < u < 1.$$

22. (a) Let $X_1, \ldots, X_n$ be i.i.d. with a d.f. given by $F(x) = 0$ if $x \le 0, = x$ if $0 < x < 1$, and $= 1$ if $x \ge 1$, called the *uniform* d.f. If $X_1^*, \ldots, X_n^*$ are the order statistics, show that a conditional density of $X_1^*, \ldots, X_k^*$ given $X_i^* = c_i, 0 < c_i < 1, i = k+1, \ldots, n$, is given by

$$\begin{aligned} &f(x_1,\ldots, x_k | c_{k+1}, \ldots, c_n) \\ &\quad = (k-1)!/c_k^{k-1} \quad \text{if } 0 < x_1 < \cdots < x_k < c_k, \end{aligned}$$

and $= 0$ otherwise, and hence that $\{X_k^*, 1 \le k \le n\}$ forms a Markov process. Deduce that, if $X_1, \ldots, X_n$ are i.i.d. with a continuous strictly increasing d.f., then their order statistics also form a Markov process.

(b)We have seen in the proof of Theorem 3.9 that $Z_k = \sum_{i=1}^k U_i$, where the U_i are independent exponentially distributed random variables. Changing the notation slightly, suppose that $S_n = \sum_{i=1}^n X_i$ where the X_i are independent, each exponentially distributed, i.e., $P(X_i < x) = 1 - e^{-x}, x = 0$, so that S_n is increasing (in the proof Z_n has a reverse labelling!) Show that, using the change of variables technique of elementary probability, the variables $Y_k = \frac{S_k}{S_{n+1}}, k = 1, \ldots, n$ have a *multivariate beta* (also called a *Dirichlet*) *distribution*. Thus, Y_k and X_k^* of part (a) above have the same (joint) distribution, the latter being obtained using a random sample from the uniform distribution on the unit interval.

23. Complete the details of proof of Theorem 4.8 on the existence of continuous parameter Markov processes, and also the details of Theorem 4.7 to use it there.

24. By Definition 3.1, an ordered set of random variables (or vectors) $\{X_t, t \in I\}$ is Markovian if the σ-algebras $\{\sigma(X_t), t \in I\}$ form a Markovian family. Let $\{X_n, n \geq 1\}$ be a family of r.v.s, and $Y_n = (X_{n+1}, \ldots, X_{n+k}) : \Omega \to \mathbb{R}^k$ be the sequence formed of the X_n. Show that $\{Y_n, n \geq 0\}$ is Markovian iff the X_n satisfy the condition

$$P^{\mathcal{B}_{n+k}}(A) = P^{\sigma(Y_n)}(A) \quad \text{a.e.,} \quad A \in \sigma(X_{n+k+1})$$

where $\mathcal{B}_n = \sigma(X_i, 1 \leq i \leq n)$. Thus if $k = 1$, we have the ordinary Markovian condition, and if $k > 1$, the X_n are said to form a kth order (or *multiple*) *Markov process*. (Follow the proof of Proposition 3.2.) The preceding relation shows that many properties of multiple Markov processes are obtainable from ordinary (vector) Markovian results. Show that any kth-order Markovian family is also $(k + l)$th-order Markovian for all integers $l > 0$ but *not* if $l < 0$. (This is an important distinction to note, since in the theory of higher-order "stochastic differential equations" of the standard type, the solutions turn out to be multiple Markov processes, but *not* of lower order ones.)

25. (a) As an application of Theorem 4.5 (or 4.7), show that there exist multiple Markov processes (on suitable probability spaces) of any given order k.

(b) Using Theorem 4.11, extend Problem 5(c) of Chapter 2 as follows. Let (Ω, Σ, P) be a complete probability space, and $\{B_x^t, x \in \mathbb{R}, t \in I\} \subset \Sigma$, where I is any index set and $B_x^t \subset B_{x'}^t$, a.e. for any $x < x'$. For each $t \in I$, there is a (unique) r.v. $X_t : \Omega \to \mathbb{R}$ such that $[X_t \leq x] \subset B_x^t$ a.e. and $[X_t > x] \subset (B_x^t)^c$ a.e. Show that for each $t_1, \ldots, t_n$, if $P[B_{x_1}^{t_1} \cap B_{x_2}^{t_2} \cap \ldots \cap B_{x_n}^{t_n}] = F_{t_1,\ldots,t_n}(x_1, \ldots, x_n), (+)$, then $\{F_{t_1,\ldots,t_n}, t_i \in I, n \geq 1\}$ is a compatible family of d.f.'s and $\{X_t, t \in I\}$ is a process with the F's as its d.f.'s. [Thus the process is determined solely by (Ω, Σ, P).] Conversely, any compatible family of d.f.'s as in Theorem 4.11, determines a probability space and a family $\{B_x^t, x \in \mathbb{R}, t \in I\}$ satisfying $(+)$. Thus we have an adjunct of Theorem 4.11.

26. Extend Proposition 3.2 to kth-order Markovian sequences.

27. As an example of multiple Markov sequences, let $\{X_n, n \geq 1\}$ be a Markov process on (Ω, Σ) and define $S_n = \sum_{k=1}^n X_k$. Show that $\{S_n, n \geq 1\}$ is a second-order Markov sequence. [*Hints*: it suffices to verify that $\{f(S_n), n \geq 1\}$ has the desired property for each bounded Borel f on $\mathbb{R}$. In particular, take

$f_t(x) = e^{itx}, t \in \mathbb{R}$, and verify that $E^{\mathcal{F}_n}(f_t(S_{n+1})) = E^{\sigma(S_n,S_{n-1})}(f_t(S_{n+1}))$ a.e., where $\mathcal{F}_n = \sigma(S_1, \ldots, S_n)$. Argue that this suffices.]

28. Let $\{X_n, n \geq 1\}$ be a sequence of r.v.s on (Ω, Σ, P) with a finite state space (=range), say, $T = (1, 2, \ldots, s)$. If the sequence is a kth-order Markov process (=*a chain* in this case), it can be regarded as an ordinary Markov chain with values in $T^k \subset \mathbb{R}^k$, by Problem 22. Let $Y_n = (X_{n+1}, \ldots, X_{n+k}), n \geq 1$, be this chain. For each $j, l \in T^k$, let $p_{jl}^{(n)} = P([Y_n = l] | Y_{n-1} = j)$ be the transition probability. Suppose that the one-step transitions are independent of n (i.e., the chain has constant or *stationary* transition probabilities.) Find the Chapman-Kolmogorov equations in this case. (Note that, of the possible s^{2k} values of the transitions, all but s^{k+1}-values vanish in this representation.)

29. Let $\{X_n, n \geq 1\}$ be a Markov chain on (Ω, Σ, P) with a countable set of states, denoted conveniently by the natural numbers $\{1, 2, \ldots\}$ and suppose that the chain has one-step stationary transitions, i.e.

$$p_{ij} = P([X_n = j] | X_{n-1} = i)$$

is independent of $n \geq 1$. The structure of the chain is then defined by the matrix $(p_{ij}, i, i \geq 1)$, where $p_{ij} \geq 0, \sum_{j\geq 1} p_{ij} = 1$. The chain is *irreducible* if any pair of states $i_1 \neq i_2$ can be reached from one to the other in a finite number of steps, in that if $(p_{ij}^{(n)}) = (p_{ij})^n$, then $p_{i_1 i_2}^{(n)} > 0$ for some $n > 0$. A sequence $\{m_j, j \geq 1\}$ of positive numbers is a *positive subinvariant (invariant) measure* of the chain if $\sum_{i\geq 1} m_i p_{ij} \leq m_j (= m_j), j \geq 1$. It is known that each irreducible Markov chain admits a positive subinvariant measure. A Markov chain is *path reversible* if for any $r \geq 1$, and states $i_1, \ldots, i_r$, we have the following condition of Kolmogorov satisfied:

$$p_{i_1 i_2} p_{i_2 i_3} \cdots p_{i_r i_1} = p_{i_1 i_r} p_{i_r i_{r-1}} \cdots p_{i_2 i_1}.$$

Show that an irreducible Markov chain is path reversible iff the chain admits a positive invariant measure $\{m_j, j \geq 1\}$ such that for all j, k we have $m_j p_{jk} = m_k p_{kj}$. In this case the invariant measure is also unique except for a constant multiplicative factor. [Further properties of such chains can be found in Feller (1957) and Kendall (1959). Note that in our definition of a Markov process (cf. Definition 3.1), it is symmetric with regard to the ordering of the index set, and here the symmetry is about the motion of the chain relative to its states.]

30. We say that any adapted sequence $\{X_n, \mathcal{B}_n, n \geq 1\}$ of integrable random variables on (Ω, Σ, P) is a (sub-) martingale if for all $A \in \mathcal{B}_n$,

$$E(\chi_A X_{n+1}) = (\leq) E(\chi_A X_n),$$

without reference to conditional expectations and thus to the Radon-Nikodým theorem. (We now show their existence.) The convergence theory can be developed with this definition. If the (sub-) martingale convergence is given, then the (Lebesgue-)Radon-Nikodým theorem can be derived as follows: (a) Let $\nu : \Sigma \to \mathbb{R}^+$ be a measure, and Σ be countably generated. Set $\mu = \nu + P : \Sigma \to \mathbb{R}^+$, so that ν is μ-continuous. If $\{B_n, n \geq 1\}$ generates Σ, let $\mathcal{F}_n = \sigma(B_k, 1 \leq k \leq n)$. If $\{B_{n,i}, 1 \leq i \leq k_n\}$ is an $\mathcal{F}_n$-partition of Ω generating $\mathcal{F}_n$, define

$$X_n = \sum_{i=1}^{k_n} \frac{\nu(B_{n,i})}{\mu(B_{n,i})} \chi_{B_{n,i}}, \qquad n \geq 1.$$

Since $\nu \ll \mu$, this is well defined. Show that $\{X_n, \mathcal{F}_n, n \geq 1\}$ is a positive martingale, so that $X_n \to X_\infty$ a.e. Verify that the sequence is also uniformly integrable. (Note $\int_{[X_n > a]} X_n d\mu = \nu[X_n > a]$.) Hence deduce that $\nu(A) = \int_A X_\infty d\mu, A \in \Sigma$, and that $d\nu/d\mu = X_\infty$, a.e.$[\mu]$. If $N = [X_\infty = 1]$, then $P(N) = 0$. If $g = X_\infty/(1 - X_\infty)$, and $\nu^c = \nu(N^c \cap \cdot)$, then $\nu^c \ll P$, and $g = d\nu^c/dP$ a.e. is the Radon-Nikodým derivative. If $\nu^s = \nu(N \cap \cdot)$, then $\nu(A) = \int_A g dP + \nu^s(A), A \in \Sigma$, gives the Lebesgue-Radon-Nikodým theorem, where $\nu^s : \Sigma \to \mathbb{R}^+$ becomes the P-singular part of ν and ν^c its P-continuous part.

(b) If now (a) is given, and $g_n = d\nu_n^c/dP_n$, where $\nu_n = \nu|\mathcal{F}_n, P_n = P|\mathcal{F}_n$, show that $\{g_n, \mathcal{F}_n, n \geq 1\}$ is a supermartingale on (Ω, Σ, P) such that $g_n \to g$ a.e. [*Remark*: Part (a), and hence (b), can be extended to the general case that Σ is not countably generated as follows. Consider $\mathcal{A} \subset \Sigma$, a countably generated σ-algebra, and let $X_\mathcal{A}$ be the corresponding $d\nu_\mathcal{A}/d\mu_\mathcal{A}$. The collection of all such σ-algebras can be directed (into I) under inclusion. Then $\{X_\mathcal{A}, \mathcal{A}, \mathcal{A} \in I\}$ forms a uniformly integrable positive martingale. Such a martingale can be shown to converge in $L^1(\mu)$ to X (but not a.e.). This is sufficient to conclude that $X = d\nu/d\mu$ and then $\tilde{g} = X/(1-X)$ on $\tilde{N}^c$, where $\tilde{N} = [X = 1]$, gives $\tilde{g} = d\nu^c/dP$. Thus the general result is obtainable from the martingale theory. See, e.g., theorem 5.3.3. in Rao (1987, 2004), for more details.]

31. Show by an example that Theorem 5.8 is not valid if the moment condition is omitted; i.e., a non-$L^1(P)$-bounded martingale need not converge a.e. (Let $\{X_n, n \geq 1\}$ be independent, $P[X_n = +n] = P[X_n = -n] = \frac{1}{2}$ and consider $\{S_n = \sum_{k=1}^n a_k X_k, n \geq 1\}$, and follow the discussion in Section 4C of Chapter 2, with suitable a_k, the interarrival times.)

32. Let $\{V_k, \mathcal{F}_{k-1}, k \geq 1\}$ be an adapted uniformly bounded sequence of r.v.s and $\{X_n, \mathcal{F}_n, n \geq 1\}$ be an $L^2(P)$-bounded martingale on a probability space (Ω, Σ, P). Show that the sequence $\{S_n = \sum_{k=1}^n V_k \phi_k, n \geq 1\}$ converges a.e. and in $L^2(P)$, where $\phi_k = X_k - X_{k-1}$ is the martingale increment or difference, and $X_0 = 0$ a.e. (Such a V_k-sequence is termed *predictable* and S_n a *predictable transform* of the X_n-process. Note that the ϕ_k are orthogonal.)

33. Let $\{X_n, n \geq 1\}$ be i.i.d. and $\mathcal{B}_n = \sigma(X_k, 1 \leq k \leq n), \mathcal{B} = \sigma(\bigcup_{n\geq 1} \mathcal{B}_n)$, on (Ω, Σ, P). If X is a $\mathcal{B}$-measurable r.v. with $E(X) = 0, E(X^2) < \infty$, show that $E^{\mathcal{B}_n}(X) \to X$ a.e. [and in $L^1(P)$] as $n \to \infty$, and that $E^{\sigma(X_n)}(X) = f(X_n), n \geq 1$, for a fixed Borel function f on $\mathbb{R}, f(X_n) \in L^2(P)$. Verify that $\{f(X_n), n \geq 1\}$ forms an independent sequence with $E(f(X_n)) = 0$, and $\sigma^2 = E(f(X_n)^2) \leq E(X^2) < \infty$, all n. (Use the image law of probabilities.) If $a_n = E(Xf(X_n))$, then $\{\sum_{k=1}^{n} a_k f(X_k), \mathcal{B}_n, n \geq 1\}$ is a martingale which converges a.e. and in $L^2(P)$. (Use Bessel's inequality.) Deduce that $E^{\sigma(X_n)}(X) = 0, n \geq 1$, then $E^{\mathcal{B}_n}(X) = 0$, and finally $X = 0$ a.e. Show that this gives a form of the Hewitt-Savage law.

34. Let $\{X_n, \mathcal{B}_n, n \geq 1\}$ be a supermartingale on (Ω, Σ, P) and $g : \mathbb{R} \to \mathbb{R}^+$ be an increasing concave function such that $g(tX_n)$ is integrable for each $n \geq 1$ and $t > 0$. Show that for any $\lambda > 0$,

$$P\left[\sup_{n\geq 1} X_n \geq \lambda\right] \leq E(g(tX_1))/g(t\lambda).$$

In case $\{X_n, \mathcal{B}_n, n \geq 1\}$ is a submartingale, then one has for any λ,

$$P\left[\sup_{n\geq 1} X_n \geq \lambda\right] \leq \lim_{n\to\infty} E(e^{t(X_n-\lambda)}), \qquad t > 0.$$

35. Let $\{X_n, \mathcal{B}_n, n \geq 1\}$ be an $L^1(P)$-bounded martingale, i.e.,

$$\sup_n E(|X_n|) < \infty.$$

If $h : \mathbb{R} \to \mathbb{R}^+$ is a continuous function such that $h(t)/t^2 \to 0$ as $t \to 0$, show that the martingale increments sequence $\{\phi_n = X_n - X_{n-1}, n \geq 1\}$ with $X_0 = 0$ satisfies $\sum_{n\geq 1} h(\phi_n) < \infty$ a.e. [This is not as simple as one might expect. First note that $\{Y_n = \sum_{k=1}^{n} \hat{X}_{k-1}(\hat{X}_k - \hat{X}_{k-1}), \mathcal{B}_n, n \geq 1\}$ with

$$\hat{X}_k = X_k \chi_{[\max_{i\leq k}, |X_i| < \lambda]}$$

is an $L^2(P)$-bounded martingale, and $E(\sup_n |Y_{n+1} - Y_n|^2) \leq 4\lambda^4 < \infty$. Next, verify that on $A^\lambda = [\sup_n |X_n| \leq \lambda], Y_n \to Y_\infty$ a.e., implying

$$\left(X_n^2 - \sum_{k=1}^{n} \phi_k^2\right) \chi_{A^\lambda} \to 0 \qquad \text{a.e.}$$

This yields the result, due originally to D.G. Austin that $\sum_{k\geq 1} \phi_k^2 < \infty$ a.e., and then our assertion follows.

36. Let $\{X_n, n \geq 1\}$ be a sequence of independent r.v.s on (Ω, Σ, P) with means zero, and $S_n = \sum_{k=1}^{n} X_k$. Then the martingale $\{S_n, \mathcal{F}_n, n \geq 1\}$ converges a.e. to $S \in L^1(P)$ iff it is uniformly integrable, where

$$\mathcal{F}_n = \sigma(S_k, 1 \leq k \leq n) = \sigma(X_k, 1 \leq k \leq n).$$

[For the necessity, let $S_n \to S$ a.e., and note that $S - S_n$ is independent of $X_1, \ldots, X_n$, so that $E^{\mathcal{F}_n}(S) = E(S - S_n) + S_n = E(S) + S_n$ a.e. Since the left side tends a.e. (and in mean) to $E^{\mathcal{B}_\infty}(S) = S$, implying $E(S) = 0$, we deduce that $S_n \to S$ in mean, and hence is uniformly integrable. The same result holds if $L^1(P)$ is replaced by $L^p(P), p \geq 1$. This appears to be due to J. Marcinkiewicz.]

37. (Continuation) In the above problem, under the same hypothesis, show that for any $p \geq 1$,

$$\left|\left| \sup_n |S_n| \right|\right|_p \leq \beta ||S||_p, \quad \beta = \min[(4^p + 2^{p-1})^{1/p}, 5].$$

[Note that, by the submartingale property, we have $||S_n||_p \uparrow ||S||_p$. If $p > 1$, this is a consequence of Theorem 5.6ii, with β replaced by $p/(p-1) = q$; thus if $p \geq 1.25$, this is ≤ 5. For $1 \leq p < 1.25$, we need to prove this. Let us sketch the argument which gives the result. This application is due to Doob. First suppose that X_n has a symmetric distribution; i.e., $P[X_n \in A] = P[X_n \in -A]$ for any Borel set $A \subset \mathbb{R}$. Since $P[\max_{k \leq n} S_k \geq \lambda, S_n \geq \lambda] = P[S_n \geq \lambda]$, verify with the decomposition of the event $[\max_{k \leq n} S_k \geq \lambda]$, as in Theorem 5.6i, that $P[\max_{k \leq n} S_k \geq \lambda, S_n < \lambda] \leq P[S_n \geq \lambda]$, so that by addition one has

$$P\left[\sup_{n \geq 1} S_n \geq \lambda\right] \leq 2P[S \geq \lambda] = P[|S| \geq \lambda], \quad \lambda \geq 0.$$

Then again as in the proof of Theorem 5.6ii with the above inequality,

$$\begin{aligned}\int_\Omega \sup_n |S_n|^p \, dP &= \int_0^\infty P\left[\sup_n |S_n|^p > \lambda\right] d\lambda \leq 2 \int_0^\infty P[|S|^p > \lambda] \, d\lambda \\ &= 2 \int_\Omega |S|^p \, dP.\end{aligned}$$

In the unsymmetric case, let (Ω', Σ', P') be a copy of the given probability space, and $\{X'_n, n \geq 1\}$ be independent r.v.s distributed as the X_n. Then on the product space $(\Omega, \Sigma, P) \times (\Omega', \Sigma', P')$, $\{X_n - X'_n, n \geq 1\}$ is a symmetrically distributed sequence, and the above result applies to give

$$\left|\left| \sup_n |S_n - S'_n| \right|\right|_p \leq 2^{1/p} ||S - S'||_p \leq 2^{1+p^{-1}} ||S||_p.$$

But $|\lambda|^p$ is convex, and so $|\lambda|^p \leq 2^{p-1}[|\lambda - \lambda'|^p + |\lambda'|^p]$. Hence

$$E\left(\sup_n |S_n|^p\right) \leq 2^{p-1}\int_{\Omega\times\Omega'} \sup_n |S_n - S_n'|^p \, dP\, dP' + 2^{p-1}\sup_n \int_\Omega |S_n|^p \, dP$$
$$\leq 2^{2p}\|S\|_p^p + 2^{p-1}\|S\|_p^p.]$$

38. (Continuation) Under the same hypothesis as in the above problem, if $S_n \to S$ a.e., and $S \in L^p(P), p \geq 1$, show that $X_n \in L^p(P)$ and $S_n \to S$ in $L^p(P)$-mean. (Thus if the martingale is of this form, the convergence statement can be improved.)

39. Let $(T, \mathcal{T}, \lambda)$ be the Lebesgue unit interval and $f : T \to \mathbb{R}$ be an integrable r.v. on it. Consider the partition $\{I_k, 0 \leq k \leq 2^n - 1\}$, where $I_k = [k/2^n, (k+1)/2^n)$. If

$$f_n = \sum_{k=0}^{2^n-1} \frac{1}{\lambda(I_k)} \left(\int_{I_k} f(t)\, dt\right) \chi_{I_k},$$

and $\mathcal{B}_n$ is the σ-algebra generated by the partition, show that $\{f_n, \mathcal{B}_n, n \geq 1\}$ is a uniformly integrable martingale and that $f_n \to f$ a.e. and in $L^1(\lambda)$. [*Thus every element of* $L^1(\lambda)$ *can be written as the a.e. and in mean limit of a martingale of simple functions.*]

40. (Generalization of the Above) Let (Ω, Σ, P) be a probability space, $(T, \mathcal{T}, \lambda)$ be the Lebesgue unit interval, and $(W, \mathcal{A}, \mu)$ be their product. Let $\xi : \Omega \to \Omega$ be a one-to-one and onto maping such that it is measurable and measure preserving, so that $P = P \circ \xi^{-1}$. Let $\tau_n : T \to T$ be a "shift," so that $\tau_n(t) = t + 2^{-n}, \tau_n^k(t) = t + k2^{-n}, 0 \leq k < 2^n, \tau_n^{2^n} =$ identity $= \tau_n^0$. If $A_0 = E \times I, E \in \Sigma, I \in \mathcal{T}$, define $A_k = \xi^k(E) \times \tau_n^k(I), 0 \leq k < 2^n$, and $I = [0, 2^{-n})$, and $\mathcal{F}_n$ as the σ-algebra in $\mathcal{A}$, generated by sets of $\tilde{A}$ of the form $\tilde{A} = \bigcup_{k=0}^{2^n-1} A_k$ with A_k defined above. If $f \in L^1(P)$, let $h(\omega, t) = f(\omega), \omega \in \Omega, t \in T$. Show that $\{h_n, \mathcal{F}_n, n \geq 1\}$ is a decreasing martingale, where

$$h_n(s) = \sum_{k=0}^{2^n-1} f(\xi^{k-j_n}(\omega)), \quad \frac{j_n}{2^n} \leq t < \frac{j_n+1}{2^n}, \quad s = (\omega, t).$$

Deduce that $h_n(s) \to \tilde{h}(s)$ a.e. $[\mu]$ as $n \to \infty$, where $\{j_n, n \geq 1\}$ is a subsequence of integers going to infinity satisfying the above restriction for t. [*Hint*: h_n is a constant on the generators of $\mathcal{F}_n$, and if $\tilde{A} \in \mathcal{F}_n$ is a generator, verify that $\int_{\tilde{A}} h_n \, d\mu = \int_{\tilde{A}} h d\mu$ after noting that $\int_{A_k} g(\omega, t)\, d\mu = \int_{A_0} g(\xi^k\omega, \tau_n^k t)\, d\mu$ for any $g \in L^1(\mu)$. Since $\int_W |h|\, d\mu = \int_\Omega |f|\, dP$, the convergence statement follows from Theorem 5.14 for all j_n such that $j_n 2^{-n} \leq t < (j_n+1)2^{-n}$ for each n. But there are infinitely many such sequences, and so the result is a weak one. This formulation is due to M. Jerison.]

41. Many sequences of r.v.s can be expressed as functions of martingales. It is their convergence that is the difficult question. We illustrate it for two important processes in this and the next problem. Let $\{X_n, \mathcal{B}_n, n \geq 1\}$ be any quasi-martingale bounded in $L^1(\Omega, \Sigma, P)$. [See Eq. (49) of Section 5 for definition.] Show that this sequence can be expressed as a difference of two positive supermartingales relative to the same stochastic base $\{\mathcal{B}_n, n \geq 1\}$. (Use Theorem 5.9 and Lemma 5.5.)

42. Let (Ω, Σ, μ) be a σ-infinite measure space and $\nu : \Sigma \to \mathbb{R}^+$ be σ-additive (hence bounded) and μ-continuous. Let $\xi : \Omega \to \Omega$ be a measurable transformation such that μ and its image $\mu \circ \xi^{-1}$ are equivalent (i.e., have the same null sets.) Let $\nu = \sum_{k=0}^{n} \nu \circ \xi^{-k}, \mu_n = \sum_{k=0}^{n} \mu \circ \xi^{-k}$ which are signed measures $(\xi^{-(k+1)} = \xi^{-1} \circ \xi^{-k})$ on Σ and $\nu_n \ll \mu_n$. Let $X_n = d\nu_n / d\mu_n$ be the Radon-Nikodým derivative. If $(\mathbb{N}, \mathcal{N}, \zeta)$ is the counting measure space on the natural numbers $\mathbb{N}$, and $(W, \mathcal{A}, \lambda) = (\Omega, \Sigma, \mu) \times (\mathbb{N}, \mathcal{N}, \zeta)$, let $\mathcal{F}_n = \mathcal{B}_n \bigotimes \mathcal{G}_n$, where $\mathcal{B}_n = \sigma(\bigcup_{k=0}^{n-1} \xi^{-k}(A), A \in \Sigma), \mathcal{G}_n = \sigma((0, 1, \ldots, n-1), \{k\}; k \geq n)$, so that $\mathcal{F}_{n+1} \subset \mathcal{F}_n \subset \mathcal{A}$. If $f(s) = X_1(\omega), s = (\omega, k) \in W$ and $f_n(s) = X_n \circ \xi^k(\omega)$ if $0 \leq k < n$, and $= f(s)$ if $k \geq n$, show that $\{f_n, \mathcal{F}_n, n \geq 1\}$ is a decreasing martingale on the σ-finite space $(W, \mathcal{A}, \lambda)$ and that $X_n \to X$ a.e. $[\mu]$ is equivalent to the convergence of the $f_n [= E^{\mathcal{F}_n}(f) -]$ martingale. [Follow the hints as in Problem 40. Note that none of our results so far can be used to deduce this convergence statement, which is the Hurewicz-Oxtoby ergodic theorem. It is known that $X_n \to X$ a.e. In a similar way other averages can be formulated as (decreasing) martingales, but the known martingale convergence theory is inadequate for this type of application. This observation is also due to M. Jerison.]

43. If $P(\cdot, \cdot) : \Omega \times \Sigma \to \mathbb{R}^+$ is a mapping on a measurable space (Ω, Σ), it is called a *Markov kernel* if $P(\cdot, A)$ is Σ-measurable and $P(\omega, \cdot)$ is a measure, $P(\omega, \Omega) = 1, \omega \in \Omega$. [It is *sub-Markov* if $P(\omega, \Omega) \leq 1$.] If P_1, P_2 are two such kernels, then by the standard Lebesgue theory we define their composition as $Q = P_1 \circ P_2$ to mean $Q(\omega, A) = \int_\Omega P_1(\omega, d\tilde{\omega}) P_2(\tilde{\omega}, A)$. Note that this composition is a form of the Chapman-Kolmogorov equation if (Ω, Σ) is the Borelian line, when P_1 and P_2 are the one-step transition functions depending only on the number of steps taken but not the starting step. [They are also called the stationary transition probabilities.] We define

$$g : x \mapsto \int_\Omega f(y) P(x, dy)$$

if f is a positive (or real bounded) measurable function on Ω. Then g will be a positive (or bounded) measurable function. If $g \leq f$, then f is called *excessive* or *superharmonic*, and if $g = f$, then f is called *invariant* or *harmonic* relative to the kernel P. (If $g \geq f$, then f is *subinvariant* or *subharmonic*.) These play important roles in potential theory. Let $\{X_n, n \geq 0\}$ on a probability space $(\mathbb{R}, \mathcal{B}, \mu)$ be a sequence of r.v.s with a stationary transition probability function $P(\cdot,\cdot)$, so that $P(x,A) = \mu([X_n \in A] | X_{n-1} = x), x \in \mathbb{R}, A \in \mathcal{B}$. Show that for Borel $f : \mathbb{R} \to \mathbb{R}$ which is bounded and invariant relative to P, we have $E_{x_0}(f(X_n)) = \int_{\mathbb{R}} f(y) P^n(x_0, dy)$, where $P^1 = P$, and $P^n = P \circ P^{n-1}$ is the composition and $X_0 = x_0$ a.e. $[P(x_0,\cdot)]$. Deduce that $\{f(X_n), \mathcal{B}_n, n \geq 1\}$ is a uniformly integrable martingale on $(\mathbb{R}, \mathcal{B}, P(x_0,\cdot))$, where $\mathcal{B}_n = \sigma(f(X_k), 1 \leq k \leq n)$. If f here is excessive (or subinvariant) then the corresponding process is super-(or sub-)martingale.

44. (Continuation) With the setup of the preceding problem, and if the tail σ-algebra $\mathcal{T}$ of $\{X_n, n \geq 0\}$ on $(\mathbb{R}, \mathcal{B}, P(x_0,\cdot))$ is degenerate, so that each $A \in \mathcal{T}$ satisfies $P(x_0, A) = 0$ or 1, and if f is invariant, then show that f must be a constant relative to $P(x_0,\cdot)$. In particular, if $P(x_0, dy) = p(x_0, y)\, dy$, so that we can define the convolution $g(x) = (f * p(x_0,\cdot))(x)$ on $\mathbb{R}$ as

$$g(x) = \int_{\mathbb{R}} f(x+y) p(x_0, y)\, dy,$$

show that each bounded continuous invariant f depends only on x_0. [Reduce the problem to the above one by writing p for $p(x_0,\cdot)$ and considering the special Markov process $\{S_n = \sum_{k=1}^n X_k, n \geq 1\}$, where the X_n are i.i.d. This is discussed, in a more general case, for finding the solutions μ of $\mu = \mu * P$, by G. Choquet and J. Deny in *C.R. Acad. Sci., Ser. A.*, 250 (1960), 799-801.]

45. The theory of super- and subharmonic functions was primarily developed by F. Riesz in the late 1920s; among other theorems, Riesz proved an important decomposition of such functions. These results have analogs in martingale theory (by substituting "martingale" for "harmonic"), of which the above two assertions are illustrative. Thus the corresponding Riesz decomposition in our theory is as follows. A positive supermartingale $\{X_n, \mathcal{B}_n, n \geq 1\}$ is termed a *potential* if $X_n \to 0$ a.e. and in $L^1(P)$-mean. Let $\{X_n, \mathcal{B}_n, n \geq 1\}$ be an arbitrary supermartingale. Then it admits a unique Riesz decomposition $X_n = Y_n + Z_n$, where $\{Y_n, \mathcal{B}_n, n \geq 1\}$ is a martingale and $\{Z_n, \mathcal{B}_n, n \geq 1\}$ is a potential iff it dominates a martingale, i.e., iff there exists a martingale $\{W_n, \mathcal{B}_n, n \geq 1\}$ such that $X_n \geq W_n$ a.e. for all $n \geq 1$, or iff there is an $\alpha > -\infty$ such that $\lim_n E(X_n) \geq \alpha$. [*Hint:* If X_n dominates a martingale, then $E(X_n) \geq E(X_{n+1}) \geq E(W_n) = \alpha > -\infty$, since the expectation of a martingale is a constant. This implies $\sup_n |E(X_n)| < \infty$. If $X_n = X_n' - A_n$ is the decomposition of Theorem 5.9, where $\{A_{n+1}, \mathcal{B}_n, n \geq 1\}$ is the increasing adapted process, $A_1 = 0$, and $\{X_n', \mathcal{B}_n, n \geq 1\}$ is a martingale, so that $E(A_n) \leq E(X_1') + \sup_n |E(X_n)| < \infty$, let $A_\infty = \lim_n A_n$. Then $E(A_\infty) < \infty$. If we let $Y_n = X_n' - E^{\mathcal{B}_n}(A_\infty)$, and $Z_n = E^{\mathcal{B}_n}(A_\infty) - A_n \geq 0$, then we have $X_n = Y_n + Z_n$ as the desired decomposition. For uniqueness if $\tilde{Y}_n + \tilde{Z}_n = X_n$ is another decomposition, then $Y_n - \tilde{Y}_n = \tilde{Z}_n - Z_n$, so that

$$E(|Y_n - \tilde{Y}_n|) \leq E(Z_n) + E(\tilde{Z}_n) \to 0.$$

Since $\{|Y_n - \tilde{Y}_n|, \mathcal{B}_n, n \geq 1\}$ is a submartingale whose expectations are nondecreasing, we must have $E(|Y_n - \tilde{Y}_n|) = 0$, so that $Y_n = \tilde{Y}_n$ and then $Z_n = \tilde{Z}_n$. The converse is immediate.]

46. Consider the Haar system of functions on [0,1]. These are defined on [0,1] as $H_0(\cdot) = 1, H_{2^k}(\cdot) = 2^{k/2}(\chi_{[0,2^{-k-1})} - \chi_{[2^{-k-1},2^{-k})}), k = 0, 1, \ldots,$ and if $1 \leq j < 2^k, k \geq 1, H_{2^k+j}(\cdot) = H_{2^k}(\cdot - j2^{-k})\chi_{[j2^{-k},(j+1)2^{-k})}$. Then $\{H_n, n \geq 1\}$ forms a complete orthonormal system in $L^2(0,1)$. (Completeness means that only the zero function is orthogonal to all $H_n, n \geq 1$.) If $f \in L^2(0,1)$, and $a_n = \int_0^1 f(x)H_n(x)\,dx$, let $S_n(f) = \sum_{k=0}^n a_k H_k$. If

$$\mathcal{B}_n = \sigma(H_1, \ldots, H_n),$$

show that $\{S_n(f), \mathcal{B}_n, n \geq 1\}$ is a uniformly integrable martingale, and, in fact, $S_n(f) = E^{\mathcal{B}_n}(f) \to f$ a.e. [Note that $\mathcal{B}_n$ has $(n+1)$ atoms. If $n = 2^k + j$, then the intervals $[l2^{-k}, (l+1)2^{-k-1}), 0 \leq l < 2j+1$, and $[m2^{-k}, (m+1)2^{-k}), j+1 \leq m < 2^k, 0 \leq j < 2^k$, are these atoms and that $E^{\mathcal{B}_n}(H_{n+1}) = 0$. This construction will be elaborated upon early in Section 8.1. Also compare with Problem 39 above.]

47. The preceding case is abstracted as follows. An orthonormal system $\{\phi_n, n \geq 1\}$ in $L^2(P)$ on a probability space (Ω, Σ, P) is an H-system if (i) each ϕ_n takes at most two nonzero values with positive probability, (ii) $\mathcal{B}_n = \sigma(\phi_1, \ldots, \phi_n)$ has exactly n atoms, and (iii) $E^{\mathcal{B}_n}(\phi_{n+1}) = 0$ a.e. Show that an orthonormal system $\{\tilde{\phi}_n\}_1^\infty$ is an H-system iff for each $f \in L^2(P), E^{\mathcal{B}_n}(f) = \sum_{k=1}^n a_k\tilde{\phi}_k, a_k = \int_\Omega f\tilde{\phi}_k\,dP$.

48. (Continuation) Let $\{X_n, \mathcal{B}_n, n \geq 1\}$ be a square integrable martingale on (Ω, Σ, P), and $\phi_n = X_n - X_{n-1}$, with $\phi_1 = X_1$, be the martingale [difference or] increments sequence. Verify that if $\{\phi_n, n \geq 1\} \subset L^2(P)$ is complete, then it is an H-system.

49. (Continuation) If $f : [0,1] \to \mathbb{R}$ is any Lebesgue measurable function, and $\{\phi_n, n \geq 1\}$ is a complete H-system in $L^2(0,1)$, then it can be expressed as $f = \sum_{k \geq 1} a_k\phi_k$, for a suitable set of $a_k \in \mathbb{R}$, the series converging a.e. Further, there exists a complete H-system on [0,1] such that every measurable (not necessarily integrable in either case) function has a series representation as in the preceding statement. [This result in its general form is due to Gundy (1966). It needs a careful analysis using the work of the last two problems.]

50. The preceding three problems admit the following further extension. Let (Ω, Σ, P) be a probability space and Σ be countably generated. Then there exists a (universal) martingale $\{X_n, \mathcal{B}_n, n \geq 1\}$ on (Ω, Σ, P) such that $\Sigma = \sigma(\bigcup_n \mathcal{B}_n)$, and every extended real-valued measurable (for Σ), but not necessarily integrable, f on Ω is a pointwise a.e. limit of a subsequence

$\{X_{n_k}, k \geq 1\}$ of the $X_n, n \geq 1$ sequence. [This result is involved, and it abstracts the preceding work, and is due to Lamb (1974). However, not all a.e. convergence results of Fourier series may be established by martingale methods. For instance, if $f \in L^p(0,1), 1 < p < \infty$, and $\phi_k(x) = e^{2\pi i k x}$, then $f = \sum_{k \geq 1} a_k \phi_k, a_k = \int_0^1 f(x)\phi_k(x)\,dx$, converges a.e. (and in L^p-norm), but the partial sums do not form a martingale, and as yet no martingale proof of this statement is available. A relatively "simple" proof of this result, which is originally due to L. Carleson and R.A. Hunt, in the middle 1960s, is given by C.L. Fefferman. *Ann. Math.* 98 (1973), 551-571.]

Part II Analytical Theory

The fine properties of probability are often obtainable by using the sharp Fourier analysis techniques They are called characteristic function methods, and Chapter 4 is utilized in establishing various important results including the Lèvy-Bochner-Cramèr representation theorems applicable for uniqueness, derivation of distributions of ratios of random variables and special properties of sums of independent random elements. Then the longest chapter of the book, Chapter 5, is devoted to the central limit theory with error estimations, to stable distributions as well as to invariance or functional limit theorems. Here the Kolmogorov law of the iterated logarithm and certain m-dependent theorems are also included. Several important complements are considered in both the chapters and should be of interest for students as well as researchers.

Chapter 4

Probability Distributions and Characteristic Functions

Structural properties of probability distributions and of characteristic functions are treated here in some detail. These include the selection principle, the Bochner and Lévy theorems on characteristic functions together with their essential equivalence, Cramér's theorem on the Fourier-Stieltjes transform of signed measures, and some multivariate extensions. The equivalence of pointwise a.e. convergence and convergence in distribution for sums of independent r.v.s is also established. Many additional results are sketched as exercises.

4.1 Distribution Functions and the Selection Principle

In the theory of the preceding chapters we have already seen the use of distribution functions at several places. We now undertake a systematic study of the structural properties of these functions and of their Fourier transforms for use in the further development of probability theory.

Recall that a distribution function of a random variable X on (Ω, Σ, P) is the image law F_X given by $F_X(x) = P[X < x]$, so that it is a nondecreasing left continuous function on $\mathbb{R}$ satisfying $F_X(-\infty) = \lim_{x \to -\infty} F_X(x) = 0$ and $F_X(+\infty) = \lim_{x \to +\infty} F_X(x) = 1$. Here we study the properties of such functions F on $\mathbb{R}$ without reference to r.v.s. In view of the Kolmogorov-Bochner-Jessen theory of the preceding chapter, the collection of such distribution functions (generalized to multidimensions) determines a probability space and a family of r.v.s. on it with these (finite dimensional) distributions for the r.v.s. thus determined. Hence a study of distributions *per se* becomes an integral part of probability theory; in fact, for most of the analytical work they even occupy a preeminent position in the subject. Consequently, we start with this analysis, through a use of Fourier transforms for a deeper insight into their analytical structure.

In Section 2.2 we introduced the concept of the convergence of a sequence of distribution functions and noted that this is the weakest of the convergences considered there. The following fundamental selection property of distributions, discovered in 1912 by E. Helly, plays an important role in our study.

Theorem 1 (Helly's Selection Principle) *Let $\{F_n, n \geq 1\}$ be a sequence of distribution functions on $\mathbb{R}$. Then there exists a nondecreasing left continuous function F (not necessarily a distribution), $0 \leq F(x) \leq 1$, $x \in \mathbb{R}$, and a subsequence $\{F_{n_k}, k \geq 1\}$ of the given sequence such that $F_{n_k}(x) \to \mathrm{F}(x)$ at all continuity points x of F.*

Proof We first establish the convergence at a dense denumerable set of points of $\mathbb{R}$ and then extend the result using the fact that the continuity points of a monotone function on $\mathbb{R}$ form an everywhere dense set. Thus, because rationals are dense and denumerable in $\mathbb{R}$, we consider them here for definiteness, and let $r_1, r_2, \ldots$ be an enumeration of this set. Since $0 \leq F_n(x) \leq 1, \{F_n(r_1), n \geq 1\}$ is a bounded sequence, so that by the Bolzano-Weierstrass theorem it has a convergent subsequence, $\{F_{n1}(r_1), n \geq 1\}$, such that $F_{n1}(r_1) \to G_1(r_1)$, as $n \to \infty$. Continuing this procedure, we get a sequence $F_{nk}(r_k) \to G_k(r_k)$ and $\{F_{nk}, n \geq 1\} \subset \{F_{n(k-1)}, n \geq 1\}$. Consider the diagonal sequence $\{F_{nn}, n \geq 1\}$. This converges at $x = r_1, r_2, \ldots$. Let $\lim_{n\to\infty} F_{nn}(r_i) = \alpha_i, i = 1, 2, \ldots$; thus $\alpha_i = G_i(r_i)$. Since the F_n are increasing, it follows that for $r_i < r_j, \alpha_i \leq \alpha_j$. For each $x \in \mathbb{R}$, define $G(x) = \inf\{\alpha_n : r_n > x\}$. Since $\alpha_n \leq \alpha_{n+1}$, it is clear that $G(\cdot)$ is nondecreasing. We now show that $F_{nn}(x) \to G(x)$ at each continuity point x of G.

If $\varepsilon > 0$, and a is a continuity point of G, choose $h > 0$ such that

$$G(a + h) - G(a - h) < \varepsilon/2.$$

This is clearly possible by the continuity of G at a. Let r_i, r_j be rationals from our enumeration such that (by density) $a - h < r_i < a < r_j < a + h$. Then

$$G(a - h) \leq \alpha_i \leq G(a) \leq \alpha_j \leq G(a + h).$$

Choose $N(= N_\epsilon)$ such that $n \geq N$ implies

$$|F_{nn}(r_i) - \alpha_i| < \varepsilon/2, \qquad |F_{nn}(r_j) - \alpha_j| < \varepsilon/2.$$

Then for all $n \geq N$ we have by the monotonicity of the F_n and G and the above inequalities:

$$\begin{aligned} F_{nn}(a) \leq F_{nn}(r_j) \leq \alpha_j + \varepsilon/2 &\leq G(a + h) + \varepsilon/2 \\ &\leq G(a) + \varepsilon/2 + \varepsilon/2 = G(a) + \varepsilon. \end{aligned}$$

Similarly

$$\begin{aligned} F_{nn}(a) \geq F_{nn}(r_i) \geq \alpha_i - \varepsilon/2 &\geq G(a - h) - \varepsilon/2 \\ &\geq G(a) - \epsilon/2 - \varepsilon/2 = G(a) - \varepsilon. \end{aligned}$$

From these two inequalities one gets

$$|F_{nn}(a) - G(a)| < \varepsilon,$$

so that $F_{nn}(x) \to G(x)$ at all continuity points x of G. Now define F on $\mathbb{R}$ by $F(x) = G(x-0)$, so that $F(x) = G(x)$ if x is a continuity point, and $F(x-0) = F(x)$ if x is a discontinuity point of G. Thus $F_{nn}(x) \to F(x)$ at all x which are continuity points of F. This completes the proof.

Remarks (1) The fact that the set $\{r_i, i \geq 1\} \subset \mathbb{R}$ is the set of rationals played no part in the above proof. Any dense denumerable set will do. Consider $F_n(x) = 0$ for $x < n, = 1$ for $x \geq n$; we see that $F_n(x) \to 0$ as $n \to \infty$ for all $x \in \mathbb{R}$, so that the limit F satisfies $F(x) \equiv 0$. Thus such an F is not necessarily a distribution function (d.f.).

(2) If $+\infty$ and $-\infty$ are continuity points of each F_n, then $F_n(-\infty) \to F(-\infty), F_n(+\infty) \to F(+\infty)$, so that $F_n(+0) - F_n(-0) = 1$ implies in this case that F is a distribution. In particular, if $\{F_n, n \geq 1\}$ is a sequence of d.f.s on a compact interval $[a,b] \subset \mathbb{R}$, then the limit of any convergent subsequence is necessarily a d.f., since {a,b} may be included in the set $\{r_i, i \geq 1\}$.

(3) The preceding theorem can be stated as follows: A uniformly bounded sequence of nondecreasing functions on $\mathbb{R}$ is *weakly sequentially compact* in the sense that it has a convergent subsequence whose limit is a bounded nondecreasing function.

The next theorem supplements the above result and is very useful in our study. It should be contrasted with the Lebesgue limit theorems, for which the integrands vary and the measure space is fixed whereas the opposite is true in the following. It is due to E. Helly in a special case, and the general case to H.E. Bray in 1919. The connection between these two viewpoints is clarified in an alternative proof below.

Theorem 2 (Helly-Bray) *Let $\{G_n, n \geq 1\}$ be a sequence of individually bounded nondecreasing functions on $\mathbb{R}$. If there exists a bounded nondecreasing function G on $\mathbb{R}$ such that*

(i) $\lim_{n\to\infty} G_n(x) = G(x)$ *at all continuity points x of G,*
(ii) $\lim_{n\to\infty} G_n(\pm\infty) = G(\pm\infty)$*in the sense that*

$$\lim_{n\to+\infty} \lim_{x\to+\infty} G_n(x) = \lim_{x\to\infty} G(x)$$

and similarly for $x \to -\infty$, then for any bounded continuous function $f : \mathbb{R} \to \mathbb{R}$, we have

$$\lim_{n\to\infty} \int_{\mathbb{R}} f(x)\, dG_n(x) = \int_{\mathbb{R}} f(x)\, dG(x). \tag{1}$$

Proof Since G is nondecreasing and bounded, given a $\delta > 0$, there is an $N_\delta \geq 1$ such that $n \geq N_\delta$ implies by hypothesis (ii),

$$G(-\infty) - \delta < G_n(-\infty) \leq G_n(x) \leq G_n(+\infty) \leq G(+\infty) + \delta.$$

Hence $\{G_n, n \geq 1\}$ is uniformly bounded. Thus there is an $M < \infty$, with $|G_n(x)| \leq M$, for all $x \in \mathbb{R}$. Next, for any $a > 0$, we consider

$$\begin{aligned} I_n = \int_{\mathbb{R}} f(x)\,d(G_n - G)(x) &= \int_{-\infty}^{-a} + \int_{-a}^{a} + \int_{a}^{\infty} f(x)\,d(G_n - G)(x) \\ &= I_n' + I_n'' + I_n''', \text{ (say.)} \end{aligned}$$

It is to be shown that $|I_n| \to 0$ as $n \to \infty$. This is accomplished by estimating the right side terms and showing that each goes to zero.

Since f is bounded, there is a $c > 0$ with $|f(x)| \leq c, x \in \mathbb{R}$. If $\varepsilon > 0$ is given, we choose a as before, so that $\pm a$ are continuity points of G and

$$\int_{[|x| \geq a]} dG(x) < \varepsilon/8c. \tag{2}$$

This is possible since G is bounded and its continuity points are dense in $\mathbb{R}$. By (i) and (ii), we may choose an $N_1(\varepsilon)$, such that $n \geq N_1(\varepsilon)$ implies

$$\begin{aligned} |G_n(a) - G(a) < \varepsilon/16c, \quad & |G_n(-a) - G(-a)| < \varepsilon/16c, \\ |G_n(+\infty) - G(+\infty)| < \varepsilon/16c, \quad & |G_n(-\infty) - G(-\infty)| < \varepsilon/16c. \end{aligned} \tag{3}$$

Then

$$\begin{aligned} |I_n'| &\leq \left| \int_{-\infty}^{-a} f(x)\,dG_n(x) \right| + \left| \int_{-\infty}^{-a} f(x)\,dG(x) \right| \\ &\leq c[G_n(-a) - G_n(-\infty) + G(-a) - G(-\infty)]. \end{aligned}$$

Similarly

$$|I_n'''| \leq c[G_n(+\infty) - G_n(a) + G(+\infty) - G(a)].$$

Adding these two and using (2) and (3), we get

$$\begin{aligned} |I_n'| + |I_n'''| &\leq c[(G_n(-a) - G_n(-\infty)) + (G_n(+\infty) - G_n(a))] + (\varepsilon/8) \\ &\leq c[G(-a) - G(-\infty) + G(+\infty) - G(a)] + (4\varepsilon/16) + (\varepsilon/8) \\ &\leq (\varepsilon/8) + (\varepsilon/4) + (\varepsilon/8) = \varepsilon/2. \end{aligned} \tag{4}$$

For $|I_n''|$, since $|G_n| \leq M$ and $[-a, a]$ is a compact interval, divide $[-a, a]$ at the x_i into m subintervals such that the oscillation of f on each is bounded by $\varepsilon/16M$ where $-a = x_0 < x_1 < \cdots < x_m = a$, the x_i also being continuity points of G. All this is clearly possible. Hence

$$\begin{aligned} |I_n''| &= \left| \sum_{i=1}^{m} \int_{x_{i-1}}^{x_i} f(x)\,d(G_n - G)(x) \right| \\ &\leq \sum_{i=1}^{m} \int_{x_{i-1}}^{x_i} |f(x) - f(x_i)|\,dG_n(x) + \sum_{i=1}^{m} \int_{x_{i-1}}^{x_i} |f(x) - f(x_i)|\,dG(x) \end{aligned}$$

$$
\begin{aligned}
&+\max_{x\in[-a,a]}|f(x)|\sum_{i=1}^{m}\left|\int_{x_{i-1}}^{x_i} d(G_n-G)(x)\right| \\
&\leq \frac{\varepsilon}{16M}\sum_{i=1}^{m}\int_{x_{i-1}}^{x_i} d(G_n+G)(x) \\
&\quad +c\sum_{i=1}^{m}[|G_n(x_i)-G(x_i)|+|G_n(x_{i-1})-G(x_{i-1})|].
\end{aligned}
$$

Let $N_2(\varepsilon)$ be chosen so that $n \geq N_2(\varepsilon) \Rightarrow |G_n(x_i) - G(x_i) < \varepsilon/8mc, i = 1, \ldots, m$. Then the above inequality becomes

$$|I_n''| \leq \frac{\varepsilon}{16M}4M + cm\frac{2\varepsilon}{8mc} = \varepsilon/2. \tag{5}$$

Thus (4) and (5) imply for $n \geq \max(N_1(\varepsilon), N_2(\varepsilon))$,

$$|I_n| \leq \varepsilon/2 + \varepsilon/2 = \varepsilon.$$

This completes the proof of the theorem.

Note that, as the example in the first part of the above remark shows, condition (ii) of the hypothesis in the Helly-Bray theorem is essential for the conclusion of (1). In that example, F_n is the d.f. of a discrete r.v. X_n, and $P[X_n > a] \to 1$ as $n \to \infty$ for each $a > 0$. Thus the probability "escapes to infinity," and condition (ii) is simply to prevent this phenomenon from happening, so that (1) is true.

We present an alternative sketch of the important Helly-Bray theorem by reducing it to the Lebesgue bounded convergence through the image probability law (cf. 1.4.1) *and* the representation in *Problem 5b of Chapter 2.* [Readers should complete the details given as hints there, if they have not already done so. A more general case will be proved in Proposition 5.4.2.]

Alternative Proof For simplicity we take G_n, G as d.f.s. (The general case, which can be reduced to this, is left to the reader.) Thus $G_n(x) \to G(x)$ as $n \to \infty$ for all continuity points x of G. Hence by the above noted problem, there exists a probability space (Ω, Σ, P) and a sequence of r.v.s X_n, X on it such that

$$P[X_n < x] = G_n(x), \qquad P[X < x] = G(x), \qquad x \in \mathbb{R},$$

and $X_n \to X$ a.e. [In fact, (Ω, Σ, P) can be taken as the Lebesgue unit interval, and $X_n(\omega) = G_n^{-1}(\omega), \omega \in (0,1)$ where G_n^{-1} is the (generalized) inverse of G_n, i.e., $G_n^{-1}(\omega) = \inf\{y \in \mathbb{R} : G_n(y) > \omega\}$, and similarly $X(\omega) = G^{-1}(\omega), \omega \in \Omega$.] Now if f is as given, $f(X_n) \to f(X)$ a.e. and $f(X)$ is an r.v. By the image law 1.4.1, we have

$$\int_{\mathbb{R}} f(x)\, dG_n(x)$$
$$= \int_{\Omega} f(X_n)\, dP \to \int_{\Omega} f(X)\, dP$$
(by the Lebesgue bounded convergence theorem)
$$= \int_{\mathbb{R}} f(x) dG(x) \qquad \text{(by the same image law).}$$

This is (1) and terminates the proof.

The technique of the second proof above will be used to establish the following result on the convergence of moments.

Proposition 3 *Let $\{X_n, n \geq 1\}$ be a sequence of r.v.s on (Ω, Σ, P) such that $E(|X_n|^s) \leq K_0 < \infty$ for all $n \geq 1$ and some $s > 0$. If $X_n \to X$ in distribution, then $E(|X_n|^r) \to E(|X|^r)$ for each $0 < r < s$, so that the rth-order absolute moments also converge.*

Proof Let $F_n(x) = P[X_n < x], F(x) = P[X < x], x \in \mathbb{R}$. Then $F_n(x) \to F(x)$ at all continuity points x of F, by hypothesis. Hence by the technique used in the last proof, there exists an auxiliary probability space (Ω', Σ', P') and a sequence $\{Y_n, n \geq 1\}$ of r.v.s on it such that $Y_n \to Y$ a.e., and $P'[Y_n < x] = F_n(x), P'[Y < x] = F(x), x \in \mathbb{R}$. Thus by the image law, we also have

$$E(|X_n|^r) = \int_{\mathbb{R}} |x|^r dF_n(x) = \int_{\Omega'} |Y_n|^r dP', \qquad n \geq 1, \tag{6}$$

and

$$E(|X|^r) = \int_{\mathbb{R}} |x|^r dF(x) = \int_{\Omega'} |Y|^r dP'. \tag{7}$$

Since $0 < r < s$, the second condition implies the uniform integrability of $\{|Y_n|^r, n \geq 1\}$. Indeed, we have for any $\alpha > 0$,

$$\int_{[|Y_n|^r > \alpha]} |Y_n|^r dP' \leq \alpha^{1-s/r} \int_{[|Y_n|^r > \alpha]} |Y_n|^s dP'$$
$$\leq K_0/\alpha^{(s/r)-1} \to 0 \qquad \text{as } \alpha \to \infty,$$

uniformly in n. Hence, since $|Y_n|^r \to |Y|^r$ a.e., by the Vitali convergence theorem we deduce that

$$\int_{\Omega'} |Y_n|^r dP' \to \int_{\Omega'} |Y|^r dP' \qquad \text{as } n \to \infty. \tag{8}$$

From (6)–(8), it follows that $E(|X_n|^r) \to E(|X|^r)$, as asserted.

Using the same ideas as in the above proof, we can also deduce the following result, which complements Theorem 2 in some respects.

Proposition 4 *Let $\{F_n, F, n \geq 1\}$ be d.f.s such that $F_n \to F$ at all continuity points of F. If $f_n : \mathbb{R} \to \mathbb{R}$ are bounded continuous functions such that $f_n \to f$ uniformly, then we can conclude that*

$$\lim_{n\to\infty} \int_{\mathbb{R}} f_n(x)\, dF_n(x) = \int_{\mathbb{R}} f(x)\, dF(x). \tag{9}$$

Proof Using the same representation as in the preceding proof, we note that there is a probability space (Ω, Σ, P) and r.v.s X_n and X on it such that $X_n \to X$ a.e. and $P[X_n < x] = F_n(x), P[X < x] = F(x), x \in \mathbb{R}$. But then (9), by the image law, is equivalent to

$$\lim_{n\to\infty} \int_{\Omega} f_n(X_n)\, dP = \int_{\Omega} f(X)\, dP. \tag{10}$$

Since $f_n \to f$ uniformly, so that f is also bounded and continuous, we deduce that $f_n(X_n) \to f(X)$ a.e. and by the bounded convergence (10) holds. Again by the image law theorem, (9) follows.

A direct proof of (9), without the representation, to get (10) is possible, and it is similar to the Helly-Bray theorem. But that is not as elegant as the above one.

Actually, the converse to (1), and hence a characterization of the convergence in distribution, is also true. If F and G are two d.f.s on $\mathbb{R}$, then we define the *Lévy metric* between them as

$$d(F, G) = \inf\{\varepsilon > 0 : F(x - \varepsilon) - \varepsilon \leq G(x) \leq F(x + \varepsilon) + \varepsilon, x \in \mathbb{R}\}. \tag{11}$$

It is not difficult to verify that $d(\cdot, \cdot)$ is a distance function on the space $\mathcal{M}$ of all d.f.s on $\mathbb{R}$. A verification of the metric axioms will be left to the reader. We have several characterizations of the concept in the following:

Theorem 5 *Let $\{F_n, F, n \geq 1\}$ be distribution functions on $\mathbb{R}$. Then the following statements are mutually equivalent:*

(i) *$F_n \to F$ at all continuity points of the latter.*

(ii) $\lim_{n\to\infty} \int_{\mathbb{R}} f(x)\, dF_n(x) = \int_{\mathbb{R}} f(x)\, dF(x)$ *for all bounded continuous $f : \mathbb{R} \to \mathbb{R}$.*

(iii) $d(F_n, F) \to 0$ *as $n \to \infty$.*

(iv) *If P_n and P are the Lebesgue-Stieltjes measures determined by F_n and F, then $\limsup_n P_n(C) \leq P(C)$ for all closed sets $C \subset \mathbb{R}$.*

(v) *If P_n and P are as in* (iv),*then $\liminf_n P_n(D) \geq P(D)$ for all open sets $D \subset \mathbb{R}$.*

Proof The method of proof here is to show that each part is equivalent to (i) or (ii). Now the Helly-Bray theorem already established that (i)⇒(ii). For the converse (cf. also the remark after the proof) let $\varepsilon > 0$ be given. If x_0 is a point of continuity of F, there is a $\delta[= \delta(x_0, \varepsilon) > 0]$ such that $|x - x_0| < \delta \Rightarrow |F(x) - F(x_0)| < \varepsilon/2$. We construct two bounded continuous functions f_1, f_2 on $\mathbb{R}$, $f_1 \leq f_2$, as follows:

$$f_1(x) = \begin{cases} 1 & x \leq x_0 - \delta \\ (x_0 - x)/\delta, & x_0 - \delta \leq x \leq x_0 \\ 0, & x \geq x_0, \end{cases}$$

$$f_2(x) = \begin{cases} 1, & x \leq x_0 \\ (x_0 + \delta - x)/\delta, & x_0 \leq x \leq x_0 + \delta \\ 0, & x \geq x_0 + \delta. \end{cases}$$

By hypothesis, for these f_1, f_2, there exists an $N[= N(\varepsilon, f_1, f_2)]$ such that $n \geq N$ implies

$$\left| \int_{\mathbb{R}} f_i(x)\, dF(x) - \int_{\mathbb{R}} f_i(x)\, dF_n(x) \right| < \varepsilon/2, \qquad i = 1, 2. \tag{12}$$

Hence,

$$\begin{aligned} \varepsilon/2 > \int_{\mathbb{R}} f_1(x)\, dF(x) - \int_{\mathbb{R}} f_1(x)\, dF_n(x) &\geq F(x_0 - \delta) - F_n(x_0) \\ &\geq F(x_0) - \varepsilon/2 - F_n(x_0). \end{aligned} \tag{13}$$

Similarly

$$\begin{aligned} -\varepsilon/2 < \int_{\mathbb{R}} f_2(x)\, dF(x) - \int_{\mathbb{R}} f_2(x)\, dF_n(x) &\leq F(x_0 + \delta) - F_n(x_0) \\ &\leq F(x_0) + \varepsilon/2 - F_n(x_0). \end{aligned} \tag{14}$$

From(13) and (14) we get

$$-\varepsilon < F(x_0) - F_n(x_0) < \varepsilon. \tag{15}$$

Since $x_0 \in \mathbb{R}$ is an arbitrary continuity point of F, (15) implies that $F_n \to F$ at all continuity points of F, and (i)⇔(ii) is established.

(i)⇔(iii) If $F_n \to F$, for each $\varepsilon > 0$, choose $\pm a$ as continuity points of F such that $F(-a) < \varepsilon/2, 1 - F(a) < \varepsilon/2$. Partition the compact interval $[-a, a]$ as $-a = a_0 < a_1 < \cdots < a_m = +a$, with $a_i - a_{i-1} < \varepsilon, i = 1, \ldots, m$, and also a_i as continuity points of F. Let N be chosen such that for $n \geq N$, we have by (i),

$$|F_n(a_i) - F(a_i)| < \varepsilon/2, \qquad i = 1, \ldots, m \tag{16}$$

To show that the F_n, F satisfy (11), let $x \in \mathbb{R}$ be arbitrary. If $x \leq -a = a_0$, then by monotonicity of F_n and F, and inequalities (16), we have

$$F_n(x) \le F_n(a_0) \le F(a_0) + \varepsilon/2 \le \varepsilon + F(x),$$
$$F_n(x) \ge 0 \ge F(a_0) - \varepsilon/2 \ge F(x) - \varepsilon \quad [\text{since } F(a_0) \le \varepsilon/2]. \tag{17}$$

If $a_{j-1} \le x \le a_j$, then by (16), and $a_j - a_{j-1} < \varepsilon, j = 1, \ldots, m,$

$$F_n(x) \le F_n(a_j) \le F(a_j) + \varepsilon/2 \le F(x+\varepsilon) + \varepsilon,$$
$$F_n(x) \ge F_n(a_{j-1}) \ge F(a_{j-1}) - \varepsilon/2 \ge F(x-\varepsilon) - \varepsilon, \tag{18}$$

Similarly if $x \ge a_m$, one gets

$$F(x) - \varepsilon \le F_n(x) \le F(x) + \varepsilon. \tag{19}$$

By (17)–(19) and (11) we have $d(F_n, F) \le \varepsilon$. Thus (iii) holds. [1]

Conversely, if (iii) is given, let $\varepsilon > 0$ and x_0 be a continuity point of F. Then there is a $\delta_1[= \delta_1(x_0, \varepsilon) > 0]$ such that for $0 < \delta < \delta_1, |x - x_0| < \delta \Rightarrow |F(x) - F(x_0)| < \varepsilon/2$. If $\eta = \min(\varepsilon/2, \delta) > 0$, then by (iii) there exits N_0 such that $n \ge N_0 \Rightarrow d(F_n, F) < \eta$, and from (11) we have

$$F_n(x_0) \le F(x_0 + \eta) + \eta \le F(x_0) + 2\eta \le F(x_0) + \varepsilon,$$
$$F_n(x_0) \ge F(x_0 - \eta) - \eta \ge F(x_0) - 2\eta \ge F(x_0) - \varepsilon.$$

Hence,$|F_n(x_0) - F(x_0)| < \varepsilon$ and (i) follows.

(ii)⇒(v) If P and P_n are as given, then (ii) may be written as

$$\lim_{n\to\infty} \int_{\mathbb{R}} f\, dP_n = \int_{\mathbb{R}} f\, dP, \qquad f : \mathbb{R} \to \mathbb{R} \text{ is bounded and continous.} \tag{20}$$

Let $A \subset \mathbb{R}$ be an open set. Then χ_A is a lower semicontinuous (l.s.c.) function. Recall that a function $h : \mathbb{R} \to \mathbb{R}$ is l.s.c. if $h(y) \le \liminf_{x\to y} h(x)$ for each $y \in \mathbb{R}$. A classical result from advanced calculus says that h is l.s.c. iff there exists a sequence of continuous functions $h_n (\ge 0 \text{ if } h \ge 0)$ such that $h_n(x) \uparrow h(x)$ for each x. We use this here. Thus let $0 \le h_n \uparrow \chi_A$ pointwise, where h_n is continuous on $\mathbb{R}$. Let k be a fixed integer, and by (ii) for each ε, there is an $n_0[= n_0(\varepsilon, h_k) \ge 1]$ such that $n \ge n_0$ implies

$$\int_{\mathbb{R}} h_k(x) P(dx) < \int_{\mathbb{R}} h_k(x) P_n(dx) + \epsilon. \tag{21}$$

Since $0 \le h_k \le \chi_A$, we also have

[1] (i)⇒(iii) can be quickly proved using Exercise 5(b) of Chapter 2 again. Thus $F_n \to F$, a d.f. $\Rightarrow X_n \to X$ a.e. on $(\Omega', \Sigma', P'), F_n = F_{X_n}, F = F_X$. So for $\varepsilon > 0$, there is an $n_0, n \ge n_0 \Rightarrow P'[|X_n - X| < \varepsilon] \ge 1 - \varepsilon$. If $\Omega^0 = [|X_n - X| < \varepsilon]$, then on $\Omega_0, X_n - \varepsilon < X < X_n + \varepsilon \Rightarrow F_n(x - \varepsilon) - \varepsilon \le F(x) \le F_n(x+\varepsilon) + \varepsilon$. Hence $d(F_n, F) \le \varepsilon$.

$$\begin{aligned}\liminf_n P_n(A) = \liminf_n \int_{\mathbb{R}} \chi_A \, dP_n &\geq \liminf_n \int_{\mathbb{R}} h_k(x) P_n(dx) \\ &\geq \int_{\mathbb{R}} h_k(x) P(dx) - \varepsilon \qquad \text{[by (21)].}\end{aligned}$$

Letting $k \to \infty$ in this sequence, we get by the monotone convergence theorem,

$$\liminf_n P_n(A) \geq \int_{\mathbb{R}} \chi_A P(dx) - \varepsilon = P(A) - \varepsilon.$$

Since $\varepsilon > 0$ is arbitrary, (v) is established, i.e., (ii)$\Rightarrow$(v).

If $C \subset \mathbb{R}$ is any closed set, then (v) implies

$$\liminf_n P_n(\mathbb{R} - C) \geq P(\mathbb{R} - C) = 1 - P(C),$$

and the left side is equal to $1 - \limsup_n P_n(C)$. Hence (iv) is true. Conversely, if (iv) holds, then considering the complements as here, we get (v). Thus (iv)$\Leftrightarrow$(v) is always true.

(v)$\Leftrightarrow$(iv) together imply (i). Indeed, let $A \subset \mathbb{R}$ be a Borel set whose boundary has P-measure zero. [For example, if $A = (a, b)$, then $\{a\}, \{b\}$ have P-measure zero, which is equivalent to saying that a, b are continuity points of F.] Thus $P(\overline{A} - \text{int}(A)) = 0$, where $\overline{A}$ is the closure of A and $\text{int}(A)$ is the interior of A. Thus by (v) and its equivalence with (iv) we have

$$\begin{aligned}P(\overline{A}) \geq \limsup_n P_n(\overline{A}) \geq \limsup_n P_n(A) &\geq \liminf_n P_n(A) \\ &\geq \liminf_n P_n(\text{int}(A)) \geq P(\text{int}(A)) \qquad \text{[by (v)].}\end{aligned}$$

But the extremes are equal. Thus $\lim_n P_n(A) = P(A)$ for every Borel A whose boundary has P-measure zero. In particular, if $A_x = (-\infty, x)$ and noting that $F_n(x) = P_n(A_x), F(x) = P(A_x)$, this yields (i). Thus the proof is complete.

Remark Since the proof is given as (i)$\Rightarrow$(ii)$\Rightarrow$(iv)$\Leftrightarrow$(v)$\Rightarrow$(i) and (i)$\Leftrightarrow$(iii), there is redundancy in showing separately that (ii)$\Rightarrow$(i). However, the separate argument shows that this implication is true if we assume (ii) only for the (subclass of) uniformly continuous bounded functions f. This insight is useful in addition to the fact that (i)$\Leftrightarrow$(ii) is the most important and often used part of the above result.

We now present a complement to the above theorem. This is a partial converse to Proposition 3. An annoying problem here is that moments of a sequence of d.f.s may converge without the d.f.s themselves converging. Further, a sequence of numbers need not be moments of a d.f.; and even if they happen to be moments, it is possible that two different d.f.s can have the same set of moments. (We give two examples in Problem 5.) Thus with restrictions, to exclude all these difficulties, we can present the following relatively simple result.

Proposition 6 *Let $\{F_n, n \geq 1\}$ be a sequence of d.f.s having moments of all orders $\{M_n^{(k)}, k \geq 1\}$. Thus we have*

$$M_n^{(k)} = \int_{\mathbb{R}} x^k \, dF_n(x), \qquad k \geq 1, \quad n \geq 1.$$

If $\lim_{n\to\infty} M_n^{(k)} = \alpha^{(k)}$ exists for each $k \geq 1$, and if $\{\alpha^{(k)}, k \geq 1\}$ determines a distribution F uniquely, then $F_n(x) \to F(x)$, as $n \to \infty$, holds for each x which is a continuity point of F, so that $F_n \to F$ as $n \to \infty$.

Proof By Theorem 1, $\{F_n, n \geq 1\}$ has a convergent subsequence. Thus there exists a nondecreasing left continuous $F, 0 \leq F(\cdot) \leq 1$, such that for some $\{F_{n_k}, k \geq 1\}$, we have $F_{n_k} \to F$. To see that F is a d.f., we need to show that $F(+\infty) - F(-\infty) = 1$. Indeed, given $\varepsilon > 0$, choose a large enough so that $\pm a$ are continuity points of F and $\alpha^{(2)}/a^2 < \varepsilon$. This is clearly possible. Then

$$\begin{aligned}
F(a) - F(-a) &= \lim_{k\to\infty} [F_{n_k}(a) - F_{n_k}(-a)] \\
&\geq \lim_{k\to\infty} \left[1 - 1/a^2 \int_{[|x|> a]} x^2 \, dF_{n_k}(x)\right] \\
&\geq \lim_{k\to\infty} \left[1 - 1/a^2 \int_{\mathbb{R}} x^2 \, dF_{n_k}(x)\right] \\
&= 1 - \alpha^{(2)}/a^2 \qquad [\text{since } M_n^{(2)} \to \alpha^{(2)}, \text{by hypothesis,}] \\
&\geq 1 - \varepsilon.
\end{aligned}$$

Since $\varepsilon > 0$ is arbitrary, we deduce that $F(\cdot)$ is a d.f., and $F_{n_k} \to F$. Hence there is a probability space (Ω, Σ, P)(as in the proof of Proposition 4) and a sequence of r.v.s $\{X_{n_k}, Y, k \geq 1\}$ such that $X_{n_k} \to Y$ a.e. and $P[X_{n_k} < x] = F_{n_k}(x), P[Y < x] = F(x), x \in \mathbb{R}$. Also for each integer $p \geq 1, 0 \leq M_{n_k}^{(2p)} \to \alpha^{(2p)}$, so that $\{M_{n_k}^{(2p)}, k \geq 1\}$ is bounded. By Proposition 3, $E(|X_{n_k}|^q) \to E(|Y|^q), 0 < q < 2p$. Hence by Proposition 1.4.6, $\{|X_{n_k}|^q, k \geq 1\}$ is uniformly integrable. But this implies that $\{X_{n_k}^q, k \geq 1\}, q \geq 1$ integer, is also uniformly integrable. Consequently by Theorem 1.4.4, $E(X_{n_k}^q) \to E(Y^q)$, and the $\alpha^{(q)}, 1 \leq q < 2p$, are the qth moments of F. Since p is arbitrary, it follows that $\{\alpha^{(q)}, q \geq 1\}$ are all the moments of F, and by hypothesis these determine F uniquely.

If $\{F_{n'}, n' \geq 1\} \subset \{F_n, n \geq 1\}$ is any other convergent subsequence, then the preceding paragraph shows that $F_{n'} \to F'$, a d.f. Since F' also has $\{\alpha^{(q)}, q \geq 1\}$ as its moments, by the uniqueness hypothesis $F = F'$. But by Theorem 5iii, the set of distribution functions on $\mathbb{R}$ is a metric space under convergence in d.f. topology, and in a metric space a sequence converges iff each of its convergent subsequences has the same limit. Thus the full sequence $\{F_n, n \geq 1\}$ converges to F, completing the proof.

This result is of use in applications only if we have some criteria for the unique determination of d.f.s by their moments. The question involved here is nontrivial and there has been a considerable amount of research on what is called the "moment problem." For an account, see Shohat and Tamarkin (1950). For instance, if $S = \{x : F(x) > 0\} \subset \mathbb{R}$ is bounded, then F is uniquely determined by its moments. This and certain other easily verifiable sufficient conditions can be obtained from the work on characteristic functions of distribution functions. We now turn to a detailed analysis of these functions and devote the rest of the chapter to this topic as it is one of the most effective tools in the subject.

4.2 Characteristic Functions, Inversion, and Lévy's Continuity Theorem

In the preceding section some properties of distribution functions (d.f.s) have been given. For a finer analysis however, we need to use the full structure of the range space, namely, $\mathbb{R}$, and turn to Fourier transforms of d.f.s. These are called *characteristic functions* (ch.f.s) in probability theory and their special properties that are of immediate interest will be considered here. Thus if $F : \mathbb{R} \to \mathbb{R}$ is a d.f., then we define its ch.f. by the Lebesgue-Stieltjes integral:

$$\phi(t) = \int_{\mathbb{R}} e^{itx}\, dF(x) = \int_{\mathbb{R}} \cos tx\, dF(x) + i \int_{\mathbb{R}} \sin tx\, dF(x). \tag{1}$$

This concept was already introduced for, and the uniform continuity of $\phi : \mathbb{R} \to C$ was established in, Proposition 1.4.2. It is clear that complex analysis plays a role, since generally ϕ is complex valued. Note that by the image probability law (Theorem 1.4.1), (1) is equivalent to saying that, if X is an r.v. on (Ω, Σ, P) with F as its d.f., then

$$\phi(t) = \phi_X(t) = E(e^{itX}) = \int_{\mathbb{R}} e^{itx} dF(x), \quad |\phi(t)| \leq \phi(0) = 1, \quad t \in \mathbb{R}. \tag{2}$$

Hence if X_1, X_2 are independent r.v.s on (Ω, Σ, P) then

$$\begin{aligned}\phi_{X_1,X_2}(t_1, t_2) &= E(e^{it_1X_1 + it_2X_2}) \\ &= E(e^{it_1X_1})E(e^{it_2X_2}) \\ &= \phi_{X_1}(t_1)\phi_{X_2}(t_2) \quad \text{for all } t_j \in \mathbb{R},\ j = 1, 2.\end{aligned} \tag{3}$$

In particular, if $X = X_1 + X_2$, then

$$\phi_X(t) = E(e^{itX}) = \phi_{X_1,X_2}(t,t) = \phi_{X_1}(t)\phi_{X_2}(t). \tag{4}$$

On the other hand, by the image law theorem one has

$$\begin{aligned} F_X(x) = P[X_1 + X_2 < x] &= \int\int_{[x_1+x_2<x]} dF_{X_1}(x_1)\, dF_{X_2}(x_2) \\ &= \int_{\mathbb{R}} dF_{X_2}(x_2) \int_{-\infty}^{x-x_2} dF_{X_1}(x_1) \\ &= \int_{\mathbb{R}} F_{X_1}(x - x_2)\, dF_{X_2}(x_2) \\ &= \int_{\mathbb{R}} F_{X_2}(x - x_1)\, dF_{X_1}(x_1). \end{aligned} \tag{5}$$

Thus F_X is the *convolution* of F_{X_1} and F_{X_2} (already seen in Problem 6(b) of Chapter 2). Also it is a commutative operation. Equations (4) and (5) together imply that the ch.f. of the convolution of a pair of d.f.s is the product of their ch.f.s. It is also clear from (5) that, if F_{X_1} and F_{X_2} have densities f_1 and f_2 relative to the Lebesgue measure, then the convolution $F_X = F_{X_1} * F_{X_2}$ becomes

$$F_X(x) = \int_{\mathbb{R}} F_{X_1}(x-t) f_2(t)\, dt = \int_{\mathbb{R}} \int_{-\infty}^{(x-u)} f_1(t)\, dt\, f_2(u)\, du, \tag{6}$$

and hence by the fundamental theorem of calculus (or the Radon-Nikodým theorem in the general case) we conclude that F_X again has a density f and from (6)

$$f(x) = \int_{\mathbb{R}} f_1(x-u) f_2(u)\, du \qquad [= (f_1 * f_2(x)]. \tag{7}$$

Note also that (5) implies F_X is absolutely continuous if either F_{X_1} or F_{X_2} has this property.

To get some feeling for these functions, we list a set of basic d.f.s that occur frequently in the theory, and then their ch.f.s will be given.

1. Gaussian or normal [often denoted $N(\mu, \sigma^2)$]:

$$F(x) = \left(\frac{1}{2\pi\sigma^2}\right)^{1/2} \int_{-\infty}^{x} e^{-(t-\mu)^2/2\sigma^2}\, dt, \qquad -\infty < \mu < \infty,$$
$$0 < \sigma^2 < \infty, \quad x \in \mathbb{R}.$$

2. Poisson:

$$F(x) = \begin{cases} \sum_{\substack{0 \le k < x \\ k \in \mathbb{Z}}} e^{-\lambda} \frac{\lambda^k}{k!}, \; x > 0, & \\ e^{-\lambda} & x = 0, \text{(we take } \lambda^0 = 1 \text{ for } \lambda \ge 0), \\ 0, & \text{if } x < 0, \text{ or } 0 < \lambda < \infty. \end{cases}$$

3. Cauchy:

$$F(x) = \frac{\lambda}{\pi} \int_{-\infty}^{x} \frac{du}{\lambda^2 + u^2}, \qquad \lambda > 0, \quad x \in \mathbb{R}.$$

4. Gamma:

$$F(x) = \begin{cases} \frac{1}{\Gamma(p)} \int_0^x e^{-u} u^{p-1} \, du, & p > 0, \quad x > 0 \\ 0 & \text{otherwise.} \end{cases}$$

5. Uniform:

$$F(x) = \begin{cases} 0, & x < a \qquad a \in \mathbb{R} \\ \frac{x-a}{b-a}, & a \le x < b, \, b \in \mathbb{R} \\ 1, & x \ge b. \end{cases}$$

6. Bernoulli:

$$F(x) = \begin{cases} 0, & x < 0 \\ q, & 0 \le x < 1 \\ 1, & x \ge 1, \end{cases} \qquad 0 < p = 1 - q \le 1.$$

7. Unitary (or degenerate):

$$F(x) = \begin{cases} 0, & x < a, \quad a \in \mathbb{R} \\ 1, & x \ge a. \end{cases}$$

8. Binomial:

$$F(x) = \begin{cases} \sum_{\substack{0 \le k < x \\ k \in \mathbb{Z}}} \binom{n}{k} p^k q^{n-k}, & 0 \le p = 1 - q \le 1, \quad 0 \le x \le n, \\ 0 & \text{otherwise.} \end{cases}$$

The respective ch.f.s are given by

1′. $\phi(t) = \exp(i\mu t - (\sigma^2 t^2/2))$,
2′. $\phi(t) = \exp(\lambda(e^{it} - 1))$,
3′. $\phi(t) = \exp(-\lambda|t|)$,
4′. $\phi(t) = (1 - it)^{-p}$,
5′. $\phi(t) = (e^{itb} - e^{ita})/(b - a)it$,
6′. $\phi(t) = (q + pe^{it})$,
7′. $\phi(t) = e^{iat}$,
8′. $\phi(t) = (q + pe^{it})^n$.

We leave a verification of these formulas to the reader with a reminder that one can use, for convenience, the calculus of residues for some of these evaluations, (such is the case with the Cauchy and Gaussian distributions).

The interest in ch.f.s stems from the fact that there is a one-to-one relation between the d.f.s and their ch.f.s, and since the latter are (uniformly) continuous, they are more suitable for a finer analysis using the well-developed results from Fourier transform theory. The uniqueness statement is a consequence of the following important result.

Theorem 1 (Lévy's Inversion Formula, 1925) *Let F be a d.f. with ch.f. ϕ. If $h > 0$ and $a \pm h$ are continuity points of F, then we have*

$$F(a+h) - F(a-h) = \lim_{T\to\infty} \frac{1}{\pi} \int_{-T}^{T} \frac{\sin ht}{t} e^{-ita} \phi(t)\, dt, \qquad a \in \mathbb{R}. \tag{8}$$

Moreover, if ϕ is Lebesgue integrable, then F has a continuous bounded density, and (8) reduces to

$$F'(x) = f(x) = \frac{1}{2\pi} \int_{\mathbb{R}} e^{-itx} \phi(t)\, dt. \tag{9}$$

Proof The importance of the result lies in the discovery of the formula (8). Once it is given, its truth can be ascertained by substitution as follows.

$$\begin{aligned}
&\frac{1}{\pi} \int_{-T}^{T} \frac{\sin ht}{t} e^{-ita} \phi(t)\, dt \\
&\quad = \frac{1}{\pi} \int_{-T}^{T} \frac{\sin ht}{t} e^{-ita} \int_{\mathbb{R}} e^{itx}\, dF(x)\, dt \\
&\quad = \frac{1}{\pi} \int_{\mathbb{R}} \int_{-T}^{T} \frac{\sin ht}{t} e^{it(x-a)} dt\, dF(x) \quad \text{(by Fubini's theorem)} \\
&\quad = \int_{\mathbb{R}} G_T(x)\, dF(x),
\end{aligned} \tag{10}$$

where

$$\begin{aligned}
G_T(x) &= \frac{1}{\pi} \int_{-T}^{T} \frac{\sin ht}{t} e^{-t(x-a)}\, dt \\
&= \frac{2}{\pi} \int_{0}^{T} \frac{\sin ht \cos t(x-a)}{t}\, dt \\
&\qquad \Big[\text{since } \frac{\sin ht}{t} \cos t(x-a) \text{ is an even function of } t \\
&\qquad \text{while } \frac{\sin ht}{t} \sin t(x-a) \text{ is an odd function of } t.\Big] \\
&= \frac{1}{\pi} \int_{0}^{T} \frac{\sin(x-a+h)t - \sin(x-a-h)t}{t}\, dt \\
&= \frac{1}{\pi} \int_{0}^{(x-a+h)T} \frac{\sin u}{u}\, du - \frac{1}{\pi} \int_{0}^{(x-a-h)T} \frac{\sin u}{u}\, du.
\end{aligned}$$

We recall from advanced calculus that $\int_0^\infty (\sin u/u)du = \pi/2$. Then

$$\lim_{T\to\infty} G_T(x) = \frac{1}{2}[\operatorname{sgn}(x-a+h) - \operatorname{sgn}(x-a-h)]$$
$$= \begin{cases} 0, \ x < a-h \\ \frac{1}{2}, \ x = a-h \\ 1, \ a-h < x < a+h \\ \frac{1}{2}, \ x = a+h \\ 0, \ x > a+h \end{cases}$$

Here "sgn" is the signum function:

$$\operatorname{sgn} x = \begin{cases} 1 & \text{if } \ x > 0 \\ 0 & \text{if } \ x = 0 \\ -1 & \text{if } \ x < 0. \end{cases}$$

Substituting this in (10), we get by the bounded convergence theorem

$$\lim_{T\to\infty} \frac{1}{\pi} \int_{-T}^{T} \frac{\sin ht}{t} e^{-ita} \phi(t)\, dt = \int_{a-h}^{a+h} 1 \cdot dF(x) = F(a+h) - F(a-h),$$

since $(dF)(a \pm h) = 0$, by hypothesis. This establishes (8).

For the second part, let ϕ be integrable. Dividing both sides of (8) by $2h$, and noting that $(\sin ht)/ht$ is bounded (by 1) for all t, we can first let $T \to \infty$ and then $h \to 0$, both by the dominated convergence theorem, since $|\phi(\cdot)|$ is the dominating integrable function. It follows that the right side has a limit and so must the left side; i.e., $F'(a)$ exists.[2] Since

$$\lim_{h\to\infty} \frac{\sin ht}{ht} = 1,$$

(8) reduces to (9). Further, for all x, it is clear that $|f(x)| \leq \int_{\mathbb{R}} |\phi(t)|\, dt$, and f is bounded. Also expressing ϕ by its real and imaginary parts and the latter by the positive and negative parts, we deduce that ϕ is a linear combination of four nonnegative integrable functions (since $|\phi|$ is), and (9) implies that f is a sum of four terms each of which is a Fourier transform of a nonnegative integrable function. By Proposition 1.4.2, it follows that each of the terms is continuous, and hence so is f. This proves the theorem completely.

We can now give the desired uniqueness assertion.

[2] Note that the existence of a symmetric derivative is equivalent to the existence of ordinary derivative for d.f.s. [Use the Lebesgue decomposition and write $F = F_a + F_s + F_d$ and observe that the singular and discrete parts F_s, F_d have no contribution and then $F' = F'_a$, the absolutely continuous part. We leave the details to the reader.]

Corollary 2 (Uniqueness Theorem) *A d.f. is uniquely determined by its ch.f.*

Proof By definition, every d.f. associates a ch.f. with it. If two d.f.s F_1 and F_2 have the same ch.f. ϕ, we need to show that $F_1 = F_2$. To this end, since each F_i has at most a countable set of discontinuities, the collection of continuity points for both F_1 and F_2, say, C_0, is the complement of a countable set, and hence is everywhere dense in $\mathbb{R}$. Let $a_i \in C_0, i = 1, 2, a_1 < a_2$. Then by (8),

$$F_1(a_2) - F_1(a_1) = F_2(a_2) - F_2(a_1), \tag{11}$$

since their right sides are equal. If P_i is the Lebesgue-Stieltjes probability determined by F_i on $\mathbb{R}$, then (11) implies that $P_1(A) = P_2(A)$ for all intervals $A \subset \mathbb{R}$ with end points in C_0. Consequently, P_1 and P_2 agree on the semiring generated by such intervals. Since C_0 is dense in $\mathbb{R}$, the σ-algebra generated by this semiring is the Borel σ-algebra of $\mathbb{R}$. But the P_i are σ-additive on the semiring, and agree there. By the Hahn extension theorem, they have unique extensions to the Borel σ-algebra of $\mathbb{R}$ and agree there. Thus $P_1 = P_2$ on this algebra, so that if $A = (-\infty, x), x \in \mathbb{R}$, which is a Borel set, we get

$$F_1(x) = P_1(A) = P_2(A) = F_2(x),$$

and the d.f.s are identical. (A direct proof of this for d.f.s. is also easy.)

In view of the preceding uniqueness theorem, it is quite desirable to have various properties of ch.f.s at our disposal both for further work on the subject and for a better understanding of their structure. The following formula is used for this purpose.

Proposition 3 *Let F be a d.f. and ϕ its ch.f. Then for any $b > 0$, we have*

$$\int_0^b [F(\alpha + x) - F(\alpha - x)]\, dx = \frac{1}{\pi} \int_{\mathbb{R}} \frac{1 - \cos bt}{t^2} e^{-it\alpha} \phi(t)\, dt, \quad \alpha \in \mathbb{R}. \tag{12}$$

Proof Replacing ϕ by its definition and simplifying the right side, exactly as in the proof of Theorem 1, we get the left side. However, for variety we present an alternative argument, following H. Cramér, and deduce the result from Theorem 1.

Let $h > 0$ be arbitrarily fixed and consider $G(x) = \int_x^{x+h} F(y)\, dy/h$. Then G is a continuous d.f. In fact, if $\tilde{F}$ is the uniform d.f. on $(-h, 0)$, as defined in the list above, then $G(x) = \int_{\mathbb{R}} F(x - y)\, d\tilde{F}(y) = (F * \tilde{F})(x)$ is the convolution. Since $\tilde{F}$ is continuous, G is also. Let ψ be the ch.f. of G. Then $\psi(t) = \phi(t)\tilde{\phi}(t), t \in \mathbb{R}$, where $\tilde{\phi}(t) = (1 - e^{ith})/ith, \tilde{\phi}(\cdot)$ being the ch.f. of $\tilde{F}$. Hence by Theorem 1,

$$
\begin{aligned}
G(a+h)-G(a) &= \lim_{T\to\infty}\frac{1}{2\pi}\int_{-T}^{T}\frac{e^{-ita}-e^{-it(a+h)}}{it}\psi(t)\,dt \\
&= \lim_{T\to\infty}\frac{1}{2\pi}\int_{-T}^{T}\left(\frac{1-e^{-ith}}{it}\right)^2\frac{e^{-ita}}{h}\phi(t)\,dt \\
&= \lim_{T\to\infty}\frac{2}{\pi h}\int_{-T}^{T}\left(\frac{\sin(ht/2)}{t}\right)^2 e^{-it(a+h)}\phi(t)\,dt. \qquad (13)
\end{aligned}
$$

Since $[(\sin ht)/t]^2$ is integrable on $\mathbb{R}$, we can take the limit as $T\to\infty$ here, and then, on substitution for G in terms of F, we get from (13)

$$
\int_{a+h}^{a+2h}F(x)\,dx-\int_{a}^{a+h}F(y)\,dy=\frac{2}{\pi}\int_{\mathbb{R}}\left(\frac{\sin(ht/2)}{t}\right)^2 e^{-it(a+h)}\phi(t)\,dt.
$$

Replacing x by $a+h+t, y$ by $a+u$ in the left-side integrands, one obtains

$$
\begin{aligned}
&\int_0^h F(a+h+t)\,dt-\int_0^h F(a+u)\,du \\
&\quad=\frac{1}{\pi}\int_{\mathbb{R}}\frac{1-\cos ht}{t^2}e^{-it(a+h)}\phi(t)\,dt. \qquad (14)
\end{aligned}
$$

Let $a+h=\alpha$ and finally $h=b$ in (14); it reduces to (12), as claimed.

We are now ready to establish the fundamental continuity theorem due to P. Lévy who discovered it in 1925.

Theorem 4 (Continuity Theorem for ch.f.s) *Let $\{F_n, n\geq 1\}$ be a sequence of d.f.s and $\{\phi_n, n\geq 1\}$ be their respective ch.f.s. Then there exists a d.f. F on $\mathbb{R}$ such that $F_n(x)\to F(x)$ at all continuity points of the latter iff $\phi_n(t)\to\phi(t), t\in\mathbb{R}$ where ϕ is continuous at $t=0$. When the last condition holds, then ϕ is the ch.f. of F.*

Proof The necessity follows from our previous work. Indeed, let $F_n\to F$, a d.f. Then by the Helly-Bray theorem (cf. Theorem 1.2, which is valid for complex functions also, by treating separately the real and imaginary parts), with $f(x)=e^{itx}$,

$$
\phi_n(t)=\int_{\mathbb{R}}e^{itx}\,dF_n(x)\to\int_{\mathbb{R}}e^{itx}\,dF(x)=\phi(t),\qquad t\in\mathbb{R}.
$$

Thus $\phi_n(t)\to\phi(t), t\in\mathbb{R}$, and ϕ is a ch.f. of F, and hence is continuous on $\mathbb{R}$.

The converse requires more detail and uses the preceding technical result. Since $\{F_n, n\geq 1\}$ is uniformly bounded, by the Helly selection (Theorem 1.1), there is a subsequence $F_{n_k}\to F$, where $0\leq F\leq 1$ and F is a left continuous nondecreasing function. We first claim that F is a d.f., using the hypothesis on ϕ. For this it suffices to show that $F(+\infty) - F(-\infty) = 1$, and later we verify that the whole sequence converges to F.

Let $b > 0$ and consider Proposition 3 with $\alpha = 0$ there. Then

$$\int_0^b [F_{n_k}(x) - F_{n_k}(-x)]\, dx = \frac{1}{\pi} \int_{\mathbb{R}} \frac{1 - \cos bt}{t^2} \phi_{n_k}(t)\, dt. \tag{15}$$

Since

$$\frac{1 - \cos bt}{t^2} = \frac{2\sin^2(bt/2)}{t^2}$$

is integrable on $\mathbb{R}$, by the dominated convergence theorem we can let $k \to \infty$ on both sides of (15) to get

$$\begin{aligned} \frac{1}{b} \int_0^b [F(x) - F(-x)]\, dx &= \frac{1}{\pi} \int_{\mathbb{R}} \frac{1 - \cos bt}{bt^2} \phi(t)\, dt \\ &= \frac{1}{\pi} \int_{\mathbb{R}} \frac{1 - \cos u}{u^2} \phi\left(\frac{u}{b}\right) du. \end{aligned}$$

Letting $b \to +\infty$ and using L'Hôpital's rule on the left and the dominated convergence theorem on the right, we get

$$\begin{aligned} F(+\infty) - F(-\infty) &= \frac{1}{\pi} \int_{\mathbb{R}} \frac{1 - \cos u}{u^2} \phi(0)\, du \\ &= \frac{2}{\pi} \int_0^{\infty} \frac{\sin^2 u}{u^2}\, du = 1, \end{aligned}$$

since, by hypothesis, ϕ is continuous at $t = 0$, and $\phi_n(0) = 1$, so that $\phi_{n_k}(0) \to \phi(0) = 1$. Here we also used again the fact from calculus that

$$\int_0^{\infty} \frac{\sin^2 u}{u^2}\, du = \frac{\pi}{2}.$$

Thus F is a d.f., and by the necessity proof we can now conclude that ϕ is the ch.f. of F.

Let $\{F_{n'}, n' \geq 1\}$ be any other convergent subsequence of $\{F_n, n \geq 1\}$, with limit $\tilde{F}$. Then by the preceding paragraph $\tilde{F}$ is a d.f. with ch.f. ϕ again (since $\phi_n \to \phi$ implying that every convergent subsequence has the same limit). By the uniqueness theorem (Corollary 2) $F = \tilde{F}$. Hence all convergent subsequences of $\{F_n, n \geq 1\}$ have the same limit d.f. F, so that the whole sequence F_n converges to the d.f. F with ch.f. ϕ. This completes the proof.

Remarks (1) The continuity of ϕ at $t = 0$ is essential for the truth of the sufficiency of the above theorem. For a simple counterexample, consider F_n defined by

$$F_n(x) = \begin{cases} 0, & x < -n \\ (x+n)/2n, & -n \leq x < n \\ 1, & x \geq n. \end{cases}$$

Then $F_n(x) \to \frac{1}{2} = F(x), x \in \mathbb{R}$, and F is not a d.f. If ϕ_n is the ch.f. of F_n, then it is seen that $\phi_n(t) = (\sin nt)/nt$, so that $\phi_n(t) \to \phi(t)$, where $\phi(t) = 0$ if $t \neq 0; = 1$ if $t = 0$. Thus $\phi(\cdot)$ is not continuous at $t = 0$ and it is not a ch.f., and F is not a d.f.

(2) It should also be noted that $\phi_n(t) \to \phi(t), t \in \mathbb{R}$, in the theorem cannot be replaced by $\phi_n(t) \to \phi(t)$, for $|t| \leq \alpha, \alpha > 0$, since two ch.f.s can agree on such a finite interval without being identical, as the following example (due to A. Khintchine) shows. Let F_1 have the density f_1 defined by $f_1(x) = (1 - \cos x)/\pi x^2$, and F_2 be discrete with jumps at $x = n\pi$ of sizes $2/n^2\pi^2, n = \pm 1, \pm 3, \ldots$, and $\frac{1}{2}$ at 0. Then using Proposition 3 with $b = 1$ and F as unitary d.f., one finds that the ch.f. of F_1 is $\phi_1(t) = 1 - |t|$ for $|t| \leq 1; = 0$ for $|t| > 1$ (cf. Exercises 8 and 9). On the other hand, a direct calculation, with the resulting trigonometric series, for $\phi_2(\cdot)$ gives

$$\phi_2(t) = \frac{1}{2} + (4/\pi^2) \sum_{k \geq 1} [\cos(2k+1)\pi t]/(2k+1)^2,$$

and so $\phi_2(t) = 1 - |t|$ for $|t| \leq 1; \neq 0$ for $|t| > 1$. Thus $\phi_1(t) = \phi_2(t)$ for $|t| \leq 1; \neq \phi_2(t)$ for $|t| > 1$. (If we expand $1 - |t|$ in the Fourier series, then the above expression results.)

The preceding remark and the theorem itself heighten the interest in the structure of ch.f.s. First, how does one recognize a uniformly continuous bounded function to be a ch.f., and second, how extensive and constructible are they? (These are nontrivial.) Regarding the second problem, note that since the product of two (or a finite number of) ch.f.s is a ch.f. [cf. (3) and (4)], we may construct new ones from a given set. In fact, if $\{\phi_n, n \geq 1\}$ is a sequence of ch.f.s and $\alpha_i \geq 0$ with $\sum_{i \geq 1} \alpha_i = 1$, then

$$\phi = \sum_{i \geq 1} \alpha_i \phi_i \tag{16}$$

is also a ch.f. Indeed, if F_n is the d.f. corresponding to the ch.f. ϕ_n, then $F = \sum_{n \geq 1} \alpha_n F_n$ is clearly a d.f., and its ch.f. is given by (16).

The preceding example admits the following extension.

Proposition 5 *Let $h : G \times \mathbb{R} \to \mathbb{C}$ be a mapping and $H : G \to [0,1]$ be a d.f. If $h(\cdot, t)$ is continuous for each $t \in \mathbb{R}$ and $h(s, \cdot)$ is a ch.f. for each $s \in G$, then*

$$\phi : t \mapsto \int_G h(s,t)\, dH(s), \qquad t \in \mathbb{R} \tag{17}$$

is a ch.f. for $G = \mathbb{Z}$ or $\mathbb{R}$. In particular, if $\psi : \mathbb{R} \to \mathbb{C}$ is a ch.f., then for each $\lambda \geq 0, \phi : t \mapsto \exp(\lambda(\psi(t) - 1))$ is a ch.f. [Here $G \subset \mathbb{R}$ is any subset.]

Proof Let $G = \mathbb{R}$. The first part of the hypothesis implies that $|h(s,t)| \leq h(s,0) = 1, s \in \mathbb{R}$ and $t \in \mathbb{R}$, so that the integral in (17) exists. By the representation used before, there is a probability space (Ω, Σ, P) and an r.v. $X : \Omega \to \mathbb{R}$ with H as its d.f. But the structure theorem of measurable functions gives the existence of a sequence of simple functions $X_n : \Omega \to \mathbb{R}$ such that $X_n \to X$ pointwise everywhere. If $H_n(x) = P[X_n < x]$, then $H_n \to H$, and by the Helly-Bray theorem

$$\lim_{n\to\infty} \int_{\mathbb{R}} h(s,t)\, dH_n(s) = \int_{\mathbb{R}} h(s,t)\, dH(s), \qquad t \in \mathbb{R}. \tag{18}$$

But H_n is a discrete d.f., since X_n is a discrete r.v. Hence the left-side integral, for each n, is of the form (16) with a finite sum, so that it is a ch.f., say, $\phi_n(\cdot)$. By (18) $\phi_n(t) \to \phi(t), t \in \mathbb{R}$, as $n \to \infty$. Since $h(s, \cdot)$ is continuous, it follows immediately that the right-side integral of (18) is continuous on $\mathbb{R}$. Theorem 4 then implies that ϕ is a ch.f. The case when $G = \mathbb{Z}$ is simpler and is left to the reader.

The last part is immediate. In fact, expanding the exponential,

$$\phi(t) = \sum_{n=0}^{\infty} e^{-\lambda} \frac{\lambda^n}{n!} \psi(t)^n = \sum_{n\geq 0} \alpha_n(\lambda)\psi(t)^n \qquad \text{(say,)} \quad t \in \mathbb{R},$$

where $\alpha_n(\lambda) > 0, \sum_{n\geq 0} \alpha_n(\lambda) = 1$, and $\psi(t)^n$ is a ch.f. for each n. This is thus of the form (16), so that $\phi(\cdot)$ is a ch.f. as shown there. The proof is completed.

Using this proposition, it is clear that we can generate a great many ch.f.s from those of the list given at the beginning of this section. The first question is treated in Section 4. To gain further insight into this powerful tool, we consider some differentiability properties.

Proposition 6 *If ϕ is a ch.f. of some d.f. F, and has p derivatives at $t = 0$, then F has $2[p/2]$ moments, where $[x]$ is the largest integer not exceeding x. On the other hand, if F has p moments [i.e., $\int_{\mathbb{R}} |x|^p\, dF(x) < \infty$], then ϕ is p times continuously differentiable. Here $p \geq 1$ is an integer.*

Proof Recall that, for a function $f : \mathbb{R} \to \mathbb{R}$, the symmetric derivative at x is defined as

$$(D_1 f)(x) = \lim_{h\to 0} [f(x+h) - f(x-h)]/2h$$

whenever this limit exists. Similarly the pth symmetric derivative, if it exists, is given by the expression

$$(D_p f)(x) = \lim_{h\to 0} \frac{\Delta_h^{(p)} f(x)}{(2h)^p}, \qquad \Delta_h^{(p)} = \Delta_h^{(1)} \Delta_h^{(p-1)},$$

with $(\Delta_h^{(1)} f)(x) = f(x+h) - f(x-h)$. It may be verified that, if f has a pth $(p \geq 1)$ ordinary derivative, then it also has the pth symmetric derivative and they are equal. But the converse is false, cf. the example after the proof. (This is also a standard fact of differentiation theory.) In our case, $f = \phi, x = 0$. If $p = 2m+1$, or $= 2m$, then $2[p/2] = 2m, m \geq 1$. Since $\phi^{(p)}(0)$ exists by hypothesis, its symmetric derivative also exists and is the same as $\phi^{(p)}(0)$. Thus we have

$$\phi^{(p)}(0) = \lim_{t \to 0} \int_{\mathbb{R}} [(e^{itx} - e^{-itx})/2t]^p \, dF(x).$$

Hence substituting for $2[p/2]$,

$$\begin{aligned} \infty > |\phi^{(2m)}(0)| &= \left| \lim_{t \to 0} i^{2m} \int_{\mathbb{R}} \left(\frac{\sin tx}{t} \right)^{2m} dF(x) \right| \\ &\geq \int_{\mathbb{R}} \liminf_{t \to 0} \left(\frac{\sin tx}{tx} \right)^{2m} x^{2m} \, dF(x) \qquad \text{(by Fatou's lemma)} \\ &= \int_{\mathbb{R}} x^{2m} \, dF(x). \end{aligned}$$

This proves the first part.

For the second part, if $p = 1$,

$$\begin{aligned} \frac{\phi(t+h) - \phi(t)}{h} &= \int_{\mathbb{R}} e^{itx} \frac{e^{ixh} - 1}{h} \, dF(x) \\ &= \int_{\mathbb{R}} e^{itx} \left(i \int_0^x e^{iuh} du \right) dF(x). \end{aligned}$$

The integrand is dominated by $|x|$, which is integrable by hypothesis. Thus by the dominated convergence theorem we may let $h \to 0$ under the integral. This shows

$$\phi'(t) = i \int_{\mathbb{R}} x e^{itx} \, dF(x), \qquad t \in \mathbb{R},$$

exists. The general case for $p > 1$ follows by induction, or by a similar direct argument. Note that $\phi^{(p)}(0) = i^p \int_{\mathbb{R}} x^p \, dF(x) = i^p \alpha_p$, where α_p is the pth moment of F, if α_p exists. This completes the proof.

To see that the first part cannot be strengthened, consider the example of a symmetric density f given by

$$f(x) = \begin{cases} 0, & |x| < 2 \\ K|x|^{-4}[\log|x|]^{-1}, & |x| \geq 2, \end{cases}$$

where

$$K = 2 \int_2^{\infty} \frac{dx}{x^4 \log x}.$$

Then it is not difficult to verify that $\phi^{(3)}(0)$ exists, and $\phi^{(3)}(0) = 0$. But it is clear that $E(|X|^3) = \int_{\mathbb{R}} |x|^3 f(x) \, dx = +\infty$. The details are left to the reader.

As a consequence of the above proposition, if a d.f. F has p finite moments then its ch.f. ϕ is p times continuously differentiable. Thus we can expand ϕ in a Taylor series around $t = 0$, and obtain the following:

Corollary 7 *Suppose F is a d.f. with p moments finite. If $\alpha_1, \ldots, \alpha_p$ are these moments, then the ch.f. ϕ of F can be expanded around $t = 0$, with either of the following two forms of its remainder:*

$$\phi(t) = 1 + \sum_{k=1}^{p} \alpha_k \frac{(it)^k}{k!} + o(|t|^p) \qquad (p \geq 1) \tag{19}$$

$$= 1 + \sum_{k=1}^{p-1} \alpha_k \frac{(it)^k}{k!} + \beta_p \theta_p \frac{|t|^p}{p!} \qquad (p > 1), \tag{19$'$}$$

where $\beta_p = \int_{\mathbb{R}} |x|^p \, dF(x), |\theta_p| \leq 1$, and $o(|t|) \to 0$ as $|t| \to 0$.

These expansions will be useful in some calculations for the weak limit laws in the next chapter. An immediate simple but useful observation is that if $\psi : \mathbb{R} \to \mathbb{R}$ is any continuous function which can be expanded in the form $\psi(t) = 1 + O(|t|^{2+\varepsilon}), \varepsilon > 0$, it will not be a ch.f. unless $\psi(t) = 1, t \in \mathbb{R}$, so that it is the ch.f. of the unitary distribution at the origin. Indeed, if $\psi(\cdot)$ is a ch.f. with such an expansion, by the above corollary, $\alpha_1 = \alpha_2 = 0$. If the second (even, or absolute) moment is zero, then the d.f. concentrates at the origin, proving our claim. In Section 4, we characterize functions which are ch.f.s, but this is a nontrivial problem, and in general there is no easy recipe for recognizing them.

To illustrate the power of ch.f.s we present an application characterizing the Gaussian law. If X_1, X_2 are two independent r.v.s with means μ_1, μ_2 and variances $\sigma_1^2, \sigma_2^2 > 0$, then $Y_i = (X_i - \mu_i)/\sigma_i, i = 1, 2$, are said to have a "reduced law," since $E(Y_i) = 0$ and $\operatorname{Var} Y_i = 1, i = 1, 2$. Note that Y_1, Y_2 are still independent. Now if X_i is Gaussian, $N(\mu_i, \sigma_i^2)$, then it is very easy to find that $E(X_i) = \mu_i$ and $\operatorname{Var} X_i = \sigma_i^2, i = 1, 2$. Hence the sum $X = X_1 + X_2$ is seen to be $N(\mu, \sigma^2)$, where $\mu = \mu_1 + \mu_2$ and $\sigma^2 = \sigma_1^2 + \sigma_2^2$. Indeed, using the ch.f.s we have

$$\begin{aligned}
\phi(t) &= E(e^{it(X_1+X_2)}) = E(e^{itX_1})E(e^{itX_2}) \quad \text{(by independence)} \\
&= \exp\left(i\mu_1 t - \frac{\sigma_1^2 t^2}{2}\right) \exp\left(i\mu_2 t - \frac{\sigma_2^2 t^2}{2}\right) \\
&= \exp\left(i\mu t - \frac{\sigma^2 t^2}{t}\right), \qquad t \in \mathbb{R}.
\end{aligned} \tag{20}$$

Hence $Z = (X - \mu)/\sigma$ is $N(0, 1)$. Thus if X_i are $N(\mu_i, \sigma_i^2)$, then they have the same reduced law as their sum. Does this property characterize the Gaussian law? The problem was posed and solved by G. Pólya. The effectiveness of ch.f.s will now be illustrated by the following result, the first part is based upon Pólya's above noted work and the second one is noted by Ibraginov and Linnik (1971), p. 32.

Proposition 8 (i) *Let X_1, X_2 be independent r.v.s with two moments. Then their reduced law is the same as that of their sum iff each X_i is Gaussian, $N(\mu_i, \sigma_i^2), i = 1, 2$.*

(ii) *Let X, Y be r.v.s with ch.f.s ϕ_X, ϕ_Y. If Y is normal $N(\mu, \sigma^2)$, and $\phi_X(u) = \phi_Y(u)$ for a bounded countable distinct set of real values $u_n, n \geq 1$ and $\phi_X(u_n) = \phi_Y(u_n), n \geq 1$, then $\phi_X(u) = \phi_Y(u)$ for all $u \in \mathbb{R}$ so that X is also $N(\mu, \sigma^2)$.*

Proof (i) If they are Gaussian, then (20) shows that the reduced law of the sum is of the same form. Thus only the converse is new. This is trivial if $\sigma_i^2 = 0$. Thus let $\sigma_i^2 > 0, i = 1, 2$.

Let Y_1, Y_2 be the reduced r.v.s from X_1, X_2 and Z be that of $X_1 + X_2$. By hypothesis, Y_1, Y_2 and Z have the same d.f., or equivalently the same ch.f., $= \phi$ (say). If ϕ_1, ϕ_2, and ϕ_3 are the ch.f.s of X_1, X_2, and $X_1 + X_2$, then we have the relations [since $X_i = \sigma_i Y_i + \mu_i, X_1 + X_2 = (\sigma_1^2 + \sigma_2^2)^{1/2} Z + (\mu_1 + \mu_2)$] :

$$\phi_1(t) = e^{i\mu_1 t}\phi(\sigma_1 t), \qquad \phi_2(t) = e^{i\mu_2 t}\phi(\sigma_2 t),$$

$$\phi_3(t) = e^{i(\mu_1+\mu_2)t}\phi((\sigma_1^2 + \sigma_2^2)^{1/2} t) = \phi_1(t)\phi_2(t), \qquad t \in \mathbb{R}.$$

This equation simplifies on substitution for $\phi_j(t)$ to

$$\phi(\sigma_1 t)\phi(\sigma_2 t) = \phi((\sigma_1^2 + \sigma_2^2)^{1/2} t), \qquad t \in \mathbb{R}. \tag{21}$$

It is the solution of this functional equation which answers our question.

To solve (21), let

$$\alpha = \sigma_1(\sigma_1^2 + \sigma_2^2)^{-1/2}, \qquad \beta = \sigma_2(\sigma_1^2 + \sigma_2^2)^{-1/2}.$$

Thus (21) becomes

$$\phi(t) = \phi(\alpha t)\phi(\beta t), \qquad t \in \mathbb{R}, \quad \alpha^2 + \beta^2 = 1, \qquad \alpha > 0, \quad \beta > 0. \tag{22}$$

Replacing t by αt and βt and iterating, we get

$$\begin{aligned}\phi(t) &= \phi(\alpha^2 t)[\phi(\alpha\beta t)]^2\phi(\beta^2 t) \\ &= \phi(\alpha^3 t)[\phi(\alpha^2\beta t)]^3[\phi(\alpha\beta^2 t)]^3\phi(\beta^3 t) \quad \text{[by using (22) again]}.\end{aligned}$$

Repeating the procedure for each term, we get at the nth stage

$$\phi(t) = \phi(\alpha^n t)^{p_0}[\phi(\alpha^{n-1}\beta t)]^{p_1} \cdots [\phi(\beta^n t)]^{p_n}, \tag{23}$$

where $p_0, p_1, \ldots, p_n$ are the coefficients in the binomial expansion of $(1+x)^n = \sum_{i=0}^n p_i x^i$, which are integers. Because $0 < \alpha < 1, 0 < \beta < 1$, it is clear that if

$u_0 = \alpha^n t, u_1 = \alpha^{n-1}\beta t, \ldots, u_n = \beta^n t$, then $u_k = \alpha^{n-k}\beta^k t \to 0$ uniformly in t (on compact sets) as $n \to \infty$. Also, since the d.f. F, whose ch.f. is ϕ, has two moments, it has mean zero, and variance one. Thus by (19) in a neighborhood of $t = 0$,

$$\phi(t) = 1 + 0 \cdot (it) + \frac{1}{2}(it)^2 + o(t^2). \tag{24}$$

Substituting (24) in (23)(with $t = u_k$ there), we get

$$\phi(t) = \left[1 - \frac{\alpha^{2n}t^2}{2} + o(\alpha^{2n}t^2)\right]^{p_0} \cdots \left[1 - \frac{\beta^{2n}t^2}{2} + o(\beta^{2n}t^2)\right]^{p_n}. \tag{25}$$

To simplify (25) further, we recall a standard fact from complex analysis, namely, in a simply connected set D, not containing 0, $\log z$ has a branch [$= f(z)$, say] and any other branch is of the form $f(z) + 2k\pi i$ (k an integer), where $f : z \mapsto f(z)$ is a continuous function. Here if $D = \phi([-t_0, t_0])$, it is connected, $0 \notin D$, and $z = \phi(t)$. For definiteness we take the principal branch (with $k = 0$). In this case, ϕ is also differentiable and we can expand it in a Taylor series. (We include a proof of a slightly more general version of the "complex logarithm" below, since that will be needed for other applications.)

Hence (25) becomes

$$\begin{aligned}\log \phi(t) &= \sum_{k=0}^{n} p_k \log[1 + \frac{1}{2}(\alpha^{n-k}\beta^k it)^2 + o(\alpha^{n-k}\beta^k t^2)^2] \\ &= -\frac{1}{2}t^2[p_0\alpha^{2n} + p_1\alpha^{2n-2}\beta^2 + \cdots + p_n\beta^{2n}] + R_n \\ &= -\frac{1}{2}t^2(\alpha^2 + \beta^2)^n + R_n,\end{aligned} \tag{26}$$

where

$$R_n = t^2[p_0\delta_0 u_0\alpha^{2n} + p_1\delta_1 u_1\alpha^{2n-2}\beta^2 + \cdots + p_n\delta_n u_n\beta^{2n}],$$

$|\delta_j| < 1$. Now for each $\varepsilon > 0$, we can choose an N such that $n > N \Rightarrow |\delta_j u_j| < \varepsilon/t^2$ for given $|t| > 0$. Since $\alpha^2 + \beta^2 = 1$, we get $|R_n| < \varepsilon$. Hence (26) for $n \geq N$ becomes

$$|\log \phi(t) + t^2/2| < |R_n| < \varepsilon. \tag{27}$$

Since the left side is independent of n, and $\varepsilon > 0$ is arbitrary,

$$\phi(t) = e^{-t^2/2}. \tag{28}$$

But (23) shows that (25), (26), and hence (27), are valid for any t, since for n large enough $\alpha^{n-k}\beta^k t$ is in a neighborhood of the origin. Thus the result holds for all $t \in \mathbb{R}$, and by the uniqueness theorem, (28) implies that ϕ is a Gaussian ch.f.

(ii) By the Bolzano-Weirstrass property of bounded sets in $\mathbb{R}$, there is a subsequence of u_n, denoted by the same symbols, with limit t. By subtracting this for the sequence, we can assume that $t = 0$, i.e., $u_n \to 0$ as $n \to \infty$ and $u_n \neq 0$. Since $\phi_Y(u) = e^{i\mu u - \frac{u^2}{2}\sigma^2}$, we consider two cases: (a) $\sigma = 0$ and (b) $\sigma > 0$, the degenerate and the general cases of Y. For (a), $\phi_X(u_n) = e^{i\mu u_n}, n \geq 1$, so that if F_X is the d.f. of X,

$$1 = e^{-i\mu u_n} \int_{\mathbb{R}} e^{iu_n x} dF_X(x) = \int_{\mathbb{R}} e^{iu_n(x-\mu)} dF_X(x), n \geq 1.$$

By the uniqueness theorem, F_X must have a jump at $x = \mu$, and it must be unitary by the fact that the u_n are distinct. So we only need to consider the case (b): $\sigma > 0$.

The idea now is to show that ϕ_X is an entire function which agrees with the exponential ϕ_Y at a sequence of distinct points u_n (with limit 0) and hence by complex function theory, the ch.f.s must be identical, which proves the result. Since $\psi(t) = \phi_X(t)\overline{\phi_X(t)}$ is a (real) symmetric ch.f. with values $\phi_Y(u_n)\overline{\phi(u_n)} = e^{-\frac{u_n^2}{2}} = (\phi_X(u_n))^2$, we may restrict $\phi_X(u)$ to be $= |\phi_Y(u)|$ at $u = u_n$, so that $\mu = 0$ and $\sigma^2 = 1$ by rescaling to establish $|\phi_X(t)| = e^{-\frac{t^2}{2}}$. This implies $\phi_X(t) = e^{-\frac{t^2}{2}} e^{i\theta t}$ for some $\theta \in \mathbb{R}$, and establishes the result.

Thus ϕ_X is (real and) symmetric, to use induction, consider

$$\phi_X(u_n) = \int_{\mathbb{R}} \cos u_n x dF_X(x) = e^{-\frac{u_n^2}{2}} = 1 - \frac{u_n^2}{2} + \frac{(\frac{u_n^2}{2})^2}{2!} - \cdots$$

This shows

$$\lim_{u_n \to 0} \frac{1 - \phi_X(u_n)}{u_n^2} = \frac{1}{2}, \text{ or } \int_{\mathbb{R}} u_n^2 x^2 \left(\frac{\sin \frac{u_n x}{2}}{\frac{u_n x}{2}} \right)^2 dF_X(x) \leq K_0 u_n^2, \quad (29)$$

for some $0 < K_0 < \infty$, and $|u_n| \leq \varepsilon$, $n \geq n_0$, K_0 being an absolute constant. Since $\frac{\sin x}{x}$ is bounded, and tends to 1 as $x \to 0$, we have for each $a > 0$

$$\int_{-a}^{a} x^2 \left(\frac{\sin \frac{u_n x}{2}}{\frac{u_n x}{2}} \right)^2 dF_X(x) \leq K_0, \quad n \geq n_0,$$

uniformly in $a > 0$ (for $|u_n| \leq \varepsilon$). This implies $\int_{-a}^{a} x^2 dF_X(x) < \infty$ and as $a \to \infty$, we get $\int_{-\infty}^{\infty} x^2 dF_X(x) < \infty$. But then by Proposition 6, $\phi_X''(\cdot)$ exists and (by symmetry of ϕ_X), $\phi_X'(0) = 0$. Moreover $\phi_X''(0) = \phi_Y''(0)$ holds.

Now using induction, let the result hold for all $m < r$ integers. Then

$$\phi_X^{2(r-1)}(0) - \phi_X^{2(r-1)}(u_n) = 2(-1)^{r-1} \int_{\mathbb{R}} x^{2(r-1)} \sin^2 \frac{u_n x}{2} dF_X(x)$$
$$= \phi_Y^{2(r-1)}(0) - \phi_Y^{2(r-1)}(u_n) = K_r u_n^2$$

where $K_r > 0$ is a finite number depending only on r. By the earlier reasoning this implies that $\int_{\mathbb{R}} v^{2r} dF_X(v) \leq K_r < \infty$, and hence by Proposition 6, $\phi_X^{(2r)}(\cdot)$ exists, and $\phi_X^{(2r)}(0) = \phi_Y^{(2r)}(0) = K_r$, and $|\phi_X^{(2r)}(u)| \leq |\phi_X^{(2r)}(0)| = |\phi_Y^{(2r)}(0)| \leq K_r$, $u \in \mathbb{R}$. Since $r \geq 1$ is arbitrary, and $\phi_Y(\cdot)$ is an entire function, this implies that $\phi_X(\cdot)$ is also an entire function and agrees with $\phi_Y(\cdot)$ on $u_n, n \geq 1$, $\phi_X^{(r)}(0) = \phi_Y^{(r)}(0)$. Hence by the classical theory $\phi_X = \phi_Y$ and the result follows. This completes the proof.

The above characterization shows how ch.f.s allow us to apply refined methods of complex analysis as well as differential calculus to solve problems of interest. Several others of a similar nature are treated later on.

In order not to interrupt our future treatment, we present the result for complex logarithms here in a form suitable for our work. This is really a simple exercise in complex analysis. [See Cartan (1963) for an extended treatment of the topic.] When the problem was discussed briefly by Loève (cf. his book (1963) p. 291 ff) some reviewers were upset and expressed that the comments were inadequate. Here we present a more detailed view, essentially following Tucker ((1967), p. 93) who witnessed Loève's lectures. An alternative method will also be included.

Proposition 9 *Let $f : \mathbb{R} \to \mathbb{C} - \{0\}$ be a continuous function so that $f(t) \neq 0, t \in \mathbb{R}$ and suppose $f(0) = 1$. Then there exists a unique continuous $g : \mathbb{R} \to \mathbb{R}$ such that $f(t) = |f(t)| \exp(ig(t)) = e^{h(t)}$ where $h(t) = \text{Log } f(t) = \log|f(t)| + i \arg f(t)$ and $g(t) = \arg f(t)$ (or $h(t)$) is continuous in t. For instance, let $f(0) = 1$, so that one can demand $g(0) = 0$ and then Log $f = \log|f(t)| + ig(t)$ is uniquely defined.*

Proof We present the argument in the form $f(t) = |f(t)|e^{ig(t)}$ for a unique continuous $g : \mathbb{R} \to \mathbb{R}, g(0) = 0$. Observe that if such a $g(\cdot)$ exists, then it is unique. Indeed, if g_1, g_2 are two functions here they differ by an integer multiple of 2π so that $g_1(t) = g_2(t) + 2\pi k(t), k : \mathbb{R} \to \mathbb{R}$ is continuous. But $g_1(0) = g_2(0) = 0$ implies $k(0) = 0$, and being integer-valued and continuous, this forces $k(t)$ to vanish and hence $g_1 = g_2 (= g$, say).

To show that such a g exists, consider $a > 0$ and since $f(t) \neq 0$, by hypothesis, $|f| : A \to \mathbb{R}^+$ is bounded and continuous on each compact set $A \subset \mathbb{R}$, and is strictly positive. Replacing f by $\frac{f}{|f|}$, we may assume that $|f(t)| = 1$. Taking $A = [0, a]$, we find $\min\{|f(t)| : 0 \leq t \leq a\} = b$ and by this reduction, $b = 1$. Since f is uniformly continuous on A, for $\varepsilon = 1$ there is a $\delta > 0$ such that for $t_1, t_2 \in A = [0, a], |t_1 - t_2| < \delta \Rightarrow |f(t_1) - f(t_2)| < 1$. Since $|f(t)| = 1$, we make the key observation that $\left|\arg\left(\frac{f(t_1)}{f(t_2)}\right)\right| < \frac{\pi}{2}$, where $\arg \rho$ stands for argument of ρ (i.e., $\tan^{-1}\left(\frac{\text{Im}(\rho)}{\text{Re}(\rho)}\right)$ for $\rho \in \mathbb{C} - \{0\}$). Consider a partition $0 = t_0 < t_1 < \cdots < t_n = a$ of $[0, a]$ such that $\max_i(t_i - t_{i-1}) < \delta$. Set $g(t) = \arg f(t), 0 = t_0 \leq t \leq t_1$. Then g is continuous on $[0, t_1]$ since $\text{Im}|f|$

and $\text{Re}(f)$ are continuous on $[0, t_1]$ and $f(t) \neq 0$ on $[0, t_1]$. Now extend g by defining inductively as:

$$g(t) = g(t_i) + \tan^{-1}\left(\frac{\text{Im}(f(t))}{\text{Re}(f(t_i))}\right), \quad i = 1, \ldots, n.$$

Then $g(0) = 0$, g is continuous and is the required function. Since $\mathbb{R}$ is σ-compact we may extend it first to $[-m, m]$ and then to $\bigcup_{m>0}[-m, m]$ continuously. If we let $h(t) = \log|f(t)| + ig(t)$ then $\text{Log} f(t) = h(t)$ is the uniquely defined continuous logarithm, as desired. [Note that $h : \mathbb{R} \to \mathbb{C}$ is a function of t and is determined by f (not a function of f which is complex valued!)]

Alternative Proof (à là K.L. Chung (1968), p. 241) The argument is based on the MacLaurin expansion of the log function $\log z$ about 1, and is slightly longer, and is as follows.

$$\text{Log } z = \sum_{j=1}^{\infty} \frac{(-1)^{j-1}}{j}(z-1)^j, \quad |z-1| < 1$$

at $z = 1$. If $h(t) = \text{Log} f(t)$ here, then $h(0) = 0$ and $h(\cdot)$ is continuous. Let $a > 0$, and for $\varepsilon = \frac{1}{2}$, we find a $\delta > 0$ and a partition of $[-a, a]$, $-a = t_m < t_{-m+1} < \cdots < t_0 < t_1 < \cdots < t_m = a, t_{j+1} - t_j = t_1 - t_0 < \delta$, such that $|f(t_1) - f(t)| \leq \frac{1}{2}$ for $t \in [t_{-1}, t_1]$, and $|f(t) - 1| = |f(t) - f(0)| \leq \frac{1}{2}$. Hence $\text{Log } f = h$ is well-defined by the series with $z = f(t), t \in [t_{-1}, t_1], h(0) = 0$. As a power series, $h(\cdot)$ is continuous and $|f(t) - f(t_i)| = |f(t_i)||\frac{f(t)}{f(t_i)} - 1| \leq \frac{1}{2}, t \in [t_i, t_{i+1}]$, since $|f(t_i)| = 1$. We may extend h onto $[t_i, t_{i+1}]$ by setting

$$h(t) = h(t_i) + \text{Log}\left(\frac{f(t)}{f(t_i)}\right),$$

and by iteration for all $t \in [-a, a]$. Then we have $[f(t_0) = 1$ and $t_i \leq t \leq t_{i+1}]$

$$e^{h(t)} = e^{\text{Log}\frac{f(t)}{f(t_i)} + h(t_i)} = e^{\text{Log}\frac{f(t)}{f(t_i)} + \sum_{j=0}^{i-1} \text{Log}\frac{f(t_{j+1})}{f(t_j)}} = f(t).$$

In the same way it can be extended to the left so that $h(\cdot)$ is defined on $[-a, a]$. As before, we can iterate the procedure to $\mathbb{R} = \bigcup_{n \geq 0}[-n, n]$, by σ-compactness. The uniqueness is, as before, immediate and the result (the unique continuous representation of $\text{Log } f$) follows.

Remark In the first edition of this book, the proof given is analogous to the first one above. But an anonymous reader indicated that the series argument is superior to the first one. So we included both the methods of proof that might appeal to a wider audience.

A useful consequence of the above proposition is given by

Corollary 10 *Let $\{f, f_n, n \geq\}$ be a sequence of ch.f.s such that $f_n(t) \to f(t)$ as $n \to \infty, t \in \mathbb{R}$. If f and f_n do not vanish anywhere, then* $\mathrm{Log}\, f_n(t) \to \mathrm{Log}\, f(t), t \in \mathbb{R}$.

Proof Since f_n, f do not vanish, by the above proposition Log f_n and Log f exist and

$$\mathrm{Log}\, f_n = \log|f_n| + i \arg f_n, \qquad \mathrm{Log}\, f = \log|f| + i \arg f, \qquad n \geq 1.$$

The hypothesis implies $|f_n|(t) \to |f|(t), t \in \mathbb{R}$. Since these are never zero, and on their ranges log is continuous, $\log|f_n| \to \log|f|$. Similarly, using the fact that $\arg(\cdot)$ is a continuous function and $f_n \to f$, the composition $\arg f_n(t) \to \arg f(t), t \in \mathbb{R}$. Hence $\mathrm{Log}\, f_n(t) \to \mathrm{Log}\, f(t), t \in \mathbb{R}$, as asserted.

Remarks (1) The ch.f.s $f_n \to f$ pointwise implies that the convergence is also uniform on compact sets of $\mathbb{R}$. Since these functions are assumed nonvanishing and $f_n(A), f(A)$ are compact sets for any compact $A \subset \mathbb{R}$, and since the Log function is uniformly continuous on compact sets of its domain, we can strengthen the conclusion of the corollary to Log $f_n \to$ Log f uniformly on compact subsets of $\mathbb{R}$. The details of this statement are left to the reader.

(2) Hereafter the unique logarithm given by the last part will be termed a *distinguished logarithm* of f denoted Log f. The same argument also shows that $\mathrm{Log}\, \phi_1\phi_2 = \mathrm{Log}\, \phi_1 + \mathrm{Log}\, \phi_2$ and $\mathrm{Log}\, \phi_1\phi_2^{-1} = \mathrm{Log}\, \phi_1 - \mathrm{Log}\, \phi_2$ for nonvanishing ϕ_1, ϕ_2 as in Proposition 9. This fact will be used without comment.

As the above two results indicate, the combination of Fourier transform theory, complex analysis, and the special ideas of probability make it a very fertile area for numerous specializations. We do not go into these here; a few of these results are given in the problems section.

4.3 Cramér's Theorem on Fourier Transforms of Signed Measures

If F_1 and F_2 are two d.f.s, then $G = F_1 - F_2$ is of bounded variation and determines a signed measure. In a number of applications it is useful to have an extension of the inversion formula (cf. Theorem 2.1) for such functions as G above. An interesting result of this nature was given by H. Cramér (1970). We present it here and include some consequences.

Theorem 1 *Let $G : \mathbb{R} \to \mathbb{R}$ be a function of bounded variation such that $\lim_{|x|\to\infty} G(x) = 0$ and $\int_{\mathbb{R}} |x||dG(x)| < \infty$. If $g(t) = \int_{\mathbb{R}} e^{itx}\, dG(x)$ and $0 < \alpha < 1$, then for each $h > 0$ and $x \in \mathbb{R}$ we have*

$$\int_x^{x+h} (u-x)^{\alpha-1} G(u)\, du = -\frac{1}{2\pi i} \int_{\mathbb{R}} \frac{g(t)}{t} e^{-itx} \int_0^h u^{\alpha-1}\, du\, dt. \tag{1}$$

If in addition $\int_{\mathbb{R}} |g(t)/t|\, dt < \infty$, then it follows that

$$G(x) = -\frac{1}{2\pi i} \int_{\mathbb{R}} \frac{g(t)}{t} e^{-itx}\, dt. \tag{2}$$

Proof We establish (1) for $x \in \mathbb{R}, h > 0$ such that x and $x + h$ are points of continuity of G. The latter set is dense in $\mathbb{R}$. This will prove (1) since both sides of (1) are continuous in h and x for each $0 < \alpha < 1$. To see this, let $(u - x)/h = v$ in the left side of (1). It becomes

$$\int_0^1 v^{\alpha-1} G(x + vh) h^\alpha\, dv \to \int_0^1 v^{\alpha-1} G(x_0 + vh) h^\alpha\, dv$$

as $x \to x_0$, since $G(x + vh) \to G(x_0 + vh)$ a.e. (Lebesgue) as $x \to x_0$. Also, the left side tends to zero as $h \to 0$, by the bounded convergence. Hence it is continuous in x and h. For the right side of (1), since $G = G_1 - G_2$, where G_i are bounded and nondecreasing with one moment existing, their Fourier transforms are differentiable by Proposition 2.8, and hence so is $g(t)$. Since $g(0) = 0$, this implies $g(t) = tg'(0) + o(t)$, so that $g(t)/t = O(1)$ as $t \to 0$. Regarding the last integral,

$$\left| \int_0^h u^{\alpha-1} e^{-itu}\, du \right| \le \frac{1}{|t|^\alpha} \int_0^{h|t|} |v^{\alpha-1} e^{-iv}|\, dv = \frac{h^\alpha}{\alpha}.$$

Hence the integral on the right of (1) as a function of t is absolutely convergent uniformly relative to x and h, and is a continuous function of x and h.

Thus let $x, x+h$ be continuity points of G. Proceeding as in Theorem 2.1, consider, for any $T > 0$, a simplification of the right side by substitution:

$$\begin{aligned} &-\frac{1}{2\pi i} \int_{-T}^T \frac{g(t)}{t} e^{-itx} \int_0^h u^{\alpha-1} e^{-itu}\, dt \\ &= \frac{-1}{2\pi i} \int_{-T}^T \frac{e^{-itx}}{t} \int_{\mathbb{R}} e^{ity}\, dG(y) \int_0^h u^{\alpha-1} e^{-itu}\, du\, dt \\ &= \frac{-1}{2\pi} \int_{\mathbb{R}} dG(y) \int_0^h u^{\alpha-1}\, du \int_{-T}^T \frac{\sin t(y-x-u)}{t}\, dt \\ &\text{[by Fubini's theorem].} \end{aligned} \tag{3}$$

With the bounded convergence theorem applied to (3) one gets

$$\begin{aligned}
&\lim_{T\to\infty} \text{ (left side of (3))} \\
&= -(1/2)\int_{\mathbb{R}} dG(y)\int_0^h u^{\alpha-1}\,\mathrm{sgn}(y-x-u)\,du \\
&\quad\left(\text{since}(2/\pi)\int_0^\infty (\sin t\alpha)/t\,dt = \mathrm{sgn}\ \alpha\right) \\
&= -\frac{1}{2}\left[\int_{-\infty}^{x} + \int_x^{x+h} + \int_{x+h}^{\infty} dG(y)\int_0^h u^{\alpha-1}\,\mathrm{sgn}(y-x-u)\,du\right] \\
&= -(1/\alpha)\int_x^{x+h}(y-x)^\alpha\,dG(y) + (h^\alpha/\alpha)G(x+h) \\
&\quad\text{(after simplification)} \\
&= \int_x^{x+h}(y-x)^{\alpha-1}G(y)\,dy \qquad \text{(with integration by parts).}
\end{aligned}$$

This establishes (1) using initial simplification from the first paragraph.

For the last part, we may differentiate both sides of (1) relative to h and get, by the fundamental theorem of calculus,

$$h^{\alpha-1}G(x+h) = -\frac{1}{2\pi i}\int_{\mathbb{R}}\frac{g(t)}{t}e^{-it(x+h)}h^{\alpha-1}\,dt,$$

the interchange of differential and integral being easily justified. Cancelling $h^\alpha > 0$, and replacing $x+h$ by x in the above, we get (2), and the theorem is proved.

The above result is of interest in calculating the distributions of ratios of r.v.s and asymptotic expansions (cf. Theorem 5.1.5). The former, also due to Cramér, is as follows.

Theorem 2 *Let X_1, X_2 be two r.v.s with finite expectations. If $P[X_2 > 0] = 1$ and ϕ is the joint ch.f., so that $\phi(t,u) = E(e^{itX_1+iuX_2})$, suppose that $\phi(t,u) = O((|t|+|u|)^{-\delta})$ for some $\delta > 0$ as $|t|+|u| \to \infty$. Then the distribution F of the ratio X_1/X_2 is given by*

$$F(x) = \frac{1}{2\pi i}\int_{\mathbb{R}}\frac{\phi_2(t)-\phi(t,-tx)}{t}\,dt, \qquad x\in\mathbb{R}, \tag{4}$$

where ϕ_2 is the ch.f. of X_2. If, further,

$$\int_{\mathbb{R}}\left|\frac{\partial\phi}{\partial u}(y,-tx)\right|\,dt < \infty$$

uniformly in x, then the density F' exists and is given by

$$f(x) = F'(x) = \frac{1}{2\pi i} \int_{\mathbb{R}} \frac{\partial \phi}{\partial u}(t, -tx)\, dt, \qquad x \in \mathbb{R}. \tag{5}$$

Proof If $F_2(x) = P[X_2 < x]$, x fixed, then by hypothesis $F_2(0) = 0$, and

$$F(x) = P[X_1 - xX_2 < 0].$$

Let $Y_x = X_1 - xX_2$ and $H(y) = P[Y_x < y]$, so that $F(x) = H(0) - F_2(0)$. If $G(y) = H(y) - F_2(y)$, then G satisfies the hypothesis of Theorem 1. The further condition on ϕ ensures that for $M > 0$,

$$\begin{aligned} \phi(t, -tx) - 1 &= i(E(X_1) - xE(X_2)t + o(t), \qquad & t \to 0, \\ \phi_2(t) - 1 &= iE(X_2)t + o(t), & t \to 0, \end{aligned} \tag{6}$$

and as $|t| \to \infty, |\phi(t, -tx)| \leq M|t|^{-\delta}$ for $x \in \mathbb{R}$. These rates of growth are utilized as follows.

If $h(t) = E(e^{itY_x})$, then $h(t) = \phi(t, -tx)$, and the Fourier transform g of G satisfies $g(t) = h(t) - \phi_2(t)$. Hence Eq. (6) shows that

$$\int_{[|t| \leq M]} |g(t)/t|\, dt < \infty,$$

and the given asymptotic rate growth implies $\int_{[|t|>M]} |g(t)/t|\, dt < \infty$. Thus we can apply the second part of Theorem 1 to get

$$G(y) = \frac{1}{2\pi i} \int_{\mathbb{R}} \frac{e^{-ity}}{t} [\phi_2(t) - \phi(t, -tx)]\, dt, \qquad x \in \mathbb{R}, \quad y \in \mathbb{R}. \tag{7}$$

Since $G(0) = F(x)$, (7) gives (4) for $y = 0$. From this (5) is obtained by differentiation using the additional hypothesis. This completes the proof.

In case X_1 and X_2 are independent r.v.s, the above result simplifies as follows.

Corollary 3 *Let X_1, X_2 be independent r.v.s with finite means and ch.f.s ϕ_1 and ϕ_2. If $P[X_2 > 0] = 1$ and $\int_{[|t| \geq M]} |\phi_2(t)/t| dt < \infty$ for some $M > 0$, then we have*

$$F(x) = P[(X_1/X_2) < x] = \frac{1}{2\pi i} \int_{\mathbb{R}} [\phi_2(t) - \phi_1(t)\phi_2(-tx)] \frac{dt}{t}, \quad x \in \mathbb{R}. \tag{8}$$

Moreover, (8) has a density f given by

$$f(x) = \frac{1}{2\pi i} \int_{\mathbb{R}} \phi_1(t) \frac{d\phi_2}{dt}(-tx)\, dt, \tag{9}$$

provided the integral exists uniformly relative to x in compact intervals.

Proof Since $\phi(t,u) = \phi_1(t)\phi_2(u)$, the result follows from (4) because the integrability conditions of (6) are satisfied as $t \to 0$, and for $|t| \geq M > 0$, we have

$$\left|\frac{\phi_2(t) - \phi_1(t)\phi_2(-tx)}{t}\right| \leq 2\left|\frac{\phi_2(t)}{t}\right|.$$

The current hypothesis implies that this is integrable on $|t| \geq M$. The rest is exactly as in the theorem, and the result holds as stated.

Let us present two applications of these results to show their utility.

Example 4 Let X, Y be independent $N(0,1)$ r.v.s and $Z = |Y|$. Then

$$\begin{aligned} P[Z < z] &= P[-z < Y < z] \\ &= (1/\sqrt{2\pi})\left[\int_{-\infty}^{z} \exp\{-x^2/2\}\,dx - \int_{-\infty}^{-z} \exp\{-x^2/2\}\,dx\right], \quad z > 0, \end{aligned}$$

and vanishes otherwise. Thus its density function is $f_z(z) = \sqrt{(2/\pi)}e^{-z^2/2}$. Since X, Z are independent, consider the distribution of the ratio X/Z. The hypothesis of Corollary 3 is easily seen to hold here. Now if ϕ_1, ϕ_2 are ch.f.s of X and Z, then $\phi_1(t) = e^{-t^2/2}$ and

$$\phi_2(t) = \int_0^\infty e^{itz} f_Z(z)\,dz, \quad \phi_2'(t) = \int_0^\infty ize^{itz} f_Z(z)\,dz.$$

Thus (9) gives the desired density, and we have

$$\begin{aligned} f(x) &= \frac{1}{2\pi i}\int_{\mathbb{R}} e^{-t^2/2} \int_0^\infty ize^{-itxz}\sqrt{\frac{2}{\pi}}\,e^{-z^2/2}\,dz\,dt \\ &= \frac{1}{\pi}\int_0^\infty ze^{-z^2/2}\int_{\mathbb{R}} e^{-itxz-t^2/2}\frac{dt}{\sqrt{2\pi}}\,dz \quad \text{(by Fubini's theorem)} \\ &= \frac{1}{\pi}\int_0^\infty ze^{-z^2/2}e^{-x^2z^2/2}dz = \frac{1}{\pi(1+x^2)}, \quad x \in \mathbb{R}. \end{aligned}$$

Thus X/Z has a Cauchy distribution. [Alternatively, $Z = X/\sqrt{Y^2}$ and Y^2 has the chi-square distribution with parameter =1 (i.e., of one degree of freedom). Hence Z is distributed as student's t—with one degree of freedom, (as in Exercise 8), which is Cauchy. This argument did not use Corollary 3. With change of variables, one can also show that X/Y has a Cauchy distribution to which Corollary 3 is not applicable!]

Example 5 Consider a pair of dependent r.v.s X_1, X_2 whose joint d.f. has the ch.f. ϕ given by ($P[X_2 > 0] = 1, X_2$ being a gamma r.v.):

$$\phi(t,u) = (1 - 2iu + t^2)^{-1/2}. \tag{10}$$

It is not evident that ϕ is a ch.f. But it arises as a limit of a sequence of ch.f.s and is continuous. [See Equation (5.6.21) latter.] Now it is easily verified that this satisfies the hypothesis of the last part of Theorem 2, and hence the density of the ratio of X_1, X_2 is given as

$$\begin{aligned} f(x) &= \frac{1}{2\pi i}\int_{\mathbb{R}} \frac{\partial\phi}{\partial u}(t, -tx)\, dt \\ &= \frac{1}{2\pi i}\int_{\mathbb{R}} \frac{i\, dt}{[1+t^2+2itx]^{3/2}} \\ &= \frac{1}{2\pi}\int_{\mathbb{R}} d\left(\frac{t+ix}{(1+x^2)\sqrt{1+t^2+2itx}}\right) \end{aligned}$$

(since the integrand above has this primitive and so no contour integration is needed)

$$\begin{aligned} &= \frac{1}{2\pi}\lim_{T\to\infty}\left[\frac{1+(ix/T)}{(1+x^2)([1+T^2+2iTx]/T^2)^{1/2}}\right. \\ &\qquad \left. - \frac{1-(ix/T)}{(1+x^2)(-1)([1+T^2-2iTx]/T^2)^{1/2}}\right] \\ &= \frac{1}{2\pi}\frac{2}{1+x^2} = \frac{1}{\pi}\frac{1}{1+x^2}, \qquad x \in \mathbb{R}. \end{aligned}$$

Thus the distribution of the ratio of two dependent r.v.s, neither of which is $N(0,1)$, is again Cauchy.

Other applications and extensions of the above theorems will be seen to have interest in our work.

4.4 Bochner's Theorem on Positive Definite Functions

We present here a fundamental characterization of ch.f.s due to S. Bochner who established it in 1932. There are at least three different methods of proof of this result. One is to base the argument on the continuity theorem for ch.f.s. The second is to obtain the result by a careful extension of an earlier (special) result of Herglotz for the discrete distributions. The third one is first to establish the result for functions in $L^2(\mathbb{R})$ and then to use it to obtain the general case for all d.f.s. Of these the first two methods are probabilistic in nature and the third one is more Fourier analytic in content. We therefore give only a probabilistic proof, which is due to Cramér.

Definition 1 Let $f : \mathbb{R} \to \mathbb{C}$ be a mapping and $t_1, \ldots, t_n$ be points in $\mathbb{R}$. Then f is said to be *positive definite* if for any $a_i \in \mathbb{C}, i = 1, \ldots, n$, we have

$$\sum_{i=1}^{n} \sum_{j=1}^{n} a_i \bar{a}_j f(t_i - t_j) \geq 0, \tag{1}$$

where $\overline{a}_j$ is the complex conjugate of a_j.

The fundamental result alluded to above is the following:

Theorem 2 (Bochner) *A continuous $\phi : \mathbb{R} \to \mathbb{C}$ with $\phi(0) = 1$ is a ch.f. iff it is positive definite.*

Proof The necessity is simple and classical, due to M. Mathias who observed it in 1923. Thus if ϕ is the ch.f. of a d.f. F, then ϕ is continuous, $\phi(0) = 1$, and

$$\begin{aligned}\sum_{j=1}^{n} \sum_{k=1}^{n} a_j \bar{a}_k \phi(t_j - t_k) &= \int_{\mathbb{R}} \sum_{j=1}^{n} \sum_{k=1}^{n} a_j \bar{a}_k e^{i(t_j - t_k)\lambda} \, dF(\lambda) \\ &= \int_{\mathbb{R}} \left| \sum_{j=1}^{n} a_j e^{it_j\lambda} \right|^2 dF(\lambda) \geq 0.\end{aligned}$$

Thus ϕ is positive definite.

For the converse, let ϕ be positive definite, continuous, and $\phi(0) = 1$. If $f : \mathbb{R} \to \mathbb{C}$ is continuous, then for any $T > 0$, (known as the integral form of positive definiteness)

$$\int_0^T \int_0^T \phi(s - t) f(s) \overline{f(t)} \, ds \, dt \geq 0, \tag{2}$$

since the integral exists on the compact set $[0, T] \times [0, T]$, and is a limit of the finite (Riemann) sums $\sum_i \sum_j \phi(t_i - t_j) f(t_i) f(t_j) \Delta t_i \Delta t_j$. But the latter is nonnegative by the fact that ϕ is positive definite. Now let $f(u) = e^{-iux}$. We define

$$p_T(x) = (1/2\pi T) \int_0^T \int_0^T \phi(u - v) e^{-i(u-v)x} \, du \, dv \geq 0 \quad [\text{by } (2)].$$

Make the change of variables $t = u - v, \tau = v$ in the above. Then after a slight simplification one gets

$$\begin{aligned} p_T(x) &= \frac{1}{2\pi T} \int_{-T}^{T} \int_{|t|}^{T} e^{-itx} \phi(t)\, d\tau\, dt \\ &= \frac{1}{2\pi} \int_{-T}^{T} e^{-itx} \left(1 - \frac{|t|}{T}\right) \phi(t)\, dt \geq 0, \end{aligned} \tag{3}$$

so that

$$p_T(x) = \int_{\mathbb{R}} e^{-itx} \phi_T(t)\, dt,$$

where $\phi_T(t) = [1 - (|t|/T)]\phi(t)$ for $|t| \leq T$, and $= 0$ for $|t| > T$. We now claim (i) p_T is a probability density on $\mathbb{R}$ and (ii) ϕ_T is its ch.f., for each T. These two points establish the result since $\phi_T(t) \to \phi(t)$ as $T \to \infty$ for each $t \in \mathbb{R}$ and $\phi(\cdot)$ is continuous on $\mathbb{R}$ by hypothesis. Thus by the continuity theorem for ch.f.s, ϕ must also be a ch.f. We thus need to establish (i) and (ii).

(i) The verification of this point involves an interesting trick. Consider the increasing sequence of functions ψ_N defined by $\psi_N(x) = [1 - (|x|/N)]$ if $|x| \leq N$, and $= 0$ for $|x| > N$. Then $\psi_N p_T : \mathbb{R} \to \mathbb{R}^+$ is continuous, and $\psi_N(x) \nearrow 1$ as $N \to \infty$, for each $x \in \mathbb{R}$. Hence

$$\begin{aligned} \int_{\mathbb{R}} p_T(x)\, dx &= \lim_{N\to\infty} \int_{\mathbb{R}} \psi_N(x) p_T(x)\, dx \\ &\quad \text{(by the monotone convergence theorem)} \\ &= \lim_{N\to\infty} \int_{\mathbb{R}} \psi_N(x) \frac{1}{2\pi} \int_{\mathbb{R}} e^{-itx} \phi_T(t)\, dt\, dx \\ &= \lim_{N\to\infty} \int_{\mathbb{R}} \phi_T(t)\, dt \frac{1}{2\pi} \int_{-N}^{N} \left(1 - \frac{|x|}{N}\right) e^{-itx}\, dx \quad \text{[by Fubini]} \\ &= \lim_{N\to\infty} \int_{\mathbb{R}} \phi_T(t) \frac{1}{2\pi} \frac{(\sin(tN/2))^2}{t^2 N/4}\, dt \\ &= \lim_{N\to\infty} \frac{1}{\pi} \int_{\mathbb{R}} \phi_T\left(\frac{2v}{N}\right) \frac{\sin^2 v}{v^2}\, dv \\ &= \frac{1}{\pi} \int_{\mathbb{R}} \phi_T(0) \frac{\sin^2 v}{v^2}\, dv = 1, \qquad \left[\text{since} \int_{\mathbb{R}} \frac{sin^2 v}{v^2}\, dv = \pi\right] \end{aligned} \tag{4}$$

where we used the dominated convergence theorem to move the limit inside the integral [since $(\sin v/v)^2$ is integrable] and then the continuity of ϕ at $t = 0$, plus the fact that $\phi(0) = 1$, implying $\phi_T(0) = 1$, and $|\phi(t)| \leq 1$ (cf. Proposition 3 below.)

(ii) To see that ϕ_T is the ch.f. of p_T, we use the just established integrability of p_T, and the dominated convergence to conclude

$$\begin{aligned}
&\int_{\mathbb{R}} e^{itx} p_T(x)\,dx \\
&\quad = \lim_{N\to\infty} \int_{\mathbb{R}} e^{itx}\psi_N(x) p_T(x)\,dx \quad [\psi_N(\cdot) \text{ is defined above for (4)}] \\
&\quad = \lim_{N\to\infty} \frac{1}{2\pi} \int_{\mathbb{R}} \psi_N(x) \int_{\mathbb{R}} e^{-i(u-t)x} \phi_T(u)\,du\,dx \\
&\quad = \lim_{N\to\infty} \frac{1}{2\pi} \int_{\mathbb{R}} \phi_T(u) \left(\frac{2\sin\frac{1}{2}(t-u)N}{(t-u)N}\right)^2 \frac{du}{N} \\
&\quad\quad \text{(by integration and Fubini's theorem)} \\
&\quad = \lim_{N\to\infty} \frac{1}{\pi} \int_{\mathbb{R}} \phi_T\left(t - \frac{2v}{N}\right)\left(\frac{\sin v}{v}\right)^2 dv. \\
&\quad = \phi_T(t) \quad \text{(by the dominated convergence, as in (i)} \\
&\quad\quad \text{and the continuity of } \phi_T \text{ at } t).
\end{aligned}$$

But the left side is a ch.f., and hence ϕ_T is the ch.f. of p_T. This proves (ii), and the theorem follows.

There is actually a redundancy in the sufficiency part of the hypothesis. The positive definiteness condition is so strong that mere continuity at $t = 0$ of ϕ implies its uniform continuity on $\mathbb{R}$. Let us establish this and some related properties.

Proposition 3 *If $\phi : \mathbb{R} \to \mathbb{C}$ is positive definite, then (i) $\phi(0) \geq 0$, (ii) $\phi(-t) = \overline{\phi}(t)$, (iii) $|\phi(t)| \leq \phi(0)$, (iv) $\overline{\phi}(\cdot)$ is positive definite, and (v) if ϕ is continuous at $t = 0$, then it is uniformly continuous on $\mathbb{R}$.*

Proof If $b_{ij} = \phi(t_i - t_j)$ and $a = (a_1, \ldots, a_n)^t$, then (1) implies that the matrix $B = (b_{ij})$ is positive definite; i.e., using the inner product notation and t for conjugate transposition of a vector or matrix and B^* for the adjoint of B, we get by positive definiteness:

$$(Ba, a) = (a, B^*a) = \overline{(B^*a, a)} = (B^*a, a) \geq 0, \tag{5}$$

But by the polarization identity of a complex inner product, we have for any pair of n-vectors, a, b

$$\begin{aligned}
4(Ba, b) &= (B(a+b), a+b) - (B(a-b), a-b) \\
&\quad + i(B(a+ib), a+ib) - i(B(a-ib), a-ib), \\
&= 4(B^*a, b) \quad [\text{by (5)}].
\end{aligned}$$

Hence $(Ba, b) = (B^*a, b)$ for all a, b, so that $B = B^*$. Taking $a = b = (1, 0, \ldots, 0)^t$, we get $\phi(0) \geq 0$. This gives (i) and (ii). The positive definiteness of B implies that each of its principal minors has a positive determinant. Thus $|\phi(t)|^2 \leq \phi(0)^2$, which is (iii). That (iv) is true follows from (1) itself (or also from (5)).

For (v) consider the third order principal minor B_3 of B. Excluding the trivial case that $\phi(0) = 0$, we may and do normalize: $\phi(0) = 1$ [since otherwise $\tilde{\phi}(t) = \phi(t)/\phi(0)$ will satisfy the conditions]. Then the determinant of B_3 is

$$\det(B_3) = 1 - |b_{12}|^2 - |b_{13}|^2 - |b_{23}|^2 + 2 \operatorname{Re}(\bar{b}_{12} b_{13} b_{23}) > 0. \tag{6}$$

Taking $t_1 = 0, t_2 = t$, and $t_3 = t'$, so that $b_{12} = \phi(-t), b_{13} = \phi(-t')$, and $b_{23} = \phi(t - t')$ in (6), we get

$$\begin{aligned} |\phi(t) - \phi(t')|^2 &= |\phi(t)|^2 + |\phi(t')|^2 - 2 \operatorname{Re}(\phi(t)\overline{\phi(t')}) \\ &\leq 1 - |\phi(t - t')|^2 + 2 \operatorname{Re}[\overline{\phi(-t)}\phi(-t')\phi(t - t') \\ &\quad - \phi(t)\overline{\phi(t')}] \quad [\text{by } (6)] \\ &\leq 1 - |\phi(t - t')|^2 + 2|\phi(t)\overline{\phi(t')}|\,|\phi(t - t') - 1|, \end{aligned}$$

which tends to zero uniformly as $t \to t'$. This establishes (v), and the proof is complete.

In general, it is not easy to determine whether a bounded continuous function on $\mathbb{R}$ is a ch.f. The proof of Theorem 2 contains information for the following result due to Cramér, which may be easier to verify in some cases.

Proposition 4 *A bounded continuous function $\phi : \mathbb{R} \to \mathbb{C}$, with $\phi(0) = 1$, is a ch.f. iff for all $T > 0$,*

$$p_T(x) = \int_0^T \int_0^T \phi(u - v) e^{i(u-v)x}\, du\, dv \geq 0, \qquad x \in \mathbb{R}.$$

The proof is essentially the same as that of Theorem 2 with simple alterations and is left to the reader.

In view of Proposition 3, it is natural to ask whether the continuity hypothesis can be eliminated from Theorem 2. F. Riesz in 1933 has shown that this is essentially possible, but we get a slightly weaker conclusion. A precise version of this statement is as follows.

Theorem 5 (Riesz) *Let $\phi : \mathbb{R} \to \mathbb{C}$ be a Lebesgue measurable mapping satisfying $\phi(0) = 1$. Then ϕ is positive definite iff it coincides with a ch.f. on $\mathbb{R}$ outside of a Lebesgue null set.*

Proof Suppose ϕ is positive definite, Lebesgue measurable, and $\phi(0) = 1$. We reduce the argument to that of Theorem 2, by the following device due to Riesz himself. Let $t_1, \ldots, t_n$ be n arbitrary points from $\mathbb{R}$. By hypothesis

$$\sum_{j=1}^{n}\sum_{k=1}^{n}\phi(t_j - t_k)e^{it_jx}e^{-it_kx} \geq 0, \qquad x \in \mathbb{R}. \tag{7}$$

This inequality holds for any vector $(t_1, \ldots, t_n) \in \mathbb{R}^n$. Hence integrating the expressions on both sides of (7) relative to the n-dimensional Lebesgue measure on the compact n-rectangle $[0, N]^n \subset \mathbb{R}^n$ and using Proposition 3, we get on the left for the diagonal terms $n\phi(0)N^n = nN^n$, and for the non-diagonal terms [there are $n(n-1)$ of them, and ϕ is a bounded measurable function]

$$n(n-1)N^{n-2}\int_0^N\int_0^N \phi(t-t')e^{ix(t-t')}\,dt\,dt'.$$

Consequently (7) becomes

$$nN^n + n(n-1)N^{n-2}\int_0^N\int_0^N \phi(t-u)e^{ix(t-u)}\,dt\,du \geq 0, \qquad x \in \mathbb{R}. \tag{8}$$

Dividing (8) by $n(n-1)N^{n-2}$ and noting that $n \geq 1$ is arbitrary, one obtains

$$\int_0^N\int_0^N \phi(t-u)e^{ix(t-u)}\,dt\,du \geq 0. \tag{9}$$

But this is $2\pi N p_N(x)$ of (3), and we can use the argument there.

Thus if ϕ_N and p_N are defined as in that theorem, then $p_N(x) \geq 0$ and $\phi_N(t) = [1 - (|t|/N)]\phi(t)$, where $[1 - (|t|/N)]$ actually defines the ch.f. of the probability density $(1 - \cos Nx)/\pi N x^2, x \in \mathbb{R}$, as we have seen in Section 1. Now the work leading to (4) implies that $0 \leq \int_{\mathbb{R}} p_N(x)dx \leq M_0 < \infty$, uniformly in N. Next consider for any $u \in \mathbb{R}$

$$\begin{aligned}
\int_0^u\int_{\mathbb{R}} & e^{itx}p_N(x)\,dx\,dt \\
&= \lim_{K\to\infty}\frac{1}{\pi}\int_{\mathbb{R}}\int_0^u \phi_N\left(t + \frac{v}{K}\right)\left(\frac{\sin v}{v}\right)^2 dt\,dv \quad \text{[by step (ii) after (4)]} \\
&= \lim_{K\to\infty}\frac{1}{\pi}\int_{\mathbb{R}}\int_{v/K}^{u+v/K} \phi_N(r)\,dr\left(\frac{\sin v}{v}\right)^2 dv \\
&= \int_0^u \phi_N(r)\,dr \\
&\quad \text{(by the dominated convergence} \\
&\quad \text{and then the second integral is unity).}
\end{aligned}$$

But the left side is $\int_0^u \hat{p}_N(t)\,dt$, where $\hat{p}_N$ is the Fourier transform of p_N (a "ch.f."), and since $\phi_N(r) \to \phi(r)$ for each $r \in \mathbb{R}$ as $N \to \infty$, by the bounded convergence theorem we conclude that

$$\lim_{N\to\infty}\int_0^N \hat{p}_N(t)\,dt = \int_0^u \phi(t)\,dt, \qquad u \in \mathbb{R}, \tag{10}$$

exists. If $G_N(x) = \int_{-\infty}^{x} p_N(v)\,dv$, then $\{G_N, N \geq 1\}$ is a uniformly bounded (by M_0) nondecreasing class, so that by the Helly selection principle we can find a subsequence G_{N_k} with $G_{N_k} \to G$ as $N_k \to \infty$ at all continuity points of G, where G is a bounded (by M_0) nondecreasing (and nonnegative) function. By the Helly-Bray theorem, we then have

$$\int_0^u \hat{p}_{N_k}(t)\,dt = \int_{\mathbb{R}} \frac{e^{iux}-1}{ix}\,dG_{N_k}(x) \to \int_{\mathbb{R}} \frac{e^{iux}-1}{ix}\,dG(x), \qquad u \in \mathbb{R}, \tag{11}$$

as $N_k \to \infty$. Actually we apply that theorem with the following modified form using the fact that $g_u(x) = (e^{iux}-1)/ix \to 0$ as $|x| \to +\infty$. Let a, b be continuity points of G, and consider

$$\left|\int_{\mathbb{R}} g_u(x)\,dG_{N_k}(x) - \int_{\mathbb{R}} g_u(x)\,dG(x)\right|$$

$$\leq \left|\int_a^b g_u(x)\,d(G_{N_k} - G)(x)\right| + 2M_0 \sup\{|g_u(x)| : x \notin (a,b)\}. \tag{12}$$

The last term can be made small if a, b are chosen suitably large, and then the integral on the compact set $[a, b]$ goes to zero by the Helly-Bray theorem. But the right side of (11) is $\int_0^u \phi(t)\,dt$. By this argument we deduce that each convergent subsequence of $\{G_N, N \geq 1\}$ has the same limit $= \int_0^u \phi(t)\,dt$, which is absolutely continuous. It follows that the limit functions of G_N differ by at most a constant, i.e., $dG_N \to dG$ for a unique G, and hence their Fourier transforms converge and if $\int_{\mathbb{R}} e^{itx}\,dG(x) = \psi(t)$, then $\int_0^u \psi(t)\,dt = \int_0^u \phi(t)\,dt, u \in \mathbb{R}$. Thus $\phi(t) = \psi(t)$a.e. But ψ is continuous and $\phi(0) = 1 = \psi(0)$, so that ψ is a ch.f. by Proposition 3. This proves the main part of the theorem.

For the converse, let $\phi = \tilde{\phi}$ a.e., where $\tilde{\phi}$ is a ch.f. and $\phi(0) = 1$. We again form $p_T(x)$ as in (3) with ϕ. Since the Lebesgue integral is unaltered if we replace the integrand by an a.e. equivalent function, p_T remains the same if ϕ is replaced $\tilde{\phi}$. Consequently (3) implies $p_T \geq 0$, and (4) shows p_T is a probability density. Also, $\tilde{\phi}_T$ is the ch.f. of p_T, and hence is positive definite. The same is true if ϕ_T is used in place of $\tilde{\phi}_T$, and ϕ_T is continuous, being the product of the bounded function ϕ and the ch.f. ψ_T defined there. Thus ϕ_T is positive definite and $\phi_T \to \phi$ pointwise as $T \to \infty$. Since a pointwise limit of a sequence of positive definite functions is clearly positive definite, it follows that ϕ is positive definite. Since ϕ is not necessarily continuous at $t = 0, \phi$ is not generally a ch.f., as simple counterexamples show. This proves the converse and the theorem is established.

Even though we used the continuity theorem for ch.f.s as a key result in the proof of Theorem 2, one can establish that the continuity theorem is a consequence of Bochner's theorem. Thus there is a certain *equivalence* between these results. Let us now establish this statement.

Theorem 6 *The continuity theorem for ch.f.s is a consequence of Bochner's theorem. More explicitly, let* $\{\phi_n, n \geq 1\}$ *be a sequence of ch.f.s and* $\{F_n, n \geq 1\}$ *be the corresponding sequence of d.f.s. If* $\phi_n(t) \to \phi(t), t \in \mathbb{R}$, *and* ϕ *is continuous at* $t = 0$, *then there is a d.f.* F *and* $F_n \to F$ *at all continuity points of* F. *Further,* ϕ *is the ch.f. of* F.

Proof Since ϕ is the pointwise limit of positive definite functions, it is positive definite; and it is continuous at $t = 0$ by hypothesis. Thus by Proposition 3 and Theorem 2, ϕ is a ch.f. Let F be its d.f. We now show that $F_n \to F$, or equivalently, if P_n and P are their Lebesgue-Stieltjes probability measures on $\mathbb{R}$, then $\limsup_n P_n(A) \leq P(A)$ for all closed sets A. (Here the truth of Theorem 2 is assumed.)

For this, we shall verify one of the equivalent hypotheses of Theorem 1.5. Let $\{t_n, n \geq 1\} \subset \mathbb{R}$ be an arbitrarily fixed dense denumerable set (e.g., rationals.) Write $K = \{e^{ix} : x \in \mathbb{R}\}$, and let K^∞ be the countable cartesian product of K with itself. Then K (identified as the unit circle in $\mathbb{C}$) is compact, and hence so is K^∞, by the Tychonov theorem, and is separable in the product topology. Consider the mapping $\tau : \mathbb{R} \to K^\infty$, defined by

$$\tau : x \mapsto (e^{it_j x}, j \geq 1). \tag{13}$$

Since for any $x_1 \neq x_2$ we can find a pair of t_j, t_k such that $e^{it_j x_1} \neq e^{it_k x_2}$, it follows that τ is a one-to-one mapping of $\mathbb{R}$ into K^∞. Also, it is continuous. Further, if $\tau_n = \tau|_{[n,n+1]}$, then τ_n^{-1}(closed set)$\subset$ closed subset of $[n, n+1]$. Since $n \geq 1$ is arbitrary, this shows that $\tau^{-1} : K^\infty \to \mathbb{R}$ and $\tau : \mathbb{R} \to K^\infty$ are both measurable. We may say, therefore, that τ is a Borel isomorphism between $\mathbb{R}$ and $\tau(\mathbb{R})$ which is countably compact. With this mapping, consider $\mu_n = P_n \circ \tau^{-1}$. Then $\mu_n(\tau(\mathbb{R})) = 1, n \geq 1$, and the same is true of $\mu = P \circ \tau^{-1}$. Moreover, since τ^{-1} is a function,

$$\int_{K^\infty} e^{it\tau^{-1}(k)} \mu_n(dk) = \int_{\mathbb{R}} e^{itx} P_n(dx) = \int_{\mathbb{R}} e^{itx}\, dF_n(x) = \phi_n(t), \tag{14}$$

and similarly ϕ is the ch.f. of μ. Since K^∞ is separable, we can use the same reasoning as in Helly's selection principle and conclude that there is a subsequence $\mu_{n_i} \to \tilde{\mu}$ and the compactness of K^∞ implies $\tilde{\mu}$ is a probability measure (no mass "escapes" to infinity). By the Helly-Bray theorem (also applicable here), $\phi_{n_i}(t) \to \tilde{\phi}(t)$, the ch.f. of $\tilde{\mu}$. But then $\tilde{\phi} = \phi$, and so by the uniqueness theorem $\mu = \tilde{\mu}$. Repeating the argument for each convergent subsequence, we deduce that $\mu_n \to \mu$. (If we had extended Theorem 1.5 somewhat, this could have been immediately deduced from it.) We still have to prove the result for the P_n-sequence.

Let $\mathbb{R} = \bigcup_n A_n$, where A_n are disjoint bounded Borel sets whose boundaries have zero P-measure. If $C \subset \mathbb{R}$ is any closed set, then ($\overline{A}_k$ denoting the closure of A_k)

$$\limsup_n P_n(C) = \limsup_n \sum_{k \geq 1} P_n(C \cap A_k) \leq \sum_{k \geq 1} \limsup_n P_n(C \cap \overline{A}_k). \tag{15}$$

We now use the result that $\mu_n \to \mu$ proved above. Thus

$$\begin{aligned}\limsup_n P_n(C \cap \overline{A}_k) &= \limsup_n P_n \circ \tau^{-1}(\tau(C \cap \overline{A}_k)) \\ &= \limsup_n \mu_n(\tau(C \cap \overline{A}_k)) \\ &\leq \mu(\tau(C \cap \overline{A}_k)) \quad \text{(by Theorem 1.5iv)} \\ &= P \circ \tau^{-1}(\tau(C \cap \overline{A}_k)) = P(C \cap \overline{A}_k). \end{aligned} \tag{16}$$

Substituting (16) in (15), we get

$$\begin{aligned}\limsup_n P_n(C) &\leq \sum_{k\geq 1} P(C \cap \overline{A}_k) \\ &= \sum_{k\geq 1} P(C \cap A_k) \quad [\text{since } P(\overline{A}_k - A_k) = 0] \\ &= P(C).\end{aligned}$$

Hence by Theorem 1.5iv again, $P_n \to P$ or $F_n \to F$, which ends the proof.

Remark This result is significant only when we present an independent proof of Bochner's theorem. Indeed, the other two proofs mentioned in the introduction of this section are of this type. It is also possible to establish Bochner's theorem using the projective limit results of Bochner and Kolmogorov (cf. Theorem 3.4.10). We give one of these versions in the next section for variety and also because these considerations illuminate the subject. It shows how several of these apparently different results are closely related to each other.

The condition of positive definiteness of Theorems 2 and 5 is thus essential for characterizing ch.f.s. But it is not very easy to verify in applications. The following *sufficient* condition can be used almost by inspection, and hence we include it. This result was obtained by G. Pólya in 1923.

Proposition 7 *A continuous symmetric nonnegative function ϕ on $\mathbb{R}$ with $\phi(0) = 1$ is a ch.f. if it is nonincreasing and convex on the positive line.*

Proof We exclude the simple case that $\phi(t) = 1, t \in \mathbb{R}$. If $\lim_{t\to\infty} \phi(t) = \alpha < 1$, then $\tilde{\phi}(t) = (\phi(t) - \alpha)/(1 - \alpha)$ satisfies the hypothesis. Thus we may assume that $\alpha = 0$. As noted in the proof of Proposition 1.3.7, ϕ can be expressed as

$$\phi(t) = \phi(0) + \int_0^t f(u)\, du, \quad t > 0, \tag{17}$$

and $\phi(0) = 1$. Here f is the right derivative of ϕ, which exists and is nondecreasing. Since ϕ is decreasing by hypothesis, f must be negative on $\mathbb{R}^+$

and $\lim_{u\to\infty} f(u) = 0$ [$\lim_{t\to\infty} \phi(t) = 0$ being the present condition]. We now complete the proof (using a remark of K. L. Chung) by reducing the result to Proposition 2.5.

As noted in Remark 2, after Theorem 2.4, we know that $h_1(t) = 1 - |t|$ for $|t| \leq 1$, and $= 0$ for $|t| > 1$, gives a ch.f. [of $(1 - \cos x)/\pi x^2, x \in \mathbb{R}$], and hence if h is defined as

$$h(s,t) = \begin{cases} 1 - (|t|/s), & |s| > |t| \\ 0, & \text{otherwise}, \end{cases} \tag{18}$$

then for each $t, h(\cdot, t)$ is also a ch.f., and $h(s, \cdot)$ is continuous for each s. We now produce a d.f. H on $\mathbb{R}$ using (17), and then get a mixture of h and H to represent ϕ and complete the argument.

Since f in (17) is increasing, consider $H(s) = \int_0^s t\,df(t) = 1 - \phi(s) + sf(s)$. Then $H(\cdot) \nearrow \geq 0$ is a d.f. because $\phi(s) \to 0$, as $s \to \infty$ so that $f(s) = o(s^{-1})$, implying $\lim_{s\to\infty} sf(s) = 0$. Hence for $t \geq 0$, if

$$\begin{aligned} g(t) &= \int_{\mathbb{R}} h(s,t)\,dH(s) = \int_t^{\infty} \left(1 - \frac{|t|}{s}\right) dH(s) \\ &= \int_t^{\infty} (s-t)\,df(s), \end{aligned}$$

we get $g'(t) = f(t)$ (by the fundamental theorem of calculus, or even by considering

$$\frac{g(t+\Delta t) - g(t)}{\Delta t} = \frac{1}{\Delta t}\int_t^{t+\Delta t} (t-s)\,df(s) + f(t+\Delta t) \to f(t), \text{a.e.},$$

as $\Delta t \to 0$ because of Lebesgue's theorem on differentiation), so that $g(t) = \phi(t) + C$. But $g(0) = \phi(0) = 1$, so that $C = 0$, and hence $\phi(t) = \int_{\mathbb{R}} h(s,t)\,dH(s)$, $t \in \mathbb{R}$, since $g(-t) = g(t) = \phi(t) = \phi(-t)$. Thus by Proposition 2.5, ϕ is a ch.f. This completes the proof.

4.5 Some Multidimensional Extensions

The preceding results for ch.f.s are all on $\mathbb{R}$ or the (discrete case) integers. Some of these extend immediately to k dimensions, but others, such as Bochner's theorem, are more involved. In this section we indicate these possibilities, but prove a generalization of Bochner's theorem using the projective limit method. This section may be skipped on a first reading.

If $X_1, \ldots, X_n$ are r.v.s on (Ω, Σ, P), then their joint d.f. $F_{X_1,\ldots,X_n}$, or F_n for short, is given by

$$F_n(x_1, \ldots, x_n) = P[X_1 < x_1, \ldots, X_n < x_n], \quad x_i \in \mathbb{R}. \tag{1}$$

Thus any nondecreasing (in each component) left continuous nonnegative function which satisfies

$$\lim_{x_i \to -\infty} F_n(x_1, \ldots, x_n) = 0, \quad \lim_{x_n \to \infty} F_n(x_1, \ldots, x_n) = F_{n-1}(x_1, \ldots, x_{n-1}),$$
$$\Delta F_n \geq 0, \quad \int \cdots \int_{\mathbb{R}^n} dF_n(x_1, \ldots, x_n) = 1$$

is a d.f., where ΔF_n is the (n-dimensional) increment of F_n. The ch.f. is defined as usual as

$$\phi_n(t_1, \ldots, t_n) = \int \cdots \int_{\mathbb{R}^n} e^{it_1x_1 + \cdots + it_nx_n} \, dF_n(x_1, \ldots, x_n), \quad t_i \in \mathbb{R}. \tag{2}$$

If μ is the Lebesgue-Stieltjes probability measure on $\mathbb{R}^n$ defined by F_n, then for any Borel set $A \subset \mathbb{R}^n$ with its boundary ∂A, measurable, we say A is a *continuity set* of μ or of F_n if $\mu(\partial A) = 0$. [Thus $\mu(\overline{A}) = \mu(\text{int}(A))$.] If A is a rectangle, then ∂A is always measurable, and if $\mu(\partial A) = 0$, then it is simply termed a *continuity interval.* The inversion formula and the uniqueness theorem are extended without difficulty. We state it as follows.

Theorem 1 *Let F_n be an n-dimensional d.f. and ϕ_n be its ch.f. (on $\mathbb{R}^n$). If $A = \times_{i=1}^n [a_i, a_j + h_i)$ is a continuity interval of $F_n, h_i > 0$, and P is its Lebesgue-Stieltjes probability, then*

$$P(A) = \lim_{T \to \infty} \frac{1}{(2\pi)^n} \int_{-T}^{T} \cdots \int_{-T}^{T} \left[\prod_{i=1}^{n} \left(\frac{1 - e^{it_jh_j}}{it_j} \right) \right] e^{-i\sum_{j=1}^n t_ja_j}$$
$$\times \phi_n(t_1, \ldots, t_n) \, dt_1 \cdots dt_n. \tag{3}$$

Hence P or the F_n is uniquely determined by its ch.f., ϕ_n.

The straightforward extension of proof of the one-dimensional case is left to the reader.

The next result contains a technique, introduced by H. Cramér and H. Wold in 1936, that allows us to reduce some multidimensional considerations to the one-dimensional case. If $(X_1, \ldots, X_n)$ is a random vector with values in $\mathbb{R}^n$, we introduce the (measurable) set $S_{A,x}$, with $\alpha = (a_1, \ldots, a_n) \in \mathbb{R}^n$, as

$$S_{\alpha,x} = \left[y = \sum_{j=1}^{n} a_j x_j < x \right]. \tag{4}$$

We now establish the above stated technique in the form of

Proposition 2 *If P_1 and P_2 are two Lebesgue-Stieltjes probabilities on $\mathbb{R}^n$ such that $P_1(S_{\alpha,x}) = P_2(S_{\alpha,x})$ for $x \in \mathbb{R}$ and vector $\alpha \in \mathbb{R}^n$, then $P_1 = P_2$ and the common measure gives the distribution of $(X_1, \ldots, X_n)$.*

Proof By hypothesis P_1 and P_2 determine the same d.f. of Y:

$$F_Y(x) = P_1(S_{\alpha,x}) = P_2(S_{\alpha,x}). \tag{5}$$

Hence

$$\begin{aligned}\phi_Y(t) = E(e^{itY}) &= E(e^{ita_1X_1+\cdots+ita_nX_n}) \\ &= \phi_n(ta_1, \ldots, ta_n),\end{aligned}$$

where ϕ_n is the joint ch.f. of $(X_1, \ldots, X_n)$. Thus if $t = 1$, this shows that P_1 and P_2 have the same ch.f.s for $\alpha \neq 0 \in \mathbb{R}^n$. If $\alpha = 0$, then $\phi_n(0) = 1$. Hence P_1 and P_2 have the same ch.f. ϕ_n. By the preceding theorem (the uniqueness part) $P_1 = P_2$ on all the Borel sets of $\mathbb{R}^n$. The last statement is immediate.

Using this result, for instance, the multidimensional continuity theorem for ch.f.s can be reduced to the one-dimensional case. We sketch this argument here. If $P_n \to P$, where P_n, P are the Lebesgue-Stieltjes probabilities on $\mathbb{R}^k$, then by the k-dimensional Helly-Bray theorem (same proof as in the one-dimensional case) the corresponding ch.f.s converge to that of P. Thus for the sufficiency, the above (reduction) technique can be applied.

If ϕ_n is the ch.f. of a k-dimensional distribution (= image probability) P_n, then, by the (multidimensional analog of) Helly selection principle, there exists a σ-additive bounded measure $\tilde{P}$ on $\mathbb{R}^k$ such that $P_n(S) \to \tilde{P}(S)$ for all Borel sets $S \subset \mathbb{R}^k$ such that $\tilde{P}(\partial S) = 0, \partial S$ being the boundary of S. On the other hand, for each fixed $t_1, \ldots, t_k (\neq 0 \text{ in } \mathbb{R}^k), \tilde{\phi}_n(t) = \phi_n(tt_1, \ldots, tt_k) \to \phi(tt_1, \ldots, tt_k) = \tilde{\phi}(t), t \in \mathbb{R}$, and $\tilde{\phi}$ is continuous at $t = 0$. Hence by the one-dimensional continuity theorem $\tilde{\phi}(\cdot)$ is a characteristic function. If $S_{\alpha,x}$ is given by (4) as a subset of $\mathbb{R}^k$ (with $a_j = t_j$ here), then $F_n^\alpha(x) = P_n(S_{\alpha,x}) \to \tilde{P}(S_{\alpha,x})$, and $F_n^\alpha(x) \to F^\alpha(x)$ at all continuity points x of F^α, a d.f. with ϕ as its ch.f. Now let $x \to \infty$ (because $F^\alpha(+\infty) = 1$), it follows that

$$\tilde{P}(\mathbb{R}^k) = \lim_{x\to\infty} \tilde{P}(S_{\alpha,x}) = F^\alpha(+\infty) = 1.$$

Hence $\tilde{P}$ is a probability function, and then ϕ will be its ch.f. Next, by the familiar argument, with Theorem 1, we conclude that each convergent subsequence of $\{P_n, n \geq 1\}$ has the same limit $\tilde{P}$, and thus the whole sequence converges to $\tilde{P}$. This gives the multidimensional continuity theorem for ch.f.s

Let us record another version of the above proposition which is very useful in some applications. In particular, we can deduce a classical result of J.

Radon from it. The desired form is the following

Proposition 2′ *A probability distribution in* $\mathbb{R}^k$ *is determined completely by its projections on a set of subspaces of dimensions* $1, 2, \ldots, k-1$ *that together exhaust the whole space.*

Proof Since the set $\{S_{\alpha,x} : x \in \mathbb{R}, \alpha \in \mathbb{R}^k\}$ of subspaces of Proposition 2 cover $\mathbb{R}^k$, it is clear that both propositions are equivalent. We sketch the argument, with $k = 2$, for simplicity and emphasis. Thus we take the ch.f. ϕ of $F(\cdot, \cdot)$:

$$\phi(t_1, t_2) = \int_{\mathbb{R}^2} e^{it_1 x + it_2 y}\, dF(x, y). \tag{6}$$

Consider a line ℓ_θ through the origin making an angle θ with the x axis. Then the projection of (x, y) on this line is given by

$$x' = x \cos\theta + y \sin\theta, \qquad y' = y \cos\theta - x \sin\theta.$$

Hence if the projection of F on ℓ_θ is F_{ℓ_θ}, its ch.f. is obtained as follows. Let x, y be the values of the r.v.s X, Y, and x' be that of $X' = X \cos\theta + Y \sin\theta$. Then

$$\phi_\theta(t) = E(e^{itX'}) = \int_{\mathbb{R}} e^{itr}\, dF_{X'}(r)$$

is assumed known for each $0 \le \theta < \pi$ and $t \in \mathbb{R}$. But

$$\phi_\theta(t) = E(e^{it(X\cos\theta + Y \sin\theta)}) = \phi(t\cos\theta, t\sin\theta). \tag{7}$$

Consequently if $\phi_\theta(t)$ is known for each θ and t, then so is ϕ, since

$$\phi(t_1, t_2) = \phi_{\tan^{-1}(t_2/t_1)}((t_1^2 + t_2^2)^{1/2}), \quad (t_1, t_2) \in \mathbb{R}^2. \tag{8}$$

Thus F is completely determined, by Theorem 1. This completes the proof.

The following consequence of the above result is due to J. Radon, proved in 1917, and plays an important role in applications. In fact, it played a crucial role in the work of A. M. Cormack, who derived a particular case of it independently in the late 1950's. The work is so useful in tomography and (brain) scanning that he was awarded a Nobel prize in medicine in 1979 for this. It was ultimately based on the Radon theorem. [A historical account and medical applications are detailed in his Nobel lecture, which is printed in *Science* **209**(1980), 1482-1486.]

Corollary 3 (Radon's Theorem) *If* K *is a bounded open set of the plane, and if the integral of a continuous function* $f : K \to \mathbb{R}$ *vanishes along every chord of* K*, then* $f = 0$.

Proof Let $f_1 = \max(f, 0)$ and $f_2 = f_1 - f$. Then f_1, f_2 are nonnegative and continuous. By hypothesis $\int_\ell f\, dx\, dy = 0$ for every chord ℓ of K, so that $\int_\ell f_1\, dx\, dy = \int_\ell f_2\, dx\, dy$ for each such ℓ. Since the f_i are integrable, we may identify them as "densities" (i.e., that their integral should equal unity is unimportant). But by Proposition 2 (or 2′) these integrals determine f_i completely. Hence $f_1 = f_2$, so that $f = 0$, as asserted.

Another consequence of some interest is contained in

Corollary 4 (Rényi) *Let $F : \mathbb{R}^2 \to \mathbb{R}^+$ be the d.f. of a bounded random vector $(X, Y) : \Omega \to \mathbb{R}^2$ of a probability space (Ω, Σ, P). [Thus the range of (X, Y) is contained in some ball of radius $0 < \rho < \infty$.] If F_{ℓ_θ}, the projection of F on a line $\ell_\theta : \ell_\theta(x, y) = x \cos\theta + y \sin\theta$ through the origin, is given for an infinite set of distinct lines [= for an infinite collection of distinct $\theta (\bmod\ \pi)$], then F is uniquely determined.*

Proof Let $\theta_1, \theta_2, \ldots$ be a distinct set as in the statement. Then this is bounded and there exists a convergence subsequence $\{\theta_{n'}\}_{n' \geq 1}$ with limit $\tilde{\theta}$. By Proposition 2 or 2′, if ϕ is the ch.f. of F and ϕ_θ is the ch.f. of F_{ℓ_θ}, then

$$\phi_\theta(t) = \phi(t \cos\theta, t \sin\theta), \qquad t \in \mathbb{R}, \tag{9}$$

is known. Since the vector (X, Y) has a bounded range, the measure determined by the d.f. F has a compact support. Thus its Fourier transform ϕ is an analytic function (since ϕ is continuously differentiable; also see Problem 6). But this transform is known for a convergent subsequence $\{\theta_{n'}\}_{n' \geq 1}$, so that (9) implies, by the method of analytic continuation, that ϕ is defined uniquely on all of the complex plane $\mathbb{C}$. This shows that $\{\phi_\theta(\cdot), \theta \in (\theta_n, n \geq 1)\}$ determines ϕ, and hence F, as asserted.

Let us now present an extension of Bochner's theorem with a proof that does *not* use the continuity theorem. The latter can thus be obtained from the former, as shown before. Our argument is founded on the Kolmogorov-Bochner Theorem 3.4.10, exhibiting the deep interrelations between three of the major results of the subject.

The first step here is based on the following special result.

Proposition 5 (Herglotz Lemma) *Let $\phi : \mathbb{Z}^k \to \mathbb{C}, k \geq 1$, be a function where $\mathbb{Z}^k$ is the set of all points of $\mathbb{R}^k$ with integer coordinates. Then for each $j = (j_1, \ldots, j_k) \in \mathbb{Z}^k$,*

$$\phi(j) = \int_{T^k} \exp\left\{ i \sum_{r=1}^{k} j_r \theta_r \right\} \mu(d\theta_1, \ldots, d\theta_k), \tag{10}$$

where T^k is the cube with a side $[0, 2\pi]$ in $\mathbb{R}^k$ and μ is a probability on the Borel sets of T^k, iff ϕ is positive definite and $\phi(0) = 1$, i.e., iff ϕ is a ch.f. on

$\mathbb{Z}^k$. *The representation (10) is unique.*

Proof If ϕ is given by (10), then clearly $\phi(0) = 1$ and it is positive definite as in the one-dimensional case. Only the converse is nontrivial.

Thus let ϕ be positive definite and $\phi(0) = 1$. If $j' = (j'_1, \ldots, j'_k)$ and $j'' = (j''_1, \ldots, j''_k)$ are two points of $\mathbb{Z}^k$, then, the sum and difference of such vectors being taken componentwise, let

$$f_N(\theta_1, \ldots, \theta_k) = \frac{1}{N^k} \sum_{\substack{1 \le j'_r, j''_r \le N \\ r = 1, \ldots, k}} \phi(j' - j'') \exp\left\{ i \sum_{r=1}^{k} (j'_r - j''_r)\theta_r \right\}. \quad (11)$$

The positive definiteness of ϕ implies $f_N \ge 0$. If we define

$$\mu_N(A) = (2\pi)^{-k} \int \cdots \int_A f_N(\theta_1, \ldots, \theta_k) d\theta_1 \cdots d\theta_k \quad (12)$$

for each Borel set A of T^k, then μ_N is a probability, since $\phi(0) = 1$. Moreover, the orthogonality relations of trigonometric functions on the circle T, by identifying 0 and 2π, give

$$\int \cdots \int_{T^k} \exp\left\{ i \sum_{r=1}^{k} j_r \theta_r \right\} d\mu_N(\theta)$$
$$= \begin{cases} \phi(j) \prod_{r=1}^{k} \left(1 - \frac{|j_r|}{N}\right) & \text{if } |j_r| < N, \ r = 1, \ldots, k \\ 0 & \text{otherwise.} \end{cases} \quad (13)$$

Here $j = (j_1, \ldots, j_k)$. But for the sequence $\{\mu_N, N \ge 1\}$ on the Borel sets of T^k, we can apply the k-dimensional version of Theorem 1.1, and extract a convergent subsequence $\{\mu_{N_s}, s \ge 1\}$; thus there is a probability μ (since T^k is compact in the product topology) which is a limit of this subsequence. Now by the corresponding Helly-Bray theorem we get on letting $N_s \to \infty$ in (13),

$$\int \cdots \int_{T^k} \exp\left\{ i \sum_{r=1}^{k} j_r \theta_r \right\} d\mu(\theta) = \lim_{N_s \to \infty} \int \cdots \int_{T^k} \exp\left\{ i \sum_{r=1}^{k} j_r \theta_r \right\} d\mu_{N_\ell}(\theta)$$
$$= \phi(j).$$

This is (10).

The uniqueness can be obtained from Theorem 1. Alternatively, since T^k is compact, we note that the set of functions $\{\exp(i \sum_{r=1}^{k} j_r \theta_r), \theta_r \in T, j \in \mathbb{Z}^k\}$ is uniformly dense in $C(T^k)$, the space of continuous complex functions on T^k, by the Stone-Weierstrass theorem. This implies the uniqueness of μ in (10) at once, completing the proof.

The preceding result will now be given in a general form using Theorem 3.4.10 in lieu of the Lévy continuity theorem for ch.f.s. If S is a set, then $T^S(= \times_{t\in S}T_t, T_t \equiv T)$ also denotes the space of all functions defined on S with values in T. Since $T = [0, 2\pi]$ is compact, T^S is then compact under the product topology and, with componentwise addition modulo 2π, it becomes a compact abelian group. We can thus express $\theta \in T^S$ as $\theta = (\theta(s), s \in S)$, so that the coordinate projection $p_s : \theta \mapsto \theta(s)$ is a mapping of T^S onto T. Let $\mathcal{B}$ be the smallest σ-algebra (=cylinder algebra in the earlier terminology) with respect to which each $p_s, s \in S$, is measurable when T is given its Borel σ-algebra $\mathcal{T}$. Thus $\mathcal{B} = \sigma(\bigcup_s p_s^{-1}(\mathcal{T}))$. Let $\mathcal{I}$ be the set of mappings $n : S \to \mathbb{Z}$ such that $n(s) = 0$ for all but a finite number of $s \in S$. With componentwise addition, $\mathcal{I}$ becomes an abelian group (the "dual" of T^S.).

Using this terminology, we have the following simple but crucial extension of the preceding result:

Proposition 6 *Let $\phi : \mathcal{I} \to \mathbb{C}$ be positive definite and $\phi(0) = 1$, where 0 is the identically zero function of $\mathcal{I}$. Then there is a unique probability P on $\mathcal{B}$ such that for each $n \in \mathcal{I}$,*

$$\int_{T^S} \exp\left\{i\sum_{s\in S} n(s)\theta(s)\right\} dP(\theta) = \phi(n). \tag{14}$$

Proof First note that the integral in (14) is really a finite-dimensional one since $n(s) = 0$ for all but a *finite* subset of S. Let us define for each *finite* set $F \subset S$,

$$\mathcal{I}_F = \{n \in \mathcal{I} : n(S - F) = 0\}.$$

Clearly $\mathcal{I}_{\mathcal{F}}$ is a group (isomorphic to $\mathbb{R}^F$). Hence by Proposition 5 applied to T^F, there is a unique probability P_F on the Borel σ-algebra of T^F such that by (10),

$$\int_{T^F} \exp\left\{i\sum_{s\in S} n(s)\theta(s)\right\} dP_F(\theta) = \phi(n), \quad n \in \mathcal{I}_{\mathcal{F}}. \tag{15}$$

If $\mathcal{F}$ denotes the directed (by inclusion) family of all finite subsets of S, then the uniqueness of P_F in (15) implies that $\{P_F, F \in \mathcal{F}\}$ is a compatible family of Borel (or Radon) probability measures on $\{T^F, F \in \mathcal{F}\}$. Now by Theorem 3.4.10 there exists a unique probability $P : \mathcal{B} \to [0, 1]$ such that $P \circ p_F^{-1} = P_F$, where $p_F : T^S \to T^F$ is the corresponding coordinate projection. This means we can replace T^F and P_F in (15) by T^S and P, so that (14) holds. This completes the proof.

To obtain the n-dimensional version of Bochner's theorem from the above result, we need to identify T^S with a simpler object. Here the structure theory (actually the Pontryagin duality theorem) intervenes. The special case for $\mathbb{R}^k$

will be given here, as this is easy. If the "duality" is assumed, then the proof carries over to the general locally compact abelian groups. However, for simplicity we restrict our treatment to $\mathbb{R}^k$ here. [A similar idea will be employed in Section 8.4 in discussing a subclass of strictly stationary processes.]

Recall that if G is a locally compact abelian group, then a continuous homomorphism α of G into the multiplicative group of complex numbers with absolute value one is called a *character* of G. Thus $\alpha : G \to \mathbb{C}$ satisfies (i) $|\alpha(x)| = 1$, (ii) $\alpha(x+y) = \alpha(x)\alpha(y), x, y \in G$ (with $+$ as group operation), and (iii) α is continuous. If $G = \mathbb{R}^k$, the additive group of k-tuples of reals, then we can give the following simple and explicit description of these characters. In addition, the set of all characters on G, denoted $\hat{G}(\subset T^G$ endowed with the product topology), will be identified.

Proposition 7 *If $G = \mathbb{R}^k$, then each character $\alpha(\in \hat{G})$ is of the form $\alpha(x) = e^{i(x,y)}$, where $x = (x_1, \ldots, x_k) \in \mathbb{R}^k$; similarly $y \in \mathbb{R}^k$ and $(x, y) = \sum_{j=1}^{k} x_j y_j$. Moreover, $\hat{G}(= (\mathbb{R}^k)\hat{})$ is isomorphic and homeomorphic to $\mathbb{R}^k$ under the identification $\alpha \leftrightarrow y$, so that $\alpha = e^{i(\cdot,y)}$ corresponds uniquely to y.*

Proof The well-known (measurable) solution of the Cauchy equation $f(x+y) = f(x)+f(y)$ is $f(x) = (x, c)$, where $c \in \mathbb{C}^k$. Setting $f(x) = \text{Log}\,\alpha(x)$ here, one gets $\alpha(x) = e^{(x,c)}$, and since $|\alpha(x)| = 1$ for all $x \in G, c$ must be a pure imaginary, $c = -iy$ for some $y \in \mathbb{R}^k$. Thus $\alpha(x) = e^{i(x,y)}$. On the other hand, clearly every $y \in \mathbb{R}^k$ defines $e^{i(\cdot,y)} \in \hat{G}$. It is also evident that $y \leftrightarrow e^{i(\cdot,y)}$ is one-to-one. Further,

$$N(1 : \varepsilon, n) = \{\alpha : |\alpha(x) - 1| < \varepsilon, |x| \leq n\}, \quad |x|^2 = \sum_{j=1}^{k} x_j^2,$$

is a neighborhood of the identity character one, and varying ε and n, these form a neighborhood basis of the identity of $\hat{G}$, and with the group property of $\hat{G}$, it determines the topology of $\hat{G}$, which is clearly the same as the induced topology of the product space T^G noted earlier. However, using the form $\alpha(x) = e^{i(x,y)}$, we see that $|\,e^{i(x,y)} - 1| < \varepsilon$ is equivalent to

$$|(x, y) < 2 \arcsin(\varepsilon/2),$$

or equivalently, $|\,y| < \delta$ with $\delta = (2/n)\text{arc}\sin(\varepsilon/2)$. Since this defines the neighborhood basis of the identity of $\mathbb{R}^k$, we conclude that the mapping between $\mathbb{R}^k$ and $(\mathbb{R}^k)\hat{}$ is bicontinuous at the identity, and since both are groups, the same must be true everywhere. This completes the proof.

A consequence of the preceding two results is the following. Since the space T can be identified with $0 \leq \theta \leq 2\pi$ (with group operation addition modulo 2π), or equivalently with the group of all complex numbers of absolute one, if $S = G = \mathbb{R}^k$, then $\hat{G} = T^G$ is isomorphic and homeomorphic to $G = \mathbb{R}^k$. Thus if $h : \hat{G} \to G$ is this mapping, then $h(\hat{G}) = \mathbb{R}^k$ and h is one-to-one, etc. Also this identification shows $(\mathbb{R}^k)\hat{} \cong \mathbb{R}^k \cong (\mathbb{R}^k)\hat{\hat{}}$. Bochner's theorem can now be established quickly as follows. We write $\mathbb{R}^k = G$ whenever it is convenient.

Theorem 8 *If $\phi : \hat{G} \to \mathbb{C}$ is a continuous mapping with $\phi(0) = 1$, then there exists a unique probability measure P on the Borel σ-algebra $\mathcal{B}$ of $\mathbb{R}^k = G$ such that*

$$\phi(y) = \int_G e^{i(x,y)} P(dx), \quad y \in \hat{G}, \tag{16}$$

iff ϕ is positive definite, or equivalently ϕ is a ch.f. on $\hat{G} \cong \mathbb{R}^k$.

Proof As before, if ϕ is given by (16), then it is immediate that ϕ is positive definite, continuous, and $\phi(0) = 1$. So we only need to establish the converse if ϕ satisfies these conditions.

Consider, to use Proposition 6, the function $\psi : \mathcal{I} \to \mathbb{C}$ defined by

$$\psi(n) = \phi\left(\sum_{y \in \hat{G}} n(y)y\right), \quad n \in \mathcal{I}. \tag{17}$$

Since $\mathcal{I}$ is a group, it is clear that ψ is positive definite and $\psi(0) = 1$. Then by Proposition 6, there is a probability μ on the cylinder σ-algebra of $T^{\hat{G}}$ such that

$$\int_{T^{\hat{G}}} \exp\left\{i \sum_{s \in \hat{G}} n(s)\theta(s)\right\} d\mu(\theta) = \psi(n), \quad n \in \mathcal{I}. \tag{18}$$

Note that θ is a complex homomorphism of absolute value one, and that $\theta(s + s') = \theta(s) \cdot \theta(s')$. Being continuous, it is a character of $\hat{G}$, and the mapping $(\theta, s) \mapsto \theta(s)$ is jointly continuous in θ and s by Proposition 7. Since $h : T^{\hat{G}} \to G = \mathbb{R}^k$ is a homeomorphism, preserving all algebraic operations, we have, on taking $n = \delta_s(\cdot)$(a delta function) in (18) and using the image law theorem, $\psi(\delta_s) = \phi(s)$ by (17), so that

$$\phi(s) = \psi(\delta_s) = \int_{T^{\hat{G}}} e^{i\theta(s)} d\mu(\theta) = \int_G e^{i(x_\theta, s)} dP(x_\theta), \quad s \in G, \tag{19}$$

where $P = \mu \circ h^{-1}$ and $\theta \leftrightarrow x_\theta \in G = \mathbb{R}^k$. Since $\hat{G} \cong \mathbb{R}^k$, (19) reduces to (16). The uniqueness of representation is a consequence of Theorem 1. This completes the proof.

Remarks (1) The above argument holds if G is any locally compact abelian group. However, we need a more delicate discussion for a determination of characters as well as the full Pontryagin duality theorem for this. The method employed above is a specialization of the general case considered by M. S. Bingham and K. R. Parthasarathy (*J. London Math. Soc.* **43** (1968), 626-632).

(2) The uniqueness part can also be established independently of Theorem 1, using an argument of the above authors. What is interesting here is that Proposition 6 (a consequence of Theorem 3.4.10) and the image law of probabilities replaced the continuity theorem for ch.f.s in this independent proof. The latter result is now a consequence of Theorem 8, as shown in the previous section.

4.6 Equivalence of Convergences for Sums of Independent Random Variables

After introducing the three central convergence concepts without regard to moments, (namely a.e., in probability, and in distribution) we established a simple general relation between them in Proposition 2.2.2. Then we stated after its proof that for sums of independent r.v.s these three convergences can be shown to be equivalent if more tools are available. The equivalence of the first two was sketched in Problem 16 of Chapter 2. However, the necessary results are now at hand, and the general assertion can be obtained, without reference to the above noted problem, as follows.

Theorem 1 *Let $X_1, X_2, \ldots$ be independent r.v.s on $(\Omega, \Sigma, P), S_n = \sum_{k=1}^{n} X_k$, and S be an r.v. Then the following convergence statements (as $n \to \infty$) are equivalent:*

(a) $S_n \to S$ *a.e.,*

(b) $S_n \to S$ *in probability,*

(c) $S_n \to S$ *in distribution.*

Proof By Proposition 2.2.2.,(*a*) $\Rightarrow$ (*b*) $\Rightarrow$ (*c*) always. Hence it suffices to prove that (*c*) $\Rightarrow$ (*a*).

Thus assume (c) and let $\phi_n(t) = E(e^{itX_n})$. By independence, if ψ_n is the ch.f. of S_n, then $\psi_n(t) = \prod_{k=1}^{n} \phi_k(t), t \in \mathbb{R}$. Since $S_n \to S$ in distribution, the continuity theorem for ch.f.s implies $\psi_n(t) \to \psi(t), t \in \mathbb{R}$, and ψ is the ch.f. of S. Hence there is an interval $I: -a \leq t \leq a, a > 0$, such that $\psi(t) \neq 0$. Let t be arbitrarily fixed in this interval. Then for each $\varepsilon > 0$, there is an $n_0 = n_0(\varepsilon, t)$ such that $n \geq n_0 \Rightarrow |\psi_n(t) - \psi(t)| < \varepsilon$, and by compactness of $[-a, a]$, we can even choose n_0 as a function of ε alone (but this fact will not be used below). Thus $\psi_n(t) \neq 0$ for all $n \geq n_0$. The idea of the proof here is to consider $\{e^{itS_n}, n \geq n_0\}$ and show that it converges a.e., and then deduce, by an exclusion of a suitable null set, that $S_n \to S$ pointwise on the complement of the null set. The most convenient tool here turns out to be the martingale convergence theorem, though the result can also be proved purely by the (finer) properties of ch.f.s. We use the martingale method.

Let $Y_n = e^{itS_n}/\psi_n(t)$ and $\mathcal{F}_n = \sigma(S_1, \ldots, S_n)$. Then we assert that $\{Y_n, \mathcal{F}_n, n \geq 1\}$ is a uniformly bounded martingale. In fact $|Y_n| = 1/|\psi_n(t)| \leq (|\psi(t)| - \varepsilon)^{-1} < \infty$, where ε may and will be taken smaller than min $\{|\psi(s)| : -a \leq s \leq a\} > 0$. Next consider

$$\begin{aligned} E^{\mathcal{F}_n}(Y_{n+1}) &= E^{\mathcal{F}_n}\left(\frac{e^{it(S_n+X_{n+1})}}{\psi_n(t)\cdot\phi_{n+1}(t)}\right) \\ &= [e^{itS_n}/\psi_n(t)]E(e^{itX_{n+1}}/\phi_{n+1}(t)) \\ &\quad \text{(by the independence of } X_{n+1} \text{ and } \mathcal{F}_n) \\ &= Y_n \quad \text{a.e.} \end{aligned}$$

This establishes our assertion. Hence by Theorem 3.5.7 or 3.5.8, $Y_n \to Y_\infty$ a.e., so that there is a set $N_t \subset \Omega$ such that $P(N_t) = 0$ and $Y_n(\omega) = e^{itS_n(\omega)}/\psi_n(t) \to Y_\infty(\omega), \omega \in \Omega - N_t$. Since $\psi_n(t) \to \psi(t) \neq 0$, and these are constants, we deduce that $e^{itS_n(\omega)} \to \tilde{Y}_\infty(\omega)$ for each $\omega \in \Omega - N_t$, where $\tilde{Y}_\infty(\omega) = Y_\infty(\omega)\psi(t)$. From this one obtains the desired convergence as follows.

Consider the mapping $(t, \omega) \mapsto e^{itS_n(\omega)}$. This is continuous in t for each ω and measurable in ω for each t. We assert that

(i) the mapping is jointly measurable in (t, ω) relative to the product σ-algebra $\mathcal{B} \otimes \Sigma$, where $(I, \mathcal{B}, \mu)$ is the Lebesgue interval and (Ω, Σ, P) is our basic probability space, and

(ii) there exists a set $\Omega_0 \in \Sigma, P(\Omega_0) = 1$, such that for each $\omega \in \Omega_0$, there is a subset $I_\omega \subset I$ satisfying $\mu(I_\omega) = \mu(I) = 2a$, and if $t \in I_\omega$, then $e^{itS_n(\omega)}$ converges to a limit $f_\omega(t)$, say, as $n \to \infty$.

These properties imply the result. Indeed, if they are granted, then from the form of the exponential function one has $f_\omega(t + t') = f_\omega(t)f_\omega(t')$. Since this is true for each t, t' in I_ω for which $t + t' \in I_\omega$, it follows that f_ω satisfies the classical Cauchy functional equation (cf. Proposition 6.7 or Problem 23 for another argument), and since $|f_\omega(t)| = \lim_n |e^{itS_n(\omega)}| = 1$, the solution is $f_\omega(t) = e^{it\alpha(\omega)}$ for some $\alpha(\omega) \in \mathbb{R}$. Hence $f_\omega(t) \neq 0$ for $t \in I_\omega$ and it is continuous for all $t \in I_\omega$ (whence at $t = 0$), so that $e^{itS_n(\omega)} = f_{\omega,n}(t) \to f_\omega(t)$ as $n \to \infty$ for all $t \in \mathbb{R}$, and f_ω is a ch.f. (of a unitary d.f.) for each $\omega \in \Omega_0$. Therefore the hypothesis of Corollary 2.10 is satisfied, and so $tS_n(\omega) = -i \operatorname{Log} f_{\omega,n}(t) \to -i \operatorname{Log} f_\omega(t) = t\alpha(\omega), t \in \mathbb{R}$. It follows that $S_n(\omega) \to \alpha(\omega), \omega \in \Omega_0$, and so $S_n \to \alpha$ a.e. But then $S_n \to \alpha$ in distribution and by hypothesis $S_n \to S$ in distribution. The limits being unique (in the Lévy metric), we must have $S = \alpha$ a.e., and (a) follows. Let us then establish (i) and (ii).

The joint measurability does not generally follow from sectional measurability. Fortunately, in the present case, this is quite easy. By hypothesis,

$S_n : \Omega \to \mathbb{R}$ is measurable, and if $g(t) = t$, the identity mapping of $I \to I$, then it is clearly measurable (for $\mathcal{B}$). Thus the product $g(S_n) : I \times \Omega \to \mathbb{R}$ is jointly $\mathcal{B} \otimes \Sigma$-measurable. Since e^{ix} is continuous in x and since a continuous function of a (real) measurable function is measurable, we conclude that $e^{ig(S_n)} : I \times \Omega \to \mathbb{C}$ is jointly measurable, proving (i).

For (ii), the set $A = \{(t, \omega) : \lim_{n\to\infty} e^{itS_n(\omega)} \text{ exists}\}$ is $\mathcal{B} \otimes \Sigma$-measurable, and so each t-section $A(t) = \Omega - N_t$ satisfies $P(A(t)) = 1$ by the martingale convergence established above. Since $\mu \otimes P$ is a finite measure, we have by the Fubini theorem applied to the bounded function χ_A,

$$\begin{aligned}\mu \otimes P(A) &= \int_I \int_\Omega \chi_A \, dP \, d\mu = \int_I d\mu(t) \int_\Omega \chi_{A(t)}(\omega) \, dP(\omega) \\ &= \int_\Omega dP(\omega) \int_I \chi_{A(\omega)}(t) \, d\mu(t),\end{aligned}$$

where $A(\omega)$ is the ω-section of A. This is $2a$ since $P(A(t)) = 1$ and $\mu(I) = 2a$. It follows that $\mu(A(\omega)) = 2a$ for almost all ω. Hence there exists an $\Omega_0 \in \Sigma, P(\Omega_0) = 1$, such that for each $\omega \in \Omega_0, \mu(A(\omega)) = 2a$. Consequently, if $\omega \in \Omega_0$, then for each $t \in A(\omega), \lim_{n\to\infty} e^{itS_n(\omega)}$ exists, and (ii) follows. This completes the proof.

Now that the basic technical tools are available, we proceed to develop the key results on distributional convergence in the next chapter. Several useful adjuncts are given as problems below.

Exercises

1. Let $\mathcal{M}$ be the set of all d.f.s on $\mathbb{R}$, and $d(\cdot, \cdot)$ be the Lévy distance, as defined in (11) of Section 1. Show that $d(\cdot, \cdot)$ is a metric on $\mathcal{M} \times \mathcal{M} \to \mathbb{R}^+$ and that ($\mathcal{M}$,d) is a complete metric space.

2. If (Ω, Σ, P) is a probability space, X, Y are a pair of r.v.s on it, let d_1, d_2 be defined as

$$d_1(X,Y) = E\left(\frac{|X - Y|}{1 + |X + Y|}\right), \quad d_2(X,Y) = d(F_X, F_Y),$$

where d is the Lévy distance. Verify that the metric d_1 is stronger than the metric d_2 in the sense that if $X_n \to X$ in d_1, then the same is true in d_2. Give an example to show that the converse is false, [i.e., convergence in distribution doesn't imply convergence in probability].

3. If $(\mathcal{M},d)$ is as in Problem 1, prove that the set of discrete d.f.s is everywhere dense in $\mathcal{M}$. [*Hint*: If $F \in \mathcal{M}$, let X be an r.v. on some (Ω, Σ, P) with F as its d.f. Then there exists a sequence of simple r.v.s $X_n \to X$ pointwise on this space, and apply Problem 2.]

4. The alternative proof of the Helly-Bray theorem given for Theorem 1.5 extends to the following case. Let Σ be the algebra generated by the open sets of $\mathbb{R}$, and let $\{P, P_n, n \geq 1\}$ be finitely additive "probabilities" on Σ.

Suppose that for each open set A such that $P(\partial A) = 0$ we have $P_n(A) \to P(A)$, where ∂A is the boundary of A. Then show that for each real bounded continuous f on $\mathbb{R}$

$$\lim_{n\to\infty} \int_{\mathbb{R}} f(x)\, dP_n(x) = \int_{\mathbb{R}} f(x)\, dP(x). \qquad (*)$$

Here integrals with respect to a finitely additive measure μ are defined as the obvious sums for step functions, and if $f_n \to f$ in μ-measure and $\int_{\mathbb{R}} |f_n - f_m|\, d\mu \to 0$, then $\int_{\mathbb{R}} f\, d\mu = \lim_n \int_{\mathbb{R}} f_n\, d\mu$, by definition. This integral has all the usual properties except that the Lebesgue limit theorems do not hold under the familiar hypotheses. The converse of the above (Helly-Bray) statement of (*) holds if P_n and P are also "regular," (i.e., μ is regular on Σ, if $\mu(A) = \inf\{\mu(B) : B \in \Sigma, A \subset \text{int}(B)\} = \sup\{\mu(C) : C \in \Sigma, \overline{C} \subset A\}$ for each $a \in \Sigma$.) [This is a specialization of a more general classical result of A. D. Alexandroff, and our proof works here.] In the above, let P, P_n be σ-additive, $P_n(A) \to P(A)$, and $f : \mathbb{R} \to \mathbb{R}^+$ be continuous. Then show that

$$\lim_{n\to\infty} \int_{\mathbb{R}} f(x) P_n(dx) = \int_{\mathbb{R}} f(x) P(dx) \text{ iff } \lim_{\lambda\to\infty} \sup_n \int_{[f>\lambda]} f(x) P_n(dx) = 0.$$

5. In proving Proposition 1.6, we noted that moments need not determine a d.f. uniquely. Here are two examples of this phenomenon:

(a) Let f_1, f_2 be two functions defined for $x > 0$ by

$$f_1(x) = (1/48) \exp[-(1/2)x^{1/3}],$$
$$f_2(x) = f_1(x)[1 + \sin(\sqrt{3}/2)x^{1/3}],$$

and $f_i(x) = 0$ for $x \leq 0, i = 1, 2$. Show that, with a calculus of residues computation, $\int_{\mathbb{R}^+} x^n \sin[(\sqrt{3}/2)x^{1/3}] f_1(x)\, dx = 0$ for all integers $n \geq 0$. Deduce that f_1, f_2 are densities having the same moments of all orders, even though $f_1 \neq f_2$ on $\mathbb{R}^+$.

(b) If the r.v. X is normally distributed, $N(0, 1)$, and $Y = e^X$, then Y has a d.f., called the *log-normal*, and the densities are

$$f_X(x) = (1/\sqrt{2\pi}) \exp\{-(1/2)x^2\}, \quad x \in \mathbb{R},$$
$$f_Y(y) = (1/\sqrt{2\pi}) y^{-1} \exp\{-(1/2)(\log y)^2\}, \quad y > 0,$$

and $f_Y(y) = 0$ for $y \leq 0$. Show that $\int_{\mathbb{R}} y^n \sin(2\pi \log y) f_Y(y)\, dy = 0$ for all integers $n \geq 0$. Deduce that f_Y and g_Y, defined by

$$g_Y(y) = f_Y(y)(1 + \sin(2\pi \log y)),\ y \in \mathbb{R},$$

are both densities with the same moments of all orders even though $f_Y \neq g_Y$ on $\mathbb{R}^+$.

6. Let F be a d.f., for which the *moment-generating function* (m.g.f.) $M(\cdot)$ exists, where by definition, $M(t) = \int_{\mathbb{R}} e^{tx}\, dF(x)$ for $|t| < \varepsilon$, for some $\varepsilon > 0$. Then verify that F has all moments finite, and F is uniquely determined by its moments. [The Taylor expansion of $M(t)$ shows that $M(\cdot)$ is analytic in the disc at the origin of $\mathbb{C}$ and of radius ε. Then the ch.f. ϕ of F and the m.g.f. M satisfy $\phi(z) = M(iz)$ on this disc and both are analytic. Thus the uniqueness follows.] This result implies, in particular, that the d.f. of a bounded r.v. is uniquely determined by its moments. Also deduce that a set $\{\alpha_n, n \geq 1\} \subset \mathbb{R}, \alpha_0 = 1$, forms a moment sequence of a unique d.f. if the series

$$\sum_{n \geq 0} (\alpha_n / n!) t^n = M(t)$$

is absolutely convergent for some $t > 0$ and if $M : t \mapsto M(t)$ satisfies $M(it) = \phi(t), t \in \mathbb{R}$, where ϕ is a ch.f.

7. Calculate the ch.f.s of the eight standard d.f.s given in Section 2.

8. Prove Proposition 2.3 by the substitution method of Theorem 2.1. Using this proposition, deduce that the ch.f. ϕ of the density $p, p(x) = (1 - \cos x)/\pi x^2, x \in \mathbb{R}$, is given by $\phi(t) = (1 - |t|)$ for $|t| \leq 1, = 0$ for $|t| > 1$.

9. Complete the details of Khintchine's example: Let f_1 be a probability density given by $f_1(x) = (1 - \cos x)/\pi x^2, x \in \mathbb{R}$, and F_2 be a discrete d.f. with jumps at $x = n\pi$ of sizes $2/n^2\pi^2, n = \pm 1, \pm 3, \ldots$, and $\frac{1}{2}$ at 0. Show that the ch.f.s coincide on the interval [-1,1] but are different elsewhere. Deduce that if $\phi_n(t) \to \phi(t), t \in [-a, a], a > 0$, where ϕ_n and ϕ are ch.f.s, then the convergence need not hold on all of $\mathbb{R}$. (Also, see Exercise 34 below.)

10. If $f(x) = K(|x|^2 \log |x|)^{-1}$ for $|x| \geq 2$, and $= 0$ for $|x| < 2$, where $K > 0$ is chosen such that $\int_{\mathbb{R}} f(x)\, dx = 1$, then verify that its ch.f ϕ is differentiable at $t = 0$ with $(d\phi/dt)(0) = 0$, even though the mean of f does not exist. [By the comment at the end of the proof of Theorem 2.1, the symmetric derivative of the ch.f. of a d.f. is equivalent to its ordinary derivative, and this may be used here.]

11. Let $X_1, X_2, \ldots$ be independent r.v.s each having the same distribution given by

$$P[X_i = -1] = P[X_i = +1] = \frac{1}{2}.$$

If $Y_n = \sum_{k=1}^{n}(X_k/2^k)$, find its ch.f. and show that $Y_n \to Y$, a.e., where Y is uniformly distributed on the interval (-1,+1). (Use ch.f.s and the continuity theorem for finding the d.f. of Y.)

12. Strengthen the conclusion of Corollary 2.10 as follows: If $\{\phi, \phi_n, n \geq 1\}$ is a sequence of ch.f.s such that $\phi_n(t) \to \phi(t), t \in \mathbb{R}$, then the convergence is uniform on each compact set of $\mathbb{R}$. [First show that the set of functions is equicontinuous; i.e., given $\varepsilon > 0$, there is a $\delta_\varepsilon > 0$ such that $|t - t'| < \delta_\varepsilon \Rightarrow |f(t) - f(t')| < \varepsilon$ for all $f \in \{\phi, \phi_n, n \geq 1\}$.] Suppose that none of the ϕ in this set vanishes. Show that Log $\phi_n(t) \to$ Log $\phi(t)$ iff $\phi_n(t) \to \phi(t), t \in \mathbb{R}$, and the convergence is uniform on compact intervals in both directions.

13. This problem contains a simple extension of the Lévy inversion formula with the same argument. Thus let F be a d.f. with ϕ as its ch.f. If $g : \mathbb{R} \to \mathbb{R}$ is an absolutely (i.e., $g, |g|$ are) Riemann integrable function such that for each $x \in \mathbb{R}$,

$$\lim_{h \to 0} g(x \pm h) = g(x \pm 0)$$

exists. Show that we then have

$$\int_{\mathbb{R}} [g(x+0) + g(x-0)]\, dF(x) = \lim_{T \to \infty} \frac{1}{\pi} \int_{-T}^{T} \int_{\mathbb{R}} g(u) e^{-itu} \phi(t)\, du\, dt.$$

Deduce Theorem 2.1 from this.

14. We give an adjunct to the selection theorem. Let $\{F_n, n \geq 1\}$ be a sequence of d.f.s. with $\{\phi_n, n \geq 1\}$ as the corresponding ch.f.s. If $F_n \to G$ (at all continuity points of G), a necessarily nondecreasing function with $0 \leq G \leq 1$, let g be its Fourier transform. Show that $\phi_n \to g$ on $\mathbb{R}$. If on the other hand, $\lim_{n\to\infty} \int_0^u \phi_n(t)\, dt$ exists for each $u \in \mathbb{R}$, show that $F_n \to$ a limit ($= H$, say), at all continuity points of H, and if h is the Fourier transform of H, then

$$\lim_{n \to \infty} \int_0^u \phi_n(t)\, dt = \int_0^u h(t)\, dt, u \in \mathbb{R}.$$

In particular, deduce that $\phi_n \to g$ a.e. [Lebesgue] on $\mathbb{R}$ implies $F_n \to G$ (at continuity points of G) and the Fourier transform of the limit agrees with g outside of a Lebesgue null set. (The argument is the same as that used for the last half of Theorem 4.5.)

15. Let $X_1, \ldots, X_n$ be r.v.s with ch.f.s defined by $\phi_k(t) = E(e^{itX_k})$ and

$$\phi_{1,\ldots,n}(t_1, \ldots, t_n) = E\left(\exp\left\{i \sum_{j=1}^{n} t_j X_j\right\}\right).$$

Show that the X_k are mutually independent if and only if

$$\phi_{1,\dots,n}(t_1,\dots,t_n) = \prod_{k=1}^{n} \phi_k(t_k).$$

Show by an example that the result is false if $(t_1,\dots,t_n)$ is replaced by only the diagonal $(t,\dots,t) \in \mathbb{R}^n$. [For a counterexample, consider $n = 2, |X_i| \le 1$, with a density given by $f(x_1, x_2) = \frac{1}{4}\{1 + x_1x_2(x_1^2 - x_2^2)\}, -1 < x_i < 1; = 0$ otherwise.]

16. We present three important facts in this exercise along with sketches of their proofs.

(a) Let $X_1, \dots, X_n$ be i.i.d. $N(0,1)$ r.v.s and set $\overline{X} = (1/n)\sum_{i=1}^n X_i$. Show that $\overline{X}$ and $\{(X_i - \overline{X}), i = 1, \dots, n\}$ are independent. [Use Problem 15, and the algebraic identity

$$\sum_{i=1}^{n} t_i(X_i - \overline{X}) = \sum_{i=1}^{n} \left[\frac{a}{n} + (t_i - \overline{t})\right] X_i - a\overline{X}, \qquad \text{all } t_i,\ a \text{ in } \mathbb{R},$$

and $\overline{t} = (1/n)\sum_{i=1}^n t_i$.] Show that the r.v. $V = \sum_{i=1}^n (X_i - \overline{X})^2$ has a gamma distribution whose ch.f. ϕ is given by $\phi(t) = (1-2it)^{-(n-1)/2}$. [Use the identity

$$\sum_{i=1}^{n} X_i^2 = \sum_{i=1}^{n} (X_i - \overline{X})^2 + n\overline{X}^2$$

and the fact that the left side has a ch.f. $t \mapsto (1 - 2it)^{-n/2}$. This result is important in statistical inference where $V/(n-1)$ is called the *sample variance* and $\overline{X}$, the *sample mean*. Just as in Proposition 2.8, it can be shown (with further work) that the independence property of $\overline{X}$ and V characterizes a normal distribution. See below.]

(b) It suffices to establish the above converse for $n = 2$. Thus let X_1, X_2 be i.i.d. with F as the common d.f., having two moments. Let $Y = X_1 + X_2$ and $Z = X_1 - X_2$. Show that if Y and Z are independent, then F is $N(\mu, \sigma^2)$ and this essentially gives the last statement of (a). [Let ϕ_Y, ϕ_Z and ψ be ch.f.'s of r.v.s Y, Z and d.f. F. Then the independence of $Y, Z \Rightarrow \phi_Y(s)\phi_Z(s) = \psi(s+t)\psi(s-t)$ and so

$$\phi(s) = \psi^2(s) \text{ and } \phi_Z(s) = \psi(s)\psi(-s).$$

These relations imply the key functional equation for ψ

$$\psi(s+t)\psi(s-t) = \psi^2(s)|\psi(t)|^2 \qquad (*)$$

since $\psi(-t) = \overline{\psi(t)}$. Put $s = t$ in (*) to get $|\psi(2s)| = |\psi(s)|^4$ so that

$$|\psi(2^2 s)| = |\psi(2s)|^4 = |\psi(s)|^{4^2}.$$

By iteration, conclude that $|\psi(2^n s)| = |\psi(s)|^{4^n}$, or $|\psi(s)| = |\psi(2^{-n}s)|^{4^n}$. Hence conclude that $\psi(\cdot)$ never vanishes. So by Propostion 2.9, $s \mapsto f(s) = \mathrm{Log}\psi(s)$, called the *cummulant* function of ψ, is well-defined. Thus (*) gives, upon taking logs, the cummulant equation

$$f(s+t) + f(s-t) = 2f(s). \qquad (**)$$

Now use the fact that F has two moments, so that by Proposition 2.6, ψ and hence f, is twice differentiable. Thus differentiate (**) twice relative to t and set $t = 0$ to get $f''(s) = -\sigma^2$ where $f''(0) = \sigma^2$, the variance of F. Since $f'(0) = i\mu$, the mean of F, the solution of this differential equation is $f(t) = \sigma^2 t^2/2 + \mu t$, so that $\psi(s) = \exp(i\mu - \sigma^2 t^2/2)$ as asserted.

Remark: The result is true without assuming any moments. But then we need a different method. The conclusion holds even if X_1, X_2 do not have the same distribution. Then $\psi_{X_1}(s) = e^{\mu_1 s + \sigma^2 s^2/2}$ and $\psi_{X_2}(s) = e^{-\mu_2 s + \sigma^2 s^2/2}$. See Stromberg (1994), p. 104. The above argument is a nice application of Propostions 2.6 and 2.9.]

(c) The independence concept is so special for probability theory, even for Gaussian families it has distinct properties. Thus let $X_1, \ldots, X_n$ be $N(0,1)$ random variables. Then they can be uncorrelated without being independent, or pairwise (or "m-wise, $m < n$) independent without mutual independence. Verify these statements by the following examples with $n = 2$ and 3.

(i) Let X_1 be $N(0,1)$ and $X_2 = X_1\chi_I - X_1\chi_{I^c}$ where I is an interval $I = [-a, a]$ such that $P[X_1 \in I] = 1/2$ and so $P[X_1 \in I^c] = P[-X_1 \in I^c] = 1/2$ since $-X_1$ is also $N(0,1)$. For any open set $J \subset \mathbb{R}$, observe that $P[X_2 \in J] = P[X_2 \in J \cap I] + P[X_2 \in J \cap I^c] = P[X_1 \in I \cap J] + P[-X_1 \in I^c \cap J] = P[X_1 \in I \cap J] + P[X_1 \in I^c \cap J] = P[X_1 \in J]$, since X_1 and $-X_1$ are identically distributed. From the arbitrariness of J, conclude that X_2 is also $N(0,1)$ and

$$E(X_1X_2) = E(X_1^2\chi_I) - E(X_1^2\chi_{I^c}) = 0$$

so that they are uncorrelated. Verify that

$$\phi_{X_1,X_2}(t_1,t_2) = E(e^{it_1X_1 + it_2X_2}) \neq e^{-\frac{t_1^2+t_2^2}{2}},$$

so that they are not jointly normal or Gaussian and are not independent.

(ii) Let X_1, X_2, X_3 be each $N(0,1)$, but with a joint density

$$\begin{aligned} f_{X_1,X_2,X_3}(x_1,x_2,x_3) = \;& g_{X_1}(x_1)g_{X_2}(x_2)g_{X_3}(x_3) \\ & \times(1 + \tilde{x}_1\tilde{x}_2\tilde{x}_3 g_{X_1}(x_1)g_{X_2}(x_2)g_{X_3}(x_3)) \end{aligned}$$

where g_{X_i} is the standard normal density $N(0,1)$ for $i = 1,2,3$, and $\tilde{x}_i = x_i\chi_{[|x_i| \le 1]}$ then $f_{X_1,X_2,X_3} \ge 0$ and is a nonfactorizable density so that X_1, X_2 and X_3 are not mutually independent, but pairwise independent $N(0,1)$. Clearly, a similar example can be given for any subset m $(< n)$ variables. [Note that if $(X_1, \ldots, X_n)$ is *jointly* normal, then these difficulties disappear.]

17. Let U, V be two r.v.s whose joint ch.f. $\phi_{U,V}(\cdot,\cdot)$ is given by

$$\phi_{U,V}(s,t) = \phi(s,t) = (1 - 2it + s^2)^{-1/2},$$

as in Example 5 of Section 3. Show that $P[V > 0] = 1$ and that the d.f. of $UV^{-1/2}$ is $N(0,1)$. [This is another consequence of Theorem 3.2. First verify that

$$\begin{aligned}\psi(s,t) = \phi_{U,\sqrt{V}}(s,t) &= \int_{\mathbb{R}^2} e^{isx+it\sqrt{y}}\,dx\,dy \int_{\mathbb{R}^2} e^{irx-iuy}\phi(r,u)\,dr\,du \\ &= \sqrt{\frac{2}{\pi}}\int_0^\infty \exp\{itv - v^2(1+s^2)/2\}\,dv.\end{aligned}$$

Next use Theorem 3.2 to get

$$\frac{1}{2\pi i}\int_{\mathbb{R}} \frac{\partial\psi}{\partial t}(s, -sx)\,ds = (2\pi)^{-1/2}\exp\{-x^2/2\}.]$$

18. Let X and Y be independent r.v.s with X as $N(0,1)$ and Y a gamma, so that their densities f_1, f_2 are given by

$$f_1(x) = (2\pi)^{-1/2}\exp\{-x^2/2\}, \quad f_2(y) = [\alpha^\lambda/\Gamma(\lambda)]y^{\lambda-1}e^{-\alpha y},$$

where $\lambda > 0, \alpha > 0, x \in \mathbb{R}, y \in \mathbb{R}^+$. Using Corollary 3.3, show that the distribution of $XY^{-1/2}$ has a density f given by

$$f(x) = \frac{1}{\sqrt{2\pi\alpha}}\frac{\Gamma(\lambda+1/2)}{\Gamma(\lambda)}\left(1+\frac{x^2}{2\alpha}\right)^{-\lambda-(1/2)}$$

If $\alpha = \lambda = n/2$, this f is called a "Student's density" with n degrees of freedom, and is of importance in statistical testing theory.

19. Let $\{X_n, n = 0, \pm 1, \ldots\}$ be a sequence of complex r.v.s with means zero and variances one. Suppose that $\tilde{r}(m,n) = E(X_m\overline{X}_n)$, and the covariance is of the form $\tilde{r}(m,n) = r(m-n)$; i.e., it depends only on the difference $m-n$. Such a sequence is called *weakly stationary*. Show that there is a unique probability function P on the Borel sets of $(0, 2\pi]$ such that

$$r(n) = \int_0^{2\pi} e^{in\lambda}\,dP(\lambda);$$

i.e., r is the "ch.f." of P. (Consider Herglotz's lemma. See Section 8.5 where the use of such r is discussed further.)

20. Suppose that $\{X_t, t \in \mathbb{R}\}$ is a set of complex r.v.s each with mean zero and variance one. Let $\tilde{r}(s,t) = E(X_s\overline{X}_t)$ and be of the form $\tilde{r}(s,t) = r(s-t)$. This is the continuous parameter analog of the above problem. If $r(\cdot)$

is Lebesgue measurable, show that there is a unique probability function P on the Borel sets of $\mathbb{R}$ such that

$$r(t) = \int_{\mathbb{R}} e^{it\lambda}\, dP(\lambda), \qquad t \in \mathbb{R} - \Lambda,$$

where Λ has Lebesgue measure zero. [Consider the Riesz extension of Bochner's theorem. In both problems, finite variances suffice, in which case $P(\cdot)$ is a finite (Borel) measure, but not necessarily a probability.]

21. Let X_n be an r.v. that has a log-normal distribution with mean μ_n and variance σ_n^2. This means that the density of X_n is defined by (compare with Problem 5b)

$$f_n(x) = \begin{cases} (x^2\alpha_n^2 2\pi)^{-1/2}\exp[-\frac{1}{2}\alpha_n^{-2}(\log x - \beta_n)^2], & x > 0, \ \beta_n \in \mathbb{R}, \ \alpha_n > 0 \\ 0, & \text{otherwise,} \end{cases}$$

and $\mu_n = \exp(\beta_n + \frac{1}{2}\alpha_n^2), \sigma_n^2 = \mu_n^2[e^{\alpha_n^2} - 1]$. Let $Z_n = (X_n - \mu_n)/\sigma_n$. Show that

$$\lim_{(\sigma_n/\mu_n)\to 0} P[Z_n < x] = \frac{1}{\sqrt{2\pi}}\int_{-\infty}^{x} e^{-t^2/2}\, dt.$$

What happens if $\sigma_n/\mu_n \not\to 0$ [Since a normal distribution is uniquely determined by its moments, by Problem 6, we can apply Proposition 1.6. (Calculation of the ch.f. of Z_n is clearly difficult.) Therefore compute the moments of Z_n and find their limits. This is not entirely simple. For instance, if $\gamma_n = e^{\alpha_n^2} - 1$, then

$$\begin{aligned} E(Z_n^k) &= (-1)^k\mu_n^k \\ &\quad \times \left[\sum_{i=1}^{\binom{k}{2}}\left\{\sum_{i=r}^{k}(-1)^i\binom{k}{i}\binom{\binom{i}{2}}{j}\right\}\gamma_n^j : (r-1)(r-2) < 2j \le r^2 - r\right] \\ &= (-1)^k\mu_n^k\sum_{j=1}^{\binom{k}{2}} C_{k,j}\gamma_n^j, \end{aligned}$$

where $C_{2k,k} = 1\cdot 3\cdots(2k-1)$. Regarding this simplification of the binomial coefficient identity, see *Canadian Math. Bulletin,* **14** (1971), 471-472.]

22. Let $\{X_n, n \ge 1\}$ be a sequence of r.v.s such that $X_n \xrightarrow{D} X$, where X is not a constant, and $a_nX_n + b_n \xrightarrow{D} \tilde{X}$, where $a_n > 0, b_n \in \mathbb{R}$, and $\tilde{X}$ is also not a constant r.v. Then show that $a_n \to a, b_n \to b$, and $\tilde{X} = aX + b$ a.e. (This assertion, due to A. Khintchine, can be proved quickly by the second method of proof given for Theorem 1.2. The result says that the convergent sequences of the form $\{a_nX_n + b_n, n \ge 1\}$ have the same *type* of limit laws.)

23. Let $\{\phi_n, n \geq 1\}$ be a sequence of ch.f.s such that $\phi_n(t) \to 1, t \in A$, where A is a set of positive Lebesgue measure in $\mathbb{R}$. Then verify that $\phi_n(t) \to 1, t \in \mathbb{R}$. [*Hints*: Since $\overline{\phi}_n(t) \to 1$ also, and $\overline{\phi}_n(t) = \phi_n(-t) \Rightarrow \phi_n(t-t') \to 0, t, t' \in A$ (CBS inequality)$\Rightarrow$ it holds on $B = A - A$ (algebraic difference). But by a classical result of H. Steinhaus, such a set B includes an open interval $(-a, a) \subset B, a > 0$. Thus $\phi_n(t) \to 1$ for $|t| < a$. By the CBS inequality, for all $|t| < a$ we get

$$\begin{aligned}|\phi_n(t) - \phi_n(2t)|^2 &\leq \int_{\mathbb{R}} dF_n(x) \int_{\mathbb{R}} |1 - e^{itx}|^2 \, dF_n(x) \\ &= 2(1 - \mathrm{Re}(\phi_n(t))) \to 0 \quad \text{as } n \to \infty.\end{aligned}$$

Hence $\phi_n(t) \to 1$ for $|t| < 2a$, and since $\mathbb{R} \subset \bigcup_{n \geq 1}(-na, na)$, the result holds first for $|t| < na, n > 1$, by induction, and then for all t.

Remark The above result is true somewhat more generally, namely: If $\phi_n(t) \to \phi(t)$ for all $|t| < a$, and ϕ is the ch.f. of a d.f. for which the m.g.f. also exists, then $\phi_n(t) \to \phi(t)$ on all of $\mathbb{R}$. See Problem 34 below for a more analytical statement.]

24. Let $X_1, \ldots, X_r$ be independent r.v.s such that X_i is $N(\tilde{\mu}, 1)$. If $Y_r = \sum_{i=1}^r X_i^2$, then Y_r is said to have a *noncentral chi-square* distribution with r and $\theta = (\sum_{i=1}^r \tilde{\mu}_i^2)$ as its parameters, called the degrees of freedom and noncentrality parameter, respectively. Verify that the ch.f. of Y_r is given by ϕ_r, where

$$\phi_r(t) = (1 - 2it)^{-r/2} \exp\{it\theta(1 - 2it)^{-1}\}.$$

Using the inversion (and calculus of residues), the density h_r of Y_r can be shown to be (verify this!)

$$\begin{aligned}h_r(x) = &\pi^{-1/2} 2^{-r/2} \exp\{-(1/2)(x + \theta)\} x^{(r/2)-1} \\ &\times \sum_{j=0}^{\infty} \frac{(\theta x)^j \Gamma(j + 1/2)}{(2j)! \Gamma(j + n/2)}, \quad x > 0,\end{aligned}$$

and $h_r(x) = 0$ for $x \leq 0$. If μ_r and σ_r^2 are the mean and variance of Y_r, show then that, as an application of the continuity theorem for ch.f.s,

$$\lim_{(\sigma_r/\mu_r) \to 0} P\left[\frac{Y_r - \mu_r}{\sigma_r} < x\right] = \frac{1}{\sqrt{2\pi}} \int_{-\infty}^{x} e^{-u^2/2} \, du.$$

What happens if $\sigma_r/\mu_r \not\to 0$ (The ratio σ_r/μ_r is called the *coefficient of variation* of Y_r and in statistics it is sometimes used to indicate the spread of probability in relation to the mean. In contrast to Problem 21, ch.f.s can be directly used here, and by Proposition 2.9, Log ϕ_r is well defined.)

25. If $\phi : \mathbb{R} \to \mathbb{C}$ is uniformly continuous, then for it to be a ch.f., it is necessary that (i) $\phi(0) = 1$, (ii) $|\phi(t)| \leq 1$, and (iii) $\phi(-t) = \overline{\phi(t)}$. But these are not sufficient.

(α) Show that the following are ch.f.s:

(a) $\phi(t) = \exp\{-|t|^r\}, 0 \le r \le 2$, (see Example 5.3.11 (ii) later),
(b) $\phi(t) = (1 + |t|)^{-1}$,
(c) $\phi(t) = 1 - |t|$ if $|t| \le \frac{1}{2}; = \frac{1}{4}|t|^{-1}$ if $|t| \ge \frac{1}{2}$.

[Use Pólya's criterion. In (a) if $1 < r \le 2$ this is not applicable. But $r = 2$ is obvious and $1 < r < 2$ can be deduced from Example 5.3.11, case (ii) as detailed there.]

(β) On the other hand, show that the following are *not* ch.f.s, but each satisfies the necessary conditions:

(a') $\psi(t) = \exp\{-|t|^k\}, k > 2$,
(b') $\psi(t) = (1 + t^4)^{-1}$,
(c') $\psi(t) = 1 - |t|^3$ if $|t| \le 1; = 0$ otherwise.

[If $\psi(t) = 1 + o(t^2)$ as $t \to 0$, then it is a ch.f. iff $\psi(t) \equiv 1$. Also $g : \mathbb{R}^+ \to \mathbb{R}^+$ is convex iff $g(x) = g(0) + \int_0^x h(t)\,dt$, where $h(\cdot)$ is nondecreasing. Use this in part (α). In (αa), if $1 < r < 2$, show that $\phi(t) = \lim_n [\psi(t/n^{1/r})]^n$, where ψ is the ch.f. of the so-called symmetric *Pareto density*, $p : x \mapsto (r/2)|x|^{-(r+1)}$ if $|x| > 1; = 0$ if $|x| \le 1$; and so $\psi(t) = 1 - c_r|t|^r + O(t^2)$ as $t \to 0$, but Pólya's result does not give us enough information!]

(γ) Show, however, that the following statements hold, and these complement the above two assertions: Let ϕ be a ch.f. and $\psi : \mathbb{R} \to \mathbb{C}$ be a function such that if $\{a_n, n \ge 1\}$ is any sequence with $a_n \nearrow \infty$, then $\phi(t)\psi(a_n t)$ defines a ch.f. for each n. Verify that ψ must in fact be a bounded continuous function and deduce (by the continuity theorem) that ψ must be actually a ch.f.

(δ) If ϕ is a ch.f., then $\phi\bar{\phi} = |\phi|^2$ is a ch.f. But the absolute value (and thus a square root) need not be a ch.f. For instance, if X is an r.v., $P[X = -1] = \frac{1}{2} = P[X = +1]$, then $|\phi(t)| = |E(e^{itX})|$ is not a ch.f. (If it were, then we may invert it and get its density. It may be seen that there is no such density by a simple calculation. For a general case, see Problem 35.)

26. This problem deals with a calculation of probabilities in certain nonparametric statistical limit distributions, and seems to have already been noted by B. V. Gnedenko in the early 1950s. Let X, Y be jointly normally distributed with means zero, unit vaiances, and correlation ρ, so that their density f is given by

$$f(x,y) = \frac{1}{2\pi(1-\rho^2)^{1/2}} \cdot \exp\left\{-\frac{1}{2(1-\rho^2)}(x^2 - 2\rho xy + y^2)\right\}, \quad (x,y) \in \mathbb{R}^2.$$

Show that, by the image law theorem,

$$P[X > 0, Y > 0] = \frac{1}{4} + \frac{1}{2\pi}\int_{x=0}^{\infty}\int_{y=0}^{\alpha x} \exp\left(-\frac{x^2+y^2}{2}\right) dy\, dx,$$

where $\alpha = \rho(1-\rho^2)^{1/2}$. [Since $\exp\{-\frac{1}{2}(x^2+y^2)\}$ is symmetric in x and y, the second term can be interpreted as the probability that (X,Y) takes its values in the sector $\{(x,y) : 0 < x < +\infty, 0 < y < \alpha x\}$ if $\rho > 0$, and replace α by $-\alpha$ if $\rho \leq 0$. If θ is the angle made by this sector, then the probability mass being uniform in each such sector, it must be $\theta/2\pi$. Since the slope of the line making an angle θ with the x axis is $\alpha = \tan\theta$, we get that the second term is $(1/2\pi)\text{arc } \sin\rho$.] From this deduce that

$$P[XY < 0] = 1 - 2P[X > 0, Y > 0] = (1/\pi)\text{arc } \cos\rho.$$

27. (a) Let $X_n = (X_{n1}, \ldots, X_{nk})$ be a k-vector of r.v.s and $X = (X_1, \ldots, X_k)$ be a random vector. If $a = (a_1, \ldots, a_k) \in \mathbb{R}^k$, show that $a \cdot X_n = \sum_{i=1}^k a_i X_{ni} \to a \cdot X$ in distribution for each $a \in \mathbb{R}^k$ as $n \to \infty$ iff the vectors $X_n \to X$ in distribution, i.e., iff the d.f.s $F_{X_n}(x) \to F_X(x)$ at all continuity points $x = (x_1, \ldots, x_k) \in \mathbb{R}^k$ of F_X. Deduce that if $X_n \xrightarrow{D} X$, then $X_{ni} \xrightarrow{D} X_i$ for each $i = 1, \ldots, k$. However, does the componentwise convergence in distribution imply its vector convergence?

(b) Suppose (X,Y) is a random vector on (Ω, Σ, P). If F_X, F_Y are (the marginal) d.f.s of X and Y, define the joint d.f. of the vector (X,Y) as F^a,

$$\begin{aligned} F^a(x,y) =& F_X(x)F_Y(y) + a(F_X(x) \\ &-F_X(x)^2)(F_Y(y) - F_Y(y)^2), \quad |a| \leq 1. \end{aligned}$$

Show that for each such a, F^a is a d.f. with the same marginals F_X, F_Y. [It should be verified also that the increment $\Delta F^a(x,y) \geq 0$ for all $(x,y) \in \mathbb{R}^2$.] Thus the marginals generally do not determine a unique joint d.f.

28. Let $\{X_n, n \geq 1\}$ be a sequence of i.i.d nonnegative r.v.s and $S_n = \sum_{k=1}^n X_k, S_0 = 0$. Then X_n may be regarded as the lifetime of, for example, the nth bulb, or of the nth machine before breakdown, so that S_n denotes the time until the next replacement or *renewal*. Let $N(t)$ be the number of renewals in the interval $[0,t]$, or equivalently $N(t)$ is the largest n such that $S_n \leq t$. Hence $N(t) = 0$ if $S_1 > t$, and $N(t) = n$ if $S_n \leq t < S_{n+1}$. Show that $N(t)$ is an r.v., and if F is the d.f. of X_1, then

$$P[N(t) = n] = F^{(n)}(t) - F^{(n+1)}(t), \quad n \geq 1, \quad t \geq 0,$$

where $F^{(1)} = F, F^{(2)} = F * F$ (the convolution), and $F^{(n)} = F * F^{(n-1)}$. If $F(x) = 1 - e^{-\lambda x}, \lambda > 0, x \geq 0$ ($= 0$ otherwise), show that

$$\lim_{t\to\infty} P\left[\frac{N(t) - \lambda t}{\sqrt{\lambda t}} \leq y\right] = (2\pi)^{-1/2} \int_{-\infty}^{y} e^{-u^2/2}\, du.$$

[The $N(t)$ process will be analyzed further in Section 8.4.]

29. Complete the details of proof of Proposition 4.4.

30. Let $X_1, X_2, \ldots$ be i.i.d. random variables each with mean zero and a finite variance $0 < \sigma^2 < \infty$. Let $S_n = \sum_{k=1}^n X_k$. Suppose that the common ch.f. ϕ of the X_i is integrable on $\mathbb{R}$. If $a \in \mathbb{R}$ and $\delta > 0$, show that, by use of the Lévy inversion formula,

$$\lim_{n\to\infty} \sqrt{n} P[a - \delta < S_n < a + \delta] = \left(\frac{2\delta^2}{\pi\sigma^2}\right)^{1/2}.$$

[*Hints*: If F_n is the d.f. of $S_n, a \pm \delta$ are continuity points of F_n, then by Theorem 2.1,

$$F_n(a+\delta) - F_n(a-\delta) = \lim_{T\to\infty} \frac{1}{\pi} \int_{-T}^{T} \frac{\sin \delta t}{t} e^{-ita} \phi(t)^n \, dt. \qquad (+)$$

Next $\phi(t) = [1 - (t^2/2) + o(t^2)]$, and so

$$\phi\left(\frac{t}{\sigma\sqrt{n}}\right)^n = E\left(\exp\left\{\frac{itS_n}{\sigma\sqrt{n}}\right\}\right) \to e^{-t^2/2}.$$

Since $|\phi(t)^n| \le |\phi(t)|$, which is integrable, replace t by $(t/\sigma\sqrt{n})$ in the above formula (+), and take limits with appropriate justification. Finally note that $a \pm \delta$ can be arbitrary continuity points, as in the original statement.]

31. Let $\{X_n, n \ge 1\}$ be i.i.d. with $E(X_1) = 0, E(X_1^2) = \sigma^2 < \infty$, and $S_n = \sum_{k=1}^n X_k$. Then $\sum_{n\ge1} P[|S_n| > n\varepsilon] < \infty$, and conversely, if this series converges, we can conclude that the moment conditions hold. [*Sketch*: The converse part was outlined in Problem 19 in Chapter 2. The direct part is also involved, and is due to Hsu and Robbins (1947). If ϕ is the common ch.f. of the X_i, then $|1-\phi(t)| \le a_1 t^2, |\phi'(t)| < a_2|t|$, and $|\phi''(t)| \le a_3$ for t near 0. Let Z be an r.v. *independent* of an X having two moments and whose ch.f. ψ vanishes outside (-4,4). [For example, $f_z(x) = 3(2\pi)^{-1} x^{-4} \sin^4 x$, and we use this density.] Let $|t| \le 4\delta, \delta > 0$, and note that

$$\sum_{n\ge1} P[|Z| > 4n\delta] \le \frac{3}{\pi} \sum_{n\ge1} \int_{n\delta}^{\infty} \frac{dx}{x^4} < \infty,$$

and

$$\sum_{n\ge1} P[|S_n| > 2n] \le \sum_{n\ge1} (P[|S_n + Z\delta^{-1}| > n] + P[|Z| > n\delta]),$$

and it suffices to verify that

$$\sum_{n=1}^{N} (P[|Z| \le n\delta] - P[|S_n + Z\delta^{-1}| \le n]) \qquad (*)$$

is bounded in N. The ch.f. of $S_n + Z\delta^{-1}$ vanishes outside of $(-4n\delta, 4n\delta)$. Thus by the inversion formula,

$$P[|S_n + Z\delta^{-1}| \leq n] = \frac{1}{\pi}\int_{-4\delta}^{4\delta} \phi^n(t)\psi\left(\frac{t}{\delta}\right)\frac{\sin nt}{t}\,dt,$$

$$P[|Z| \leq n\delta] = \frac{1}{\pi}\int_{-4\delta}^{4\delta} \psi\left(\frac{t}{\delta}\right)\frac{\sin nt}{t}\,dt.$$

Hence on subtraction we get

$$(*) = \frac{1}{\pi}\int_{-4\delta}^{4\delta} \psi\left(\frac{t}{\delta}\right)\sum_{n=1}^{N}(1 - \phi^n(t))\frac{\sin nt}{t}\,dt = \frac{1}{\pi}A_N \quad \text{(say)}.$$

Use the above inequalities for $\phi(t)$, and choose $\delta > 0$ so that for some constants B, C,

$$|\sin(t/2)| \geq B|t|, \quad |(1-\phi(t))^2 - 4(1-\phi(t))\sin^2(t/2) + 4\sin^2(t/2)| \geq Ct^2,$$

and simplify this to show

$$A_N = \int_{-\delta}^{\delta}(1-\phi(t))\phi(t)^N\frac{\sin Nt}{t^3}\,dt$$
$$- \int_{-\delta}^{\delta} 2(1-\phi(t)^{N-1})\frac{\sin t/2\cos(N-1/2)t}{t^3}\,dt + O(1).$$

By a careful estimation, show that these integrals are bounded. There is some delicate estimation here, which proves that (*) is bounded, and thus the result follows.]

32. Different forms of Theorem 2.1 can be used to calculate probabilities at the discontinuity points of the d.f. of an r.v. (a) If X is an r.v. whose d.f. has a jump at a, and if ϕ is the ch.f. of X, then, using the method of proof of the just noted theorem, show that

$$\lim_{T\to\infty}\frac{1}{2T}\int_{-T}^{T} e^{-iat}\phi(t)\,dt = P[X = a].$$

(b) If $\{a_n, n \geq 1\}$ is the set of all discontinuity points of the d.f. of X, then show that we have an extension of the above result as

$$\lim_{T\to\infty}\frac{1}{2T}\int_{-T}^{T}|\phi(t)|^2\,dt = \sum_{n\geq 1}(P[X = a_n])^2.$$

[*Hints*: Let Y be another r.v. which is independent of X and which has the same d.f. F as X. Then $\phi(t)\cdot\overline{\phi(t)} = E(e^{it(X-Y)})$. But the d.f. G of $X - Y$ is the convolution of that of X and $-Y$ given by

$$G(x) = \int_{\mathbb{R}} F(x-y)\, d\tilde{F}(y), \quad x \in \mathbb{R}$$

where $\tilde{F}(y) = P[-Y < y]$. By (a), the left side (above) gives the discontinuity of G at $x = 0$, and for this value show that the integral for G is $\sum_n (P[X = a_n])^2$, since a_n is a discontinuity of F iff $-a_n$ is such for $\tilde{F}$.]

33. This exercise extends Proposition 2.6. Let X_1 and X_2 be independent r.v.s on (Ω, Σ, P) and $X = X_1 + X_2$. If X has $2n$ moments, then the same is true of X_1 and X_2. Moreover, there exist real numbers K_j and α_j such that

$$E(X_j^{2k}) \le K_j E(\alpha_j + X)^{2k}, \quad j = 1, 2 \quad \text{and} \quad 0 \le k \le n.$$

[*Hints*: Let ϕ_j and ϕ be the ch.f.s of X_j and X, so that $\phi = \phi_1\phi_2$. Replacing X_2 by $\beta + X_2$ if necessary, we may assume that X_2 takes values in both $\mathbb{R}^+$ and $\mathbb{R}^-$ with positive probability. Since $E(X^{2n}) < \infty$, by Proposition 2.6, $\phi^{(2n)}(0)$ exists; expressing it as the symmetric derivative, we get

$$(-1)^n \phi^{(2n)}(0) = \int\int_{\mathbb{R}^2} (x+y)^{2n}\, dF_{X_1}(x)\, dF_{X_2}(y).$$

Hence for $0 \le k \le 2n$,

$$(-i)^k \phi^{(k)}(0) = \int\int_{\mathbb{R}^2} (x+y)^k\, dF_{X_1}(x)\, dF_{X_2}(y)$$

and for $0 \le x, y \le a$, since $0 \le x^{2n} \le (x+y)^{2n}$, we get from this

$$\begin{aligned}\lim_{a\to\infty} \int_0^a dF_{X_2}(y) \int_0^a x^{2k}\, dF_{X_1}(x)& \\ \le \lim_{a\to\infty} \int_0^a \int_0^a (x+y)^{2k}\, dF_{X_1}(x)\, dF_{X_2}(y) &\le |\phi^{(2k)}(0)|.\end{aligned}$$

Interchanging F_{X_1}, F_{X_2}, here, the second inequality holds since

$$|E(e^{it(\alpha+X)})| = |E(e^{itX})|.$$

This is a form of a result of A. Devinatz and extends a classical result due to P. Lévy and independently to D. A. Raikov.]

34. If ϕ_1, ϕ_2 are ch.f.s on $\mathbb{R}$, which agree on $(-a, a), a > 0$, and if one of them is regular, then they agree everywhere. (Compare with Exercise 9. This is a form of a classical result due to J. Marcinkiewicz, and the proof depends on complex function theory. Compare with Proposition 2.8 (ii).)

35. An r.v. X is of *lattice type* if its range is of the form $\{\alpha + k\beta : k = 0, \pm 1, \pm 2, \ldots, \beta > 0 \text{ and } \alpha \text{ real}\}$. Show that its ch.f. ϕ_X is periodic of period β. If $\beta = 2\pi$, then

$$P[X = k] = \frac{1}{2\pi} \int_{-\pi}^{\pi} \phi_X(t) e^{-ikt} \, dt.$$

In particular, if X is a symmetric lattice r.v., deduce that $|\phi_X(\cdot)|$ cannot be a ch.f. This extends Exercise 25(δ). (For a stronger negative statement, with ϕ never vanishing, see Exercise 9 of Chapter 5.)

Chapter 5

Weak Limit Laws

The strong (or pointwise a.e.) limit theory of Chapter 2 naturally leads to the distributional convergence of random sequences. Such a shift in viewpoint enabled an enormous growth of probability theory. This chapter contains a general outline of this picture. It starts with the classical central limit theorems of Lévy and Liapounov and contains their modern versions as well as an error estimate of Berry and Esséen. Some aspects of infinite divisibility together with the Lévy-Khintchine representation and stability are treated. The invariance principles of Donsker and Prokhorov are discussed, and two important applications are included. Further, Kolmogorov's law of the iterated logarithm and related results are given. Applications and extensions to m-dependent sequences establish the generality and limitations of invariance principles. The tools developed in Chapter 4 are essential here. The material in this chapter represents a central aspect of analytical probability theory, some of which will be essential for Part III of this book dealing with some important applications. In fact, much of additive process analysis in Chapter 8 depends upon the work of this chapter.

5.1 Classical Central Limit Theorems

In all the results of Chapter 2, we demanded pointwise convergence of the sequences of various r.v.s (either partial sums or averages). The conclusions are the strongest possible in this setup, and all considerations relate to the given underlying probability space. However, if we lower our demands and settle for somewhat weaker conclusions, such as convergence in probability or in distribution (hence the appellation "weak limit laws"), then many new results can be proved. In this, the intermediate versions with "in probability" are not very illuminating. Really new areas are opened for investigation when we go to the convergence theory on the image spaces, i.e., to the distributional convergence. In this case, we can employ the new tools developed in the preceding chapter, thereby bringing in the well-known machinery of classical Fourier analysis. Moreover, as Theorem 4.6.1 shows, in some important cases the weak and strong convergence statements coincide. With these as motivation, we shall concentrate henceforth on distributional convergence, inequalities, and the like. Note that by the Kolmogorov-Bochner theorem (see 3.4.10), with a consistent family of distribution functions, we can manufacture a probability space and a set of r.v.s on it having the given d.f.s as its finite-dimensional distributions. Thus with a probability space and r.v.s on it, we can go to the image space with d.f.s satisfying the compatibility conditions, and given the latter family, we can invent a probability space that is, in a well-defined sense, measure-theoretically indistinguishable from the original space. This type of change of spaces to suit our needs is a distinguishing feature of probability theory. In this vein, we prove some key results that play a central role in theory and applications; these results are called *central limit theorems*, especially when the limit distribution is normal or Gaussian (or more generally, one that is "infinitely divisible," a term to be defined later).

Let us start with a very simple result, the weak law of large numbers, due to A. Khintchine, which was given as Theorem 2.3.2. But we can slightly improve it here. [We use modern methods and simplifications afforded by ch.f.s.]

Proposition 1 *Let $\{X_n, n \geq 1\}$ be independent r.v.s with the same distribution. If their common ch.f. has merely a derivative at the origin of $\mathbb{R}$, with value $i\mu$, then $(1/n)\sum_{k=1}^{n} X_k \to \mu$ in probability.*

Remark It was noted in the last chapter (Problem 10) that a ch.f. can have a derivative at the origin without the d.f. having a finite mean. Thus the hypothesis here is weaker than that of Theorem 2.3.2. That proof does not apply. (However, the proof there illustrates the truncation technique which is useful for other results.)

Proof If $\phi(t) = E(e^{itX_k})$, then by hypothesis $\phi(t) = 1 + i\mu t + o(t)$ (as $t \to 0$), since ϕ has a derivative $= i\mu$ at 0. If $S_n = \sum_{k=1}^{n} X_k$, then

$$\tilde{\phi}_n(t) = E(e^{itS_n/n}) = \prod_{k=1}^{n} E(e^{itX_k/n}) \quad \text{(by independence)}$$

$$= \phi\left(\frac{t}{n}\right)^n = \left[1 + \frac{i\mu}{n}t + o\left(\frac{t}{n}\right)\right]^n \to e^{i\mu t} \qquad \text{as } n \to \infty.$$

By the continuity theorem, this implies $(S_n/n) \to \mu$ in distribution, hence in probability, by Proposition 2.2.2. This completes the proof.

The following result was established by A. de Moivre in about 1730 if the r.v.s are Bernoulli, taking values 1 and 0, each with probability $\frac{1}{2}$, and it was extended by P. S. Laplace nearly a century later if the probabilities are p, $1-p$ $(0 < p < 1)$, usually called the Laplace-DeMoivre central limit theorem. The result was generalized by J. W. Lindeberg in the early 1920's, and to the present form by P. Lévy a little later.

Theorem 2 *Let $\{X_n, n \geq 1\}$ be i.i.d. random variables each with mean μ and variance $\sigma^2 > 0$. If $S_n = \sum_{k=1}^n X_k$, then*

$$\lim_{n\to\infty} P\left[\frac{S_n - n\mu}{\sqrt{n\sigma^2}} < x\right] = (2\pi)^{-1/2} \int_{-\infty}^{x} e^{-u^2/2}\, du, \qquad x \in \mathbb{R}. \tag{1}$$

Proof Let ϕ_n be the ch.f. defined by

$$\begin{aligned}
\phi_n(t) &= E(\exp[it(S_n - n\mu)/\sqrt{n\sigma^2}]) \\
&= \prod_{j=1}^{n} E\left(\exp\left\{\frac{it}{\sqrt{n\sigma^2}}(X_j - \mu)\right\}\right) \qquad \text{(by independence)} \\
&= \left[\phi\left(\frac{t}{\sqrt{n\sigma^2}}\right)\right]^n \qquad [\phi(\cdot) \text{ being the common ch.f. of } X_i - \mu], \\
&= \left[1 + 0 + \frac{(it)^2\sigma^2}{2n\sigma^2} + o(t)^2\right]^n \qquad \text{(by Corollary 4.2.7)} \\
&\to e^{-t^2/2} \quad \text{as} \quad n \to \infty.
\end{aligned} \tag{2}$$

Since $t \mapsto e^{-t^2/2}$ is the ch.f. of the standard normal $N(0,1)$ d.f., the result follows by the continuity theorem for ch.f.s, completing the proof.

A natural question is to consider the case that the X_n are not identically distributed; if this is solved one tries to extend the results for sequences which form certain dependence classes. We consider the first problem here in some detail and aspects of the second (dependent) case later. It is noted that the above theorem fails if the i.i.d. hypothesis is omitted. (See Example 4 below.) The early treatment of a nonidentically distributed case was due to A. Liapounov in 1901, and his results were generalized in the works of J. W. Lindeberg and W. Feller. This circle of results attained the most precise and definitive treatment in the theory of infinitely divisible distributions. First we present Liapounov's theorem, since his method of proof has not only admitted a solution of the central limit problem, as indicated above, but contained a calculation for the speed of convergence in the limit theory as well. Here is the convergence result.

Theorem 3 (Liapounov) *Let $\{X_n, n \geq 1\}$ be a sequence of independent r.v.s with means $\{\mu_n, n \geq 1\}$, variances $\{\sigma_n^2, n \geq 1\}$, and finite third absolute central moments $\rho_k^3 = E(|X_k - \mu_k|^3)$. If $S_n = \sum_{k=1}^n X_k$, and we write $\rho^3(S_n) = \sum_{k=1}^n \rho_k^3$ (not* the third absolute moment of S_n*), $\sigma^2(S_n) = \sum_{k=1}^n \sigma_k^2$, then*

$$\lim_{n\to\infty} P\left[\frac{S_n - E(S_n)}{\sigma(S_n)} < x\right] = (2\pi)^{-1/2} \int_{-\infty}^{x} e^{-t^2/2}\, dt, \qquad x \in \mathbb{R}, \tag{3}$$

whenever $\lim_{n\to\infty}[\rho(S_n)/\sigma(S_n)] = 0$.

Proof Let $\phi_k(t) = E(\exp\{it(X_k - \mu_k)\})$ and

$$\psi_n(t) = E\left(\exp\left\{\frac{it}{\sigma(S_n)}(S_n - E(S_n))\right\}\right),$$

so that

$$\psi_n(t) = \prod_{k=1}^{n} \phi_k\left(\frac{t}{\sigma(S_n)}\right),$$

by independence. By Corollary 4.2.7, [see Eq. (19′) there] we have

$$\phi_k\left(\frac{t}{\sigma(S_n)}\right) = 1 - \frac{\sigma_k^2 t^2}{2\sigma^2(S_n)} + \theta_1 \frac{\rho_k^3 t^3}{6\sigma^3(S_n)} = 1 + y_k \quad \text{(say)}, \quad |\theta_1| \le 1. \tag{4}$$

Hence

$$|y_k| \le \frac{\sigma_k^2 t^2}{2\sigma^2(S_n)} + \frac{\rho_k^3 |t|^3}{6\sigma^3(S_n)} \le \frac{\rho_k^2}{\sigma^2(S_n)}\left(\frac{t^2}{2} + \frac{\rho_k}{\sigma(S_n)}\frac{|t|^3}{6}\right),$$

since $\sigma_k \le \rho_k$ by Liapounov's inequality (cf. Corollary 1.3.5 with $r = 2, s = 3$, and $\mu = 0$ there). But $\rho_k/\sigma(S_n) \le \rho(S_n)/\sigma(S_n)$, which tends to zero as $n \to \infty$. Thus there exists an n_0 such that $n \ge n_0 \Rightarrow \rho_k/\sigma(S_n) < 1$ and for fixed but arbitrary t, $|y_k| \le \frac{1}{2}$ since $|y_k| \to 0$ as $n \to \infty$ for each k. Now writing

$$y_k = \theta_2 \rho_k^2 \left(\frac{t^2}{2} + \frac{|t|^3}{6}\right) / \sigma^2(S_n), \qquad |\theta_2| \le 1,$$

we have

$$|(\text{Log } (1 + y_k) - y_k)/y_k^2| \le \frac{1}{2} + \frac{1}{4} + \cdots = 1.$$

Consequently for some $|\theta_3| \le 1$, [using (4)],

$$\begin{aligned}
\text{Log } (1 + y_k) &= y_k + \theta_3 y_k^2 \\
&= \frac{-\sigma_k^2 t^2}{2\sigma^2(S_n)} + \frac{\theta_1 \rho_k^3 t^3}{6\sigma^3(S_n)} + \theta_3 \left[\frac{\theta_2 \rho_k^2}{\sigma^2(S_n)}\left(\frac{t^2}{2} + \frac{|t|^3}{6}\right)\right]^2 \\
&= -\frac{\sigma_k^2 t^2}{2\sigma^2(S_n)} + \theta_4 \frac{\rho_k^3}{\sigma^3(S_n)}\left[\frac{|t|^3}{6} + \left(\frac{t^2}{2} + \frac{|t|^3}{6}\right)\right],
\end{aligned}$$

where $\theta_4 = \theta_{4,k}$ and $|\theta_4| \le 1$. Hence we can take the logarithms of ψ_n and ϕ_k (cf. Proposition 4.2.9) to get

$$\begin{aligned}
\text{Log } \psi_n(t) &= \sum_{k=1}^{n} \phi_k\left(\frac{t}{\sigma(S_n)}\right) \\
&= -\frac{t^2}{2} + \theta_5 \frac{\rho^3(S_n)}{\sigma^3(S_n)}\left[\frac{|t|^3}{6} + \left(\frac{t^2}{2} + \frac{|t|^3}{6}\right)^2\right] \\
&\to -t^2/2 \quad \text{as} \quad n \to \infty,
\end{aligned} \tag{5}$$

since $[\rho(S_n)/\sigma(S_n)] \to 0$ by hypothesis. Here we have set

$$\left|\sum_{k=1}^{n} \theta_{4,k}\rho_k^3\right| \leq \sum_{k=1}^{n} \rho_k^3 = \rho^3(S_n),$$

so that for some $\theta_5(=\theta_{5,n}), |\theta_5| \leq 1$,

$$\sum_{k=1}^{n} \theta_4 \rho_k^3 = \theta_5 \rho^3(S_n).$$

Hence $\lim_{n\to\infty} \psi_n(t) = \exp(-t^2/2)$, and the proof is completed by the continuity theorem.

Remark In contrast to Theorem 2, in the nonidentically distributed case, the conclusion of the above theorem does not hold if an additional condition on the relative growth of the moments higher than 2 of r.v.s is not assumed.

Let us amplify the significance of this remark by the following example:

Example 4 Let $\{X_n, n \geq 1\}$ be independent r.v.s such that

$$P[X_k = k^{1/2}] = 1/2k = P[X_k = -k^{1/2}]$$

and

$$P[X_k = 0] = 1 - (1/k).$$

Then $E(X_k) = 0, \sigma^2(X_k) = 1$, so that $\sigma^2(S_n) = n$, where $S_n = \sum_{k=1}^n X_k$ and $\rho^3(S_n) = \sum_{k=1}^n k^{3/2}$. Thus $\rho^3(S_n)$ is asymptotically of the order $n^{5/2}$ (by Euler's formula on such expressions). Hence $\rho^3(S_n)/\sigma^3(S_k) \not\to 0$. On the other hand,

$$\psi_n(t) = E(\exp\{itS_n/\sqrt{n}\}) = \prod_{j=1}^{n} \left(1 - \frac{2}{j}\sin^2[(t/2)\sqrt{j/n}]\right).$$

Therefore

$$\log\psi_n(t) = \sum_{j=1}^{n} \log\left(1 - \frac{2}{j}\sin^2\frac{t}{2}\sqrt{j/n}\right) \to -2\int_0^1 \left(\frac{\sin[(t/2)\sqrt{u}]}{\sqrt{u}}\right)^2 du$$

$$= f(t)(\neq -\frac{t^2}{2}),$$

where we used the Riemann approximation to the integral. Thus

$$\lim_{n\to\infty} \psi_n(t) = \exp\{f(t)\} \neq \exp(-t^2/2).$$

Hence the limit distribution of $S_n/\sqrt{n}$ exists, but it is not normal, even though all the X_k have the same means and the same variances, but not identically distributed, and the hypothesis of Theorem 3 is not satisfied.

For the validity of Liapounov's theorem it suffices to have $2+\delta, \delta > 0$ moments for the r.v.s. Then define $\rho_k^{2+\delta} = E(|X_t|^{2+\delta}), \rho^{2+\delta}(S_n) = \sum_{k=1}^n \rho_k^{2+\delta}$. The sufficient condition of the theorem becomes

$$\lim_{n\to\infty} [\rho^{2+\delta}(S_n)/\sigma^{2+\delta}(S_n)] = 0.$$

The demonstration is a slight modification of that given above, and we shall not consider it here (but will indicate the result in problem 2). See the computation following Theorem 3.6 below for another method.

In the proof of the preceding theorem [cf. especially Equation (5)], there is more information than that utilized for the conclusion. The argument gives a crude upper bound for the error at the nth stage. This is clearly useful in applications, since one needs to know how large n should be in employing the limit theory. We now present one such result, due independently to A.C. Berry in 1941 and C.G. Esséen in 1945, which is a generalization of the original work of Liapounov as well as an improvement of that of Cramér's, (cf. the latter's monograph (1970)).

Theorem 5 *Let $\{X_n, n \geq 1\}$ be independent r.v.s as in Theorem 3, i.e., $E(X_n) = \mu_n, Var\ X_n = \sigma_n^2(> 0)$, and $\rho_n^3 = E(|X_n - \mu|^3)$ with $\rho^3(S_n) = \sum_{k=1}^n \rho_k^3$, where $S_n = \sum_{k=1}^n X_k$. Then there exists an absolute constant $0 < C_0 < \infty$ such that*

$$\sup_{x\in\mathbb{R}} \left| P\left[\frac{S_n - E(S_n)}{\sigma(S_n)} < x\right] - (2\pi)^{-1/2}\int_{-\infty}^{x} e^{-u^2/2}\,du\right| < C_0 \frac{\rho^3(S_n)}{\sigma^3(S_n)}. \quad (6)$$

If the X_n are i.i.d., and the rest of the hypothesis is satisfied, let Var $X_1 = \sigma^2, \rho^3 = E(|X_1 - \mu_1|^3)$, so that $\rho^3(S_n)/\sigma^3(S_n) = (\rho/\sigma)^3(1/\sqrt{n})$. Under these conditions we deduce from (6) the following:

Corollary 6 *Let $\{X_n, n \geq 1\}$ be i.i.d. with three moments finite. Then there is an absolute constant $0 < C_1 < \infty$ such that*

$$\sup_{x\in\mathbb{R}} \left| P\left[\frac{S_n - n\mu}{\sigma\sqrt{n}} < x\right] - (2\pi)^{-1/2}\int_{-\infty}^{x} e^{-u^2/2}\,du\right| < \frac{C_1}{\sqrt{n}} \cdot \left(\frac{\rho}{\sigma}\right)^3. \quad (7)$$

Remark In this case A.C. Berry also indicated a numerical value of C_1. Carrying out the details carefully, V.M. Zolotarev in 1966 showed that $C_1 \leq 1.32132\cdots$. The best possible value of C_1 is not known. For (6), H. Bergström in 1949 indicated that $C_0 < 4.8$, which perhaps can be improved.

The proof of Theorem 5 is involved, as it depends on many estimates. The basic idea, however, is the same as that of Liapounov's theorem. Thus we present these estimates separately as two key lemmas, and then complete the proof of the theorem thereafter. If F_n is the d.f. of $[S_n - E(S_n)]/\sigma(S_n)$, and Φ is the standard normal d.f., so that $F_n \to \Phi$ under our assumptions, we should get uniform (in x) bounds on $|F_n(x) - \Phi(x)|$ for each n. Thus if F and G are two d.f.s, G is continuous with a bounded density G', let $H = F - G$. (We later set $G = \Phi$ and $F = F_n$.) For nontriviality let $H \neq 0$. Since H is a function of bounded variation, if we add some integrability conditions (which will hold in our context) then Theorem 4.3.1 suggests the method to be employed here. In fact, that theorem was devised primarily for the (present) error estimation needs by Cramér, and the analysis with it can be carried forward much further. (One can give an asymptotic expansion: $F = G + H + \cdots$, where the H_i are certain functions of the Čebyšev-Hermite polynomials, but we do not consider that aspect here.)

Lemma 7 *Let* $H = F - G$ *and* $M = \sup_{x\in\mathbb{R}} |G'(x)|$. *If* H *is Lebesgue integrable on* $\mathbb{R}$, *and* h *is its Fourier-Stieltjes transform, then for every* $T > 0$ *we have*

$$\sup_{x\in\mathbb{R}} |H(x)| \leq \frac{2}{\pi}\int_0^T \left|\frac{h(t)}{t}\right| dt + \frac{24M}{\pi T}. \tag{8}$$

Proof Since the result is clearly true if the integral is infinite on the right side of (8), let it be finite. Then by definition

$$h(t) = \int_{\mathbb{R}} e^{ixt} dH(x) = -it\int_{\mathbb{R}} H(x)e^{ixt}\,dx,$$

since $H(\pm\infty) = 0$, and integration by parts is used. Hence for any fixed but arbitrary $a_0 \in \mathbb{R}\,(t \neq 0)$,

$$-\frac{h(t)}{it}e^{-ita_0} = \int_{\mathbb{R}} H(x + a_0)e^{itx}\,dx. \tag{9}$$

Since $[-h(t)/it]$ is the Fourier transform of $H(\cdot)$, we would like to convert it into a convolution by multiplying with a *suitable* ch.f. with compact support. This is the key trick in the proof. So consider the "triangular" ch.f. $\phi_T(t) = [1 - |t|/T]$ for $|t| \leq T$, and $=0$ for $|t| > T$. This is the ch.f. of the symmetric probability density $(1 - \cos Tx)/\pi T x^2$, and we have already used it in proving Pólya's theorem on ch.f.s.

Thus multiplying both sides of (9) by ϕ_T and integrating, we get

$$\begin{aligned}
&-\int_{-T}^{T} \frac{h(t)}{it}e^{-ita_0}\phi_T(t)\,dt \\
&\quad= \int_{-T}^{T}\int_{\mathbb{R}} H(x + a_0)e^{itx}\phi_T(t)\,dx\,dt \\
&\quad= 2\pi\int_{\mathbb{R}} \frac{1}{2\pi}\int_{-T}^{T} e^{itx}\phi_T(t)\,dt\,H(x + a_0)\,dx
\end{aligned}$$

$$\text{(by Fubini's Theorem)}$$
$$= \int_{\mathbb{R}} H(x + a_0) \frac{2(1 - \cos Tx)}{Tx^2}\, dx$$
$$\text{(by the inversion Theorem 4.2.1 applied to } \phi_T)$$
$$= 2 \int_{\mathbb{R}} \frac{2\sin^2(Tx/2)}{Tx^2} H(x + a_0)\, dx$$
$$= 2 \int_{\mathbb{R}} \left(\frac{\sin v}{v}\right)^2 H_{a_0}\left(\frac{2v}{T}\right) dv, \tag{10}$$

where $H_{a_0}(x) = H(x + a_0)$. Hence (10) gives

$$\left| \int_{\mathbb{R}} \frac{\sin^2 v}{v^2} H_{a_0}\left(\frac{2v}{T}\right) dv \right| \leq \frac{1}{2} \int_{-T}^{T} \left| \frac{h(t)}{t} \right| |\phi_T(t)|\, dt$$
$$\leq \int_0^T \left| \frac{h(t)}{t} \right| dt \quad [\text{since } |\phi_T(t)| \leq 1]. \tag{11}$$

It is now necessary to find a lower estimate of the left side of the integral of (11), after an analysis of the behavior of H. This involves some computation, and we now set down the details.

If $\alpha = \sup_{x\in\mathbb{R}} |H(x)|$, then $\pm\alpha$ is attained at some $a \in \mathbb{R}$, i.e., $H(a) = \pm\alpha$ (by left continuity), or $H(a + 0) = \alpha$. Indeed, since $0 < \alpha \leq 2$, there exists a sequence $\{x_n, n \geq 1\} \subset \mathbb{R} \subset \overline{\mathbb{R}}$ with a convergent subsequence $x_{n_i} \to a \in \overline{\mathbb{R}}$. But $H(x) \to 0$ as $x \to \pm\infty$, and $\alpha > 0$, so that $a \in \mathbb{R}$, i.e., a must be finite. Thus there is a subsequence $\{x_{n_j}, j \geq 1\} \subset \{x_{n_i}, i \geq 1\}$ such that $x_{n_j} \to a$ and $H(x_{n_j}) \to \alpha$ or $H(x_{n_j}) \to -\alpha$. Consider the first case: $H(x_{n_j}) \to \alpha$. Now $\{x_{n_j}, j \geq 1\}$ must have a further subsequence which converges to a from the left or from the right. In the former case, by left continuity $H(a) = +\alpha$. In the latter case $H(a+0) = \alpha$. Also, by the Fatou inequality, and the continuity of G,

$$\alpha \leq H(a) \leq F(a + 0) - G(a) = F(a + 0) - G(a + 0) = H(a + 0) = \alpha.$$

Thus $H(a) = +\alpha$ holds. The case that $H(a) = -\alpha$ is similar.

Let $\beta = \alpha/2M, b = a+\beta$, and consider $H(x+b)$ for $|x| < \beta$. Then $b+x \geq a$ and, since $|G'(x)| \leq M$ by hypothesis,

$$\begin{aligned} H_b(x) &= F(x + b) - G(x + a + \beta) \\ &\geq F(a) - G(a + x + \beta) \\ &= F(a) - [G(a) - (x + \beta)G'(\theta)] \\ &\quad \text{(say)} \quad \text{(by the Taylor expansion of } G \text{ about } a) \\ &= H(a) + (x + \beta)G'(\theta) \geq \alpha - (x + \beta)M \\ &= 2M\beta - (x + \beta)M = M(\beta - x). \end{aligned} \tag{12}$$

If $H(a) = -\alpha$, then this bound would be $-M(\beta + x)$ with the inequality reversed. Now consider for (10)

$$\int_{[|x|<\beta]} H_b(x)\frac{1-\cos Tx}{x^2}\,dx$$
$$\geq M\int_{[|x|<\beta]} (\beta - x)\frac{1-\cos Tx}{x^2}\,dx \quad \text{[by (12)]}$$
$$= M\beta\int_{[|x|<\beta]} \frac{1-\cos Tx}{x^2}\,dx$$

(since the second integral vanishes)

$$= 2M\beta\int_0^\beta \frac{2\sin^2(Tx/2)}{x^2}\,dx$$
$$= 2M\beta T\left(\pi/2 - \int_{\beta T/2}^\infty \frac{\sin^2 v}{v^2}\,dv\right). \tag{13}$$

On the other hand,

$$\int_{[|x|\geq\beta]} H_b(x)\frac{1-\cos Tx}{x^2}\,dx \leq \alpha\int_{[|x|\geq\beta]} \frac{1-\cos Tx}{x^2}\,dx$$
$$= 2\alpha T\int_{\beta T/2}^\infty \frac{\sin^2 v}{v^2}\,dv. \tag{14}$$

But then the left side of (11) simplifies, using (13) and (14) (set $a_0 = b$ there), as follows:

$$\left|\int_{\mathbb{R}} \frac{\sin^2 v}{v^2} H_b\left(\frac{2v}{T}\right)\,dv\right|$$
$$\geq \int_{\mathbb{R}} \frac{\sin^2 v}{v^2} H_b\left(\frac{2v}{T}\right)\,dv, \text{ since } H_b \text{ can be negative,}$$
$$= \int_{\mathbb{R}} H_b(x)\frac{1-\cos Tx}{Tx^2}\,dx$$
$$= \int_{[|x|<\beta]} H_b(x)\frac{1-\cos Tx}{Tx^2}\,dx + \int_{[|x|\geq\beta]} H_b(x)\frac{1-\cos Tx}{Tx^2}\,dx$$
$$\geq 2M\beta\left(\frac{\pi}{2} - \int_{\beta T/2}^\infty \frac{\sin^2 v}{v^2}\,dv\right) - \left|\int_{[|x|\geq\beta]} H_b(x)\frac{1-\cos Tx}{Tx^2}\,dx\right|$$
$$\geq \alpha\left(\frac{\pi}{2} - 3\int_{\beta T/2}^\infty \frac{\sin^2 v}{v^2}\,dv\right) \qquad \text{[by (13) and (14)]}$$
$$\geq \alpha\left(\frac{\pi}{2} - \frac{6}{\beta T}\right) \quad \left(\text{since } \frac{\sin^2 v}{v^2} \leq \frac{1}{v^2}\right).$$

Putting this in (11) and transposing the terms, we get (8). In case $H(a) = -\alpha$, with the reverse inequality of (12), we get the same result after an analogous computation. This completes the proof of the lemma.

Let us specialize this lemma to the case where $G = \Phi$ and F is the d.f. F_n of our normalized partial sum $(S_n - E(S_n))/\sigma(S_n)$. To use the result of (8), one needs to find an upper bound for the right-side integral involving $h(\cdot)$, which is the difference of two ch.f.s. This is obtained in the next lemma.

Lemma 8 *Let $\{X_n, n \geq 1\}$ be independent r.v.s with three moments: $E(X_n) = \mu_n$, $\operatorname{Var} X_n = \sigma_n^2$, and $\rho_n^3 = E(|X_n - \mu|^3)$. If $S_n = \sum_{k=1}^n X_k$, $\rho^3(S_n) = \sum_{k=1}^n \rho_k^3$, and $\psi_n : t \mapsto E(\exp\{it(S_n - E(S_n))/\sigma(S_n)\})$, then there exists an absolute constant $0 < C_2 < \infty$ such that ($C_2 = 16$ is permissible below)*

$$|\psi_n(t) - \exp\{-t^2/2\}| \leq C_2 \frac{\rho^3(S_n)}{\sigma^3(S_n)} |t|^3 e^{-t^2/6} \tag{15}$$

for $|t| < \sigma^3(S_n)/2\rho^3(S_n)$.

Proof It may be assumed that the means μ_n are zero. Let $\phi_k(t) = E(e^{itX_k})$. Then by Corollary 4.2.7, we can again use the computations of Liapounov's proof, i.e. of (4) and (5), if, with the notations there, we write

$$\phi_k\left(\frac{t}{\sigma(S_n)}\right) = 1 + y_k,$$

where

$$|y_k| < \rho_k^2 \left(\frac{t^2}{2} + \frac{\rho_k}{\sigma(S_n)} \frac{|t|^3}{6}\right) / \sigma^2(S_n),$$

and if $|t| < \sigma(S_n)/2\rho(S_n)$, we get $|y_k| < (1/8) + (1/48) < 1/2$. Thus for t in this range with the $|\theta_i| \leq 1$, one gets on using $\sigma_k \leq \rho_k$

$$\begin{aligned}
\operatorname{Log}(1 + y_k) &= y_k + \theta_3 y_k^2 \\
&= -\frac{\sigma_k^2 t^2}{2\sigma^2(S_n)} + \frac{\theta_1 \rho_k^3 |t|^3}{6\sigma^3(S_n)} + \theta_3 \left(\frac{\rho_k^2 t^2}{2\sigma^2(S_n)} + \theta_1 \frac{\rho_k^3 |t|^3}{6\sigma^3(S_n)}\right)^2 \\
&= -\frac{\sigma_k^2 t^2}{2\sigma^2(S_n)} + \frac{\theta_4 \rho_k^3 |t|^3}{\sigma^3(S_n)} [\frac{1}{6} + \frac{1}{2}(\frac{1}{2} + \frac{1}{12})^2].
\end{aligned}$$

Hence, summing on k,

$$\operatorname{Log} \psi_n(t) = -\frac{t^2}{2} + \theta_5 \frac{97}{288} \frac{\rho^3(S_n)}{\sigma^3(S_n)} |t|^2. \tag{16}$$

Let us now simplify the left side of (15) with the estimate (16). Here we use the trivial inequality $|e^x - 1| \leq |x| e^{|x|}$. Thus

$$\begin{aligned}
&|\psi_n(t) - e^{-t^2/2}| \\
&= \left| \exp\left\{\theta_5 \frac{97}{288} \frac{\rho^3(S_n)}{\sigma^3(S_n)} |t|^3\right\} - 1 \right| e^{-t^2/2} \\
&\leq \frac{97}{288} e^{-t^2/2} \left(\frac{\rho^3(S_n)}{\sigma^3(S_n)} |t|^3\right) \exp\left\{\frac{97}{288} \frac{\rho^3(S_n)|t|^3}{\sigma^3(S_n)}\right\} \\
&< \frac{1}{2} e^{-t^2/2} e^{1/16} \left(\frac{\rho^3(S_n)}{\sigma^3(S_n)} |t|^3\right) \quad \left[\text{since} \frac{|t|\rho(S_n)}{\sigma(S_n)} < \frac{1}{2}\right] \\
&< e^{-t^2/2} \frac{\rho^3(S_n)}{\sigma^3(S_n)} |t|^3 \quad (\text{since } e^{1/16} < 2) \qquad (17)
\end{aligned}$$

Next we extend the range of t as given in (15). For this we symmetrize the d.f. Thus let X_k' be an r.v. independent of X_k but with the same d.f. Then $X_k - X_k'$ has for its ch.f. $|\phi_k|^2$, and since (by convexity of $|x|^3$)

$$E(|X_k - X_k'|^3) \leq 8\rho_k^3, \qquad \operatorname{Var}(X_k - X_k') = 2\sigma_k^2$$

(with zero means), we have on writing $|\phi_k|^2$ in place of ϕ_k in (4), and using the above estimates for the variance and third absolute moment there (this is simpler than squaring the value of $|\phi_k|$)

$$\begin{aligned}
&\left|\phi_k\left(\frac{t}{\sigma(S_n)}\right)\right|^2 \\
&= 1 - \frac{t^2(2\sigma_k^2)}{2\sigma^2(S_n)} + \tilde{\theta}_1 \frac{8\rho_k^3|t|^3}{\sigma^3(S_n)6} \quad (|\tilde{\theta}_1| \leq 1) \\
&\leq 1 - t^2 \frac{\sigma_k^2}{\sigma^2(S_n)} + 4\frac{|t|^3}{3} \frac{\rho_k^3}{\sigma^3(S_n)} \\
&\leq \exp\left\{-\frac{t^2\sigma_k^2}{\sigma^2(S_n)} + 4\frac{|t|^3}{3} \frac{\rho_k^3}{\sigma^3(S_n)}\right\} \quad (\text{since } 1 + x \leq e^x, x > 0).
\end{aligned}$$

Multiplying over $k = 1, \ldots, n$, we get

$$\begin{aligned}
|\psi_n(t)|^2 &\leq \exp\{-t^2 + 4|t|^3\rho^3(S_n)/3\sigma^3(S_n)\} \\
&\leq \exp\{-t^2 + 2t^2/3\} = e^{-t^2/3} \\
&\quad [\text{since } |t| \leq \sigma^3(S_n)/2\rho^3(S_n) \quad \text{in (15)}]. \qquad (18)
\end{aligned}$$

Now to extend (17), if $|t| \geq \sigma(S_n)/2\rho(S_n)$ [but satisfying the range condition of (15)],

$$\begin{aligned}
|\psi_n(t) - e^{-t^2/2}| &\leq |\psi_n(t)| + e^{-t^2/2} \\
&\leq e^{-t^2/6} + e^{-t^2/2} \leq 2e^{-t^2/6} \quad [\text{by (18)}] \\
&\leq 8\left(\frac{\rho(S_n)}{\sigma(S_n)}|t|\right)^3 2e^{-t^2/6}. \qquad (19)
\end{aligned}$$

Thus (17) and (19) together establish (15), with $C_2 = 16$, and the lemma follows.

With these estimates we are now ready to complete our proof.

Proof of Theorem 5 We set $G(x) = \Phi(x)$, so that $G'(x) = (2\pi)^{-1/2} e^{-x^2/2}$ and $M = (2\pi)^{-1/2}$ in Lemma 7. Also, if

$$F_n(x) = P\left[\frac{S_n - E(S_n)}{\sigma(S_n)} < x\right],$$

then by Čebyšev's inequality $F_n(x) \leq 1/x^2$ if $x < 0$ and $\geq 1 - (1/x^2)$ if $x > 0$. The same is true for $\Phi(x) = P[X < x]$. Hence $H = F_n - \Phi$ is Lebesgue integrable first on $[|x| > \varepsilon]$ and, being bounded, also on $[-\varepsilon, \varepsilon]$ for $\varepsilon > 0$, thus on $\mathbb{R}$ itself. Let $T = [\sigma(S_n)/\rho(S_n)]^3$ in Lemma 8. Since the hypothesis of Lemma 7 is also satisfied, we get

$$\begin{aligned}
&\sup_{x \in \mathbb{R}} |\, F_n(x) - \Phi(x)| \\
&\leq \frac{2}{\pi} \int_0^T \left| \frac{\psi_n(t) - e^{-t^2/2}}{t} \right| dt + \frac{48}{(2\pi)^{3/2} T} \quad \text{[by (8)]} \\
&\leq \frac{2}{\pi} \int_0^T 16 \frac{\rho^3(S_n)}{\sigma^3(S_n)} t^2 e^{-t^2/6}\, dt + \frac{48}{(2\pi)^{3/2}} \left[\frac{\rho(S_n)}{\sigma(S_n)}\right]^3 \\
&\leq \left(\frac{\rho(S_n)}{\rho(S_n)}\right)^3 \left[\frac{32}{\pi} \int_0^\infty t^2 e^{-t^2/6}\, dt + \frac{48}{(2\pi)^{3/2}}\right], \\
&= C_0 \left(\frac{\rho(S_n)}{\sigma(S_n)}\right)^3,
\end{aligned}$$

where C_0 is the above constant. This is (6) and the proof is complete.

Remark It should be noted that the various "standard" tricks used in the above estimates have their clear origins in Liapounov's proof of this theorem, and thus the important problem of error estimation is considered there for the first time. Taking the X_k as Bernoulli r.v.s in the Berry-Esséen theorem, one may note that the order of $(\rho^3/\sigma^3)(S_n)$ cannot be smaller than what we have obtained in (6). Under the given hypothesis, it is "the best" *order* of magnitude, though C_0 can (with care) be improved. Also, if the limit distribution is different (but continuous with a bounded density such as gamma), Lemma 7 can still be used. For an up-to-date treatment of this subject, see R. N. Bhattacharya and R. R. Rao (1976). For another account including the multidimensional problem, see Sazonov (1981), Springer Lecture Notes in Math. No. 879.)

In the central limit theorems of Lévy and Liapounov, we considered the independent sequences $\{X_k/\sigma(S_n), 1 \leq k \leq n\}, n = 1, 2, \ldots$, and their partial sums. Abstracting this, one may consider the double sequences $\{X_{nk}, 1 \le k \le n\}$ of independent r.v.s in each row. These results can appear as interesting problems in their own right; S. D. Poisson already considered such a question in 1832. We can establish it with our present tools quite easily, but as seen in the next section, this turns out to be an important new step in the development of our subject. Its striking applications will appear later in Chapter 8.

Theorem 9 (Poisson) *Let $\{X_{nk}, 1 \le k \le n, n \ge 1\}$ be a sequence of finite sequences of Bernoulli r.v.s which are i.i.d. in each row:*

$$P[X_{nk} = 1] = \lambda/n, \qquad P[X_{nk} = 0] = 1 - \lambda/n, \qquad k = 1, \ldots, n, \ \lambda > 0.$$

If $S_n = \sum_{k=1}^n X_{nk}$, then

$$\lim_{n\to\infty} P[S_n = k] = e^{-\lambda}(\lambda^k/k!), \qquad k = 0, 1, 2, \ldots$$

Proof Let $\phi_{nk}(t) = E(e^{itX_{nk}}), \phi_n(t) = E(e^{itS_n})$. Then

$$\begin{aligned}\phi_n(t) &= \prod_{k=1}^n \phi_{nk}(t) \qquad \text{(by independence in each row)}\\ &= \left[1 + \frac{\lambda}{n}(e^{it} - 1)\right]^n \to \exp\{\lambda(e^{it} - 1)\} \quad \text{as } n \to \infty.\end{aligned}$$

Since the limit is a Poisson ch.f., the result follows from the continuity theorem.

The interest in this proposition stems from the fact that the limit d.f. is not Gaussian. Unfortunately, the significance of the result was not recognized until the late 1920s, and the theorem remained a curio for all those years. Its status, on a par with the Gaussian d.f., was realized only after the formulation and study of infinitely divisible distributions as a culmination of the central limit problem. We take it up in the next section, presenting several key results in that theory, and then deduce certain important theorems, which were originally proved by different methods. For instance, we obtain, among others, the Lindeberg-Feller theorem, which generalizes both Theorems 2 and 3, later in Section 5.3. Its key part is considered further in the continuous parameter case in Section 8.4. As is clear from the above work, the use of ch.f.s in the analysis is crucial. The basic probability space will be in the background. For that reason, this aspect of the subject is often referred to as the analytical theory of probability.

5.2 Infinite Divisibility and the Lévy-Khintchine Formula

One of the important points raised by the Poisson limit theorem is that the asymptotic d.f.s for partial sums of a sequence of sequences of r.v.s independent in each row can be different from the Gaussian family. This is a generalized version of the classical central limit problem, and the end result here is different–it is not normal. The only thing common to both these d.f.s is that their ch.f.s are exponentials. An abstraction of these facts leads to the introduction of one of the most important classes as follows:

Definition 1 A d.f. F with ch.f. ϕ is called *infinitely divisible* if for each integer $n \geq 1$, there is a ch.f. ϕ_n such that $\phi = (\phi_n)^n$.

Since a product of ch.f.s corresponds to the sum of independent r.v.s, the above definition can be translated into saying that an r.v. has an infinitely divisible d.f. iff it can be written as a sum of n i.i.d. variables for each $n \geq 1$, or equivalently, the d.f. is the n-fold convolution of some d.f. for each $n \geq 1$. Clearly the normal and Poisson ch.f.s are infinitely divisible.

Two immediate properties of real interest are given in the following:

Proposition 2 (a) *An infinitely divisible ch.f. ϕ never vanishes on $\mathbb{R}$. (b) If X is an infinitely divisible r.v., and X is nonconstant, then X takes infinitely many values, which form an unbounded set.*

Proof (a) By definition, for each integer $n \geq 1$ there is a ch.f. ϕ_n such that $\phi = (\phi_n)^n$, so that $|\phi_n|^2 = |\phi|^{2/n}$. But $\phi_n\overline{\phi}_n = |\phi_n|^2$, and since ϕ_n and $\overline{\phi}_n$ are ch.f.s, their product is a ch.f. (It is the ch.f. of $X - X'$, where X, X' are i.i.d. with ch.f. ϕ_n.) Similarly, $|\phi|^2$ is a ch.f. Since $|\phi(t)| \leq 1$,

$$\lim_{n\to\infty} |\phi_n(t)|^2 = \lim_{n\to\infty} |\phi(t)|^{2/n} = g(t) \quad (\text{say}), \quad t \in \mathbb{R}, \tag{1}$$

exists, and $g(t) = 0$ on the set $\{t : \phi(t) = 0\}$, and $=1$ on the set $\{t : |\phi(t)| > 0\}$. But $\phi(0) = 1$ and ϕ is continuous at $t = 0$. Thus there exists a $T > 0$ such that $|\phi(t)| > 0$ for $|t| < T$. Hence $g(0) = 1$ and g is continuous for $-T < t < T$. Since each $|\phi_n|^2$ is a ch.f., and $|\phi_n|^2 \to g$, which is continuous at $t = 0$, by the continuity theorem, g must be a ch.f., and hence is continuous on $\mathbb{R}$. But g takes only two values: 0, 1. Thus we can conclude that $g(t) = 1$ for all $t \in \mathbb{R}$ which yields $|\phi(t)| > 0$ for all $t \in \mathbb{R}$. This proves (a).

(b) If X concentrates at a single point $\mu \in \mathbb{R}$, then its ch.f. $\phi : t \mapsto e^{it\mu}$ is clearly infinitely divisible. If X takes finitely many values, then it is a bounded r.v. Thus, more generally, suppose X is a bounded r.v. We assert that it must be degenerate. Indeed, if $|X| \leq M < \infty$ a.e., and if X is an infinitely divisible r.v., then by definition, for each integer $n \geq 1$, there exist n i.i.d. variables $X_{n1}, \ldots, X_{nn}$ such that $X = \sum_{k=1}^{n} X_{nk}$. The fact that $|X| \leq M$ a.e. implies $|X_{nk}| \leq M/n$ a.e. for all $1 \leq k \leq n$. If not, $A_1 = [X_{nk} > M/n]$ or

$A_2 = [X_{nk} < -M/n], P(A_i) > 0$, for $i = 1$ or 2, and at least one (hence all by i.i.d.)$k, P[X_{nk} > M/n] > 0$ or $P[X_{nk} < -M/n] > 0$, then on $A_1, [X > M]$, or on $A_2, [X < -M]$, which is impossible. Thus $P(A_1 \cup A_2) = 0$. Hence

$$\text{Var } X = \sum_{k=1}^{n} \text{Var } X_{nk} \leq \frac{M^2}{n^2} \cdot n = \frac{M^2}{n}.$$

Since n is arbitrary, Var X=0, so that $X = E(X)$ a.e., as asserted. (See Problem 8 which asks for another argument.)

An immediate consequence is that the binomial, multinomial, or uniform d.f.s are *not* infinitely divisible. On the other hand, the r.v. whose d.f. has a density f given by

$$f(x) = (1 - \cos x)/\pi x^2, \qquad x \in \mathbb{R},$$

is unbounded. Its ch.f. ϕ, which we found before, is given by $\phi(t) = 1 - |t|$ if $|t| \leq 1$, and $=0$ if $|t| > 1$. Since $\phi(t) = 0$ on $[|t| > 1]$, by the proposition this ch.f. cannot be infinitely divisible. In addition to normal and Poisson, we shall see that the gamma ch.f. is also infinitely divisible (cf. Problem 14).

Remark (Important) Since by (a) of the proposition, an infinitely divisible ch.f. ϕ never vanishes, by Proposition 4.2.9 there exists a unique continuous function $f : \mathbb{R} \to \mathbb{C}$ such that $f(0) = 0$ and $\phi(t) = e^{f(t)}$, i.e., $f(t) = \text{Log } \phi(t), t \in \mathbb{R}$, the *distinguished logarithm*. We call $e^{(1/n)f(t)}$ *the (distinguished)* nth *root* of $\phi(t)$, and denote it $(\phi(t))^{1/n}$.

Hereafter, the nth root of ϕ is always meant to be the distinguished one as defined above. This appears often in the present work. Thus, we have the following fact on the extent of these ch.f.s.

Proposition 3 *Let $\mathcal{F}$ be the family of infinitely divisible ch.f.s on $\mathbb{R}$. Then $\mathcal{F}$ is closed under multiplication and passages to the (pointwise) limits in that $\phi_n \in \mathcal{F}, \phi_n \to \phi$ on $\mathbb{R}, \phi$ is a ch.f. $\Rightarrow \phi \in \mathcal{F}$.*

Proof Let $\phi_i \in \mathcal{F}, i = 1, \ldots, k$. Then by definition for each integer $n \geq 1$, there exist ψ_{in} such that $\phi_i = (\psi_{in})^n$. Thus $\phi = \prod_{i=1}^k \phi_i = (\prod_{i=1}^k \psi_{in})^n$ and $\prod_{i=1}^k \psi_{in}$ is a ch.f. Hence ϕ is infinitely divisible and $\phi \in \mathcal{F}$. Next suppose that $\phi_n \in \mathcal{F}$ and $\phi_n \to \tilde{\phi}$ a ch.f., as $n \to \infty$. Then for each integer $m \geq 1$, we have that $|\phi_n|^{2/m}$ is a ch.f., as seen before, and the hypothesis implies that $|\phi_n|^{2/m} \to |\tilde{\phi}|^{2/m}$ and the limit is continuous at $t = 0$. Thus by the continuity theorem $|\tilde{\phi}|^{2/m}$ is a ch.f., and $|\tilde{\phi}|^2$ is infinitely divisible. Now by Proposition 2a, $\tilde{\phi}$ never vanishes and is continuous, so that Log $\tilde{\phi}$ is well defined. The same is true of Log ϕ_n. Hence on $\mathbb{R}$

$$\phi_n^{1/m} = \exp\left\{\frac{1}{m} \text{Log } \phi_n\right\} \to \exp\left\{\frac{1}{m} \text{Log } \tilde{\phi}\right\} = (\tilde{\phi})^{1/m}, \tag{2}$$

and $(\tilde{\phi})^{1/m}$ is continuous at $t = 0$. Thus $(\tilde{\phi})^{1/m}$ is a ch.f., implying that $\tilde{\phi}$ is infinitely divisible. Thus $\tilde{\phi} \in \mathcal{F}$, and the proof is complete. [Note that $\phi_n \in \mathcal{F}, \phi_n \to \phi$ on $\mathbb{R} \not\Rightarrow \phi \in \mathcal{F}$, as $\phi_n(t) = e^{-nt^2/2}$ shows.]

The argument used for (2) implies that, if $\phi \in \mathcal{F}$, then ϕ^r is a ch.f. for each rational $r \geq 0$. Then the continuity theorem and the fact that each real $\lambda \geq 0$ can be approximated by a sequence of rationals $r_n \geq 0$ give the result for all $\lambda \geq 0$. A similar argument (left to the reader) gives the second part of the following, since $(|\phi|^2)^{1/n}$ is a ch.f., $n \geq 1$.

Corollary 4 *Let ϕ be an infinitely divisible ch.f. Then for each $\lambda \geq 0, \phi^\lambda$ is also a ch.f. Further, $|\phi|$ is an infinitely divisible ch.f.*

But there are ch.f.s ϕ such that ϕ^λ is not a ch.f. for some $\lambda > 0$. (See Problem 9.) These properties already show the special nature and intricate structure of infinitely divisible ch.f.s. There exist pairs of noninfinitely divisible ch.f.s whose product is infinitely divisible; and also pairs one member of which is infinitely divisible and the other one not, and the product is not infinitely divisible. We illustrate this later (see Example 8.)

The problem of characterizing these ch.f.s was proposed, and a first solution given, by B. de Finetti in 1929, and a general solution, if the r.v.s have finite variances, was given by A. Kolmogorov in 1932. Since, as is easily seen, the Cauchy distribution is infinitely divisible and it has no moments, a further extension was needed. Later P. Lévy succeeded in obtaining the general formula, which includes all these cases, but it was not yet in canonical form. The final formula was derived by A. Khintchine in 1937 in a more compact form from Lévy's work. We now present this *fundamental result* and then obtain the original formulas of Lévy and of Kolmogorov from it.

Theorem 5 (Lévy-Khintchine Representation) *Let $\phi : \mathbb{R} \to \mathbb{C}$ be a mapping. Then ϕ is an infinitely divisible ch.f. iff it can be represented as*

$$\phi(t) = \exp\left\{i\gamma t + \int_{\mathbb{R}} \left(e^{ixt} - 1 - \frac{itx}{1+x^2}\right) \frac{1+x^2}{x^2} dG(x)\right\}, \tag{3}$$

where $\gamma \in \mathbb{R}$, and G is a bounded nondecreasing function such that $G(-\infty) = 0$. With G taken as left continuous, the representation is unique.

Proof We first prove the sufficiency and uniqueness of the representation, since the argument here is more probabilistic than the converse part. Thus let ϕ be given by (3). The idea of proof is to show that ϕ is a limit of a sequence of ch.f.s each of which is infinitely divisible, so that by Proposition 3 the limit function ϕ, which is continuous on $\mathbb{R}$, will be infinitely divisible. We exclude the true and trivial case that $G \equiv 0$.

Let $\varepsilon > 0$ and $T > 0$ be given. For $|t| \leq T$, choose $a_\varepsilon > 0$ such that

$$[G(\infty) - G(a_\varepsilon) + G(-a_\varepsilon) - G(-\infty)] < \varepsilon/2K, \tag{4}$$

where $K(= K_T) = \sup\{|f_t(x)| : |t| \le T, x \in \mathbb{R}\}$ with

$$f_t(x) = (e^{itx} - 1 - itx(1+x^2)^{-1})(1+x^2)/x^2,$$

and so $\lim_{x\to 0} f_t(x) = -t^2/2$. On the other hand, $\overline{\lim}_{x\to\infty}|f_t(x)| \le 2$ for each t. Hence we deduce that for $|t| \le T, K < \infty$. Now choose an integer N_ε and an n-partition, of $[-a_\varepsilon, a_\varepsilon]$ with $n \ge N_\varepsilon, -a_\varepsilon = x_0 < x_1 < \cdots < x_n = a_\varepsilon$, such that

$$|f_t(x) - f_t(x_{j-1})| < (\varepsilon/2)[G(a_\varepsilon) - G(-a_\varepsilon)]^{-1}, \qquad x_{j-1} \le x \le x_j. \tag{5}$$

This is possible since f_t is uniformly continuous on $[-a_\varepsilon, a_\varepsilon]$. But ϕ, given by (3), is never zero; thus Log ϕ is defined by Proposition 4.2.9, and for $|t| \le T$ we get

$$\begin{aligned}\left|\operatorname{Log}\phi(t) - i\gamma t - \int_{[|x|\le a_\varepsilon]} f_t(x)\,dG(x)\right| &= \left|\int_{[|x|>a_\varepsilon]} f_t(x)\,dG(x)\right| \\ &\le K\int_{[|x|>a_\varepsilon]} dG(x) < \varepsilon/2 \quad \text{[by (4)]}.\end{aligned} \tag{6}$$

Using the Riemann-Stieltjes approximating sums for the integral on the left in (6), one has, by (5),

$$\left|\int_{[|x|\le a_\varepsilon]} f_t(x)\,dG(x) - \sum_{j=1}^{n} f_t(x_{j-1})[G(x_j) - G(x_{j-1})]\right| < \frac{\varepsilon}{2}. \tag{7}$$

Hence (6) and (7) imply

$$\left|\operatorname{Log}\phi(t) - \left\{i\gamma t + \sum_{j=1}^{n} f_t(x_{j-1})[G(x_j) - G(x_{j-1})]\right\}\right| < \varepsilon/2 + \varepsilon/2 = \varepsilon. \tag{8}$$

This may be simplified by writing out the values of f_t. Let

$$\mu_n = \gamma - \sum_{\substack{j=1 \\ x_j \ne 0}}^{n} [G(x_j) - G(x_{j-1})]/x_j,$$

$$\lambda_j = (1+x_j^2)[G(x_j) - G(x_{j-1})]/x_j^2,$$

if $x_j \ne 0$, and at $x_j = 0$, if G has a jump of size α^2 at 0, then

$$\lim_{x\to 0} f_t(x)[G(x) - G(0)] = -\alpha^2 t^2/2.$$

Hence (8) shows that $\phi(t)$ is approximated by the following in which we substitute $-\alpha^2 t^2/2$ at $x_j = 0$:

$$\exp\left\{it\mu_n - \frac{\alpha^2 t^2}{2}\right\} \prod_{j=1, x_j \neq 0}^{n} \exp\{\lambda_j(e^{itx_j} - 1)\}.$$

But this is a finite product of the ch.f.s of Gaussian and Poisson d.f.s. Thus it is infinitely divisible for each n, and by Proposition 3, the limit ϕ is also infinitely divisible as ϕ is clearly continuous at $t = 0$.

Let us establish uniqueness before turning to the proof of necessity. As noted above, Log ϕ is a well-defined continuous complex function on $\mathbb{R}$ by Proposition 4.2.9. Consider

$$w(t) = -(1/2)\int_{t-1}^{t+1} \text{Log } \phi(u)\, du + \text{Log } \phi(t), \quad t \in \mathbb{R}.$$

Using the representation (3), this can be simplified immediately to get

$$\begin{aligned} w(t) &= -\frac{1}{2}\int_{t-1}^{t+1} i\gamma u\, du - \frac{1}{2}\int_{t-1}^{t+1}\int_{\mathbb{R}} f_u(x)\, dG(x)\, du + i\gamma t + \int_{\mathbb{R}} f_t(x)\, dG(x) \\ &= \int_{\mathbb{R}} (e^{itx} - 1)\frac{1+x^2}{x^2} dG(x) \\ &\quad -\frac{1}{2}\int_{\mathbb{R}} \left(e^{itx}\left[\frac{e^{ix} - e^{-ix}}{ix}\right] - 2\right)\frac{1+x^2}{x^2} dG(x) \\ &= \int_{\mathbb{R}} e^{itx}\left(1 - \frac{\sin x}{x}\right)\frac{1+x^2}{x^2} dG(x). \end{aligned}$$

Thus if (G being left continuous also) we set

$$W(x) = \int_{-\infty}^{x} \left(1 - \frac{\sin u}{u}\right)\frac{1+u^2}{u^2} dG(u),$$

then $W(\cdot)$ is a bounded nondecreasing left continuous function and $W(-\infty) = 0$. Hence it is a d.f. except for a normalization, and $w(\cdot)$ is its ch.f. Thus W is uniquely determined by w, which in turn is uniquely determined by ϕ. It then follows that the left continuous G is uniquely determined by W, and hence so is ϕ. Since ϕ and G determine γ in (3) uniquely, the representation in (3) is unique.

We now establish the representation (3) if ϕ is an infinitely divisible ch.f., i.e., the necessity part of (3). (Because of its importance, an alternative proof of this part based on a result of Gnedenko's is also given later.) Thus for each $n \geq 1$ there exists a ch.f. ψ_n such that $\phi = (\psi_n)^n$, and since Log ϕ exists, consider

$$n(\psi_n(t) - 1) = n\left\{\exp\left[\frac{1}{n}\text{Log } \phi(t)\right] - 1\right\}$$

$$= \left[\text{Log } \phi(t) + \frac{1}{2n}(\text{Log } \phi(t))^2 + \cdots\right]$$

$$\to \text{Log } \phi(t) \quad \text{as } n \to \infty. \tag{9}$$

If F_n is the d.f. of ψ_n, then (9) may be written as

$$\lim_{n\to\infty} n \int_{\mathbb{R}} (e^{itx} - 1)\, dF_n(x) = \text{Log } \phi(t). \tag{10}$$

Let

$$G_n : x \mapsto n \int_{-\infty}^{x} \frac{u^2}{1+u^2} dF_n(u), \quad x \in \mathbb{R}.$$

Then $G_n(-\infty) = 0, G_n$ is nondecreasing, left continuous, and $G_n(+\infty) < \infty$. We show that (i) $\{G_n, n \geq 1\}$ is uniformly bounded, so by the Helly selection principle, $G_{n_k}(x) \to G(x)$, a bounded nondecreasing left continuous function (for a subsequence), and (ii) this G and ϕ determine γ and (3) obtains.

To establish (i) consider the following integral for each fixed t and n:

$$I_n(t) = \int_{\mathbb{R}} (e^{itx} - 1) \frac{1+x^2}{x^2} dG_n(x)$$

$$= n \int_{\mathbb{R}} (e^{itx} - 1)\, dF_n(x) \to \text{Log } \phi(t) \quad [\text{by } (10)]. \tag{11}$$

Hence the real and imaginary parts of $I_n(t)$ converge to the corresponding parts of Log $\phi(t)$, so that one has, from the real parts,

$$\lim_{n\to\infty} \int_{\mathbb{R}} [(\cos tx - 1)(1 + x^2)/x^2]\, dG_n(x) = \log|\phi(t)|.$$

If $A_n = \int_{[|x|\leq 1]} dG_n(x), B_n = \int_{[|x|>1]} dG_n(x)$, so that $G_n(+\infty) = A_n + B_n$, for each $\varepsilon > 0$ there is an n_ε such that $n \geq n_\varepsilon$ implies

$$-\log|\phi(t)| + \varepsilon \geq \int_{[|x|\leq 1]} (1 - \cos tx) \frac{1+x^2}{x^2} dG_n(x) \geq 0. \tag{12a}$$

$$-\log|\phi(t)| + \varepsilon \geq \int_{[|x|>1]} (1 - \cos tx) \frac{1+x^2}{x^2} dG_n(x). \tag{12b}$$

If $|x| \leq 1$,

$$\frac{1 - \cos x}{x^2} = \frac{1}{2} - \frac{x^2}{4!} + \frac{x^4}{6!} - \cdots \geq \frac{1}{2} - \left[\frac{1}{4!} + \frac{1}{6!} + \cdots\right]$$

$$= 2 - \frac{e + e^{-1}}{2} \approx 0.4569 > \frac{1}{3},$$

and so (12a) gives

$$-\log|\phi(1)|+\varepsilon \geq (1/3)\int_{[|x|\leq 1]}(1+x^2)\,dG_n(x) \geq (1/3)A_n.$$

Thus $\{A_n, n \geq 1\}$ is bounded. If $0 \leq t \leq 2$, (12b) gives

$$\begin{aligned}-\frac{1}{2}\int_0^2 \log|\phi(t)|\,dt+\varepsilon &\geq \int_{[|x|>1]}\left(1-\frac{\sin 2x}{2x}\right)\frac{1+x^2}{x^2}dG_n(x)\\ &\geq \frac{1}{2}\int_{[|x|>1]} dG_n(x) = \frac{1}{2}B_n,\end{aligned}$$

since

$$1-\frac{\sin 2x}{2x} \geq 1-\frac{|\sin 2x|}{2|x|} \geq 1-\frac{1}{2}=\frac{1}{2} \qquad \text{for } |x|>1.$$

Thus $\{B_n, n \geq 1\}$ is bounded, and hence $\{G_n(+\infty), n \geq 1\}$ is bounded.

Now by the Helly selection principle there is a subsequence $\{G_{n_k}\}_{k\geq 1}$ which converges to G, a bounded nondecreasing left continuous function, at all points of continuity of G. Clearly $G(-\infty)=0$. To see that $G_{n_k}(+\infty) \to G(+\infty)$, let $\varepsilon > 0$ be given and choose $a_0 > 1$, a continuity point of G, such that $G(+\infty)-G(a_0) < \varepsilon/3$ and for $|t| \leq 2/a_0, |\log|\phi(t)|| \leq \varepsilon/12$. Choose $N_\varepsilon > 1$ such that $n \geq N_\varepsilon \Rightarrow |G_{n_k}(a_0)-G(a_0)| < \varepsilon/3$. Then

$$\begin{aligned}|G_{n_k}(+\infty)-G(+\infty)| &\leq |G_{n_k}(+\infty)-G_{n_k}(a_0)|+|G_{n_k}(a_0)-G(a_0)|\\ &\quad +|G(a_0)-G(+\infty)|\\ &< \int_{[|x|>a_0]} dG_{n_k}(x)+\varepsilon/3+\varepsilon/3. \qquad (13)\end{aligned}$$

Now if $N_1 \geq n_\varepsilon$, then (12b) holds if $|x| > a_0 > 1$, so that for $n \geq N_1$ we get

$$-\log|\phi(t)|+\varepsilon/12 \geq \int_{[|x|>a_0]}(1-\cos tx)\,dG_n(x). \qquad (14)$$

Integrating (14) for $0 \leq t \leq 2/a_0$, and dividing by the length of the interval,

$$\begin{aligned}-\frac{a_0}{2}\int_0^{2/a_0}\log|\phi(t)|\,dt+\frac{\varepsilon}{12} &\geq \int_{[|x|>a_0]}\left[1-\left(\sin\frac{2x}{a_0}\right)\frac{a_0}{2x}\right]dG_n(x)\\ &\geq \frac{1}{2}\int_{[|x|>a_0]} dG_n(x).\end{aligned}$$

Thus

$$\int_{[|x|>a_0]} dG_n(x) \leq \varepsilon/6+2\sup_{|t|\leq 2a_0^{-1}}|\log|\phi(t)|| \leq \varepsilon/6+2\cdot\varepsilon/12=\varepsilon/3, \quad (15)$$

by the choice of a_0. Hence if $n \geq N_2 = \max(N_\varepsilon, N_1)$, then (13) and (15) yield

$$|G_{n_k}(+\infty)-G(+\infty)| < \varepsilon. \qquad (16)$$

Finally, let

$$\gamma_{n_k} = n_k \int_{\mathbb{R}} (x/(1+x^2))\, dF_{n_k}(x) = \int_{\mathbb{R}} (1/x)\, dG_{n_k}(x).$$

Then $|\gamma_{n_k}| < \infty$, and I_n of (11) becomes

$$\begin{aligned} I_{n_k}(t) &= \int_{\mathbb{R}} (e^{itx} - 1)\frac{1+x^2}{x^2} dG_{n_k}(x) \\ &= \int_{\mathbb{R}} \left(e^{itx} - 1 - \frac{itx}{1+x^2}\right) \frac{1+x^2}{x^2} dG_{n_k}(x) + it\gamma_{n_k}. \end{aligned} \tag{17}$$

Since the integrand in (17) is bounded and continuous for each t, and $G_{n_k} \to G$ as shown by (16), we may apply the Helly-Bray theorem for it and interchange the limits. But by (11) $I_{n_k}(t) \to \operatorname{Log} \phi(t)$, so that (17) implies γ_{n_k} must converge to some number γ. Hence we get

$$\operatorname{Log} \phi(t) = it\gamma + \int_{\mathbb{R}} \left(e^{itx} - 1 - \frac{itx}{1+x^2}\right) \frac{1+x^2}{x^2}\, dG(x),$$

which is (3). This gives the necessity, and with it the theorem is completely proved.

An alternative proof of necessity will be given after we record the special but also useful forms of Lévy and Kolmogorov, as consequences of formula (3).

Define $M : \mathbb{R}^- \to \mathbb{R}^+$ and $N : \mathbb{R}^+ \to \mathbb{R}^-$ by the equations

$$M(x) = \int_{-\infty}^{x} \frac{1+u^2}{u^2} dG(u), \qquad N(x) = -\int_{x}^{\infty} \frac{1+u^2}{u^2} dG(u). \tag{18}$$

If $\sigma^2 = G(0+) - G(0-) \geq 0$ is the jump of G at $x = 0$, then (i) M, N are both nondecreasing, (ii) $M(-\infty) = N(+\infty) = 0$, (iii) M and $G|\mathbb{R}^-, N$ and $G|\mathbb{R}^+$ have the same points of continuity, and (iv) for each $\varepsilon > 0$,

$$\begin{aligned} &\int_{-\varepsilon}^{0-} u^2 dM(u) + \int_{0+}^{\varepsilon} u^2\, dN(u) \\ &\quad = \int_{-\varepsilon}^{0-} (1+u^2)\, dG(u) + \int_{0+}^{\varepsilon} (1+u^2)\, dG(u) < \infty. \end{aligned} \tag{19}$$

Given G, (18) and (19) determine M and N, and conversely, if M, N are given to satisfy (18) and (19) and the conditions (i)–(iv), then G is determined in terms of which (3) becomes, with a $\gamma \in \mathbb{R}$,

$$\begin{aligned} \phi(t) = \exp\Bigg\{ &i\gamma t - \frac{\sigma^2 t^2}{2} + \int_{\mathbb{R}^-} \left(e^{ixt} - 1 - \frac{ixt}{1+x^2}\right) dM(x) \\ &+ \int_{\mathbb{R}^+} \left(e^{ixt} - 1 - \frac{ixt}{1+x^2}\right) dN(x) \Bigg\}, \qquad t \in \mathbb{R}. \end{aligned} \tag{20}$$

The collection (γ, σ^2, M, N) is the *Lévy (spectral) set* for the infinitely divisible ch.f. ϕ, and the pair (M, N) the *Lévy measures*. These will be needed in Section 8.4 as well.

If the infinitely divisible d.f. has a finite variance (equivalently, its ch.f. ϕ is twice differentiable), then we can also get the Kolmogorov formula from (3) as follows.

Define $K : \mathbb{R} \to \mathbb{R}^+$, called the *Kolmogorov function*, by

$$K(x) = \int_{-\infty}^{x} (1 + u^2)\, dG(u), \qquad x \in \mathbb{R}. \tag{21}$$

Clearly $K(x) \geq 0$ for $x \in \mathbb{R}$, and the following definition of ϕ is formally correct for suitable $\tilde{\gamma}$:

$$\phi(t) = \exp\left\{ i\tilde{\gamma}t + \int_{\mathbb{R}} (e^{itx} - 1 - itx) \frac{dK(x)}{x^2} \right\}. \tag{22}$$

We have to show that $K(+\infty) < \infty$ if the d.f. has finite variance. In fact, from (22) by differentiation of the holomorphic function Log ϕ at $t = 0$, we get

$$\begin{aligned} \frac{d}{dt}(\text{Log } \phi(t))\,|_{t=0} &= \phi'(0) = iE(X) = i\tilde{\gamma}, \\ \frac{d^2}{dt^2}(\text{Log } \phi(t))\,|_{t=0} &= -\text{Var } X = -\int_{\mathbb{R}} dK(x) = -K(+\infty), \end{aligned}$$

where X is the r.v. with ϕ as its ch.f. Thus $K(+\infty) < \infty$, and then

$$\tilde{\gamma} = \gamma + \int_{\mathbb{R}} (u/(1 + u^2))\, dK(u).$$

Thus (22) is obtained from (3) and is rigorously correct. We state this result as follows:

Theorem 6 *Let $\phi : \mathbb{R} \to \mathbb{C}$ be a mapping. Then it is an infinitely divisible ch.f. iff it admits a representation (20) for a Lévy set (γ, σ^2, M, N). On the other hand, if $\phi : \mathbb{R} \to \mathbb{C}$ is a ch.f. of a d.f. with finite variance, then it is infinitely divisible iff ϕ admits a representation (22) for a Kolmogorov pair $(\tilde{\gamma}, K)$. If (M, N) and K are taken left continuous, then these representations are unique.*

It may be remarked that (22) can also be obtained directly using the argument of (3) with slight simplifications, though it is unnecessary to reproduce them here (cf. Problem 12).

To present the alternative proof of the necessity part of Theorem 5 noted earlier, we need the following result on the convergence of infinitely divisible d.f.s, which is of independent interest. It is due to B.V. Gnedenko.

Proposition 7 *Let $\{F_n, n \geq 1\}$ be a sequence of infinitely divisible d.f.s with the Lévy-Khintchine pairs $\{(\gamma_n, G_n), n \geq 1\}$. Then F_n tends to a d.f. F (necessarily infinitely divisible) iff there exists a pair (γ, G) as above such that*

(i) $\lim_{n\to\infty} G_n(x) = G(x)$ *at all continuity points x of G,*
(ii) $\lim_{n\to\infty} G_n(\pm\infty) = G(\pm\infty)$ *and*
(iii) $\lim_{n\to\infty} \gamma_n = \gamma (\in \mathbb{R})$.

The pair (γ, G) determines the ch.f. of F by the (Lévy-Khintchine) formula (3).

Proof Let F_n be infinitely divisible and ϕ_n be its ch.f. Then for each n,

$$\phi_n(t) = \exp\left\{ i\gamma_n t + \int_{\mathbb{R}} \left(e^{itx} - 1 - \frac{itx}{1+x^2} \right) \frac{1+x^2}{x^2} dG_n(x) \right\}, \quad t \in \mathbb{R}. \tag{23}$$

If (i)–(iii) hold, then $\{G_n(+\infty), n \geq 1\}$ is convergent, so that it is bounded and $|G_n(x)| \leq \sup_n G_n(+\infty) < \infty$. Thus the $G_n, n \geq 1$, are uniformly bounded. The integrand in (23) is a bounded continuous function in x for each t. Hence by the Helly-Bray theorem [with (iii)] $\phi_n(t) \to \phi(t)$ and ϕ is given by (23) with (γ_n, G_n) replaced by (γ, G), and is continuous. Thus ϕ is an infinitely divisible ch.f. by the sufficiency part of Theorem 5. By the continuity theorem $F_n \to F$ and ϕ is the ch.f. of F, which is infinitely divisible.

The converse can be proved using the arguments of the necessity part of Theorem 5. But we present an alternative method which does not depend on that result, so that the present result can be used to obtain a simpler proof of its necessity. The idea here is to use the same trick employed for the uniqueness part of the proof of Theorem 5. Thus let $\psi_n = \operatorname{Log} \phi_n$ and

$$\begin{aligned} w_n(t) &= \psi_n(t) - (1/2) \int_{t-1}^{t+1} \psi_n(u)\, du \\ &= \int_{\mathbb{R}} e^{itx} \left(1 - \frac{\sin x}{x} \right) \frac{1+x^2}{x^2}\, dG_n(x) \end{aligned}$$

(by substitution for ψ_n from (23) and simplification).

By hypothesis of this part, $\phi_n \to \phi$, a ch.f. Hence $\psi_n \to \psi$ and then $w_n(t) \to w(t), t \in \mathbb{R}$, and since ϕ is continuous, so is w. But

$$\left(1 - \frac{\sin x}{x} \right) \frac{1+x^2}{x^2} > 0$$

and G_n is increasing, $G_n(-\infty) = 0$. Thus $w_n(\cdot)$ is the ch.f. of the "d.f." H_n, where

$$H_n(x) = \int_{-\infty}^{x} \left(1 - \frac{\sin u}{u} \right) \frac{1+u^2}{u^2}\, dG_n(u) = \int_{-\infty}^{x} h(u)\, dG_n(u) \quad \text{(say)}. \tag{24}$$

Then by the continuity theorem, $H_n \to H$ at all continuity points of H, and $H_n(\pm\infty) \to H(\pm\infty)$. But h is a positive bounded continuous function such that $\lim_{u\to 0} h(u) = \frac{1}{6}$, $\lim_{u\to\infty} h(u) = 1$. It follows, by the Radon-Nikodým theorem, that

$$G_n(x) = \int_{-\infty}^{x} \frac{1}{h(u)}\, dH_n(u). \tag{25}$$

Since $1/h(u)$ is also bounded and continuous, one concludes from (25) and the Helly-Bray theorem that

$$G_n(x) \to G(x) = \int_{-\infty}^{x} \frac{1}{h(u)}\, dH(u)$$

at all continuity points x of G (equivalently of H) and $G_n(\pm\infty) \to G(\pm\infty)$. Since $\phi_n \to \phi$ by hypothesis, (23) and this imply that $\gamma_n \to \gamma$, and hence (i)–(iii) hold; therefore ϕ is given by (23) with (γ, G) in place of (γ_n, G_n). This completes the proof.

Remark The corresponding limit theorems hold for the Lévy and Kolmogorov representations, although each statement needs a separate but entirely similar argument.

We are now ready to present the following:

Alternative Proof of (the Necessity of) Theorem 5 By hypothesis $\phi = (\psi_n)^n$ for each n, where ψ_n is a ch.f. Then by (9)

$$n(\psi_n(t) - 1) \to \operatorname{Log} \phi(t), \quad n \to \infty, \quad t \in \mathbb{R}. \tag{9'}$$

Let F_n be the d.f. of ψ_n. Hence

$$\begin{aligned} n(\psi_n(t) - 1) &= n \int_{\mathbb{R}} (e^{itx} - 1)\, dF_n(x) \\ &= nit \int_{\mathbb{R}} \frac{x}{1+x^2}\, dF_n(x) + n \int_{\mathbb{R}} \left(e^{itx} - 1 - \frac{itx}{1+x^2} \right) dF_n(x) \\ &= it\gamma_n + \int_{\mathbb{R}} \left(e^{itx} - 1 - \frac{itx}{1+x^2} \right) \frac{1+x^2}{x^2}\, dG_n(x), \end{aligned}$$

where

$$\gamma_n = n \int_{\mathbb{R}} \frac{x}{1+x^2}\, dF_n(x), \quad G_n(x) = n \int_{-\infty}^{x} \frac{u^2}{1+u^2}\, dF_n(u).$$

If ϕ_n is defined by (3) with (γ_n, G_n), then it is clear (by the sufficiency) that ϕ_n is infinitely divisible. The hypothesis and (9′) imply that $\phi_n(t) \to \phi(t), t \in \mathbb{R}$. Thus the necessity part of Proposition 7 implies that $\gamma_n \to \gamma, G_n \to G$ satisfying the stated conditions and that ϕ is given by (3) for the pair (γ, G). This proves the result.

In general, the infinite divisibility is verified by an explicit construction of the pair (γ, G) and the formula (3), (20), or (22). There is no other quick and easy test. Let us present an example to show how this may be accomplished.

Example 8 Let $0 < \alpha \leq \beta < 1$ and X, Y be independent r.v.s on (Ω, Σ, P) whose distributions are specified as

$$P[X = n] = (1 - \beta)\beta^n, \qquad n = 0, 1, 2, \ldots,$$

and

$$P[Y = 0] = 1/(1 + \alpha) = 1 - P[Y = -1].$$

Let $Z = X + Y$. Then X is an infinitely divisible r.v., but Y and Z are not. However, if $\tilde{Z}$ is independent of Z and has the same d.f. as Z, then $V = Z - \tilde{Z}$ is an infinitely divisible r.v.

Proof We establish these facts as follows. Here Y is a Bernoulli r.v., and X is an r.v. representing the 1st head on the $(n+1)$th trial in a sequence of tosses of a coin whose probability of obtaining a tail is β (then X is said to have a *geometric* d.f.):

$$\phi(t) = E(e^{itX}) = \sum_{n \geq 0} e^{int} \beta^n (1 - \beta) = (1 - \beta)(1 - \beta e^{it})^{-1}.$$

Thus ϕ never vanishes, and we have by 4.2.9,

$$\begin{aligned} \operatorname{Log} \phi(t) &= \log(1 - \beta) - \operatorname{Log}(1 - \beta e^{it}) \\ &= \sum_{n \geq 1} \frac{\beta^n}{n} + \sum_{n \geq 1} \frac{\beta^n e^{int}}{n} = \sum_{n \geq 1} (e^{int} - 1)\beta^n / n. \end{aligned} \tag{26}$$

Hence

$$\phi(t) = \lim_{n \to \infty} \prod_{k=1}^{n} \exp\left\{ \frac{\beta^k}{k} (e^{ikt} - 1) \right\}.$$

Since the product is the ch.f. of a sum of n independent Poisson r.v.s, it is infinitely divisible, and by Proposition 3, ϕ is infinitely divisible. Also by Proposition 2b, the bounded r.v. Y is not infinitely divisible.

Regarding Z, consider its ch.f.

$$\psi(t) = E(e^{itZ}) = \phi(t) \cdot E(e^{itY}) = (1 - \beta)(1 - \beta e^{it})^{-1} \left(\frac{1}{1 + \alpha} + \frac{\alpha}{1 + \alpha} e^{-it} \right).$$

Since the ch.f. of Y in absolute value lies between $(1 - \alpha)/(1 + \alpha)$ and 1, it also never vanishes. Thus $\psi(t)$ never vanishes, and again by 4.2.9,

$$\begin{aligned}
\text{Log } \psi(t) &= \text{Log } \phi(t) + \text{Log}\left(\frac{1+\alpha e^{-it}}{1+\alpha}\right) \\
&= \sum_{n\geq 1}(e^{int}-1)\beta^n/n \\
&\quad + \text{Log}(1+\alpha e^{-it}) - \log(1+\alpha) \quad [\text{by } (26)] \\
&= \sum_{n\geq 1}(e^{int}-1)\beta^n/n + \sum_{n\geq 1}(-1)^{n-1}(e^{-int}-1)\alpha^n/n, \\
&= i\gamma t + \int_{\mathbb{R}}\left(e^{itx} - 1 - \frac{itx}{1+x^2}\right)\frac{1+x^2}{x^2}\,dG(x),
\end{aligned} \tag{27}$$

where $\gamma = \sum_{n\geq 1}(\beta^n + (-1)^n\alpha^n)/(1+n^2)$ and G is a function on the integers of bounded variation with jumps of size $n\beta^n/(1+n^2)$ and $(-1)^{n-1}[n\alpha^n/(1+n^2)]$ at positive and negative integers, respectively. It has no jump at 0. This formula is like (3); but G is *not* monotone increasing. Hence by Theorem 5, ψ cannot be infinitely divisible.

If $\zeta(t) = E(e^{itV}) = \psi(t)\psi(-t) = |\psi(t)|^2$, then using (27) we get

$$\begin{aligned}
\zeta(t) &= \exp\left\{Re\int_{\mathbb{R}}\left(e^{itx} - 1 - \frac{itx}{1+x^2}\right)\frac{1+x^2}{x^2}\,dG(x)\right\} \\
&= \exp\left\{\sum_{n\geq 1}(\cos tn - 1)[\beta^n + (-1)^{n-1}\alpha^n]/n\right\} \\
&= \exp\left\{0\cdot it + \int_{\mathbb{R}}\left(e^{itx} - 1 - \frac{itx}{1+x^2}\right)\frac{1+x^2}{x^2}\,d\tilde{G}(x)\right\},
\end{aligned} \tag{28}$$

where $\tilde{G}$ is now monotone nondecreasing and bounded with positive jumps at $\pm 1, \pm 2, \ldots$, of sizes $[|n|/(1+n^2)](\beta^{|n|} + (-1)^{|n|-1}\alpha^{|n|})$, since $\alpha \leq \beta$, and no jump at 0. Hence (28) is the same as (3) with "$\gamma = 0$" and "$G = \tilde{G}$," so that V is an infinitely divisible r.v.

This last part implies that $\zeta = \psi\overline{\psi}$ is an infinitely divisible ch.f. even though neither ψ nor $\overline{\psi}$ is such. Also $\zeta(t) = |\phi|^2(t)|E(e^{itY})|^2$ is a product of an infinitely divisible $|\phi|^2$ and a noninfinitely divisible one. Thus the intricate structure of these ch.f.s as well as the beauty and depth of the Lévy-Khintchine formula are exhibited by this example.

The next result is motivational for the work of the following section.

Example 9 Let $X_{nk}, 1 \leq k \leq n, n \geq 1$, be independent and have the d.f.s defined by $P[X_{nk} = k/n] = 1/n = P[X_{nk} = -k/n]$ and $P[X_{nk} = 0] = 1 - (2/n), 1 \leq k \leq n$. Then $S_n = \sum_{k=1}^n X_{nk}$ is not infinitely divisible for any $n \geq 1$, but $S_n \xrightarrow{D} S$ and S is infinitely divisible; it is not Gaussian.

Proof Since each S_n is a bounded nondegenerate r.v., it cannot be infinitely divisible by Proposition 2b. Let ϕ_n be the ch.f. of S_n, so that

$$\phi_n(t) = E(e^{itS_n}) = \prod_{k=1}^{n} E(e^{itX_{nk}}) \qquad \text{(by independence)}$$
$$= \prod_{k=1}^{n} \{1 + (2/n)[\cos(kt/n) - 1]\}.$$

Thus ϕ_n is real and for each $n(\geq 5)$, $\phi_n(t)$ is never zero $[> (\frac{1}{5})^n]$. Consequently

$$\log \phi_n(t) = \sum_{k=1}^{n} \log \left[1 + \frac{2}{n}\left(\cos \frac{kt}{n} - 1\right)\right]$$
$$= \frac{2}{n}\sum_{k=1}^{n}[(\cos kt/n) - 1] - o(1/n).$$

Hence the Riemann integral approximation gives

$$\lim_{n\to\infty} \log \phi_n(t) = 2\int_0^1 (\cos tx - 1)\, dx, \qquad t \in \mathbb{R}. \tag{29}$$

Since the limit is continuous at $t = 0$,

$$\lim_{n\to\infty} \phi_n(t) = \phi(t) = \exp\left\{\int_{-1}^{1} (\cos tx - 1)\, dx\right\}$$

is a ch.f., which, by the continuity theorem, implies $S_n \xrightarrow{D} S$ and ϕ is the ch.f. of S. Clearly S is not a normal (=Gaussian) r.v.

To see that ϕ is infinitely divisible, note that

$$\phi(t) = \exp\left\{\int_{-1}^{1} (e^{itx} - 1)\, dx\right\}$$
$$= \exp\left\{\int_{-1}^{1} \left(e^{itx} - 1 - \frac{itx}{1+x^2}\right) \frac{1+x^2}{x^2}\, dG(x)\right\}, \tag{30}$$

where G is a nonnegative nondecreasing bounded (by 2) function. It may be written as

$$G(x) = \begin{cases} 0, & x < -1 \\ \int_{-1}^{x} (u^2/(1+u^2))\, du, & -1 \leq x < 1 \\ 2, & x \geq 1. \end{cases}$$

By Theorem 5, (0, G) is a Lévy-Khintchine pair and ϕ is infinitely divisible, as asserted.

The properties of the sequence $\{X_{nk}, 1 \leq k \leq n\}$ can be expressed as follows: $E(X_{nk}) = 0$;

$$\text{Var } X_{nk} = \frac{2k^2}{n^3},$$

thus

$$\text{Var } S_n = 2\sum_{k=1}^{n} \frac{k^2}{n^3} \to \frac{2}{3} \quad \text{as } n \to \infty;$$

and for $\varepsilon > 0$,

$$P[|X_{nk} - E(X_{nk})| \geq \varepsilon] \leq \frac{2}{n}$$

and

$$\lim_{n\to\infty} \max_{1\leq k\leq n} P[|X_{nk} - E(X_{nk})| \geq \varepsilon] = 0.$$

Note that the variances of the partial sums are bounded, and the independent components X_{nk} are uniformly absolutely negligible.

Abstracting these two properties, we present some limit theorems and some important results, such as the Lindeberg-Feller theorem, conditions for the limit law to be Poisson, and the like.

5.3 General Limit Laws, Including Stability

Let us now abstract the properties of the X_{nk}-sequence appearing in Example 9 of the preceding section. The "smallness" of X_{nk} is made precise in

Definition 1 (a) Let $\{X_{nk}, 1 \leq k \leq k_n, n \geq 1\}$ be a sequence of sequences of r.v.s on a probability space (Ω, Σ, P). Then the X_{nk} are called *infinitesimal* if for each $\varepsilon > 0$

$$\lim_{n\to\infty} \max_{1\leq k\leq k_n} P[|X_{nk}| \geq \varepsilon] = 0. \tag{1}$$

(b) More generally, the X_{nk} are called *asymptotically constant* if there exist constants a_{nk} [which typically are $E(X_{nk})$ if these exist, and are medians in general] such that in $\{X'_{nk} = X_{nk} - a_{nk}, 1 \leq k \leq k_n, n \geq 1\}$, the X'_{nk} are infinitesimal.

For computational convenience, it is better to have alternative forms of condition (1). These are given by

Proposition 2 *Let $\{X_{nk}, 1 \leq k \leq k_n, n \geq 1\}$ be a sequence of r.v.s. Then the following are equivalent:*

(i) *The X_{nk} are infinitesimal.*
(ii) *If ϕ_{nk} is the ch.f. of X_{nk}, then*

$$\lim_{n\to\infty} \max_{1\leq k\leq k_n} |\phi_{nk}(t) - 1| = 0$$

uniformly in $|t| \leq T$ *for each* $T > 0$.

(iii) *If* F_{nk} *is the d.f. of* X_{nk}, *then*

$$\lim_{n\to\infty} \max_{1\leq k\leq k_n} \int_{\mathbb{R}} \frac{x^2}{1+x^2}\, dF_{nk}(x) = 0.$$

Proof (i)⇒(ii) Let $T > 0$ be arbitrarily fixed. Since (1) holds, given $\varepsilon > 0$, there exists an N_ε such that $n \geq N_\varepsilon$ implies

$$\max_{1\leq k\leq k_n} \int_{[|x|\geq \varepsilon/2T]} dF_{nk} < \varepsilon/4. \tag{2}$$

Consider

$$\begin{aligned}
|\phi_{nk}(t) - 1| &\leq \left| \int_{[|x|<\varepsilon/2T]} (e^{itx} - 1)\, dF_{nk}(x) \right| + \int_{[|x|\geq\varepsilon/2T]} |e^{itx} - 1|\, dF_{nk}(x) \\
&\leq \int_{[|x|<\varepsilon/2T]} \left| \int_0^{tx} e^{iu}\, du \right| dF_{nk}(x) + 2 \int_{[|x|\geq\varepsilon/2T]} dF_{nk}(x) \\
&\leq \int_{[|x|<\varepsilon/2T]} |tx|\, dF_{nk}(x) + 2 \cdot \frac{\varepsilon}{4}, \qquad \text{by (2)} \\
&\leq \frac{\varepsilon}{2T} T + \frac{\varepsilon}{2} = \varepsilon \qquad \text{for all } |t| \leq T.
\end{aligned}$$

Thus (ii) holds. This implication is the one often used.

(ii)⇒(i) By hypothesis $\lim_{n\to\infty} \phi_{nk}(t) = 1$ uniformly in k for $|t| \leq T$, $T > 0$. Thus $X_{nk} \to 0$ in distribution, hence in probability, uniformly in k. This means, in symbols, (1) is true, so that (i) follows.

(i)⇒(iii) If $0 < \varepsilon < \frac{1}{2}$, then there is an N_ε such that

$$n \geq N_\varepsilon \Rightarrow \max_{1\leq k\leq k_n} P[|X_{nk}| \geq \varepsilon] < \varepsilon/2.$$

Hence

$$\begin{aligned}
&\max_{1\leq k\leq k_n} \int_{\mathbb{R}} \frac{x^2}{1+x^2}\, dF_{nk}(x) \\
&\leq \max_{1\leq k\leq k_n} \int_{[|x|<\varepsilon]} x^2\, dF_{nk}(x) + \max_{1\leq k\leq k_n} \int_{[|x|\geq\varepsilon]} dF_{nk}(x) \\
&\leq \varepsilon^2 + \varepsilon/2 < (\varepsilon/2) + (\varepsilon/2) = \varepsilon.
\end{aligned}$$

Thus (iii) holds.

On the other hand, if (iii) is given, for each $\varepsilon > 0$,

$$\max_{1\le k\le k_n}\int_{\mathbb{R}}\frac{x^2}{1+x^2}\,dF_{nk}(x)$$
$$\ge \max_{1\le k\le k_n}\int_{|x|\ge\varepsilon}\frac{\varepsilon^2}{1+\varepsilon^2}\,dF_{nk}(x), \qquad \text{because } \frac{x^2}{1+x^2}\uparrow \text{ on } \mathbb{R}^+,$$
$$= \frac{\varepsilon^2}{1+\varepsilon^2}\max_{1\le k\le k_n} P[|X_{nk}|\ge\varepsilon].$$

Taking limits as $n\to\infty$, this implies (i), and the proof is complete.

If we are given a sequence of sequences of independent r.v.s $\{X_{nk}, 1\le k\le k_n\}$, conditions should be found such that the sequence $\{(S_n - A_n)/B_n, n\ge 1\}$ converges in distribution to an r.v., where $S_n = \sum_{k=1}^{k_n} X_{nk}$, $B_n > 0$, $A_n\in\mathbb{R}$ are suitable constants. The key idea now is to express the ch.f. ϕ_n of $(S_n - A_n)/B_n$ in terms of a ch.f. ψ_n, as in Eq. (2.3), for some suitable pair (γ_n, G_n) with an error term tending to zero, as $n\to\infty$ uniformly on compact t-sets. Then the given sequence converges iff the ψ_n sequence does, and for the latter sequence Proposition 2.7 tells us how to proceed. The remarkable first steps were taken in 1936 by G. Bawly even before the above result was discovered. This simple, yet important proposition, which suffices for many applications, will now be presented. (Of course, the Kolmogorov representation was known by 1932.) For nontriviality, henceforth we *assume that* $k_n\to\infty$ *as* $n\to\infty$ without further mention.

Theorem 3 (Bawly) *Let $\{X_{nk}, 1\le k\le k_n, n\ge 1\}$ be a sequence of sequences of rowwise independent r.v.s with finite variances. Suppose that the r.v.s $X_{nk} - E(X_{nk})$ are infinitesimal and $\sup_{n\ge1}\operatorname{Var} S_n < \infty$, where $S_n = \sum_{k=1}^{k_n} X_{nk}$. Then for any sequence $\{A_n, n\ge 1\}\subset\mathbb{R}$, $S_n - A_n\to\tilde{S}$ in distribution iff the following sequence of infinitely divisible ch.f.s ψ_n (or their d.f.s) converges to a (necessarily infinitely divisible) ch.f. ψ as $n\to\infty$, uniformly on compact subsets of $\mathbb{R}$, where*

$$\psi_n(t) = \exp\left\{i\alpha_n t + \int_{\mathbb{R}}(e^{itx} - 1 - itx)\frac{1}{x^2}\,dK_n(x)\right\}. \tag{3}$$

Here

$$\alpha_n = \sum_{k=1}^{k_n} E(X_{nk}) - A_n,$$
$$K_n(x) = \sum_{k=1}^{k_n}\int_{-\infty}^{x} u^2\,dF_{nk}(u + E(X_{nk})),$$

F_{nk} being the d.f. of X_{nk}. The limit d.f. in both cases is the same, and it is infinitely divisible.

Remark The associated d.f.s with the ch.f.s ψ_n are called the *accompanying laws* of the given sequence $\{S_n - A_n, n \geq 1\}$.

Proof Let

$$\phi_{nk}(t) = E(\exp\{it(X_{nk} - E(X_{nk}))\}), \qquad \phi_n(t) = E(\exp\{it(S_n - A_n)\}), \quad t \in \mathbb{R}.$$

Then by independence

$$\phi_n(t) = \exp\left\{-itA_n + it\sum_{k=1}^{k_n} E(X_{nk})\right\} \cdot \prod_{k=1}^{k_n} \phi_{nk}(t). \tag{4}$$

We now associate an infinitely divisible ch.f. with the right side on using the infinitesimality of $X'_{nk} = X_{nk} - E(X_{nk})$.

Let F'_{nk} be the d.f. of X'_{nk}, so that $F'_{nk} = F_{nk}(x + E(X_{nk})), F_{nk}$ being the d.f. of X_{nk}. Consider

$$a_{nk} = \phi_{nk}(t) - 1 = \int_{\mathbb{R}} (e^{itx} - 1)\, dF'_{nk}(x), \qquad t \in \mathbb{R}.$$

By Proposition 2 [(i)⇒(ii)],

$$\lim_{n\to\infty} \max_{1\leq k\leq k_n} |a_{nk}| = 0, \qquad |t| \leq T; \ T > 0.$$

Hence there exists an N_1 such that $n \geq N_1 \Rightarrow \max_{1\leq k\leq k_n} |a_{nk}| < \frac{1}{2}$. Because for $|t| \leq T, \phi_{nk}(t) = 1 + a_{nk}, \phi_{nk}$ does not vanish for $n \geq N_1$ and, by Proposition 4.2.9, Log ϕ_{nk} is well defined. Thus

$$\begin{aligned} |\text{Log}\, \phi_{nk}(t) - a_{nk}| &\leq \sum_{r\geq 2} |a_{nk}|^r / r \\ &\leq \frac{1}{2}\sum_{r\geq 2} |a_{nk}|^r = \frac{1}{2}\frac{|a_{nk}|^2}{1 - |a_{nk}|} < |a_{nk}|^2. \end{aligned} \tag{5}$$

Also, $E(X'_{nk}) = 0$, whence

$$a_{nk} = \int_{\mathbb{R}} (e^{itx} - 1 - itx)\, dF'_{nk}(x). \tag{6}$$

Note that (6) implies

$$\begin{aligned} |a_{nk}| &\leq \int_{\mathbb{R}} \left| x^2 \int_0^t \left(\int_0^u e^{itx}\, d\tau\right) du \right| dF'_{nk}(x) \\ &\leq \frac{t^2}{2}\int_{\mathbb{R}} x^2\, dF'_{nk}(x) = \frac{t^2}{2} \text{Var}\, X_{nk}. \end{aligned} \tag{7}$$

If ψ_n is as defined in (3) and ϕ_n is given by (4), then one has the following simplification in which $n \geq N_1$, so that Log ϕ_n is well defined for $|t| \leq T$:

$$\begin{aligned}
&\text{Log}\,\psi_n(t) - \text{Log}\,\phi_n(t)\\
&= it\alpha_n + \int_{\mathbb{R}} \frac{(e^{itx} - 1 - itx)}{x^2}\,dK_n(x) - \left[it\alpha_n + \sum_{k=1}^{k_n} \text{Log}\,\phi_{nk}(t)\right]\\
&= \sum_{k=1}^{k_n} \left[\int_{\mathbb{R}} (e^{itx} - 1 - itx)\,dF'_{nk}(x) - \text{Log}\,\phi_{nk}(t)\right], \text{[by definition of } K_n\text{]},\\
&= \sum_{k=1}^{k_n} [a_{nk} - \text{Log}\,\phi_{nk}(t)]. \qquad (8)
\end{aligned}$$

Hence with (5) and (7), (8) becomes

$$\begin{aligned}
|\text{Log}\,\psi_n(t) - \text{Log}\,\phi_n(t)| &\leq \sum_{k=1}^{k_n} |a_{nk}|^2 \leq \max_{1\leq k\leq k_n} |a_{nk}| \frac{t^2}{2} \text{Var}(S_n)\\
&\leq \frac{t^2}{2} \cdot \sup_{n\geq 1} \text{Var}\,S_n \max_{1\leq k\leq k_n} |a_{nk}| \to 0
\end{aligned}$$

as $n \to \infty$, by Proposition 2. Thus $(\phi_n/\psi_n)(t) \to 1 \Rightarrow |\phi_n(t) - \psi_n(t)| \to 0$ as $n \to \infty$ uniformly in $|t| \leq T$. Then if $\psi_n(t) \to \psi(t), |t| \leq T$, it is clear that $\phi_n(t) \to \psi(t)$ also, $|t| \leq T$ and conversely. The proof of the theorem is complete.

It is now easy to present conditions for a prescribed infinitely divisible limit distribution to be the limit d.f. of the sequences $\{X_{nk}, 1 \leq k \leq k_n, n \geq 1\}$ of Theorem 3, using Proposition 2.7. The following result, due to Gnedenko, is an adaptation of the former.

Theorem 4 *Let $\{X_{nk}, 1 \leq k \leq k_n, n \geq 1\}$ be a sequence of sequences of rowwise independent r.v.s with finite variances and each $X_{nk} - E(X_{nk})$ be infinitesimal. Then the d.f.s of the sequence of sums $S_n = \sum_{k=1}^{k_n} X_{nk} - A_n$, for a given constant sequence $\{A_n, n \geq 1\} \subset \mathbb{R}$ converges to a limit d.f. F, and their variances tend to the variance of F iff there exist an $\alpha \in \mathbb{R}$ and a bounded nondecreasing function $K : \mathbb{R} \to \mathbb{R}^+$ with $K(-\infty) = 0$ such that*

(i) $$\lim_{n\to\infty} \sum_{k=1}^{k_n} \int_{-\infty}^{x} u^2\,dF_{nk}(u + E(X_{nk})) = K(x)$$

at all continuity points x of $K(x)$,

(ii) $$\lim_{n\to\infty} \sum_{k=1}^{k_n} \int_{-\infty}^{\infty} u^2\,dF_{nk}(u + E(X_{nk})) = K(+\infty),$$

(iii)
$$\lim_{n\to\infty}\left(\sum_{k=1}^{k_n} E(X_{nk}) - A_n\right) = \alpha,$$

where F_{nk} is the d.f. of the X_{nk}. The ch.f. of F is given by (22) of Section 2, with (α, K) as the Kolmogorov pair determined by (i)–(iii) above.

Proof Let $\alpha_n = \sum_{k=1}^{k_n} E(X_{nk}) - A_n$, and

$$K_n(x) = \sum_{k=1}^{k_n} \int_{-\infty}^{x} u^2\, dF_{nk}(u + E(X_{nk})). \tag{9}$$

Then $K_n(-\infty) = 0, K_n$ is nondecreasing, and by (i)–(iii), $K_n(x) \to K(x)$ at K-continuity points x and $K_n(+\infty) \to K(+\infty), \alpha_n \to \alpha$. Since Var $S_n = K_n(+\infty)$, and a convergent (real) sequence is bounded, the "if" part of Bawly's theorem is satisfied; and if ψ_n is defined by (3) with this α_n and K_n, then ψ_n is an infinitely divisible ch.f., and by (an analog for the Kolmogorov representation of) Proposition 2.7, $\psi_n(t) \to \psi(t), t \in \mathbb{R}$, where ψ is given by (3) with the above α and K. Hence $S_n - A_n \xrightarrow{D} \tilde{S}$ and, further, the proposition implies that the variances $\sigma^2(S_n)$ converge to the variance of $\tilde{S}$ because in the Kolmogorov formula $K(+\infty) = \text{Var } \tilde{S}$. Thus the result holds in this direction.

Conversely, let $S_n - A_n \xrightarrow{D} \tilde{S}$ and $\sigma^2(S_n - A_n) = \sigma^2(S_n) \to \sigma^2(\tilde{S})$. Then the variances are bounded and $X_{nk} - E(X_{nk})$ are infinitesimal in both directions, by hypothesis. Hence by the necessity part of Bawly's theorem (and Proposition 2.7 or the Kolmogorov representation), α_n and $K_n(\cdot)$ defined above must converge to α and $K(\cdot)$ (always taken left continuous), and the latter pair determines ψ uniquely, which is then a ch.f. of an infinitely divisible d.f. Thus conditions (i)–(iii) above hold. This completes the proof.

The result enables us to obtain conditions for distributional convergence of partial sums of independent r.v.s with finite variances to any infinitely divisible d.f., as soon as we are able to calculate the Kolmogorov pair (α, K).

Let us now present a specialization of the above result if the desired limit d.f. is the standard normal $N(0,1)$. Since its ch.f. is $t \mapsto e^{-t^2/2}$, in the Kolmogorov formula [see (22) in Section 2], it is seen that $\gamma = 0$ and K must have a jump of size 1 at $x = 0$; i.e., $K(x) = 0$ for $x < 0, = 1$ for $x \geq 0$. With this knowledge of (α, K) the following normal convergence criterion holds.

Theorem 5 *Let $\{X_{nk}, 1 \leq k \leq k_n\}$ be a sequence of sequences of rowwise independent r.v.s with two moments finite. Then for some sequence of constants $\{A_n, n \geq 1\}$ the sequence $S_n^* = \sum_{k=1}^{k_n} X_{nk} - A_n \xrightarrow{D} \tilde{S}$, which is distributed as $N(0,1), \sigma^2(S_n^*) \to 1$, and $X_{nk} - E(X_{nk})$ are infinitesimal, iff for each $\varepsilon > 0, F_{nk}$ being the d.f. of X_{nk}, we have*

(i) $$\lim_{n\to\infty}\sum_{k=1}^{k_n}\int_{[|x|\geq\varepsilon]} x^2\,dF_{nk}(x+E(X_{nk}))=0,$$

(ii) $$\lim_{n\to\infty}\sum_{k=1}^{k_n}\int_{[|x|<\varepsilon]} x^2\,dF_{nk}(x+E(X_{nk}))=1,$$

(iii) $$\lim_{n\to\infty}\left(\sum_{k=1}^{k_n}E(X_{nk})-A_n\right)=0.$$

Proof Suppose that the conditions hold. Then the $(X_{nk}-E(X_{nk}))$ are infinitesimal since for each $\varepsilon>0$,

$$\begin{aligned}\max_{1\leq k\leq k_n} P[|X_{nk}-E(X_{nk})|\geq\varepsilon] &= \max_{1\leq k\leq k_n}\int_{[|x|\geq\varepsilon]} dF_{nk}(x+E(X_{nk}))\\ &\leq \frac{1}{\varepsilon^2}\sum_{k=1}^{k_n}\int_{[|x|\geq\varepsilon]} x^2\,dF_{nk}(x+E(X_{nk}))\\ &\to 0 \qquad \text{[by (i)]}.\end{aligned}$$

The variances $\sigma^2(S_n^*)$ are bounded because by adding (i) and (ii), we get

$$\lim_{n\to\infty}\sum_{k=1}^{k_n}\int_{\mathbb{R}} u^2\,dF_{nk}(u+E(X_{nk}))=1$$

and

$$\lim_{n\to\infty}\sum_{k=1}^{k_n}\int_{-\infty}^{x} u^2\,dF_{nk}(u+E(X_{nk}))=\begin{cases}0 & \text{if } x<0\\ 1 & \text{if } x\geq 0.\end{cases}$$

Thus if

$$K(x)=\begin{cases}0, & x<0\\ 1, & x\geq 0,\end{cases}$$

then conditions (i)–(iii) of Theorem 4 are satisfied for the infinitesimal sequence $\{X_{nk}-E(X_{nk}), 1\leq k\leq k_n, n\geq 1\}$. Hence, by that theorem, the limit distribution of the S_n^* is $N(0,1)$.

Conversely, if the d.f.s F_n of S_n^* converge to $N(0,1)$ and the variances of S_n^* tend to 1, the r.v.s $X_{nk}-E(X_{nk})$ being infinitesimal, then by Theorem 4 again, its conditions (i)–(iii) must hold. These are now equivalent to the present conditions, since (iii), with $\alpha=0$, is the same in both cases. Next for all $\varepsilon>0$ with the two-valued K going with $N(0,1)$, we have as $n\to\infty$,

$$\sum_{k=1}^{k_n}\int_{-\infty}^{-\varepsilon} x^2\,dF_{nk}(x+E(X_{nk}))\to 0,$$
$$\sum_{k=1}^{k_n}\int_{-\infty}^{+\varepsilon} x^2\,dF_{nk}(x+E(X_{nk}))\to 1.$$

These imply as $n \to \infty$,

$$\sum_{k=1}^{k_n} \int_{[|x|<\varepsilon]} x^2 \, dF_{nk}(x + E(X_{nk})) \to 1,$$

$$\sum_{k=1}^{k_n} \int_{[|x|\geq\varepsilon]} x^2 \, dF_{nk}(x + E(X_{nk})) \to 0.$$

This completes the proof.

An immediate consequence of this result is the celebrated Lindeberg-Feller theorem. Here the $X_n, n \geq 1$, are independent r.v.s with $E(X_n) = 0$, Var $X_n = \sigma^2$, and $X_{nk} = X_k/\sigma(S_n)$, where $S_n = \sum_{k=1}^n X_k$. If

$$\max_{1\leq k\leq n} (\sigma_k^2/\sigma^2(S_n)) \to 0,$$

then the X_{nk} are infinitesimal and $\mathrm{Var}(\sum_{k=1}^n X_{nk}) = 1$, so that conditions (ii) and (iii) of Theorem 4 are automatic. Now for the convergence to $N(0,1)$, only (i) of Theorem 4, which became (i) of Theorem 5, need be satisfied. Note that if F_k and F_{nk} are the d.f.s of X_k and X_{nk}, then

$$F_{nk}(x) = P[X_{nk} < x] = P[X_k < x\sigma(S_n)] = F_k(x\sigma(S_n)),$$

and hence for any $\varepsilon > 0$,

$$\int_{[|x|\geq\varepsilon]} x^2 \, dF_{nk}(x) = \int_{[|x|\geq\varepsilon]} x^2 \, dF_k(x\sigma(S_n)) = \frac{1}{\sigma^2(S_n)} \int_{[|y|\geq\varepsilon\sigma(S_n)]} y^2 \, dF_k(y). \tag{10}$$

Thus we have the following result as a consequence of Theorem 5. The sufficiency is due to Lindeberg and the necessity to Feller.

Theorem 6 (Lindeberg and Feller) *Let $\{X_n, n \geq 1\}$ be independent r.v.s with finite variances and zero means. Let F_k be the d.f. of X_k, and $S_n = \sum_{k=1}^n X_k$. Then $(S_n/\sigma(S_n)) \xrightarrow{D} S$, which is $N(0,1)$, and the $X_k/\sigma(S_n)$ are infinitesimal iff the following* CONDITION OF LINDEBERG *is satisfied for each $\varepsilon > 0$:*

$$\lim_{n\to\infty} \sum_{k=1}^{n} \frac{1}{\sigma^2(S_n)} \int_{[|x|>\varepsilon\sigma(S_n)]} x^2 \, dF_k(x) = 0. \tag{11}$$

It is useful to note that if the r.v.s have three moments finite, then the Liapounov condition $\rho^3(S_n)/\sigma^3(S_n) \to 0$ implies (11), so that this is an important generalization of the Liapounov theorem. To see this implication, consider for $\varepsilon > 0$,

$$
\begin{aligned}
0 &\leq \sum_{k=1}^{n} \frac{1}{\sigma^2(S_n)} \int_{[|x|>\varepsilon\sigma(S_n)]} x^2 \, dF_k(x) \\
&\leq \sum_{k=1}^{n} \frac{\varepsilon^{-1}}{\sigma^3(S_n)} \int_{[|x|>\varepsilon\sigma(S_n)]} |x^3| \, dF_k(x) \\
&\leq \frac{\rho^3(S_n)}{\varepsilon\sigma^3(S_n)} \to 0 \quad \text{as} \quad n \to \infty.
\end{aligned}
$$

The same computation holds if only $2+\delta, \delta > 0$, moments exist and moreover $\sum_{k=1}^{n} E(|X_k|^{2+\delta})/\sigma^{2+\delta}(S_n)$ goes to zero as $n \to \infty$.

Observing that, for a Poisson distribution with parameter $\lambda > 0$, the Kolmogorov pair (α, K) is given by $\alpha = \lambda$,

$$
K(x) = \begin{cases} 0, & x < 1 \\ \lambda, & x \geq 1, \end{cases}
$$

we can present analogous conditions for sums of independent r.v.s for convergence to a Poisson limit. This will extend Proposition 1.9. However, it is again an immediate consequence of Theorem 4. The easy verification is left to the reader.

Theorem 7 *Let $\{X_{nk}, 1 \leq k \leq k_n, n \geq 1\}$ be a sequence of rowwise independent sequences of infinitesimal r.v.s with finite variances. Then for some sequences $\{A_n, n \geq 1\}$ of constants, $S_n^* = \sum_{k=1}^{k_n} X_{nk} - A_n$ converges in distribution to a Poisson r.v. with parameter $\lambda > 0$, and $\sigma^2(S_n^*) \to \lambda$, iff for each $\varepsilon > 0$ we have*

(i) $$\lim_{n\to\infty} \sum_{k=1}^{k_n} \int_{[|x-1|\geq\varepsilon]} x^2 \, dF_{nk}(x + E(X_{nk})) = 0,$$

(ii) $$\lim_{n\to\infty} \sum_{k=1}^{k_n} \int_{[|x-1|<\varepsilon]} x^2 \, dF_{nk}(x + E(X_{nk})) = \lambda,$$

(iii) $$\lim_{n\to\infty} \left(A_n - \sum_{k=1}^{k_n} E(X_{nk}) \right) = \lambda.$$

It is now natural to ask whether there are analogous results for the sequences of partial sums $\{S_n, n \geq 1\}$, $S_n = \sum_{k=1}^{k_n} X_{nk}$, if the infinitesimal X_{nk} do not have finite moments. Indeed, the answer is yes, and with the Lévy or Lévy-Khintchine representations and Proposition 2.7, such results have also been obtained, primarily by B.V. Gnedenko. Now a more delicate estimation of various integrals is needed. We state the main result–a generalized Bawly theorem. Then one can understand the type of conditions that replace those of Theorem 4. For a proof of the result, we direct the reader to the classic by Gnedenko and Kolmogorov (1954) where references to the original sources and other details are given.

Theorem 8 (Generalized Bawly) *Let* $\{X_{nk}, 1 \leq k \leq k_n, n \geq 1\}$ *be a sequence of rowwise independent sequences of infinitesimal r.v.s and* $S_n^* = \sum_{k=1}^{k_n} X_{nk} - A_n$ *for some constants* $\{A_n, n \geq 1\}$. *Then* $S_n^* \xrightarrow{D} S$ *iff the following accompanying sequence* $\{Y_n, n \geq 1\}$ *of infinitely divisible r.v.s converges in distribution to* Y, *and then* $Y{=}S$ *a.e. Here* Y_n *is an r.v. with ch.f.* ψ_n *determined by*

$$E(e^{itY_n}) = \psi_n(t) = \exp\left\{ i\gamma_n t + \int_{\mathbb{R}} \left(e^{itx} - 1 - \frac{itx}{1+x^2} \right) \frac{1+x^2}{x^2} \, dG_n(x) \right\},$$

where for any fixed but arbitrary $\tau > 0$, *with* F_{nk} *as the d.f. of* X_{nk}, *we have*

$$\begin{aligned}
\alpha_{nk} &= \int_{[|x|<\tau]} x \, dF_{nk}(x), \\
\gamma_n &= \sum_{k=1}^{k_n} \alpha_{nk} - A_n + \sum_{k=1}^{k_n} \int_{\mathbb{R}} \frac{x}{1+x^2} \, dF_{nk}(x + \alpha_{nk}), \\
G_n(x) &= \sum_{k=1}^{k_n} \int_{-\infty}^{x} \frac{u}{1+u^2} \, dF_{nk}(u + \alpha_{nk}).
\end{aligned}$$

(Even though the Y_n *depend on* $\tau > 0$, *the limit r.v.* Y *does not.)*

The other, equally useful, forms of the limit distributions use the Lévy representation. All such results are discussed in detail, with a beautiful presentation, in the above monograph. In all these theorems the r.v.s X_{nk} were assumed asymptotically constant. If this hypothesis is dropped, the methods of proof undergo drastic changes. No general theory is available. But the following extension of the Lindeberg-Feller theorem, obtained by V.M. Zolotarev in 1967, indicates these possibilities.

Let $X_{nk}, k \geq 1, n \geq 1$, be rowwise independent r.v.s with $E(X_{nk}) = 0$, Var $X_{nk} = \sigma_{nk}^2$, and $\sum_{k \geq 1} \sigma_{nk}^2 = 1$. Let $S_n = \sum_{k \geq 1} X_{nk}$, which converges by Theorem 2.2.6. If $\tilde{G}$ is the normal d.f., $N(0,1)$, let G_{nk} be defined by $G_{nk}(x) = \tilde{G}(x/\sigma_{nk})$. If F, H are two d.f.s, let $d(F, H)$ denote the Lévy metric defined in Eq. (4.1.11). We then have

Theorem 9 (Zolotarev) *Let* $\{X_{nk}, k \geq 1, n \geq 1\}$ *be a sequence of rowwise independent sequences of r.v.s with means zero, variances* σ_{nk}^2, *such that*

$$\sum_{k=1}^{\infty} \sigma_{nk}^2 = 1.$$

Let $\tilde{G}$ be the normal d.f., $N(0,1)$, and F_n the d.f. of $S_n = \sum_{k\geq 1} X_{nk}$. Then $F_n(x) \to \tilde{G}(x)$ for all $x \in \mathbb{R}$ [or $d(F_n, \tilde{G}) \to 0$] as $n \to \infty$ iff

(i) $\alpha_n = \sup_k d(F_{nk}, G_{nk}) \to 0$, *where F_{nk} is the d.f. of X_{nk} and G_{nk} is defined above, and*

(ii) *for each $\varepsilon > 0$, with $\mathcal{A}_n = \{k : \sigma_{nk}^2 < \sqrt{\alpha_n}\}$*

$$\sum_{k\in\mathcal{A}_n} \int_{[|x|\geq\varepsilon]} x^2 \, dF_{nk}(x) \to 0, \qquad \text{as } n \to \infty.$$

Again we omit the proof of this interesting result to avoid the digression. In this vein, we state an important consequence of Theorem 8. Its proof involves an alternative form of the conditions of Theorem 8, and they are still nontrivial.

Theorem 10 (Khintchine) *Let $S_n = \sum_{k=1}^{k_n} X_{nk}$, the $X_{nk}, 1 \leq k \leq k_n$, being independent and infinitesimal. Suppose $S_n \xrightarrow{D} S$ as $n \to \infty$. Then S is normal $N(0,1)$ iff for each $\varepsilon > 0$*

$$\lim_{n\to\infty} \sum_{k=1}^{k_n} P[|X_{nk}| \geq \varepsilon] = 0, \tag{12}$$

which is equivalent to the condition

$$\lim_{n\to\infty} P\left[\max_{1\leq k\leq k_n} |X_{nk}| \geq \varepsilon\right] = 0. \tag{13}$$

There are other interesting specializations for Poisson and degenerate convergence and then to single sequences. For details, we refer the reader to the Gnedenko-Kolmogorov fundamental monograph noted above.

The preceding theory shows that any infinitely divisible d.f. can be a limit d.f. of the sequence of sums of rowwise independent infinitesimal r.v.s. The classical central limit theory leads to the normal d.f. as the limit element. However, it is of interest to look for a subclass of the infinitely divisible laws which can arise having a "simpler" form than the general family. This turns out to be a family, called stable laws. Let us introduce this class by the following motivational example, which gives some concreteness to the general concept to be discussed.

Example 11 Let $\{X_n, n \geq 1\}$ be i.i.d. random variables with the common d.f. F_p given by its density $F_p' = f_p$, (Pareto density, c.f., Exercise 4.25)

$$f_p(x) = \begin{cases} 0, & |x| \leq 1 \\ (p/2)|x|^{-1-p}, & |x| > 1, \ p > 0. \end{cases}$$

If $S_n = \sum_{k=1}^n X_k$, we find the numbers $b_n(p) > 0$ (called "normalizing constants") such that $S_n/b_n(p) \xrightarrow{D} \tilde{S}$, where $\tilde{S}$ is an infinitely divisible r.v.

(i) It is clear that, if $p > 2$, then $\sigma^2 = \text{Var } X_1 = (p-2)^{-1} < \infty$, and the classical central limit theorem applies with $b_n(p) = \sqrt{n\sigma^2}$, so that

$$\lim_{n\to\infty} P\left[\frac{S_n}{\sqrt{n\sigma^2}} < x\right] = \frac{1}{\sqrt{2\pi}} \int_{-\infty}^{x} e^{-u^2/2}\, du, \qquad x \in \mathbb{R}.$$

Thus we only need to consider $0 < p \le 2$. This has to be treated in two parts: $0 < p < 2$ and $p = 2$. The variances do not exist in both cases.

(ii) $0 < p < 2$. We cannot directly try to verify (12) here since it is first necessary to find the $b_n(p)$ such that $S_n/b_n(p) \xrightarrow{D} \tilde{S}$. Only then (12) gives conditions for $\tilde{S}$ to be $N(0,1)$. However, this is the key part. Let us try, by analogy with (i), n^α, for some $\alpha > 0$ as the normalizing factor. A heuristic reason for this will become clear a little later. We use the technique of ch.f.s. Thus

$$\begin{aligned}
\phi_n(t) &= E(e^{itS_n/n^\alpha}) \\
&= \prod_{k=1}^{n} E(e^{itX_k/n^\alpha}) \\
&= \left[p\int_1^\infty \frac{\cos(tx/n^\alpha)}{x^{1+p}}\, dx\right]^n, \\
&= \left[1 - p\int_1^\infty \frac{1-\cos(tx/n^\alpha)}{x^{1+p}}\, dx\right]^n \\
&= [1 - I_n(t)]^n \quad \text{(say)}.
\end{aligned} \tag{14}$$

It is clear that ϕ_n is real and never zero on $\mathbb{R}$. Thus we have, on expansion of (14),

$$\log \phi_n(t) = n[-I_n - I_n^2/2 - \cdots] \tag{15}$$

if $|I_n| < 1$. We now find α such that $I_n = O(1/n)$. In fact,

$$nI_n = 2np\int_1^\infty \frac{\sin^2(tx/2n^\alpha)}{x^{1+p}}\, dx = \frac{p|t|^p}{2^{p-1}n^{\alpha p-1}} \int_{|t|/(2n^\alpha)}^\infty \frac{\sin^2 u}{u^{1+p}}\, du.$$

Taking $\alpha p = 1$, we can then let $n \to \infty$. But note that the integral on the right converges only if $p < 2$. (Its singularity is at the lower end point if $p = 2$.) Hence for $0 < p < 2$, letting $\alpha = 1/p$, we have

$$\lim_{n\to\infty} nI_n = \frac{p|t|^p}{2^{p-1}} \int_0^\infty \frac{\sin^2 u}{u^{1+p}}\, du = c_p \frac{1}{2^{p-1}} |t|^p \quad \text{(say)}. \tag{16}$$

From (15) and (16), it follows that

$$\lim_{n\to\infty} \phi_n(t) = \exp\left(-\frac{c_p}{2^{p-1}}|\,t|^p\right), \qquad 0 < p < 2, \tag{17}$$

and since the right side is continuous, it is a ch.f., by the continuity theorem. If $p = 1$, then the limit is the Cauchy ch.f. Thus for $0 < p < 2, b_n(p) = n^{1/p}$ is the correct normalizing constant for S_n.

(iii) $p = 2$. Since $b_n = \sqrt{n}$ is not enough to control the growth of S_n, as seen from the divergence of (16), we need b_n to grow somewhat faster. Let us try $b_n = (n \log n)^{1/2}$. Then (15) becomes with this new normalization

$$\log \phi_n(t) = n[-J_n - J_n^2/2 - \cdots] \tag{17'}$$

if $|\,J_n| < 1$. Here

$$\begin{aligned} nJ_n &= 4n \int_1^\infty \frac{\sin^2(tx/2[n\log n]^{1/2})}{x^3}\,dx \\ &= \frac{t^2}{\log n}\int_{|\,t|/2(n\log n)^{1/2}}^\infty \frac{\sin^2 u}{u^3}\,du. \end{aligned}$$

It is clear that the integral on (ε, ∞) converges for each $\varepsilon > 0$, and hence the right side goes to zero as $n \to \infty$ on this interval. Therefore its value, as $n \to \infty$, is the same as the limit of

$$J_n' = \frac{t^2}{\log n}\int_{|\,t|/2(n\log n)^{1/2}}^{\varepsilon} u^{-3}\sin^2 u\,du.$$

We now estimate this. First, choose $\varepsilon > 0$ such that given $\eta > 0, 1 - \eta < \sin^2 u/u^2 < 1$ for $0 \le u \le \varepsilon$. Then

$$\frac{(1-\eta)t^2}{\log n}\int_{|\,t|/2(n\log n)^{1/2}}^{\varepsilon} \frac{du}{u} < J_n' < \frac{t^2}{\log n}\int_{|\,t|/2(n\log n)^{1/2}}^{\varepsilon} \frac{du}{u}.$$

Hence for each $t \neq 0$,

$$\begin{aligned} &\frac{(1-\eta)t^2}{\log n}[\log\varepsilon - \log(|\,t|/2(n\log n)^{1/2})] \\ &\qquad < J_n' < \frac{t^2}{\log n}[\log\varepsilon - \log(|\,t|/2(n\log n)^{1/2})]. \end{aligned}$$

First letting $n \to \infty$ on both sides and then $\eta \to 0$ we see that the extremes have the limit $t^2/2$. Consequently $\lim_{n\to\infty} J_n' = \lim_{n\to\infty} nJ_n = t^2/2$. Substituting this in (17) we get

$$\lim_{n\to\infty} \phi_n(t) = e^{-t^2/2}. \tag{18}$$

Now that $b_n = (n\log n)^{1/2}$ is seen as the correct normalizing constant, so that $S_n/b_n \xrightarrow{D} \tilde{S}$, we could also immediately verify (12), so that $\tilde{S}$ is $N(0,1)$, which agrees with (18).

Thus if $b_n(p) = n^{1/p}$ for $0 < p < 2; = (n \log n)^{1/2}$ for $p = 2$; or $= \sqrt{n}$ if $p > 2$, we see that

$$\lim_{n\to\infty} \phi_n(t) = \exp(-c'_p|t|^{\tilde{p}}), \qquad t \in \mathbb{R}, \tag{19}$$

where $\tilde{p} = p$ on $[0, 2]$ and $\tilde{p} = 2$ when $p > 2$. A ch.f. of the type on the right side is clearly the nth power of a similar ch.f. for each $n \geq 1$, so that it is infinitely divisible. Such ch.f.s define *symmetric stable distributions*. We therefore introduce the concept following P. Lévy.

Definition 12 An r.v. X is said to have a *stable* d.f. F if, when X_1, X_2 are i.i.d. with the d.f. F, then for each $a_1 > 0, a_2 > 0$ and $b_i \in \mathbb{R}, i = 1, 2$, there exist a pair $a > 0, b \in \mathbb{R}$, such that $(a_1X_1 + b_1) + (a_2X_2 + b_2)$ and $aX+b$ have the same d.f. Equivalently, $F(a_1^{-1}((\cdot)-b_1)) * F(a_2^{-1}((\cdot)-b_2))(x) = F(a^{-1}(x-b)), x \in \mathbb{R}$, where $*$ denotes convolution.

Stated in terms of ch.f.s, the above becomes: if ϕ is the ch.f. of X, then

$$\phi(a_1t)\phi(a_2t) = \phi(at)\exp\{it(b - b_1 - b_2)\}, \qquad t \in \mathbb{R}. \tag{20}$$

From this it follows that normal, Cauchy, and degenerate d.f.s are stable. But there are others. Their structure is quite interesting, though somewhat intricate, as we shall see. [Some important applications of this class will be considered in Section 8.4.]

Iterating (20) n times, one gets with $a_1 \ldots, a_n \in \mathbb{R}^+$, an $a \in \mathbb{R}^+$ and a $b \in \mathbb{R}$ such that

$$\prod_{i=1}^{n} \phi(a_it) = \phi(at)e^{itb}, \qquad t \in \mathbb{R}.$$

In particular, setting $a_1 = a_2 = \cdots = a_n = 1$, there exists an $a'_n > 0$ and a $b' \in \mathbb{R}$ such that

$$(\phi(t))^n = \phi(a'_nt)e^{itb'}, \tag{21}$$

so that

$$\phi(t) = \left[\phi\left(\frac{t}{a'_n}\right)\exp\left(-it\frac{b'}{na'_n}\right)\right]^n, \qquad n \geq 1. \tag{21'}$$

Since the factor in [ ] in (21′) is a ch.f. for each n, we conclude that ϕ is infinitely divisible, and so the *stable class is a subset of the infinitely divisible family*. Hence a stable ch.f. never vanishes.

Remark Evidently (21) is derived from the definition of stability. It says that if X is an r.v. and $X_1, \ldots, X_n$ are i.i.d. with the d.f. of X, then

$S_n = \sum_{i=1}^n X_i$ and $a_n' X + b'$ have the same d.f. for each n and for some $a_n' > 0, b' \in \mathbb{R}$. [Equivalently, $(S_n - b_n)/a_n$ has the same d.f. as X.] However, the converse of this statement is also true; i.e., if ϕ is a ch.f. which satisfies (21) for each n, then ϕ is stable in the sense of Definition 12, or (20). This follows from the representation (or characterization) below of the class of stable ch.f.s. In (21), the a_n are called *norming constants.* One says that the d.f.s F, G are of the *same type* if $F(x) = G(ax + b)$ for all $x \in \mathbb{R}$ and some $a > 0, b \in \mathbb{R}$. In words, F, G differ only by the scale and location parameters. From this point of view, if F is stable, then so is G, where $G(x) = F(ax + b)$ for some $a > 0, b \in \mathbb{R}$, and all $x \in \mathbb{R}$. Thus we have *stable types* of laws.

Regarding a heuristic reason for the normalizing factors used in Example 11 (however, the r.v.s there are not stable, only the limit is), or in (21), may be compared with the following definitive statement.

Proposition 13 *The norming constants $a_n > 0$ in (21) are always of the form $a_n = n^{1/\alpha}, \alpha > 0$. If, moreover, ϕ is nondegenerate, then $\alpha \leq 2$.*

Proof To simplify the argument, we first reduce it to the symmetric case and complete the demonstration essentially following Feller. If X_1, X_2 are i.i.d. stable r.v.s, then $Y = X_1 - X_2$ is a symmetric stable r.v and the X_i, Y have the same norming constants. Indeed, if ϕ is the (common) ch.f. of the X_i and ψ that of Y, we have by (21)

$$\begin{aligned}(\psi(t))^n &= (\phi(t)\phi(-t))^n \\ &= \phi(a_n t)\phi(-a_n t)\exp\{it(b_n' - b_n')\} = \psi(a_n t), \qquad n \geq 1\end{aligned}$$

and $\psi(t) \geq 0$. Thus ψ and ϕ have the same norming constants $a_n > 0$. Consequently we may (and do) assume that X is a symmetric stable r.v. with $a_n > 0$ as its norming constant for the rest of the proof. Also let $X \not\equiv 0$ (to avoid trivialities).

As noted in the remark before the statement of the proposition, the stability hypothesis on X implies that if $X_1, \ldots, X_n$ are i.i.d. as X, then $S_n = \sum_{i=1}^n X_i$ and $a_n X + b_n'$ are identically distributed for each n, where a_n is the norming constant and $n \geq 1$ is arbitrary. Replacing n by $m + n$ ($m, n \geq 1$ integers), we first note that $(S_{m+n} - S_m)$ and S_m are independent r.v.s, and the stability hypothesis implies the following set of equations, since $(S_{m+n} - S_m)$ and S_n are identically distributed:

$$S_{m+n} \stackrel{D}{=} a_{m+n}X, \qquad S_m \stackrel{D}{=} a_m X', \qquad S_{m+n} - S_m \stackrel{D}{=} a_n X'', \tag{22}$$

where X', X'' are i.i.d. as X. The symmetry of X implies that $b' = 0$ in this representation of the S_n. Since $S_{m+n} = (S_{m+n} - S_m) + S_m$, (22) yields

$$a_{m+n}X \stackrel{D}{=} a_m X' + a_n X'', \qquad m \geq 1, \quad n \geq 1. \tag{23}$$

In terms of ch.f.s, this becomes

$$\phi(ta_{m+n}) = \phi(a_m t)\phi(a_n t), \qquad t \in \mathbb{R}, \tag{24}$$

since X', X'' are i.i.d. and have the same ch.f. as X. Replacing t by t/a_n and setting $m = n$ in (24), we obtain

$$\begin{aligned}\phi(ta_{2n}/a_n) &= \phi(t)^2 \\ &= \phi(a_2 t), \qquad t \in \mathbb{R}, \qquad [\text{by } (21)].\end{aligned} \tag{25}$$

Since $t \in \mathbb{R}$ is arbitrary, we see (why?) from (25) that $a_{2n} = a_2 a_n$. In a similar manner considering r blocks of n terms each for S_{rn}, we get $a_{rn} = a_r a_n, r \geq 1, n \geq 1$. If now $n = r^p$, so that $a_{r^{k+1}} = a_r a_{r^k}, k = 1, 2, \ldots, p$, multiply them to get $a_n = (a_r)^p$. We next obtain a few other arithmetical properties of $\{a_n, n \geq 1\}$. Since the result is true and trivial in the degenerate case for ϕ (any a_n works), we exclude this case in the following argument.

The sequence $\{a_n, n \geq 1\}$ is monotone increasing and tends to infinity. Indeed, let $u > 0$ be arbitrarily fixed. Consider with (23) and the symmetry of the r.v. X,

$$\begin{aligned}P[a_{m+n}X > a_m u] &= P[a_m X' + a_n X'' > a_m u] \\ &\geq P[a_m X' > a_m u, a_n X'' \geq 0] \\ &= P[a_m X' > a_m u] P[a_n X'' \geq 0] \quad \text{(by independence)} \\ &\geq \frac{1}{2} P[X' > u] \quad \text{(by symmetry of the d.f. of } X''\text{)}.\end{aligned}$$

Thus

$$P[X > (a_m/a_{m+n})u] \geq \frac{1}{2} P[X' > u] = \frac{1}{2} P[X > u]. \tag{26}$$

Now the right side is a fixed positive constant for a suitable $u > 0$. If a_m/a_{m+n} is not bounded as $m \to \infty$, and $n \to \infty$, then the left side goes to zero, contradicting the inequality of (26). Hence

$$\overline{\lim}_{m\to\infty, n\to\infty} \frac{a_m}{a_{m+n}} \leq \beta_0 < \infty.$$

In particular, if $m = r^k$ and $n = (r+1)^k - m$, where $r > 0$ is fixed, we get for large m, n,

$$\frac{a_m}{a_{m+n}} = \frac{a_{r^k}}{a_{(r+1)^k}} = \left(\frac{a_r}{a_{r+1}}\right)^k \leq \beta_0 < \infty$$

by the preceding paragraph. Letting $k \to \infty$, this implies $a_r/a_{r+1} \leq 1$ so that $\{a_r, r \geq 1\}$ is monotone.

Next we assert that $a_r = r^\beta$, for some $\beta > 0$, so that a_r tends to infinity and proves the main part. In fact, if $k, p \geq 1$ are integers and $q > 1$,then we can find a unique integer $j \geq 0$ such that

$$p^j \leq k^q < p^{j+1} \Rightarrow a_{pj} = (a_p)^j \leq (a_k)^q < (a_p)^{j+1}. \tag{27}$$

This implies $a_p > 1$ and then on taking "logs" for these inequalities and dividing, one gets (all $a_k = 1 \Rightarrow X = 0$ a.e. by (21), so that $a_k \neq 1$):

$$\frac{j \log p}{(j+1) \log a_p} < \frac{\log k}{\log a_k} \leq \frac{(j+1) \log p}{j \log a_p}. \tag{28}$$

This is independent of q, and so by taking q large, we can make j large, so that the extremes can be made arbitrarily close to each other. This implies the middle ratio is independent of k. Letting $(\log k)/\log a_k = \alpha > 0$, we get $a_k = k^{1/\alpha}$ and $\beta = 1/\alpha$. It only remains to show that if $\alpha > 2$, then ϕ is degenerate.

Again consider (21) for $|\phi(\cdot)|$. Since ϕ is infinitely divisible, $|\phi(t)|$ is never zero. Thus

$$n \log |\phi(t)| = \log |\phi(a_n t)|, \qquad t \in \mathbb{R}.$$

But we just proved that $a_n = n^{1/\alpha}$. Hence replacing t by $\tau/n^{\alpha^{-1}}$ in the above we get

$$\log |\phi(\tau)| = n \log |\phi(\tau/n^{\alpha^{-1}})|,$$

and therefore $\log |\phi(1/n^{\alpha^{-1}})| = O(1/n)$. Consequently, $n^{2/\alpha} \log |\phi(1/n^{1/\alpha})| = O(n^{(2/\alpha)-1})$, and this tends to zero as $n \to \infty$ if $\alpha > 2$. It means that $|\phi(t)| = 1 + o(t^2)$ as $t \to 0$. This implies that the r.v. X with ϕ as its ch.f. has finite second moment, and then it is zero, so that the r.v. is a constant a.e. To see this, consider $Y = X - \tilde{X}$, where $\tilde{X}$ is i.i.d. as X. Thus Y has two moments iff X has the same property (cf. Problem 33 of Chapter 4), and its ch.f. is $|\phi|^2 \geq 0$. If F is the d.f. of Y, then

$$\begin{aligned}
\int_{\mathbb{R}} x^2 \, dF(x) &= 2 \int_{\mathbb{R}} \lim_{t \to 0} \frac{1 - \cos tx}{t^2} \, dF(x) \\
&\leq 2 \liminf_{t \to 0} \int_{\mathbb{R}} \frac{1 - \cos tx}{t^2} \, dF(x) \qquad \text{(by Fatou's lemma)} \\
&= 2 \liminf_{t \to 0} \frac{1 - |\phi(t)|^2}{t^2} \leq 2 \liminf_{t \to 0} \frac{1 - e^{-Ct^2}}{t^2} = C < \infty.
\end{aligned}$$

Here C is a constant satisfying $|[\log |\phi(t)|^2]/t^2| \leq C$ as $t \to 0 \Rightarrow |\phi(t)|^2 \geq e^{-Ct^2}$. But $|\phi(t)| = 1 + o(t^2), t \to 0$. Thus the second moment of Y must vanish. The proof is finished.

Remark If $\alpha = 2$, the above computation implies that $\log |\phi(t)| = O(t^2)$ as $t \to 0$, so that $\phi''(0)$ exists and X has two moments. Then (21′) shows that ϕ is the ch.f. of $(S_n - nb')/\sqrt{n}$ for all $n \geq 1$, where $S_n = \sum_{k=1}^n X_k, X_k$ are i.i.d. as X, and $b = E(X)$. The classical central limit law (Theorem 1.2) shows that ϕ must be a normal ch.f. We leave the formal statement to the reader (see Problem 23.)

The preceding result has another consequence. We also term the constant $\alpha > 0$ of the above proposition the *characteristic exponent* of (the d.f. of) X.

Corollary 14 *If X is a nondegenerate symmetric stable r.v. with a characteristic exponent $\alpha > 0$, and X_1, X_2 are i.i.d. as X, then for any positive numbers a, b we have*

$$(a+b)^{1/\alpha} X \stackrel{D}{=} a^{1/\alpha} X_1 + b^{1/\alpha} X_2. \tag{29}$$

Proof If X, X_1, X_2 are as in the statement, and ϕ is the ch.f. of X, then by (24)

$$\phi(m+n)^{1/\alpha} t) = \phi(a_{m+n} t) = \phi(a_n t)\phi(a_m t) = \phi(n^{1/\alpha} t)\phi(m^{1/\alpha} t) \tag{30}$$

for any positive integers m, n. Replacing m, n by np and $mq (p, q \geq 1$ integers) and t by $t/(nq)^{1/\alpha}$ in (30), we get

$$\phi((p/q + m/n)^{1/\alpha} t) = \phi((p/q)^{1/\alpha} t)\phi((m/n)^{1/\alpha} t), \quad t \in \mathbb{R}. \tag{31}$$

Hence (29) is true if $a = p/q, b = m/n$, i.e., all positive rationals. If $a, b > 0$ are real, then they can be approximated by sequences of rationals for which (31) holds. Since ϕ is continuous, (31) implies

$$\phi((a+b)^{1/\alpha} t) = \phi(a^{1/\alpha} t)\phi(b^{1/\alpha} t).$$

This is (29), and the result follows.

The interest in stable laws is enhanced by the fact that they are the only laws that can arise as limit distributions of the normalized sums of i.i.d. sequences of r.v.s as originally noted by P. Lévy. That is how he first introduced this concept. Let us present this result precisely.

Proposition 15 (Lévy) *Let $\{X_n, n \geq 1\}$ be i.i.d. and $S_n = \sum_{k=1}^n X_k$. Then for some constants $A_n > 0, B_n \in \mathbb{R}, (1/A_n)S_n - B_n \xrightarrow{D} \tilde{S}$ iff $\tilde{S}$ is a stable r.v.*

Proof That every stable law is a limit of the described type is immediate from definition. In fact, by the remark following Definition 12, if X is a stable r.v. and $X_1, \ldots, X_n$ are i.i.d. as X, then $S_n = \sum_{i=1}^n X_i \stackrel{D}{=} a_n X + b_n, a_n > 0$, so that $(1/a_n)S_n - \tilde{b}_n \stackrel{D}{=} X$, where $\tilde{b}_n = b_n/a_n$. Thus only the converse is nontrivial. The true and trivial case that $\tilde{S}$ is degenerate will again be eliminated in the rest of the proof.

Suppose then $(1/A_n)S_n - B_n \xrightarrow{D} \tilde{S}$, as given. Hence every convergent subsequence on the left has the same limit $\tilde{S}$. Consider the following blocks of i.i.d. sequences: $S_{1n} = \sum_{i=1}^n X_i, S_{2n} = \sum_{i=n+1}^{2n} X_i, \ldots, S_{kn} = \sum_{i=(k-1)n+1}^{kn} X_i$. By hypothesis, for each $k \geq 1$,

$$(1/A_n)S_{kn} - B_n \xrightarrow{D} \tilde{S}_k \quad \text{as } n \to \infty, \tag{32}$$

and $\{\tilde{S}_k, k \geq 1\}$ are i.i.d. Let k be arbitrarily fixed. Consider

$$Y_{kn} = \frac{1}{A_{kn}} \sum_{i=1}^{kn} X_i - B_{kn} \quad \left(= \frac{1}{A_{kn}} \sum_{j=1}^{k} S_{jn} - B_{kn} \right) \xrightarrow{D} \tilde{S}, \tag{33}$$

as $n \to \infty$, since $Y_{1n} \xrightarrow{D} \tilde{S}$ by hypothesis and $\{kn, n \geq 1\}$ is a cofinal subsequence of the integers. (This is immediate if we use the image laws and go to another probability space on which the corresponding sequence converges a.e., as in the second proof of Theorem 4.1.2.) From (32) and the definition of Y_{kn} in (33), one gets

$$\begin{aligned}\sum_{j=1}^{k} \tilde{S}_j \xleftarrow{D} \sum_{j=1}^{k} \left[\frac{1}{A_n} S_{jn} - B_n \right] &= \frac{A_{kn}}{A_n} Y_{kn} + \left(\frac{B_{kn} A_{kn}}{A_n} - kB_n \right) \\ &= a_{kn} Y_{kn} + b_{kn} \quad (say).\end{aligned} \tag{34}$$

But (33) and (34) imply that $Y_{kn} \xrightarrow{D} \tilde{S}$ and $a_{kn} Y_{kn} + b_{kn} \xrightarrow{D} \tilde{Y}$. Since $\tilde{S}$ is nondegenerate by assumption, we can apply the result of Problem 22 in Chapter 4 (and the reader should verify it now, if it was not already done), $a_{kn} \to a_k > 0, b_{kn} \to b_k$, so that (34) implies

$$\sum_{j=1}^{k} \tilde{S}_j \stackrel{D}{=} a_k \tilde{S} + b_k.$$

Since $k \geq 1$ is arbitrary, this implies $\tilde{S}$ is stable, which ends the proof.

The preceding work heightens interest in the stable laws and it is thus natural to study and determine this subset of the infinitely divisible class. Such a characterization has been obtained again jointly by A. Khintchine and P. Lévy in 1938, and we present it now with a somewhat simpler proof. The original one depended on the canonical representation of infinitely divisible laws given in Theorem 2.5.

Theorem 16 *Let $\phi : \mathbb{R} \to \mathbb{C}$ be a mapping. Then ϕ is a stable ch.f. iff it admits the representation,*

$$\phi(t) = \exp\{i\gamma t - c|t|^\alpha(1 - i\beta \operatorname{sgn} t \cdot \varpi(t, \alpha))\}, \tag{35}$$

where $\gamma \in \mathbb{R}, -1 \leq \beta \leq 1, c \geq 0, 0 < \alpha \leq 2$, and

$$\varpi(t, \alpha) = \begin{cases} \tan(\pi\alpha/2) & if \ \alpha \neq 1 \\ -(2/\pi) \log |t| & if \ \alpha = 1. \end{cases}$$

Remark The constant $\alpha > 0$ here will be seen to be the same as that of Proposition 13, and hence it is just the *characteristic exponent* of ϕ. The case that $\alpha > 2$ of Proposition 13 corresponds to $c = 0$ here. Also $\alpha = 2$ gives the normal, and $\alpha = 1, \beta = 0, \gamma = 0$ gives the Cauchy ch.f.s. If ϕ is nondegenerate (thus $c > 0$), then $|\phi|$ is Lebesgue integrable on $\mathbb{R}$, and so every nondegenerate stable d.f. is absolutely continuous with a continuous density (by Theorem 4.2.1). However, an explicit calculation of most of these densities is not simple. [Some asymptotic expansions for such densities have been presented in 1954 by A.V. Skorokhod.] We derive the representation (35), but the complete proof will be given *only for the symmetric stable case* and comment on the omitted part (shifted to the problem section, see Exercises 24 and 25). The argument uses some number theoretical techniques of great value.

Proof Let ϕ be a stable ch.f. in the sense of Definition 12. Then (21) holds by Proposition 13, with $a_n = n^{1/\alpha}$ for some $\alpha > 0$, so that, setting $\delta = 1/\alpha$ for convenience, we have

$$(\phi(t))^n = \phi(n^\delta t)e^{itb_n}, \qquad b_n \in \mathbb{R}. \tag{36}$$

Since ϕ is also infinitely divisible by (21′), it never vanishes by Proposition 2.2a. (This elementary property is the only one taken from the theory of Section 2.) By the remark following Proposition 13, $\alpha = 2$ implies ϕ is normal and (35) is true. We exclude this and the (true and) trivial degenerate cases from the following discussion. So assume that $0 < \alpha < 2$ (or $\delta > \frac{1}{2}$). Hence by Proposition 4.2.9,

$$\operatorname{Log} \phi(t) = \log|\phi(t)| + ig(t), \tag{37}$$

where $g(0) = 0$ and $g(\cdot)$ is continuous on $\mathbb{R}$. Let $h(t) = |\phi(t)|$. Then (36) implies

$$(h(t))^n = h(n^\delta t), \qquad n \geq 1, \qquad t \in \mathbb{R}. \tag{38}$$

Since (ϕ and so) h is continuous, we deduce from (38) for integers $m, n \geq 1$,

$$h(n^\delta) = (h(1))^n \quad \text{and} \quad h(m^\delta) = h(n^\delta(m/n)^\delta) = h((m/n)^\delta))^n,$$

so that

$$h((m/n)^\delta) = (h(m^\delta))^{1/n} = (h(1))^{m/n}.$$

By the continuity of h, we get $h(t^\delta) = (h(1))^t$ for $t > 0$. Replacing t by t^δ in the above, one obtains

$$h(t) = h(1)^{t^\alpha} = \exp\{-c|t|^\alpha\}, \tag{39}$$

where $c = -\log h(1) > 0$. Clearly (39) is true for $t = 0$ also, and then for all $t \in \mathbb{R}$.

Let us next consider $g(\cdot)$. From (36) and (37) (considering the imaginary parts), one has

$$ng(t) = g(n^\delta t) + tb_n. \tag{40}$$

Hence replacing n by mn gives

$$\begin{aligned} mng(t) &= g(m^\delta n^\delta t) + tb_{mn} \\ &= mg(n^\delta t) - b_m n^\delta t + tb_{mn} \quad \text{[by (40)]} \\ &= m(ng(t) - tb_n) - b_m n^\delta t + tb_{mn}. \end{aligned}$$

Rewriting this, one has (set $t = 1$)

$$b_{mn} = mb_n + n^\delta b_m = nb_m + m^\delta b_n \quad \text{(because } b_{mn} = b_{nm}\text{)}.$$

Thus

$$b_m(n - n^\delta) = b_n(m - m^\delta), \quad n, m \geq 1 \text{ integers}. \tag{41}$$

If now $\alpha \neq 1$, so that $\delta \neq 1$, we get a solution of (41), for some $a_0 \in \mathbb{R} : b_n = a_0(n - n^\delta)$. Setting $f(t) = g(t) - a_0 t$, then with (40) one has

$$nf(t) = f(n^\delta t), \quad t \in \mathbb{R}, \quad n \geq 1. \tag{42}$$

For this functional equation we can apply the same argument as in (38). Thus

$$f(n^\delta) = nf(1) \quad \text{and} \quad f(m^\delta) = f(n^\delta(m/n)^\delta) = nf((m/n)^\delta).$$

Hence

$$f(m^\delta/n^\delta) = (1/n)f(m^\delta) = (m/n)f(1), \quad m, n \geq 1.$$

Since $f(\cdot)$ is continuous, this gives

$$f(t^\delta) = tf(1), \quad t > 0.$$

Next replacing t by t^α, one gets (because $\delta = 1/\alpha$)

$$f(t) = f(1) \cdot t^\alpha = g(t) - a_0 t. \tag{43}$$

Substituting (39) and (43) in (37), we have for $\alpha \neq 1$,

$$\begin{aligned} \operatorname{Log} \phi(t) &= -c|t|^\alpha + ia_0 t + if(1)t^\alpha \qquad (t > 0) \\ &= ia_0 t - ct^\alpha(1 - i\beta_0) \qquad (t > 0) \end{aligned} \tag{44}$$

where $\beta_0 = f(1)/c$. Since $\phi(-t) = \overline{\phi(t)}$, the result for $t < 0$ also is obtained from the above. This becomes an identity for $t = 0$. Hence (44) yields

$$\operatorname{Log} \phi(t) = ia_0 t - c|t|^\alpha(1 - i\beta \operatorname{sgn} t \cdot \theta), \quad t \in \mathbb{R}, \tag{45}$$

where $\theta = \beta_0/\beta$ if $\beta_0 \neq 0, = 0$, if $\beta_0 = 0$, with $-1 \leq \beta \leq 1$. Now (45) is the same as (35) *if we can identify* θ as $\tan(\pi\alpha/2)$. This is discussed further later on.

Let us consider the case $\alpha = 1$ [so that $\delta = 1$ in (41)]. There is no problem for (39). Now (40) becomes $ng(t) = g(nt) + tb_n$. From this we deduce, on replacing t by mt and eliminating b_n between these equations,

$$\begin{aligned} ng(mt) &= g(nmt) + mtb_n \\ &= g(nmt) + m[ng(t) - g(nt)]. \end{aligned} \tag{46}$$

To solve the functional equation (46), let $w(u) = e^{-u}g(e^u), u \in \mathbb{R}$. Then if $u = \log t, a_n = \log n, n \geq 1, t > 0$, we get from (46)

$$w(u + a_m + a_n) - w(u + a_m) - w(u + a_n) + w(u) = 0. \tag{47}$$

For each fixed m, let $v(r) = w(r + a_m) - w(r)$. Then (47) implies that $v(r + a_n) = v(r), n \geq 1$ and $r \in \mathbb{R}$. This periodicity behavior in turn gives for any integers p, q, m, n with $m \geq 1, n \geq 1$

$$\begin{aligned} v(pa_m + qa_n) &= v(pa_m + (q-1)a_n) \\ &\vdots \\ &= v(pa_m) \\ &\vdots \\ &= v(0 + (p-1)a_m) \\ &\vdots \\ &= v(0). \end{aligned} \tag{48}$$

Choose $m_0 \geq 1, n_0 \geq 1$ such that a_{m_0}/a_{n_0} is irrational. [This is possible. Indeed let m_0, n_0 be relatively prime. For the continuous function $f : x \mapsto f(x) = n_0^x, f(x) = m_0$ has a solution x_0 (by the intermediate value theorem) which is irrational.] Then the set $\{pa_{m_0} + qa_{n_0} : p, q \text{ all integers}\}$ is dense in $\mathbb{R}$. Hence (48) implies $v(u) = v(0)$, by continuity of $v(\cdot)$. This means in terms of $w(\cdot)$.

$$w(u + a_m) - w(u) = w(a_m) - w(0), \quad u \in \mathbb{R}, \quad m \geq 1.$$

Let $\zeta(u) = w(u) - w(0)$. Then the above equation becomes

$$\zeta(u + a_m) = \zeta(u) + \zeta(a_m), \quad u \in \mathbb{R}, \quad m \geq 1. \tag{49}$$

Replacing a_m by $pa_m + qa_n$ in (49), we get

$$\zeta(u + pa_m + qa_n) = \zeta(u) + \zeta(pa_m + qa_n), m \geq 1, \quad n \geq 1.$$

If m, n are replaced by m_0, n_0, then, using the density argument, we can deduce that

$$\zeta(u + r) = \zeta(u) + \zeta(r), \quad u, r \in \mathbb{R}. \tag{50}$$

But $\zeta(\cdot)$ is continuous, and thus (50) is the classical Cauchy functional equation. So $\zeta(u) = a_0 u$, or $w(u) = a_0 u + b_0$ for some constants a_0, b_0. Hence for $t > 0$, and $u = \log t$,

$$g(t) = g(e^u) = w(u) \cdot e^u = (a_0 \log t + b_0)t. \tag{51}$$

Substituting (39) and (51) in (37) with $\alpha = 1$ [and using $\phi(-t) = \overline{\phi(t)}$] gives

$$\text{Log}\, \phi(t) = ib_0 t - c|t|\{1 + i(a_0/c) \operatorname{sgn} t \cdot \log|t|\}, \qquad t \in \mathbb{R}. \tag{52}$$

If $\beta = -\pi a_0/2c$, then (52) is of the form (35), provided that we show $|\beta| \le 1$. In other words, (45) and (52) together give only the *form* of (35). To show that the expressions for ϕ given by (45) and (52) do in fact determine ch.f.s, we have to verify that they are positive definite to use Bochner's theorem. This is quite difficult. There is the following alternative (detailed, but elementary) method for $0 < \alpha < 2, \alpha \ne 1$, i.e., for (45). Since ϕ is integrable one considers its inverse Fourier transform and shows that it determines a positive (continuous) integrable function (=density of a d.f.) iff $\theta = \tan(\pi\alpha/2)$. The annoying case $\alpha = 1$ needs a separate proof to show that $|a_0/c| \le 2/\pi$. Here we omit this (nontrivial) work. (But a detailed sketch of the argument for $0 < \alpha \le 2, \alpha \ne 1$, is given as Problems 24 and 25.)

If ϕ is the ch.f. of a *symmetric* stable law, then ϕ is real and the above (unproved) case disappears, and one gets

$$\phi(t) = \exp\{-c|t|^\alpha\}, \qquad 0 < \alpha \le 2, \quad c > 0, \tag{53}$$

which proves (35). Note that the α in (35) is *the same constant as that in Proposition* 13. For $\alpha > 2, \phi$ is a ch.f. only if $c = 0$.

Conversely, if ϕ is given by (35), in the symmetric case it reduces to (53). The latter is a ch.f. by Pólya's criterion (cf. Problem 25 of Chapter 4). To see that ϕ is then a stable ch.f., it suffices to verify the relation (20)(i.e., Definition 12). Thus for $a_1 > 0, a_2 > 0$ we see that

$$\phi(a_1 t)\phi(a_2 t) = \phi(at),$$

where $a = (a_1^\alpha + a_2^\alpha)^{1/2}$. Hence a function ϕ defined by (53) is always a symmetric stable ch.f. Actually if we first verify that ϕ given by (35) is a ch.f. (indicated in the problem), then the above simple argument implies that it is a stable ch.f. (Of course, the "if" here involves a nontrivial amount of work.) This finishes the proof.

Remark Let X be a symmetric stable r.v. so that its ch.f. $\phi_{X,c}(\cdot)$ is given by (53). Then Schilder (1970) has observed that $\|\cdot\|_\alpha : X \to |c|^{1\wedge\alpha^{-1}}, 0 < \alpha \le 2$, defines a metric on the class of independent symmetric r.v.s in $L^\alpha(P)$. Thus $\|X\|_\alpha = \log \phi_{X,|c|^{1\wedge\alpha^{-1}}}(1)$, and with a small computation it is also seen that for independent r.v.s X_1, X_2 with the same characteristic exponent α,

$||X_1 + X_2||_\alpha = ||X_1||_\alpha + ||X_2||_\alpha$. This statement is also true if X_1, X_2 and $X_1 + X_2$ are all symmetric stable with the same $\alpha > 0$, but here the joint (or multivariate) stability concept is needed. (See Problem 25 (c) on this notion.)

The simple argument concerning the representation (35) presented above follows essentially that of Ramaswamy et al, (1976). A related treatment by S. Bochner is given as Problems 26 and 27. An example of (35) with $\alpha = \frac{1}{2}, \beta = 1, \gamma = 0, c = 1$, due independently to P. Lévy and N.V. Smirnov, is as follows:

$$f(x) = \begin{cases} 0 & \text{if } x \leq 0 \\ (2\pi x^3)^{-1/2} \exp\{-1/(2x)\} & \text{if } x > 0. \end{cases}$$

Several striking properties of stable laws are known. An excellent account of these may be found in the monograph of Gnedenko and Kolmogorov (1954). For more recent accounts, one may see Zolotarev (1986) and Samorodnitsky and Taqqu (1994). We give them no further considerations here, but strongly urge the reader to review the material carefully. The most interesting aspect of this subject here is that the whole analysis depends only on the structure of the real line $\mathbb{R}$ and the key concept of (statistical) independence. Replacing $\mathbb{R}$ by an algebra of matrices defined on a (Hilbert) space, and introducing a new concept called "free independence" on the new space relative to an expectation like functional (a trace operation on matrices), it is possible to extend most of the above analysis to this new setting. This is being done by D. Voiculescu (see his recent CRM monograph, 1992). All the preceding work is necessary to understand this extension which has theoretical consequences. The matrix algebra actually goes over to C^*-algebras and von Neumann algebras! We briefly consider an application of the (classical) stable class in Section 8.4 and will see how an important and very interesting new chapter of the subject emerges.

5.4 Invariance Principles

Let us take another look at the classical central limit theorem. If $\{X_n, n \geq 1\}$ is an i.i.d. sequence of r.v.s with means zero and unit variances, and $S_n = \sum_{k=1}^n X_k$, then Theorem 1.2 says that $(S_n/\sqrt{n}) \xrightarrow{D} Y$, where Y is $N(0,1)$. From some early (1931) results of A. Kolmogorov, and of P. Erdös and M. Kac in the middle 1940s, it is possible to look at the problem in the following novel way. If $I = \{t : 0 \leq t \leq 1\}$, then define a mapping $Y_n(\cdot,\cdot) : I \times \Omega \to \mathbb{R}$ by the equation

$$Y_n(t,\omega) = \frac{1}{\sqrt{n}}[S_{[nt]}(\omega) + (nt - [nt])X_{[nt]+1}(\omega)], \quad \omega \in \Omega, \quad t \in I, \quad (1)$$

where $[nt]$ is the integral part of nt, so that for $t = 1, Y_n(1,\omega) = S_n(\omega)/\sqrt{n}, \omega \in \Omega$. Thus if we set $S_0 = 0, Y_n(\cdot,\omega)$ is a polygonal path, joining (0,0) and

$(t, Y_n(t,\omega))$; and hence for each $n \geq 1$, and $\omega \in \Omega$, the curve $Y_n(\cdot,\omega)$ starts at 0 and is continuous on I. The central limit theorem, slightly extended (to be discussed later), shows not only that $Y_n(1,\cdot) \xrightarrow{D} Y(1,\cdot)$, which is $N(0,1)$, but that $Y_n(t,\cdot) \xrightarrow{D} Y(t,\cdot)$ and$Y(t,\cdot)$ is $N(0,t)$. Moreover, for $0 < t_1 < t_2 < 1, Y(1,\cdot) - Y(t_2,\cdot)$ and $Y(t_2,\cdot) - Y(t_1,\cdot)$ are independent, $N(0, 1-t_2)$ and $N(0, t_2 - t_1)$, respectively. This led M.D. Donsker to look at $Z_n(\cdot) = Y_n(\cdot,\cdot) : \Omega \to C[0,1]$, the space of real continuous functions as the range space of $\{Z_n, n \geq 1\}$, and if $\mu_n = P \circ Z_n^{-1}$ is the image law, then to investigate the convergence of μ_n as well as to determine the limit. Thus it is desired to show, in general, that, under reasonable conditions, one can assert

$$\lim_{n\to\infty} \int_S f(x)\, d\mu_n(x) = \int_S f(x)\, d\mu(x), \qquad f \in C(S), \tag{2}$$

$C(S)$ being the space of scalar continuous functions on a metric space S. Here $S = C[0,1]$. This is equivalent to saying that $\mu_n(A) \to \mu(A)$ for all Borel sets $A \subset S$ such that the boundary ∂A satisfies $\mu(\partial A) = 0$ (essentially the same proof of Theorem 4.1.5 given there for $S = \mathbb{R}$.) In 1951 Donsker was able to establish this result for the space $S = C[0,1]$ and identify μ as the Wiener measure on S. Of course, this includes the Lindeberg-Lévy theorem, and opened up a whole new area of research in probability theory. These ideas have been extended and perfected by Prokhorov (1956), and there has been much research activity thereafter.

Since μ_n is the image measure of the Z_n in $C[0,1]$, and $\mu_n \to \mu$ in the above sense, one can consider the corresponding theorems if S is taken to be other interesting metric spaces. This new development is called the weak convergence of probability measures in metric spaces. The work is still being pursued in the current research. The second possibility is to note that $Z_n \xrightarrow{D} Y$ [to mean that for each $0 < t_1 < \cdots < t_k \leq 1, (Z_n(t_1,\cdot), \ldots, Z_n(t_k,\cdot)) \xrightarrow{D} (Y(t_1,\cdot), \ldots, Y(t_k,\cdot))$ is equivalent to showing $h(Z_n) \xrightarrow{D} h(Y)$ for each bounded continuous mapping $h : C[0,1] \to C[0,1]$ and calculating the distributions of $h(Y)$ for several interesting hs]. But this is not simple in general. However, μ_n is determined by the distribution of Z_n or of $Y_n(\cdot,\cdot)$ and this in turn is determined by the i.i.d. sequence $\{X_n, n \geq 1\}$. The classical limit theorem says (cf. Theorem 1.2) that for all distributions satisfying these moment conditions the limit d.f. remains the same. Hence in the general case the measure determined by the finite dimensional d.f.s of Y is the Wiener measure for all the initial $\{X_n, n \geq 1\}$-measures. Thus choose some simple and convenient d.f. for the X_n-sequence, calculate the d.f. of $h(Y_n(\cdot,\cdot))$, and then find its limit by letting $n \to \infty$. This will give the d.f. of $h(Y)$. The underlying idea is then called the *invariance principle*. Since it is based on weak convergence it is sometimes also referred to as weak invariance principle or a *functional central limit theorem*. In other cases as in the first SLLN (cf. Theorem 2.3.4 or 2.3.6) the convergence of the averages is a.e., and the corresponding ideas lead to a "strong" invariance principle. We present a few of the results of Donsker and Prokhorov in this section because of their importance and great interest in applications.

The preceding discussion clearly implies that we need to consider new technical problems before establishing any general results. The first one is the definition and existence of Wiener measure and process, which can be stated as follows. (From now on an r.v. $X(t,\cdot)$ is also written as X_t, for convenience.)

Definition 1 An indexed family $\{X_t, 0 \le t \le 1\}$ of r.v.s on a probability space (Ω, Σ, P) is called a *Brownian motion* (or a *Wiener process*) if each X_t is a Gaussian (or normal) r.v. $N(0, \sigma^2 t)$ and for each $0 \le t_1 < t_2 < \cdots < t_n \le 1, n \ge 1$, the r.v.s $X_{t_1}, X_{t_2} - X_{t_1}, \ldots, X_{t_n} - X_{t_{n-1}}$ are mutually independent with $E(|X_{t_i} - X_{t_{i-1}}|^2) = \sigma^2(t_i - t_{i-1})$, where $X_0 = 0$ a.e. Thus

$$P[X_{t_i} - X_{t_{i-1}} < u_i, i = 1, \ldots, n] = \prod_{i=1}^{n} [2\pi(t_i - t_{i-1})]^{-1/2} \int_{-\infty}^{u_i} e^{-v^2/2(t_i - t_{i-1})}\, dv. \tag{3}$$

(Here the index [0, 1] is taken only for convenience. The concept holds if the index is any subset of $\mathbb{R}^+$, or even $\mathbb{R}$ with simple modifications.)

It is not obvious that such a process exists. Indeed, if $F_{t_1,\ldots,t_n}$ is the joint d.f. of $X_{t_1}, \ldots, X_{t_n}$, then from (3) we can immediately note that $\{F_{t_1,\ldots,t_n}, n \ge 1\}$ is a compatible family. In terms of ch.f.s this is immediate, since

$$\begin{aligned}
\phi_{t_1,\ldots,t_n}(u_1, \ldots, u_n)
&= E(\exp\{iu_1 X_{t_1} + \cdots + iu_n X_{t_n}\}) \\
&= E\left(\exp\left\{i\left(\sum_{j=1}^{n} u_j\right) X_{t_1} + i\left(\sum_{j=2}^{n} u_j\right)(X_{t_2} - X_{t_1}) + \cdots \right.\right. \\
&\qquad \left.\left. + iu_n(X_{t_n} - X_{t_{n-1}})\right\}\right) \\
&= \phi_{X_{t_1}}\left(\sum_{j=1}^{n} u_j\right) \phi_{X_{t_2} - X_{t_1}}\left(\sum_{j=2}^{n} u_j\right) \cdots \phi_{X_{t_n} - X_{t_{n-1}}}(u_n),
\end{aligned}$$

and the compatibility conditions on the F become [cf. Eqs. (3.4.2), (3.4.3)]

(i) $\lim_{u_n \to 0} \phi_{t_1,\ldots,t_n}(u_1, \ldots, u_n) = \phi_{t_1,\ldots,t_{n-1}}(u_1, \ldots, u_{n-1})$,

(ii) $\phi_{t_{i_1},\ldots,t_{i_n}}(u_{i_1}, \ldots, u_{i_n}) = \phi_{t_1,\ldots,t_n}(u_1, \ldots, u_n), [(i_1, \ldots, i_n) \to (1, 2, \ldots, n)]$.

Since $\phi_t(u) = \exp\{-\frac{1}{2}u^2\sigma^2 t\}$, this is clearly true for the d.f.s given by (3). Hence by Theorem 3.4.11, it follows that there exists a probability space (Ω, Σ, P) and a process $\{X_t, t \in [0,1]\}$ on it with the given finite-dimensional distributions (3). In fact $\Omega = \mathbb{R}^{[0,1]}, \Sigma =$ the σ-algebra generated by the cylinder sets of Ω, and if $\omega \in \Omega$, then $X_t(\omega) = \omega(t)$, the coordinate function. However, we need to know more precisely the range of the r.v.s. In other words, since for each $\omega, X_{(\cdot)}(\omega) = \omega(\cdot)$ is a t-function (called the *sample function*), what is the subspace of these ω that satisfy (3)? For instance, (3) implies for each $0 < s < t < 1, X_t - X_s$ is $N(0, \sigma^2(t-s))$, so that we can find all its moments. In particular, taking $\sigma^2 = 1$ for convenience,

$$E(|X_t - X_s|^4) = 3|t-s|. \tag{4}$$

However, a classical result of Kolmogorov's asserts that any process X_t for which (K_0 being a constant)

$$E(|X_t - X_s|^\alpha) \leq K_0|t-s|^{1+\delta}, \quad \delta > 0, \quad \alpha > 0, \quad t, s \in \mathbb{R}, \tag{5}$$

must necessarily have almost all its sample functions continuous. In other words, the P-measure is not supported on all of $\mathbb{R}^{[0,1]}$, but it concentrates (except for a set of P-measure zero) on the subset $C[0,1] \subset \mathbb{R}^{[0,1]}$. On the other hand, for the probability space (Ω, Σ, P) constructed above, the only measurable sets (of Σ) are those which are determined by at most countably many points. This implies that $C[0,1] \notin \Sigma$, but it has P-outer measure one and P-inner measure zero (this needs a computation). Consequently one has to extend P to $\tilde{P}$ and expand Σ to $\tilde{\Sigma}$, so that the new σ-algebra is determined by all $\{\omega : X_t(\omega) < u\}, t \in [0,1], \omega \in C[0,1], u \in \mathbb{R}$, and

$$\begin{aligned} &\tilde{P}\{\omega \in C[0,1] : X_{t_1}(\omega) < u_1, \ldots, X_{t_n}(\omega) < u_n\} \\ &\qquad = P\{\omega \in \Omega : X_{t_1}(\omega) < u_1, \ldots, X_{t_n}(\omega) < u_n\}. \end{aligned}$$

The right side is given by (3). Fortunately this is all possible, and not too difficult. We omit the proof here, since it is not essential for this discussion. (It may be found, for instance in the first author's (1979) monograph, pp. 186-191.) One then notes from the work in real analysis, that because $C[0,1]$ is a separable metric space [under the sup norm as metric, $\|X\| = \sup_{0 \leq t \leq 1} |X(t)|$], its Borel σ-algebra (i.e., the one determined by the open sets) and $\tilde{\Sigma}$ are the same and that any finite measure on such a Borel σ-algebra is automatically regular, [i.e., $P(A) = \sup\{P(K) : K \subset A, \text{ compact }\}, A \in \tilde{\Sigma}$]. This regular probability measure P on the Borel σ-algebra $\mathcal{B}$ of $C[0,1]$, is called the *Wiener measure*, and is also denoted $W(\cdot)$. Thus $(C[0,1], \mathcal{B}, W)$ is the Wiener space and $\{X_t, 0 \leq t \leq 1\}$ can be regarded as a process on (Ω, Σ, P) with its sample functions in $C[0,1]$. It is the *Wiener (or Brownian motion) process.*

There are other ways of constructing this process. We establish one such construction in the last chapter and present some deep results. N. Wiener was the first in 1923 to demonstrate rigorously the existence of this process (hence the name Wiener process), even though R. Brown, an English botanist, observed the process experimentally, i.e., the erratic behavior of its sample paths (or functions), as early as 1826 (hence Brownian motion). Now it can be approximated by a random walk; and other methods, such as Wiener's original construction, are available. However, there is no really very "simple" proof, and the method outlined above seems to be the "bottom line."

For further work, it is useful to have the Skorokhod mapping theorem in its general form. We had given a special case of it as Problem 5b in Chapter 2, and utilized it in deriving alternative (and simpler) proofs in Chapter 4 (e.g., a second proof of the Helly-Bray theorem). Since $C[0,1]$ is not finite dimensional, our special case extends only by a nontrivial argument, which we present here for a separable metric space. (Cf., also Problem 5(c)–(d) of Chapter 2 to understand its place here as a consequence of earlier ideas.)

Proposition 2 (Skorokhod) *Let S be a complete separable metric space with $\mathcal{B}$ as its Borel σ-algebra. If P_n, P are probability measures on $\mathcal{B}$ such that $P_n \to P$, in the sense that $P_n(A) \to P(A)$ for all $A \in \mathcal{B}$ with $P(\partial A) = 0$ (such sets are also termed P-**continuity sets**), then there exist r.v.s X_n, X on the Lebesgue unit internal $[0, 1]$ with values in S [so that $X_n^{-1}(\mathcal{B}), X^{-1}(\mathcal{B})$ are Lebesgue measurable classes of sets] such that $X_n \to X$ a.e. $[\mu]$, and*

$$P_n(A) = \mu\{\omega : X_n(\omega) \in A\}, \quad P(A) = \mu\{\omega : X(\omega) \in A\}, \quad A \in \mathcal{B} \qquad (6)$$

μ being the Lebesgue measure. Thus P_n, P are the distributions of X_n, X in S.

Proof Let ρ denote the metric function on S. We construct measurable P-continuous partitions of S and analogous partitions of $[0, 1)$ having equal Lebesgue measure. Then define countably valued r.v.s on $[0, 1)$ into the above partitions of S and show that by the refinement order these converge to the desired r.v.s relative to P_n- and P-measures. Here are the details. (see also Billingsley (1995), p.333.)

For each integer $k \geq 1$, let $\{B_n^k\}_{n\geq 1}$ be balls of diameter less than 2^{-k} which cover S and such that $P(\partial B_n^k) = 0$. Since P is a finite measure, there are only countably many balls whose boundaries can have positive measure, so that by changing the radius slightly the above can be achieved. Since for any balls $B_1, B_2, \partial(B_1 \cap B_2) \subset \partial(B_1) \cup \partial(B_2)$, we may disjunctify the above B_n^k and still retain them as P-continuous sets. Thus let $A_1^k = B_1^k, A_2^k = B_2^k - B_1^k$, and $A_n^k = B_n^k - \bigcup_{i=1}^{n-1} B_i^k$, and let us continue this procedure for each $k \geq 1$. If we let $S_{i_1,\ldots,i_k} = \bigcap_{j=1}^{k} A_{i_j}^j$, then for each k-tuple of integers $(i_1, \ldots, i_k)$, the collection $\{S_{i_1,\ldots,i_k}, k \geq 1\}$ are disjoint P-continuity sets such that for each $k, \bigcup_{i_k \geq 1} S_{i_1,\ldots,i_k} = S_{i_1,\ldots,i_{k-1}}$, the diameters satisfy diam $(S_{i_1,\ldots,i_k}) \leq 2^{-k}$ and together they cover S. (Verification is left to the reader.)

Next obtain for the interval $[0, 1)$ the corresponding decompositions such that $I_{i_1,\ldots,i_k}$ and $I_{i_1,\ldots,i_k}^n$ are chosen to satisfy $\mu(I_{i_1,\ldots,i_k}) = P(S_{i_1,\ldots,i_k})$ and $\mu(I_{i_1,\ldots,i_k}^n) = P_n(S_{i_1,\ldots,i_k})$, if $(i_1, \ldots, i_k) \prec (i_1', i_2', \ldots, i_k')$, (lexicographic order), take $I_{i_1,\ldots,i_k}$ to the left of $I_{i_1',\ldots,i_k'}$. Here we use the order property of the real line. Similarly we order $I_{i_1,\ldots,i_k}^n$ to the left of $I_{i_1',\ldots,i_k'}^n$ and both of these cover the unit interval. With such a decomposition we construct the desired r.v.s as follows: Choose a point $x_{i_1,\ldots,i_k} \in S_{i_1,\ldots,i_k}$ and set

$$X_n^k(\omega) = x_{i_1,\ldots,i_k} \quad \text{for} \quad \omega \in I^n_{i_1,\ldots,i_k},$$
$$X^k(\omega) = x_{i_1,\ldots,i_k} \quad \text{for} \quad \omega \in I_{i_1,\ldots,i_k}. \tag{7}$$

(Omit the empty I's.) This defines X_n^k and X^k on $[0, 1)$ as r.v.s into S, since $S_{i_1,\ldots,i_k} \in \mathcal{B}$ for each $k \geq 1$. Furthermore, $\rho(X_n^k(\omega), X_n^{k+1}(\omega)) \leq 2^{-k+1}, l \geq 1$, and the same is true of $X^k(\omega)$. Thus for each n, these are Cauchy sequences, and by the completeness of S, there exist mappings X_n and X such that $X_n^k(\omega) \to X_n(\omega), X^k(\omega) \to X(\omega)$, as $k \to \infty$, for all but at most a countable set of ω that are boundary points of these I-sets. Hence defining them arbitrarily at these points (which have μ-measure zero) we see that X_n, X are measurable mappings on $[0, 1)$ into S, i.e., r.v.s. Also,

$$\mu(I^n_{i_1,\ldots,i_k}) = P_n(S_{i_1,\ldots,i_k}) \to P(S_{i_1,\ldots,i_k}) = \mu(I_{i_1,\ldots,i_k}),$$

by hypothesis, since $S_{i_1,\ldots,i_k}$ are P-continuity sets. Thus for each $\omega \in I_{i_1,\ldots,i_k}$, an $n_0(\omega, k)$ can be found such that $n \geq n_0(\omega, k) \Rightarrow X_n^k(\omega) = X^k(\omega)$, and so

$$\begin{aligned}
&\rho(X_n(\omega),\ X(\omega)) \\
&\qquad \leq \rho(X_n(\omega), X_n^k(\omega)) + \rho(X_n^k(\omega), X^k(\omega)) + \rho(X^k(\omega), X(\omega)) \\
&\qquad \leq 2^{-k+1} + 2^{-k+1} = 2^{-k+2} \to 0, \quad \text{as } k \text{ (hence } n) \to \infty.
\end{aligned}$$

Thus $X_n(\omega) \to X(\omega)$ for each $\omega \in [0,1)$. It remains to establish (6).

Let $A \in \mathcal{B}$, and let $\bigcup' S_{i_1,\ldots,i_k}$ be the union of all $S_{i_1,\ldots,i_k}$ that meet with A. If A^ε is the ε-neighborhood of A [i.e., $A^\varepsilon = \{y : \rho(y, A) < \varepsilon\}$], then by definition taking $\varepsilon > 2^{-k+1}$, we see that $\bigcup' S_{i_1,\ldots,i_k} \subset A^\varepsilon$ and that as $\varepsilon \downarrow 0$ we get $\overline{A}^\varepsilon \searrow \overline{A}$ (the overbar denotes closure). Consequently

$$\begin{aligned}
\mu \circ (X^k)^{-1}(A) &\leq \mu \circ (X^k)^{-1}(\textstyle\bigcup' S_{i_1,\ldots,i_k}) = \textstyle\sum' \mu \circ (X^k)^{-1}(S_{i_1,\ldots,i_k}) \\
&= \textstyle\sum' P(S_{i_1,\ldots,i_k}) \leq P(\overline{A}^\varepsilon).
\end{aligned}$$

If now A is a closed set, then on letting $k \to \infty$ and then $\varepsilon \downarrow 0$, we get

$$\limsup_k \mu \circ (X^k)^{-1}(A) \leq P(A). \tag{8}$$

This implies $\mu \circ (X^k)^{-1} \to \mu \circ X^{-1} = P$ by the analog of Theorem 4.1.5, which holds here with essentially no change in its proof. Hence $\mu \circ X^{-1} = P$. Replacing X by X_n and X^k by X_n^k in the above, the result implies $\mu \circ X_n^{-1} = P_n$. [In the above-noted theorem we constructed a function in showing that $P_n \to P \Leftrightarrow \overline{\lim}_n P_n(C) \leq P(C)$. The following function g may be used for the same purpose: $g(u) = f((1/\varepsilon)\rho(u, C))$, where $f(u) = 1$ if $u \geq 0; = 1 - u$ if $0 \leq u \leq 1; = 0$ if $u \geq 1$. Then $g = 0$ outside of C^ε, and

$$\limsup_n P_n(C) \leq \lim_n \int_S g\, dP_n = \int_S g\, dP = P(C^\varepsilon) < P(C) + \varepsilon.]$$

This completes the proof.

We have the following interesting consequence, to be used below.

Corollary 3 *Let $(S, \mathcal{B})$ and $(\tilde{S}, \tilde{\mathcal{B}})$ be two metric spaces as in the proposition. If P_n, P are probabilities on $\mathcal{B}$ such that $P_n \to P$ in the sense of the proposition, and if $f : S \to \tilde{S}$ is a $(\mathcal{B}, \tilde{\mathcal{B}})$-measurable mapping such that the discontinuity points $D_f \in \mathcal{B}$ satisfy $P(D_f) = 0$, then $P_n \circ f^{-1} \to P \circ f^{-1}$. In particular, if (Ω, Σ, Q) is a probability space, Y_n, Y are r.v.s from Ω into S such that $Y_n \xrightarrow{D} Y$ (i.e., $Q \circ Y_n^{-1} \to Q \circ Y^{-1}$), then $f(Y_n) \xrightarrow{D} f(Y)$ for each continuous $f : S \to \tilde{S}$.*

Proof By the above proposition, there exist r.v.s, X_n, X on the Lebesgue unit interval $([0, 1), \mathcal{L}, \mu)$ into $(S, \mathcal{B})$ such that $X_n \to X$ a.e., and $P_n = \mu \circ X_n^{-1}, P = \mu \circ X^{-1}$. Also, $f(X_n)(\omega) \to f(X)(\omega)$ for all ω for which f is continuous at $X(\omega)$. The set of discontinuities of f is contained in $X^{-1}(D_f)$, which is μ-null. Hence $\mu \circ f(X_n)^{-1} \to \mu \circ f(X)^{-1}$, or equivalently

$$\mu \circ X_n^{-1} \circ f^{-1} = P_n \circ f^{-1} \to \mu \circ X^{-1} \circ f^{-1} = P \circ f^{-1}.$$

Setting $P_n = Q \circ Y_n^{-1}, P = Q \circ Y^{-1}$ in the above, and since $P_n \circ f^{-1} = Q \circ f(Y)^{-1}, P \circ f^{-1} = Q \circ f^{-1}(Y)^{-1}$, the main result implies the last part. This finishes the proof.

Note that both the above proposition and corollary reduce to what we have seen before if $S = \mathbb{R}$. Moreover, the calculus of "in probability" results of Mann and Wald given as Problems 9–11 in Chapter 2 extend to the case considered here.

We are ready to prove Donsker's (1951) invariance principle discussed earlier in this section, and it still needs many details, to be given here.

Theorem 4 *Let $\{X_n, n \geq 1\}$ be a sequence of i.i.d. random variables on (Ω, Σ, P) with zero means and variances $\sigma^2 > 0$. If Y_n is defined by (1) as a polygonal function on $\Omega \to C[0, 1]$, which is a random element with distribution $P_n(= P \circ Y_n^{-1})$ on the Borel sets of the separable metric space $C[0, 1]$ (with uniform norm as its metric), then $P_n \to W$, where W is the Wiener measure. Equivalently, $Y_n \xrightarrow{D} Z$, where $\{Z(t, \cdot), t \in [0, 1]\}$ is the Wiener or Brownian motion process with $Z(t, \cdot)$ as $N(0, \sigma^2 t)$. Hence for each $h : C[0, 1] \to C[0, 1]$ which is measurable and whose discontinuities form a set of Wiener measure zero, we have $h(Y_n) \xrightarrow{D} h(Z)$ or $P_n \circ h^{-1} \to W \circ h^{-1}$.*

Proof By definition of Y_n in (1), the central limit theorem implies $Y_n(1, \cdot) \xrightarrow{D} Y(1, \cdot)$ where $\{Y(t, \cdot), 0 \leq t \leq 1\}$ is the Wiener process in which we may and do take $\sigma^2 = 1$ for convenience. Note that $Y_n(t, \cdot) \xrightarrow{D} Y(t, \cdot)$ also, as $n \to \infty$. Indeed, for $\varepsilon > 0$,

$$P[|Y_n(t,\cdot) - n^{-1/2}S_{[nt]}| \geq \varepsilon] \leq P[n^{-1/2}|X_{[nt]+1}| \geq \varepsilon] \quad [\text{ by}(1)]$$

$$\leq \frac{1}{n\varepsilon^2} \to 0 \quad \text{as} \quad n \to \infty \quad \text{(Čebyšev's inequality)}. \tag{9}$$

On the other hand, since $[nt]/n \to t$ as $n \to \infty$, from Lévy's central limit theorem we get with $\phi(u) = E(e^{iuX_1})$,

$$E(e^{iuS_{[nt]}/\sqrt{n}}) = \left(\phi\left(\frac{u}{\sqrt{n}}\right)^n\right)^{[nt]/n} \to \exp\left(-\frac{u^2t}{2}\right),$$

so that $(1/\sqrt{n})S_{[nt]} \xrightarrow{D} Y(t,\cdot)$. But the mapping $h : (u_1, \ldots, u_k) \mapsto (u_1, u_2 - u_1, \ldots, u_k - u_{k-1})$ is a homeomorphism of $\mathbb{R}^k$, and thus if P_n is the probability measure of $(1/\sqrt{n})(S_{[nt_1]}, \ldots, S_{[nt_k]}), 0 \leq t_1 < \cdots < t_k \leq 1$, then it converges (weakly or distributionally) to $W(\cdot)$ iff $P_n \circ h^{-1} \to W \circ h^{-1}$ for each fixed k. Taking $k = 2, 0 < t_1 < t_2 < 1$, the general case being similar, we get by independence of $S_{[nt_1]}$ and $S_{[nt_2]} - S_{[nt_1]}$,

$$\frac{1}{\sqrt{n}}(S_{[nt_1]}, S_{[nt_2]} - S_{[nt_1]}) \xrightarrow{D} (Y(t_1), Y(t_2) - Y(t_1)). \tag{10}$$

Thus the finite-dimensional distributions of $\{Y_n(t,\cdot), 0 \leq t \leq 1\}$ converge at all points to the corresponding ones of the Brownian motion process $\{Y(t), 0 \leq t \leq 1\}$ by (9) and (10). Using this we shall show that the probability measure P_n induced by Y_n on $C[0,1]$ converges to the Wiener measure W on $C[0,1]$. For this, we need to establish a stronger assertion: for each Borel set A satisfying $W(\partial A) = 0$, one has $P_n(A) \to W(A)$. This is the crux of the result and we present it in two parts, as steps for clarity.

I. Since the Borel σ-algebra $\mathcal{B}$ is also generated by the (open) balls of $C[0,1]$, it is sufficient to prove the convergence result for sets A which are finite intersections of such balls satisfying $W(\partial A) = 0$ because sets of the latter kind again generate $\mathcal{B}$. Now let $f_1, \ldots, f_m$ be elements of $C[0,1]$ and A be the set which is the intersection of m-balls centered at these points and radii $r_1, \ldots, r_m$ having the boundary of W-measure zero. Thus A is of the form

$$A = \Big\{g : \tilde{f}(t) = \max_{1 \leq i \leq m}(f_i(t) - r_i) < g(t) < \tilde{\tilde{f}}(t) = \min_{1 \leq i \leq m}(f_i(t) + r_i), t \in [0,1]\Big\}.$$

Let $k \geq 1$ be fixed and $t = j/2^k, j = 0, 1, \ldots, 2^k$, a dyadic rational. Replacing t by these numbers in the definition of A, the new set A_k (say) approximates A from above as $k \to \infty$. Given $\delta > 0$, we can find a $k \geq 1$ such that [we assume $W(\partial A) = 0$], $W(A_k) - W(A) < \delta$. But by the first paragraph, (the multi-dimensional central limit theorem) if $\tilde{B}$ is a k-cylinder with B as its base (so $\pi_k^{-1}(B) = \tilde{B}$ and $\tilde{B}$ is spelled out below in $P_n[\tilde{B}]$)

$$\lim_{n\to\infty} P_n[\tilde{B}] = \lim_{n\to\infty} P_n[(Y(t_1),\dots,Y(t_k)) \in B] = W(\tilde{B}),$$

where the base B is a k-dimensional Borel set and $W(\partial\tilde{B}) = 0$. In particular, taking $\tilde{B} = A_k$, one has $\lim_{n\to\infty} P_n(A_k) = W(A_k)$, and since $P_n(A_k) \geq P_n(A)$, we get

$$\limsup_n P_n(A) \leq \limsup_n P_n(A_k) = W(A_k) \leq W(A) + \delta. \tag{11}$$

It follows from the arbitrariness of $\delta > 0$ that $\overline{\lim}_{n\to\infty} P_n(A) \leq W(A)$. Thus the next step is to establish the opposite inequality, $\liminf_n P_n(A) \geq W(A)$, which needs more work than that of establishing (11).

II. Let $\varepsilon > 0, \eta > 0$ be given. Choose $\alpha > 0$ such that if

$$H_\alpha = \{g : \tilde{f}(t) + \alpha < g(t) < \tilde{\tilde{f}}(t) - \alpha, 0 \leq t \leq 1\},$$

which increases to A as $\alpha \downarrow 0$, then $W(H_\alpha) > W(A) - \varepsilon$. One can approximate H_α from above if t is replaced by the rationals as in the last paragraph. If $t = i/k, i = 0, 1, \dots, k$, and

$$H_\alpha^k = \{g : \tilde{f}(i/k) + \alpha < g(i/k) < \tilde{\tilde{f}}(i/k) - \alpha, i = 0, 1, \dots, k\},$$

then $W(A) < W(H_\alpha) + \varepsilon < W(H_\alpha^k) + \varepsilon$ for each $k \geq 1$. But $\tilde{f}, \tilde{\tilde{f}} \in C[0,1]$. Hence they are uniformly continuous. Thus we can find $k > 1$ such that if $|t - s| < 1/k$, then $|\tilde{f}(t) - \tilde{f}(s)| < \alpha/3, |\tilde{\tilde{f}}(t) - \tilde{\tilde{f}}(s)| < \alpha/3$. In particular, if $n \geq k$, and $C_n \subset C[0,1]$ is the set of functions which are piecewise linear on $[(i-1)/n, i/n], i = 1, \dots, n$, then by the uniform continuity of these functions (and by the Weierstrass approximation theorem) we see that $P_n(C_n) = 1$ for each n (every $1/n$-neighborhood of C_n in $C[0,1]$). Next let

$$F_n = \{g \in C[0,1] : g(i/n) < \tilde{f}(i/n) + \alpha/3 \text{ or } g(i/n) > \tilde{\tilde{f}}(i/n) - \alpha/3 \text{ for some } 0 \leq i \leq n\}.$$

If $\mu > 1/\alpha^2 k$, then $C_n \cap F_n^c \subset A$ [each $g \in C_n$ not in F_n must satisfy $\tilde{f}(t) + \alpha/3 \leq g(t) \leq \tilde{\tilde{f}}(t) - \alpha/3$.] We want to estimate the probabilities of A for the P_n and show that $\overline{\lim}_n P_n(A^c) \leq W(A^c)$ to complete the proof. By the above inclusion [and the fact that $P(C_n) = 1$] one has

$$P_n(A^c) \leq P(F_n) + P(C_n^c) = P(F_n) = \sum_{r=1}^{n} P(F_{n,r}), \tag{12}$$

where $F_{n,r}$ are the disjoint sets ($F_n = \bigcup_{r=1}^n F_{n,r}$) defined by

$$F_{n,r} = \{g : \tilde{f}(i/n) + \alpha/3 \leq g(i/n) \leq \tilde{\tilde{f}}(i/n) - \alpha/3, i = 0, \ldots, r-1,$$
$$\text{but } \textit{not} \text{ for } i = r \geq 1\}.$$

If $q_{n,r}$ is that q satisfying $(q-1)/k < r/n \leq q/k, 1 \leq q \leq k \leq n$ (set $q_{n,0} = 0$), then

$$\sum_{r=1}^{n} P(F_{n,r}) = \sum_{r=1}^{n} P(F_{n,r} \cap \{g : |g(r/n) - g(q_{n,r}/k)| < \alpha/3\})$$
$$+ \sum_{r=1}^{n} P(F_{n,r} \cap \{g : |g(r/n) - g(q_{n,r}/k)| \geq \alpha/3\}). \quad (13)$$

By definition of $F_{n,r}$,

$$g \in F_{n,r} \Rightarrow \tilde{f}(q_{n,r}/k) + \alpha \leq g(q_{n,r}/k) \leq \tilde{\tilde{f}}(q_{n,r}/k) - \alpha$$

is false and hence $g \notin H_\alpha$ or is in H_α^c. Hence the first term on the right side of (13) is dominated by $P_n((H_\alpha^k)^c)$, since $n \geq k$.

Consider the second term of (13). Since $Y_n(t, \cdot)$- and $Y(t, \cdot)$-processes have independent increments, we have for $0 \leq t, t' \leq 1$ and $S = C[0, 1]$,

$$\left[\int_S |Y(t) - Y(t')|^2 \, dP_n\right]^{1/2}$$

$$\leq \left(\left[\int_S (Y(t) - \frac{1}{\sqrt{n}} S_{[nt]})^2 \, dP_n\right]^{1/2} + \left(\int_S \left(\frac{1}{\sqrt{n}} (S_{[nt]} - S_{[nt']})\right)^2 dP_n\right)^{1/2}\right.$$
$$\left. + \left(\int_S (Y(t') - \frac{1}{\sqrt{n}} S_{[nt']})^2 \, dP_n\right)^{1/2}\right), \quad (14)$$

by the triangle inequality. But by the first paragraph [cf. (10)] the first and last integrals of (14) tend to 0. The middle term $\leq (|[nt] - [nt']|/n)^{1/2}$. Hence for large enough n, the right side of (14) $\leq 3\sqrt{|t - t'|}$. Consequently

$$P_n(F_{n,r} \cap \{g : |g(r/n) - g(q_{n,r}/k)| \geq \alpha/3\})$$
$$= P_n(F_{n,r}) P_n(\{g : |g(r/n) - g(q_{n,r}/k) \geq \alpha/3\})$$
(by independence of the X_i defining S_n and hence P_n)
$$\leq P_n(F_{n,r}) \frac{1}{\alpha^2} \left|\frac{r}{n} - \frac{q_{n,r}}{k}\right|$$
(by Čebyšev's inequality and the above estimate).

But $(r/n) - (q_{n,r}/k) \leq 1/k$, by definition of $q_{n,r}$. Hence the right-side term is at most $P_n(F_{n,r})(1/\alpha^2 k) \leq P_n(F_{n,r})\eta$. This gives for the second term of (13) a bound $\leq \eta$, since $\sum_{r=1}^{n} P_n(F_{n,r}) = P_n(F_n) \leq 1$. Substituting this in (13) and then putting together all the estimates in (12), one has

$$P_n(A^c) \leq P_n((H_\alpha^k)^c) + \eta, \qquad n \geq k. \tag{15}$$

But by (10), $\lim_{n\to\infty} P_n((H_\alpha^k)^c) = W((H_\alpha^k)^c)$ for each $k \geq 1$. Hence (15) yields

$$\limsup_n P_n(A^c) \leq W((H_\alpha^k)^c) + \eta \leq W(A^c) + \eta + \varepsilon. \tag{15'}$$

From this we deduce that ($\varepsilon > 0, \eta > 0$ being arbitrary) $\liminf_n P_n(A) \geq W(A)$. This and the inequality (11) above imply $\lim_n P_n(A) = W(A)$, and the proof of the theorem is complete since the last part is an immediate consequence of Corollary 3.

Discussion 5 The above proof is essentially based on Donsker's argument. There is another method of proving the same result. Since by (10) the finite-dimensional distributions converge to the corresponding distributions of the Brownian motion process, the result follows if one can show that the sequence $\{P_n, n \geq 1\}$ is weakly compact. What we proved is actually a particular case of the latter. (This follows by Helly's selection principle if the space is finite dimensional.) The general compactness criterion, of independent interest, was established by Prokhorov for all complete separable metric spaces in 1956, and then he deduced Donsker's theorem from a more general result. When such a compactness criterion is available, the rest of the generalizations are, in principle, not difficult, though the computations of individual estimates need special care. For a detailed treatment of these and related matters, one may refer to the books by Billingsley (1968), Parthasarathy (1967) and Gikhman and Skorokhod (1969); and the former is followed in the above demonstration.

Let us state Prokhorov's result, which illuminates the structure of the invariance principle as well as the above theorem.

Theorem 6 (Prokhorov) *Let $\{X_{nk}, 1 \leq k \leq k_n\}_{n\geq 1}$ be a sequence of rowwise independent sequences of r.v.s which are infinitesimal and which have two moments finite such that $E(X_{nk}) = 0$, Var $S_n = 1$, where $S_{nr} = \sum_{k=1}^r X_{nk}$ and $S_n = S_{nk_n}$. Let*

$$Y_n(t) = S_{nr} + (t - t_{nr})(S_{n(r+1)} - S_{nr})/(t_{n(r+1)} - t_{nr}),$$

$t \in [t_{nr}, t_{n(r+1)}]$, $S_{n0} = 0, t_{n0} = 0$, and $t_{nr} =$ Var S_{nr}. Let P_n be the distribution of Y_n in $C[0,1]$. Then $P_n \to W$, the Wiener measure in $C[0,1]$, iff the X_{nk}-sequence satisfies the Lindeberg condition:

$$\lim_n \sum_{k=1}^{k_n} \int_{[|x|>\lambda]} x^2 \, dF_{nk}(x) = 0 \tag{16}$$

for each $\lambda > 0$, where F_{nk} is the d.f. of X_{nk}. [Compare this form with the Lindeberg-Feller form given in Theorem 3.6.]

For the i.i.d. sequence of Theorem 4, the Lindeberg condition is automatic, and hence this is a considerable extension of the previous result. To see this implication, let $X_{nk} = X_k/n^{1/2}$, so that $F_{nk}(x) = F(x\sqrt{n})$, where F is the common d.f. of the X_i there. Since $k_n = n$, we have

$$\sum_{k=1}^{n} \int_{[|x|>\lambda]} x^2\, dF_{nk}(x) = \sum_{k=1}^{n} \int_{[|x|>\lambda]} x^2\, dF(x\sqrt{n}) = \int_{[|x|>\lambda\sqrt{n}]} x^2\, dF(x)$$

by a change of variable. Since Var $X = 1$, the right side $\to 0$ as $n \to \infty$ for each $\lambda > 0$. Thus (16) holds.

We shall omit a proof of Prokhorov's theorem, which can be found in the above references. However, we illustrate its idea by a pair of very important applications.

Two Applications Let $\{X_{nk}, 1 \le k \le k_n, n \ge 1\}$ be a sequence of rowwise independent infinitesimal r.v.s with means zero and finite variances. Let $S_{nr} = \sum_{k=1}^{r} X_{nk}$, $t_{nr} = \text{Var } S_{nr} = \sum_{k=1}^{r} \text{Var } X_{nk}$. We then assert

Theorem 7 *Suppose that the X_{nk} satisfy the Lindeberg condition* (16). *Thus if $t_n = t_{nk_n}$, then with the above notation and assumptions,* (16) *becomes*

$$\lim_{n\to\infty} \frac{1}{t_n} \sum_{k=1}^{k_n} \int_{[|x|>\lambda t_n^{1/2}]} x^2\, dF_{nk}(x) = 0 \tag{16'}$$

implying that

$$\lim_{n\to\infty} P\left[\max_{1\le r\le k_n} S_{nr}/t_n^{1/2} < x\right] = \sqrt{\frac{2}{\pi}} \int_0^x e^{-u^2/2}\, du, \tag{17}$$

if $x > 0$; $= 0$ if $x \le 0$.

Remark According to Theorem 6, the limit d.f. of $M_n = \max_{1\le r\le k_n} S_{nr}/t_n^{1/2}$ is the same as that of $V = \sup_{0\le t\le 1} B(t)$, where $\{B(t), 0 \le t \le 1\}$ is the standard Brownian motion process. Thus the right side of (17) is $P[V < x]$. However, there is no simple method of evaluating the latter probability. By the invariance principle this limit d.f. does not depend on the particular d.f.s F_{nk}, and we take advantage of this point in choosing simple enough d.f.s in calculating the limit. For variety, we prove the result without invoking the preceding theorem, whose proof we have not included.

Proof The result is established in two stages. First we show that the left-side limit of (17) exists and is independent of the F_{nk}. (This would be immediate from Theorem 4 if the X_n were i.i.d., since then $h(f) = \sup_{0\le t\le 1} f(t), f \in C[0,1]$, is a continuous functional on $C[0,1]$.) Next we choose X_{nk} to be symmetric i.i.d. Bernoulli r.v.s and evaluate the limit. These are nontrivial computations.

I. For the first point, let $\{Y_n, n \geq 1\}$ be a sequence of independent $N(0,1)$ r.v.s on (Ω, Σ, P) on which the X_{nk} are defined. The probability space may be assumed rich enough to support all these r.v. families, by enlarging it if necessary. Let $Z_r = \sum_{k=1}^{r} Y_k$ and if $F_n(x)$ is the left side of (17) before taking the limit, then we assert that for each $\varepsilon > 0$ and integer $r \geq 1$,

$$\liminf_n F_n(x) \geq P\left[\max_{1\leq k\leq r} Z_k/r^{1/2} < x - \varepsilon\right] - \frac{1}{\varepsilon^2 r}, \tag{18}$$

and

$$\limsup_n F_n(x) \leq P\left[\max_{1\leq k\leq r} Z_k/r^{1/2} < x\right]. \tag{19}$$

These two inequalities are analogous to those of steps I and II of the proof of Theorem 4, and the following argument proceeds on the same lines but is tailored to the case under consideration.

Let $i_1, \ldots, i_r$ be integers chosen such that for each $1 \leq j \leq r$, we have

$$\sum_{k=1}^{i_j} \text{Var } X_{nk} \geq (j/r)t_n, \qquad \sum_{k=1}^{i_j-1} \text{Var } X_{nk} < (j/k)t_n, \tag{19$'$}$$

where $t_n = \text{Var } S_{nk_n}$, as in the statement. Thus i_j is the smallest integer such that $t_{ni_j} \geq (j/r)t_n$, and $1 \leq i_1 \leq \cdots \leq i_r = k_n$. Consider $U_{n1} = S_{ni_1}$ and $U_{nj} = S_{ni_j} - S_{ni_{j-1}}, j = 2, \ldots, r$. We note that as $n \to \infty$ the U_{nj} satisfy Lindeberg's condition. In fact, by (16$'$) one has

$$1 = \lim_{n\to\infty} \sum_{k=1}^{k_n} (1/t_n) \int_{[|x|<\lambda t_n^{1/2}]} x^2 \, dF_{nk}(x), \qquad \lambda > 0. \tag{16$''$}$$

Hence for each $\varepsilon > 0$,

$$\begin{aligned}\max_{1\leq k\leq k_n} \text{Var } X_{nk}/t_n \leq& \max_{1\leq k\leq k_n} (1/t_n) \int_{[|x|\leq \varepsilon t_n^{1/2}]} x^2 \, dF_{nk}(x) \\ &+ (1/t_n) \sum_{k=1}^{k_n} \int_{[|x|>\varepsilon t_n^{1/2}]} x^2 \, dF_{nk}(x) \\ &\leq \varepsilon^2 \; [\text{ as } n \to \infty \text{ by } (16') \text{ and } (16'')].\end{aligned} \tag{19$''$}$$

Since $\varepsilon > 0$ is arbitrary, (19$''$) implies $\text{Var } X_{nk}/t_n \to 0$ uniformly in k, as $n \to \infty$. Now if $a_{nj} = \text{Var}(U_{nj})$, (19$'$) and (19$''$) imply $1/r \leq a_{n1}/t_n < 1/r -$ Var $X_{ni_1}/t_n = 1/r - o(1), 2/r \leq a_{n1} + a_{n_2}/t_n < 2/r -$ Var $X_{ni_1}/t_n = 2/r - o(1)$, so that $a_{n_2}/t_n = 1/r + o(1)$, and similarly $a_{nj}/t_n = 1/r + o(1), 1 \leq j \leq r$. Thus for $0 < \eta < 1$, there is an $n_0[= n_0(\eta)]$ such that $n \geq n_0$ implies

$$t_n(1-\eta)/r \leq a_{nj} \leq t_n(1+\eta)/r. \tag{20}$$

Hence by (20) on setting $B_{nr} = [|x| > \lambda\sqrt{(1-\eta)t_n/r}]$,

$$\frac{1}{a_{nj}} \sum_{k=i_{j-1}+1}^{i_j} \int_{B_{nr}} x^2 \, dF_{nk}(x) \le \frac{r}{1-\eta} \frac{1}{t_n} \sum_{k=1}^{k_n} \int_{B_{nr}} x^2 \, dF_{nk}(x) \to 0,$$

as $n \to \infty$, using (16′). Thus the U_{nj} also satisfy Lindeberg's condition for each j. By Theorem 3.5, $U_{nj}/a_{nj}^{1/2} \xrightarrow{D}$ to an $N(0,1)$ r.v., for each $j = 1, \ldots, r$, as $n \to \infty$. Since the $U_{nj}, j = 1, \ldots, r$, are independent, so are these limit r.v.s. But by (20) $\lim_n (a_{nj}/t_n) = 1/r$. Consequently, $r^{1/2} U_{nj}/t_n^{1/2} \xrightarrow{D}$ to an $N(0,1)$ r.v. as $n \to \infty, j = 1, \ldots, r$, these being independent. This means

$$\lim_{n\to\infty} P[U_{nj} < x_j \sqrt{t_n/r}, j = 1, \ldots, r] = (1/2\pi)^{r/2} \prod_{j=1}^{r} \int_{-\infty}^{x_j} e^{-u^2/2} \, du. \quad (21)$$

Now going to an auxiliary probability space and using Corollary 3 with $h(\tilde{x}) = \max_{1\le k\le r}(\Sigma_{j=1}^{k} \tilde{x}_j), \tilde{x} = (\tilde{x}_1, \ldots, \tilde{x}_r) \in \mathbb{R}^r$, we get from (20) that

$$\lim_{n\to\infty} P\left[\max_{1\le j\le r} S_{ni_j} < x t_n^{1/2}\right] = P\left[\max_{1\le j\le r} Z_j < x r^{1/2}\right], \qquad x \in \mathbb{R}. \quad (22)$$

II. It will be shown presently that (22) implies (19) as well as (18). For this let $G_{nr}(x) = P[\max_{1\le j\le r} S_{ni_j} < x t_n^{1/2}]$. We express the event $[\max_{1\le j\le r} S_{ni_j} < x t_n^{1/2}]$ here as a disjoint union in much the same way as in (13) (as was done many times before, e.g. in martingale convergence). Thus, let

$$H_{n1}(x) = [S_{n1} \ge x t_n^{1/2}],$$

and for $j > 1, H_{nj}(x) = [S_{nj} \ge x t_n^{1/2}, S_{ni} < x t_n^{1/2}$ for $1 \le i \le j-1]$. These are disjoint, and if $Q_{nj}(x) = P(H_{nj}(x))$, we have

$$\sum_{j=1}^{k_n} Q_{nj}(x) = 1 - F_n(x). \quad (23)$$

If $i_{j-1} < i_j$, then for each $i_{j-1} < k \le i_j$, set

$$Q'_{nk}(x) = P\left[S_{nk} \ge x t_n^{1/2}, \max_{1\le i\le k-1} S_{ni} < x t_n^{1/2}, |S_{ni_j} - S_{nk}| \ge \varepsilon t_n^{1/2}\right]$$

and

$$Q''_{nk}(x) = P\left[S_{nk} \ge x t_n^{1/2}, \max_{1\le i\le k-1} S_{ni} < x t_n^{1/2}, |S_{ni_j} - S_{nk}| < \varepsilon t_n^{1/2}\right].$$

Thus $Q_{nk}(x) = Q'_{nk}(x) + Q''_{nk}(x)$, and by independence

$$Q'_{nk}(x) = Q_{nk}(x) P[|S_{ni_j} - S_{nk}| \ge \varepsilon t_n^{1/2}] \le Q_{nk}(x)[a_{nj}/(\varepsilon^2 t_n)]$$

by Čebyšev's inequality for the last one. By (20), this becomes

$$Q'_{nk}(x) \le Q_{nk}(x)[(1+\eta)/(\varepsilon^2 r)]$$

This and (23) give

$$\begin{aligned} 1 - F_n(x) &= \sum_{j=1}^{k_n} Q'_{nj}(x) + \sum_{j=1}^{k_n} Q''_{nj}(x) \\ &< \frac{1+\eta}{\varepsilon^2 r} \cdot 1 + \sum_{j=1}^{k_n} Q''_{nj}(x). \end{aligned} \tag{24}$$

However, on each set appearing in Q''_{nk}, we have $S_{nk} \ge x t_n^{1/2}$ and $|S_{ni_j} - S_{nk}| < \varepsilon t_n^{1/2}$, so that $S_{ni_j} > S_{nk} - \varepsilon t_n^{1/2} \ge (x-\varepsilon) t_n^{1/2}$. Hence

$$\sum_{j=1}^{k_n} Q''_{nj}(x) \le 1 - G_{nr}(x-\varepsilon). \tag{25}$$

Thus (24) and (25) give

$$1 - F_n(x) \le [(1+\eta)/(\varepsilon^2 r)] + 1 - G_{nr}(x-\varepsilon).$$

Clearly $F_n(x) \le G_{nr}(x)$ for all $r \ge 1$. Hence the above inequality becomes

$$G_{nr}(x-\varepsilon) - [(1+\eta)/(\varepsilon^2 r)] \le F_n(x) \le G_{nr}(x). \tag{26}$$

Thus with (22) one has

$$\limsup_n F_n(x) \le \lim_n G_{nr}(x) \le P\left[\max_{1\le j\le r} Z_j < x r^{1/2}\right],$$

which is (19). Taking lim inf of the first two terms and using (22), we get (18), since $\eta > 0$ is arbitrary.

From (18) and (19) it follows that

$$\begin{aligned} P\left[\max_{1\le j\le r} Z_j < (x-\varepsilon) r^{1/2}\right] - \frac{1}{\varepsilon^2 r} &\le \underline{\lim}_n F_n(x) \le \overline{\lim_n} F_n(x) \\ &\le P\left[\max_{1\le j\le r} Z_j < x r^{1/2}\right]. \end{aligned} \tag{27}$$

If $Q_r(x) = P[\max_{1\le i\le r} Z_j / r^{1/2} < x]$, then by the Helly selection principle for a subsequence $r_i, Q_{r_i}(x) \to Q(x)$ at all continuity points x of Q, a nondecreasing function, $0 \le Q \le 1$. Since F_n does not depend on r_i, we get from (27) on letting $r_i \to \infty$,

$$Q(x-\varepsilon) \le \underline{\lim}_n F_n(x) \le \overline{\lim_n} F_n(x) \le Q(x). \tag{27$'$}$$

Letting $\varepsilon \searrow 0$ so that $x - \varepsilon$ is a continuity point of Q, we see that $\lim_{n\to\infty} F_n(x)$ exists and $= Q(x)$ where x is a continuity point of Q. [Also note that (27) and

(27′) ⇒ Q is a d.f. But this is again independently obtained in step III.] This proves our first assertion, as well as the fact that the limit does not depend on the d.f.s of X_{nk}. It only remains to calculate this limit.

III. Consider $\{Y'_n, n \geq 1\}$, where the Y'_n are independent and $P[Y'_k = 1] = P[Y'_k = -1] = \frac{1}{2}, k \geq 1$. Now we may take $X_{nk} = Y'_k, 1 \leq k \leq k_n = n$ (or $X_{nk} = Y'_k/n^{1/2}, 1 \leq k \leq n$, so that the X_{nk} are infinitesimal, but this is not used at this point) and set $S_n = \sum_{k=1}^n Y'_k$.

Let $S_n^* = \max_{1 \leq k \leq n} S_k$ and $x > 0$. Now if $N = [x\sqrt{n}]$, the integral part, then since S_n^* takes only integer values, we have (with largest $N < x\sqrt{n}$)

$$\begin{aligned} F_n(x) &= P[S_n^* < x\sqrt{n}] \\ &= P[S_n^* \leq N] \\ &= 1 - P[S_n^* \geq N+1] \\ &= 1 - P[S_n^* \geq N+1, S_n < N+1] - P[S_n^* \geq N+1, S_n \geq N+1] \\ &= 1 - P[S_n^* \geq N+1, S_n < N+1] - P[S_n \geq N+1]. \end{aligned} \tag{28}$$

To simplify the middle term consider for any integer $J \geq 1$,

$$\begin{aligned} P[S_n^* \geq J, S_n < J] &= \sum_{j=1}^{n-1} P[S_j \geq J, S_i < J, 1 \leq i \leq j-1, S_n < J] \\ &\quad \text{(the first set being } [S_1 \geq J, S_n < J] \text{ and the } n\text{th term is zero)} \\ &= \sum_{j=1}^{n-1} P[S_j \geq J, S_i < J, 1 \leq i \leq j-1, S_n - S_j < 0] \\ &= \sum_{j=1}^{n-1} P[S_j \geq J, S_i < J, 1 \leq i \leq j-1] P[S_n - S_j < 0], \\ &\quad \text{(by independence)} \\ &= \sum_{j=1}^{n-1} P[S_j \geq J, S_i < J, 1 \leq i \leq j-1] P[S_n - S_j > 0], \\ &\quad \text{(by symmetry of the } Y_i) \\ &= \sum_{j=1}^{n-1} P[S_j \geq J, S_i < J, 1 \leq i \leq j-1, S_n - S_j > 0], \\ &\quad \text{(by independence)} \\ &= \sum_{j=1}^{n-1} P[S_j \geq J, S_i < J, 1 \leq i \leq j-1, S_n > J] \\ &= P[S_n^* \geq J, S_n > J] \\ &= P[S_n > J]. \end{aligned}$$

Substituting this in (28) with $J = N + 1$, one gets

$$\begin{aligned}F_n(x) &= 1 - P[S_n > N+1] - P[S_n \geq N+1] \\ &= 1 - 2P[S_n \geq N+1] + P[S_n = N+1] \\ &= 1 - 2(1 - P[S_n < N+1]) + P[S_n = N+1] \\ &= 2P[S_n/\sqrt{n} < ([x\sqrt{n}]+1)/\sqrt{n}] - 1 + P[S_n/\sqrt{n} = (N+1)/\sqrt{n}].\end{aligned} \tag{29}$$

But

$$\lim_n P[S_n/\sqrt{n} < x] = \left(\frac{1}{2\pi}\right)^{1/2} \int_{-\infty}^{x} e^{-u^2/2}\, du, \qquad x \in \mathbb{R},$$

by the central limit theorem. Since the limit is continuous, the last term in (29) goes to zero, and hence for all $x > 0$

$$\lim_{n\to\infty} F_n(x) = 2\frac{1}{\sqrt{2\pi}} \int_{-\infty}^{x} e^{-u^2/2}\, du - 1 = \sqrt{\frac{2}{\pi}} \int_0^x e^{-u^2/2}\, du,$$

and $=0$ for $x \leq 0$, since $F_n(x) = 0$ for all $n \geq 1, x \leq 0$. Substituting this in (27′), we see that (17) holds, and the theorem is completely proved.

In the special calculation of probabilities following (28) we have used the symmetry in deducing $S_n - S_j > 0$ has the same probability as $S_n - S_j < 0$ (and, of course, also the independence). This is called the *reflection principle* due to D. André. It is thus clear that in all these problems involving invariance principles a considerable amount of special ingenuity is needed to obtain specific results. For instance $\max_{j\leq n} |S_j|, \min_{j\leq n} S_j$, etc., are all useful problems with applicational potential and admit a similar analysis.

The next illustration shows how another class of problems called empiric distributional processes can be treated, and new insight gained. We use the preceding result in the following work.

If $\{X_n, n \geq 1\}$ is a sequence of i.i.d. random variables with F as their common d.f., and, for each n, if $F_n(x) = (1/n)\sum_{j=1}^n \chi_{[X_i < x]}$, called the empiric distribution, then we have shown in 2.4.1 (the Glivenko-Cantelli theorem) that $F_n(x) \to F(x)$ as $n \to \infty$, uniformly in $x(\in \mathbb{R})$, with probability one. Can we get, by a proper normalization, a limit distribution of the errors $(F_n(x) - F(x))$? If F is a continuous d.f., then $F(X_1)$ is uniformly distributed on the unit interval. By the classical central limit theorem, $\sqrt{n}(F_n(x) - F(x)) \xrightarrow{D} Z_x$, an $N(0,1)$ r.v. Consequently, one should consider processes of the type $Y_n(x) = \sqrt{n}(F_n(x) - F(x))$ and investigate whether $Y_n(t) \xrightarrow{D} Y(t)$, and whether $\{Y(t), t \in \mathbb{R}\}$ is somehow related to the Brownian motion process. This is our motivation. But there are obstacles at the very beginning. Even if F is continuous, $Y_n(\cdot)$ [unlike (1)] has jump discontinuities. Thus $Y_n : \Omega \to D(\mathbb{R})$, the space of real functions on $\mathbb{R}$ without discontinuities of the second kind. If, for convenience, we restrict our attention to the uniform distribution $F(t) = t$ (cf. Theorem 3.3.9, where such a transformation for strictly increasing functions was discussed in detail), even then $Y_n(\Omega) \subset D[0,1]$ and $C[0,1] \subset D[0,1]$. This and similar problems show that one has to introduce a topology in the larger space $D[0,1]$ to make it a complete *separable* metric space so that the above inclusion or embedding is continuous, and if P_n is the induced measure of Y_n in $D[0,1]$, then one needs to find conditions for $P_n \to P$, and determine P. A suitable topology here was introduced by A.V. Skorokhod, perfected by Kolmogorov, and soon thereafter the corresponding extension of Theorem 6 was obtained by Prokhorov (1956). It turns out that P concentrates on the subspace $C[0,1]$ of $D[0,1]$, and P is related to W in a simple way. [It is the "tied down" Wiener process $X(t) : X(0) = 0, X(1) = 0$.] We do not need this general theory here. The next application for the empiric distribution processes can be proved using Theorem 7 above and Theorem 3.3.9 in a reasonably simple manner. We do this following Rényi (1953).

The result to be established is on the limit d.f. of the *relative errors* instead of the actual errors $F_n(x) - F(x)$. We assume that F is continuous, and consider $-\infty < x_a < \infty$ such that $0 < F(x_a) = a < 1$. Then one has the following result, which complements an application detailed in Theorem 3.3.9. However we need to use the latter theorem in the present proof.

Theorem 8 *Let F be a continuous strictly increasing d.f. and $x_a, 0 < a < 1$, be as above. If F_n is the empiric d.f. of a set of n independent r.v.s with F as their common d.f., then*

$$\lim_{n\to\infty} P\left[\sup_{x_a \le x < \infty} \frac{\sqrt{n}(F_n(x) - F(x))}{F(x)} < y\right]$$

$$= \begin{cases} 0 & if \quad y \le 0 \\ \sqrt{\frac{2}{\pi}} \int_0^{y(a)} e^{-u^2/2}\, du & if \quad y > 0, \end{cases} \tag{30}$$

where $y(a) = y(a/(1-a))^{1/2}$. If, moreover, $F(x_b) = b$, with $0 < a < b < 1$, then we have

$$\lim_{n\to\infty} P\left[\sup_{x_a \le x \le x_b} \frac{\sqrt{n}(F_n(x) - F(x))}{F(x)} < y\right] = G(y), \quad y \in \mathbb{R}, \tag{31}$$

where

$$G(y) = \frac{1}{\pi} \int_{-\infty}^{y(b)} e^{-u^2/2}\, du \int_0^{(y(b)-u)c(a,b)} e^{-t^2/2}\, dt,$$

with $c(a,b) = [a(1-b)/(b-a)]^{1/2}$. In particular,

$$G(0) = \frac{1}{\pi} \text{arc sin} \frac{a(1-b)}{b(1-a)^{1/2}}$$

Proof First note that if we let $b \to 1$, so that $X_b \to +\infty$, then $G(y)$ becomes the right side of (30), as it should. Thus it suffices to establish (31).

However, (31) is also an easy extension of (30). Thus we obtain (30) for simplicity, and then modify it to deduce (31). Because of the separability of $\mathbb{R}$, there are no measurability problems for the events in (30) and (31) (cf. the proof of the Gilvenko-Cantelli Theorem 2.4.1). We present the proof again in steps, for clarity. The first step deals with a reduction of the problem.

I. Since $Y[= F(X)]$ is uniformly distributed on (0, 1), if $u = F^{-1}(x)$, then $H_n(x) = F_n(u) = F_n(F^{-1}(x))$ is the empiric distribution of the observations $Y_i = F(X_i), i = 1, \ldots, n$, from the uniform d.f. Hence

$$\sup_{x_a \leq x < \infty} \frac{[F_n(x) - F(x)]}{F(x)} = \sup_{a \leq x < 1} \frac{[H_n(x) - x]}{x}.$$

Consequently (30) is equivalent to the following:

$$\lim_{n\to\infty} P\left[\sup_{a \leq x < 1} \sqrt{n}(H_n(x) - x)/x < y\right] = \begin{cases} 0 & y \leq 0 \\ \sqrt{\frac{2}{\pi}} \int_0^{y(a)} e^{-u^2/2}\, du, & y > 0. \end{cases} \tag{32}$$

We claim that, for (32), it suffices to establish

$$\lim_{n\to\infty} P\left[\sup_{a \leq H_n(x) \leq 1} \frac{\sqrt{n}(H_n(x) - x)}{x} < y\right] = \sqrt{\frac{2}{\pi}} \int_0^{y(a)} e^{-u^2/2}\, du, \quad y > 0. \tag{33}$$

To see this, let

$$A_n = \left[\sup_{a \leq x} [H_n(x) - x]/x < y/\sqrt{n}\right],$$

$$\tilde{A}_n = \left[\sup_{H_n(x) \geq a+\varepsilon} [H_n(x) - x]/x < y/\sqrt{n}\right],$$

where $\varepsilon > 0$ is arbitrarily fixed. If $B_n = [| H_n(x) - x| \leq \varepsilon]$, then on B_n for $H_n(x) \geq a + \varepsilon$ or $a \leq H_n(x) - \varepsilon \leq x$, one has

$$\sup_{a \leq x}(H_n(x) - x)/x \geq \sup_{a+\varepsilon \leq H_n(x)} (H_n(x) - x)/x.$$

It follows that $A_n \cap B_n \subset \tilde{A}_n \cap B_n$. Hence from

$$\begin{aligned} P(A_n) &= P(A_n \cap B_n) + P(A_n \cap B_n^c) \\ &\leq P(\tilde{A}_n \cap B_n) + P(A_n \cap B_n^c) \leq P(\tilde{A}_n) + P(B_n^c), \end{aligned}$$

we get, on noting that as $n \to \infty$, $H_n(x) \to x$ a.e., uniformly in x (Glivenko-Cantelli), that if (33) is true,

$$\overline{\lim_n} P(A_n) \leq \lim_n P(\tilde{A}_n) = \sqrt{2/\pi} \int_0^{y(a+\varepsilon)} e^{-u^2/2}\, du. \tag{34}$$

Similarly, starting with $H_n(x) \geq a - \varepsilon$, on B_n we have $a - \varepsilon \leq H_n(x) \geq x - \varepsilon$, and if

$$\tilde{\tilde{A}}_n = \left[\sup_{H_n(x) \geq a-\varepsilon} [H_n(x) - x]/x < y/\sqrt{n} \right],$$

then $A_n \cap B_n \supset \tilde{\tilde{A}}_n \cap B_n$, and since $a \leq x$, we get with (33)

$$\underline{\lim}_n P(A_n) \geq \lim_n P(\tilde{A}_n) = \sqrt{2/\pi} \int_0^{y(a-\varepsilon)} e^{-u^2/2}\, du. \qquad (35)$$

Since the right sides of (34) and (35) are continuous functions of $y(a)$, and $y(a \pm \varepsilon) \to y(a)$ as $\varepsilon \to 0$, (34) and (35) imply (32) if (33) is true.

II. Let us therefore establish (33). Here we use the properties of order statistics, especially Theorem 3.3.9. If $Y_1^* < Y_2^* < \cdots < Y_n^*$ are the order statistics of the uniform r.v.s $Y_1, \ldots, Y_n$ $[Y_k = F(X_k)]$, then clearly the empiric d.f., H_n can also be defined as

$$H_n(x) = \begin{cases} 0 & \text{if } \; Y_1^* \geq x \\ k/n & \text{if } \; Y_k^* < x \leq Y_{k+1}^* \\ 1 & \text{if } \; Y_n^* < x. \end{cases}$$

(This is obviously true for any d.f., not necessarily uniform.) Since H_n is a step function which is constant between Y_k^* and Y_{k+1}^*, it follows that

$$\sup_{Y_k^* < x < Y_{k+1}^*} \left(\frac{H_n(x) - x}{x} \right) = \frac{H_n(Y_k^* + 0)}{Y_k^*} - 1 = \frac{k}{nY_k^*} - 1. \qquad (36)$$

But the Glivenko-Cantelli theorem implies $(H_n(x) - x)/x \to 0$ a.e. uniformly in x, so that in (36) $(k/(nY_k^*)) - 1 \to 0$, a.e. uniformly for $a \leq k/n < 1$ as $n \to \infty$. Thus for each $\varepsilon > 0$, there is an n_0 $[= n_0(\varepsilon)]$ such that $n \geq n_0$ implies $|(k/(nY_k^*)) - 1| < \varepsilon/n$ a.e. for all $a \leq k/n \leq 1$. Hence

$$\sqrt{n} \log \frac{k}{nY_k^*} = \sqrt{n} \log \left[\left(\frac{k}{nY_k^*} - 1 \right) + 1 \right] = \sqrt{n} \left(\frac{k}{nY_k^*} - 1 \right) + o_p(1/\sqrt{n}), \qquad (37)$$

where $o_p(1/\sqrt{n}) \to 0$ a.e. as $n \to \infty$ for $a \leq k/n \leq 1$. If follows thereafter that

$$\lim_{n \to \infty} P\left[\max_{an \leq k \leq n} \sqrt{n} \left(\frac{k}{nY_k^*} - 1 \right) < y \right] = \lim_{n \to \infty} P\left[\max_{an \leq k \leq n} \sqrt{n} \log \frac{k}{nY_n^*} < y \right]. \qquad (38)$$

Thus (33) will be proved, because of (36) and (38), if we show

$$\lim_{n \to \infty} P\left[\max_{an \leq k \leq n} \sqrt{n} \log \frac{k}{nY_n^*} < y \right] = \sqrt{\frac{2}{\pi}} \int_0^{y(a)} e^{-u^2/2}\, du. \qquad (39)$$

III. Now $\{Y_k^*, 1 \le k \le n\}$ is a set of order statistics from the uniform d.f. But from Theorem 3.3.9 (or even directly here) $Z_k = -\log Y_{n+1-k}^*$ is an order statistic from the exponential d.f., and then (from the theorem) $\{Z_k, 1 \le k \le n\}$ has independent increments. In fact, as we saw in the proof of that theorem, if $U_k = Z_k - Z_{k-1} (Z_0 = 0)$, then $U_1, \ldots, U_n$ are independent; the density of U_k is given by

$$g_{U_k}(x) = \begin{cases} \text{const. } e^{-(n-k+1)x}, & x > 0 \\ 0, & x \le 0 \end{cases}$$

In other words, if $U_k = V_k/(n-k+1)$, then V_k has the standard exponential density with mean 1, i.e., the density g_{V_k} is

$$g_{V_k}(x) = \begin{cases} e^{-x}, & x > 0 \\ 0, & x \le 0. \end{cases} \tag{40}$$

Thus

$$Z_k = \sum_{i=1}^{n} V_i/(n-1+1) = \sum_{j=1}^{k} U_j;$$

the V_i are i.i.d. with density given by (40). But $-\log Y_k^* = Z_{n-k+1} = \sum_{j=1}^{n-k+1} U_j$, and the U_j satisfy the Lindeberg condition. In fact they satisfy the stronger Liapounov condition $\lim_{n\to\infty} \rho(Z_n)/\sigma(Z_n) = 0$, where

$$\rho^3(Z_n) = \sum_{k=1}^{n} E|U_k - E(U_k)|^3, \quad \sigma^2(Z_n) = \sum_{k=1}^{n} \text{Var } U_k = \sum_{k=1}^{n} (n-k+1)^{-2}.$$

But a simple computation shows that $\rho^3(Z_n) < 17 \sum_{k=1}^{n} (n-k+1)^{-3}$. Hence the above condition holds. Thus Theorem 7 applies, and we get the following on setting $s^2(n) = \sum_{an \le k \le n} \text{Var } U_{n-k+1}$:

$$\lim_{n\to\infty} P\left[\max_{an \le k \le n} (Z_{n+1-k} - E(Z_{n+1-k})) < xs(n)\right] = \sqrt{\frac{2}{\pi}} \int_0^x e^{-u^2/2}\, du, \tag{41}$$

where $x > 0$. But the left-handed side (LHS) of (41) can be written as

$$\text{LHS}(41) = \lim_{n\to\infty} P\left[\max_{an \le k \le n} \left(\log \frac{1}{Y_k^*} - \sum_{j=k}^{n} j^{-1}\right) < x \left(\sum_{an \le j \le n} j^{-2}\right)^{1/2}\right]. \tag{42}$$

Since $0 < a < 1$ and $j \ge an$, we have with the standard calculus approximations

$$\sum_{j=k}^{n} \frac{1}{j} = \log\left(\frac{n}{k}\right) + O\left(\frac{1}{n}\right), \quad \sum_{j=[an]}^{n} \frac{1}{j^2} = \frac{1-a}{an} + O\left(\frac{1}{n^2}\right).$$

Hence the right-hand side (RHS) of (42) simplifies to

$$\mathrm{RHS}(42) = \lim_{n\to\infty} P\left[\max_{an\le k\le n}\left(\log\frac{1}{Y_k^*} - \log\frac{n}{k}\right) < x\left(\frac{1-a}{an}\right)^{1/2}\right]. \qquad (43)$$

Finally, (39) follows from (41) and (43) if we set

$$y = x[(1-a)/a]^{1/2} \qquad \text{or} \qquad x = y(a)$$

in (43). This establishes (30) in complete detail.

IV. It is now clear how to extend this for (31). The only changes are in the limits for the maximum. Hence consider for $0 < a < b < 1$,

$$\begin{aligned}
\zeta_n &= \max_{an\le k\le bn} \sqrt{n}\left(\log\frac{1}{Y_k^*} - \sum_{j=k}^{n}\frac{1}{j}\right) \\
&= \max_{an\le k\le bn} \sqrt{n}\sum_{j=1}^{n+1-k}(U_{n-j+1} - E(U_{n-j+1})) \\
&= \max_{n-bn\le k\le n-an} \sqrt{n}\left(\sum_{1\le j\le n+1-bn} U'_{n-j+1} + \sum_{j=k}^{n-bn} U'_j\right) \\
&\qquad \text{(the primed variables being centered at means)} \\
&= \sqrt{n}\sum_{1\le j\le n+1-bn} U'_{n-j+1} + \max_{n-bn\le k\le n-an}\sum_{j=k}^{n-bn} U'_j \\
&\qquad \text{(since the first term does not depend on } k\text{)} \\
&= \zeta_n^{(1)} + \zeta_n^{(2)} \quad \text{(say)}.
\end{aligned} \qquad (44)$$

Thus $\zeta_n^{(1)}$ and $\zeta_n^{(2)}$ are independent r.v.s and $\zeta_n^{(1)} \xrightarrow{D}$ to an r.v., $N(0,(1-b)/b)$ as $n \to \infty$, and by the first part

$$\lim_{n\to\infty} P[\zeta_n^{(2)} < x] = \sqrt{2/\pi}\int_0^{x\sqrt{a/(b-a)}} e^{-u^2/2}du.$$

The limit d.f. of ζ_n is thus a convolution of these two, so that

$$\lim_{n\to\infty} P[\zeta_n < y] = \frac{1}{\pi}\left(\frac{b}{1-b}\right)^{1/2}\int_{-\infty}^{y} e^{-bu^2/2(1-b)}\,du \times \int_0^{(y-u)[ab/(b-a)]^{1/2}} e^{-v^2/2}\,dv.$$

The right side reduces to $G(y)$ and (31) is obtained.

Finally, for $G(0)$, note that it can be written as

$$G(0) = \lim_{n\to\infty} P\left[\sup_{x_a \le x \le x_b} (F_n(x) - F(x)) < 0\right] = \frac{1}{\pi}\int_{\mathbb{R}^+} e^{-u^2/2}\, du$$
$$\times \int_{[0<t<uc(a,b)]} e^{-t^2/2}\, dt.$$

The right-side expression was shown in Problem 26 in Chapter 4 to be the desired quantity. This proves the theorem completely.

We have included here all the (brutal) details because of the importance of this result in statistical hypothesis testing. It is related to classical theorems of Kolmogorov and Smirnov, and is used to test the hypothesis that a sample came from a specified continuous (strictly increasing) d.f. Various alterations [e.g., with absolute deviations in (30) and (31)] are possible. We do not consider them here, but indicate some in a problem (see Problem 32).

Before ending this section, we add some remarks on two points raised before. Processes $\{X_t, 0 \le t \le 1\}$ more general than Brownian motion are those which have independent increments and are stochastically continuous, i.e., $\lim_{t\to s} P[|X_t - X_s| \ge \varepsilon] = 0$ for $t, s \in [0,1]$ and a given $\varepsilon > 0$. Such processes can be shown to have no discontinuities of the second kind, so that their sample paths belong to $D[0,1]$. Consequently one can seek conditions for the convergence of P_n of $Y_n(t,\cdot)$—the random polygon obtained as in Theorem 6 for sequences of rowwise independent asymptotically constant r.v.s. Here the theory of infinitely divisible distributions (as a generalization of the classical central limit problem) enters. The corresponding characterization has been obtained by Prokhorov (1956). The second point is about the "strong" invariance principle. This is about a statement that the random polygonal processes, obtained as in (1), converge to the Brownian motion process with probability one, if both can be defined on the same probability space. Can this always be done or only sometimes? A first positive solution was given by V. Strassen in the early 1960s. To describe his formulation, let $\{X_n, n \ge 1\}$ be i.i.d. with zero means and unit variances. If $S_0 = 0, S_n = \sum_{k=1}^n X_k$, let $\tilde{Y}_n(t,\cdot)$ be the polygonal process defined similarly (but we can not use the central limit theorem and so must have a different normalization). Thus let us define on the probability space (Ω, Σ, P) of the X_n

$$\tilde{Y}(t) = ([t] + 1 - t)S_{[t]} + (t - [t])S_{[t]+1}, \quad t \ge 0,$$

where $[t]$ is the integral part. Extending the ideas from the embedding method of Proposition 2, and using other tools (such as the iterated logarithm law–see the next section) one can obtain the following result.

Theorem 9 (Strassen) *There exists a probability space (Ω', Σ', P') (actually the Lebesgue unit interval will be the candidate) and a Brownian motion* *process* $\{B(t), t \geq 0\}$ *and another process* $\{\hat{Y}(t), t \geq 0\}$ *on it such that*

(i) $\{\tilde{Y}(t), t \geq 0\}$ *and* $\{\hat{Y}(t), t \geq 0\}$ *have the same finite-dimensional distributions, and*

(ii) $P'[\lim_{t\to\infty} \sup_{s\leq t} |\hat{Y}(t) - B(t)|/(2t \log\log t)^{1/2} = 0] = 1.$

Even though we discussed the existence of Brownian motion on [0, 1], the general case of $\mathbb{R}^+$ is similar. A number of other "old" results have been extended to this setting. A survey of these and other possible extensions with references have been given by Csörgö and Révész (1981), to which we refer the interested reader for information on this line of investigation. As is clear from the statement of Theorem 9, one needs to use several properties of Brownian motion, (some of these will be studied in Chapter 8) and will lead us tangentially in our treatment. So the proof will not be detailed here.

It is clear that the results of this section indicate that a study of limit theorems in suitable metric spaces more general than those of $C[0,1]$ and $D[0,1]$ can be useful in applications. These may clarify the structure of the concrete problems we described above. Relative compactness of families of measures and their equivalent usable forms, called *tightness* conditions from Prokhorov's work, have been the focus of much recent research. One of the main points of this section is that an essential part of probability theory merges with the study of measures on (infinite-dimensional) function spaces. We have to leave these specializations at this stage.

5.5 Kolmogorov's Law of the Iterated Logarithm

The preceding result, due to Strassen, contains a factor involving an iterated logarithm. In fact the law of the iterated logarithm (LIL) is about the growth of partial sums of (at first) independent r.v.s which is a far-reaching generalization of the SLLN and is also needed for a proof of Theorem 4.9. Here we demonstrate the first basic LIL result given in 1929 by A. Kolmogorov. Similar to the other theorems, this one also has been extended in many directions, but for now we shall be content with the presentation of a complete proof of this fundamental result. We include a historical motivation.

The problem originated in trying to sharpen the statements about the normal numbers (cf. Problem 7 in Chapter 2). If $0 < x < 1$, and in the dyadic expansion of x, $S_n(x)$ denotes the number of 1's in the first n digits, then $S_n(x) \to \infty$ a.e. as $n \to \infty$. (A similar problem also arises for decimal expansions.) But what is the rate of growth of S_n? Since $S_n/n \to 1/2$ a.e. by the SLLN, one should find the rates of growth of $|S_n - (n/2)|$. The first result in 1913, due to F. Hausdorff, gives the bound as $O(n^{(1/2)+\varepsilon})$, $\varepsilon > 0$. Then in the following year G.H. Hardy and J.E. Littlewood, who were the masters of the "$\varepsilon - \delta$" estimations of "hard" analysis, improved the result to $O(\sqrt{n \log n})$. Almost 10 years later A. Khintchine, using probabilistic analysis, was able to prove the best result $O(\sqrt{n \log \log n})$, and more precisely,

$$\limsup_n \frac{|S_n - n/2|}{\sqrt{n \log \log n}} = 2^{-1/2} \quad \text{a.e. (Lebesgue).} \tag{1}$$

Thus the law of the iterated logarithm was born. In the current terminology, if S_n is the partial sum of i.i.d. symmetric Bernoulli r.v.s, then (1) is equivalent to stating that

$$\limsup_n (|S_n| / \sqrt{n \log \log n}) = 1/\sqrt{2} \quad \text{a.e. (Lebesgue).} \tag{2}$$

This result showed the power of the probabilistic methods and represented a great achievement in the subject. That was enhanced by the next result of Kolmogorov's, when he generalized the above for an arbitrary independent sequence of bounded r.v.s. We establish this here. Because it is a pointwise convergence statement, the result is a strong limit theorem. However, error estimates in the proof depend critically on the weak limit theory, and so the result is placed in this chapter. (Thus the strong statement is really based on the "weak statement.") Actually in the modern development of probability, the strong and weak limit theories intertwine; and this greatly enriches the subject.

The desired result can be stated as follows.

Theorem 1 (Kolmogorov) *Let $\{X_n, n \geq 1\}$ be a sequence of independent individually bounded r.v.s such that $E(X_n) = 0$, $s_n^2 = \mathrm{Var}(S_n) \nearrow \infty$, where $S_n = \sum_{k=1}^n X_k$. If $|X_n| = o(s_n(\log \log s_n)^{-1/2})$ a.e., $n \geq n_0$, then*

$$P\left[\overline{\lim_{n\to\infty}} (S_n / \sqrt{2 s_n^2 \log \log s_n}) = +1\right] = 1, \tag{3a}$$

and

$$P\left[\underline{\lim}_{n\to\infty} (S_n / \sqrt{2 s_n^2 \log \log s_n}) = -1\right] = 1. \tag{3b}$$

Hence we also have

$$P\left[\overline{\lim_{n\to\infty}} (|S_n| / \sqrt{2 s_n^2 \log \log s_n} = 1\right] = 1. \tag{3c}$$

For a proof of (3a), we need to establish with the weak limit theory certain exponential bounds on probabilities. Thus we first present them separately for clarity. Note that if (3a) is established, then considering the sequence $\{-X_n, n \geq 1\}$ which satisfies the same hypothesis, one gets (3b), and then (3c) is a consequence. Hence it suffices to establish (3a). This is done in two stages. First one proves that

$$\overline{\lim_{n\to\infty}}\,(S_n/\sqrt{2s_n^2 \log\log s_n}\,) \leq 1 \quad \text{a.e.,} \tag{4}$$

and then for each $\varepsilon > 0$, one shows that the left side is $\geq 1 - \varepsilon$ a.e. These two parts need different methods. Kolmogorov obtained some exponential upper and lower bounds for $P[\max_{1\leq k\leq n} S_k \geq \lambda c_n]$. Then using the (relatively easy) first Borel-Cantelli lemma, (4) is proved with the upper estimate. The lower one [for the opposite inequality of (4)] is more involved. It should be calculated for certain subsequences by bringing in the independent property so as to use the second Borel-Cantelli lemma. This is why a proof of Theorem 1 has always been difficult. A relatively simple computation of the bounds, essentially following H. Teicher (*Z. Wahrs.* **48** (1979), 293-307), is given here. These are stated in the next two technical lemmas for convenience.

It is useful to consider a property of the numerical function

$$h(x) = (e^x - 1 - x)/x^2 = \frac{1}{2!} + \frac{x}{3!} + \frac{x^2}{4!} + \cdots. \tag{5}$$

For $x > 0$, $h(x) > 0$, $h'(x) = (dh/dx)(x) > 0$, and $h''(x) = (d^2h/dx^2)(x) > 0$. Hence $(h(\cdot)$ is a positive increasing convex function. The same is also true if $x < 0\,[h(0) = 1/2]$. To see this, let $x = -y, y > 0$, so that

$$h(-y) = g(y) = (e^{-y} - 1 + y)/y^2 = \frac{1}{y^2}\int_0^y \int_0^t e^{-v}\, dv\, dt. \tag{6}$$

Hence $h(x) > 0$ for $x < 0$, and $\lim_{y\to 0} g(y) = 1/2, \lim_{y\to\infty} g(y) = 0$. Also, $g'(y) = [1 - (y+2)g(y)]/y$. Thus $g'(y) < 0$ iff $g(y) > (y+2)^{-1}$. And $g''(y) = [g(y)(y^2+4y+6) - (y+3)]/y$, so that $g''(y) > 0$ iff $g(y) > (y+3)(y^2+4y+6)^{-1}$. For us it suffices to establish the first inequality. Since $(y+2)^{-1} < (y+3)(y^2+4y+6)^{-1}, y > 0$, if we verify the second, both inequalities follow. Thus consider the latter. Now

$$e^y - 1 - y - (y^2/2) = \frac{y^3}{3!}\left[1 + \frac{y}{4} + \frac{y^2}{4\cdot 5} + \cdots\right] < \frac{y^3}{3!}\left(1 + \frac{y}{4} + \frac{y^2}{4^2} + \cdots\right)$$

$$= \begin{cases} y^3/6[1-(y/4)] & \text{if } 0 < y < 4 \\ < y^3/(6-2y) & \text{if } 0 < y < 3. \end{cases}$$

Thus

$$e^y < 1 + y + y^2/2 + y^3/(6-2y) = (y^2+4y+6)/(6-2y), \quad \text{if } 0 < y < 3.$$

Hence for $0 < y < 3$,

$$\begin{aligned} g(y) = (e^{-y} - 1 + y)/y^2 &> ([(y^2+4y+6)/(6-2y)]^{-1} - 1 + y)/y^2 \\ &= (y+3)/(y^2+4y+6). \end{aligned}$$

If $y \geq 3$, $g(y) > (y-1)/y^2 \geq (y+3)/(y^2+4y+6)$. Hence in all cases h given by (5) is a nonnegative increasing convex function on $\mathbb{R}$. This function h plays a key role in the next two lemmas, giving the exponential bounds alluded to before. (Only $0 \leq h \nearrow$ is used in the next lemma, and X_k's are as in Theorem 1.)

Lemma 2 *Let $\{X_n, n \geq 1\}$ be independent r.v.s with means zero and finite variances. Let $S_n = \sum_{k=1}^n X_k$, $s_n^2 = \text{Var } S_n$ $(= \sum_{k=1}^n \sigma_k^2, \sigma_k^2 = \text{Var } X_k)$, and $c_n > 0$ be such that $0 < c_n s_n$ and increasing. Let $h(\cdot)$ be given by (5). Then*

(i) $P[X_k \leq c_k s_k, 1 \leq k \leq n] = 1$ *implies*

$$E(\exp\{tS_n/s_n\}) \leq \exp\{t^2 h(c_n t)\}, \qquad t > 0, \tag{7}$$

and for $\lambda_n > 0, b > 0, x_n > 0$, we have

$$P\left[\max_{1 \leq k \leq n} S_k > \lambda_n x_n s_n\right] \leq \exp\{-x_n^2(\lambda_n b - b^2 h(c_n x_n b))\}. \tag{8}$$

(ii) $P[X_k \geq -c_k s_k, 1 \leq k \leq n] = 1$ *implies*

$$E(\exp\{tS_n/s_n\}) \geq \exp\left\{t^2 h(-c_n t)\left(1 - t^2 h(-c_n t)/s_n^{-4} \sum_{k=1}^n \sigma_k^4\right)\right\}, \qquad t \geq 0, \tag{9}$$

and if also $\sigma_k \leq c_k s_k, k \geq 1$, one has

$$E(\exp\{tS_n/s_n\}) \geq \exp\{t^2 h(-c_n t)(1 - c_n^2 t^2 h(-c_n t))\}, \quad t > 0. \tag{10}$$

Proof (i) Consider for $1 \leq k \leq n$

$$\begin{aligned} E(\exp\{tX_k/s_n\}) &\\ &= 1 + E(\exp\{tX_k/s_n\} - 1 - tX_k/s_n) \quad [\text{since } E(X_k) = 0] \\ &= 1 + E(h(tX_k/s_n)(t^2 X_k^2/s_n^2)) \qquad (11) \\ &\leq 1 + t^2 E(h(tc_n)(X_k^2/s_n^2)) \\ &\quad (\text{by hypothesis and the monotonicity of } h) \\ &= 1 + (t^2\sigma_k^2/s_n^2)h(tc_n) \leq \exp\{(t^2\sigma_k^2/s_n^2)h(tc_n)\} \\ &\quad (\text{since } 1 + x < e^x \text{ for } x > 0). \qquad (12) \end{aligned}$$

Multiplying over $1 \leq k \leq n$, we get (7) by independence.

Also, $\{S_k, \sigma(X_1, \ldots, X_k), k \geq 1\}$ is a martingale sequence, and since $x \mapsto \exp\{tx\}, t > 0$, is convex, $\{\exp\{tS_k\}, \sigma(X_1, \ldots, X_k), k \geq 1\}$ is a positive bounded submartingale (by the bound condition on the X_k). Hence the maximal inequality gives (cf. Theorem 3.5.6i)

$$\begin{aligned} P\left[\max_{1 \leq k \leq n} S_k > \lambda_n x_n s_n\right] &= P\left[\max_{1 \leq k \leq n} \exp\{tS_k\} > \exp\{\lambda_n t x_n s_n\}\right] \\ &\leq \exp\{-\lambda_n t x_n s_n\}\, E(\exp\{tS_n\}) \\ &\leq \exp\{-\lambda_n t x_n s_n + t^2 s_n^2 h(c_n s_n t)\} \quad [\text{by } (7)] \\ &= \exp\{-x_n^2(\lambda_n b - b^2 h(c_n b x_n)\} \quad (\text{if } t = b x_n/s_n). \end{aligned}$$

This is (8), and (i) follows.

(ii) For (9), we use the inequality $1+x \geq \exp(x-x^2)$ for $x \geq 0$. In fact, this is obvious if $x \geq 1$ and for $x=0$. If $0<x<1$, then this is equivalent to showing that $(1+x)\exp(x^2-x)>1$, or that

$$\log(1+x)+(x^2-x)=f(x)>0. \tag{13}$$

Since $f(0)=0$ and $f'(x)=x(1+2x)(1+x)^{-1}>0$, $f(\cdot)$ is strictly increasing, so that (13) is true. We use this with the expansion (11). Thus by the condition of (ii) and the monotonicity of h on $\mathbb{R}$,

$$\begin{aligned} E(\exp\{tX_k/s_n\}) &\geq 1+h(-c_nt)t^2\sigma_k^2/s_n^2 \\ &\geq \exp\left\{h(-c_nt)\frac{t^2\sigma_k^2}{s_n^2}-\frac{t^4\sigma_k^4}{s_n^4}h^2(-c_nt)\right\} \quad \text{[by (13)]}. \end{aligned}$$

Multiplying over $1 \leq k \leq n$, this yields (9) since $s_n^2 = \sum_{k=1}^n \sigma_k^2$. If also $\sigma_k^2 \leq c_n^2 s_n^2$, so that

$$\sum_{k=1}^n \sigma_k^4 \leq c_n^2 s_n^2 \sum_{k=1}^n \sigma_k^2 = c_n^2 s_n^4,$$

we get (10) from (9) on substituting the bound c_n^2 for $\sum_{k=1}^n \sigma_k^4/s_n^4$. This completes the proof of the lemma.

Next we proceed to derive the lower bound using (10). This is somewhat involved, as remarked before.

Lemma 3 *Let $\{X_n, n \geq 1\}$ be an independent sequence of individually bounded centered r.v.s. If $0<\sigma_k^2 = \operatorname{Var} X_k, |X_k| \leq d_k$ a.e., where $0<d_k \uparrow$ and $x_n > x_0 > 0$, satisfying $\lim_{n\to\infty}(d_n x_n/s_n)=0$, with $s_n = \operatorname{Var} S_n, S_n = \sum_{k=1}^n X_k$, then for each $0<\varepsilon<1$, there exists a $0<C_\varepsilon<1/2$ such that*

$$P[S_n > 2^{1/2}(1-\varepsilon)^2 s_n x_n] \geq C_\varepsilon \exp\{-x_n^2(1+\varepsilon)(1-\varepsilon)^2\} \tag{14}$$

for all $n \geq n_0(=n_\varepsilon)$.

Proof Since X_k is a bounded r.v., its moment-generating function (m.g.f.) exists. Thus if $\phi_k(t)=E(e^{tX_k})$, then $0<\phi_k(t), \phi_k(0)=1$, and ϕ_k is continuous. Hence there is a $t_0>0$ such that $0<\phi_k(t)<\infty$ for all $0 \leq t < t_0$, since the set $J=\{t : \phi(t)<\infty\}$ is an interval containing 0. This is because the function $t \mapsto \exp\{tx\}$ is convex, and hence so is ϕ_k, which implies that J is a convex set. If $a_0 = \inf\{t : t \in J\}, b_0 = \sup\{t : t \in J\}$, then $-\infty \leq a_0 \leq 0 < b_0 \leq +\infty$, so that J is the interval (a_0, b_0) with or without the end points. Thus $\psi_k(t) = \log\phi_k(t), 0 \leq t < t_0$, is finite, so that $\phi_k(t) = \exp\psi_k(t)$.

The idea of the proof is (i) to obtain lower and upper bounds for ψ_k (so we get exponential bounds for ϕ_k), and (ii) then use an ancient transformation, due to F. Esscher in 1932, to relate the result to the desired inequality (14). There is no motivation for this computation, except that it was successfully used by Feller (1943) in an extension of a result of Cramér and then in his generalization of Kolmogorov's LIL. We employ it here in the same manner.

To proceed with point (i), let $c_n = d_n/s_n$ and consider $\phi_{kn}(t) = \phi_k(t/s_n)$. Since ϕ_k is actually holomorphic (on $\mathbb{C}$), we have on differentiation

$$\begin{aligned}\phi'_{kn}(t) &= \frac{d}{dt}E(e^{tX_k/s_n}) = E\left(\frac{X_k}{s_n}(e^{tX_k/s_n}-1)\right) \qquad [\text{since} E(X_k)=0]\\ &= E(th_1(tX_k/s_n)X_k^2/s_n^2),\end{aligned} \tag{15}$$

where, as in Lemma 2, $h_1(x) = (e^x - 1)/x$. Clearly h_1 is increasing on $\mathbb{R}^+$. But on $\mathbb{R}^-$, let $h_2(y) = h_1(-y), y > 0$. Because $h_1(0) = 1, h_2(y) = (1 - e^{-y})/y \to 0$ as $y \to \infty$, it is decreasing for $y \nearrow$ because $h'_2(y) = [(y+1)e^{-y} - 1]/y^2$ and $e^y > 1 + y \Rightarrow h'_2(y) < 0$, so that $h_1(x)$ is increasing on all of $\mathbb{R}$ as $x \nearrow$. Consequently for $t > 0$, because $|X_k/s_n| \le c_n = d_n/s_n$ a.e. for $k \le n$,

$$th_1(-tc_n)\sigma_k^2/s_n^2 \le \phi'_{kn}(t) \le th_1(tc_n)\sigma_k^2/s_n^2. \tag{16}$$

Similarly e^x is increasing so that

$$e^{-tc_n}(\sigma_k^2/s_n^2) \le E[X_k^2 e^{tX_k/s_n}/s_n^2] = \phi''_{kn}(t) \le e^{tc_n}(\sigma_k^2/s_n^2). \tag{17}$$

However, $\sigma_k^2 \le d_n^2 = c_n^2 s_n^2, k \le n$, and for $t > 0, \phi_k(t/s_n) \ge 1$. Thus the inequalities (16), (17), and (11) yield, with h of that lemma, $[\psi_{kn} = \log \phi_{kn}]$

$$\frac{th_1(-tc_n)(\sigma_k^2/s_n^2)}{1 + t^2c_n^2 h(tc_n)} \le \frac{\phi'_{kn}(t)}{\phi_{kn}(t)} = \psi'_{kn}(t) \le th_1(tc_n)(\sigma_k^2/s_n^2), \tag{18}$$

and similarly

$$\psi''_{kn}(t) = \frac{\phi''_{kn}(t)}{\phi_{kn}(t)} - [\psi'_{kn}(t)]^2 \le \phi''_{kn}(t) \le e^{tc_n}(\sigma_k^2/s_n^2) \quad [\text{by (17)}]. \tag{19}$$

A lower bound is obtained by using (12), (17), and (18):

$$\psi''_{kn}(t) \ge (\sigma_k^2/s_n^2)[\exp\{-tc_n - t^2c_n^2 h(tc_n)\} - t^2h_1^2(tc_n)]. \tag{19'}$$

Consequently, if $\psi_n(t) = \sum_{k=1}^n \psi_{kn}(t)$, then (18) implies

$$\frac{th_1(-tc_n)}{1 + t^2c_n^2 h(tc_n)} \le \psi'_n(t) = \sum_{k=1}^n \psi'_{kn}(t) \le th_1(tc_n), \qquad t > 0. \tag{20}$$

For point (ii) we proceed to the key Esscher transformation and use these bounds in its simplification. Let $0 < t < t_0$ be fixed, and if F_k is the d.f. of

X_k, so that F_{kn} is the d.f. of $X_k/s_n, 1 \le k \le n$, define a new d.f. F^t_{kn} by the equation

$$dF^t_{kn}(x) = [\phi_{kn}(t)]^{-1} e^{tx}\, dF_{kn}(x). \tag{21}$$

Let $\{X_{kn}(t), 1 \le k \le n\}$ be independent r.v.s each with d.f. F^t_{kn}. It may be assumed that these r.v.s are defined on the same probability space as the original X_k, by enlarging the underlying space if necessary. Let $S_n(t) = \sum_{k=1}^n X_{kn}(t)$ and F^t_n be its d.f. Noting that if the ch.f. $\tilde{\phi}^t_{kn}$ of F^t_{kn} is calculated, then the ch.f. $\tilde{\phi}^t_n$ of $S_n(t)$ is given by

$$\begin{aligned}
\tilde{\phi}^t_n(u) &= \prod_{k=1}^n E(e^{iuX_{kn}(t)}) = \prod_{k=1}^n [\phi_{kn}(t)]^{-1} \int_{\mathbb{R}} e^{(iu+t)y} dF_{kn}(y) \\
&= \prod_{k=1}^n \frac{\phi_{kn}(t+iu)}{\phi_{kn}(t)} = \exp\left\{\sum_{k=1}^n (\psi_{kn}(t+iu) - \psi_{kn}(t))\right\}
\end{aligned}$$

[by the fact that $\psi_{kn}(t) = \log \phi_{kn}(t)$]

$$= \exp\{\psi_n(t+iu) - \psi_n(t)\} \qquad \left(\text{since } \psi_n = \sum_{k=1}^n \psi_{kn}\right). \tag{22}$$

Thus the mean and variance of $S_n(t)$ are

$$\frac{1}{i}\frac{d}{du} \log \tilde{\phi}^t_n(u)|_{u=0} \qquad \text{and} \qquad -\frac{d^2}{du^2} \log \tilde{\phi}^t_n(u)|_{u=0},$$

respectively, so that they are $\psi'_n(t)$ and $\psi''_n(t)$. If $\tilde{F}^t_n$ is the d.f. of $\tilde{S}_n(t) = (S_n(t) - \psi'_n(t))/\sqrt{\psi''_n(t)}$, then it is clear that

$$\tilde{F}^t_n(x) = F^t_n(\psi'_n(t) + x\sqrt{\psi''_n(t)}).$$

We use this transformation to connect the probabilities of S_n/s_n and $\tilde{S}_n(t)$ for each $0 \le t < t_0$, since then by a suitable choice of t ($t = 0$ corresponds to S_n) we will be able to get the desired lower bound. But the ch.f. of S_n/s_n is given by $\exp \psi_n$, and (22) implies (by the uniqueness theorem) that the d.f. F^t_n of $S_n(t)$, with $\tilde{F}_n$ as the d.f. of S_n/s_n, is

$$F^t_n(x) = e^{-\psi_n(t)} \int_{-\infty}^x e^{ty}\, d\tilde{F}_n(y), \qquad x \in \mathbb{R},\ t \in \mathbb{R}. \tag{23}$$

Here we use

$$\phi_n(u) = E(\exp\{uS_n/s_n\}) = \prod_{k=1}^n \phi_k(u/s_n) \Rightarrow \psi_n(u) = \log \phi_n(u).$$

Now,

$$\begin{aligned}
P[(S_n/s_n) > w] &= \int_w^\infty d\tilde{F}_n(x) \\
&= e^{\psi_n(t)} \int_w^\infty e^{-tx}\, dF_n^t(x) \quad \text{[by (23)]} \\
&= e^{\psi_n(t)} \int_{(w-\psi_n'(t))/\sqrt{\psi_n''(t)}}^\infty \exp[-t(\psi_n'(t) + x\sqrt{\psi_n''(t)})] \\
&\quad \times dF_n^t(\psi_n'(t) + x\sqrt{\psi_n''(t)}) \quad \text{(by change of variables).}
\end{aligned} \tag{24}$$

To get a lower bound, note that if t is replaced by a sequence t_n in (19) and (19′) such that $t_n c_n \to 0$, implying $e^{t_n c_n} = 1 + O(t_n c_n)$, we have

$$\psi_n''(t_n) = \sum_{k=1}^n \psi_{kn}''(t_n) = 1 + O(t_n c_n), \tag{25}$$

since $\sum_{k=1}^n \sigma_k^2 = s_n^2$. To get a lower estimate for $\exp\{\psi_n(t) - t\psi_n'(t)\}$ in (24), consider

$$\begin{aligned}
\psi_n(t_n) - t_n\psi_n'(t_n) &\geq t_n^2 h(-c_n t_n) - c_n^2 t_n^4 h^2(-c_n t) - t_n^2 h_1(t_n c_n) \\
&\qquad \text{[by (10) and (20)]} \\
&\geq t_n^2[\tfrac{1}{2} + o(t_n c_n) - o(t_n^2 c_n^2) - (1 + o(t_n c_n))] \\
&\qquad \text{[since } h(x) = \tfrac{1}{2} + o(x), h_1(x) = 1 + o(x) \text{ as } x = t_n c_n \to 0] \\
&\geq -(t_n^2/2)(1+\varepsilon) \qquad \text{where } 0 < \varepsilon < 1.
\end{aligned} \tag{26}$$

Define

$$\tilde{v}_n = [(1-\varepsilon)t_n - \psi_n'(t_n)]/\sqrt{\psi_n''(t_n)} = -\varepsilon t_n + o(t_n c_n)$$

as $t_n c_n \to 0$ by (20) and (25). Hence $\tilde{v}_n \leq -\varepsilon t_n/2$ as $t_n c_n \to 0$. Setting $w = (1-\varepsilon)t_n$ in (24), we get with (26)

$$P[S_n > (1-\varepsilon)t_n s_n] \geq \exp\left\{-\frac{t_n^2}{2}(1+\varepsilon)\right\} \int_{-\varepsilon t_n/2}^\infty dF_n^{t_n}(\psi_n'(t_n) + x\sqrt{\psi_n''(t_n)}). \tag{27}$$

However, if

$$Z_{kn} = [X_{kn}(t_n) - E(X_{kn}(t_n))]/\sqrt{\psi_n''(t_n)},$$

then $E(Z_{kn}) = 0, \sum_{k=1}^n Z_{kn} = (S_n(t_n) - \psi_n'(t_n))/\sqrt{\psi_n''(t_n)}$, the $Z_{kn}, 1 \leq k \leq n$, are independent and infinitesimal, because by hypothesis

$$|Z_{kn}| \leq 2c_n(\psi_n''(t_n))^{-1/2} = o(1) \qquad \text{a.e.} \quad \text{uniformly in } k.$$

Also by hypothesis

$$\lim_{n\to\infty} \frac{d_n x_0}{s_n} \leq \lim_{n\to\infty} \frac{d_n x_n}{s_n} = 0,$$

so that $s_n \to \infty$ faster than d_n.

Since X_k, and hence X_{kn} [and $X_{kn}(t)$], take values in the interval $[-d_k, d_k]$, it follows that for large enough n [note that $\phi_k(t/s_n) \geq 1$]

$$\sum_{k=1}^{n} \int_{[|x|>\varepsilon]} x^2\, dF_{kn}^{tn}(x + E(X_{kn}(t_n))) \leq \sum_{k=1}^{n} \frac{1}{s_n^2} \int_{[|y|>\varepsilon s_n]} y^2 e^{t_n y/s_n}\, dF_k(y)$$
$$= o(1),$$

because $dF_k(x) = 0$ for $|x| > d_k$ and $\varepsilon s_n > d_n$ for large enough n, by the above noted condition. But $\mathrm{Var}(S_n(t_n)/\sqrt{\psi_n''(t)}) = 1$. It follows from Theorem 3.5 that $(S_n(t_n) - \psi_n'(t_n))/\sqrt{\psi_n''(t_n)} \to$ an r.v. which is $N(0,1)$. Consequently the right-side integral of (27) is $\geq 1/2$. Hence (27) becomes for large enough n

$$P[S_n > (1-\varepsilon)s_n t_n] \geq C_\varepsilon \exp\{-t_n^2(1+\varepsilon)/2\}, \qquad 0 < C_\varepsilon < 1/2,$$
$$0 < \varepsilon < 1. \tag{28}$$

This is (14) if $t_n = x_n(1-\varepsilon)2^{1/2}$, and the proof of the lemma is complete.

Note how significantly the central limit theorem (for rowwise independent sequences) enters into the argument in addition to all the other computations for the lower bound. We are now ready to complete the

Proof of Theorem 1 Let $d_n = o(s_n(\log\log s_n)^{-1/2})$ and so $d_n \uparrow$, $|X_n| \leq d_n$, a.e. If $c_n = d_n/s_n$, then $c_n = o(1)$ and $\sigma_n^2/s_n^2 \leq c_n^2$. Also, $c_n(\log\log s_n)^{-1/2} \to 0$ as $n \to \infty$. For any $a > 0, b > 0$, consider $\alpha > b^{-1} + bh(ab)$, where $h(x) = (e^x - 1 - x)/x^2$. Choose $\beta > 1$ such that $\alpha/\beta^2 > b^{-1} + bh(ab)$. Since $s_{n+1}^2 = s_n^2 + \sigma_n^2$, so that $s_{n+1}/s_n \to 1$ as $n \to \infty$ (because $\sigma_n^2/s_n^2 \leq c_n^2 \to 0$), we deduce that there exist $n_k < n_{k+1} < \cdots$ such that $s_{n_k} \leq \beta^k < s_{n_{k+1}}$. For otherwise there will be no s_n in (β^k, β^{k+1}) such that $\lim(s_{n+1}/s_n) \geq \beta > 1$, contradicting the preceding sentence. This implies also that $s_{n_k} \sim \beta^k$, the symbol $\sim$ indicating that the ratio $\to 1$ as $k \to \infty$. Let $x_n^2 = \log\log s_n$, so that for large enough $n, c_n x_n \leq a$, since c_n goes to zero. We can now use Lemma 2. Taking $\lambda_n = \alpha/\beta^2 > 0, b > 0$, and $x_n > 0$ as here, we then get by (8),

$$P\left[\max_{1\leq k\leq n_k} S_k > \alpha\beta^{-2} s_{n_k}(\log\log s_{n_k})^{1/2}\right]$$
$$\leq \exp\{-[(\alpha b/\beta^2) - b^2 h(ab)]\log\log s_{n_k}\}. \tag{29}$$

But $(\alpha b/\beta^2) - b^2 h(ab) > 1$, by the choice of β. Hence there is an $\eta > 0$ such that the above probability in (29) is not larger than

$$\exp\{-(1+\eta)\log\log s_{n_k}\} \leq \{k \log\beta)^{-1-(n/2)}$$

for all large enough n, because of $s_{n_k} \sim \beta^k$. Thus by the first Borel-Cantelli lemma, since $\sum_{k\geq 1}(k\log\beta)^{-1-(n/2)} < \infty$, we have

$$
\begin{aligned}
&P[S_n > \alpha s_n(\log\log s_n)^{-1/2}, \text{i.o.}] \\
&\quad \leq P\left[\max_{n_k \leq n < n_{k+1}} S_n > \alpha s_{n_k}(\log\log s_{n_k})^{1/2}, \text{i.o.}\right] \\
&\quad \leq P\left[\max_{1 \leq n \leq n_{k+1}} S_n > \alpha\beta^{-2} s_{n_k+1}(\log\log s_{n_{k+1}})^{1/2}, \text{i.o.}\right] = 0.
\end{aligned}
$$

It follows that

$$\overline{\lim_{n\to\infty}} \frac{S_n}{s_n(\log\log s_n)^{-1/2}} \leq \alpha \quad \text{a.e.} \tag{30}$$

Since $\alpha > b^{-1} + bh(ab)$ for all $a > 0, b > 0$, we see, on letting $a \to 0$, so that $h(0) = 1/2$, that (30) is true for all $\alpha > b^{-1} + b/2$. The least value of the right side is $\sqrt{2}$, and so (30) holds a.e. if $\alpha = \sqrt{2}$. This establishes (4). Note that by applying this result to $\{-X_n, n \geq 1\}$, we deduce that

$$\overline{\lim_{n\to\infty}} \frac{|S_n|}{s_n(\log\log s_n)^{1/2}} \leq \sqrt{2} \quad \text{a.e.} \tag{31}$$

We now prove the opposite inequality to (4). Again choose the $\{n_k, k \geq 1\}$ as before, and let $0 < \varepsilon < 1$. To use Lemma 3 and the second Borel-Cantelli lemma (cf. Theorem 2.1.9ii), it is necessary first to consider at least pairwise independent events. We actually can define mutually independent ones as follows. Let

$$A_k = [S_{n_k} - S_{n_{k-1}} > (1-\varepsilon)^{3/2} a_k b_k],$$

where, since $s_{n_k} \sim \beta^k, k \geq 1, \beta > 1$ we let

$$
\begin{aligned}
a_k^2 &= s_{n_k}^2 - s_{n_{k-1}}^2 \sim s_{n_k}^2(1-\beta^{-2}), \\
b_k^2 &= 2\log\log a_k \sim 2\log\log s_{n_k} < 2(1+\varepsilon)\log k,
\end{aligned}
$$

for all large enough n_k (or k). Thus $d_{n_k} b_k a_k^{-1} = o(s_{n_k}^{-1}) = o(1)$.

Let $x_{n_k} = b_k$ in Lemma 3. Then the above definitions of a_k, b_k yield with (14), and $S_{n_k} - S_{n_{k-1}}$ for S_n there (all the conditions are now satisfied),

$$
\begin{aligned}
P(A_k) &\geq C_\varepsilon \exp\{-\tfrac{1}{2} b_k^2 (1+\varepsilon)(1-\varepsilon)^2\} \geq C_\varepsilon \exp\{-(1+\varepsilon)^2(1-\varepsilon)^2 \log k\} \\
&= C_\varepsilon k^{-(1-\varepsilon^2)^2}.
\end{aligned}
$$

Since $\sum_{k\geq 1} k^{-(1-\varepsilon^2)^2} = \infty$, Theorem 2.1.9 implies $P(A_k$, i.o.$)=1$.

It is now necessary to embed this A_k-sequence in a larger, but dependent, sequence $D_k = [S_{n_k} > \delta_k]$ for a suitable δ_k and show that $\overline{\lim}_k D_k \supset \overline{\lim}_k A_k$ to deduce the result. The crucial work is over, and this is a matter of adjustment. For simplicity, let $v_n^2 = 2\log\log s_n$. For large enough n_k, consider

$$
\begin{aligned}
(1-\varepsilon)^2 a_k b_k - 2 s_{n_{k-1}} v_{n_{k-1}} &\sim (1-\varepsilon)^2 (1-\beta^{-2})^{1/2} s_{n_k} v_{n_k} - 2\beta^{-1} s_{n_k} v_{n_k} \\
&= [(1-\varepsilon^2)(1-\beta^{-2})^{1/2} - 2\beta^{-1}] s_{n_k} v_{n_k} > (1-\varepsilon)^3 s_{n_k} v_{n_k}
\end{aligned} \tag{32}
$$

where $\beta > 1$ is chosen large enough so that $(1-\varepsilon)^2(1-\beta^{-1})^{1/2} > (1-\varepsilon)^3 + 2/\beta$ holds. This can be done. With this choice, consider the events $B_k = [|S_{n_{k-1}}| \leq 2s_{n_{k-1}}v_{n_{k-1}}]$. By (31) $P[B_k^c$, i.o.$]=0$. But we have

$$\begin{aligned} A_k \cap B_k &\subset [S_{n_k} > (1-\varepsilon)^2 a_k b_k - 2s_{n_{k-1}}v_{n_{k-1}}] \\ &\subset [S_{n_k} > (1-\varepsilon)^3 s_{n_k} v_{n_k}] \\ &\quad \text{(by the above choice for all large enough } k). \end{aligned}$$

Since $P[A_k$, i.o.$]=1$ and $P[B_k^c$,i.o.$]=0$, we get $P[A_k \cap B_k$, i.o.$]=1$. Thus

$$P[S_{n_k} > (1-\varepsilon)^3 s_{n_k} v_{n_k}, \text{i.o.}] \geq P[A_k \cap B_k, \text{i.o.}] = 1.$$

Consequently, we have

$$\overline{\lim_{n\to\infty}} \frac{S_n}{s_n v_n} \geq \overline{\lim_{n\to\infty}} \frac{S_{n_k}}{s_{n_k} v_{n_k}} \geq (1-\varepsilon)^3 \quad \text{a.e.} \tag{33}$$

Since $\varepsilon > 0$ is arbitrary, by letting $\varepsilon \searrow 0$ through a sequence in (33) we get the opposite inequality of (4) with probability one. These two together imply the truth of (3a). As noted before, this gives (3) itself, and thus the theorem is completely proved.

This important theorem answers some crucial questions but raises others. The first one of the latter kind is this: Since many of the standard applications involving i.i.d sequences of r.v.s are not necessarily bounded but will have some finite moments, how does one apply the above result? Naturally one should try the truncation method. As remarked by Feller (1943), if the $X_n, n \geq 1$ are i.i.d., with slightly more than two moments, then they obey the LIL. In fact, let $Y_n = X_n \chi_{A_n}$, where $A_n = [|X_n| \leq n^{1/2} \log\log n]$ and $E(X_1^2(\log|X_1|)^{1+\varepsilon}) < \infty$ for some $\varepsilon > 0$. Then one can verify with the Borel-Cantelli lemma that $\sum_{n\geq 1} P[X_n \neq Y_n] < \infty$, so that $X_n = Y_n$ a.e. for all large n. But the Y_n are bounded and the reader may verify that the Y_n-sequence satisfies the hypothesis of Theorem 1. Hence that result can be applied to the i.i.d. case with this moment condition. However, this is not the best possible result. The following actually holds. The sufficiency was proved in 1941 by P. Hartman and A. Wintner, and finally in 1966 the necessity by V. Strassen. We state the result without proof. Note that, since "$\limsup_n S_n$" defines a tail event, the probability of that event is either 0 or 1, by the "0-1 law."

Proposition 4 *Let $\{X_n, n \geq 1\}$ be a sequence of i.i.d. random variables with zero means. Let $S_n = \sum_{k=1}^n X_k$. Then* LIL *holds for the S_n-sequence, in the sense that*

$$P\left[\overline{\lim_{n\to\infty}} \frac{S_n}{(n \log\log n)^{1/2}} = \sqrt{2}\right] = 1 \tag{34}$$

iff $E(X_1^2) < \infty$, and then $E(X_1^2) = 1$.

In a deep analysis of the case of bounded r.v.s, Feller (1943) has shown that for any increasing positive sequence $\{a_n, n \geq 1\}$ and $S_n = \sum_{k=1}^{n} X_k$ with X_k as (bounded) and independent r.v.s, one has with $s_n^2 = \text{Var } S_n, E(X_k) = 0$,

$$P[S_n > s_n a_n, \text{i.o.}] = 0 \quad \text{or} \quad 1,$$

according to whether $\sum_{n \geq 1}(1/n)a_n \exp\{-a_n^2/2\}$ is finite or infinite. Moreover, the same result holds for unbounded r.v.s, provided there exist numbers $\alpha_0 > 0, \varepsilon > 0$ such that

$$E(X_k^2 | \log | X_k||^{1+\varepsilon}) < \alpha_0 \text{Var } X_k, \qquad k \geq 1. \tag{35}$$

In the i.i.d. case, (35) reduces to the sufficiency part of the previously stated result. (For the Bernoulli case, see Problem 35.)

Another point to note here is that in both Theorem 1 and Proposition 4 the $\{S_n, n \geq 1\}$-sequence satisfies the central limit theorem, in the sense that

$$\lim_{n \to \infty} P\left[\frac{S_n - E(S_n)}{\sqrt{\text{Var } S_n}} < x\right] = (2\pi)^{-1/2} \int_{-\infty}^{x} e^{-u^2/2}\, du, \qquad x \in \mathbb{R}.$$

Thus one can ask the question: Does every sequence of independent r.v.s $\{X_n, n \geq 1\}$ which obeys the central limit theorem also obey the LIL? What about the converse? In the i.i.d. case, Proposition 4 essentially gives an answer. The general case is still one of the current research interests in probability theory. The preceding work already shows how the combination of the ideas and methods of the strong and weak limit theory is essential for important investigations in the subject. In another direction, these results are being extended for dependent r.v. sequences. Some of these ideas will be presented in the next section which will also motivate the topic of the following chapter.

5.6 Application to a Stochastic Difference Equation

The preceding work on weak and strong limit theorems is illustrated here by an application. Let $\{X_n, n \geq 1\}$ be a sequence of r.v.s satisfying the first-order linear stochastic difference equation

$$X_t = \sum_{i=1}^{k} \alpha_i X_{t-i} + \varepsilon_t, \quad t = 1, 2, \ldots, \tag{1}$$

where $X_0 = 0$ for simplicity and the ε_t, $t \geq 1$, are a sequence of i.i.d. r.v.s with $P[\varepsilon_1 = 0] = 0$, means zero and variances $\sigma^2 > 0$. The constants α_i are real, but usually not known.

A problem of interest is to estimate the α_i, based upon the observations of $\{X_t, t = 1, \ldots, n\}$. Thus, if the values of (the nonindependent) $X_t, t = 1, \ldots, n$ are observed and α_i is estimated then using (1) the value of X_{n+1} can be "predicted." For this to be fruitful, the estimators $\hat{\alpha}_{in}$ which are functions of $X_1, \ldots, X_n$, should be "close" to the actual α_i in some well-defined sense. The subject can be considered as follows. This is a useful application of the Donsker invariance principle which illustrates its use as well as its limitations.

For this, one can use the familiar *principle of least squares*, due to Gauss. This states that, for $n > k$ one should find those $\hat{\alpha}_{in}(\omega)$ for which

$$\sum_{t=1}^{n} \left(X_t - \sum_{i=1}^{k} \alpha_i X_{t-i} \right)^2 (\omega) = \sum_{t=1}^{n} \varepsilon_t^2(\omega)$$

is a minimum for a.a. (ω). This expression is a quadratic form, so it has a unique minimum given by solving the set of equations (with $X_t = 0$ for $t \leq 0$)

$$\sum_{i=1}^{k} \hat{\alpha}_{in} \sum_{t=1}^{n} X_{t-i} X_{t-j} = \sum_{t=1}^{n} X_t X_{t-j}, \quad j = 1, \ldots, k, \text{ a.e.} \tag{2}$$

Assuming that the matrix $(\sum_{t=1}^{n} X_{t-i} X_{t-j}, 1 \leq i, j \leq k)$ is a.e. nonsingular, (2) gives a unique solution to $(\hat{\alpha}_{1n}, \ldots, \hat{\alpha}_{kn})$ by Cramer's rule.

In this setting, the first-order (i.e., k=1) stochastic difference equation (1) (with $\alpha_1 = \alpha$), and the least squares estimator $\hat{\alpha}_n$, illustrates the preceding analysis for obtaining the asymptotic d.f. of the (normalized) errors $(\hat{\alpha}_n - \alpha)$. It will be seen that the limiting probability distributions of these "errors" are different for the cases: $|\alpha| < 1, |\alpha| = 1$ $(\alpha = 1$ and $\alpha = -1)$, as well as $|\alpha| > 1$. There is no "continuity" in the parameter α. It illuminates the *use and limitation* of invariance principles of the preceding sections at the same time.

A comprehensive account of the asymptotic distribution problem can now be given. The solution of the problem for k=1, which is already non-trivial, is given by:

Theorem 1 *Let $\{X_n, n \geq 1\}$ be a sequence of r.v.s satisfying* (1) *with $k = 1$, and the $\{\varepsilon_n, n \geq 1\}$ be i.i.d. with means zero and (for convenience) unit variances. If $\hat{\alpha}_n$ is the least squares estimator of α based on n observations, and $g(n;\alpha)$ is defined by $g(n;\alpha) = [n/(1-\alpha^2)]^{-1/2}$ if $|\alpha| < 1; = n/\sqrt{2}$ if $|\alpha| = 1$; and $= |\alpha|^n(\alpha - 1)^{-1/2}$ if $|\alpha| > 1$, then*

$$\lim_{n \to \infty} P[g(n;\alpha)(\hat{\alpha}_n - \alpha) < x] = \int_{-\infty}^{x} f(u)\, du, \qquad x \in \mathbb{R}, \tag{3}$$

exists, where (i) $f(u) = (1/\sqrt{2\pi})\exp(-u^2/2)$ *if* $|\alpha| < 1$ *[so that the limit is the $N(0,1)$ d.f.],* (ii)$f(u) = [\pi(1+u^2)]^{-1}$ *for* $|\alpha| > 1$ *when the ε_n are also normal (so that the limit is the Cauchy d.f.), and* (iii) *for* $|\alpha| = 1$ *[the ε_n being as in* (ii)*]*

$$f(x) = (8\pi^2)^{-1/2} \int_{\mathbb{R}} \frac{\rho(x,t)}{[r(x,t)]^{3/2}} \cos(\delta(x,t) - \frac{3}{2}\theta(x,t)) \times (\chi_{\mathbb{R}_+^2}(x,t) + \chi_{\mathbb{R}_+^2}(-x,-t)) \frac{dt}{\sqrt{tx}}, \tag{4}$$

where ρ, r, δ, θ are defined by the (complicated) expressions as follows:

$$\rho^2(x,t) = 2\left(1 - \frac{\alpha}{\sqrt{8x^2}}\right)^2 (\sinh^2\sqrt{2tx} + \sin^2\sqrt{2tx}) + \frac{t}{x}(\sinh^2\sqrt{2tx} + \cos^2\sqrt{2tx}) - \alpha\sqrt{t/x}\left(1 - \frac{\alpha}{\sqrt{8x^2}}\right)(\sin\sqrt{8tx} + \sinh\sqrt{8tx}),$$

$$r^2(x,t) = \sinh^2\sqrt{2tx} + \cos^2\sqrt{2tx} + \frac{t}{2x}(\sinh^2\sqrt{2tx} + \sin^2\sqrt{2tx}) + \frac{\alpha}{2}\sqrt{t/x}(\sin\sqrt{8tx} - \sinh\sqrt{8tx}),$$

$$\theta(x,t) = \text{arc } \tan\left\{\frac{1 - (\alpha/2)\sqrt{t/x}\,(\coth\sqrt{2tx} + \cot\sqrt{2tx})}{1 - (\alpha/2)\sqrt{t/x}\,(\tanh\sqrt{2tx} - \tan\sqrt{2tx})} \times \tan\sqrt{2tx}\,\tanh\sqrt{2tx}\right\},$$

$$\delta(x,t) = \text{arc } \tan\left(\frac{C\cos(\alpha t/\sqrt{2}) + D\sin(\alpha t/\sqrt{2})}{D\cos(\alpha t/\sqrt{2}) - C\sin(\alpha t/\sqrt{2})}\right).$$

Here C, D are also functions of x, t, α and are given by:

$$C = \left(1 - \frac{\alpha}{\sqrt{8x^2}}\right)(\sinh\sqrt{2tx}\cos\sqrt{2tx} - \cosh\sqrt{2tx}\sin\sqrt{2tx}) + \alpha\sqrt{t/x}\sinh\sqrt{2tx}\sin\sqrt{2tx},$$

$$D = \left(1 - \frac{\alpha}{\sqrt{8x^2}}\right)(\sinh\sqrt{2tx}\cos\sqrt{2tx} + \cosh\sqrt{2tx}\sin\sqrt{2tx}) - \alpha\sqrt{t/x}\cosh\sqrt{2tx}\cos\sqrt{2tx}.$$

Furthermore, if $|\alpha| > 1$ and the ε_n are also normal, or $|\alpha| < 1$ and the ε_n satisfy the only conditions of (i)(so that they need not be normal) then we have

$$\lim_{n\to\infty} P\left[\sqrt{\left(\sum_{k=1}^{n} X_{k-1}^2\right)}(\hat{\alpha}_n - \alpha) < x\right] = \frac{1}{\sqrt{2\pi}}\int_{-\infty}^{x} e^{-u^2/2}\,du. \tag{4'}$$

In fact, if $|\alpha| > 1$, the limit d.f. of the $(\hat{\alpha}_n - \alpha)$ depends on the d.f. of the ε_n—the "noise," in the model (1) (with k=1) and the invariance principle is inapplicable.

The proof of this result is long and must be considered separately in the three cases: $|\alpha| < 1, |\alpha| = 1$, and $|\alpha| > 1$. This result exemplifies how real life applications demand serious mathematical analysis. Thus it is useful to present some auxiliary results on the way to the proof. The next statement is an extension of the *central limit theorem for "m-dependent" r.v.s* used in proving (i), and the result has independent interest.

A sequence $\{U_n, n \geq 1\}$ of r.v.s is called *m-dependent* ($m \geq 0$ an integer) if $U_1, U_2, \ldots, U_k$ is independent of $U_{\ell+1}, U_{\ell+2}, \ldots$ whenever $\ell - k \geq m$. If $m = 0$, then one has the usual (mutual) independence. We should remark that this unmotivated concept is introduced here, since in many estimation problems of the type considered above such sequences appear. Let us establish a typical result which admits various generalizations. (Cf. Problems 41 and 43.)

Proposition 2 *Let $\{U_n, n \geq 1\}$ be a sequence of m-dependent r.v.s with means zero and variances σ_n^2, but $\sup_n \sigma_n^2 = \sigma^2 < \infty$. Let $S_n = \sum_{k=1}^n U_k$. Then*

$$\lim_{n \to \infty} P[S_n/\sigma(S_n) < x] = (1/\sqrt{2\pi}) \int_{-\infty}^{x} e^{-v^2/2}\, dv, \qquad x \in \mathbb{R}, \tag{5}$$

whenever $\sigma^3(S_n)/n \to +\infty$ as $n \to \infty$.

Proof The condition $\sigma^3(S_n)/n \to \infty$ is always satisfied in the i.i.d. case, since $\sigma^3(S_n) = n^{3/2}[E(U_1^2)]^{3/2}$ with $E(U_1^2) > 0$. Let $k \geq 1$ be an integer to be chosen later, and set $n_j = [jn/k]$, the integral part, $0 \leq j \leq k$. Define two sequences $\{X_j, j \geq 0\}$ and $\{Y_j, j \geq 0\}$ as follows ($j \geq 0$):

$$\begin{aligned} X_j &= U_{n_j+1} + U_{n_j+2} + \cdots + U_{n_{j+1}-m}, \\ Y_j &= U_{n_{j+1}-m+1} + \cdots + U_{n_{j+1}}. \end{aligned}$$

The m-dependence hypothesis implies that $X_1, X_2, \ldots$ are independent r.v.s, and if n is large enough [*e.g.*, if $n > k(2m-1)$], then $Y_1, Y_2, \ldots$ are also independent. Since $\sigma^2(S_n) \to \infty$, one sees that the initial segment of $k(2m-1)$ r.v.s has no influence on the problem because $(X_1 + \cdots + X_{k(2m-1)})/\sigma(S_n) \xrightarrow{P} 0$ even if k is made to depend on n, but it grows "slowly enough". Consider the decomposition

$$S_n = \sum_{j=0}^{k-1} X_j + \sum_{j=0}^{k-1} Y_j = S_n' + S_n'' \quad \text{(say)}. \tag{6}$$

If $U_{i'}$ and $U_{i''}$ are parts of S_n' and S_n'', respectively, then they are uncorrelated unless $|i' - i''| < m$, and thus at most m terms are correlated. Also,

$$E((S_n')^2) = \sum_{j=0}^{k-1} E(X_j^2) \leq km^2\sigma^2,$$

since the X_j are independent and $E(X_j^2) \leq m^2\sigma^2$. Similarly (since the Y_j are also independent) one gets the following estimate:

$$E((S_n'')^2) \le km^2\sigma^2.$$

Hence

$$\begin{aligned}|E(S_n'S_n'')| &\le [E((S_n')^2)E((S_n'')^2)]^{1/2} \qquad \text{(by the CBS inequality)}\\ &\le km^2\sigma^2. \end{aligned} \tag{7}$$

But from (6)

$$\begin{aligned}|E(S_n^2) - E((S_n')^2)| &\le E((S_n'')^2) + 2|E(S_n'S_n'')|\\ &\le km^2\sigma^2 + 2km^2\sigma^2 = 3km^2\sigma^2 \qquad \text{[by (7)]}.\end{aligned} \tag{8}$$

Thus far $k \ge 1$ has been arbitrary, and the additional growth on $\sigma(S_n)$ has not been used. Let $k = k_n = [n^{2/3}]$, so that $\sigma^2(S_n)/k_n \to \infty$. Hence with this choice, (8) gives

$$|1 - \sigma^2(S_n')/\sigma^2(S_n)| \le 3m^2\sigma^2 k_n/\sigma^2(S_n) \to 0 \quad \text{as } n \to \infty.$$

Also [using (7)], $\sigma^2(S_n'')/\sigma^2(S_n) \to 0, E(S_n'S_n'')/\sigma^2(S_n) \to 0$, and then $S_n''/\sigma(S_n) \xrightarrow{P} 0$ (by Proposition 2.3.1), so that (cf. Problem 11—the Cramér-Slutsky theorem—in Chapter 2)

$$\frac{S_n}{\sigma(S_n)} = \frac{\sigma(S_n')}{\sigma(S_n)}\frac{S_n'}{\sigma(S_n')} + \frac{S_n''}{\sigma(S_n)} \stackrel{D}{=} \frac{S_n'}{\sigma(S_n')}. \tag{9}$$

Thus we have reduced the problem to finding the limit d.f. of $S_n'/\sigma(S_n')$ of independent summands which, however, are not identically distributed. For this we now verify Lindeberg's condition (cf. Theorem 5.3.6). If F_j is the d.f. of X_j, then the following should tend to zero. Indeed,

$$\frac{1}{\sigma^2(S_n')}\sum_{j=1}^{k_n}\int_{[|x|>\varepsilon\sigma(S_n')]} x^2\, dF_j(x) \le \frac{1}{\sigma^2(S_n')}\sum_{j=1}^{k_n} \operatorname{Var} X_j \le \frac{k_n m^2\sigma^2}{\sigma^2(S_n)} \to 0$$

as $n \to \infty$, since $\sigma^2(S_n)/k_n \to \infty$ by hypothesis and the choice of k_n. Consequently, by the Lindeberg-Feller theorem, $S_n'/\sigma(S_n') \xrightarrow{D}$ to an r.v. which is $N(0,1)$. This and (9) establish the proposition.

Remark In a note in *Math. Nachr.* 140 (1989), 249-250, P. Schatte has given an example to show that the central limit theorem does not hold for "weakly m-dependent" random sequences. However, his calculations based on conditional moments with evaluations, not satisfying the Conditional Analysis of Chapter 3, do not apply to the situation considered here (or in Chung (1974), Sec. 7.3), contrary to his assertions there. The difficulty may be in his manipulation on conditioning sets of probability zero as explained in our Chapter 3 in detail.

We now turn to the proof of Theorem 1, in stages. First the case $|\alpha| < 1$ is considered.

Proposition 3 *If* $|\alpha| < 1$, *then Theorem 1 is true as stated.*

Proof Consider

$$g(n;\alpha)(\hat{\alpha}_n - \alpha) = \sqrt{n}\sum_{t=1}^{n} \varepsilon_t X_{t-1} \Big/ \sum_{t=1}^{n} X_{t-1}^2 (1-\alpha^2)$$

$$= n^{-1/2}\sum_{t=1}^{n} \varepsilon_t X_{t-1} \Big/ \left(\frac{1}{n}\sum_{t=1}^{n} X_{t-1}^2\right) (1-\alpha^2)$$

$$\stackrel{D}{=} n^{-1/2}\sum_{t=1}^{n} \varepsilon_t X_{t-1}/\sigma^2 \qquad \text{(cf. Problem 38).} \tag{10}$$

[Actually in this calculation we use the result of Problem 11 of Chapter 2, especially the Cramér-Slutsky calculus. The reader should review it now if not already done so.] We now assert that the d.f. of this r.v. tends to $N(0,1)$. In fact, let

$$A_n = \sum_{t=1}^{n} \varepsilon_t X_{t-1} = \sum_{t=2}^{n} \varepsilon_t \varepsilon_{t-1} + \alpha \sum_{t=3}^{n} \varepsilon_t \varepsilon_{t-2} + \cdots + \alpha^{n-2}\varepsilon_n \varepsilon_1. \tag{11}$$

Given $\eta > 0$, choose $m \geq 1$ such that $\alpha^{2(m+1)} < \eta(1-\alpha^2)\sigma^{-4}$. We fix this m and produce an m-dependent sequence whose partial sums have the same limit behavior as that of the A_n. If an r.v. $B_{m,n}$ is defined by

$$B_{m,n} = \sum_{t=2}^{n} \varepsilon_t \varepsilon_{t-1} + \cdots + \alpha^m \sum_{t=m+2}^{n} \varepsilon_t \varepsilon_{t-m-1}, \qquad m < n,$$

then $\mathrm{Var}(A_n - B_{m,n}) \leq (n-m-2)\sigma^4 \alpha^{2(m+1)}/(1-\alpha^2)$. Hence for $n > m$,

$$\mathrm{Var}((A_n - B_{m,n})/\sqrt{n}) \leq \sigma^4 \alpha^{2(m+1)}/(1-\alpha^2) < \eta.$$

Thus $(A_n/\sqrt{n}) \stackrel{D}{=} (B_{m,n}/\sqrt{n})$. Consider $Y_j = \varepsilon_j \sum_{k=0}^{m} \alpha^k \varepsilon_{j-k-1}$. Then

$$n^{-1/2} C_{m,n} = \frac{1}{\sqrt{n}} \sum_{j=m+2}^{n} Y_j \stackrel{D}{=} \frac{B_{m,n}}{\sqrt{n}}, \tag{12}$$

since $E(C_{m,n}) = 0 = E(B_{m,n})$ and

$$\mathrm{Var}\left(\frac{B_{m,n} - C_{m,n}}{\sqrt{n}}\right) = \frac{\sigma^4}{n}(m + \alpha^2(m-1) + \cdots + \alpha^{2(m-1)} \cdot 1) \to 0$$

as $n \to \infty$. Also, in (12) $\mathrm{Var}(C_{m,n}) = n\sigma^4(1-\alpha^{2(m+1)})(1-\alpha^2)^{-1}$, so that $\sigma(C_{m,n})/n^{1/3} \to +\infty$. Thus $\{Y_j, j \geq 1\}$ is an m-dependent sequence satisfying the hypothesis of Proposition 2. Hence (10) and (12) imply that

$$A_n/\sigma^2(1-\alpha^2)n^{1/2} \xrightarrow{D} Z,$$

which is $N(0,1)$. This completes the proof, on setting $\sigma^2 = 1$.

For the cases $|\alpha| \geq 1$, by hypothesis the ε_i are $N(0,1)$. The exact d.f. of $g(n;\alpha)(\hat{\alpha}_n - \alpha)$ can be derived in principle. This is accomplished by calculating the moment-generating function (m.g.f.) of $(X_1, \ldots, X_n)$, and then we shall be able to find the limit m.g.f. of this quantity. By inverting the latter, the limit d.f. of $g(n;\alpha)(\hat{\alpha}_n - \alpha)$ is obtained. Let us fill in the details.

Because the ε_i are independent $N(0,1)$ and $\varepsilon_k = X_k - \alpha X_{k-1}$, the transformation from the ε_i to the X_i is one-to-one with Jacobian unity, and we find that the density of the X_k to be:

$$\begin{aligned} f_{X_1,\ldots,X_n}(x_1,\ldots,x_n) &= (2\pi)^{-n/2}\exp\left\{-(1/2)\sum_{k=1}^{n}(x_k - \alpha x_{k-1})^2\right\} \quad (x_0 = 0) \\ &= (2\pi)^{-n/2}\exp\{-(1/2)(x'Ax)\}, \end{aligned} \tag{13}$$

where A is the n-by-n symmetric positive definite matrix given by $A = (a_{ij})$ with

$$\begin{aligned} &a_{ii} = 1 + \alpha^2, && i = 1, \ldots, n-1 \\ &a_{nn} = 1, \quad a_{i(i+1)} = -\alpha, && i = 1, \ldots, n-1. \end{aligned}$$

and $a_{ij} = 0$ otherwise. Now $x' = (x_1, \ldots, x_n)$ is the row vector (prime for transpose).

On the other hand, $(\hat{\alpha}_n - \alpha)$ can be written (c.f. (1) and (2) with k = 1) as

$$\begin{aligned} \hat{\alpha}_n - \alpha &= \sum_{t=1}^{n}\varepsilon_t X_{t-1}\Big/\sum_{t=1}^{n}X_{t-1}^2 = \sum_{t=1}^{n}(X_t X_{t-1} - \alpha X_{t-1}^2\Big/\sum_{t=1}^{n}X_{t-1}^2 \\ &= (X'BX)/(X'CX), \end{aligned} \tag{14}$$

with $X' = (X_1, \ldots, X_n)$ and B a symmetric n-by-n matrix having 2α on the diagonal except for the nth element, which is zero, -1 for the first line parallel to the diagonal and zeros elsewhere. Here C is an n-by-n identity matrix except for the nth element, which is zero. The joint m.g.f. of the numerator and denominator r.v.s in (14) is given by

$$\begin{aligned} m(u,v) &= E(\exp\{u(X'BX) + v(X'CX)\}) \\ &= (2\pi)^{-n/2}\int_{\mathbb{R}^n}\exp\{-(1/2)(x'Ax) + u(x'Bx) + v(x'Cx)\}\,dx_1\cdots dx_n \\ &= (2\pi)^{-n/2}\int_{\mathbb{R}^n}\exp\{-(1/2)(x'Dx)\}\,dx_1\cdots dx_n, \end{aligned} \tag{15}$$

where $D = A - 2uB - 2vC$, which is positive definite if u, v are sufficiently small, so that the integral in (13) exists. Since D is symmetric, it can be diagonalized: $D = Q'F_\lambda Q$, where Q is an orthogonal matrix, and F_λ is diagonal with eigenvalues $\lambda_j \geq 0$ of D. Setting $y = Q'x$ and noting that the Jacobian in absolute value is $=1$, we get for (15)

$$m(u,v) = (2\pi)^{-n/2} \int_{\mathbb{R}^n} \exp\{-(1/2)(y'F_\lambda y)\} \, dy_1 \cdots dy_n$$

$$= \left(\prod_{j=1}^{n} \lambda_j \right)^{-1/2} = (\det D)^{-1/2}. \tag{16}$$

But if we let $p = 1 + \alpha^2 - 2v + 2\alpha u, q = -(\alpha + u)$, then writing D_n for the D which depends on n, we get by expansion (of $D_n = A_n - 2uB_n - 2vC_n$)

$$D_n = pD_{n-1} - q^2 D_{n-2}, \qquad n > 2, \tag{17}$$

with boundary values $D_1 = 1$ and $D_2 = p - q^2$. Let μ_1, μ_2 be the roots of the characteristic equation of the difference equation (17):

$$\mu^2 - p\mu + q = 0, \qquad \text{so that} \quad \mu_1, \mu_2 = \frac{1}{2}p \pm \frac{1}{2}(p^2 - 4q^2)^{1/2}. \tag{18}$$

Hence

$$D_n = \frac{1 - \mu_2}{\mu_1 - \mu_2} \mu_1^n + \frac{1 - \mu_1}{\mu_2 - \mu_1} \mu_2^n. \tag{19}$$

Substituting (19) in (16), we get the m.g.f. of the r.v.s for each n. As we shall see later, one can employ Cramér's theorem (Theorem 4.3.2) to obtain the *exact distribution* of $\hat{\alpha}_n - \alpha$ without inverting $m(\cdot, \cdot)$ itself. This is still involved. So we turn to the asymptotic result.

For the limiting case, first observe that

$$g(n;\alpha)(\hat{\alpha}_n - \alpha) = \left(\frac{X'BX}{g(n;\alpha)} \right) \Big/ \left(\frac{X'CX}{g^2(n;\alpha)} \right),$$

and so consider the joint m.g.f. obtained from $m(u, v)$ by setting

$$M_n(t_1, t_2) = m\left(\frac{t_1}{g(n;\alpha)}, \frac{t_2}{g^2(n;\alpha)} \right) = (\det \tilde{D}_n)^{-1/2}, \tag{20}$$

where $\tilde{D}_n$ is D with u, v replaced by $t_1/g(n;\alpha)$ and $t_2/g^2(n;\alpha)$, respectively.

We now calculate $\lim_{n\to\infty} M_n(t_1, t_2)$ using the conditions $|\alpha| < 1, = 1$, or > 1 in (20). Recalling the values of $g(n;\alpha)$ and expanding the radical in μ_1, μ_2 of (18), we get the following expressions with an easy computation:

Case 1 $|\alpha| < 1$:

$$\mu_1 = 1 - \frac{t_1^2 + 2t_2}{n} + O(n^{-3/2}),$$

$$\mu_2 = \alpha^2 + 2\alpha t_1[(1-\alpha^2)/n]^{1/2} + \frac{t_1^2 + 2\alpha^2 t_2}{n} + O(n^{-3/2}).$$

Case 2 $|\alpha| > 1$:

$$\mu_1 = 1 + \frac{(t_1^2 + 2t_2)(\alpha^2 - 1)}{\alpha^{2n}} + O(|\alpha|^{-3n}),$$

$$\mu_2 = \alpha^2 + \frac{2\alpha t_1(\alpha^2 - 1)}{|\alpha|^n} - \frac{(t_1^2 + 2\alpha^2 t_2)(\alpha^2 - 1)}{\alpha^{2n}} + O(|\alpha|^{-3n}).$$

Case 3 $|\alpha| = 1$:

$$\mu_1 = 1 + \frac{\sqrt{2}\alpha t_1}{n} + \frac{2i\sqrt{t_2}}{n} + O(n^{-2}),$$

$$\mu_2 = 1 + \frac{\sqrt{2}\alpha t_1}{n} - \frac{2i\sqrt{t_2}}{n} + O(n^{-2}).$$

Substituting these values in (20) and simplifying, one gets

$$\lim_{n\to\infty} M_n(t_1, t_2) = \begin{cases} \exp(t_2 + t_1^2/2) & \text{if } |\alpha| < 1 \\ (1 - 2t_2 - t_1^2)^{-1/2} & \text{if } |\alpha| > 1 \\ \exp\left(-\frac{\alpha t_1}{\sqrt{2}}\right)\left(\cos 2\sqrt{t_2} - \frac{\sqrt{2}\alpha t_1}{2\sqrt{t_2}} \sin 2\sqrt{t_2}\right)^{-1/2} & \text{if } |\alpha| = 1. \end{cases} \tag{21}$$

[The calculations from (17) leading to (21) are due to John S. White.] Replacing t_1, t_2 by it_1 and it_2, we see that the limit ch.f. ϕ is given by

$$\phi(t_1, t_2) = \lim_{n\to\infty} M_n(it_1, it_2). \tag{22}$$

Incidentally, this shows that (corresponding to $|\alpha| > 1$),

$$(t_1, t_2) \mapsto (1 - 2it_2 + t_1^2)^{-1/2}$$

is a ch.f., which we stated (without proof) in Example 4.3.5.

To get the desired limit d.f. of $g(n;\alpha)(\hat{\alpha}_n - \alpha)$, consider

$$\begin{aligned} F(x) &= \lim_{n\to\infty} P[g(n;\alpha)(\hat{\alpha}_n - \alpha) < x] \\ &= \lim_{n\to\infty} P\left[\frac{X'B_nX}{g(n;\alpha)} < x\frac{X'C_nX}{g^2(n;\alpha)}\right] \\ &= \lim_{n\to\infty} P\left[\frac{X'B_nX}{g(n;\alpha)} - x\frac{X'C_nX}{g^2(n;\alpha)} < 0\right] \\ &= P[U - xV < 0] = P[U/V < x], \end{aligned} \tag{23}$$

where

$$\left(\frac{X'B_nX}{g(n;\alpha)}, \frac{X'C_nX}{g^2(n;\alpha)}\right) \xrightarrow{P} (U,V).$$

Hence knowing the ch.f. of (U,V) by (22), we need to find the d.f. of U/V which gives the desired result. Here one invokes Theorem 4.3.2. With that result, we get the density f of F, the d.f. of U/V, as follows:

Case 1 $|\alpha| < 1$:

$$\begin{aligned} f(x) &= \frac{1}{2\pi i}\int_{\mathbb{R}} \frac{\partial \phi}{\partial u}(v, -xv)\, dv \\ &= \frac{1}{2\pi i}\int_{\mathbb{R}} ie^{-(ivx+(v^2/2))}\, dv \quad \text{[by (21) and (22)]} \\ &= \frac{1}{\sqrt{2\pi}}\int_{\mathbb{R}} e^{-ivx-(v^2/2)}\frac{dv}{\sqrt{2\pi}} = \frac{1}{\sqrt{2\pi}}e^{-x^2/2}. \end{aligned}$$

This was seen in Proposition 3, even without the normality of the ε_i. The above simple calculation verifies the result with this specialization, and is included only for illustration. It also shows how invariance principle is at work.

Case 2 $|\alpha| > 1$: In this form, we have already evaluated the integral in Example 4.3.5, and it gives the assertion of the theorem, with ε as $N(0,1)$. Here the limit distribution depends on that of ε_t's and the invariance principle does *not* apply!

Case 3 $|\alpha| = 1$: In this case, one uses Theorem 4.3.2 and evaluates the resulting integrals. It involves similar techniques, though a careful (and a tedious) calculation using some tricks with trigonometric and hyperbolic functions [and $\sqrt{i} = (1+i)/\sqrt{2}$, etc.] is required. We shall omit this here. (It is given as Problem 40.) This then yields the desired result of (4). The invariance principle is applicable here.

It remains to prove (4'). Again note that

$$\lim_{n\to\infty} P\left[\sqrt{\sum_{t=1}^{n} X_{t-1}^2}(\hat{\alpha}_n - \alpha) < x\right] = P[U/\sqrt{V} < x], \tag{24}$$

using the notation of (23). Hence the asserted density is obtainable from Theorem 4.3.2 after getting the ch.f. ψ of $(U/\sqrt{V})$. This is seen by using the inversion formula for (U,V) and simplifying the following where

$$\begin{aligned} \psi(u,v) &= \int\int_{\mathbb{R}^2} e^{iux+iv\sqrt{y}}\left(\frac{1}{(2\pi)^2}\int\int_{\mathbb{R}^2} e^{-irx-isy}\phi(r,s)\,dr\,ds\right)dx\,dy \\ &= \left(\frac{2}{\pi}\right)^{1/2}\int_0^\infty e^{ivt-(t^2/2)(1+u^2)}\,dt. \end{aligned} \tag{25}$$

Here, on substituting for ϕ from (21) [or(22)], one recognizes the ch.f. of the gamma density in the simplification. The easy computation is left to the reader. Then by a straightforward evaluation, one gets

$$g(x) = (1/2\pi i) \int_{\mathbb{R}} (\partial\psi/\partial v)(u, -ux)\, du = (1/\sqrt{2\pi})e^{-x^2/2},$$

which is (4').

In the case that $|\alpha| < 1$, (cf. Problem 38) we have

$$[(1-\alpha^2)/n] \sum_{t=1}^{n} X_{t-1}^2 \to \sigma^2 = 1 \quad \text{in probability.}$$

However, if $|\alpha| > 1$, then a slightly more involved calculation of a similar nature shows that

$$g(n;\alpha)^{-1} \left(\sum_{t=1}^{n} \varepsilon_t X_{t-1} \right) \xrightarrow{D} UV \quad \text{and} \quad g(n;\alpha)^{-2} \sum_{t=1}^{n} X_{t-1}^2 \xrightarrow{P} V^2,$$

where $V > 0$ a.e. Hence $g(n;\alpha)(\hat{\alpha}_n - \alpha) \xrightarrow{D} U/V$; and since V is a positive r.v. (which is not a constant), its d.f. is determined by that of the ε_t. Consequently the limit d.f. of the estimators in this case depends on the initial d.f. of the i.i.d. "errors." This is the substance of the last comment. With this the proof of Theorem 1 is finally finished.

Remarks 4 The result of Theorem 1 for $|\alpha| < 1$ holds for all ε_i as stated there, without specification of their d.f. This is a case of an invariance principle for dependent r.v.s which we have not proved. Also, in the case that $\alpha = 1$ one can apply Donsker's theorem itself, since the functional

$$h(f) = \frac{\int_0^1 f(t)f'(t)dt}{\int_0^1 f^2(t)dt}, \quad f \in C(0,1)$$

and ($f' = df/dx$ is to exist), has for its set of discontinuities zero Wiener measure, and we may apply the corollary of Skorokhod's theorem. If the ε_i have a symmetric distribution, then $\alpha^{t-i}\varepsilon_i = \pm\varepsilon_i$ for $|\alpha| = 1$, and the same reasoning holds. However, if $|\alpha| > 1$, then no invariance principle applies. Even in the best circumstances, the Lindeberg condition fails, and, as Theorem 4.6 (Prokhorov) implies, an invariance principle cannot be expected. The last part of Theorem 1 is an explicit recognition of this situation. An extension of Theorem 1 if X_0 is a constant ($\neq 0$) is possible and is not difficult.

We have not considered the case that $k > 1$ in (1) here. It is not a simple extension of the above work, but needs additional new ideas and work. A brief discussion was included in the earlier edition. We omit its consideration now.

Exercises

1. Let $X_1, X_2, \ldots$ be independent r.v.s such that $P[X_k = k^{3/2}] = P[X_k = -k^{3/2}] = 1/2k$ and $P[X_k = 0] = 1 - k^{-1}$. If $S_n = \sum_{k=1}^{n} X_k$, verify that the sufficiency condition $\rho(S_n)/\sigma(S_n) \to 0$ as $n \to \infty$ is not satisfied in Theorem 1.3. Show, however, that $S_n/\sigma(S_n) \xrightarrow{D} S$, which is *not* $N(0,1)$.

2. For the validity of Liapounov's theorem existence of $2+\delta$ moments is sufficient, where $0 < \delta \leq 1$. Establish Theorem 1.3 by modifying the given proof, with the condition that $(\sum_{k=1}^{n} E(|X_k|^{2+\delta})/\sigma(S_n)^{2+\delta}) \to 0$ as $n \to \infty$. [*Hint*: Note that $|e^{ix} - 1| \leq 2^{1-\delta}|x|$, and hence for the ch.f. ϕ_k of X_k with $E(X_k) = 0$, one has

$$\phi_k(t) = 1 - \frac{t^2}{2}E(X_k^2) + \frac{2^{1-\delta}\theta |t|^{2+\delta} E(|X_k|^{2+\delta})}{(1+\delta)(2+\delta)}, \qquad |\theta| \leq 1.]$$

3. Let $\{X_n, n \geq 1\}$ be independent r.v.s such that $P[X_k = k^{\alpha}] = \frac{1}{2} = P[X_k = -k^{\alpha}]$. Let $S_n = \sum_{k=1}^{n} X_k$. Show that if $\alpha > -\frac{1}{2}$, then $S_n/n^{\alpha+(1/2)} \xrightarrow{D} Y$, and if $\alpha = \frac{1}{2}, P[|Y| < y] = (4/\pi)^{1/2}\int_0^y e^{-u^2}\,du, y > 0$. Deduce that if $\alpha = \frac{1}{2}$, WLLN does not hold for this sequence X_k. [Use Euler's formula for the asymptotic expressions, $\sum_{k=1}^{n} k^m \sim n^{m+1}/(m+1)$ for $m > 0$.]

4. Just as in Problem 2, show that the Berry-Esséen result can be given the following form if only $2+\delta, 0 < \delta \leq 1$, moments exist (and the corresponding $\rho/\sigma \to 0$): If the $X_n, n \geq 1$, have $2+\delta$ moments and the r.v.s are independent, let $S_n = \sum_{k=1}^{n} X_k, \rho^{2+\delta}(S_n) = \sum_{k=1}^{n} E(|X_k|^{2+\delta})$ and $\sigma^2(S_n) = \text{Var } S_n > 0$. Then there exists an absolute constant $C_0[= C_0(\delta)]$ such that if $G(\cdot)$ is $N(0,1)$,

$$\sup_{x \in \mathbb{R}} |P[(S_n - E(S_n))/\sigma(S_n) < x] - G(x)| \leq C_0(\rho(S_n)/\sigma(S_n))^{2+\delta}.$$

5. Let $\{X_n, n \geq 1\}$ be i.i.d. Bernoulli r.v.s, $P[X_1 = +1] = P[X_1 = -1] = \frac{1}{2}$, so that $P[S_n = k] = \binom{n}{k}(1/2)^n, S_n = \sum_{k=1}^{n} X_k$. Using Stirling's approximation, $n! = \sqrt{2\pi} n^{n+(1/2)} \cdot e^{-n+\theta_n}$, where $|\theta_n| < 1/12n$, and letting $x_k = (k - n/2)/\sqrt{n/4}$, show that, for each $-\infty < a \leq x_k \leq b < \infty$, we have as $n \to \infty$,

$$|P[a\sqrt{n/4} \leq S_n < b\sqrt{n/4}] - 1/\sqrt{2\pi}\int_a^b e^{-x^2/2}\,dx| = O(n^{-1/2}),$$

so that the *order* of the error in Theorem 1.5 cannot be improved.

6. Let $\{X_n, n \geq 1\}$ be i.i.d. random variables with a common d.f. whose density f is given by

$$f(x) = \begin{cases} 0 & \text{if } |x| < 2 \\ K(x^2 \log |x|)^{-1} & \text{if } |x| \geq 2, \end{cases}$$

where

$$\frac{1}{2K} = \int_2^\infty \frac{dx}{x^2 \log x}.$$

If $S_n = \sum_{i=1}^n X_i$, show that $(\log n/n)S_n \xrightarrow{D} S$ as $n \to \infty$, and find the d.f. of S. [Hint: Use ch.f.'s and the continuity theorem.]

7. If in the above example the common d.f. has the following density:

$$g(x) = K(1 + x^2)^{-m}, \qquad m \geq 2,$$

find the normalizing sequence $a_n > 0$ such that $S_n/a_n \xrightarrow{D} S$ as $n \to \infty$. Does $a_n = [n/(2m-1)]^{1/2}$ work? What is the d.f. of S? Discuss the situation when $m = 1$ (the Cauchy distribution).

8. Give a proof, based on ch.f.s, of Proposition 2.2(b): If X is a bounded nondegenerate r.v., then it cannot be infinitely divisible.

9. In contrast to the assertion of Corollary 2.4, if ϕ is a ch.f. which is not infinitely divisible (but nonvanishing), then $\phi^\lambda [= \exp(\lambda \text{ Log } \phi)]$ need not be a ch.f. for $\lambda > 0$, where λ is not an integer but an arbitrary positive real number. The following two ch.f.s illustrate this. (a) Let $\phi(t; n) = (q + pe^{it})^n, p + q = 1, \frac{1}{2} < q < 1$. Show that for some real $\lambda > 0, \phi(\cdot; \lambda)$ is not a ch.f. (b) A stronger assertion is as follows. Let $\phi(\cdot, \cdot, m, n)$ be a bivariate ch.f., given for $|\rho_{12}| < 1$ by

$$\begin{aligned} \phi(t_1, t_2; m, n) = {} & (1 - it_1)^{-mn/2}(1 - it_2)^{-m/2} \\ & \times [1 - \rho_{12}^2 t_1 t_2/(1 - it_1)(1 - it_2)]^{-m/2}. \end{aligned}$$

This is known to be a ch.f. Consider (for $\rho_{12} \neq 0$)

$$\begin{aligned} \tilde{\phi}(t_1, t_2; M, N, R) = {} & (1 - it_1)^{-M}(1 - it_2)^{-N} \\ & \times [1 - \rho_{12}^2 t_1 t_2/(1 - it_1)(1 - it_2)]^{-R}, \end{aligned}$$

where $M, N, R > 0$ are real numbers. Verify, on assuming that it is in fact a ch.f. and inverting, that its "density function" exists for $|\rho_{12}| < 1$, but is negative on a set of positive Lebesgue measure, if $R\rho_{12}^2 > \min(M, N)$. Thus $\tilde{\phi}$ is not a ch.f. for a continuum of values of M, N, R. [This is not a trivial problem, and it is first observed by W. F. Kibble. Note that $\tilde{\phi}$ would have been a ch.f. if ϕ were infinitely divisible, and the conclusion implies that it cannot be infinitely divisible.]

10. Let $\{X_{nk}, 1 \leq k \leq k_n, n \geq 1\}$ be rowwise independent r.v.s with $S_n = \sum_{k=1}^{k_n} X_{nk}, E(X_{nk}) = 0$. Suppose that $E(|S_n|^{4+\delta}) \leq K_0 < \infty$ for a $0 < \delta \leq 1$ (so that each X_{nk} has $4+\delta$ moments by Problem 2 in Chapter 2). Suppose also that Var $X_{nk} \to 0$ uniformly in k as $n \to \infty$ (essentially infinitesimal) and Var $S_n \to \sigma^2 < \infty$. Show that $S_n \xrightarrow{D} S$, which is $N(0, \sigma^2)$ iff $E(S_n^4) - 3[E(S_n^2)]^2 \to 0$ as $n \to \infty$. [*Hints*: Use Proposition 4.1.3 in one direction; for the converse, infinitesimality implies that when $S_{n_k} \xrightarrow{D} S$ for any subsequence, S must be infinitely divisible. Then the given moment condition implies that in the Kolmogorov pair $(\alpha, K), \alpha = 0$, and K must have a jump at $x = 0$ of size σ^2 and be constant elsewhere.]

11. Let $\{S_n, n \geq 1\}$ be partial sums as in the preceding problem and with 4 moments, but assume that each S_n is infinitely divisible. (This is the tradeoff for $\delta = 0$ in the above.) Then $S_n \xrightarrow{D} S$ and S in $N(0, \sigma^2)$ iff $E(S_n^2) \to \sigma^2$ and $E(S_n^4) - 3[E(S_n^2)]^2 \to 0$ as $n \to \infty$. (Since S is necessarily infinitely divisible now, one can proceed as in the last problem again. The point of these two problems is that the conditions are only on the moments and not on the d.f.s themselves. These observations are due to P. A. Pierre (1971).)

12. Establish the Lévy form of the representation of an infinitely divisible ch.f. ϕ; i.e., prove formula (20) of Section 2, or Theorem 2.6, in complete detail.

13. (a) Let ϕ be an infinitely divisible ch.f. and (γ, G) be its Lévy-Khintchine pair. Then show that $\phi^{(2k)}$, the (2k)th derivative, exists iff $\int_{-\infty}^{\infty} x^{2k}$ $dG(x) < \infty$. (This is a specialization of Proposition 4.2.6.) (b) Let ψ be the ch.f. of a d.f. Show that $\phi = \exp(\psi - 1)$ is an infinitely divisible ch.f. Hence, using the remark after Theorem 4.2.4, conclude that $\phi_n \to \tilde{\phi} \neq 0, \phi_n$ infinitely divisible ch.f. $\not\Rightarrow \tilde{\phi}$ is a ch.f. [But if $\tilde{\phi}$ is a ch.f. then it must necessarily be infinitely divisible, by Proposition 2.3.]

14. Let X be a gamma r.v. with density $f_{\alpha\beta}$ given by

$$f_{\alpha\beta}(x) = \beta^{\alpha} x^{\alpha-1} e^{-\beta x} / \Gamma(\alpha) \quad \text{if } x > 0, \quad \alpha > 0, \quad \beta > 0,$$

and =0 otherwise. Show that X is infinitely divisible and the Lévy-Khintchine pair (γ, G) of its ch.f. is given by

$$\gamma = \alpha \int_0^{\infty} e^{-\beta x} \frac{dx}{1+x^2}, \quad G(x) = \begin{cases} \alpha \int_0^x \frac{u}{1+u^2} e^{-\beta u} \, du & \text{for } x > 0 \\ 0 & \text{for } x \leq 0. \end{cases}$$

(First define γ_n, G_n as in the proof of Theorem 2.5 and then obtain γ, G by a limit process.)

15. For the same d.f. as in the preceding problem, show that the corresponding Kolmogorov and Lévy pairs are (γ, K) and (γ, σ^2, M, N) where

$$\text{(i) } \gamma = \alpha/\beta, \quad K(x) = \begin{cases} \alpha \int_0^x ue^{-\beta u}\, du & \text{for } x > 0 \\ 0 & \text{for } x \leq 0. \end{cases}$$

$$\text{(ii) } \gamma = \alpha \int_0^\infty e^{-\beta x} \frac{dx}{1+x^2}, \qquad M(x) \equiv 0, \quad x \in \mathbb{R},$$

$$N(x) = -\alpha \int_x^\infty e^{-\beta u} \frac{du}{u}, \quad x > 0.$$

16. The following interesting observation is due to Khintchine on a relation between the Riemann zeta function and the theory of infinitely divisible ch.f.s. Recall that the zeta function $\zeta(\cdot)$ is defined by the following sum (and the Euler product):

$$\zeta(s) = \sum_{n=1}^\infty n^{-s} = \prod_{i \geq 1} (1 - p_i^{-s})^{-1}, \qquad s = \sigma + iu, \quad \sigma > 1,$$

where the product is extended over all prime numbers $p_i > 1$. If ϕ is defined as $\phi(t) = \zeta(\sigma + it)/\zeta(\sigma)$, show that $\phi(\cdot)$ is an infinitely divisible ch.f. (Note that ϕ never vanishes, and Log ϕ can be expressed, with the Euler product, as a limit of suitable Poisson ch.f.s)

17. The definition of infinite divisibility may be stated in an apparently weaker form. Since X is infinitely divisible iff for each $n, X = \sum_{k=1}^n X_{nk}$, where X_{nk} are i.i.d. for $1 \leq k \leq n$, we may relax the identically distributedness as follows. Say X is infinitely divisible in the "generalized sense" iff for each $\varepsilon > 0$, there exists an n $[=n(\varepsilon)]$ and independent r.v.s $X_k, 1 \leq k \leq n$, such that $X = \sum_{k=1}^n X_k$ and $P[|X_k| \geq \varepsilon] \leq \varepsilon, 1 \leq k \leq n$. Alternatively, if ϕ is the ch.f. of X, then $\phi = \prod_{k=1}^n \phi_k$, where ϕ_k is the ch.f. of X_k with $|1 - \phi_k(t)| < \varepsilon$ for $|t| \leq 1/\varepsilon$, so that both n and the ϕ_k depend on ε. We then have the following assertion: The r.v. X (or its ch.f. ϕ) is infinitely divisible in the generalized sense iff it is infinitely divisible in the sense of Definition 2.1 [The method of proof uses several estimates on ch.f.s. That the ordinary $\Rightarrow$ generalized sense is easy. For the reverse implication, note that ϕ never vanishes, and the proof proceeds by centering the r.v.s at their medians and writing:

$$\text{Log } \phi(t) = i\gamma_\varepsilon t + \sum_{j=1}^n [\tilde{\phi}_j(t) - 1] + o(1),$$

$$\tilde{\phi}_j(t) = E(\exp\{it(X_j - m_j)\}) \quad \text{with } m_j \text{ as a median of } X_j,$$

so that

$$\begin{aligned}\text{Log } \phi(t) &= i\gamma_\varepsilon t + \int_{\mathbb{R}} (e^{itx} - 1) \sum_{j=1}^{n} d\tilde{F}_j(x) + o(1) \\ &= it\left(\gamma_\varepsilon + \int_{\mathbb{R}} \frac{x}{1+x^2}\right) d\tilde{F}_\varepsilon(x) \\ &\quad + \int_{\mathbb{R}} \left(e^{itx} - 1 - \frac{itx}{1+x^2}\right) \frac{1+x^2}{x^2}\, dG_\varepsilon(x) + o(1)\end{aligned}$$

for suitable $\gamma_\varepsilon, \tilde{F}_\varepsilon, G_\varepsilon$. Since $\varepsilon > 0$ is arbitrary, let $\varepsilon \searrow 0$, and show that the right side gives the Lévy-Khintchine formula, and hence the converse. The details need care as in the necessity proof of Theorem 1.5. This result is due to Doob (1953). Because of this result, "in the generalized sense" is omitted.]

18. Let X be an r.v. with density function f_n given by

$$f_n(x) = \begin{cases} |x|^{-3} n^{-3}, & x < -1/n \\ (2n-3)/4, & |x| \le 1/n \\ 0, & 1/n < x < 1 - (1/n) \\ 1/4, & 1 - (1/n) \le x < 1 + (1/n) \\ (x-1)^{-3} n^{-3}, & x \ge 1 + (1/n). \end{cases}$$

If X_{nk} is distributed as X for each $1 \le k \le n$, show that, with Proposition 3.2 or otherwise, the $X_{nk}, 1 \le k \le n$, are infinitesimal. Let $S_n = \sum_{k=1}^{n} X_{nk}$. Ascertain the truth of the following: $S_n \xrightarrow{D} S$ as $n \to \infty$ and S is a Poisson r.v. with parameter $\lambda = 1$. [Even though the result can be established by finding the limit of the ch.f. of S_n by a careful estimation of various integrals, it is better to use Theorem 3.8 after verifying that the Lévy-Khintchine pair (γ, G) for a Poisson d.f. is given by $\gamma = \lambda/2 : G(x) = \lambda/2$ if $x \ge 1, = 0$ if $x < 1$.]

19. The theory of infinitely divisible d.f.s in higher dimensions proceeds by essentially the same arguments as in $\mathbb{R}$. Thus the (integral) representation of the ch.f. ϕ of a p-dimensional r.v. is given by Lévy's form (by P. Lévy himself) as

$$\phi(t) = \exp\left\{ i\gamma' t - \frac{1}{2} t' \Gamma t + \int_{\mathbb{R}^p} \left(e^{it'x} - 1 - \frac{it'x}{1+x'x}\right) \nu(dx) \right\},$$

for $t \in \mathbb{R}^p$, where $t = (t_1, \ldots, t_p)'$ (prime for transpose) is the column vector, $\gamma' = (\gamma_1, \ldots, \gamma_p)$ with $\gamma_i \in \mathbb{R}, (\gamma' t = \sum_{i=1}^{p} \gamma_i t_i), \Gamma = (\sigma_{ij})$ is a $p \times p$ positive (semi-)definite matrix of real numbers, and v is a measure on the Borel sets of $\mathbb{R}^p$ such that

$$\int_{[x'x<1]} x'x\, \nu(dx) < \infty, \qquad \int_{[x'x \ge 1]} \nu(dx) < \infty.$$

The factor $\exp\{i\gamma' t - \frac{1}{2} t' \Gamma t\}$ is called the *Gaussian component* of ϕ and the rest the *generalized Poisson component.* The corresponding Kolmogorov form, if X has two moments finite, is given by

$$\phi(t) = \exp\left\{i\tilde{\gamma}'t - \frac{1}{2}t'\Gamma t + \int_{\mathbb{R}^p}(e^{it'x} - 1 - it'x)(x'x)^{-1}K(dx)\right\}, \quad t \in \mathbb{R}^p,$$

where $K(\{0\}) = 0$, and the measure K is finite on the Borel σ-algebra of $\mathbb{R}^p$. Here the vector $\tilde{\gamma}$ is given by

$$\tilde{\gamma} = \gamma + \int_{\mathbb{R}^p} x(1 + x'x)^{-1}K(dx).$$

Verify these two forms following closely the proof in the one-dimensional case. In the finite variance case, show that the mean and covariance matrices are

$$E(X) = \tilde{\gamma}, \quad E(XX') = \Gamma + \int_{\mathbb{R}^p}(xx')(x'x)^{-1}K(dx).$$

Deduce that if X is a p-vector which is an infinitely divisible r.v. and $a \in \mathbb{R}^p$, then $Y = a'X$ is again infinitely divisible. [Many of these limit theorems admit immediate multidimensional extensions but not automatically. The next problem shows that the converse of the last statement is not always true.]

20. We now present an example, also due to P. Lévy, showing that the converse of the last statement of the preceding problem is not true. Further, this result has additional interesting information. Let X, Y be independent $N(0,1)$ r.v.s and $\xi = X^2, \eta = 2XY$ and $\zeta = Y^2$. Then show that any *two* of the three r.v.s (ξ, η, ζ) have joint d.f. which is infinitely divisible, but that the joint d.f. of (ξ, η, ζ) is not. Deduce that if each linear combination $\sum_{i=1}^{p} a_i\xi_i$ is infinitely divisible $(a_1, \ldots, a_p) \in \mathbb{R}^p$, then one cannot conclude that the vector $(\xi_1, \ldots, \xi_p), p > 1$, must be infinitely divisible. [*Sketch*: By the image law theorem, the ch.f. ϕ of (ξ, η, ζ) is

$$\begin{aligned}\phi(t_1, t_2, t_3) &= \frac{1}{2\pi}\int\int_{\mathbb{R}^2} \exp\left\{it_1x^2 + 2it_2xy + it_3y^2 - \frac{x^2}{2} - \frac{y^2}{2}\right\} dx\, dy \\ &= [(1 - 2it_1)(1 - 2it_3) + 4t_2^2]^{-1/2}.\end{aligned}$$

Since ξ, ζ are independent and each is infinitely divisible being gamma (or Chi-square variables), so is the pair (ξ, ζ). It suffices to verify that (ξ, η) is also. From the Lévy canonical form for ξ, one gets

$$-\log(1 + 4t^2) = 2\int_0^\infty (\cos 2tv - 1)\exp\{-v\}\, dv/v. \tag{1}$$

Note that

$$(1 - iu + t^2)^{-1} = \int_0^\infty e^{iux - x(1+t^2)}\, dx,$$

and integrating relative to u from 0 to x, rearranging, and using the ch.f. of $N(\mu, \sigma^2)$ suitably, we get

$$-\text{Log}\left(\frac{1+ix+t^2}{1+t^2}\right) = \frac{1}{2\sqrt{\pi}}\int_0^\infty du \int_{\mathbb{R}} (e^{i(xu+tv)} - 1)e^{-u-v^2/(4u)}u^{-3/2}\, dv$$
$$- \frac{1}{2\sqrt{\pi}}\int_{\mathbb{R}} (e^{itv} - 1)\, dv \int_0^\infty e^{-u-v^2/(4u)}u^{-3/2}\, du. \tag{2}$$

The last integral in (2) is known to be $= v^{-1}e^{-v/2}$. Substituting this in (2) and using (1), one has

$$-\text{Log}(1 - ix + 4t^2) = \int_0^\infty du \int_{\mathbb{R}} (e^{i(xu+2tv)} - 1)f(u,v)\, dv$$
$$+ \int_{\mathbb{R}} (e^{2itv} - 1)g(v)\, dv, \tag{3}$$

where

$$f(u,v) = \frac{1}{2\sqrt{\pi}} \exp\{-u - v^2/(4u)\}u^{-3/2}$$

and $g(v) = |v|^{-1}e^v$ if $v < 0, = v^{-1}(e^{-v} - e^{-v\sqrt{2}})$ if $v > 0$. Comparing this with the Lévy representation of Problem 19, we get that (ξ, η) is infinitely divisible. From this one concludes that $a_1\xi + a_2\eta + a_3\zeta$ is also infinitely divisible for any $(a_1, a_2, a_3) \in \mathbb{R}^3$.

To see that (ξ, η, ζ) itself is not infinitely divisible, suppose the contrary. Then, by the representation of ϕ (cf. Problem 19), since ξ, η, ζ depend on the infinitely divisible r.v.s without Gaussian components, Log ϕ is a sum of terms of the form $\int_{\mathbb{R}} (e^{it'x} - 1)\nu(dx)$, each a Stieltjes integral. Looking at terms for $t_1 = 0, t_2 = 0$, or $t_3 = 0$, one concludes that this is the exponent of the generalized Poisson part, and thus it must be a sum of functions of a form that depends on $(t_1, t_2), (t_2, t_3)$, and (t_1, t_3) alone. Hence, in particular, it must satisfy (ξ, η, ζ having all moments finite)

$$\frac{\partial^2 \text{ Log } \phi(t_1, t_2, t_3)}{\partial t_1\, \partial t_3} = 0.$$

But the ϕ of our example does not satisfy this, so that it cannot be infinitely divisible. Thus there are more surprises than one ordinarily expects from these r.v.s. Incidentally, this implies that the distribution of the sample covariance matrix of a random sample from a multivariate normal r.v., called the *Wishart distribution,* is not infinitely divisible. Also compare this with the pathological example given by Problem 9.]

21. Let $X = (X_1, \ldots, X_p)$ be a p-vector which is infinitely divisible. If X has four moments finite, $E(X) = 0$, and ϕ is its joint ch.f., let

$$q(X_i, X_j) = \frac{\partial^4 \text{ Log } \phi}{\partial t_i^2\, \partial t_j^2}\Big|_{t_i=0=t_j}, \quad 1 \le i,\ j \le p.$$

Show that $q(X_i, X_j) \geq 0$. Verify that if all the X_i are bounded below by a constant (so that its ch.f. has no Guassian component), then the X_i are independent whenever they are uncorrelated. Without the boundedness hypothesis (using the Kolmogorov representation) show that the components $X_1, \ldots, X_p$ are independent iff $X_1^2, \ldots, X_p^2$ are uncorrelated. [*Hints*: Since $q(X_i, X_j) \geq 0$, the uncorrelatedness of the X_i^2 implies that $\sum_{i \neq j} q(X_i, X_j) = 0$. Thus the measure $K(\cdot)$ must concentrate on the axes $x_i = 0, i = 1, \ldots p$, and implies the uncorrelatedness of X_i and $X_j, i \neq j$. The converse is clear. Several special properties of such r.v.s with four moments have been discussed by Pierre (1971).]

22. If $X = (X_1, \ldots, X_p)$ is an infinitely divisible random vector, show, by using the Lévy form of Problem 19, that $X_1, \ldots, X_p$ are mutually independent iff they are pairwise independent. [This is also due to Pierre (*loc.cit.*) with finite variances so that Kolmogorov's form is applicable. The idea is essentially the same even here since one verifies that the Lévy measure must concentrate on the axes. In general the finiteness of various integrals must be verified as in the proof of the Lévy (or the Lévy-Khintchine) representation.]

23. Complete details of the remark after Proposition 3.13 regarding the case of the characteristic exponent $\alpha = 2$ (namely, ϕ is nondegenerate, $\alpha = 2 \Rightarrow \phi$ is a normal ch.f.)

24. This problem and the next show that the bounds on α, β in Theorem 3.16 are necessary and complete the proof there (at least for the case $0 < \alpha \leq 2, \alpha \neq 1$). Let us write the ch.f. ϕ of Eq. (44) of Section 3 as follows:

$$\phi_{\alpha,\beta}(t) = \exp\{iat - ct^{\alpha}(1 - i\beta)\},$$

where $t > 0, 0 < \alpha \leq 2, \alpha \neq 1$, and $c > 0$. It is to be shown that $\phi_{\alpha,\beta}$ is a ch.f. only if $0 < \alpha \leq 2$ and $|\beta| < |\tan(\pi\alpha/2)|$. For convenience, express $\beta = \tan \pi\gamma/2, |\gamma| < 1$, and thus for the above it suffices to establish that (since $\phi_{\alpha,\beta}$ is a ch.f. iff $\psi_{\alpha,\gamma}$ defined below is) the following is a ch.f.:

$$\psi_{\alpha,\gamma}(t) = \exp\{-t^{\alpha} e^{i\pi\gamma/2}\}, \qquad t > 0.$$

Let $\psi_{\alpha,\gamma}$ be a ch.f., and consider the density

$$p(x : \alpha, \gamma) = \int_{\mathbb{R}} e^{-itx} \psi_{\alpha,\gamma}(t)\, dt, \qquad \alpha > 0, \quad |\gamma| < 1;$$

using $\phi_{\alpha,\beta}$, consider the density

$$g(x; \alpha, \beta) = \int_{\mathbb{R}} e^{-itx - |t|^{\alpha}(1 \pm i\beta)}\, dt, \qquad \alpha > 0,$$

and $\pm\beta$ depending on whether $t > 0$ or $t < 0$.

(a) Verify that $g(x;\alpha,\beta) = g(-x;\alpha,-\beta)$, and if $\frac{1}{3} \le \alpha < 1$ and $\beta > \tan(\pi\alpha/2)$, then $g(0;\alpha,\beta) < 0$. For this note that

$$g(0;\alpha,\beta) = 2\int_0^\infty e^{-t^\alpha}\cos(\beta t^\alpha)\,dt = (2/\alpha)\Gamma(1/\alpha)\,\mathrm{Re}(1-i\beta)^{-1/\alpha},$$

by a known integral formula. Writing $1 - i\beta = re^{-i\theta}, 0 < \theta < \pi/2$, so that $g(0;\alpha,\beta) = (2/\alpha)\Gamma(1/\alpha)r^{-1/\alpha}\cos(\theta/\alpha)$, deduce that $\beta > \tan(\pi\alpha/2)$ implies the assertion.

(b) If $0 < \alpha \le 1/3, \tan(\pi\alpha/2) < \beta < \tan(3\pi\alpha/2)$, then again $g(0;\alpha,\beta) < 0$. Indeed, in the last expression for $g(0;\alpha,\beta)$ use the fact that the limits for β imply that $\pi\alpha/2 < \theta < 3\pi\alpha/2$, and hence $\cos(\theta/\alpha) < 0$. Thus the assertion follows.

(c) Using (a) and (b) and the fact that g is a continuous function, deduce that it cannot be a density, so that $\phi_{\alpha,\beta}$ for $0 < \alpha < 1, |\beta| > \tan(\pi\alpha/2)$, is not a ch.f.

25. (Continuation) The result for $1 < \alpha < 2$ is somewhat more involved. First verify that $p(x:\alpha,\gamma) = p(-x;\alpha,-\gamma)$ as before, since $\overline{\psi}_{\alpha,\gamma}(t) = \psi_{\alpha,\gamma}(-t)$.

(a) Let $0 < \alpha, -1 < \gamma < \min(2\alpha - 1, 1)$; then for each $x > 0$

$$p(x:\alpha,\gamma) = x^{-(1+\alpha)}p(x^{-\alpha}:1/\alpha,\gamma^*), \qquad \gamma^* = (\gamma + 1 - \alpha)/\alpha,$$

for, on integrating by parts, note that [with a change of variable $u = (tx)^\alpha b$]

$$p(x:\alpha,\gamma) = -2x^{-(1+\alpha)}\mathrm{Im}\left(\int_0^\infty f(u)\,du\right),$$

where $f(z) = \exp\{i\pi\gamma/2 - iz^{1/\alpha} - zx^{-\alpha}e^{\pi i\gamma/2}\}$, with $z = u + iv$. Evaluate the integral by the residue theorem along a sector of angle θ such that $0 < \theta < (\pi/2)(\gamma+1)$ and $\theta/\alpha < \pi$. Since $-1 < \gamma < \min(2\alpha - 1, 1)$, this sector is nonempty. Setting $z = re^{-i\theta}, 0 < r < \infty$, and noting that

$$\lim_{R\to\infty} = \int_{\Gamma_R} f(z)\,dz = 0,$$

where $\Gamma_R = \{z = Re^{-iu} : 0 \le u \le \theta\}$, we get

$$\int_0^\infty f(u)\,du = -i\int_0^\infty \exp\{ix^{-\alpha}r - r^{1/\alpha}e^{i\pi(1-2\theta/\pi\alpha)/2}\}\,dr.$$

From this the desired relation follows after a computation.

(b) With the result of (a), show that for $1 < \alpha < 2, \psi_{\alpha,\gamma}$ or $\phi_{\alpha,\beta}$ is not a ch.f. when $|\beta| > |\tan(\pi\alpha/2)|$, for, letting $\beta = \tan(\pi\gamma/2)$, with $|\gamma| < 1$, this is equivalent to $|\gamma| > (2 - \alpha)$. If $-1 < \gamma < 2 - \alpha < 1$, then (a) is applicable. If $\beta^* = \tan\pi\gamma^*/2$, where γ^* is as in (a), then

$|\beta^*| > \tan(\pi/2\alpha)$ iff $|\gamma^*| > \alpha^{-1} > 2^{-1}$. But if $|\gamma^*| > \alpha^{-1}$, then, by Problem 24a, $g(0; \alpha^{-1}, \beta^*) < 0$. Thus $p(0 : \alpha^{-1}, \gamma^*) < 0$. Hence if $\beta < -|\tan(\pi\alpha/2)|$, then $\gamma < \alpha - 2 \Rightarrow \gamma^* < -1/\alpha$, and so $p(x : \alpha, \gamma) < 0$ as $x \to \infty$. By the continuity of p, this again shows that $\phi_{\alpha,\beta}$ is not a ch.f. However, the set $\{\beta : \phi_{\alpha,\beta} = \text{a ch.f.}\}$ is convex, and symmetric around the origin. This shows that if $\beta > |\tan(\pi\alpha/2)|$, then $\phi_{\alpha,\beta}$ is not a ch.f., and completes the necessity ($\alpha = 2$ being immediate, since then $\beta = 0$). [This argument, which did not use the theory of infinitely divisible d.f.s, is due to Dharmadhikari and Sreehari (1976).]

(c) We also have a multivariate formulation of stability again due to P. Lévy (1937). A random vector $\boldsymbol{X} = (X_1, \ldots, X_n)$ is symmetric stable of type $\alpha, 0 < \alpha \leq 2$, if its ch.f. is expressible as:

$$\begin{aligned}\varphi_{\boldsymbol{X}}(t_1, \ldots, t_n) &= E(e^{i\sum_{i=1}^{n} t_j X_j}) \\ &= \exp\left\{-\int_S |t_1\lambda_1 + \ldots + t_n\lambda_n|^\alpha dG(\lambda_1, \ldots, \lambda_n)\right\}, t_i \in \mathbb{R},\end{aligned}$$

where G is a (σ-finite) measure on $(\mathbb{R}^n, \mathcal{B})$ whose support is on the surface of the unit sphere S of $\mathbb{R}^n$. Verify that this reduces to Theorem 3.16 for $n = 1$. If we define a functional $||\cdot||_\alpha$ as

$$||X_1 + \ldots + X_n||_\alpha = \left[\int_S |\lambda_1 + \ldots + \lambda_n|^\alpha dG(\boldsymbol{\lambda})\right]^{1\wedge\frac{1}{\alpha}}, \quad 0 < \alpha \leq 2,$$

then show that it is a metric on all such random variables (and a norm if $\alpha \geq 1$). [This is a standard Minkowski metric, cf. e.q. Rao (1987, or 2004), Theorem 4.5.4 and Proposition 4.5.6.] The result shows that each linear combination of X_j ($\sum_{i=1}^{u} a_j X_j$) is a symmetric α-stable random variable (e.g., take $a_i = b_i u$ in the above). Show that if one defines on the measure space $(\mathbb{R}, \mathcal{B}, G)$, [$\mathcal{B}$ being Borel σ-algebra] a probability measure R whose Fourier transform $\hat{R}$ is given by

$$\hat{R}_{A_1,\ldots,A_u}(\boldsymbol{u}) = \exp\left\{-\left|\left|\sum_{j=1}^{n} u_j \mathcal{X}_{A_j}\right|\right|_\alpha\right\}, \quad \boldsymbol{u} = (u_1, \ldots, u_n) \in R^n, A_j \in \mathcal{B}_0$$

where $\mathcal{B}_0 = \{B \in \mathcal{B}, G(B) < \infty\}$, then we have

$$\hat{R}_A(t) = \exp\{-|t|^\alpha G(A)\}, \quad A \in \mathcal{B}_0, \quad t \in \mathbb{R}.$$

Show that for disjoint $A_k \in \mathcal{B}_0$, $A = \bigcup_{k=1}^{n} A_k$, $R_{A_1,\ldots,A_n} = R_{A_1} \otimes \ldots \otimes R_{A_n}$, a product measure. Using Theorem 3.4.10 (or 11), conclude that there is a probability space (Ω, Σ, P) and a (random) mapping $\mu : \mathcal{B}_0 \to L^0(P)$ such that the ch.f. of $\mu(A), A \in \mathcal{B}_0$, is given by

$$\varphi_{\mu(A)}(t) = \exp(-|t|^\alpha G(A)), \quad t \in \mathbb{R},$$

and moreover $R_{A_1,\dots,A_n} = P \circ \pi^{-1}_{A_1,\dots,A_n}, \pi_{A_1,\dots,A_n} : \Omega \to \mathbb{R}^n$ being the coordinate projection, guaranteed by Theorem 3.4.10. Finally, conclude that the family of finite dimensional distributions (or image measures) of $\{\mu(A), A \in \mathcal{B}_0\}$ is precisely $\{R_A, A \in \mathcal{B}_0\}$ which is a symmetric stable class with the property that the values of μ are pairwise independent on disjoint sets, hence, also mutually independent in this case, and $\mu(A) = \sum_{n=1}^{\infty} \mu(A_n), A_n \in \mathcal{B}_0$ with $A = \cup_{n=1}^{\infty} A_n$, a disjoint union, the series converging in probability (whence a.e. by Theorem 4.6.1). [A brief application of such random measures will be given in the last chapter, and the resulting analysis plays an important role in certain parts of probability as well as potential theories. For many extensions and generalizations of these results to Banach space valued functions, the book by Linde (1986) is a good place to turn to. Contrast the conditions here with Exercise 20 above, in obtaining a multivariate extension for the subclass under consideration.]

26. This and the next problem present an extension of the stable distribution theory and help in understanding that subject further.

Call a continuous mapping $\phi : \mathbb{R}^+ \to \mathbb{C}$ *stable* if (i)$\phi(0) = 1$, (ii)$|\phi(t)| \leq 1$, and (iii) for each $n(= 1, 2, \dots)$, there exist $a_n > 0$ and $b_n \in \mathbb{R}$ such that

$$\phi(t)^n = \phi(a_n t) e^{ib_n t}, \quad n \geq 1, \quad t \in \mathbb{R}^+.$$

Here ϕ is *not* assumed to be positive definite. Thus even if we extend the definition to $\mathbb{R}$ by setting $\phi(-t) = \overline{\phi(t)}, t \geq 0, \phi$ is not necessarily a ch.f. We now characterize such stable functions, following Bochner (1975).

Let $\phi : \mathbb{R}^+ \to \mathbb{C}$ be a continuous mapping. Then it is a stable function in the above sense iff the following two conditions hold:

(i) ϕ never vanishes, so that $\psi = \operatorname{Log} \phi$ is defined and is continuous, $\psi(0) = 0$, and

(ii) either $\psi(t) = iCt, C \in \mathbb{R}$ (degenerate case), or there exists an exponent $0 < p < \infty$ such that $a_n = n^{1/p}$ (in the definition of stability) and

$$\psi(t) = \begin{cases} (A + iB)t^p + iCt & \text{if } 0 < p < 1, \quad 1 < p < \infty \\ (A + iB \log t)t + iCt & \text{if } p = 1, \end{cases}$$

where $A \geq 0, B \in \mathbb{R}, C \in \mathbb{R}$, and $A + iB \neq 0$.

[The proof of this result proceeds in all its details similar to the argument given for Theorem 3.16. Starting from the identity $n\psi(t) = \psi(a_n t) + ib_n t$, and hence for all rationals $r > 0$, one gets

$$r\psi(t) = \psi(a(r)t) + ib(r)t, \qquad a(r) > 0, \quad b(r) \in \mathbb{R}.$$

Then one shows that $\{a(r), r \in \mathbb{R}^+\}$ is bounded; otherwise the degenerate case results. In the nondegenerate case, one shows that ultimately, if

$\rho(t) = (\psi(t)/t) - \psi(1)$, then $\rho(\alpha\beta) = \rho(\alpha) + \rho(\beta)$, and the familiar functional equations result, as in the text. For another generalization of stability, see, Ramachandran and C.R. Rao (1968).]

27. (Continuation) A function $\phi : \mathbb{R}^+ \to \mathbb{C}$ is *minimally positive definite* if for all triples $t_1 = 0, t_2 = t, t_3 = 2t, 0 < t < \infty$, we have

$$\sum_{i,j=1}^{3} \phi(t_i - t_j) z_i \bar{z}_j \geq 0, \quad z_i \in \mathbb{C}.$$

Let ϕ be a stable function (as in the preceding problem) which is minimally positive definite. Then in the representation of ϕ of the last problem, $0 < p \leq 2$, the constants A, B satisfy the restraints $A > 0, B \in \mathbb{R}$)

$$|B| \leq d_p A, \qquad 0 < p \leq 2, \qquad d_p = |1 - 2^{2-p}|^{1/2}/|1 - 2^{1-p}|$$

for $0 < p < 1, 1 < p \leq 2$, and $d_1 = (\log 2)^{-1}$. [Since the matrix $(\phi(t_i - t_j), 1 \leq i, j \leq 3)$ is positive definite, letting $u = \phi(t), v = \phi(t'), w = \phi(t + t')$, where $t', t > 0$, we get from its determinant

$$|\phi(t + t') - \phi(t)\phi(t')|^2 \leq (1 - |\phi(t)|^2)(1 - |\phi(t')|^2),$$

which gives, if $t' = t, |u|^2 - |w| \leq |w - u^2| \leq 1 - |u|^2 \Rightarrow |1 - \phi(2t)|^2 \leq 4(1 - |\phi(t)|^2)$. Substitute the value of $\phi(t)$ from Problem 26 in these expressions and simplify when $0 < p < 1, 1 < p \leq 2$, and $p = 1$; then the stated bounds result after some limiting argument. If ϕ is assumed to be fully positive definite, after extending it to $\mathbb{R}$, using the argument of Problems 24 and 25, can one get the precise bounds stated in Theorem 3.16?] It will be of interest to extend Bochner's stability analysis to the multivariate case as in Problem 25(c) above. It opens up another area of investigation extending P. Lévy's pioneering work.

28. Let $\mathcal{L}$ be the class of d.f.s F which are the d.f.s of r.v.s S such that for some sequence of independent r.v.s $X_1, X_2, \ldots, S_n = \sum_{k=1}^{n} X_k$, then $(S_n - b_n)/a_n \xrightarrow{D} S$, where $b_n \in \mathbb{R}$ and $0 < a_n \to \infty$, with $a_{n+1}/a_n \to 1$. The problem of characterization of $\mathcal{L}$ was proposed by A. Khintchine in 1936 and P. Lévy has given a complete solution in 1937. Thus it is called the $\mathcal{L}$ (or *Lévy) class.* Show that $F \in \mathcal{L}$ iff the ch.f. ϕ of F has the property that for each $0 < \alpha < 1, \tilde{\phi}_\alpha$ is a ch.f., where $\tilde{\phi}_\alpha(t) = \phi(t)/\phi(\alpha t)$, and hence ϕ is infinitely divisible. [Note first that ϕ cannot vanish. Indeed, if $\phi(2a) = 0$ for some $a > 0$ such that $\phi(t) \neq 0, 0 \leq t < 2a$, then $\tilde{\phi}_\alpha(2a) = 0$ and

$$1 = 1 - |\tilde{\phi}_\alpha(2a)|^2 \leq 4\{1 - |\phi_\alpha(a)|^2\}$$

(see the inequality in the last problem for ϕ satisfying [even minimal] positive definiteness). But $\phi_\alpha(a) \to 1$ as $\alpha \to 1$, for each $a \in \mathbb{R}$, by the continuity of

ϕ, and this contradicts the above inequality.

Sketch of Proof Using

$$\phi(nt) = \prod_{k=1}^{n} \frac{\phi(kt)}{\phi((k-1)t)},$$

and each factor is a ch.f., by hypothesis, of an r.v. X_k, and replacing t by τ/n, letting $n \to \infty$, the right-side product converges to ϕ, which is a ch.f. This means $(1/n)\sum_{k=1}^{n} X_k \xrightarrow{D} S$, where the ch.f. of S is ϕ. Conversely, consider $S_n^* = (1/a_n)\sum_{k=1}^{n} X_k - b_n$. The ch.f. ϕ_n of S_n^* can be expressed as

$$\phi_n(t) = \phi_m\left(\frac{a_m}{a_n}t\right)\psi_{m,n}(t), \quad n > m,$$

where $\psi_{m,n}$ is the ch.f. of $(\sum_{k=m+1}^{n} X_k - b_n + b_m)a_m/a_n$. Letting $m, n \to \infty$ so that $a_m/a_n \to \alpha, 0 < \alpha < 1$ (which is possible), since, by hypothesis, $\phi_n(t) \to \phi(t)$ as $n \to \infty$ and ϕ never vanishes, we get

$$\phi(t)/\phi(\alpha t) = \lim_{m,n\to\infty} \psi_{m,n}(t)$$

and it is continuous at $t = 0$. Thus it is a ch.f.$=\tilde{\phi}_\alpha$ (say). This gives the necessity. The first part shows that for each n, ϕ is the product of n ch.f.s and that $\phi(kt)/\phi((k-1)t) \to 1$ as $k \to \infty$. By Problem 17, ϕ is infinitely divisible. This fact also follows from the result that the $(1/n)X_k(= X_{nk})$ are infinitesimal, and hence their partial sums can only converge to infinitely divisible r.v.s.]

29. Let X be an r.v. with a double exponential (also called Laplace) density f_X given by

$$f_X(x) = [1/(2\alpha)]\exp\{-|x-\beta|/\alpha\}, \quad \alpha > 0, \quad \beta \in \mathbb{R}, \quad x \in \mathbb{R}.$$

Find its ch.f. ϕ. Show that f_X is of class $\mathcal{L}$ but that ϕ is not a stable ch.f. [*Hint*: Observe that an exponential density is infinitely divisible, and using the result of Problem 14 with the Lévy-Khintchine or Kolmogorov canonical form, deduce that f_X is of class $\mathcal{L}$. With the help of Theorem 3.16, conclude that ϕ cannot be in the stable family, thus the latter is a proper subset.]

30. If $\{X_n, n \geq 1\}$ is an i.i.d. sequence of r.v.s with d.f. F and if $S_n = \sum_{k=1}^{n} X_k$, then F is said to belong to the *domain of attraction* of V, provided that for some normalizing constants $a_n > 0$ and numbers $b_n \in \mathbb{R}, (S_n - b_n)/a_n \xrightarrow{D} \tilde{S}$ whose d.f. is V. Thus if F has two moments, then [with $a_n = \sqrt{n \operatorname{Var} X_1}, b_n = E(S_n)$] F belongs to the domain of attraction of a normal law. Show that only stable laws have nonempty domain of attraction and that each such law belongs to its own domain of attraction (cf. Proposition 3.15).

If in the above definition the full sequence $\{(S_n - b_n)/a_n, n \geq 1\}$ does not converge, but there is a subsequence $n_1 < n_2 < \cdots$ such that $(S_{n_k} - b_{n_k})/a_{n_k} \xrightarrow{D} W$, then the d.f. H of W is said to have a *domain of partial attraction,* with F as a member of this domain. Establish the following beautiful result due to Khintchine: Every infinitely divisible law has a nonempty domain of partial attraction. [*Sketch* (after Feller): If ϕ is an infinitely divisible ch.f., then $\phi = e^{\psi}$ by the Lévy-Khintchine formula, and in fact $\phi(t) = \lim_{n\to\infty} \prod_{i=1}^{n} e^{\psi_i^n(t)}$, where the right side is a Poisson ch.f. (a "generalized Poisson") and $\psi^n(0) = 0 = \psi_i^n(0)$, the ψ_i being continuous. Each ψ_i^n is bounded. As a preliminary, let $\phi_k = e^{\zeta_k}$ be any sequence of infinitely divisible ch.f.s with each ζ_k bounded. Let $\lambda(t) = \sum_{k\geq 1} n_k^{-1}\zeta_k(a_k t)$. Choose $a_k > 0, n_k < n_{k+1} < \cdots$ such that $|n_r\lambda(t/a_r) - \zeta_r(t)| \to 0$. Indeed, choose first a sequence of integers such that $(n_k/n_{k-1}) > 2^k \sup_{t\in\mathbb{R}} |\zeta_k(t)|$. Then

$$\left| n_r\lambda\left(\frac{t}{a_r}\right) - \zeta_r(t)\right| \leq n_r \sum_{k=1}^{r-1} \left|\zeta_k\left(\frac{a_k t}{a_r}\right)\right| + \sum_{k\geq r+1} 2^{-k}.$$

Now choose $a_1 = 1$, and after $a_1, \ldots, a_{r-1}$, let a_r be so large that for $|t| < r$ [since $\zeta_k(\tau) \to 0$ as $\tau \to 0$] $|\zeta_k(t)| < (2r^2 n_r)^{-1}$. With this choice, the right side is $< r^{-1}$, so that the left side $\to 0$ as $r \to \infty$. Next, since the given ϕ is a limit of $\phi_n = \prod_{i=1}^{n} e^{\psi_i^n(t)} = e^{\xi_n(t)}$ (say), so that ξ_n are bounded and continuous, $\xi_n(t) \to 0$ as $t \to 0$. Define $\lambda(\cdot)$ with this ξ_n in place of ζ_n above. Then for a choice of n_k, we have $e^{\lambda(\cdot)}$ infinitely divisible, and

$$\lim_{r\to\infty} \exp\{n_r\lambda(t/a_r)\} = \lim_{r\to\infty} \exp\{\xi_r(t)\} = \phi(t).$$

Since $\exp\{n_r\lambda(t/a_r)\}$ is a ch.f. tending to the ch.f. ϕ, it follows that $e^{\lambda(\cdot)}$ belongs to the domain of partial attraction. The original proof is more involved.]

31. Let $\{X_{nk}, 1 \leq k \leq k_n, n \geq 1\}$ be a rowwise independent sequence of infinitesimal r.v.s such that $E(X_{nk}) = 0, t_{nk}^2 = \text{Var } S_{nk}$, where

$$S_{nk} = \sum_{i=1}^{k} X_{ni}, \qquad t_n^2 = t_{nk_n}^2.$$

Suppose the sequence satisfies the Lindeberg condition (as in Theorem 4.6). Show that, following the argument of Theorem 4.7,

$$\lim_{n\to\infty} P[-at_n \leq S_{nk} < bt_n, 1 \leq k \leq k_n] = \begin{cases} \frac{4}{\pi}\sum_{k=0}^{\infty}(2k+1)^{-1}\exp\left\{-\frac{(2k+1)^2\pi^2}{2(a+b)^2}\right\}\sin\frac{(2k+1)a\pi}{a+b} & \text{if } a \geq 0, \quad b > 0 \\ 0, & \text{otherwise.} \end{cases}$$

[*Hints*: First show that the limit exists as in Theorem 4.7. To calculate that limit, again consider the special r.v.s $P[X_{nk} = +1] = P[X_{nk} = -1] = \frac{1}{2}$. Letting $\alpha_n = [a\sqrt{n}] + 1, \beta_n = [b\sqrt{n}] + 1$, verify that

$$P[-a\sqrt{n} < S_{nk} < b\sqrt{n}; 1 \leq k \leq k_n] = \sum_{k=-\alpha_n+1}^{\beta_n - 1} A_{kn},$$

where

$$A_{kn} = a_k^n + \sum_{j=1}^{\infty} [a_{2j(\alpha_n+\beta_n)+k}^n - a_{2j(\alpha_n+\beta_n)-2\alpha_n-k}^n + a_{-2j(\alpha_n+\beta_n)+k}^n - a_{-2j(\alpha_n+\beta_n)+2\beta_n-k}^n],$$

with

$$a_k^n = \binom{n}{(n+k)/2} \frac{1}{2^n},$$

the binomial coefficient being zero if $(n+k)/2$ is not an integer. Verify that $\sum_k A_{kn}$ tends to the desired limit as $n \to \infty$ on using the central limit theorem. The A_{kn} are essentially planar random walk probabilities.]

Deduce from the above that

$$\lim_{n\to\infty} P\left[\max_{1\leq k\leq k_n} |S_{nk}| < xt_n\right] = \begin{cases} \frac{4}{\pi}\sum_{k\geq 0} \frac{(-1)^k}{2k+1} \exp\left\{-\frac{(2k+1)^2\pi^2}{8x^2}\right\}, & x > 0 \\ 0, & \text{otherwise.} \end{cases}$$

32. Let F be a continuous strictly increasing d.f. and F_n be the empiric d.f. of n i.i.d. random variables with F as their d.f. Establish the following "relative error analog" of the Kolmogorov-Smirnov theorem for $0 < a < 1$:

$$\lim_{n\to\infty} P\left[\sup_{a\leq F(x)} \left|\frac{\sqrt{n}(F_n(x) - F(x))}{F(x)}\right| < y\right] = G(y),$$

where

$$G(y) = \begin{cases} \frac{4}{\pi}\sum_{k=0}^{\infty} \frac{(-1)^k}{2k+1} \exp\left\{-\frac{(2k+1)^2}{8}\frac{1-a}{ay^2}\right\} & \text{if } y > 0 \\ 0 & \text{if } y \leq 0. \end{cases}$$

[*Hints*: As in Theorem 4.8, take F as uniform, and $H_n(x) = F_n(F^{-1}(x))$ the empiric d.f. of the uniform r.v.s $Y_i = F(X_i)$. Note that (with $\vee = \max$)

$$\sup_{a\leq H_n(x)} \left|\frac{\sqrt{n}(H_n(x) - x)}{x}\right| = \sup_{an\leq k\leq n} \sqrt{n}\left(\left|\frac{k}{nY_k^*} - 1\right| \vee \left|\frac{k}{nY_{k-1}^*} - 1\right|\right)$$

where $Y_k^*, 1 \leq k \leq n$, are order statistics from $Y_i, 1 \leq i \leq n$. Thus it suffices to calculate the limit d.f. of

$$\sup_{an \leq k \leq n} \sqrt{n} \left| \frac{1}{nY_k^*} - 1 \right|.$$

Use the last part of the preceding exercise for this. The details are similar to those of Theorem 4.8. See Rényi (1953), in connection with both of these results, i.e., this and the preceding one.]

33. Let $\{X_n, n \geq 1\}$ be i.i.d. random variables with $S_n = \sum_{k=1}^{n} X_k, S_0 = 0$. In the study of fluctuations of the random walk sequence $\{S_n, n \geq 0\}$ in Section 2.4c, we had to analyze the behavior of $R_n = \max\{S_k : 0 \leq k \leq n\}$. In many of these random walk problems, the joint distributions of (R_n, S_n) are of interest. Let $\phi_n(u,v) = E(\exp\{iuR_n + ivS_n\})$, the ch.f. of the vector (R_n, S_n). If S_n^+, S_n^- are the positive and negative parts of S_n [i.e., $S_n^+ = (|S_n| + S_n)/2, S_n^- = S_n^+ - S_n$], let $\psi_n(u,v) = E(\exp\{iuS_n^+ + ivS_n\})$. Show that one has the identity

$$\sum_{n=0}^{\infty} t^n \phi_n(u,v) = \exp\left\{\sum_{n=1}^{\infty} t^n \psi_n(u,v)/n\right\}, \qquad t \in \mathbb{R}. \tag{$*$}$$

If $c_n(u,v) = E(\exp\{iuR_n + iv(R_n - S_n)\})$, $p_n(u) = E(\exp\{iuS_n^+\})$, and finally $q_n(v) = E(\exp\{ivS_n^-\})$, then verify that $(*)$ is equivalent to the identity

$$\sum_{n=0}^{\infty} t^n c_n(u,v) = \exp\left\{\sum_{n=1}^{\infty} \frac{t^n}{n}(p_n(u) + q_n(v) - 1)\right\}. \tag{$+$}$$

In particular, if $c_n(u) = c_n(u,0) = E(\exp\{iuR_n\})$, then $(+)$ gives

$$\sum_{n=0}^{\infty} t^n c_n(u) = \exp\left\{\sum_{n=1}^{\infty} t^n p_n(u)/n\right\}. \tag{$*+$}$$

The important identity $(+)$ [or equivalently $(*)$], and hence $(*+)$, was obtained by Spitzer (1956), and is sometimes referred to as *Spitzer's identity.* He established it by first proving a combinatorial lemma, and using it in the argument. Here we outline an alternative algebraic method of $(*)$, which is simple, elegant, and short, due to J. G. Wendel, who has obtained it after Spitzer's original work. Since $\phi_n(u,v) = c_n(u+v, -v)$, we get $(*) \Leftrightarrow (+)$.

Proof Let $G_n(x,y) = P[R_n < x, S_n < y]$, and note that

$R_{n+1} = \max(R_n, S_n + X_{n+1})$ and $S_{n+1} = S_n + X_{n+1}$, so that $\{(R_n, S_n), n \geq 0\}$ is a Markov process. Also if $A_n^{x,y} = [R_n < x, S_n < y]$, then by a simple property of conditioning (cf. Proposition 3.1.2), we have

$$P(A^{x,y}_{n+1}) = E(E(\chi_{A^{x,y}_{n+1}}|X_{n+1}))$$
$$= \int_{\Omega} P(A^{x,y}_{n+1}|X_{n+1})(\omega)P(d\omega).$$

Thus, going to the image spaces, this becomes, if F is the common d.f. of the X_n, (using a manipulation as though the conditioning event has positive probability, cf. Kac-Slepian paradox, Section 3.2, and thus we get the following which needs a rigorous justification, which may be provided since all the conditional measures are regular in this application, and since Spitzer's original argument does not use the conditioning! See also Example 3.3.3 (b)) We outline the argument as follows:

$$\begin{aligned} G_{n+1}(x,y) &= \int_{\mathbb{R}} P[R_{n+1} < x, S_{n+1} < y|X_{n+1} = r]P(dr) \\ &= \int_{\mathbb{R}} P[R_n < x, S_n < \min(x,y) - r]\, dF(r) \\ &= \int_{\mathbb{R}} G_n(x, \min(x,y) - r)\, dF(r), \qquad n \geq 0. \end{aligned} \tag{i}$$

To go from here to $(*)$, we introduce an algebraic construct on $\mathbb{R}^2$. Namely, let $\mathcal{M}$ be the space of all two-dimensional d.f.s on $\mathbb{R}^2$, with convolution as multiplication and linear combinations of such d.f.s. [Then $\mathcal{M}$ is essentially identifiable with all signed measures on $\mathbb{R}^2$.] With this multiplication, $\mathcal{M}$ becomes an algebra. Using the total variation as norm, $\mathcal{M}$ actually is seen to be a complete normed algebra with the property that $G, H \in \mathcal{M} \Rightarrow \|G*H\| \leq \|G\| \cdot \|H\|$, where $G * H$ is the convolution and $\|G\|$ is the total variation norm of G. [Thus $(\mathcal{M}, \|\cdot\|)$ is a "Banach algebra."] If δ is the degenerate distribution at the origin, then $\delta * G = G * \delta = G$ and $\delta \in \mathcal{M}$ is the identity of this algebra. If $G \in \mathcal{M}$, $\|G\| < 1$, then $\sum_{n\geq 0} G^n/n! \in \mathcal{M}$ (by completeness), where $G^n = G*G*\cdots*G$ (n times). This is denoted $\exp G (\in \mathcal{M})$ (which holds for all $G \in \mathcal{M}$) and similarly $\log(\delta - G) = -\sum_{n\geq 1} G^n/n \in \mathcal{M}$ (for $\|G\| < 1$.) One can now verify that $\exp\{\log(\delta - G)\} = \delta - G$ for G with $\|G\| < 1$. Next define two linear operators L, M on $\mathcal{M}$ into itself by the equations $LG = \tilde{F} * G$, where $\tilde{F}$ is the common d.f. of $(0, X_n)$ (the X_n are i.i.d.), and $(MG)(x,y) = G(x, \min(x,y)), G \in \mathcal{M}$. It is clear that $M^2G = MG$, so that M is a projection, and $M\delta = \delta$. Also $\|MG\| \leq \|G\|$ and $\|LG\| \leq \|G\|$, so that L, M are contractions. Further $M(\mathcal{M})$ and $(I - M)(\mathcal{M})$ are closed subalgebras, $\mathcal{M} = M(\mathcal{M}) \bigoplus (I - M)(\mathcal{M})$. Since $MG \in M(\mathcal{M})$, we get $\exp\{MG\} \in M(\mathcal{M}), \exp\{(I - M)G\} \in (I - M)(\mathcal{M})$. Thus for each $G \in \mathcal{M}$ there is a $\tilde{G} \in \mathcal{M}$ such that $\exp\{(I - M)G\} - \delta = (I - M)\tilde{G}$. This completes our introduction of an abstract algebra structure.

Let us now transform the recursion relation (i) for the G_n into this abstract situation. Now (i), in the formal computation, is expressible as $G_{n+1} = MLG_n$, and (by iteration) $G_n = (ML)^n\delta$, where $G_0 = \delta$. Thus for $|t| < 1$,

$$\sum_{n\geq 0} t^n G_n = \sum_{n\geq 0} t^n (ML)^n \delta = (I - tML)^{-1}\delta = G \qquad \text{(say)}. \qquad \text{(ii)}$$

To derive $(*)$ from (ii), we first assert that

$$G = \exp\left\{\sum_{n\geq 1} t^n M(\tilde{F}^n)\right\}; \qquad \text{(iii)}$$

for (ii)$\Rightarrow G - tMLG = \delta$, so that $MG - tMLG = M\delta = \delta$, since $M^2 = M$. By subtraction, $MG = G$, and also $M[G - tLG] = M[G - t\tilde{F} * G] = \delta$. An element G satisfying the last pair of equations is unique, since if G and G' are two such elements, then $MG' = G', M[G' - tLG'] = \delta \Rightarrow G' - tMLG' = \delta \Rightarrow G' = (I - tML)^{-1}\delta = G$ by (ii). To exhibit a solution, consider $\tilde{G} = \exp\{-M(\log(\delta - t\tilde{F}))\} \in \mathcal{M}$. Then, since $M(\log(\delta - t\tilde{F})) \in M(\mathcal{M})$, we get $\tilde{G} \in M(\mathcal{M})$, so that $MG' = G'$. Also, because the convolution on $\mathbb{R}^2$ is commutative ($\tilde{F} * \tilde{G} = \tilde{G} * \tilde{F}$), we get

$$(\delta - t\tilde{F}) = \exp\{\log(\delta - t\tilde{F})\}, \qquad |t| < 1,$$

and so (multiplication between d.f.s being convolution)

$$\begin{aligned}(\delta - t\tilde{F})\tilde{G} &= \exp\{\log(\delta - t\tilde{F})\}\exp\{-M(\log(\delta - t\tilde{F}))\}\\ &= \exp\{\log(\delta - t\tilde{F}) - M(\log(\delta - t\tilde{F}))\}\\ &\quad \text{(by the commutativity noted above)}\\ &= \exp\{(I - M)\log(\delta - t\tilde{F})\}\\ &= \delta + (I - M)G_1\end{aligned}$$

for a $G_1 \in \mathcal{M}$, since $\log(\delta - t\tilde{F}) \in \mathcal{M}$. Hence

$$M[(\delta - t\tilde{F})\tilde{G}] = M\delta + M(I - M)G_1 = M\delta = \delta.$$

Thus $\tilde{G}$ is a solution and so is $G, \Rightarrow \tilde{G} = G$ in (ii). But then

$$G = \tilde{G} = \exp\{-M(\log(\delta - t\tilde{F}))\} = \exp\left\{\sum_{n\geq 1} t^n M(\tilde{F}^n)/n\right\}.$$

This gives (iii), after (i) and its consequence (ii) are rigorized.

To establish $(*)$ from (ii), let us take Fourier transforms on both sides. Note that the Fourier transform $\mathcal{F}$ of a convolution on $\mathbb{R}^2$ is the product of the transforms and also $||\mathcal{F}(G)||_\infty \leq ||G||$. Thus the left side of (ii) becomes

$$\mathcal{F}\left(\sum_{n\geq 0} t^n G_n\right)(u,v) = \sum_{n\geq 0} t^n \mathcal{F}(G_n)(u,v) = \sum_{n\geq 0} t^n \phi_n(u,v).$$

For the right side,

$$\mathcal{F}\left(\exp\left\{\sum_{n\geq 1}\frac{t^n}{n}M(\tilde{F}^n)\right\}\right)(u,v) = \exp\left\{\sum_{n\geq 1}\frac{t^n}{n}\mathcal{F}(M(\tilde{F}^n))(u,v)\right\}. \quad \text{(iv)}$$

But we see that for any bounded Borel function h on $\mathbb{R}^2$,

$$\begin{aligned}\int\int_{\mathbb{R}^2} h(x,y)(MG)(dx,dy) &= \int\int_{\mathbb{R}^2} h(x,y)\,dG(x,\min(x,y))\\ &= \int\int_{[(x,y):x\geq y]} h(x,y)\,dG(x,y)\\ &= \int\int_{\mathbb{R}^2} h(\max(x,y),y)\,dG(x,y). \quad \text{(v)}\end{aligned}$$

Taking $h(x,y) = \exp\{iux + ivy\}$ in (v), we get

$$\begin{aligned}&\mathcal{F}(M\,(\tilde{F}^n))(u,v)\\ &\quad= \int\int_{\mathbb{R}^2} \exp\{iux+ivy\}\,d(M\tilde{F}^n)(x,y)\\ &\quad= \int\int_{\mathbb{R}^2} \exp\{i(u\max(x,y)+vy)\}\,d\tilde{F}^n(x,y)\\ &\qquad\text{(by definition of } M)\\ &\quad= \int_{\Omega} \exp\{iu\max(0,S_n)+ivS_n\}\,dP\\ &\qquad[\text{since } \tilde{F} \text{ is the joint d.f. of } (0,X_1)\\ &\qquad\text{and then the image law relation is used}]\\ &\quad= \psi_n(u,v).\end{aligned}$$

Substituting this in (iv) and using it in (ii) and (iii), (*) follows. [Can this method be used to solve the integral equation of Exercise 21 in Chapter 3? In both cases the method of (i) is to be justified with the Kolmogorov definition of conditioning, without taking the manipulation for granted.]

34. Let $\{X_n, n \geq 1\}$ be an i.i.d. sequence as in the preceding problem, and $S_n = \sum_{k=1}^n X_k, S_0 = 0$, and S_n^+ be the positive part. Show that

(a) (M. Kac-G. A. Hunt)

$$E\left(\max_{1\leq k\leq n} S_k^+\right) = \sum_{k=1}^n E(S_k^+)/k,$$

and

(b) (E. S. Anderson)

$$\sum_{n\geq 0} P[S_k \geq 0, 1 \leq k \leq n]t^n = \exp\left\{\sum_{k=1}^{\infty} t^k P[S_k \geq 0]/k\right\}.$$

[*Hints*: For (a) differentiate $(*+)$ of the above problem relative to u at $u = 0$, and identify the coefficient of t^n on both sides. Similarly for (b), get the result for $\max_{k\leq n} S_k^-$ by replacing X_i with $-X_i$, then setting $u = -i\lambda$, and letting $\lambda \to +\infty$. These and other specializations and applications are given in Spitzer's paper noted above. Further applications of these results in obtaining precise conditions for a unique solution F of the Wiener-Hopf integral equation $F(x) = \int_0^\infty f(x-y)\,dF(x)$, where F is a d.f. on $\mathbb{R}^+$ and $f(\cdot)$ is a probability density on $\mathbb{R}$ can be given. In fact, this equation may be written, a little more generally, in terms of r.v.s as $Y \stackrel{D}{=} (X+Y)^+$, where X, Y are independent and F is the d.f. of Y, f being the density of X. Thus if $X_n, n \geq 1$, are i.i.d., $S_0 = 0, S_n = \sum_{k=1}^n X_k$, let $Y_0 = S_0$ and $Y_1 = \max[Y_0, S_1] = X_1^+, \ldots, Y_n = \max(S_0, S_1, \ldots, S_n)$. The existence of a solution is equivalent to proving that $Y_n \to Y$ a.e. and $Y < \infty$ a.e. This may be shown to hold if the series $(*+)$ of the above problem converges when $t = 1$, and there will be no solution if it diverges at $t = 1$.]

35. Let $S_n, n \geq 1$, be the partial sums of i.i.d. Bernoulli r.v.s X_n, with $P[X_n = 1] = p, P[X_n = 0] = 1-p = q, 0 < p < 1$. If $\{\phi(n), n \geq 1\}$ is a positive monotone increasing sequence, then show that $P[S_n > np + \sqrt{npq}\phi(n)$, i.o.$]=0$ or 1 according to whether $\sum_{n\geq 1}[\phi(n)/n]\exp\{-\frac{1}{2}\phi^2(n)\}$ converges or diverges. Verify that the latter series converges iff

$$\sum_{r\geq 1}(\phi(n_r))^{-1}\exp\{-\frac{1}{2}\phi^2(n_r)\}$$

converges, where $n_r = [\exp\{r\log r\}]$, the integral part of the number shown. If $\phi^2(n) = 2\log\log n$, then one obtains the LIL for the Bernoulli r.v.s. [*Hints*: Let $A_r = [S_n - np > \sqrt{npq}\phi(n_r)$ for some n, $n_r < n < n_{r+1}]$. Verify that $P(A_r) \leq C\exp\{-\frac{1}{2}\phi^2(n_r)\}\phi(n_r)$ for some constant $0 < C < \infty$. By hypothesis, this is the rth term of a convergent series, so that $\sum_{r\geq 1} P(A_r) < \infty$. For the second part, which is more involved, note that $n_r - n_{r-1} \sim n_r(1 - (\log r)^{-1})$, and that $\sum_{n\geq 1}[\phi(n)/n]\exp\{-\frac{1}{2}\phi^2(n)\} < \infty$ only if $\phi^2(n) > 2\log\log n$, and it diverges if $\phi^2(n) \leq 2\log\log n$. By various estimates deduce that for some positive constants C_1, C_2,

$$C_1/\phi^2(n_r) < (n_{r+1} - n_r)/n_r < C_2/\phi^2(n_r),$$

and hence there are positive constants C_3 and C_4, such that

$$\frac{C_3}{\phi^2(n_r)} \leq \frac{1}{2}\log\frac{n_{r+1}}{n_r} \leq \sum_{n=n_r+1}^{n_{r+1}} \frac{1}{n} \leq \log\frac{n_{r+1}}{n_r} \leq \frac{C_4}{\phi^2(n_r)}.$$

Define another sequence $m_r < m_{r+1} < \cdots$ such that

$$\phi(m_r)\exp\{-\frac{1}{2}\phi^2(m_r)\} = \max_{n_r < n \leq n_{r+1}} \phi(n)\exp\{-\frac{1}{2}\phi^2(n)\},$$

and show that $\phi(m_r)/\phi(n_r) \sim 1$. This gives after a careful estimation of terms

$$\sum_{n\geq 1} \frac{1}{n}\phi(n)\exp\{-\frac{1}{2}\phi^2(n)\} \leq C_5 \sum_{r\geq 1} \frac{1}{\phi(n_r)}\exp\{-\frac{1}{2}\phi^2(n_r)\}$$

and

$$\sum_{n\geq 1} \frac{1}{n}\phi(n)\exp\{-\frac{1}{2}\phi^2(n)\} \geq C_6 \sum_{r\geq 1} \frac{1}{\phi(n_{r+1})}\exp\{-\frac{1}{2}\phi(m_{r+1})\}.$$

These imply the last part, and then the probability statement obtains. This result if $p = q = \frac{1}{2}$ and $\phi^2(n) = 2\lambda \log\log n$ was first proved by P. Erdös, and the case of ϕ for more general r.v.s is due to Feller (1943), as noted in the text. Regarding this problem, see also Feller (1957).]

The following problems elaborate and complement the special application discussed in Section 6 in illustrating the invariance principle.

36. Let $\{\varepsilon_n, n \geq 1\}$ be i.i.d. random variables with means zero and variances one. If $X_n = \varepsilon_n \varepsilon_{n+1}$, show that $\{X_n, n \geq 1\}$ is a sequence of uncorrelated r.v.s with bounded variances and that $(1/n)\sum_{k=1}^n X_k \to 0$, a.e. and in L^2-mean. (Use Theorem 2.3.4.)

37. If $X_n = \sum_{k=1}^n \alpha^{n-k}\varepsilon_k, |\alpha| < 1$, and the ε_k are i.i.d., $E(\varepsilon_k) = 0, E(\varepsilon_k^2) = 1$, and if $\tilde{X}_n = \sum_{k=1}^n \alpha^{k-1}\varepsilon_k$, show that $X_n \stackrel{D}{=} \tilde{X}_n$, that $\tilde{X}_n^2 \to Y^2$ a.e., and that $E(Y^2) = (1-\alpha^2)^{-1}$; note that Y is not a constant r.v. Deduce that $(1/n)\sum_{k=1}^n \tilde{X}_{k-1}^2 \to Y^2$ a.e., and that $E[(1/n)\sum_{k=1}^n X_k^2] \to (1-\alpha^2)^{-1}$. Does the latter sequence of r.v.s also converge to Y^2 in distribution? [Recall Cramér-Slutksy Theorem from Exercise 2.11.]

38. Let $\{X_t, t \geq 0\}$ be a sequence of r.v.s which satisfy the first-order (stochastic) difference equation where $\{\varepsilon_t, t \geq o\}$ is an i.i.d. sequence with $E(\varepsilon_t) = 0, 0 < E(\varepsilon_t^2) = \sigma^2 < \infty$,

$$X_t = \alpha X_{t-1} + \varepsilon_t, \quad t = 1, 2, \ldots.$$

Using this expression, one gets by iteration:

$$X_t = \alpha(\alpha X_{t-2} + \varepsilon_{t-1}) + \varepsilon_t = \ldots = \alpha^{t-1}\varepsilon_1 + \alpha^{t-2}\varepsilon_2 + \ldots + \alpha\varepsilon_{t-1} + \varepsilon_t.$$

To determine the convergence properties one needs to consider the cases that (i) $|\alpha| < 1$, (ii) $|\alpha| = 1$, and (iii) $|\alpha| > 1$, where (for reference) the corresponding processes are termed *stable, unstable,* and *explosive*, respectively. Since

$E(X_t) = 0$ and $Var X_t = \sum_{k=1}^{t} \alpha^{2(k-1)}$, it follows that $\lim_{t\to\infty} Var X_t < \infty$ iff $|\alpha| < 1$. Since $Var X_t = (1-\alpha^{2t})/(1-\alpha^2)$ if $\alpha \neq 1$, $= t$ if $\alpha = 1$, let $g(n;\alpha) = [n/(1-\alpha^2)]^{-1/2}$ if $|\alpha| < 1$; $= n/\sqrt{2}$ if $|\alpha| = 1$; and $= |\alpha|^n(\alpha^2-1)^{-1/2}$ if $|\alpha| > 1$. Then we can establish the following useful fact: If $V_n = \sum_{t=1}^{n} X_t^2$ with X_t given the above difference equation shows that we can conclude $g(n;\alpha)^{-2}V_n = W_n \xrightarrow{D} V$ as $n \to \infty$ and $P[V = 0] = 0$. [The result is not entirely easy. It uses the SLLN, the Cramér-Slutsky calculus, and Donsker's invariant principle. The details were given in the first edition, but can be tried here.]

39. Let $\hat{\alpha}_n$ be the least squares estimator of α of the first order model as given by Eq. (2) of Section 6 under the same conditions. Show that if $|\alpha| > 1$ and $g(n;\alpha) = |\alpha|^n/(\alpha^2-1)^{1/2}$, then $g(n;\alpha)(\hat{\alpha}_n - \alpha) \xrightarrow{D} V$ and that the limit d.f. namely of V, *depends on the common distribution of the errors* ε_n. [Thus V is Cauchy distributed if the ε_n are $N(0,1)$.]

40. Use the method of proof of Theorem 6.1 (i.e., employ Theorem 4.3.2), and complete the proof of the case $|\alpha| = 1$ of that result. With a similar method find the limit d.f. of $\hat{\alpha}_n$ if $\alpha = 1$ and note that this is different from that of the case $|\alpha| = 1$ [the norming factor $g(n;\alpha)$ being the same]. A similar statement holds for $\alpha = -1$ also. [The computations are not simple and need care.]

41. Proposition 6.2 admits the following extension. Let $\{U_n, n \geq 1\}$ be a sequence of m-dependent r.v.s with $E(U_n) = 0, E(U_n^2) \leq M < \infty, n \geq 1$. If $S_n = \sum_{k=1}^{n} U_k$, then $S_n/\sigma(S_n) \xrightarrow{D} Z$, where Z is $N(0,1)$ distributed, provided that $(\sum_{k=1}^{n} \text{Var } U_k)/\sigma^2(S_n) = O(1)$ and that the Lindeberg condition holds, i.e. for each $\varepsilon > 0$

$$\frac{1}{\sigma^2(S_n)} \sum_{k=1}^{n} \int_{[|x|>\varepsilon\sigma(S_n)]} x^2\, dF_k(x) \to 0,$$

where the F_k is the d.f. of U_k. [This is a special case of a result due to S. Orey.]

42. To establish an analog of Theorem 3.1iii for a multiparameter estimation problem, it is necessary to extend Theorem 4.3.2 to higher dimensions as a first step. We present such a result here. Let (X_1, Y_1, X_2, Y_2) be integrable r.v.s with $P[Y_i > 0] = 1, i = 1,2$. Let $H(x_1, x_2) = P[X_i/Y_i < x_i, i = 1,2]$ be the joint distribution of the ratios shown. If ϕ is the joint ch.f. of the vector (X_1, Y_1, X_2, Y_2), show that the joint density $h = \partial^2 H/\partial x_1\, \partial x_2$ of the ratios $X_i/Y_i, i = 1,2$, is given, on assuming that ϕ is twice differentiable, by

$$h(x_1, x_2) = -\frac{1}{4\pi^2} \int\int_{\mathbb{R}^2} \frac{\partial^2 \phi}{\partial t_2\, \partial t_4}(\tau_1, -x_1\tau_1, \tau_2, -x_2\tau_2)\, d\tau_1\, d\tau_2$$

whenever the integral is uniformly convergent for (x_1, x_2) in open intervals of $\mathbb{R}^2$.

43. An extension of the m-dependent central limit theorem when m is not bounded is as follows. Let $\{X_{kj}, 1 \leq j \leq j_k, k \geq 1\}$ be a sequence of sequences of m_k-dependent r.v.s with means zero and $2+\delta, \delta > 0$, moments. Suppose that (a) $\sup_{j,k} E(|X_{kj}|^{2+\delta}) < \infty$, (b) $\mathrm{Var}(\sum_{j=i+1}^{\ell} X_{kj}) \leq (\ell - i)K_0$ for all $i < \ell$ and k (integers > 0), (c) $\lim_{k\to\infty}(1/j_k)\,\mathrm{Var}(\sum_{j=1}^{j_k} X_{kj}) = \alpha > 0$ exists, and (d) $\lim_{k\to\infty} m_k^{2+(2/\delta)}/j_k = 0$. Then show that $\sum_{j=1}^{j_k} X_{kj}/\sqrt{j_k} \xrightarrow{D} Z$, where Z is $N(0, \alpha)$. [The proof of this result is a careful adaption of the m-dependent case; cf. Proposition 5.6.2. Here condition (d), which is trivial if $m_k \equiv \tilde{m}$ (fixed), is essential for the truth of the result. This extension is due to K. N. Berk.]

Part III Applications

This part uses the theory developed in the preceding two parts, and presents different ideas based upon them. Chapters 6 and 7 are short but the former includes the relatively new concept of stopping times and classes of (central) limit theorems for dependent sequences. This leads to the introduction of ergodic sequences and the Birkhoff's theorem, and the strict stationarity concept. The work motivates a glimpse of stochastic processes proper. We consider the key classes related to Brownian motion (its quick existence through random Fourier transform) and the Poisson measures leading to a brief analysis of general classes of additive processes. These are used to show how various classes of families arise in applications. They include strong, strict and weak stationarities as well as the corresponding strict, strong and weak harmonizabilities. The key role of Bochner's V-boundedness principle is discussed. Numerous problems of interest, relating to queueing, birth-death processes, generalized random measures, and several new facts are discussed as applications with extended sketches. As a result, this part and particularly Chapter 8 received the largest amount of new material in comparison with the earlier presentation.

Chapter 6

Stopping Times, Martingales, and Convergence

A new tool, called a stopping time transformation, is introduced here; it plays an important role for a refined analysis of the subject. With this, some properties of stopped martingales, the Wald equation, the optional sampling theorem, and some convergence results are given. This work indicates the basic role played by the new tool in the modern developments of probability theory.

6.1 Stopping Times and Their Calculus

The concept of a random stopping originated in gambling problems. A typical example is the following. Suppose that a gambler plays a game in succession, starting with a capital X_1, and his fortunes are $X_2, X_3, \ldots$. At game n (or at time n) he decides to stop, *based on his present fortunes* $X_1, X_2, \ldots, X_n$ and on no future outcomes. Thus the n will be a function of these fortunes, which means that the r.v. n satisfies $\{\omega : n(\omega) \leq k\} \in \sigma(X_1, \ldots, X_k), k = 1, 2, \ldots$. We abstract this idea and introduce the new concept as follows.

Definition 1 Let $\{\mathcal{F}_n, n \geq 1\}$ be an increasing sequence of σ-subalgebras of Σ in a probability space (Ω, Σ, P). Then a mapping $T : \Omega \to \mathbb{N} \cup \{\infty\}$ is called a *stopping time* (or an *optional* or a *Markov time*) of the class $\{\mathcal{F}_n, n \geq 1\}$ if for each $k \in \mathbb{N}, [T = k] \in \mathcal{F}_k$, or, equivalently, $[T \leq k] \in \mathcal{F}_k$, [or $[T > k] \in \mathcal{F}_k$. A sequence $\{T_n, n \geq 1\}]$ of stopping times of the fixed class $\{\mathcal{F}_n, n \geq 1\}$ is termed a *stopping time process* if $T_n \leq T_{n+1}, n \geq 1$. The family $\mathcal{F}(T) = \{A \in \mathcal{F}_\infty = \sigma(\bigcup_n \mathcal{F}_n) : A \cap [T = k] \in \mathcal{F}_k$, all $k \geq 1\}$ is known as the class of *events prior* to T.

In this definition, if $P[T = +\infty] = 0$, then T is a *finite* stopping time, and if this probability is positive, then it is *nonfinite* (or extended real valued). In general a linear combination of stopping times is *not* a stopping time. Because of their use in our work, we detail some properties in the following:

Proposition 2 *Let $\{\mathcal{F}_n, n \geq 1\}$ be an increasing sequence of σ-subalgebras of (Ω, Σ, P) and $\{T, T_1, T_2, \ldots\}$ be a collection of stopping times of $\{\mathcal{F}_n, n \geq 1\}$. Then the following statements hold:*

(i) *$\mathcal{F}(T)$ is a σ-algebra, and if $[T = n] = \Omega$, then $\mathcal{F}(T) = \mathcal{F}_n, \mathcal{F}(T) \subset \mathcal{F}_\infty$.*
(ii) *$\max(T_1, T_2), \min(T_1, T_2)$, and $T_1 + T_2$ are $\{\mathcal{F}_n, n \geq 1\}$-stopping times.*
(iii) *T is $\mathcal{F}(T)$-measurable and if $\tau \geq T, \tau$ being $\mathcal{F}(T)$-measurable implies τ is an $\{\mathcal{F}_n, n \geq 1\}$-stopping time.*
(iv) *If $T_1 \leq T_2$, then $\mathcal{F}(T_1) \subset \mathcal{F}(T_2)$.*
(v) *If T_1, T_2 are any stopping times of $\{\mathcal{F}_n, n \geq 1\}$, then $\{[T_1 \leq T_2], [T_2 \leq T_1], [T_1 = T_2]\} \subset \mathcal{F}(T_1) \cap \mathcal{F}(T_2)$.*
(vi) *More generally, $\liminf_n T_n, \limsup_n T_n$ are stopping times of $\{T_n\}_{n \geq 1}$*

$$\{\mathcal{F}_n, n \geq 1\},$$

and if $\{T_n, n \geq 1\}$ is a monotone sequence of stopping times of the same $\mathcal{F}_n$-family, then $\lim_n \sup_n T_n = T$ is a stopping time, and $\lim_n \mathcal{F}(T_n) = \mathcal{F}(T)$.

Proof Consider (ii). Since

$$[\max(T_1, T_2) \leq n] = [T_1 \leq n] \cap [T_2 \leq n] \in \mathcal{F}_n, \quad n \geq 1,$$

and

$$[\min(T_1, T_2) \leq n] = [T_1 \leq n] \cup [T_2 \leq n] \in \mathcal{F}_n, \quad n \geq 1,$$

we deduce that $\min(T_1, T_2)$ and $\max(T_1, T_2)$ are stopping times of $\{\mathcal{F}_n, n \geq 1\}$. Note that the argument holds for $\sup_n T_n, \inf_n T_n$ also for sequences. Next

$$[T_1 + T_2 \leq n] = \bigcup_{k=0}^{n} [T_1 \leq k] \cap [T_2 \leq n - k] \in \mathcal{F}_n, \tag{1}$$

so that $T_1 + T_2$ is a stopping time of $\{\mathcal{F}_n, n \geq 1\}$; and (i) and (iii) are simple.

For (iv), let $A \in \mathcal{F}(T_1)$. Since $T_1 \leq T_2, A$ can be expressed as $A = A \cap [T_1 \leq T_2]$. Thus it suffices to show that the latter set is in $\mathcal{F}(T_2)$. But by definition, we need to verify that $A \cap [T_1 \leq T_2] \cap [T_2 \leq n] \in \mathcal{F}_n, n \geq 1$. Now for *any* stopping times T_1, T_2 of $\{\mathcal{F}_n, n \geq 1\}$ we have

$$\begin{aligned} A \cap [T_1 \leq T_2] \cap [T_2 \leq n] = A \cap [T_1 \leq n] \cap [T_2 \leq n] \\ \cap[\min(T, n) \leq \min(T_2, n)]. \end{aligned} \tag{2}$$

But by (ii), $\min(T_i, n)$ is a stopping time of $\{\mathcal{F}_n, n \geq 1\}, i = 1, 2$, and

$$[\min(T_i, n) < x] = [T_i < x]$$

if $x \leq n, = \Omega$ if $x > n$. Thus for all $x \geq 0, \min(T_i, n)$ is $\mathcal{F}_n$-measurable, so that the last set of (2) is in $\mathcal{F}_n$. Since $A \in \mathcal{F}(T_1)$, the first set is in $\mathcal{F}_n$. But T_2

is an $\{\mathcal{F}_n, n \geq 1\}$-stopping time. Thus the middle one is also in $\mathcal{F}_n$, so that $A \in \mathcal{F}(T_2)$. This proves (iv) a little more generally than asserted.

Regarding (v), the argument for (2) above with $A = \Omega$ shows that $[T_1 \leq T_2] \in \mathcal{F}(T_2)$, and hence its complement $[T_1 > T_2] \in \mathcal{F}(T_2)$. On the other hand, $\min(T_1, T_2)$ is a stopping time by (ii) and is measurable relative to

$$\mathcal{F}(\min(T_1, T_2)) \subset \mathcal{F}(T_i), i = 1, 2,$$

by (iv). Hence $[\min(T_1, T_2) < T_i] \in \mathcal{F}(T_i)$, by the first line of this paragraph, and $[\min(T_1, T_2) = T_2] \in \mathcal{F}(T_2) \Rightarrow [T_1 = T_2]$ and $[T_1 < T_2]$ belong to $\mathcal{F}(T_2)$. By interchanging 1 and 2, it follows that these events are also in $\mathcal{F}(T_1)$, and hence are in their intersection.

Finally, for (vi), since $\limsup_n T_n = \inf_k \sup_{n \geq k} T_n$ and by (ii) $\sup_{n \geq k} T_n$ is a stopping time of $\{\mathcal{F}_n, n \geq 1\}$, it follows that lim sup and lim inf and lim, if it exists, of stopping times of $\{\mathcal{F}_n, n \geq 1\}$ are again stopping times of the same family. Let $T_n(\omega) \to T(\omega), \omega \in \Omega$. If $T_n \uparrow T$, then by (iv) $\sigma(\bigcup_n \mathcal{F}(T_n)) \subset \mathcal{F}(T)$. To show there is equality, we verify that each generator of $\mathcal{F}(T)$ is in the left-side σ-algebra. Let $A \in \mathcal{F}_n$, and consider $A \cap [T > n]$, which is a generator of $\mathcal{F}(T)$, since this is just $A \cap [T \leq n]^c$. Now $A \cap [T > n] = \bigcup_{k \geq 1} A \cap [T_k > n] \in \sigma(\bigcup_k \mathcal{F}(T_k))$. Thus $\mathcal{F}(T) \subset \sigma(\bigcup_{k \geq 1} \mathcal{F}(T_k))$. Next let $T_n \downarrow T$. Then $\bigcap_{k \geq 1} \mathcal{F}(T_k) \supset \mathcal{F}(T)$. For the reverse inclusion, let $A \in \mathcal{F}(T_k)$ for all $k \geq 1$ (i.e., is in the intersection). Then we have, for $n \geq 1$,

$$\bigcup_{k=1}^{\infty} A \cap [T_k \leq n] = A \cap \left[\inf_k T_k \leq n\right] = A \cap [T \leq n] \in \mathcal{F}(T).$$

Hence $\bigcap_{k \geq 1} \mathcal{F}(T_k) \subset \mathcal{F}(T)$, and in both cases $\lim_k \mathcal{F}(T_k) = \mathcal{F}(T)$. This finishes the proof.

Remark The same concepts are meaningful if $\{\mathcal{F}_n, n \geq 1\}$ is replaced by $\{\mathcal{F}_t, t \geq 0\}$ with $\mathcal{F}_t \subset \mathcal{F}_{t'}$ for $t < t'$. Now $\mathcal{F}_{t+} = \bigcap_{s > t} \mathcal{F}_s \supset \mathcal{F}_t$ with a (possibly) strict inclusion. The equality must be *assumed*, which means a "right continuity" relative to inclusion order; similarly left continuity is defined. These problems do not arise in the discrete case, so that the above proposition is true in the continuous index case only if we assume this right order continuity. Thus the theory becomes delicate in the continuous case, and we are not going into it here.

However, the following deduction is immediate (for continuous time also):

Corollary 3 *Let $\{\mathcal{F}_t, t \geq 0\}$ be an increasing family of σ-algebras of (Ω, Σ, P) and T_1, T_2 be its stopping times. Then*

$$\mathcal{F}(T_1) \cap \mathcal{F}(T_2) = \mathcal{F}(\min(T_1, T_2)),$$

where $\mathcal{F}(T)$ denotes the σ-algebra of events prior to T, as in the proposition.

Proof Since $\min(T_1, T_2)$ is a stopping time $\leq T_i, i = 1, 2$, it is clear that $\mathcal{F}(\min(T_1, T_2)) \subset \mathcal{F}(T_1) \cap \mathcal{F}(T_2)$. For the opposite inclusion, let $A \in \mathcal{F}(T_1) \cap \mathcal{F}(T_2)$. Then

$$A \cap [\min(T_1, T_2) \leq t] = (A \cap [T_1 \leq t]) \cup (A \cap [T_2 \leq t]) \in \mathcal{F}_t,$$

since $A \in \mathcal{F}(T_i), i = 1, 2$, and the definition of $\mathcal{F}(T_i)$ implies this. Hence $A \in \mathcal{F}(\min(T_1, T_2))$, as asserted.

A standard and natural manner in which stopping times enter our analyses may be illustrated as follows. Let $X_1, X_2, \ldots$ be a sequence of r.v.s and $A \subset \mathbb{R}$ be an interval (or a Borel set). The first *time* the sequence X_n enters A is clearly

$$T_A = \inf\{n > 0 : X_n \in A\}, \tag{3}$$

where $T_A = +\infty$ if the set $\{\ \} = \emptyset$. If $\mathcal{F}_n = \sigma(X_1, \ldots, X_n)$, then we assert that T_A is a stopping time of $\{\mathcal{F}_n, n \geq 1\}$ (or of the X_n-sequence). Since

$$[T_A \leq k] = \bigcup_{m=1}^{k} [X_m \in A]$$

and

$$[T_A = k] = [X_k \in A, X_i \notin A, 1 \leq i \leq k - 1],$$

and these sets belong to $\mathcal{F}_k$, it follows that T_A is an r.v. and is an $\{\mathcal{F}_n, n \geq 1\}$-stopping time. It is called the *debut* of A. If $\{X_n, \mathcal{F}_n, n \geq 1\}$ is an arbitrary *adapted* (i.e., X_n is $\mathcal{F}_n$-measurable and $\mathcal{F}_n \subset \mathcal{F}_{n+1}$) sequence and T is an $\{\mathcal{F}_n, n \geq 1\}$-stopping time, we define $X_T : \Omega \to \bar{\mathbb{R}}$ to be that function which is given by the equation

$$(X_T)(\omega) = \begin{cases} X_{T(\omega)}(\omega), & \omega \in [T < \infty] \\ \limsup_n X_n(\omega), & \omega \in [T = +\infty]. \end{cases} \tag{4}$$

Then X_T is an extended real-valued measurable function as a composition of the two measurable functions $\{X_n, n \geq 1\}$ and T. It is always an r.v. if T is a finite stopping time (and extended valued otherwise). Similarly, if $T_n = \min(T, n)$, then $\{T_n, n \geq 1\}$ is a stopping time process of $\{\mathcal{F}_k, k \geq 1\}$ and $\{X_{T_n}, n \geq 1\}$ is a new sequence of r.v.s, obtained from the given set $\{X_n, n \geq 1\}$, by the stopping time transformations. Clearly this definition makes sense if $\{T_n, n \geq 1\}$ is, more generally, any finite stopping time process.

It is also important to note that X_T is $\mathcal{F}(T)$-measurable, since

$$[X_T < x] = \bigcup_{n \geq 1} [X_n < x] \cap [T = n] \in \mathcal{F}_\infty,$$

$$[X_n < x, T = n] \in \mathcal{F}_n, \quad n \geq 1.$$

Thus, $[X_T < x] \in \mathcal{F}(T), x \in \mathbb{R}$ (cf. Definition 1). Hence $\{X_{T_n}, \mathcal{F}(T_n), n \geq 1\}$ is an adapted sequence. It should be noted that we are fully using the linear ordering of the range space of the T. If it is only partially ordered, then the arguments get more involved. Here we consider only the simple case of $\mathbb{N}$. If the range of T is *finite*, then T is called a *simple stopping* time. [If the range is in $\overline{\mathbb{R}}^+$, and it is bounded (or finite) then T is a *bounded(simple)* stopping time.] Now we present an application which has interest in statistics.

6.2 Wald's Equation and an Application

The following result was established by A. Wald for the i.i.d. case and is useful in statistical sequential analysis (cf. Wald (1947)). As seen later, it can be deduced from the martingale theory using stopping time transformations (cf. Problem 6). However, the present account motivates the latter work, and illuminates that transformation.

Theorem 1 *Let $\{X_n, n \geq 1\}$ be independent r.v.s on (Ω, Σ, P) with a common mean,* $\sup_n E(|X_n|) < \infty, S_n = \sum_{k=1}^n X_k, \mathcal{F}_n = \sigma(X_1, \ldots, X_n)$, *and T be an $\{\mathcal{F}_n, n \geq 1\}$-stopping time. If $E(T) < \infty$ (so that T is a finite time), then*

$$E(S_T) = E(T)E(X_1). \tag{1}$$

If, further, the X_i have mean zero and a common variance, then

$$E(S_T^2) = E(T) \operatorname{Var} X_1. \tag{2}$$

Both (1) and (2) are deduced from the following observation of some independent interest. It appears in Neveu (1965) as an effective tool.

Lemma 2 *Let $\{Y_n, \mathcal{F}_n, n \geq 1\}$ be an adapted integrable sequence of r.v.s, and T be a bounded $\{\mathcal{F}_n, n \geq 1\}$- stopping time. Then setting $Y_0 = 0$, we have*

$$E(Y_T) = E\left(\sum_{0 \leq k < T} E^{\mathcal{F}_k}(Y_{k+1} - Y_k)\right). \tag{3}$$

Proof Let $T_n = \min(T, n)$, which is a stopping time of $\{\mathcal{F}_k, k \geq 1\}$ by Proposition 1.2. If $Y_n' = Y_{T_n}, T \leq n_0 < \infty$, and $0 \leq n \leq n_0$, we have

$$\begin{aligned} Y_{n+1}' &= \sum_{k=1}^{n} Y_k \chi_{[T=k]} + Y_{n+1}\chi_{[T>n]} = Y_n' + Y_{n+1}\chi_{[T>n]} - Y_n\chi_{[T>n]} \\ &= Y_n' + (Y_{n+1} - Y_n)\chi_{[T>n]}. \end{aligned} \tag{4}$$

Hence on noting that $[T > k] \in \mathcal{F}_k$, one has, from (4) by adding for $0 \le n < n_0$,

$$E(Y'_{n_0}) = \sum_{k=0}^{n_0-1} E[\chi_{[T>K]} E^{\mathcal{F}_k}(Y_{k+1} - Y_k)] = E\left(\sum_{0\le k<T} (E^{\mathcal{F}_k}(Y_{k+1}) - Y_k)\right),$$

which implies (3) since $\min(T, n_0) = T$ and $Y'_{n_0} = Y_T$, by the boundedness of T.

Proof of Theorem 1 First let $Y_n = S_n$ in the above lemma, with $\mathcal{F}_n = \sigma(X_1, \ldots, X_n)$. Then S_n is $\mathcal{F}_n$-adapted, and hence

$$\begin{aligned} E^{\mathcal{F}_n}(S_{n+1}) &= E^{\mathcal{F}_n}(S_n) + E^{\mathcal{F}_n}(X_{n+1}) \\ &= S_n + E(X_{n+1}) \quad \text{(since } \mathcal{F}_n \text{ and } X_{n+1} \text{ are independent)} \\ &= S_n + E(X_1) \quad \text{(by hypothesis).} \end{aligned} \tag{5}$$

Hence (3) becomes, if $T_n = \min(T, n)$ again,

$$\begin{aligned} E(S_{T_n}) &= E\left(\sum_{0\le k<T_n} (E^{\mathcal{F}_k}(S_{k+1}) - S_k)\right) = E\left(\sum_{0\le k<T_n} E(X_1)\right) \quad \text{[by (5)]} \\ &= E(T_n E(X_1)) = E(X_1)E(T_n). \end{aligned} \tag{6}$$

Letting $n \to \infty$, by the monotone convergence on the right and the dominated convergence on the left in (6), we get (1), since $S_{T_n} \to S_T$ a.e., $|S_{T_n}| \le |S_T|$, and S_T is integrable. In fact,

$$\begin{aligned} E(|S_T|) &\le E\left(\sum_{n=1}^{\infty} |X_n|\chi_{[T\ge n]}\right) \\ &= \sum_{n=1}^{\infty} E(|X_n|)P[T \ge n] \\ &\quad \text{(since } [T \ge n] \in \mathcal{F}_{n-1}, \text{ which is independent of } X_n) \\ &\le \sup_n E(|X_n|)E(T) < \infty. \end{aligned}$$

For (2), let us set $Y_n = S_n^2, T_n = \min(T, n)$, and $\mathcal{F}_n$ as given. Then $Y_{n+1} = S_n^2 + X_{n+1}^2 + 2X_{n+1}(X_1 + \cdots + X_n)$, so that (3) becomes

$$\begin{aligned} E(S_{T_n}^2) &= E\left(\sum_{0\le k<T_n} E^{\mathcal{F}_k}(Y_{k+1} - Y_k)\right) \\ &= E\left(\sum_{0\le k<T_n} E^{\mathcal{F}_k}(X_{k+1}^2)\right) \end{aligned}$$

$$[\text{since } E^{\mathcal{F}_k}(X_{k+1}) = E(X_{k+1}) = 0,$$
$$\text{by independence and the vanishing mean hypothesis}]$$
$$= \text{Var } X_1 \cdot E(T_n) \quad \text{(as before).} \tag{7}$$

However, S_{T_n} can be written as follows [cf. (4)]:

$$S_{T_{n+1}} = S_{T_n} + (S_{n+1} - S_n)\chi_{[T>n]} = S_{T_n} + X_{n+1}\chi_{[T \geq n+1]}.$$

By iteration, if $n > m$, we have

$$S_{T_{n+1}} - S_{T_m} = \sum_{k=m}^{n} X_{k+1}\chi_{[T \geq k+1]},$$

and hence

$$E(S_{T_{n+1}} - S_{T_m})^2 = \sum_{k=m}^{n} E(X_{k+1}^2)P[T \geq k+1] \quad \text{(by independence)}$$
$$= \text{Var } X_1 \cdot [E(T_{n+1}) - E(T_m)] \to 0 \quad \text{as } n, m \to \infty.$$

Since $S_{T_n} \to S_T$ a.e., this shows the convergence is also in $L^2(P)$. Thus $\lim_{n\to\infty} E(S_{T_n}^2) = E(S_T^2)$. Letting $n \to \infty$ in (7) and using this, we get (2). This completes the proof.

Remark If $\{X_n, n \geq 1\}$ is an i.i.d. sequence with two moments finite, then the hypothesis is satisfied for (1) and (2). Since $\{S_n, \mathcal{F}_n, n \geq 1\}$ is a martingale if $E(X_i) = 0$, part of the above proof is tailored to proving that $\{S_{T_n}, \mathcal{F}(T_n), n \geq 1\}$ is also a martingale. This result is proved in the next section. However, the above argument did not use martingale theory and is self-contained.

As a companion to this theorem, we present a *renewal application.* It deals with the behavior (or fluctuation) of the partial sums of i.i.d. positive r.v.s and corresponds to the first renewal (or replacement) of the lifetimes of objects (such as light bulbs).

Theorem 3 *Let $\{X_n, n \geq 1\}$ be a sequence of i.i.d. positive r.v.s such that $E(X_1) = \mu > 0$. If $S_n = \sum_{k=1}^{n} X_k, a > 0$, and*

$$T_a = \inf\{n \geq 1 : S_n > a\}, \tag{8}$$

then T_a is a stopping time of $\{\sigma(X_1, \ldots, X_n), n \geq 1\}, E(T_a^k) < \infty, k \geq 1$, and one has the **renewal assertion,** *$\lim_{a\to\infty} E(T_a)/a = \mu^{-1}$, where $\mu^{-1} = 0$ if $\mu = +\infty$. Further, for each $a > 0$, we have the bounds*

$$\frac{1}{2}E(T_a) \leq a/E(\min(X_1, a)) \leq E(T_a). \tag{9}$$

Proof Let $\mathcal{F}_n = \sigma(X_1, \ldots, X_n)$. Then it is evident that

$$[T_a = n] = [S_1 \le a, \ldots, S_{n-1} \le a, S_n > a] \in \mathcal{F}_n, n \ge 1, a > 0.$$

Thus T_a is a stopping time of $\{\mathcal{F}_n, n \ge 1\}$. To see that T_a has all moments finite, we proceed as follows. For each integer $m \ge 1$, consider the independent r.v.s formed from $\{X_n, n \ge 1\}$:

$$S_1' = \sum_{i=1}^{m} X_i, S_2' = \sum_{i=m+1}^{2m} X_i, \ldots, S_j' = \sum_{i=(j-1)m+1}^{jm} X_i, \ldots.$$

Since the X_i are i.i.d., so are the $S_j', j = 1, 2, \ldots$. Next let $p = P[S_j' \le a]$. Then the fact that $P[X_1 > 0] > 0$ implies that $p > 0$. Also $p < 1$ if m is chosen large enough. This is because $(1/m)\sum_{i=1}^{m} X_i \to E(X_1) > 0$ a.e. by the SLLN, and hence $\sum_{i=1}^{m} X_i \to \infty$ a.e. as $m \to \infty$. Thus for any $a > 0, P[S_j' > a] > 0$ if m is large enough. We fix such an m, so that $0 < p < 1$. Consider now for $jm \le n < (j+1)m$,

$$\begin{aligned} P[T_a \ge jm] &\le P[S_1 \le a, S_2 \le a, \ldots, S_{jm-1} \le a] \\ &\le P[S_1' \le a, S_2' \le a, \ldots, S_{j-1}' \le a] \\ &= p^{j-1} \quad \text{(by the i.i.d. property of the } S_i'\text{)} \\ &= p^{[n/m]-1} \le p^{(n/m)-2} = (p^{1/m})^n p^{-2} \\ &\quad \text{(with } [n/m] \text{ for the integral part of } n/m\text{)}. \end{aligned} \tag{10}$$

Hence for any integers $n_0 > 1$,

$$E(T_a^k) = \sum_{n=1}^{\infty} n^k P[T_a = n] \le \sum_{n=1}^{n_0} n^k P[T_a = n] + \sum_{n \ge n_0} e^{k \log n} P[T_a \ge n]. \tag{11}$$

Let $0 < t < -\log p^{1/m}$, where m and p are chosen for (10). Now we take n_0 large enough so that $k \log n < tn$ for $n \ge n_0$. Then (11) becomes

$$\begin{aligned} E(T_a^k) &\le \sum_{n=1}^{n_0} n^k P[T_a = n] + \sum_{n \ge n_0} e^{tn} p^{-2} (p^{1/m})^n \quad [\text{by } (10)] \\ &= \sum_{n=1}^{n_0} n^k P[T_a = n] + e^{tn_0} p^{(n_0/m)-2} (1 - e^t p^{1/m})^{-1} < \infty. \end{aligned} \tag{11'}$$

Since $k \ge 1$ is an arbitrary integer, (11′) shows that T_a has all moments finite. It also implies that for each $a, T_a < \infty$ a.e.

For the second part, since $\mu > 0$, let $0 < \alpha < \mu$. Set $X_t' = X_i$ if $X_i \le \alpha_0, =$ 0 otherwise. Then $\{X_n', n \ge 1\}$ are i.i.d. bounded r.v.s, with

$$E(X_1') = \int_0^{\alpha_0} x \, dF(x) \to \mu \text{ as } \alpha_0 \to \infty,$$

where F is the d.f. of X_1. Thus choose α_0 such that $\alpha < E(X_1) < \mu$. If $S'_n = \sum_{i=1}^n X'_i$, and $T'_a = \inf\{n \geq 1 : S'_n > a\}$, then by the first part T'_a is a stopping time with $E(T'_a) < \infty$. Hence by (1),

$$E(T'_a)E(X'_1) = E(S'_{T'_a}) = \sum_{n\geq 1} \int_{[T'_a=n]} S'_n \, dP$$

$$= \sum_{n\geq 1} \int_{[T'_a=n]} (S'_{n-1} + X'_n) \, dP \leq \alpha_0 + a. \tag{12}$$

Since by definition, $S'_n \leq S_n$, we get $T_a \leq T'_a$, and then (12) implies

$$E(T_a) \leq E(T'_a) \leq (\alpha_0 + a)/E(X'_1) \leq (\alpha_0 + a)/\alpha.$$

Hence

$$\limsup_{a\to\infty} E(T_a)/a \leq 1/\alpha. \tag{13}$$

Since $\alpha < \mu$ is arbitrary, $\limsup_{a\to\infty}[E(T_a)/a] \leq 1/\mu$. If $\mu = +\infty$, this gives $\lim_{a\to\infty} E(T_a/a) = 0$. If $0 < \mu < \infty$, then

$$E(S_{T_a}) = \sum_{n\geq 1} \int_{[T_a=n]} S_n \, dP \geq a \sum_{n\geq 1} P[T_a = n] = a.$$

Hence using (1), we get

$$a \leq E(S_{T_a}) = E(T_a)E(X_1) = \mu E(T_a). \tag{14}$$

Thus by (14) $\liminf_{a\to\infty} E(T_a/a) \geq 1/\mu$. This and the earlier inequality of (13) yield $\lim_{a\to\infty} E(T_a/a) = 1/\mu$.

The last statement follows from (14) on setting $X''_n = \min(X_n, a)$, and T''_a is defined with $S''_n = \sum_{k=1}^n X''_k$, the X''_k being i.i.d. [cf. (12) with $\alpha_0 = a$] :

$$a \leq E(S''_{T_a}) = E(T_a)E(X''_1) \leq 2a. \tag{15}$$

Since $T''_a \geq T_a$, (15) implies (9), and this completes the proof of the theorem.

The limit statement that $\lim_{a\to\infty} E(T_a/a) = \mu^{-1} \geq 0$ is called a *renewal theorem*. Many extensions and applications of the renewal theorem have appeared; for an account of these we refer the reader to Feller (1966).

6.3 Stopped Martingales

The availability of stopping time transformations enabled martingale analysis to enrich considerably. A few aspects of this phenomenon are discussed here.

As indicated at the beginning of Section 3.5, a martingale is a fair game and a submartingale is a favorable one. If the player (in either game) wishes to skip some bets in the series but participate at later times in an ongoing play because of boredom or some other reason (but *not* because of clairvoyance), then the instances of skipping become stopping times and it is reasonable to expect the observed or transformed process to be a (sub-) martingale if the original one is. That this expectation is generally correct can be seen from the next result, called the *optional stopping theorem*, due to Doob (1953). This is presented in two forms, one for "stopping" and the other for "sampling."

If $\{X_n, \mathcal{F}_n, n \geq 0\}$ is an adapted sequence of r.v.s and $\{V_n, n \geq 0\}$ is another sequence with $V_0 = 0$ such that for each $n \geq 1, V_n$ is $\mathcal{F}_{n-1}$-adapted and $\mathcal{F}_0 = \{\emptyset, \Omega\}$, then let

$$(V \cdot X)_n = \sum_{k=1}^{n} V_k(X_k - X_{k-1}). \tag{1}$$

It is clear that $\{(V \cdot X)_n, \mathcal{F}_n, n \geq 1\}$ is an adapted sequence. The $(V \cdot X)_n$-sequence is called the *predictable transform* of the X-sequence, and the V-sequence itself is termed *predictable*, since the present V_n is already determined (i.e., measurable relative to $\mathcal{F}_{n-1}$, the "past" σ-algebra). Thus the predictable sequence transforms the increments $(X_k - X_{k-1})$ into a new sequence $\{(V \cdot X)_n, \mathcal{F}_n, n \geq 1\}$. If the X-process is a martingale and the V-sequence is bounded, then the transformed process (1) is also called a "martingale transform," and the increment sequence $\{X_k - X_{k-1}, k \geq 1\}$, is sometimes termed a "martingale difference sequence," for the discrete time index t.

If T is a stopping time of $\{\mathcal{F}_n, n \geq 1\}$, and $V_n = \chi_{[T \geq n]}$, then

$$\{V_{n+1}, \mathcal{F}_n, n \geq 1\}$$

is a bounded (in fact $\{0,1\}$-valued) predictable sequence. On the other hand, if $\{V_{n+1}, \mathcal{F}_n, n \geq 1\}$ is *any* $\{0,1\}$-valued decreasing predictable sequence, it arises from a stopping time T of $\{\mathcal{F}_n, n \geq 1\}$. In fact, let $T = \inf\{n \geq 1\} : V_{n+1} = 0\}$, where $\inf(\emptyset) = +\infty$. Then $[T = n] \in \mathcal{F}_n$, and $[V_n = 1] = [T = n]$, so that T is the desired stopping time.

If $\{X_n, \mathcal{F}_n, n \geq 1\}$ is an adapted process, and T is a stopping time of $\{\mathcal{F}_n, n \geq 1\}$, then the adapted process $\{Y_n, \mathcal{F}(T_n), n \geq 1\}$ is called the transformed X-process by T, where $T_n = \min(T, n)$ (a stopping time by Proposition 1.2ii) and $Y_n = X_{T_n}$. If $V_n = \chi_{[T \geq n]}$, then $X \cdot V$ is written X^T, called a *stopped* process. This is a special form of (1). The problem considered here is this. If $\{X, \mathcal{F}_n\}_{n \geq 1}$ is a martingale, and T is a stopping time of $\{\mathcal{F}_n, n \geq 1\}$, when is the transformed process $\{X_{T_n}, \mathcal{F}(T_n)\}_{n \geq 1}$ also a martingale? Without further restrictions, $\{X_{T_n}, \mathcal{F}(T_n)\}_{n \geq 1}$ need not be a martingale, as the following example shows. Let $\{X_n, n \geq 1\}$ be i.i.d. with mean zero. If $S_n = \sum_{k=1}^n X_k, \mathcal{F}_n = \sigma(X_1, \ldots, X_n)$, then $\{S_n, \mathcal{F}_n, n \geq 1\}$ is a martingale. Consider

$$T = \inf\{n \geq 1 : S_n > 0\}. \tag{2}$$

Then T is an $\{\mathcal{F}_n, n \geq 1\}$-stopping time, and

$$E(S_T) = \sum_{n \geq 1} \int_{[T=n]} S_n \, dP > 0, \tag{3}$$

since $S_n > 0$ on each set $[T = n]$. But $T > 1$ because, by the queueing aspects of the random walk $\{S_n, n \geq 1\}, P[\sup_n S_n > 0] = 1$ (cf. Theorem 2.4.4). If the transformed process $\{S_1, S_T\}$ by $\{1, T\}$ is a martingale relative to $\{\mathcal{F}_1, \mathcal{F}(T)\}$, then we must have $0 = E(S_1) = E(S_T)$, and this contradicts (3). Incidentally, this shows that, in the Wald equation (1) of Section 2, $E(T) = +\infty$ must be true, so that the expected time for the random walk to reach the positive part of the real line is infinite!

We start with the following optional stopping assertion:

Proposition 1 *If $\{X_n, \mathcal{F}_n, n \geq 1\}$ is a martingale, $\{V_{n+1}, \mathcal{F}_n, n \geq 1\}$ is a predictable process such that $(+)\{(V \cdot X)_n, \mathcal{F}_n, n \geq 1\} \subset L^1(P)$, then $(+)$ is a martingale. In particular, this is true if the V_n are bounded. So for any stopping time T the stopped martingale process X^T is again a martingale.*

Proof This is immediate since by (1), $Y_n = (V \cdot X)_n$ is $\mathcal{F}_n$-measurable, and Y_n is integrable for each n, so we have

$$\begin{aligned} E^{\mathcal{F}_n}(Y_{n+1}) &= E^{\mathcal{F}_n}\left[\sum_{k=1}^n V_k(X_k - X_{k-1}) + V_{n+1}(X_{n+1} - X_n)\right] \\ &= (V \cdot X)_n + V_{n+1} E^{\mathcal{F}_n}(X_{n+1} - X_n) \quad \text{a.e.} \\ &\quad (V_{n+1} \text{being } \mathcal{F}_n\text{-measurable}) \\ &= Y_n \quad \text{a.e.} \quad (X_n \text{ being a martingale}). \end{aligned} \tag{4}$$

If the V_n are also bounded, then the integrability hypothesis is automatically satisfied, and in case $V_n = \chi_{[T \geq n]}$ this condition holds. The result follows.

Remarks 1. The above proposition is clearly true if the X-process is a sub- (or super-) martingale, with a similar argument. Further, taking expectations in (4) one has for the submartingale case [since $Y_n = X_{T_n}, T_n = \min(T, n)$]

$$\begin{aligned} E(X_1) = E(Y_1) &\leq E(Y_n) \\ &= \int_\Omega X_{\min(T,n)} \, dP \end{aligned}$$

$$= \sum_{k=1}^{n-1} \int_{[T=k]} X_k \, dP + \int_{[T \geq n]} X_n \, dP$$

$$\leq \sum_{k=1}^{n-1} \int_{[T=k]} X_n \, dP + \int_{[T \geq n]} X_n \, dP$$

(since $[T = k] \in \mathcal{F}_k$ and we used the submartingale property of the X_n)

$$= \int_{\Omega} X_n \, dP = E(X_n). \tag{5}$$

There is equality in (5) throughout in the martingale case.

2. It is of interest to express X^T more explicitly. By definition,

$$\begin{aligned}(X^T)_n = Y_n &= \sum_{k=1}^{n} \chi_{[T \geq k]}(X_k - X_{k-1}) \\ &= X_1\chi_{[T=1]} + X_2\chi_{[T=2]} + \cdots + X_{n-1}\chi_{[T=n-1]} + X_n\chi_{[T \geq n]} \\ &= X_{\min(T,n)}. \end{aligned} \tag{6}$$

We now present the famous *optional sampling theorem*, generalizing the above result.

Theorem 2 (Doob) *Let $\{X_n, \mathcal{F}_n, n \geq 1\}$ be a submartingale and $\{T_n, n \geq 1\}$ a stopping time process of $\{\mathcal{F}_n, n \geq 1\}$, where $P[T_n < \infty] = 1$. If $Y_n = X_{T_n}$, suppose that* (i) *$E(Y_n^+) < \infty$ and* (ii) *$\liminf_{n \to \infty} E(X_n^+ \chi_{[T_k > n]}) = 0, k \geq 1$. Then $\{Y_n, \mathcal{F}(T_n), n \geq 1\}$ is a submartingale. In particular, if the X_n-process is a positive (super) martingale, then* (i) *holds,* (ii) *can be omitted, so that $\{Y_n, \mathcal{F}(T_n), n \geq 1\}$ is a (super) martingale. The same conclusion obtains for the given X_n-process (not necessarily positive) if either T_n is bounded for each n or $X_n \leq E^{\mathcal{F}_n}(Z)$ for some $Z \in L^1(P), n \geq 1$.*

Proof It is sufficient to consider a pair of stopping times, say S and T, with $S \leq T$, for which (i) and (ii) are satisfied. We need to show, since X_S is $\mathcal{F}(S)$-adapted and $\mathcal{F}(S) \subset \mathcal{F}(T)$, by Proposition 1.2, that $E(|X_S|) < \infty, E(|X_T|) < \infty$ and

$$E^{\mathcal{F}(S)}(X_T) \geq X_S \quad \text{a.e.} \tag{7}$$

Indeed, let $A \in \mathcal{F}_n$ and consider, for the integrability of X_S,

$$\begin{aligned}\int_{A \cap [S \geq n]} X_n \, dP &= \int_{A \cap [S=n]} X_n \, dP + \int_{A \cap [S>n]} X_n \, dP \\ &\leq \int_{A \cap [S=n]} X_S \, dP + \int_{A \cap [S>n]} X_{n+1} \, dP\end{aligned}$$

(since $\{X_n, \mathcal{F}_n\}_{n \geq 1}$ is a submartingale and $[S > n] \in \mathcal{F}_n$)

$$\leq \int_{A\cap[S=n]} X_S\, dP + \int_{A\cap[S=n+1]} X_S\, dP$$

$$+ \int_{A\cap[S>n+1]} X_{n+2}\, dP$$

(by iteration)

$$\vdots$$

$$\leq \sum_{k=n}^{n+\ell} \int_{A\cap[S=k]} X_S\, dP + \int_{A\cap[S>n+\ell+1]} X_{n+\ell+2}\, dP. \tag{8}$$

Since $X_n \leq X_n^+$, we can use (ii) with $T_k = S$, in (8), so that (letting $\ell \to \infty$)

$$\int_{A\cap[S\geq n]} X_n\, dP \leq \int_{A\cap[S\geq n]} X_S\, dP. \tag{9}$$

Thus if $A = \Omega$, and $n = 1$ (so that $[S \geq n] = \Omega$) in (9), we get

$$-\infty < E(X_1) \leq E(X_S) \leq E(X_S^+) < \infty,$$

using (i). Since $E(\cdot)$ is a Lebesgue integral, this implies $E(|X_S|) < \infty$. Similarly $E(|X_T|) < \infty$. Also, (7) follows from (8) and (9). In fact, if $A_1 \in \mathcal{F}(S)$, then $A_1 \cap [S = k] \in \mathcal{F}_k$ and $S \leq T$ implies $[S = k] \subset [T \geq k]$. Hence letting $A = A_1 \cap [S = k]$ in (9) and replacing S by T there, we get

$$\int_{A_1\cap[S=k]} X_S\, dP = \int_{A_1\cap[S=k]} X_k\, dP = \int_{A_1\cap[S=k]\cap[T\geq k]} X_k\, dP$$

$$\leq \int_{A_1\cap[S=k]\cap[T\geq k]} X_T\, dP \qquad [\text{by } (9)]$$

$$= \int_{A_1\cap[S=k]} X_T\, dP.$$

Hence summing over $k = 1, 2, \ldots$, we get

$$\int_{A_1} X_S\, dP \leq \int_{A_1} X_T\, dP,$$

which is (7).

For the second part, if X_n is a positive martingale or supermartingale, (ii) is unnecessary [cf. (8)], and we verify (i). The positivity implies

$$0 \leq E(X_n) \leq E(X_1) < \infty,$$

with equality in the martingale case. Now for any stopping time T of $\{\mathcal{F}_n, n \geq 1\}$, $\min(T, k)$ is also one, and by Proposition 1, $\{X_1, X_{\min(T,k)}\}$ is a martingale or a supermartingale for $\{\mathcal{F}_1, \mathcal{F}(\min(T, k))\}$, so that

$$0 \leq E(X_{\min(T,k)}) \leq E(X_1). \tag{10}$$

But $\lim_{k\to\infty} X_{\min(T,k)} = X_T$ a.e. Thus by Fatou's lemma, (10) implies $E(X_T) < \infty$, which is (i).

For the last part, clearly (i) and (ii) both hold if T is bounded. On the other hand, if the submartingale is closed on the right by Z, then (by definition)

$$\{X_n, \mathcal{F}_n, 1 \leq n < \infty, E^{\mathcal{F}_\infty}(Z), \mathcal{F}_\infty\}$$

is again a submartingale. So also is $\{X_n^+, \mathcal{F}_n, 1 \leq n < \infty, E^{\mathcal{F}_\infty}(Z^+), \mathcal{F}_\infty\}$ and hence $E(X_n^+) \leq E(Z^+) \leq E(|Z|) < \infty$. Since by the preceding, if $\tilde{T}_n^k = \min(T_n, k)$, then $\tilde{T}_n^k \leq \tilde{T}_{n+1}^k$ and these are bounded stopping times so that $\{X_{\tilde{T}_n^k}, \mathcal{F}(\tilde{T}_n^k), k \geq 1\}$ is a submartingale, we have

$$\begin{aligned} E(|X_{\tilde{T}_n^k}|) &= 2E(X^+_{\tilde{T}_n^k}) - E(X_{\tilde{T}_n^k}) \\ &= 2\sum_{j=1}^{k-1} \int_{[T_n=j]} X_j^+ \, dP + 2\int_{[T_n\geq k]} X_k^+ \, dP - E(X_{\tilde{T}_n^k}) \\ &\leq 2\sum_{j=1}^{k-1} \int_{[T_n=j]} X_k^+ \, dP + 2\int_{[T_n\geq k]} X_k^+ \, dP - E(X_1^+) \\ &\quad \text{(by the submartingale property and } X_1^+ = X^+_{\tilde{T}_n^1}) \\ &\leq 2E(X_k^+) - E(X_1) \leq 2E(Z^+) - E(X_1) < \infty. \end{aligned} \tag{11}$$

Now letting $k \to \infty$ on the left, using Fatou's lemma, we get $E(|X_{T_n}|) < \infty$. But

$$\begin{aligned} 0 \leq E(X_n^+ \chi_{[T>n]}) &= E(E^{\mathcal{F}_{n-1}}(X_n^+ \chi_{[T>n]})) \\ &= E(\chi_{[T>n]} E^{\mathcal{F}_{n-1}}(X_n^+)) \\ &\leq E(\chi_{[T>n]} E^{\mathcal{F}_{n-1}}(Z^+)) \\ &= E(Z^+ \chi_{[T>n]} \to 0 \quad \text{as} \quad n \to \infty. \end{aligned} \tag{12}$$

Hence (ii) is also true. This proves the theorem.

The last part of the above proof actually gives a little more than the assertion. We record this for reference. The bound follows from (11).

Corollary 3 *Let $\{X_n, \mathcal{F}_n, n \geq 1\}$ be a submartingale. If $\sup_n E(|X_n|) < \infty$, and T is any stopping time of $\{\mathcal{F}_n, n \geq 1\}$, then $E(|X_T|) \leq 3\sup_n E(|X_n|)$. If the submartingale is uniformly integrable and $\{T_n, n \geq 1\}$ is a stopping time process of $\{\mathcal{F}_n, n \geq 1\}$, then $\{X_{T_n}, \mathcal{F}(T_n), n \geq 1\}$ is a submartingale.*

As an application of the optional stopping results we present a short derivation (but not materially different from the earlier proof, though the new viewpoint is interesting) of the martingale maximal inequality (cf. Theorem 3.5.6).

Theorem 4 *Let $\{X_k, \mathcal{F}_k, 1 \leq k \leq n\}$ be a submartingale and $\lambda \in \mathbb{R}$. Then*

$$\lambda P\left[\max_{k\leq n} X_k > \lambda\right] \leq \int_{[\max_{k\leq n} X_k > \lambda]} X_n \, dP \leq E(X_n^+), \tag{13}$$

$$\lambda P\left[\min_{k\leq n} X_k < \lambda\right] \geq E(X_1) - \int_{[\min_{k\leq n} X_k \geq \lambda]} X_n \, dP. \tag{14}$$

Proof Let $T_1 = \inf[k \geq 1 : X_k > \lambda]$, with $T_1 = n$, if the set $[\,] = \emptyset$. Let $T_2 = n$. Then $T_1 \leq n$ and both T_1, n are (bounded) stopping times of $\{\mathcal{F}_k, 1 \leq k \leq n\}$. Thus $\{X_{T_i}, \mathcal{F}(T_i)\}_1^2$ is a submartingale by Theorem 2. Hence $E(X_{T_2}|\mathcal{F}(T_1) \geq X_{T_1}$ a.e. Since $A_\lambda = [X_{T_1} > \lambda] \in \mathcal{F}(T_1)$, we have

$$\begin{aligned}\int_{A_\lambda} X_n \, dP = \int_{A_\lambda} X_{T_2} \, dP &\geq \int_{A_\lambda} X_{T_1} \, dP \\ &= \sum_{k=1}^{n} \int_{[T_1=k]} X_{T_1} \, dP \quad (\text{since } A_\lambda = \bigcup_{k=1}^{n} [T_1 = k]) \\ &= \sum_{k=1}^{n} \int_{[T_1=k]} X_k \, dP \geq \lambda \sum_{k=1}^{n} \int_{[T_1=k]} dP \\ &= \lambda P(A_\lambda).\end{aligned}$$

The second inequality is similarly proved, and the details are omitted.

We now briefly discuss improvements obtainable for the martingale convergence statements using the ideas of stopping times. Here we present a proof of the martingale convergence without the use of the maximal inequality (but with the optional stopping result instead, i.e., with Theorem 2 above). This further illuminates the theory. Here is the announced short proof of the martingale convergence.

Theorem 5 *Let $\{X_n, \mathcal{F}_n, n \geq 1\}$ be an $L^1(P)$-bounded martingale. Then $X_n \to X_\infty$ a.e. and $E(|X_\infty|) \leq \liminf_n E(|X_n|)$.*

Proof By Lemma 3.5.5, which is purely measure theoretic and does not involve any martingale convergence, $X_n = X_n^{(1)} - X_n^{(2)}$, and $\{X_n^{(i)}, \mathcal{F}_n, n \geq 1\}$ is a positive martingale, $i = 1, 2$. Thus for this proof it may be assumed that $X_n \geq 0$ itself. We give an indirect argument.

Suppose that a positive martingale does not converge a.e. Then there exist $0 < a < b < \infty$ such that the following event,

$$A = \left\{\omega : \liminf_n X_n(\omega) < a < b < \limsup_n X_n(\omega)\right\}$$

must have positive probability. Let us define a stopping time process $\{T_n, n \geq 1\}$ of $\{\mathcal{F}_n, n \geq 1\}$ such that $\{X_{T_n}, \mathcal{F}(T_n), n \geq 1\}$ is not a martingale, contradicting Theorem 2 and thereby proving the result.

Let $T_0 = 1$, and define $\{T_n, n \geq 1\}$ as follows:

$$T_1(\omega) = \inf\{n > 1 : X_n(\omega) > b\}, \quad \inf(\emptyset) = +\infty,$$

and inductively set for $k > 1$

$$T_{2k}(\omega) = \inf\{n > T_{2k-1}(\omega) : X_n(\omega) < a\},$$
$$T_{2k+1}(\omega) = \inf\{n > T_{2k}(\omega) : X_n(\omega) > b\}, \quad \omega \in \Omega.$$

But $P(A) > 0$ by assumption. Thus $\{T_n, n \geq 1\}$ is an increasing sequence of functions with $T_n(\omega) \to \infty, \omega \in \Omega$. Since $[T_k = n]$ is determined by $X_1, \ldots, X_n$ only, and the sets in braces are measurable, it follows that $[T_k = n] \in \mathcal{F}_n$, so that $\{T_n, n \geq 1\}$ is an $\{\mathcal{F}_n, n \geq 1\}$-stopping time process. Moreover, $T_{2k} < \infty$ implies $X_{T_{2k}} < a$, and similarly, $X_{T_{2k+1}} > b$ if $T_{2k+1} < \infty$. Hence

$$\begin{aligned} E(X_{T_{2k}} \chi_{[T_{2k}<\infty]}) &< aP[T_{2k} < \infty] \leq aP[T_{2k-1} < \infty] \\ &< bP[T_{2k-1} < \infty] \leq E(X_{T_{2k-1}} \chi_{[T_{2k-1}<\infty]}). \end{aligned} \tag{15}$$

However, by the optional sampling theorem (cf. Theorem 2),

$$\{X_{T_k}, \mathcal{F}(T_k), k \geq 1\}$$

is again a positive martingale, so that in particular the expectations are constant. Hence $0 \leq E(X_{T_k}) < \infty$ and is the same for all k. Moreover,

$$E(X_{T_k}) = \int_{[T_k<\infty]} X_{T_k}\, dP < \infty. \tag{16}$$

Thus

$$\begin{aligned} 0 &= E(X_{T_{2k-1}}) - E(X_{T_{2k}}) \\ &= \int_{[T_{2k-1}<\infty]} X_{T_{2k-1}}\, dP - \int_{[T_{2k}<\infty]} X_{T_{2k}}\, dP \\ &\geq bP[T_{2k-1} < \infty] - aP[T_{2k-1} < \infty] \quad \text{[by (15)]} \\ &= (b-a)P[T_{2k-1} < \infty] \geq (b-a)P(A) > 0, \end{aligned}$$

since $A \subset [T_{2k-1} < \infty]$. This contradiction shows that $P(A) = 0$, and we must have $X_n \to X_\infty$ a.e. The last statement is a consequence of Fatou's lemma. This completes the proof. The convergence of $L^1(P)$-bounded sub- (or super-) martingales follows from this, as before (c.f. Theorem 3.5.11).

The argument here is due to J. Horowitz. We can extend it, even without the optional stopping result and by weakening the integrability hypothesis to include some non-$L^1(P)$-bounded (sub-) martingales. In this case Lemma 3.5.5 is not applicable. The generalization is adapted from one due to Y. S. Chow, who obtained the result for the directed index sets satisfying a "Vitali condition." The details of the latter are spelled out in the first author's monograph [(1979), Theorem IV.4.1]. Its proof is again by contradiction. We therefore present this result without further detail, because the basic idea is similar to that of the above theorem.

Theorem 6 *Let $\{X_n, \mathcal{F}_n, n \geq 1\}$ be a [not necessarily $L^1(P)$-bounded] submartingale such that for each stopping time T of $\{\mathcal{F}_n, n \geq 1\}$ we have $E(X_T^+) < \infty$. Then $X_n \to X_\infty$ a.e., but X_∞ may take infinite values on a set of positive probability.*

We thus end this chapter with these results using stopping times. A few complements are included in the exercises, as usual.

Exercises

1. Complete the proofs of the omitted parts of Proposition 1.2.

2. Find a pair of stopping times T_1, T_2 of a stochastic base $\{\mathcal{F}_n, n \geq 1\}$ such that αT_1 and $T_1 - T_2$ are not stopping times of the base for an $\alpha \in \mathbb{R}$.

3. Let $\{X_n, n \geq 1\}$ be i.i.d. and $X_1 > 0$ a.e. If $S_n = \sum_{k=1}^n X_k$ and $c > 0$, let $\tau_c = \max\{n \geq 1 : S_n \leq c\}$. Show that τ_c is not a stopping time of $\mathcal{F}_n = \sigma(X_1, \ldots, X_n)$, but that $T_c = \tau_c + 1$ is one, and $T_c = \inf\{n \geq 1 : S_n > c\}$.

4. Let $\{X_n, n \geq 1\}$ be i.i.d. and $P[|X_1| > 0] > 0$. If $S_n = \sum_{k=1}^n X_k$ and $0 < a, b < \infty$, let $T_{ab} = \inf\{n \geq 1 : S_n \in [-a, b]\}$. Show that $P[T_{ab} < \infty] = 1$ and in fact $E(T_{ab}^k) < \infty, k \geq 1$, slightly extending part of Theorem 2.3.

5. Let $\{X_n, \mathcal{F}_n, n \geq 1\}$ be a submartingale and $\{T_n, n \geq 1\}$ be an integrable stopping time process of $\{\mathcal{F}_n, n \geq 1\}$. Suppose there exists an adapted sequence of positive r.v.s $\{Y_n, \mathcal{F}_n, n \geq 1\}$ such that $E^{\mathcal{F}_n}(Y_{n+1}\chi_{[T_j > n]})$ is uniformly bounded for all $j \geq 1$ and that $|X_n| \leq \sum_{j=1}^n Y_j$ a.e. on $[T_j \geq n]$ for $j \geq 1$. Then show that $\{X_{T_n}, \mathcal{F}(T_n), n \geq 1\}$ is a submartingale by verifying the hypothesis of Theorem 3.2. In particular, Y_n can be the absolute increments of the X_n-process.

6. Let $\{X_n, n \geq 1\}$ be an i.i.d. sequence of r.v.s and $S_n = \sum_{k=1}^n X_k$. Suppose that the moment-generating function $M(\cdot)$ of the X_k exists in some nondegenerate interval around the origin. If T is a stopping time of

$$\{\mathcal{F}_n = \sigma(S_1, \ldots, S_n), n \geq 1\}$$

such that $E^{\mathcal{F}_n}(|S_{n+1}|\chi_{[T \geq n]}) \leq K_0 < \infty$ for all n and $E(T) < \infty$, then show that $\{Y_n = e^{tS_n}/(M(t))^n, \mathcal{F}_n, n \geq 1\}$ is a martingale and if $T_0 = 1$, then $\{Y_{T_0}, Y_T\}$ is a martingale for $\{\mathcal{F}_1, \mathcal{F}(T)\}$ and the *fundamental identity of sequential analysis* obtains:

$$E(Y_T) = E(Y_{T_0}) = E(Y_1) = 1.$$

(*Hint:* Use the result of Problem 5 in showing the martingale property.) Deduce, from this result, the conclusions of Theorem 2.1 after justifying the differentiation under the integral sign.

7. Let X be an integrable r.v. on (Ω, Σ, P) and $\{\mathcal{F}_n, n \geq 1\}$ be a stochastic base with $\Sigma = \sigma(\bigcup_n \mathcal{F}_n)$. If T_1, T_2 are a pair of stopping times of $\{\mathcal{F}_n, n \geq 1\}$, show that, for the martingale $\{X_n = E^{\mathcal{F}_n}(X), \mathcal{F}_n, n \geq 1\}, X_{\min(T_1,T_2)} = E^{\mathcal{F}(T_2)}(Y) = E^{\mathcal{F}(\min(T_1,T_2))}(X)$, where $Y = E^{\mathcal{F}(T_1)}(X)$. [Use Theorem 3.2.] Deduce that

$$\text{(i)} \;\; E^{\mathcal{F}(T_1)}E^{\mathcal{F}(T_2)} = E^{\mathcal{F}(T_2)}E^{\mathcal{F}(T_1)} = E^{\mathcal{F}(\min(T_1,T_2))},$$

and

$$\text{(ii)} \;\; \mathcal{F}(T_1) \cap \mathcal{F}(T_2) = \mathcal{F}(\min(T_1, T_2)).$$

(For (ii), note that $A \in \mathcal{F}(T_1) \cap \mathcal{F}(T_2)$ implies $A \cap [\min(T_1, T_2) \leq n] \in \mathcal{F}_n, n \geq 1$. The analysis needs some thought. See, e.g., the first author's book (1979), p. 351, eq. (6).)

Chapter 7

Limit Laws for Some Dependent Sequences

This chapter is devoted to a brief account of some limit laws including the central limit theorem,and SLLN, for certain classes of dependent random variables. These cover martingale increments and stationary sequences. A limit theorem and a general problem for a random number of certain dependent random variables are also considered in some detail. Moreover, Birkhoff's ergodic theorem, its comparison with SLLN, and a motivation for strict stationarity are discussed.

7.1 Central Limit Theorems

In Section 5.6 we saw a central limit theorem for m-dependent r.v.s and its application to a limit distribution of certain estimators. Here we present a similar theorem for square integrable martingale increments sequences. It will facilitate the analysis if we first establish the following key technical result based on arguments used in a classical problem. Unfortunately, the conditions assumed are not well motivated except that they are needed for the following proofs. Again $\lim_n k_n = +\infty$ will be assumed without mention.

Proposition 1 *Let $\{X_{nk}, \mathcal{F}_{nk}, 1 \leq k \leq k_n\}, n \geq 1$, be a sequence of rowwise adapted sequences of r.v.s with $\mathcal{F}_{nk} \subset \mathcal{F}_{n(k+1)}$. Let*

$$Y_n(t) = \prod_{k=1}^{k_n}(1 + itX_{nk}), \quad t \in \mathbb{R}, \ i = \sqrt{-1}.$$

Suppose that the following conditions hold:

(i) $E(Y_n(t)) \to 1$ *as* $n \to \infty, t \in \mathbb{R}$,
(ii) $\{Y_n(t), n \geq 1\} \subset L^1(P)$ *is uniformly integrable, for each* $t \in \mathbb{R}$,
(iii) $\sum_{k=1}^{k_n} X_{nk}^2 \xrightarrow{P} 1$,

(iv) X_{nk} *are strongly infinitesimal in that* $\max_{1\le k\le k_n}|X_{nk}| \xrightarrow{P} 0$ *as* $n \to \infty$.

Then $S_n = \sum_{k=1}^{k_n} X_{nk} \xrightarrow{D}$ *to an r.v. that is* $N(0,1)$.

Proof It is to be shown that the ch.f. $t \mapsto E(e^{itS_n})$ tends to $t \mapsto e^{-t^2/2}$, as $n \to \infty$. Consider a representation of e^{ix} as (and this is the key observation):

$$e^{ix} = (1+ix)\exp\{r(x) - \frac{1}{2}x^2\}, \tag{1}$$

where $r : \mathbb{R} \to \mathbb{C}$ is defined by the above equation. We now assert that

$$|r(x)| \le \frac{|x|^3}{3} + \frac{|x|^4}{4} \le \frac{2}{3}|x|^3 \quad \text{if } |x| \le 1. \tag{2}$$

This is elementary, but it needs a little care. Thus taking logarithms and expanding (1),

$$\begin{aligned} r(x) &= ix + \frac{x^2}{2} - \left[ix - \frac{(ix)^2}{2} + \frac{(ix)^3}{3} - \cdots\right] \\ &= -i\left[\sum_{n\ge1}(-1)^n \frac{x^{2n+1}}{2n+1}\right] + \sum_{n\ge2}(-1)^n \frac{x^{2n}}{2n}. \end{aligned}$$

Hence, for the complex number $r(x)$, one has:

$$|r(x)|^2 = \left(\sum_{n\ge1}(-1)^n \frac{|x|^{2n+1}}{2n+1}\right)^2 + \left(\sum_{n\ge2}(-1)^n \frac{x^{2n}}{2n}\right)^2.$$

But each of the terms inside parentheses is positive if $|x| = \alpha < 1$, so that one has

$$|r(x)|^2 \le \left[\sum_{n\ge1}(-1)^n \frac{\alpha^{2n+1}}{2n+1} + \sum_{n\ge2}(-1)^n \frac{\alpha^{2n}}{2n}\right]^2$$

because $a, b > 0$ implies $a^2 + b^2 \le (a+b)^2$, so that

$$\begin{aligned} |r(\alpha)|^2 &\le \left[\int_0^\alpha \frac{u^2}{1+u^2}\,du + \int_0^\alpha \frac{u^3}{1+u^2}\,du\right]^2 \\ &\le \left[\frac{\alpha^3}{3} + \frac{\alpha^4}{4}\right]^2 \le \left(\frac{2}{3}\alpha^3\right)^2, \end{aligned} \tag{3}$$

since $0 \le \alpha \le 1$. This implies (2). Replacing x by tX_{nk} and multiplying over $1 \le k \le k_n$, (1) becomes

$$e^{itS_n} = Y_n(t)\exp\left\{-\frac{t^2}{2}\sum_{k=1}^{k_n} X_{nk}^2 + \sum_{k=1}^{k_n} r(tX_{nk})\right\} = Y_n(t)Z_n(t) \quad \text{(say)},$$
$$= Y_n(t)e^{-t^2/2} + Y_n(t)[Z_n(t) - e^{-t^2/2}]. \tag{4}$$

Taking expectations and letting $n \to \infty$, the left side gives a sequence of ch.f.s that tends to the desired normal $N(0,1)$ ch.f. [with the first term because of (i) of the hypothesis] if we show that the second term on the right side of (4) goes to 0 in $L^1(P)$ using conditions (ii)–(iv). This is verified as follows. By (3),

$$\left|\sum_{k=1}^{k_n} r(tX_{nk})\right| \le \frac{2}{3}\sum_{k=1}^{k_n} |t|^3 |X_{nk}|^3$$
$$\le \frac{2}{3}|t|^3 \max_{1\le k\le k_n} |X_{nk}| \sum_{k=1}^{k_n} X_{nk}^2 \xrightarrow{P} 0,$$

using (iii) and (iv). This implies [by (iii)] that

$$Z_n(t) \xrightarrow{P} e^{-t^2/2} \quad \text{as } n \to \infty.$$

Also, $\{e^{itS_n}, n \ge 1\}$, being bounded, is uniformly integrable. Now by (ii) $e^{-t^2/2}$, being a finite constant for each $t \in \mathbb{R}$, $\{Y_n(t)e^{-t^2/2}, n \ge 1\}$ is uniformly integrable. Hence

$$Y_n(t)[Z_n(t) - e^{-t^2/2}] = e^{itS_n} - Y_n(t)e^{-t^2/2}$$

gives a uniformly integrable sequence which $\xrightarrow{P} 0$. Thus it also goes to zero in $L^1(P)$ by the Vitali convergence (cf. Theorem 1.4.4), completing the proof.

This result will be used to obtain a central limit theorem for martingale increments (double) arrays. Recall that an adapted integrable process $\{X_n, \mathcal{F}_n, n \ge 1\}$ qualifies as a martingale increments sequence iff for each $n \ge 1, E^{\mathcal{F}_n}(x_{n+1}) = 0$ a.e. (cf. Proposition 3.5.2). Similarly, if $\{X_{nk}, \mathcal{F}_{nk}, 1 \le k \le k_n, n \ge 1\}$ is a double array of martingale increments sequences, and if for each $n, \mathcal{F}_{nk} \subset \mathcal{F}_{nk'}$ for $k \le k'$, then $E^{\mathcal{F}_{nk}}(X_{n(k+1)}) = 0$ a.e. For such a family the following result holds:

Theorem 2 *Let the double array $\{X_{nk}, \mathcal{F}_{nk}, 1 \le k \le k_n, n \ge 1\}$ of martingale increments satisfy the three conditions:*

(i) $E\left(\max_{k\le k_n} |X_{nk}|^2\right) \le C < \infty, \quad n \ge 1$

(ii) $\lim_n P\left[\max_{k\le k_n} |X_{nk}| \ge \varepsilon\right] = 0 \quad$ for each $\varepsilon > 0$,

(iii) $\sum_{k=1}^{k_n} X_{nk}^2 \xrightarrow{P} 1$ as $n \to \infty$.

Then $S_n = \sum_{k=1}^{k_n} X_{nk} \xrightarrow{D}$ *to an r.v. which is* $N(0,1)$.

Remark After the proof we shall indicate how (i) and (ii) are consequences of the classical Lindeberg condition. Also other forms of (iii), and some specializations of this result will be recorded.

Proof We consider a suitable predictable transform of the X_{nk}-sequence so that the new r.v.s satisfy the conditions of Proposition 1 and that the partial sums of the X_{nk} and of the transformed ones are asymptotically equal in probability. Then the result follows from the preceding one.

Because of (iii) we define $X'_{nk} = X_{nk}$ on the set $[\sum_{i=1}^{k-1} X_{ni}^2 \leq 2]$, and $=0$ otherwise, where $X'_{n1} = X_{n1}$. Because the above set is in $\mathcal{F}_{n(k-1)}$, this becomes a useful mapping, a special case of what was called a predictable transform, defined in Exercise 32 of Chapter 3, and discussed again at the beginning of Section 3. Let $S'_n = \sum_{k=1}^{k_n} X'_{nk}$ and note that $P[|S'_n - S_n| \geq \varepsilon] \to 0$ as $n \to \infty$ for any $\varepsilon > 0$. In fact,

$$\begin{aligned}
P[|S'_n - S_n| \geq \varepsilon] &\leq P\left[\sum_{k=1}^{k_n} |X'_{nk} - X_{nk}| \geq \varepsilon\right] \\
&\leq P\left[\bigcup_{k=1}^{k_n} [|X'_{nk} - X_{nk}| \geq \varepsilon/k_n]\right] \\
&\leq P\left[\bigcup_{k=1}^{k_n} [X'_{nk} \neq X_{nk}]\right] \\
&\leq P\left[\sum_{k=1}^{k_n} X_{nk}^2 > 2\right] \to 0 \quad \text{[by (iii)]}. \qquad (5)
\end{aligned}$$

In particular, the last two inequalities of (5) show that $X'_{nk} \stackrel{P}{=} X_{nk}$ as $n \to \infty$. This implies that $\{X'_{nk}, 1 \leq k \leq k_n, n \geq 1\}$ also satisfies conditions (i)–(iii) of the theorem. We now assert that this transformed sequence satisfies the hypothesis of Proposition 1.

Let $Y_n(t) = \prod_{k=1}^{k_n}(1 + itX'_{nk})$. Then

$$\begin{aligned}
E(Y_n(t)) &= E\left[\prod_{k=1}^{k_n-1}(1 + itX'_{nk}) E^{\mathcal{F}_{n(k_n-1)}}(1 + itX'_{nk_n})\right] \\
&= E\left(\prod_{k=1}^{k_n-1}(1 + itX'_{nk})\right) \qquad [\text{since } E^{\mathcal{F}_k}(X'_{n(k+1)}) = 0] \\
&\;\;\vdots
\end{aligned}$$

$$= E(1 + itX'_{n1}) = 1.$$

Next define an r.v. T_n by the relation

$$T_n = \min\left[j \geq 1 : \sum_{k=1}^{j} X_{nk}^2 > 2\right],$$

with $\min(\emptyset) = k_n$. Then T_n is an integer valued measurable function, i.e., $[T_n = k] \in \mathcal{F}_{n_k}$, (a stopping time of $\{\mathcal{F}_{nk}, 1 \leq k \leq k_n\}$.) Such mappings, introduced and discussed in Chapter 6, are useful in the following analysis. We thus have, on using $1 + x \leq e^x$ for $x \geq 0$,

$$\begin{aligned} E(|Y_n(t)|^2) &= E\left(\prod_{k=1}^{k_n}(1 + itX'_{nk})(1 - itX'_{nk})\right) \\ &= E\left(\prod_{k=1}^{k_n}(1 + t^2 X'^2_{nk})\right) \\ &\leq E\left[\exp\left\{t^2 \sum_{k=1}^{T_n - 1} X'^2_{nk}\right\}(1 + t^2 X^2_{nT_n})\right], \\ &\quad \text{(by definition of } T_n\text{)}, \\ &\leq \exp\{2t^2\} \cdot (1 + t^2 E(X^2_{nT_n})) \qquad (6) \\ &\quad \text{(by Doob's opitional sampling theorem 6.3.2).} \end{aligned}$$

But by (i) the expectation on the right side of (6) is bounded by C, and hence $\{Y_n(t), n \geq 1\}$ is uniformly integrable, since it is a bounded set of $L^2(P)$. Conditions (iii) and (ii) of this theorem are the same as conditions (iii) and (iv) of Proposition 1. Thus we have verified that all the four conditions are satisfied by the X'_{nk}-sequence, and hence $S'_n \to$ an r.v. which is $N(0,1)$ distributed, so that S_n has the same limit distribution. This completes the proof of the theorem.

Before discussing the relation of the above hypothesis (iii) to the Lindeberg condition, it is useful to show that the assumptions are nearly optimal for normal convergence. More precisely, the following supplement to the above result holds.

Proposition 3 *Let $\{X_{nk}, \mathcal{F}_{nk}, 1 \leq k \leq k_n, n \geq 1\}$ be any adapted integrable sequence of r.v.s for which the following conditions hold:*

(i), (ii), *and* (iii) *are the same as those of Theorem 2,*

(iv) $\sum_{k=1}^{k_n} E^{\mathcal{F}_{n(k-1)}}(X_{nk}) \xrightarrow{P} 0$ *as* $n \to \infty$, *and*

(v) $\sum_{k=1}^{k_n} [E^{\mathcal{F}_{n(k-1)}}(X_{nk})]^2 \xrightarrow{P} 0$ *as* $n \to \infty$.

Then $S_n = \sum_{k=1}^{k_n} X_{nk} \xrightarrow{D}$ *to an r.v. which is* $N(0,1)$ *distributed.*

Proof Let $Y_{nk} = X_{nk} - E^{\mathcal{F}_{n(k-1)}}(X_{nk})$, so that $E^{\mathcal{F}_{n(k-1)}}(Y_{nk}) = 0$. We now assert that $\{Y_{nk}, \mathcal{F}_{nk}, 1 \leq k \leq k_n\}_{n \geq 1}$ satisfies the hypothesis of Theorem 2. Then condition (iv) implies $S_n - \sum_{k=1}^{k_n} Y_{nk} \xrightarrow{P} 0$, and hence S_n has the desired limit distribution.

To see that (i)-(iii) of Theorem 2 hold for the Y_{nk}-sequence, note that

$$|Y_{nk}| \leq |X_{nk}| + E^{\mathcal{F}_{n(k-1)}}(\max|X_{nk}|). \tag{7}$$

Taking the maximum of both sides and noting, by hypothesis (ii), that $Z_n = \max_{k \leq k_n} |X_{nk}| \xrightarrow{P} 0$, we get, as $n \to \infty$,

$$\max_{k \leq k_n} |Y_{nk}| \xrightarrow{P} 0 \quad \text{if} \quad \max_{k \leq k_n} E^{\mathcal{F}_{n(k-1)}}(Z_n) \xrightarrow{P} 0. \tag{8}$$

To see that the last part of (8) is true, note that $\{E^{\mathcal{F}_{nk}}(Z_n), \mathcal{F}_{nk}, 1 \leq k \leq k_n\}$ is a positive submartingale and $Z_n \in L^2(P)$. Hence by Theorem 3.5.6ii, with $p = 3/2$, we have, if $U_{n(k-1)} = E^{\mathcal{F}_{n(k-1)}}(Z_n), q = p/(p-1)(= 3)$,

$$\begin{aligned}
E\left(\max_{k \leq k_n} U^p_{n(k-1)}\right) &\leq q^p E(U^p_{n(k_n-1)}) \\
&= q^p E(U^{1/2}_{n(k_n-1)} U_{n(k_n-1)}) \quad \text{(since } p = 3/2) \\
&\leq q^p [E(U_{n(k_n-1)}) E(U^2_{n(k_n-1)})]^{1/2} \\
&\quad \text{(by the CBS inequality)} \\
&\leq q^p [E(E^{\mathcal{F}_{n(k_n-1)}}(Z_n)) E(E^{\mathcal{F}_{n(k_n-1)}}(Z_n^2))]^{1/2} \\
&\quad \text{(by the conditional Jensen inequality)} \\
&= 3^{3/2} [E(Z_n) E(Z_n^2)]^{1/2} \to 0
\end{aligned} \tag{9}$$

as $n \to \infty$, since $Z_n \xrightarrow{P} 0$ and $\{Z_n, n \geq 1\}$, being $L^2(P)$-bounded by (i) and (ii), is uniformly integrable, so that the first factor $\to 0$ and the second one is bounded. Hence

$$E\left(\max_{k \leq k_n} U_{n(k-1)}\right) \leq E\left(\max_{k \leq k_n} U^{3/2}_{n(k-1)}\right) \to 0$$

by (9), and this shows that (8) is true. Thus condition (ii) of Theorem 2 holds for the Y_{nk} even in a stronger form. [Namely, the sequence $\to 0$ in $L^1(P)$.]

For (iii) of Theorem 2, note that by (iii) and (v) of the hypothesis,

$$\sum_{k=1}^{k_n} Y^2_{nk} = \sum_{k=1}^{k_n} X^2_{nk} + \sum_{k=1}^{k_n} [E^{\mathcal{F}_{n(k-1)}}(X_{nk})]^2 - 2 \sum_{k=1}^{k_n} X_{nk} E^{\mathcal{F}_{n(k-1)}}(X_{nk}), \tag{10}$$

so that the first term $\xrightarrow{P} 1$ and the second term $\xrightarrow{P} 0$ on the right side of (10). Thus we only need to show that the last term $\xrightarrow{P} 0$ also. This follows by the CBS inequality again, since

$$\left|\sum_{k=1}^{k_n} X_{nk} E^{\mathcal{F}_{n(k-1)}}(X_{nk})\right|^2 \le \left(\sum_{k=1}^{k_n} X_{nk}^2\right)\left(\sum_{k=1}^{k_n}[E^{\mathcal{F}_{n(k-1)}}(X_{nk})]^2\right) \xrightarrow{P} 0.$$

The Y_{nk}-sequence satisfies (iii) of Theorem 2. Finally, (i) of that theorem is verified using the argument of (8) and (9). In fact,

$$E\left(\max_{k\le k_n}|Y_{nk}|\right)^2 \le 2E\left(\max_{k\le k_n} X_{nk}^2\right) + 2E\left(\max_{k\le k_n}[E^{\mathcal{F}_{n(k-1)}}(X_{nk})]^2\right),$$

since $(a_1+a_2)^2 \le 2(a_1^2+a_2^2)$ for $a_i \in \mathbb{R}$. But the first term is bounded, by the present hypothesis (i), and the second term is majorized by $E(\max_{k\le k_n} E^{\mathcal{F}_{n(k-1)}}(Z_n))^2$, where Z_n is given in (8). The inequalities (9) now establish the desired bound, since using the notation there,

$$\begin{aligned} E\left(\max_{k\le k_n} U_{n(k-1)}^2\right) &\le 4E(U_{n(k_n-1)}^2) \\ &\le 4E(E^{\mathcal{F}_{n(k_n-1)}}(Z_n^2)) \quad \text{(by Jensen's inequality)} \\ &= 4E(Z_n^2), \end{aligned}$$

which is bounded by (i) of the present hypothesis. Thus $\{Y_{nk}, \mathcal{F}_{nk}, 1 \le k \le k_n, n \ge 1\}$ is a martingale increments sequence satisfying the hypothesis of Theorem 2. Hence, as noted at the beginning, the result follows.

Recall that the usual Lindeberg condition states for $\{X_{nk}, 1 \le k \le k_n\}_{n\ge1}$ that for all $\varepsilon > 0$, as $n \to \infty$

$$\sum_{k=1}^{k_n} \int_{[|X_{nk}|>\varepsilon]} X_{nk}^2 dP \to 0, \tag{11a}$$

$$\sum_{k=1}^{k_n} E(X_{nk}^2) \to 1. \tag{11b}$$

This implies, in particular, that

$$\sum_{k=1}^{k_n} X_{nk}^2 \chi_{[|X_{nk}|>\varepsilon]} \xrightarrow{P} 0 \quad \text{as } n \to \infty. \tag{12}$$

It can be shown by examples that (12) is strictly weaker than (11). On the other hand, condition (ii) of Theorem 2 above can be written as

$$P\left[\max_{k\le k_n} |X_{nk}| > \varepsilon\right] = P\left[\bigcup_{k=1}^{k_n} [|X_{nk}| > \varepsilon]\right]$$
$$= P\left[\sum_{k=1}^{k_n} X_{nk}^2 \chi_{[|X_{nk}|>\varepsilon]} > \varepsilon^2\right]. \quad (13)$$

Hence (12) is equivalent to condition (ii) of Theorem 2. Thus the hypotheses of the preceding results are weaker than those of similar limit theorems given in many previous studies [cf., e.g., Loève, 1963, Sec. 28]. The preceding treatment follows essentially the interesting paper by McLeish (1974).

An account with references to earlier works, and concentrating on the martingale increment (double) arrays, is given in the book by Hall and Heyde (1980), to which we refer the reader for further information and applications.

7.2 Limit Laws for a Random Number of Random Variables

The previous work on the central limit theorem has been developed around an i.i.d. sequence $\{X_n, n \ge 1\}$ of random variables. It is of interest in some applications to consider a random number of such random variables. This idea is important for the area of sequential analysis, and also plays a key role in Section 8.4 where we analyze the (compound) Poisson process. Here a central limit theorem and some extensions are given to show how a change of viewpoint from the work of the last section is desirable.

Our first assertion about the problem can be stated as follows:

Theorem 1 *Let $\{X_n, n \ge 1\}$ be an i.i.d. sequence of r.v.s on (Ω, Σ, P) with mean 0 and variance 1. Let $\{N_n, n \ge 1\}$ be a sequence of integer valued r.v.s such that $N_n/n \xrightarrow{P} Y$, where Y is a positive discrete r.v. If $S_n = \sum_{k=1}^{n} X_k$, then*

$$\lim_{n\to\infty} P[(S_{N_n}/\sqrt{N_n}) < x] = \frac{1}{\sqrt{2\pi}} \int_{-\infty}^{x} e^{-u^2/2}\, du, \qquad x \in \mathbb{R}. \quad (1)$$

Remark If the set of r.v.s $\{N_n, n \ge 1\}$ is independent of the X_n-sequence, then such a result was known before in special studies in sequential analysis and without such an independence assumption but with $Y =$ constant. It was treated by F.J. Anscombe in the early 1950s. The result in the present form is due to Rényi (1960). Note that $\{S_{N_n}, n \ge 1\}$ is *no* longer a sequence of sums of independent r.v.s.

Let us first present an auxiliary result, to be used in the proof of the above theorem, which is of independent interest. A sequence $\{A_n, n \ge 1\} \subset \Sigma$

is called (*strongly*) *mixing* with density $0 < \alpha < 1$, if $\lim_{n\to\infty} P(A_n \cap B) = \alpha P(B)$ for each $B \in \Sigma$. A consequence of this concept is given by the following:

Proposition 2 *If $\{A_n, n \geq 1\} \subset \Sigma$ is mixing with density $0 < \alpha < 1$ in (Ω, Σ, P) and $Q : \Sigma \to [0,1]$ is a probability such that $Q \ll P$, then $\lim_{n\to\infty} Q(A_n) = \alpha$ holds. On the other hand, if $P(A_n) > 0$ for each $n \geq 1$, and if for each $m \geq 1$, we have with $0 < \alpha < 1$*

$$\lim_{n\to\infty} P(A_n|A_m) = \alpha = \lim_{n\to\infty} P(A_n), \tag{2}$$

where $P(A_n|A_m)$ is the conditional probability of A_n given the event A_m, then $\{A_n, n \geq 1\}$ is mixing with density α.

Proof First note that if $\overline{Q}(\cdot) = P(B \cap \cdot)$, then $\overline{Q}(A_n) = P(B \cap A_n) \to \alpha P(B) = \alpha\overline{Q}(\Omega)$. Thus if $Q = \overline{Q}/\overline{Q}(\Omega)$, so that $Q \ll P$ and Q is a probability, we get $Q(A_n) \to \alpha$. In general, by the Radon-Nikodým theorem, let $f = dQ/dP$, so that $\int_\Omega f\,dP = 1$. There exists a sequence $0 \leq f_n \uparrow f$ of simple functions, and if $f_n = \sum_{k=1}^{k_n} a_k^n \chi_{B_k^n}$, let

$$Q_n(A) = \int_A f_n\,dP, \quad a \in \Sigma.$$

Then as $m \to \infty$ we get

$$\begin{aligned} Q_n(A_m) &= \int_{A_k} f_n\,dP = \sum_{j=1}^{k_n} a_j^n P(B_j^n \cap A_m) \\ &\to \alpha \sum_{j=1}^{k_n} a_j^n P(B_j^n) = \alpha \int_\Omega f_n\,dP = \alpha Q_n(\Omega). \end{aligned}$$

Letting $n \to \infty$, one has $Q_n(A_m) \to Q(A_m)$ so that $Q(A_m) \to \alpha Q(\Omega) = \alpha$.

For the converse direction, let $\{A_n, n \geq 1\}$ be a sequence of events for which (2) holds. Consider the simple functions

$$g_n = (1-\alpha)\chi_{A_n} - \alpha\chi_{A_n^c}, \quad n \geq 1.$$

Note that $|g_n| \leq \max(\alpha, 1-\alpha) \leq 1$. Also,

$$\begin{aligned} E(g_n g_k) &= (1-\alpha)^2 P(A_n \cap A_k) + \alpha^2 P(A_n^c \cap A_k^c) \\ &\quad -\alpha(1-\alpha)[P(A_n \cap A_k^c) + P(A_k \cap A_n^c)] \\ &= P(A_n \cap A_k) + \alpha^2 - \alpha(P(A_n) + P(A_k)) \\ &\to \alpha P(A_k) + \alpha^2 - \alpha(\alpha + P(A_k)) = 0 \end{aligned} \tag{3}$$

as $n \to \infty$ for each $k \geq 1$, by (2). Let $\mathcal{H}_0 = \mathrm{sp}\{g_n, n \geq 1\}$ be the linear span and $\mathcal{H}_1 = \overline{\mathcal{H}}_0 \subset L^2(P)$ be its closure. Thus (3) implies $\lim_n E(g_n h_0) = 0$ for each $h_0 \in \mathcal{H}_0$. If $h \in \mathcal{H}_1$ and $\varepsilon > 0$, then there exists a $g_\varepsilon \in \mathcal{H}_0$ such that $||h - g_\varepsilon||_2 < \varepsilon$, where $||f||_2^2 = E(|f|^2)$. Hence

$$|E(hg_n) - E(g_\varepsilon g_n)| \le ||h - g_\varepsilon||_2 ||g_n||_2 < \varepsilon$$

by the CBS inequality and the fact that $||g_n||_2 \le 1$. It follows that, with (3), $\limsup_n |E(hg_n)| \le \varepsilon$, so that $\lim_{n\to\infty} E(hg_n) = 0$ for all $h \in \mathcal{H}_1$. If $\mathcal{H}_2 = \mathcal{H}_1^\perp \subset L^2(P)$, the orthogonal complement, then $E(fh) = 0$ for all $f \in \mathcal{H}_2, h \in \mathcal{H}_1$, and since each $u \in L^2(P)(= \mathcal{H}_1 \bigoplus \mathcal{H}_2)$ can be uniquely expressed as $u = u_1 + u_2, u_i \in \mathcal{H}_i, i = 1, 2$, it follows that $\lim_n E(ug_n) = 0$ for all $u \in L^2(P)$. In particular, if $u = \chi_B, B \in \Sigma$, then

$$\begin{aligned} E(ug_n) &= (1 - \alpha)P(A_n \cap B) - \alpha P(A_n^c \cap B) \\ &= P(A_n \cap B) - \alpha P(B) \to 0 \quad \text{as} \quad n \to \infty. \end{aligned}$$

Hence $\{A_n, n \ge 1\} \subset \Sigma$ is mixing if (2) is true, as asserted.

The following is a useful consequence of the above result:

Proposition 3 *Let $\{X_n, n \ge 1\}$ be independent and suppose that*

$$\lim_{n\to\infty} P[(S_n - b_n)/a_n < x] = F(x) \tag{4}$$

exists at all continuity points x of the d.f. F, where $0 < a_n \uparrow \infty, b_n \in \mathbb{R}$, and $S_n = \sum_{k=1}^n X_k$. Then $S_n^* = (S_n - b_n)/a_n$ is a mixing sequence with "density" $F(x)$, in that for all $A \in \Sigma, P(A) > 0, \lim_{n\to\infty} P[S_n^* < x| A] = F(x), x$ a continuity point of F. In particular, if $Q \ll P$ is any probability, then $\lim_{n\to\infty} Q[S_n^* < x] = F(x)$ also holds at each continuity point x of F.

Proof Let x be a continuity point of F such that $F(x) > 0$, and consider $A_n = [(S_n - b_n)/a_n < x], n \ge 1$. Since $F(x) > 0$ and $\lim_{n\to\infty} P(A_n) = F(x) > 0$, there exists an n_0 such that $n \ge n_0 \Rightarrow P(A_n) > 0$. To show that $\{A_n, n \ge n_0\}$ is a mixing sequence, it suffices to establish (2) for all $m \ge n_0$. Now for each $k, S_k/a_n \xrightarrow{P} 0$ as $n \to \infty$, since $a_n \to \infty$. Hence $S_n^* \pm (S_k/a_n) \xrightarrow{D} Y$, where Y has d.f. F by the elementary part of Slutsky's result (cf. Problem 11b in Chapter 2). But

$$S_n^* - S_k/a_n = (S_n - S_k)/a_n - b_k/a_n,$$

so that this is independent of S_k^*, and since $P(A_k) > 0$,

$$P\left[S_n^* - \frac{S_k}{a_n} < x| A_k\right] = P\left[S_n^* - \frac{S_k}{a_n} < x\right] \to F(x).$$

It follows that, since $S_k/a_n \xrightarrow{P} 0$, for each $k \ge n_0$,

$$P[A_n|A_k] \to F(x) \quad \text{as } n \to \infty. \tag{5}$$

This is (2), and thus $\{A_n, n \geq n_0\}$ is (strongly) mixing for any such x by Proposition 2.

For the last part, let $\tilde{Q}(\cdot) = P(\cdot| A_k)$ for any fixed but arbitrary $k \geq n_0$. Since $P(A_k) > 0$, this elementary conditional probability is well defined. Let Q be as given. Then $Q \ll \tilde{Q} \ll P$, and by the first part of Proposition 2

$$\lim_{n\to\infty} Q(A_n) = \lim_{n\to\infty} P(A_n| A_k) = F(x).$$

This finishes the proof of the proposition.

We are now ready to present the

Proof of Theorem 1 Let $p_k = P[Y = \ell_k] > 0, k \geq 1$, the ℓ_k being the possible values of Y. If $r_n^k = [n\ell_k]$, the integral part, then $r_n^k \uparrow \infty$ as $n \to \infty$. By the central limit theorem (cf. Theorem 5.1.2) and Proposition 3, we have if $M_n = [nY], S_n^* = S_n/\sqrt{n}$, and $A_k = [Y = \ell_k]$,

$$\begin{aligned}\lim_{n\to\infty} P[S_{M_n}^* < x] &= \lim_{n\to\infty} \sum_{k=1}^{\infty} P[S_{r_n^k}^* < x| A_k] p_k \\ &= \sum_{k=1}^{\infty} \lim_{n\to\infty} P[S_{r_n^k}^* < x| A_k] p_k \\ &\quad \text{(by the dominated convergence theorem)} \\ &= \sum_{k=1}^{\infty} G(x) p_k = G(x) = \frac{1}{\sqrt{2\pi}} \int_{-\infty}^{x} e^{-u^2/2}\, du. \end{aligned} \tag{6}$$

Here with independence, we used the fact that $\{[S_{r_n^k}^* < x], n \geq 1\}$ is mixing by Proposition 3 and that $\{S_n^*, n \geq 1\}$ obeys the central limit theorem (and $\sum_{k\geq 1} p_k = 1$). On the other hand, consider the identity

$$S_{N_n}^* = S_{M_n}^* + \sqrt{\frac{M_n}{N_n}} \left(\frac{S_{N_n} - S_{M_n}}{\sqrt{M_n}}\right) + \frac{S_{M_n}}{\sqrt{M_n}} \left(\sqrt{\frac{M_n}{N_n}} - 1\right). \tag{7}$$

We note that $N_n/M_n = (N_n/n)(n/M_n) \xrightarrow{P} Y/Y = 1$. Since $S_{M_n}^* \xrightarrow{D} Z$, an $N(0,1)$ distributed r.v., it is bounded in probability. Thus the last term of (7) $\xrightarrow{P} 0$. The first term on the right converges to the desired limit r.v. Thus the theorem follows (from Slutsky's result again) if the middle term on the right side of (7)$\xrightarrow{P} 0$. Since $M_n/N_n \xrightarrow{P} 1$, it suffices to show that $(S_{N_n} - S_{M_n})M_n^{-1/2} \xrightarrow{P} 0$. This is inferred from the Kolmogorov inequality (cf. Theorem 2.2.5) as follows.

Let $\varepsilon > 0, \delta > 0$ be given. Let $A_k = [Y = \ell_k]$ as before, and set $B_n = [|N_n - [nY]| < n\eta], \eta > 0$. If $C_{nk} = [|S_{N_n} - S_{r_n^k}| > \varepsilon\sqrt{r_n^k}]$, then

$$\begin{aligned}
P[A_k \cap B_n \cap C_{nk}] &\le P[|S_{N_n} - S_{r_n^k}| > \varepsilon\sqrt{r_n^k}, |N_n - r_n^k| < n\eta] \\
&\le P\left[\max_{|j - r_n^k| < n\eta} |S_j - S_{r_n^k}| > \varepsilon\sqrt{r_n^k}\right] \\
&\le P\left[\max_{r_n^k \le j < n\eta + r_n^k} |S_j - S_{r_n^k}| > \varepsilon\sqrt{r_n^k}\right] \\
&\quad + P\left[\max_{r_n^k - n\eta < j \le r_n^k} |S_j - S_{r_n^k}| > \varepsilon\sqrt{r_n^k}\right] \\
&\le \frac{2(n\eta - 1)}{\varepsilon^2[n\ell_k]} \le \frac{2\eta}{\varepsilon^2 \ell_k} \quad \text{(by Theorem 2.2.5).} \qquad (8)
\end{aligned}$$

Let $D_k = [Y \ge \ell_k]$. Then we have

$$\begin{aligned}
&P[|S_{N_n} - S_{M_n}| > \varepsilon\sqrt{M_n}] \\
&\quad \le \sum_{k \ge 1} P(A_k \cap B_n \cap C_{nk}) + P(B_n^c) \\
&\quad \le \sum_{k=1}^{k_0 - 1} P(A_k \cap B_n \cap C_{nk} \cap [Y = \ell_k]) + P(D_{k_0}) + P(B_n^c) \\
&\quad \le \frac{2\eta}{\varepsilon^2} \sum_{k=1}^{k_0 - 1} \frac{1}{\ell_k} + P(D_{k_0}) + P(B_n^c) \quad [\text{by } (8)]. \qquad (9)
\end{aligned}$$

Now choose k_0 large enough so that $P(D_{k_0}) < \delta/3$, and then choose $\eta_0 > 0$ small enough so that $0 < \eta < \eta_0$ implies

$$(2\eta/\varepsilon^2)\left(\sum_{k=1}^{k_0 - 1} \ell_k^{-1}\right) < \delta/3. \qquad (10)$$

Finally, choose $n_0[= n_0(k_0, \eta_0(\varepsilon), \delta]$ such that $n \ge n_0 \Rightarrow P(B_n^c) < \delta/3$. Thus if $n \ge n_0$, we have from (9) that the left-side probability is bounded by δ. Since $\delta > 0$ and $\varepsilon > 0$ are arbitrary, this implies that $(S_{N_n} - S_{M_n})/\sqrt{M_n} \xrightarrow{P} 0$, so that (7) implies that $S_{N_n}^*$ has the limit d.f. which is $N(0,1)$. This completes the proof of the theorem.

Remark 4 It is clear from the above proof that the independent r.v.s should satisfy an appropriate central limit theorem (e.g., Liapounov's or the Lindeberg-Feller's) and that the i.i.d. assumption is not crucial. With a little care in the error estimates, the result can be shown to be valid if the limit r.v. Y is merely positive and not necessarily discrete, since (as noted in Problem 3 in Chapter 4) it can be approximated by a discrete r.v. to any degree of accuracy desired.

We now sketch another extension of the above ideas if the sequence $\{X_n, n \ge 1\}$ is replaced by a certain dependent double array of considerable practical as well as theoretical interest. This will also indicate how other extensions are possible.

The problem to be described arises typically in life-testing situations. If the life span of the ith object (bulb, the drug effect on an animal, etc.) of a certain population is Z_i, then it is usually found (to be a good approximation) that $Z_i, i = 1, \ldots, n$ form order statistics from an exponential population. (In electric bulb life problems this is indeed found to be an acceptable approximation.) Suppose that one wants to test just a subset k of the n objects because of cost or some other reason. Let r_j be the jth selected moment for observation using a chance mechanism. Thus $1 \leq r_1 < \cdots < r_k \leq n$ and the r_i are integer-valued r.v.s. The problem of interest is the asymptotic distribution of $\{Z_{r_i}, i = 1, \ldots, k\}$ as $n \to \infty$, where $r_i/n \xrightarrow{P}$ to an r.v. Note that each $r_i (= r_i^n)$ is also a function of n, so that this is in fact a double array of (a random number of) r.v.s. We can now assert the following generalization of the above situation. Some previous work (especially Theorem 3.3.9) is also utilized in this analysis.

We recall the following familiar concept.

Definition 5 A random vector $U = (U_1, \ldots, U_k)$ is k-variate Gaussian if for each real vector $(t_1, \ldots, t_k)$, the linear combination $V = \sum_{i=1}^k t_i U_i$ is $N(\mu_k, \sigma_k^2)$, where $\mu_k = \sum_{i=1}^k t_i E(U_i)$ and $\sigma_k^2 = \sum_{i=1}^k t_i t_j a_{ij} \geq 0$. Then the vector $v = (v_1, \ldots, v_k), v_i = E(U_i)$ is called the *mean* and $\Sigma = (a_{ij})$ the *covariance (matrix)* of U.

Using this terminology, we present the following main result of the section, which originally appeared in a compressed form in Rao (1962, Theorem 4.2).

Theorem 6 *Let $Y_1, \ldots, Y_n$ be order statistics from an absolutely continuous strictly increasing d.f. F. Let $1 \leq r_i < \cdots < r_{k+1} \leq n$ be integer-valued r.v.s such that $r_i/n \xrightarrow{P} q_i X > 0, i = 1, \ldots, k+1$, as $n \to \infty$, where X is a positive discrete r.v., $0 < q_1 < \cdots < q_{k+1} = 1, q_i = F(Q_i)$. If the $Y_{r_i}, 1 \leq i \leq k+1$, are thus optionally selected r.v.s from the order statistics $Y_1, \ldots, Y_n$ of F, then*

$$\lim_{n\to\infty} P[\sqrt{r_{k+1}}(Y_{r_1} - Q_1) < x_1, \ldots, \sqrt{r_{k+1}}(Y_{r_k} - Q_k) < x_k] = G(x_1, \ldots, x_k), \tag{11}$$

where G is a k-variate Gaussian d.f. with mean vector zero and covariance matrix $D\Sigma D$ of the following description: D is a $k \times k$ diagonal matrix $(d_{ii}), d_{ii} = q_i/F'(Q_i)$, and $\Sigma = (a_{ij})$, with $a_{ij} = a_i = (1 - q_i)/q_i, i \leq j \leq k, i = 1, \ldots, k$, where $k \geq 1$ is fixed (and $F'(x) = dF/dx$, which exists a.e.). (A k-variate Gaussian d.f. is given in Definition 5 above.)

The proof is presented in a series of propositions. The argument is typical and is useful in several other asymptotic results. The first consideration is the case that $F(x) = 1 - e^{-x}, x \geq 0$, the exponential d.f., and the r_i are nonrandom. In any event Q_i is the "ith quantile" of F and Y_{r_i} is an estimator of Q_i.

Then (11) gives the limit d.f. of the vector estimators of $\{Q_i, 1 \leq i \leq k\}$ and has independent interest in applications.

Proposition 7 *Let $Z_1 < \cdots < Z_n$ be order statistics from the exponential d.f. F_1 given by $F_1(x) = 1 - e^{-x}$ if $x \geq 0 (= 0$ if $x < 0)$. Suppose $1 \leq i_1 < \cdots < i_k \leq n$ and $\lim_{n\to\infty}(i_j/n) = q_j, 0 < q_1 < \cdots < q_k < 1 (p_i = 1 - q_i)$; then*

$$\lim_{n\to\infty} P[\sqrt{n}(Z_{i_j} + \log p_j) < x_j, 1 \leq j \leq k] = \tilde{G}(x_1, \ldots, x_k), \tag{12}$$

where $\tilde{G}$ is a k-variate normal d.f. with mean zero and covariance matrix $\tilde{\Sigma}$, which is of the same form as Σ of Theorem 6, with a_i^{-1} for a_i there.

Proof In view of the multivariate extension theory of Section 4.5, it suffices to establish, for this proof, that if $\tilde{Z}_n = \sum_{j=1}^k t_j Z_{i_j} (t_j \in \mathbb{R})$, then $(\tilde{Z}_n - \mu_n)/\sigma_n \xrightarrow{D}$ an r.v. with $N(0,1)$ as its d.f., where μ_n and σ_n are the mean and variance of $\tilde{Z}_n$, and $\mu_n \to -\sum_{j=1}^k t_j \log p_j, n\sigma_n^2 \to \sum_{i,j=1}^k t_i t_j \tilde{a}_{ij} = \tilde{\Sigma}$ of the theorem, because of the Slutsky result. We fill in the details now.

Let ψ_n be the ch.f. of $\tilde{Z}_n$. Then using the form of the density of the Z_i given in Theorem 3.3.9, we get (ψ_n never vanishes, Z_i being infinitely divisible.)

$$C_n(u) = \mathrm{Log}\,\psi_n(u) = \log[n!/(n-i_k)!] - \sum_{j=0}^{k-1} \sum_{m=i_j}^{i_{j+1}-1} \mathrm{Log}(n - iu\tau_{j+1} - m), \tag{13}$$

where $\tau_j = \sum_{v=j}^k t_v$. In view of the continuity theorem, it is enough to verify that $\lim_{n\to\infty}[C_n''(u/\sigma_n] = -1$ uniformly in some neighborhood of $u = 0$. The desired Gaussian limit assertion then follows. (Here C_n'' is the second derivative of C_n relative to u.) Thus

$$C_n''(u) = -\sum_{j=0}^{k-1} \sum_{m=i_j}^{i_{j+1}-1} \tau_{j+1}^2 (n - iu\tau_{j+1} - m)^{-2}, \tag{14}$$

and similarly,

$$\sigma_n^2 = \sum_{j=1}^{k} t_j^2 \sum_{s=0}^{i_j - 1} (n-s)^{-2} + 2 \sum_{1 \leq j < j' \leq k} t_j t_{j'} \sum_{s=0}^{i_j - 1} (n-s)^{-2}. \tag{15}$$

For any integers $h, H, 0 < h < H$, and α such that $s + \alpha \neq 0$, one has

$$\sum_{s=h}^{H} (s+\alpha)^{-2} = (h+\alpha)^{-1} - (H+\alpha)^{-1} + \theta[(h+\alpha)(H+\alpha)]^{-1},$$

where $0 < \theta < 1, \theta/h \to 0$ if $(H-h)(h+\alpha)^{-1} \to 0$ as $h \to \infty$. Substituting this in (14) and (15) and remembering that $i_j/n \to q_j$ as $n \to \infty$, we get

$$\lim_{n\to\infty} n\sigma_n^2 = \sum_{j=1}^{k} t_j^2 q_j p_j^{-1} + 2\sum_{j=1}^{k-1} t_j \tau_{j+1} p_j q_j^{-1} = M^2 \quad \text{(say)} \tag{16}$$

and, after a similar simplification,

$$\lim_{n\to\infty} nC_n''(u/\sigma_n) = -\sum_{i=1}^{k} \tau_i^2 (q_i - q_{i-1})(p_i p_{i-1})^{-1}, \tag{17}$$

where $q_0 = 0$ (so that $p_0 = 1$). Therefore it follows, since $t_j = \tau_j - \tau_{j+1}, 1 \le j \le k-1, t_k = \tau_k$, that

$$\lim_{n\to\infty} [C_n''(u/\sigma_n)/\sigma_n^2] = -1, \tag{18}$$

uniformly in $u \in \mathbb{R}$. Finally, since $\mu_n = \sum_{j=1}^{k} t_j E(Z_{i_j})$ and

$$a_j = E(Z_{i_j}) = \sum_{s=0}^{i_j-1} (n-s)^{-1} = \log \frac{n}{n-i_j} + \varepsilon_n^j, \tag{19}$$

where $|\varepsilon_n^j| = O((n-i_j)^{-1})$, one has $E(Z_{i_j}) \to -\log p_j$ and $\sqrt{n}(a_j + \log p_j) \to 0$ as $n \to \infty$. Hence $\sqrt{n}(Z_{i_j} - a_j) \stackrel{P}{=} \sqrt{n}(Z_{i_j} + \log p_j)$, and by the first paragraph, this tends to the standard normal r.v. as $n \to \infty$. The proof is finished.

The conclusion is valid in a stronger form, namely, it is true on (Ω, Σ, P_B), where $P_B = P(B \cap \cdot)/P(B)$, the conditional probability for any $B \in \Sigma, P(B) > 0$. This is also called a "stable" convergence. In the above, it is proved for $B = \Omega$. We sketch this strengthening here, since this is also needed.

Proposition 8 *Under the same hypothesis as in Proposition 7, the limit* (12) *holds if the probability measure P is replaced by $P_B(\cdot) = P(B \cap \cdot)/P(B)$, the conditional probability given B, where $P(B) > 0$; i.e., "stable" convergence holds.*

Proof The result can be established by a careful reworking of the above argument. However, it is instructive to use a version of Proposition 1.1 instead. We do this with $k = 1$, for simplicity, in (12). By the Cramér-Wold device (cf. Proposition 4.5.2) this is actually sufficient. Since also $\sqrt{n}(E(Z_{i_j}) + \log p_j) \to 0$ as $n \to \infty$, it is enough to show that

$$\sqrt{n}(Z_{i_j} - E(Z_{i_j})) \xrightarrow{D} \tilde{G} \quad \text{as } n \to \infty. \tag{20}$$

By Theorem 3.3.9, $Z_{i_j} = \sum_{k=0}^{i_j-1} V_k/(n-k+1)$, where the V_k are i.i.d. exponential or gamma distributed on $\mathbb{R}^+$ with $E(V_1) = 1$. Let $X_{ni} = \sqrt{n}[V_i - 1]/(n-i+1)$ and $S_n = \sum_{i=0}^{i_j-1} X_{ni}$, with $i_j/n \to q_j, 0 < q_j < 1$. These r.v.s have the following properties. Let $\varepsilon > 0$ be arbitrary. Then

$$P\left[\max_i |X_{ni}| \geq \varepsilon\right] \leq P\left[\frac{\sqrt{n}}{n - i_j + 1}\left(1 + \max_j V_j\right) \geq \varepsilon\right] \to 0 \qquad (21)$$

as $n \to \infty$, where we used the fact that $\varepsilon(n - i_j + 1)/\sqrt{n} \to \infty$. Also,

$$E\left(\sum_{i=0}^{i_j-1} X_{ni}^2\right) = \sum_{i=0}^{i_j-1} \frac{n}{(n-i+1)^2} \to \left(\frac{1}{1-q_j} - 1\right) = \frac{q_j}{1-q_j} > 0 \qquad (22)$$

and

$$\begin{aligned}\operatorname{Var}\left(\sum_{i=0}^{i_j-1} X_{ni}^2\right) &= \sum_{i=0}^{i_j-1} \frac{n^2}{(n-i+1)^4} \operatorname{Var} X_{ni}^2 \\ &\leq 9n^2 \sum_{i=0}^{i_j-1} (n-i+1)^{-4} \to 0 \quad \text{as } n \to \infty. \qquad (23)\end{aligned}$$

Hence

$$\sum_{i=0}^{i_j-1} X_{ni}^2 \xrightarrow{P} 2q_j(1-q_j)^{-1} = 2q_j p_j^{-1}. \qquad (24)$$

Let $Y_n = \prod_{k=0}^{i_j-1}(1 + itX_{nk})$ and $\mathcal{F}_{nk} = \sigma(X_{nj}, j \leq k)$. Then

$$\begin{aligned}E(|Y_n(t)|^2) &= E\left(\prod_{k=0}^{i_j-1}(1 + t^2X_{nk}^2)\right) \\ &= E\left(\prod_{k=0}^{i_j-2}(1 + t^2X_{nk}^2)E^{\mathcal{F}_{n(i_j-2)}}(1 + t^2X_{n(i_j-1)}^2)\right) \\ &= \left(1 + \frac{t^2 n}{(n-i_j+2)^2}\right) E\left(\prod_{k=0}^{i_j-2}(1 + t^2X_{nk}^2)\right) \\ &\qquad [\text{by independence of } \mathcal{F}_{nk} \text{ and } X_{n(k+1)}] \\ &\ \ \vdots \\ &= \prod_{k=0}^{i_j-1}\left(1 + \frac{t^2 n}{(n-k+1)^2}\right) \quad \text{(by iteration)} \\ &\leq \exp\left\{t^2 n \sum_{k=0}^{i_j-1}(n-k+1)^{-2}\right\} \to \exp\{2t^2 q_j p_j^{-1}\}, \qquad (25)\end{aligned}$$

as in (16). Here we used the inequality $1 + x \leq e^x, x > 0$. Hence $\{Y_n(t), n \geq 1\}$ is bounded in $L^2(P)$, which implies its uniform integrability. We now verify that $Y_n(t) \to$ to a constant weakly in $L^1(P)$, and this gives the desired distributional convergence, as in Proposition 1.1.

Thus let k' be fixed, $1 \leq k' \leq i_j - 1$, and $A \in \mathcal{F}_{nk'}$. Then for $n \geq k'$ we have, as in (25),

$$\begin{aligned} E(\chi_A Y_n(t)) &= E\left[\chi_A \prod_{k=0}^{k'}(1+itX_{nk})E^{\mathcal{F}_{nk'}}\left(\prod_{j=k'+1}^{i_j-1}(1+itX_{nj})\right)\right] \\ &= E\left[\chi_A \prod_{k=0}^{k'}(1+itX_{nk})\right]\cdot E\left(\prod_{j=k'+1}^{i_j-1}(1+itX_{nj})\right) \\ &\quad \text{(by independence of } X_j \text{ and } \mathcal{F}_{nk'}) \\ &= E\left[\chi_A \prod_{k=0}^{k'}(1+itX_{nk})\right]\prod_{j=k'+1}^{i_j-1}(1+itE(X_{nj})) \\ &= P(A) + r^0_{nk'} \quad \text{(say)}, \end{aligned} \tag{26}$$

since $E(X_{nj}) = 0$. Regarding $r^0_{nk'}$, it consists of a finite number $(\leq 2^{k'} - 1)$ of terms of the form $\chi_A X_{nk_1} \cdots X_{nk_s}$, where $1 \leq k_j \leq k_s \leq k'$, with $1 \leq j \leq s \leq k'$. Each of these terms satisfies the inequality

$$\begin{aligned} E(\chi_A |X_{nk_1} \cdots X_{nk_s}|) &\leq \prod_{j=1}^{k_s} E(|X_{nk_j}|) \quad \text{(by independence)} \\ &\leq \left[\frac{2\sqrt{n}}{n-k_s+1}\right]^{k'} \to 0 \quad \text{as } n \to \infty. \end{aligned} \tag{27}$$

Thus (26) and (27) show that for each $A \in \bigcup_{n\geq 1}\bigcup_{k=1}^{i_j} \mathcal{F}_{nk} = \mathcal{A}$,

$$E(\chi_A Y_n(t)) \to P(A) \quad \text{as } n \to \infty. \tag{28}$$

By linearity of the expectation, (28) gives

$$E(f\cdot(Y_n(t)-1)) \to 0 \quad \text{as } n \to \infty \tag{29}$$

for all $f = \sum_{i=1}^m a_i\chi_{A_i}, A_i \in \mathcal{A}$. If $\Sigma_0 = \sigma(\mathcal{A})$, then the above class of step functions is dense in $L^2(\Sigma_0)$, and since $Y_n(t) - 1 \in L^2(\Sigma_0)$, (29) implies that the same relation is true for all bounded $f \in L^2(\Sigma_0)$, and then for all $f \in L^2(\Sigma_0)$. This means $Y_n(t) \to 1$ weakly in $L^2(\Sigma_0)$.

We can now apply the result of Proposition 1.1. By using the same symbolism with the present $Y_n(t)$ in Eq. (1.4), one gets for $A \in \Sigma_0$,

$$\int_A e^{itS_n}\,dP = e^{-t^2/2}\int_A Y_n(t)\,dP + \int_A Y_n(t)[Z_n(t) - e^{-t^2/2}]\,dP. \tag{30}$$

As noted there, Eqs. (21)-(24) imply $(Z_n(t) - e^{-t^2/2}) \xrightarrow{P} 0$ and $\{Y_n(t), n \geq 1\}$ is uniformly integrable by (25). Hence $\{e^{itS_n} - e^{-t^2/2}Y_n(t), n \geq 1\}$ is uniformly integrable. This shows that the last integral in (30)→ 0 and by (29) the right side tends to $e^{-t^2/2}P(A)$. Thus we have for each $A \in \Sigma_0$ with $P(A) > 0$,

$$\lim_{n\to\infty} \int_{\Omega} e^{itS_n}\, dP_A = e^{-t^2/2}, \quad t \in \mathbb{R}. \tag{31}$$

Since $S_n = \sqrt{n}(Z_{i_j} - E(Z_{i_j}))$, the proposition is established.

We have included this alternative and detailed argument here because with simple modifications it can be used if the X_{ni} are not independent, but satisfy the hypothesis of Theorem 1.2 or Proposition 1.3. This work can be applied also to certain other problems involving "stable" convergence. Further, it should be noted that Proposition 7 is deducible from the proof of the above result. But the separation has some methodological interest. We now extend it to the optional sampling problem.

Proposition 9 *Let $s_j = [nX_j]$ be the integral part, with $X_j = q_jX$, as in Theorem 6. If Z_{i_j} is replaced by Z_{s_j} in Proposition 8, so that the Z_{s_j} are an optionally selected set of r.v.s, then (12) holds if n is replaced by s_{k+1} there.*

Proof As in the proof of Proposition 7, consider $\tilde{Z}_{s_{k+1}} = \sum_{j=1}^{k} t_j Z_{s_j}, \mu = -\sum_{j=1}^{k} t_j \log p_j$, with M given by the right side of (16). If X takes values $d_1, d_2, \ldots$ and $\delta_j = P[X = d_j], (\delta_j > 0)$ then

$$P\left[\sqrt{s_{k+1}}(\tilde{Z}_{s_{k+1}} - \mu) < Mx\right] = \sum_{j=1}^{\infty} P\left[\sqrt{n'_j}(\tilde{Z}'_{n'_j} - \mu) < Mx \middle| X = d_j\right] \delta_j, \tag{32}$$

where $n'_j = [nd_j], \tilde{Z}'_{n'_i} = \sum_{j=1}^{k} t_j Z_{[nq_j d_i]}$. Letting $n \to \infty$ and using Proposition 8 and the bounded convergence theorem in (32), one gets

$$\lim_{n\to\infty} P[\sqrt{s_{k+1}}(\tilde{Z}_{s_{k+1}} - \mu) < Mx] = G(x),$$

where $G(\cdot)$ is the $N(0,1)$ d.f. This implies the conclusion.

For the next extension we establish:

Proposition 10 *Let $(Z_1, \cdots, Z_n), (q_1, \ldots, q_{k+1})$ be as in the preceding proposition and $r_1, \ldots, r_{k+1}$ be as in Theorem 6. Then*

$$\lim_{n\to\infty} P[\sqrt{r_{k+1}}(Z_{r_1} + \log p_1) < x_1, \ldots, \sqrt{r_{k+1}}(Z_{r_k} + \log p_k) < x_k]$$
$$= \tilde{G}(x_1, \ldots, x_k),$$

where $\tilde{G}$ is as defined in Proposition 7.

Proof Let s_k be defined as in the above proof, and consider

$$\sqrt{r_{k+1}}\ (Z_{r_j} + \log p_j)$$
$$= \sqrt{s_{k+1}}(Z_{s_j} + \log p_j) + (r_{k+1}/s_{k+1})^{1/2} \cdot \sqrt{s_{k+1}}(Z_{r_j} - Z_{s_j})$$
$$+ \sqrt{s_{k+1}}(Z_{s_j} + \log p_j)((r_{k+1}/s_{k+1})^{1/2} - 1). \tag{33}$$

But

$$\frac{r_{k+1}}{s_{k+1}} = \frac{r_{k+1}}{n} \frac{n}{[nq_{k+1}X]} \xrightarrow{P} \frac{q_{k+1}X}{q_{k+1}X} = 1,$$

and $\sqrt{s_{k+1}}(Z_{s_j} + \log p_j)$ has a limit d.f., so that it is bounded in probability. It follows that the last term of (33)$\xrightarrow{P} 0$. The first term on the right side of (33) gives the desired result by Proposition 9. The middle term $\xrightarrow{P} 0$, and then the result follows by the Cramér-Slutsky theorem if we show that

$$\sqrt{s_{k+1}}(Z_{r_j} - Z_{s_j}) \xrightarrow{P} 0$$

for $j = 1, \ldots, k$.

For this observe that $r_j/r_k \xrightarrow{P} q_j/q_k > 0$ as $n \to \infty$, and hence r_j and r_k grow at the same rate. Let $d_1, d_2, \ldots$ be the values taken by X, and fixing j, set $n_k = [nq_j d_k]$, the integral part. The desired conclusion depends on Kolmogorov's inequality (or equivalently here, on the submartingale inequality). Given $\varepsilon > 0, \delta > 0$, consider, with r_j, s_j as before,

$$P[\sqrt{s_{k+1}}|Z_{r_j} - Z_{s_j}| > \varepsilon] \le \sum_{k=1}^{\infty} P(A_k \cap B_n^{\delta} \cap C_{nk}) + P(B_n^c), \tag{34}$$

where $B_n^c = \Omega - B_n$, and

$$A_k = [X = d_k], \qquad B_n^{\delta} = [|\, r_j - s_j| < n\delta],$$
$$C_{nk} = [\sqrt{n_k}|Z_{r_j} - Z_{n_k}| > \varepsilon].$$

But $A_k \cap B_n^{\delta} = [|\, r_j - n_k| < n\delta]$, and (for $n_k - \delta_n < \ell < n_k + \delta_n$) one has;

$$P[A_k \cap B_n^{\delta} \cap C_{nk}]$$
$$\le P\left[\max_{|\ell - n_k| < \delta n} \sqrt{n_k}|Z_\ell - Z_{n_k}| > \varepsilon\right]$$
$$\le P\left[\max_{|\ell - n_k| < \delta n} \left|\sum_{t=n_k}^{\ell} \sqrt{n_k} U_{nt}\right| > \varepsilon\right]$$

(where $U_{nt} = \frac{V_t}{n-t+1}$ and$Z_\ell = \sum_{t=1}^{\ell} U_{nt}$,

V_t being i.i.d. gamma r.v.s with unit mean)

$$\le \frac{2}{\varepsilon^2} \sum_{t=1}^{[\delta n]} \frac{n_k}{(n_k - t + 1)^2} \quad \text{(by Kolmogorov's inequality)}$$
$$\le \frac{2K_0\delta}{\varepsilon^2} \left(\frac{(d_k - 1)n_k}{n_k - d_k + 1}\right), \tag{35}$$

where $K_0 > 0$ is a constant independent of n. Now choose m_0 such that $P[\bigcup_{k>m_0} A_k] < \varepsilon/3$, and hence (34) becomes, by (35),

$$\text{LHS}(34) \le \frac{2K_0\delta}{\varepsilon^2}\sum_{k=1}^{m_0}\frac{(d_k-1)n_k}{(n_k-d_k+1)}+\frac{\varepsilon}{3}+P(B_n^c). \tag{36}$$

Since m_0 is fixed, and $n_k/(n_k-d_k+1) \to 1$, for each k, as $n \to \infty$, we can choose $\delta > 0$ small enough so that the first term on the right of (36) is at most $\varepsilon/3$. Then choose n large enough so that $P(B_n^c) < \varepsilon/3$. Thus as $n \to \infty$, (36) gives

$$\lim_{n\to\infty} P[\sqrt{s_{k+1}}|Z_{r_j} - Z_{s_j}| > \varepsilon] < \varepsilon.$$

This implies the desired result.

We now extend the above proposition from the Z_i to the original r.v. sequence; namely the X_n's:

Proposition 11 *Let $X_n = (X_{1n}, \ldots, X_{kn}), n = 1, 2, \ldots,$ be a sequence of random vectors and $\theta = (\theta_1, \ldots, \theta_k)$ be a constant vector. Let $\{h_n, n \ge 1\}$ be an increasing sequence of r.v.s such that $P[\lim_n h_n = +\infty] = 1$, and that*

$$\lim_{n\to\infty} P[h_n(X_{1n}-\theta_1) < x_1, \ldots, h_n(X_{kn}-\theta_k) < x_k] = F(x_1, \ldots, x_k), \tag{37}$$

at all continuity points $(x_1, \ldots, x_k)$ of F. If $f : B \to \mathbb{R}^k$ is a continuously differentiable function with B as a convex set containing θ in its interior, then one has

$$\lim_{n\to\infty} P[h_n(f(X_{1n}, \ldots, X_{kn}) - f(\theta_1, \ldots, \theta_k)) < x] = \tilde{F}(x) \tag{38}$$

at all continuity points x of $\tilde{F}, x = (x_1, \ldots, x_k)$ and inequality between vectors is componentwise. Here, if F of (37) denotes the d.f. of the vector $(X_1, \ldots, X_k)$, then $\tilde{F}$ of (38) is the d.f. of $(X_1, \ldots, X_k)D'$, where the k-by-k matrix $D = (\partial f_i/\partial x_j|_\theta, 1 \le i, j \le k)$ and prime stands for transposition.

Proof Expand f about θ in a Taylor series,

$$f(x) = f(\theta) + (x-\theta)D' + \eta(x-\theta),$$

where D is the (Jacobian) matrix given in the statement and the Euclidean norm of the (column) vector $|\eta(x-\theta)| = o(|x-\theta|)$ by elementary analysis. Hence

$$h_n(f(X_n) - f(\theta)) = h_n(X_n-\theta)D' + h_n\eta(X_n-\theta). \tag{39}$$

Since by (37) $h_n(X_n-\theta)$ is bounded in probability and $h_n \uparrow \infty$ a.e., it follows from the stochastic calculus that $X_n - \theta \xrightarrow{P} 0$, and then $h_n|\eta(X_n-\theta)| \xrightarrow{P} 0$.

From this and (39) we deduce that the stated limit d.f. of (38) must be true.

We can now quickly complete the proof of the main result.

Proof of Theorem 6 Let $B = \mathbb{R}_+^k, f_i(u) = F^{-1}(e^{-u}), i = 1, \ldots, k$, and $f = (f_1, \ldots, f_k)$ in Proposition 11. Let $h_n = \sqrt{r_{k+1}}, \theta_j = \log p_j, X_{jn} = Z_{r_j}$, and $q_j = 1 - p_j$. Since F is the k-dimensional normal d.f. with means zero and covariance matrix given for (12), it is seen that (11) is a consequence of Proposition 11 with $\tilde{F} = G$. Note that the ordering of the Z and Y is reversed, so that the p_j go into the q_j as given in (11). This completes the proof.

If the sampling variables $\{r_j, 1 \leq j \leq k\}$ are assumed to be independent of the Y, then the proof can be simplified considerably. However, because we have not made any such supposition, our proof was necessarily long. The details are presented since this type of result is of interest in applications. If $k = 1$ and $r_j = i_j$, $r_{k+1} = n$ (both nonrandom), then the result of Theorem 6 is essentially given in Cramér's classic [(1946), p. 369], and Rényi has also contributed to this problem. In the above results $0 < q_j < 1$ was essential and $q_j = 0$ gives a different (nonnormal) limit d.f. We indicate an instance of this as an exercise (Problem 6). On the other hand the assumption that the limit r.v. X in Theorem 6 be discrete is not essential, although it is used crucially in the above proof in invoking the conditional probability analysis, and it will be interesting to extend the result to the general case, i.e., $X > 0$ is any r.v.

7.3 Ergodic Sequences

In this section we include a considerable generalization of the sufficiency part of Kolmogorov's SLLN to certain dependent sequences of r.v.s, called ergodic sequences. The results are of fundamental importance in physics, and ergodic theory has now grown into a separate discipline. Here we present the basic Birkhoff ergodic theorem as an extension of Theorem 2.3.7 and also use it in the last chapter in analyzing some key ideas of stationary type processes. Let us motivate the concept.

If (Ω, Σ, P) is a probability space, $T : \Omega \to \Omega$ is a measurable mapping [i.e., $T^{-1}(\Sigma) \subset \Sigma$], then T is *measure preserving* if $P(T^{-1}(E)) = P(E), E \in \Sigma$. For such transformations, the following interesting recurrence phenomenon was noted by H. Poincaré in about 1890.

Proposition 1 *If $A \in \Sigma, T : \Omega \to \Omega$ is a measure-preserving transformation on (Ω, Σ, P), then for almost all $\omega \in A, T^n(\omega) \in A$ for infinitely many n, where $P(A) > 0$.*

Proof We first note that the result holds for at least one n. Indeed, let

$$B = \{\omega \in A : T^n(\omega) \notin A \quad \text{for all} \quad n \geq 1\} = \bigcap_{n=1}^{\infty} \{\omega \in A : T^n(\omega) \notin A\}$$

$$= A \cap \bigcap_{n=1}^{\infty} T^{-n}(A^c), \qquad A^c = \Omega - A. \tag{1}$$

Since T is measurable, $B \in \Sigma$. Next note that for $n \neq m, T^{-n}(B) \cap T^{-m}(B) = \emptyset$. In fact, if $n > m$, and $\omega \in T^{-n}(B) \cap T^{-m}(B) = T^{-m}(B \cap T^{-(n-m)}(B))$, then there must be a point $\omega' \in B \cap T^{-(n-m)}(B)$ such that $T^m(\omega) = \omega' \in B$. But this is impossible by definition of B above. On the other hand, $P(T^{-k}(B)) = P(B)$ for all $k \geq 1$ by the hypothesis on T. Since $\bigcup_{n=1}^{\infty} T^{-n}(B) \subset \Omega$, we have

$$\sum_{n=1}^{\infty} P(T^{-n}(B)) \leq P(\Omega) = 1.$$

This is possible only if $P(B) = 0$. Thus for each $\omega \in A - B, T^n(\omega) \in A$ for at least one n.

Now for any integer $k \geq 1$, consider

$$B_k = A \cap \bigcap_{n=1}^{\infty} T^{-kn}(A^c). \tag{2}$$

Then by the first paragraph, $P(B_k) = 0$, so that $P(\bigcup_{k=1}^{\infty} B_k) = 0$. If $\omega \in A \cap \bigcap_{k=1}^{\infty} B_k^c$, then $\omega \in A \cap B_1^c = \bigcup_{n=1}^{\infty} A \cap T^{-n}(A)$, so that $\omega \in T^{-n}(A)$ for some n. Consider $\omega \in A \cap B_k^c = \bigcup_{n=1}^{\infty} A \cap T^{-kn}(A); \omega \in T^{-kn'}(A)$ for some n', for each k. Thus $T^n(\omega) \in A$ for infinitely many n, as asserted.

The result is clearly false if P is replaced by an infinite measure here. For instance, if $\Omega = \mathbb{R}, T(\omega) = \omega + a$, and $\mu =$ Lebesgue measure on $\mathbb{R}$, then we get a counterexample.

An interesting point here is that almost every ω of $A \subset \Omega$ returns to A infinitely often. Thus $\sum_{k=0}^{n-1} \chi_A(T^k(\omega))$ is the number of times ω visits A under T during the first n instances. A natural problem of interest in statistical mechanics and in some other applications in physics is to know whether such (recurrent) points have a mean sojourn time; i.e., does the limit

$$\lim_{n \to \infty} (1/n) \sum_{k=0}^{n-1} \chi_A(T^k(\cdot)) \tag{3}$$

exist in some sense? This leads immediately to a more general question. If $f \in L^1(P)$, does the limit

$$\lim_{n \to \infty} \frac{1}{n} \sum_{k=0}^{n-1} f(T^k(\cdot)) \tag{4}$$

exist pointwise a.e. or in $L^1(P)$ mean? Thus if X is an r.v., and T is such a transformation, let $X_n = X \circ T^n$. Then the problem of interest is to consider the SLLN and WLLN for $\{X_n, n \geq 1\}$, and thus study their averages:

$$\frac{1}{n}\sum_{k=0}^{n-1} X_k, n \geq 1,$$

for various transformations T. This forms the ergodic theory. Note that the X_k are *not* independent. Here we shall prove the pointwise convergence of (4), which is a fundamental result originally obtained by G. D. Birkhoff in 1931. In many respects, the proof proceeds along the lines of the martingale convergence theorem, but cannot be deduced from the latter. As in the martingale case, a certain maximal inequality is needed. The proof of the latter has been re-worked and simplified since its first appearance, and the following short version is essentially due to A. M. Garsia who obtained it in the middle 1960s. If T is measure preserving and $Q : f \mapsto f \circ T, f \in L^1(P)$, it defines a positive linear mapping with $||Qf||_1 \leq ||f||_1$ and $Q1 = 1$. We actually can prove the result, more generally, for such "contractive" operators Q on $L^1(P)$. More precisely, one has a basic inequality as:

Proposition 2 (Maximal Ergodic Theorem) *Let $Q : L^1(P) \to L^1(P)$ be a positive linear operator such that $||Qf||_1 \leq ||f||_1$. Then for any $f \in L^1(P)$ we have*

$$\int_{A_f} f\, dP \geq 0 \tag{5}$$

where $A_f = \bigcup_{n\geq 0}\{\omega : \sum_{k=0}^{n}(Q^k f)(\omega) > 0\}$.

Proof Let $f_n = \sum_{k=0}^{n} Q^k f$ and $\tilde{f}_n = \sup_{0\leq k\leq n} f_k$. Then $\tilde{f}_n \uparrow$ and $[\tilde{f}_n > 0] \uparrow A_f$. Since Q is positive and $f = \tilde{f}_0$, we have $f \leq f + Q\tilde{f}_n^+$ and $f_{m+1} = f + \sum_{k=0}^{m}(Q^{k+1}f) = f + Qf_m \leq f + Q\tilde{f}_m^+, 0 \leq m \leq n$. Hence $\tilde{f}_n \leq \tilde{f}_{n+1} \leq f + Q\tilde{f}_n^+$. Consequently

$$\begin{aligned}\int_{[\tilde{f}_n>0]} f\, dP &\geq \int_{[\tilde{f}_n>0]} (\tilde{f}_n - Q\tilde{f}_n^+)\, dP \quad [\text{since } \tilde{f}_n \in L^1(P)] \\ &= \int_{[\tilde{f}_n>0]} \tilde{f}_n^+\, dP - \int_{[\tilde{f}_n>0]} Q\tilde{f}_n^+\, dP \\ &\geq ||\tilde{f}_n^+||_1 - ||Q\tilde{f}_n^+||_1 \geq ||\tilde{f}_n^+||_1 - ||\tilde{f}_n^+||_1 = 0 \end{aligned} \tag{6}$$

because Q is a (positive) contraction. Since $f \in L^1(P)$ and $[\tilde{f}_n > 0] \uparrow A_f$, we get (5) by letting $n \to \infty$ in (6) on using the dominated convergence theorem. This completes the proof.

Note that the finiteness of P is *not* used here. With this we can establish

Theorem 3 (Birkhoff's Ergodic Theorem) *Let (Ω, Σ, μ) be a measure space and $T : \Omega \to \Omega$ be a measurable and measure-preserving transformation. Then for each $f \in L^1(\mu)$, we have*

$$\lim_{n\to\infty} (1/n) \sum_{k=0}^{n-1} f \circ T^k = f^* \tag{7}$$

exists a.e. and f^* is invariant, in the sense that $f^* \circ T = f^*$ a.e. If, moreover, $\mu = P$, a probability, and $\mathcal{F} \subset \Sigma$ is the σ-subalgebra of invariant sets under T, so that $\mathcal{F} = \{A \in \Sigma : P(A \,\Delta\, T^{-1}(A)) = 0\}$, then $f^* = E^{\mathcal{F}}(f)$ a.e., and the convergence of (7) is also in $L^1(P)$-mean. ($A \,\Delta\, B$ is the symmetric difference of the sets, A, B.)

Proof Since $f \circ T^k \in L^1(\mu), k \geq 1$, it follows that

$$A_k = \{\omega : (f \circ T^k)(\omega) \neq 0\}$$

is σ-finite for each k, so that $A = \bigcup_{k=1}^{\infty} A_k \subset \Omega$ is σ-finite. Replacing Ω by A, if necessary, μ may be assumed σ-finite. We may (and do) take for (7) that $f \geq 0$ a.e. by considering $f^{\pm}$ separately. The proof of the existence of limit in (7) to be given here, is similar to the first proof of Theorem 3.5.7.

Let $0 < a < b < \infty, S_n(f) = (1/n) \sum_{k=0}^{n-1} f \circ T^k$, and consider the set

$$B_{ab} = \left\{\omega : \liminf_n S_n(f)(\omega) < a < b < \limsup_n S_n(f)(\omega)\right\}. \tag{8}$$

Since clearly $\liminf_n S_n(f)(T\omega) = \liminf_n S_n(f)(\omega)$ and similarly for lim sup, we conclude that $B_{ab} = T^{-1}(B_{ab})$, so that B_{ab} is invariant. If $\mu(B_{ab}) < \infty$, then the argument of Proposition 1 yields $\mu(B_{ab}) = 0$. Let us show that $\mu(B_{ab}) < \infty$.

If $A \in \Sigma, A \subset B_{ab}$ and $\mu(A) < \infty$, consider $g = f - b\chi_A \in L^1(\mu)$. Also, let B be the set on which $S_n(g) \geq 0$ for at least one n. Then by Proposition 2,

$$0 \leq \int_B g\, d\mu = \int_B f\, d\mu - b\mu(A \cap B) = \int_B f\, d\mu - b\mu(A), \tag{9}$$

since $A \subset B_{ab} \subset B$. To see the last inclusion, let $\omega \in B_{ab}$. Then

$$\begin{aligned} b < \limsup_n S_n(f)(\omega) &\Rightarrow b < S_n(f)(\omega) \quad \text{for at least one } n, \\ &\Leftrightarrow \frac{1}{n}\sum_{k=0}^{n-1}(f \circ T^k(\omega) - b\chi_\Omega(\omega)) \geq 0 \\ &\Rightarrow \frac{1}{n}\sum_{k=0}^{n-1}(f(T^k(\omega)) - b\chi_A(T^k(\omega))) \geq 0 \\ &\Leftrightarrow S_n(g)(\omega) \geq 0 \Rightarrow \omega \in B. \end{aligned}$$

Also, since μ is σ-finite, by (9) $\mu(A) \leq (1/b) \int_B |f|\, d\mu$ for all $A \in \Sigma(B_{ab})$, trace of Σ on B_{ab}, with $\mu(A) < \infty$, it follows that there exist $A_n \uparrow B_{ab}, \mu(A_n) < \infty$ and

$$\mu(B_{ab}) = \lim_{n\to\infty} \mu(A_n) \leq (1/b) \int_B |f|\, d\mu < \infty, \tag{10}$$

as needed for (8). Thus $h = f - b\chi_{B_{ab}} \in L^1(\mu)$, and applying Proposition 2 to h (and B_{ab} in place of A_f there, since $S_n(h) \geq 0$ on B_{ab}), we get

$$0 \leq \int_{B_{ab}} h\, d\mu = \int_{B_{ab}} (f - b)\, d\mu. \tag{11}$$

Consider $h' = a - f \geq 0$ so that $S_n(h') \geq 0$ on B_{ab}; then Proposition 2 again implies

$$0 \leq \int_{B_{ab}} h'\, d\mu = \int_{B_{ab}} (a - f)\, d\mu. \tag{12}$$

Adding (11) and (12), we obtain $0 \leq (a - b)\mu(B_{ab}) \leq 0$, since $a < b$. Hence $\mu(B_{ab}) = 0$. Letting $a < b$ run through the rationals, one deduces that the set $N = [\liminf_n S_n(f) < \limsup_n S_n(f)]$ is μ-null, whence $\lim_{n\to\infty} S_n(f) = f^*$ exists a.e.(μ), proving the SLLN result in (7). We take $\mu = P$, a probability measure, for the last part.

It is clear that $\mathcal{F}$ is a σ-algebra and f^* is $\mathcal{F}$-measurable. Since $f \in L^1(P), S_n(f) \in L^1(P)$, and by Fatou's inequality

$$E(|f^*|) \leq \limsup_n E(|S_n(f)|) < \infty.$$

Also, for each $A \in \mathcal{F}$

$$\begin{aligned}
\int_A S_n(f)\, dP &= (1/n) \sum_{k=0}^{n-1} \int_A f(T^k(\omega))\, dP \\
&= (1/n) \sum_{k=0}^{n-1} \int_{T^{-k}(A)} f\, dP \\
&= \int_A f\, dP \quad [\text{since } T^{-k}(A) = A \text{ a.e.}]
\end{aligned} \tag{13}$$

If we show that $\{S_n(f), n \geq 1\}$ is uniformly integrable, then $S_n(f) \to f^*$ a.e. implies, by Vitali's theorem (cf. Theorem 1.4.4), that we can take limits in (13) and get

$$\int_A f^*\, dP = \lim_{n\to\infty} \int_A S_n(f)\, dP = \int_A f\, dP = \int_A E^{\mathcal{F}}(f)\, dP, \quad A \in \mathcal{F}, \tag{14}$$

and since the extreme integrands are $\mathcal{F}$-measurable, $E^{\mathcal{F}}(f) = f^*$ a.e. follows. Now $\|S_n(f)\|_1 \leq \|f\|_1 < \infty$, so that $\{S_n(f), n \geq 1\}$ is bounded in $L^1(P)$. Also, given $\varepsilon > 0$, choose $\delta > 0$ such that $P(F) < \delta$ implies $E(|f|\chi_F) < \varepsilon$, which is possible since $f \in L^1(P)$. Hence

$$\int_F |S_n(f)|\, dP \leq (1/n) \sum_{k=0}^{n-1} \int_F | f(T^k(\omega)|\, dP$$
$$= (1/n) \sum_{k=0}^{n-1} \int_{T^{-k}(F)} | f|\, dP < \varepsilon, \tag{15}$$

because $P(T^{-k}(F)) = P(F) < \delta$, so that $\int_{T^{-k}(F)} | f|\, dP < \varepsilon$. Thus the uniform integrability holds. This also implies, by the same Vitali theorem, that $||S_n(f) - f^*||_1 \to 0$, completing the proof.

A measure preserving transformation $T : \Omega \to \Omega$ is called *ergodic* or *metrically transitive* if its invariant sets A (in $\mathcal{F}$) have probabilities 0 or 1 only. In this case, for each r.v. f, the sequence $\{f_n = f \circ T^n, n \geq 1\}$ is called an *ergodic process.* A family $\{X_n, n \geq 0\}$ (or $\{X_n, -\infty < n < \infty\}$) is called *strictly stationary* if for each integer $\ell \geq 0$ (or any integer ℓ) the finite dimensional d.f.s of $(X_{n_1}, \ldots, X_{n_k})$ and $X_{n_1+\ell}, \ldots, X_{n_k+\ell}$ are identical. Thus X_n need not have any finite moments for strict stationarity and it also need not be ergodic. If $T : \Omega \to \Omega$ is one-to-one and both T and T^{-1} are measure preserving, and X_0 is an r.v., then the sequence $\{X_n = X_0 \circ T^{-n}, n \geq 1\}$ is strictly stationary, as is easily seen. Recalling the work of Section 5.3, it may be noted that the symmetric stable processes (or sequences) form a subclass of strictly stationary family. There is a weaker notion of stationarity if the process has two (translation invariant) moments which will be discussed in the last and final chapter. Theorem 3 implies the following statement:

Corollary 4 *Let $\{X_n, n \geq 1\}$ be an integrable strictly stationary sequence of r.v.s. Then the sequence obeys the SLLN with limit as an invariant r.v., which is $E^{\mathcal{I}}(X_1)$ where $\mathcal{I}$ is the σ-algebra of invariant sets in Σ (of (Ω, Σ, P)) and $E^{\mathcal{I}}(\cdot)$ is the conditonal expecation relative to $\mathcal{I}$. If, moreover, $\{X_n = X \circ T^n, n \geq 1\}$ is an ergodic sequence, then we have*

$$(1/n) \sum_{k=0}^{n-1} X_k \to E(X)(= E^{\mathcal{I}}(X),\ \mathcal{I} = \{\phi, \Omega\}),\ \text{a.e. and in } L^1(P). \tag{16}$$

It should be remarked that if $\{X_n, n \geq 0\}$ is a sequence of independent r.v.s, then it is strictly stationary iff the X_n are identically distributed. Further, an i.i.d. sequence $\{X_n, n \geq 0\}$ is not only stationary as noted, but may be taken to be ergodic. This means that if the (Ω, Σ, P) is the canonical space given by the Kolmogorov representation (cf. Theorem 3.4.11) with $\Omega = \mathbb{R}^I (I = \text{integers})$ (this can always be arranged), then $(TX_n)(\omega) = X_{n+1}(\omega)$ is well defined, and easily shown to be ergodic in this case. With such an identification, (16) is equivalent to the sufficiency part of the SLLN given in Theorem 2.3.8. Since in the above T can be more general, it is clear how ergodic theory can be developed in different directions and also on infinite measure spaces.

Starting with the strictly stationary (and perhaps ergodic) processes, one can extend the central limit theory and other results to this dependent class. We leave the matter here; a few complements are indicated as exercises.

Exercises

1. The result of Proposition 1.1 can be stated in the following slightly more general form. Assume the conditions (ii) and (iv) there and let (i) and (iii) be replaced by (i′) $E(Y_n(t)\chi_A) \to P(A), A \in \Sigma$, and (iii′)$\sum_{k=1}^{k_n} X_{nk}^2 \xrightarrow{P} Z > 0$ a.e. Then, with simple changes in the proof (see also Proposition 2.7), show that $S_n \xrightarrow{D} V$ where the ch.f. of V is given by $\phi_V(t) = E(\exp(\frac{-t^2 Z}{2}))$.

2. Let (Ω, Σ, P) be the Lebesgue unit interval. ($\Omega = [0,1], P =$ Lebesgue measure.) Consider the r.v.s V_n defined by $V_n(\omega) = \mathrm{sgn}(\sin(2^{n+1}\pi\omega)), \omega \in \Omega, n \geq 0$. These are independent with means 0 and variances 1, known as the *Rademacher functions.* If $X_{nk} = (V_k/\sqrt{n}) + 2^{n/2}\chi_{A_n}, A_n = [0, 2^{-n}], 0 \leq k \leq n, \mathcal{F}_{nk} = \sigma(X_{nj}, j \leq k)$, then verify that the sequence

$$\{X_{nk}, 1 \leq k \leq n, n \geq 1\}$$

satisfies condition (ii) of Theorem 1.2 but not the Lindeberg condition [not even its weaker form: $\sum_{i=1}^{k_n} X_{ni}^2 \chi_{[|X_{ni}|>\varepsilon]} \xrightarrow{D} 0$ as $n \to \infty$]. Both these observations are due to McLeish (1974).

3. Complete the computational details of Propositions 2.7 and 2.11.

4. Let $\{N_n, n \geq 1\}$ be a sequence of integer-valued (positive) r.v.s such that $N_n/n \xrightarrow{P} a > 0$ as $n \to \infty$, and let $\{X_n, n \geq 1\}$ be i.i.d. variables with two moments finite. Show that $X_{N_n}/\sqrt{N_n} \xrightarrow{P} 0$ as $n \to \infty$.

5. Show that the proof of Theorem 2.1 can be simplified if the r.v.s N_n there are independent of the $\{X_n, n \geq 1\}$-sequence.

6. Let $\{Y_{r_1}, \ldots, Y_{r_{k+1}}\}$ be a set of optionally selected r.v.s from $Y_1, \ldots, Y_n$ which form order statistics from a random sample of size n with a continuous strictly increasing d.f. F on the interval $(\alpha, \beta], -\infty \leq \alpha < \beta < \infty$, $F'(\beta) > 0$ (cf. Theorem 2.6), and $1 \leq r_1 < r_2 < \cdots < r_{k+1} \leq n$, are integer-valued r.v.s. If $r_{k+1}/n \xrightarrow{P} X$, a discrete r.v., while $r_j/n \xrightarrow{P} 0, j = 1, \ldots, k$, with $P[r_j = i_j] = 1$, let $r_j^* = r_{k+1} - r_j$. Then show, using the procedure of the proof of Theorem 2.6, that

$$\lim_{n\to\infty} P[r_{k+1}F'(\beta)(\beta - Y_{r_1^*}) < x_1, \ldots, r_{k+1}F'(\beta)(\beta - Y_{r_k^*}) < x_k]$$
$$= \overline{G}(x_1, \ldots, x_k),$$

where $\overline{G}$ is the k-fold convolution of "chi-square" r.v.s with $2(i_{j+1} - i_j - 1)$ "degrees of freedom," $j = 0, \ldots, k - 1 (i_0 = 0)$. Thus the ch.f. ϕ of $\overline{G}$ is given by

$$\phi(u_1, \ldots, u_k) = \prod_{j=0}^{k-1} (1 - iv_j)^{-(i_{j+1}-i_j-1)}, \qquad v_j = \sum_{m=j}^{k} u_m.$$

[The argument is similar to (but simpler than) that of Theorem 2.6. Both these results on order statistics were sketched by the first author (1962). Also Rényi's (1953) results on order statistics are of interest.]

7. Let $L^p(P)$ be the Lebesgue space (Ω, Σ, P), $p \geq 1$, and let $T_n : L^1(P) \to L^1(P)$ be a positive linear mapping such that $||T_n f||_1 \leq ||f||_1$ and $||T_n f||_\infty \leq ||f||_\infty$. Then it can be verified that the adjoint operator T_n^* of T_n is also positive and satisfies the same norm conditions on $L^1(P)$ and $L^\infty(P)$. Let $T_{1n} = T_n T_{n-1} \ldots T_1$ and $V_n = T_{1n}^* T_{1n}$. Then V_n is well defined on $L^\infty(P)$, satisfies the same norm conditions, and is positive. If $f \in L^p(P), 1 < p < \infty$, and $g_n^f = V_n f$, show that $\{g_n^f, \mathcal{F}_n, n \geq 1\}$ is an adapted integrable sequence. Let $\mathcal{T}$ be the directed set of all bounded or simple stopping times of $\{\mathcal{F}_n, n \geq 1\}$. The sequence $\{g_n^f, \mathcal{F}_n, n \geq 1\}$ is called an *asymptotic martingale* (or amart) if the net $\{E(g_\tau^f), \tau \in \mathcal{T}\}$ of real numbers converges. It is asserted that $g_n^f \to g^f$ a.e. and in $L^p(P)$, $f \in L^p(P), 1 < p < \infty$. This is equivalent to showing that $\{g_n^f, \mathcal{F}_n, n \geq 1\}$ is an amart. [Proofs of these results are nontrivial, c.f. G.A. Edgar and L. Sucheston, (1976). We hope that a simple proof can be found!]

8. Show that the conclusion of Theorem 2.1 is valid if Y there is not necessarily discrete but just strictly positive.

9. Let $\{X_n, n \geq 1\}$ be i.i.d. variables and $Y_n = \max(X_1, \ldots, X_n)$. If N_n is an integer-valued r.v. such that $N_n/n \xrightarrow{P} X > 0$, where X is an r.v., and if $(Y_n - a_n)/b_n \xrightarrow{D} Z$ (nondegenerate) for $a_n \in \mathbb{R}, b_n > 0$, then using the method of proof of Section 2, show that the above convergence to Z is also "stable," as defined there. With this, prove that, F_X being the d.f. of X,

$$\lim_{n\to\infty} P[Y_{N_n} < a_n + b_n x] = \int_0^\infty (P[Z < x])^t F_T(dt).$$

[One has to prove several details. This result is due to O. Barndorff-Nielsen (1964).]

10. We now present an application of Birkhoff's result, which is called a *random ergodic theorem,* formulated by S. M. Ulam and J. von Neumann in 1945. Let (Ω, Σ, P) and $(S, \mathcal{I}, \mu)$ be two probability spaces. If

$(Y, \mathcal{Y}, \nu) = \times_{i=-\infty}^{\infty}(S, \mathcal{I}, \mu)^i$, the doubly infinite product measure space in the sense of Jessen (cf. Theorem 3.4.3), let $\phi_s : \Omega \to \Omega, s \in S$, be a family of one-to-one measurable and P-measure-preserving transformations. Note that the mapping $(\omega, s) \mapsto (\phi_s(\omega), s)$ is $P \times \mu$-measurable. If $X : \Omega \to \mathbb{R}$ is an integrable r.v. and for each $y \in Y, s_k(y)$ is as usual the kth coordinate of $y[s_k(y) \in S$, all $k]$, then the mapping $(\omega, y) \mapsto (\phi_{s_0(y)}(\omega), \psi(y))$ is $P \times \nu$-measurable and $P \times \nu$-measure preserving, where $\psi : Y \to Y$, defined by $s_n(\psi(y)) = s_{n+1}(y)$, is called a *shift transformation*. Then $\tilde{\phi}_n(y, s) = (\phi_{s_{n-1}(y)} \cdots \phi_{s_0(y)}(s), \psi^n(y))$ is well defined for $n = 1, 2, \ldots$. Show, using Theorem 3.3, that

$$\lim_{n\to\infty} (1/n) \sum_{k=0}^{n-1} X(\phi_{s_{k-1}(y)} \cdots \phi_{s_0(y)}(\cdot)) = X_y^*(\cdot)$$

a.e.[P], and in $L^1(P)$ also, for each $y \in Y - N, P \times \nu(N) = 0$. (The various subsidiary points noted above are useful in the proof. More on this and extensions with interesting applications to Markov processes can be found in an article by S. Kakutani (1951).)

Chapter 8

A Glimpse of Stochastic Processes

In this final chapter a brief account of continuous parameter stochastic processes is presented. It includes a direct construction of Brownian motion together with a few properties leading to the law of the iterated logarithm for this process. Also, processes with independent increments, certain other classes based on random measures and their use in integral representation of various processes are discussed. In addition, the classification of strictly, strongly, weakly stationary and harmonizable processes are touched on, so that a bird's-eye view of the existing and fast developing and deeply interesting stochastic theory can be perceived.

8.1 Brownian Motion: Definition and Construction

One of the most important continuous parameter stochastic processes and the best understood is the Brownian motion, named after the English botanist Robert Brown, who first observed it experimentally in 1826 as a continual irregular motion of small particles suspended in fluid, under the impact of the surrounding molecules. Later this motion was described in (plausible) mathematical terms by Bachelier in 1900, by Einstein in 1905, and by von Smoluchovski in 1906. However, it was Norbert Wiener who in 1923 gave a rigorous mathematical derivation of this process, and so it is also called a Wiener process. It occurs in physical phenomena (as Einstein and Smoluchovski considered) and can also be used to describe fluctuations of the stock market averages (as Bachelier noted) among other things. We now define this process and derive its existence by presenting a direct construction without using the Kolmogorov-Bochner theory of projective limits (of Chapter 3) which may also be used.

Definition 1 Let $\{X_t, t \geq 0\}$ be a family of r.v.s on a probability space (Ω, Σ, P). Then it is called a *Brownian motion* if (i) $P[X_0 = 0] = 1$ and if (ii) for any $0 \leq t_0 < \cdots < t_n$, the $X_{t_{i+1}} - X_{t_i}, i = 0, 1, \ldots, n-1$, are mutually independent, with $X_{t_{i+1}} - X_{t_i}$ distributed as $N(0, \sigma^2(t_{i+1} - t_i))$ for some constant $\sigma^2 > 0$. (For simplicity, we take $\sigma^2 = 1$. Compare with Definition 5.4.1.)

It is immediately seen that this collection of r.v.s $\{X_{t_i}, i = 1, \dots, n\}$, as n varies, has a compatible set of d.f.s in the sense of Section 3.4 (it was verified after 5.4.1). Hence by Theorem 3.4.11 such a process, as required by the above definition, exists with $\Omega = \mathbb{R}^{[0,\infty]}$, $\Sigma = \sigma$-algebra of cylinder sets of Ω, and P given by the normal d.f.s that satisfy conditions (i) and (ii). However, it turns out that the function $t \mapsto X_t(\omega)$ is continuous for almost all $\omega \in \Omega$, so that the process really concentrates on the set of continuous functions $C[0, \infty) \subset \Omega$. This and several properties of great interest in the analysis of Brownian motion need separate (and nontrivial) proofs. The present special work is facilitated by direct constructions that do not essentially use the Kolmogorov-Bochner theory, and so we present one such here. It is based on an interesting argument given by Ciesielski [(1961), especially p. 406]. For this we need the completeness property of Haar functions of abstract analysis, indicated in Exercise 46 in Chapter 3. Since this is essential, let us present it here.

The functions given in the above-noted problem can be written, for all $0 \leq x \leq 1$, as

$$h_0(x) = 1, \quad h_0^{(1)}(x) = \chi_{[0,1/2]} - \chi_{(1/2,1]},$$

and for $n \geq 1, k = 1, \dots, 2^n$,

$$h_n^{(k)}(x) = \begin{cases} \sqrt{2^n} & \text{if } (k-1)2^{-n} \leq x < (2k-1)2^{-n-1} \\ -\sqrt{2^n} & \text{if } (2k-1)2^{-n-1} < x \leq k2^{-n} \\ 0 & \text{otherwise.} \end{cases} \tag{1}$$

Writing these as a sequence in increasing values of n (and for each n in increasing k), let us relabel the set thus denoted as $\{H_n, n \geq 0\}$. Then it is a bounded orthonormal sequence in $L^2(0,1)$, so that for any $f \in L^p(0,1), p \geq 1$, its Fourier expansion can be obtained as

$$S_n(f) = \sum_{k=0}^{n} \alpha_k(f) H_k, \quad f \in L^p(0,1), \tag{2}$$

where $\alpha_k(f) = \int_0^1 f(x)H_k(x)dx$. The completeness of the Haar system may be proved in different ways. For instance, one can show that they form a basis in $L^p(0,1), 1 \leq p < \infty$. However, a short probabilistic argument runs as follows. If $\mathcal{B}_n = \sigma(H_1, \dots, H_n)$, then the atoms of $\mathcal{B}_n$ are intervals of the form $[k2^{-n}, (k+1)2^{-n})$, and $E^{\mathcal{B}_{n-1}}(H_n) = 0$ by definition of each H_n. Thus $\{S_n, \mathcal{B}_n, n \geq 1\}$ of (2) is an $L^2(0,1)$-bounded martingale for each $f \in L^2(0,1)$. Hence it converges a.e. and in $L^2(0,1)$ to f, since $\sigma(\bigcup_{n\geq 1} \mathcal{B}_n)$ is the Borel σ-algebra of (0,1). Thus $\int_0^1 (fH_n)(x)dx = 0$ for all $n \geq 0$ implies $f = 0$ a.e., which is completeness of the system by definition. As is well-known, in $L^2(0,1)$ this is equivalent to Parseval's equation. Thus $f, g \in L^2(0,1)$ implies

$$(f,g) = \int_0^1 f(x)g(x)\,dx = \sum_{n=0}^{\infty}(f,H_n)(g,H_n). \tag{3}$$

We use this form of completeness in the following work. It is of interest to note that the integrated objects $\psi_n : x \mapsto \int_0^x H_n(v)\,dv$ are called "Schauder functions," and their orthonormalized set (by the Gram-Schmidt process) is known as the "Franklin functions." These various forms are profitably used in different types of computations in harmonic analysis.

The existence result is explicitly given by the following:

Theorem 2 *Let $\{\xi_n, n \geq 0\}$ be an i.i.d. sequence of $N(0,1)$ r.v.s on a probability space (Ω, Σ, P). [The existence of this space follows at once from the product probability theorem of Jessen (cf. Theorem 3.4.3).] Then the series*

$$\sum_{n=0}^{\infty} \xi_n(\omega) \int_0^t H_n(x)\,dx = X_t(\omega), \quad 0 \leq t \leq 1, \tag{4}$$

converges uniformly in t for almost all $\omega \in \Omega$, and the family $\{X_t, 0 \leq t \leq 1\}$ is a Brownian motion process on this probability space, which can then be extended to all of $\mathbb{R}^+$.

Remark: The crucial part is the uniform convergence of the series in (4) for *all* t, and a.a. (ω). This is an important reason for using "random Fourier series" to study the sample path properties of their sums.

Proof Let us first establish the uniform convergence of (4). Using the $N(0,1)$ hypothesis, we have for $u > 0$

$$\begin{aligned} P[|\xi_n| > u] &= \frac{2}{\sqrt{2\pi}} \int_u^{\infty} e^{-v^2/2}\,dv = \sqrt{\frac{2}{\pi}} \int_{u^2/2}^{\infty} e^{-r} \frac{dr}{\sqrt{2r}} \\ &\leq \sqrt{\frac{2}{\pi}} [\exp(-u^2/2)]/u \qquad \text{(integration by parts)}. \end{aligned}$$

Hence if $u^2 = 3\log n$, we get

$$\sum_{n=2}^{\infty} P[|\xi_n| > u] \leq \sqrt{\frac{2}{3\pi}} \sum_{n=2}^{\infty} \frac{n^{-3/2}}{\sqrt{\log n}} < \infty.$$

Thus by the Borel-Cantelli lemma (cf. Theorem 2.1.9iii)

$$P[|\xi_n| > \sqrt{3\log n}, \text{i.o.}] = 0. \tag{5}$$

With this, if ψ_m is the Schauder function, then by integration we note that $\psi_m(t) \geq 0$, and if $2^n \leq m < 2^{n+1}, 0 \leq t \leq 1$, then $0 \leq \psi_m(t) \leq 2^{-n/2}/4$. But (5) implies

$$a_n = \max_{2^n \leq k < 2^{n+1}} |\xi_k| \leq [3 \log 2^{n+1}]^{1/2}$$

with probability 1 for all large enough n. Consequently (4) is dominated by the following series for a.a.(ω):

$$\sum_{n=m}^{\infty} |\xi_n|\psi_n(t) \leq \frac{1}{4}\sum_{n\geq m} \alpha_n 2^{-n/2} \leq \frac{\sqrt{3\log 2}}{4}\sum_{n\geq m}(n+1)^{1/2}2^{-n/2} < \infty. \quad (6)$$

By (6) the series defined in (4) is uniformly convergent for all t, $0 \leq t \leq 1$ with probability one. It remains to verify only that $\{X_t, 0 \leq t \leq 1\}$ is a Brownian motion process, since the continuity of $X_t(\omega)$, a.a. (ω), is immediate from the uniform convergence of continuous summands.

It is clear that $X_0 = 0$ a.e. One uses the completeness property in the form of (3) to verify that Definition 1ii is also true. Let $0 < t_1 < \cdots < t_m \leq 1$. In view of the uniqueness theorem for ch.f.s (cf. Theorem 4.5.1), it suffices to show that the joint ch.f. of $(X_{t_1}, \ldots, X_{t_m})$ is normal and that the $(X_{t_{i+1}} - X_{t_i})$ are independent. For this consider

$$\begin{aligned}
\phi(u_1, \ldots, u_m) &= E[\exp(iu_1X_{t_1} + iu_2(X_{t_2} - X_{t_1}) + \cdots + iu_m(X_{t_m} - X_{t_{m-1}}))] \\
&= E\left[\exp\left(i\sum_{k=0}^{\infty}\xi_k\{(u_1 - u_2)\psi_k(t_1) + \cdots + u_m\psi_k(t_m)\}\right)\right] \\
&\qquad \text{[by the representation (4) just established]} \\
&= \lim_{n\to\infty}\prod_{k=0}^{n}\exp\left(-\frac{1}{2}[(u_1 - u_2)\psi_k(t_1) + \cdots + u_m\psi_k(t_m)]^2\right) \\
&\qquad \text{[by the mutual independence and } N(0,1) \text{ of the } \xi_k] \\
&= \exp\left(-\frac{1}{2}\left[\sum_{j=1}^{m}(u_j - u_{j+1})^2\sum_{k=0}^{\infty}\psi_k^2(t_j)\right.\right. \\
&\qquad \left.\left. + \; 2\sum_{1\leq j<j'\leq m}(u_j - u_{j+1})(u_{j'} - u_{j'+1})\sum_{k=1}^{\infty}\psi_k(t_j)\psi_k(t_{j'})\right]\right) \\
&\qquad (\text{where } u_{m+1} = 0). \quad (7)
\end{aligned}$$

Note that if we consider $f = \chi_{[0,t_i)}, g = \chi_{[0,t_j)}, 0 \leq t_i \leq t_j \leq 1$, then (3) becomes

$$t_i = (f, g) = \sum_{n=0}^{\infty}(f, H_n)(g, H_n) = \sum_{k=0}^{\infty}\psi_k(t_i)\psi_k(t_j). \quad (8)$$

Using (8) in (7), we get, after a direct computation and a simplification,

$$\begin{aligned}
\phi(u_1, \ldots, u_m) \\
= \exp\left\{-\frac{1}{2}\left(\sum_{j=1}^{m}(u_j - u_{j+1})^2 t_j + 2\sum_{1\leq j<j'\leq m}(u_j - u_{j+1})(u_{j'} - u_{j'+1})t_j\right)\right\}
\end{aligned}$$

$$= \prod_{j=1}^{m} E(\exp(iu_j(X_{t_j} - X_{t_j-1}))) \qquad \text{(where } t_0 = 0\text{)}. \tag{9}$$

This implies that $X_{t_j} - X_{t_{j-1}}$ are independent $N(0, (t_j - t_{j-1}))$ r.v.s, and completes the proof for $0 \le t \le 1$.

To get the result for $0 \le t < \infty$, define (inductively) the process as follows. Let $X_t = X_t^{(1)}$ for $0 \le t \le 1$, and if X_t is defined on $0 \le t \le n$, then let $X_t = X_{t-n}^{(n+1)} + X_n$ for $n \le t \le n+1$, where $\{X_t^{(n)}, 0 \le t \le 1\}, n \ge 1$, are independent copies given by (4).

From Eq. (9), for $0 < t_1 < t_2$, we get $E(X_{t_1}X_{t_2}) = E(X_{t_1}^2) = t_1$, so that the covariance function $r(\cdot,\cdot)$ of the Brownian motion process is given by

$$r(s,t) = E(X_s X_t) = \min(s,t), \qquad 0 < s,t < \infty. \tag{10}$$

This completes the proof on $\mathbb{R}^+$ itself.

Remark Of the other methods of construction, one that is interesting and intuitively simple is to show that the Brownian motion process is the limiting form of a suitable sequence of random walk processes. This yields another connection of this theory with our earlier work. However, we do not present it here.

8.2 Some Properties of Brownian Motion

Even though the *sample functions* $t \mapsto X_t(\omega), t \ge 0$, of a Brownian motion are uniformly continuous for almost all $\omega \in \Omega$, their behavior is quite irregular. Almost all are non-differentiable, and do not have finite variation, on any interval of positive length, so that one cannot easily extend the classical integration for this process as an integrator. Let us establish these properties before discussing others.

First we note the following parametric characterization of a Gaussian process, a family $\{X_t, t \in T\}$ all of whose finite-dimensional d.f.s are Gaussian.

Proposition 1 *A real Gaussian process $\{X_t, t \in T\}$ is characterized by a mean function $\mu : t \mapsto \mu(t) = E(X_t)$ and a covariance function $r : (s,t) \mapsto r(s,t) = Cov(X_s, X_t)$, i.e., a function satisfying $r(s,t) = r(t,s)$ and for each finite set $\{t_1, \ldots, t_n\} \subset T$, the n-by-n matrix $(r(t_1, t_j), i, j = 1, \ldots, n)$ is positive (semi-) definite.*

Proof Let μ, r be the mean and covariance functions (parameters) of the given Gaussian process $\{X_t, t \in T\}$. Then for each finite set $(t_1, \ldots, t_n)$ from

T, the joint d.f. of $(X_{t_1}, \ldots, X_{t_n})$, is an n-dimensional Gaussian d.f. so that if $(\alpha_1, \ldots, \alpha_n) \in \mathbb{R}^n, Y_n = \sum_{i=1}^n \alpha_i X_{t_i}$, then Y_n is $N(\nu_n, \sigma_n^2)$ distributed, where $\nu_n = \sum_{i=1}^n \alpha_i \mu(t_i)$, $\sigma_n^2 = \sum_{i=1}^n \sum_{j=1}^n \alpha_i \alpha_j r(t_i, t_j)$. Thus its ch.f. ϕ_n is given by

$$\begin{aligned}\phi_n(u) &= E(\exp\{iuY_n\}) \\ &= \exp\{iu\nu_n - \frac{1}{2}u^2\sigma_n^2\} \\ &= \exp\left\{i\sum_{j=1}^n u\alpha_j\mu(t_j) - \frac{1}{2}\sum_{j=1}^n\sum_{k=1}^n u\alpha_j r(t_j, t_k)\alpha_k u\right\}.\end{aligned}$$

Hence writing $u_j = u\alpha_j$, we can deduce that the ch.f. ψ_n of $(X_{t_1}, \ldots, X_{t_n})$ is

$$\psi_n(u_1, \ldots, u_n) = \exp\left\{i\sum_{j=1}^n u_j\mu(t_j) - \frac{1}{2}\sum_{j=1}^n\sum_{k=1}^n u_j u_k r(t_j, t_k)\right\}. \tag{1}$$

Since each n-dimensional Gaussian ch.f. ψ_n given by (1) is completely determined by the parameters $(\mu(t_1), \ldots, \mu(t_n))$ and $r(t_i, t_j), i, j = 1, \ldots, n)$, that is, by the functions μ and r, it follows that a Gaussian process is determined by these two functions.

On the other hand, if $\mu : T \to \mathbb{R}$ and $r : T \times T \to \mathbb{R}$ are given with r as positive definite, then by substitution of $\mu(t_i), r(t_i, t_j)$ in (1), we get classes of n-dimensional ch.f.s. Because of the uniqueness theorem (cf. Theorem 4.5.1) and the classical Fubini theorem, we deduce at once that these ch.f.s (and hence their d.f.s) form a compatible family. Thus by Kolmogorov's theorem (cf. Theorem 3.4.11) we can construct a probability space (Ω, Σ, P), (with $\Omega = \mathbb{R}^T$, etc.) and a process $\{\tilde{X}_t, t \in T\}$ which has these n-dimensional Gaussian d.f.s as its finite dimensional d.f.s, so that the process is Gaussian with μ and r as its mean and covariance functions. Thus as far as the finite-dimensional distributions are concerned, the given process and the constructed process are indistinguishable. In this sense there is essentially a unique Gaussian process having a given mean and covariance function. This proves more than the asserted statement.

Since a Brownian motion is also a Gaussian process with $\mu(t) = E(X_t) = 0$ and covariance $r(s, t)$ given by [cf. Eq. (1.10)]

$$r(s, t) = E(X_s X_t) = \min(s, t), \tag{2}$$

a simple calculation of μ and r shows that we have the following consequence:

Corollary 2 *Let $\{X_t, t \geq 0\}$ be a Brownian motion process. Then*

(i) $\{X_{t+\alpha} - X_\alpha, t \geq 0\}$ *is a Brownian motion process for each* $\alpha > 0$;
(ii) $\{tX_{t^{-1}}, t > 0\}$ *and*

(iii) $\{\beta^{-1}X_{\beta^2 t}, t \geq 0\}$ *for* $\beta \neq 0$,

are also Brownian motion processes, (i.e., all have the same finite dimensional d.f.s).

We can now present a mathematical description of the irregular behavior of the process, observed by R. Brown under a microscope:

Theorem 3 (Wiener) *The sample functions* $t \mapsto X_t(\omega)$ *of the Brownian motion process* $\{X_t, t \geq 0\}$ *on a complete probability space* (Ω, Σ, P), *which are uniformly continuous for a.a.*(ω) *by Theorem 1.2, are nowhere differentiable for a.a.* $\omega \in \Omega$. *In particular, they are not rectifiable in any interval of positive length [for a.a.*(ω)].

Proof In view of the preceding corollary, it suffices to prove this result for $\{X_t, 0 \leq t \leq 1\}$. We present a short argument here following Dvoretzky et. al. (1961). If X_t is differentiable at a point $s \in [0,1]$, then for t, s close enough, we must have $|X_t(\omega) - X_s(\omega)| \leq C|t-s|$. One may translate this into the following statement: Let $j = [ns]+1$, the next integer to the integral part $[ns]$ for any natural number $n \geq 1$. Then the above inequality gives

$$|X_{i/n} - X_{(i-1)/n}|(\omega) < \ell/n, \quad i = j+1, j+2, j+3 \tag{3}$$

for an integer $\ell \geq 1$ and n large enough. Let $A(j,\ell,n)$ be the ω-set defined by (3) and let A be the set of ω for which $X_{(\cdot)}(\omega)$ is differentiable at some point of [0,1]. It is not obvious that $A \in \Sigma$. However, we show that it is contained in a P-null set, so that (by completeness of the space (Ω, Σ, P)), Σ contains all subsets of P-null sets which then have zero probabilities. Note that $A(j,\ell,n) \in \Sigma$.

It is seen that, by definition,

$$A \subset \bigcup_{\ell=1}^{\infty} \bigcup_{m=1}^{\infty} \bigcap_{n=m}^{\infty} \bigcup_{j=1}^{n+1} A(j,\ell,n). \tag{4}$$

Consider

$$\begin{aligned}
&P\left(\bigcap_{n\geq m} \bigcup_{j=1}^{n+1} A(j,\ell,n)\right) \\
&\leq \liminf_n P\left(\bigcup_{j=1}^{n+1} \bigcap_{i=j+1}^{j+3} \left[|X_{i/n} - X_{(i-1)/n}| < \frac{\ell}{n}\right]\right) \\
&\leq \liminf_n \sum_{j=1}^{n+1} P\left[|X_{i/n} - X_{(i-1)/n}| < \frac{\ell}{n}, i = j+1, j+2, j+3\right] \\
&\leq \liminf_n \sum_{j=1}^{n+1} [P[|X_{1/n}| < \ell/n]]^3
\end{aligned}$$

(by the i.i.d. property of "equal" increments and Corollary 2)

$$= \liminf_n (n+1)(P[|X_{1/n}| < \ell/n])^3$$

$$= \liminf_n (n+1)\left(\sqrt{\frac{2n}{\pi}} \int_0^{\ell/n} e^{-nu^2/2}\, du\right)^3$$

[since $X_{1/n}$ has the $N(0, 1/n)$ d.f.]

$$\leq \liminf_n (n+1)\left(\frac{3\ell}{\sqrt{n}}\sqrt{\frac{2}{\pi}}\, e^{-\ell^2/n}\right)^3 = 0. \tag{5}$$

From (4) and (5) it follows that A is contained in a countable union of P-null sets and hence has probability zero.

If the path $\{X_t, (\omega), 0 \leq t \leq 1\}$ is rectifiable, then it is of bounded variation. But from real variable theory we know that a function of bounded variation has a derivative at almost all points $s \in [0,1]$ (Lebesgue measure), and by the above, the set of ω-points with this property is a P-null set. Thus the last assertion follows, and the proof is complete.

Actually the above authors have shown, in the paper cited, a stronger result than this theorem, namely: the process $\{X_t + \alpha t, 0 \leq t \leq 1\}$, with a "linear trend" αt, has almost no points of increase or decrease for any $\alpha \in \mathbb{R}$. Many intricate properties are found for this process because of the analytical work available for the d.f.s $N(0, t)$ of X_t, as in (5).

By the preceding theorem, almost no Brownian path $t \mapsto X_t(\omega), 0 \leq t \leq 1$, is of bounded variation. However, P. Lévy has given the following interesting property on its "second variation":

Proposition 4 *Let $\{X_t, 0 \leq t \leq 1\}$ be a Brownian motion. Then*

$$\lim_{n\to\infty} \sum_{k=1}^{2^n} [X_{k2^{-n}} - X_{(k-1)2^{-n}}]^2 = 1 \qquad a.e. \tag{6}$$

Proof Let $Y_{kn} = X_{k2^{-n}} - X_{(k-1)2^{-n}}$, so that the Y_{kn} are $N(0, 2^{-n})$ distributed and independent. Consequently, $\{Y_{kn}^2, 1 \leq k \leq 2^n\}$ are also independent, $E(\sum_{k=1}^{2^n} Y_{kn}^2) = 1$, and

$$\mathrm{Var}\left(\sum_{k=1}^{2^n} Y_{kn}^2\right) = \sum_{k=1}^{2^n} \mathrm{Var}\ Y_{kn}^2 = \sum_{k=1}^{2^n} 2 \cdot 2^{-2n} = 2^{-n+1}.$$

Hence by the Čebyšev inequality, for each $\varepsilon > 0$, if

$$A_{n,\varepsilon} = \left[\left| \sum_{k=1}^{2^n} Y_{kn}^2 - 1 \right| \geq \varepsilon \right],$$

then

$$P(A_{n,\varepsilon}) \leq \varepsilon^{-2} 2^{-n+1}.$$

Thus

$$\sum_{n=1}^{\infty} P(A_{n,\varepsilon}) \leq \varepsilon^{-2} \sum_{n=1}^{\infty} 2^{-n+1} = 2\varepsilon^{-2} < \infty,$$

and by the Borel-Cantelli lemma (cf. Theorem 2.1.9i), $P(A_{n,\varepsilon}, \text{i.o.}) = 0$. Hence for the complementary events, $P(A_{n,\varepsilon}^c, n \geq n_0) = 1$, so that

$$P\left\{ \omega : \left| \sum_{k=1}^{2^n} Y_{kn}^2(\omega) - 1 \right| < \varepsilon, n \geq n_0(\omega) \right\} = 1. \tag{7}$$

Since $\varepsilon > 0$ is arbitrary, (7) implies (6), as asserted.

A few other properties of the process are included in the problems section.

8.3 Law of the Iterated Logarithm for Brownian Motion

We have proved the Kolmogorov LIL as Theorem 5.5.1 and the details there are considerably involved. Here we present a similar LIL for Brownian motion, due originally to A. Khintchine in 1933, and its proof is easier than the earlier result. The work is further shortened by using a martingale maximal inequality. We follow McKean (1969) for its proof.

Theorem 1 *Let $\{X_t, t \geq 0\}$ be a Brownian motion process. Then*

$$P\left[\limsup_{t \searrow 0} \frac{X_t}{(2t \log\log 1/t)^{1/2}} = 1 \right] = 1, \tag{1}$$

and also

$$P\left[\liminf_{t \searrow 0} \frac{X_t}{(2t \log\log 1/t)^{1/2}} = -1 \right] = 1. \tag{2}$$

Proof It is sufficient to prove (1), since by Corollary 2.2 $-X_t$ is also a Brownian motion process and $-\limsup(-X_t) = \liminf X_t$. (There are no measurability problems.) As in Theorem 5.5.1, we establish (1) in two parts: for ≥ 1 and for ≤ 1. The key observation here is to note that $Y_t = \exp\{\alpha X_t -$

$\alpha^2 t/2\}$ is a martingale for $\mathcal{F}_t = \sigma(X_s, 0 \le s \le t), t \ge 0$, for each $\alpha \in \mathbb{R}$. Indeed, Y_t is integrable, since

$$E(Y_t) = e^{-\alpha^2 t/2} E(e^{\alpha X_t}) \\ = e^{-\alpha^2 t/2} e^{\alpha^2 t/2} = 1, \tag{3}$$

with the formula for ch.f.s. Also, Y_t is $\mathcal{F}_t$-adapted, and for $0 < s < t$ we have

$$E^{\mathcal{F}_s}(Y_t) = Y_s E^{\mathcal{F}_s}(e^{\alpha(X_t - X_s) - \alpha^2(t-s)/2}) \\ = Y_s E^{\mathcal{F}_s}(e^{\alpha X_{t-s} - \alpha^2(t-s)/2}) \quad \text{(by Corollary 2.2)} \\ = Y_s \quad \text{a.e., [by (3)]}. \tag{4}$$

Hence $\{Y_t, \mathcal{F}_t, t \ge 0\}$ is a positive martingale.

Let $b > 0$ and consider the maximal inequality (cf. Theorem 3.5.6i)

$$P\left[\max_{s \le t}\left(X_s - \frac{a}{2}s\right) > b\right] = P\left[\max_{s \le t} Y_s \ge e^{ab}\right] \quad (a \in \mathbb{R}) \\ \le e^{-ab} E(Y_t) = e^{-ab}. \tag{5}$$

Note that since X_s, and hence Y_s, is (uniformly) continuous on $[0, t]$, the maximum exists, and there are again no measurability problems. (In fact we can consider the result when s ranges over all the rationals and by continuity the above assertion follows.) In order to apply the Borel-Cantelli lemma, one chooses a, b of (5) suitably.

Let $h(t) = (2t \log\log 1/t)^{1/2}, 0 < \delta < 1, \varepsilon > 0$ and $t > 0$ be small. Set $t_n = \delta^n, a_n = (1 + \varepsilon)\delta^{-n} h(\delta^n)$, and $b_n = \frac{1}{2} h(\delta^n)$. Then

$$a_n b_n = (1 + \varepsilon) \log\log \delta^{-n}, \qquad e^{-a_n b_n} = (\log 1/\delta)^{-(1+\varepsilon)} n^{-(1+\varepsilon)}. \tag{6}$$

By (5) and (6) we get

$$\sum_{n \ge 1} P\left[\max_{s \le t_n}\left(X_s - \frac{a_n}{2}s\right) > b_n\right] \le (\log 1/\delta)^{-(1+\varepsilon)} \sum_{n \ge 1} n^{-(1+\varepsilon)} < \infty.$$

Thus the (first) Borel-Cantelli lemma implies

$$P\left[\max_{s \le t_n}(X_s - \frac{1}{2}a_n s) \le b_n, \text{all but finitely many } n\right] = 1. \tag{7}$$

Hence for a.a.(ω), we can find $n_0(\omega)$ such that $n \ge n_0(\omega)$ and $t_{n-1} < t \le t_n$ implies

$$X_t(\omega) \le \max_{s \le t_n} X_s(\omega) < b_n + a_n t_n/2 = \frac{1}{2} h(\delta^n)[(1 + \varepsilon) + 1] \\ = \frac{1}{2} h(t_n)[2 + \varepsilon].$$

Thus

$$\limsup_{\varepsilon \searrow 0\, \delta \nearrow 1} (X_t/h(t)) \leq 1 \qquad \text{a.e.} \tag{8}$$

For the opposite inequality consider the *independent* events

$$A_n = [(X_{\delta^n} - X_{\delta^{n+1}}) \geq (1 - \sqrt{\delta})h(\delta^n)], \tag{9}$$

where the $0 < \delta < 1$ and h are as before. Let us calculate

$$\begin{aligned} P(A_n) &= P\left[\frac{X_{\delta^n} - X_{\delta^{n+1}}}{\sqrt{(\delta^n - \delta^{n+1})}} \geq c_n\right], \qquad \left[c_n = \frac{1-\sqrt{\delta}}{\sqrt{(\delta^n - \delta^{n+1})}} h(\delta^n) > 0\right] \\ &= \sqrt{\frac{1}{2\pi}} \int_{c_n}^{\infty} e^{-u^2/2}\, du \quad \text{(by Corollary 2.2).} \end{aligned} \tag{10}$$

To obtain a lower estimate, consider

$$\begin{aligned} \int_a^\infty e^{-u^2/2}\, du &> \int_a^\infty a^2 u^{-2} e^{-u^2/2}\, du \\ &= a e^{-a^2/2} - a^2 \int_a^\infty e^{-u^2/2}\, du \quad \text{(by integration by parts).} \end{aligned}$$

Hence

$$\begin{aligned} \int_a^\infty e^{-u^2/2}\, du &> \frac{a}{1+a^2} e^{-a^2/2} = \frac{1}{a}(1+a^{-2})^{-1} e^{-a^2/2} \\ &= \frac{1}{a}(1 - a^{-2} + a^{-4}(1+a^{-2})^{-1}) e^{-a^2/2} > (a^{-1} - a^{-3}) e^{-a^2/2}. \end{aligned}$$

Substituting this into (10), we get on simplification

$$\begin{aligned} P(A_n) &> (2\pi)^{-1/2} \left(\frac{1+\sqrt{\delta}}{2(1-\sqrt{\delta})}\right)^{1/2} \frac{(\log n)^{-1/2}}{1 + (\log\log \delta^{-1})/\log n} \\ &\times \left(1 - \frac{1+\sqrt{\delta}}{2(1-\sqrt{\delta})\log\log \delta^{-n}}\right)(n \log \delta^{-1})^{-(1-\sqrt{\delta})/(1+\sqrt{\delta})}. \end{aligned} \tag{11}$$

For large enough n, the right side of (11) is at least as great as

$$\text{const.}(\log n)^{-1/2} n^{-(1-\sqrt{\delta})/(1+\sqrt{\delta})},$$

and this is the general term of a divergent series. Since the A_n of (9) are independent events, we get by the second Borel-Cantelli lemma (cf. Theorem 2.1.9iii), because $\sum_{n \geq 1} P(A_n) = \infty$, that $P(A_n$, i.o.)=1. Hence for infinitely many n,

$$X_{\delta^n} \geq (1 - \sqrt{\delta})h(\delta^n) + X_{\delta^{n+1}}, \quad \text{a.e..} \tag{12}$$

On the other hand, by (8), $X_{\delta^{n+1}} < h(\delta^{n+1})$ from some n onward. But X_{δ^n} is $N(0, \delta^n)$, so that (by symmetry) $X_{\delta^{n+1}} > -h(\delta^{n+1})$ also. Thus (12) becomes

$$\begin{aligned} X_{\delta^n} &\geq (1-\sqrt{\delta})h(\delta^n) - h(\delta^{n+1}) \\ &= (1-\sqrt{\delta})h(\delta^n) - \sqrt{\delta}h(\delta^n) - \sqrt{\delta}h(\delta^n)^{-1}\delta^n \log(1+\log\delta^{-1}/\log\delta^{-n}) \\ &\quad + \cdots \\ &> (1-\sqrt{\delta}-3\sqrt{\delta})h(\delta^n) \quad \text{(for large enough } n\text{, since } o<\delta<1\text{), a.e..} \end{aligned}$$

Consequently

$$\limsup_{t \searrow 0} X_t/h(t) \geq (1-4\sqrt{\delta}) \quad \text{a.e.} \tag{13}$$

Since $0 < \delta < 1$ is arbitrary, (13) implies

$$\limsup_{t \searrow 0} X_t/h(t) \geq 1 \quad \text{a.e.} \tag{14}$$

Thus (8) and (14) imply (1), and the proof is complete.

In view of the "scale invariance" (cf. Corollary 2.2ii), one has the following consequence of (1) and (2):

Corollary 2 *For a Brownian motion* $\{X_t, t \geq 0\}$ *on* (Ω, Σ, P),

$$P\left[\limsup_{t \nearrow \infty}(X_t/(2t\log\log t)^{1/2}) = 1\right] = 1 \tag{15}$$

and

$$P\left[\liminf_{t \nearrow \infty}(X_t/2t\log\log t)^{1/2}) = -1\right] = 1. \tag{16}$$

Many other results can be obtained by such explicit calculations using the exponential form of the d.f.s. For a good account, we refer the reader to Lévy's (1948) classic. For further developments, McKean (1969) and Hida (1980) may be consulted. The analysis leads to stochastic integration with numerous applications. (See, e.g., Revuz and Yor (1999) 3rd. ed. Springer) for a recent treatment.)

8.4 Gaussian and General Additive Processes

Since a Brownian motion process is both a Gaussian process and one that has independent increments, the preceding study leads us in (at least) two directions. One is to consider analogous sample function analysis for Gaussian processes whose mean functions, for instance, are zero but whose covariance functions r are more general than (perhaps not factorizable as) the one given by $r(s,t) = \min(s,t)$—the Brownian motion covariance. The second possibility is to consider processes having independent (and perhaps strictly stationary) increments which need not be Gaussian distributed. The latter processes are sometimes called *additive processes.* We discuss these two distinct extensions briefly in this section to indicate a view of the progression of the subject.

Since by Proposition 1.2, a (real) Gaussian process is completely determined by its mean and covariance functions, many properties of the process can be studied from the behavior of these two parameters—mean and covariance. But $r : (s,t) \mapsto E(X_s X_t)$ is a symmetric positive definite function, and as such it qualifies to be a kernel of the classical Hilbert-Schmidt (and Fredholm) theory of integral equations. Thus it is to be expected that this classical theory plays a role in the present analysis. Of basic interest is the following well-known result of T. Mercer. Its proof is not included here.

Theorem 1(Mercer) *Let $r : [a,b] \times [a,b] \to \mathbb{R}$ be a continuous covariance function. Then the equation $(-\infty < a < b < \infty)$*

$$\lambda \int_a^b r(s,t)u(t)\,dt = u(s), \quad a \leq s \leq b, \tag{1}$$

admits an infinite number of values $\lambda_i > 0$ and a corresponding system of continuous solutions $\{u_n, n \geq 1\}$ which form a complete orthonormal sequence in the Hilbert space of (equivalence class of) functions on the Lebesgue interval (a,b), denoted simply as $L^2(a,b)$, such that

$$r(s,t) = \sum_{k=1}^{\infty} u_k(s)u_k(t)/\lambda_k, \tag{2}$$

the series converging absolutely and uniformly in the square $[a,b] \times [a,b]$.

This result enables one to consider Gaussian processes with continuous covariances as follows. Let $\{\xi_n, n \geq 1\}$ be a sequence of independent $N(0,1)$ r.v.s on a (product) probability space (Ω, Σ, P) and let r be a continuous covariance function on the square $[a,b] \times [a,b]$. If the u_n and λ_n are as in the Mercer theorem, let

$$X_t = \sum_{n=1}^{\infty} \xi_n u_n(t) \lambda_n^{-1/2}. \tag{3}$$

Because of (2), Theorem 2.2.6 implies that the above series converge a.e. and in $L^2(P)$-mean. Hence $E(X_t) = 0$, and by independence of the ξ_n,

$$E(X_s X_t) = \sum_{n=1}^{\infty} u_n(s)u_n(t)/\lambda_n = r(s,t), \tag{4}$$

using (2) again. Then by Proposition 2.1, since each X_t is clearly Gaussian, $\{X_t, t \in [a,b]\}$ is a Gaussian process on the space (Ω, Σ, P) of the ξ_n with mean zero and covariance r. Such a representation as (2) and (3) is useful for many computations.

To illustrate the effectiveness of the representation (3), and the need for special techniques, let us calculate the distribution of a general quadratic functional of a Gaussian process $\{X_t, 0 \leq t \leq 1\}$ with mean zero and a continuous covariance r. The quadratic functional is

$$Q(X) = \int_0^1 (X_t - \alpha q(t))^2 \, dt, \tag{5}$$

where $q \in L^2(0,1)$ and $\alpha \in \mathbb{R}$. Our assumptions imply that $X : (t, \omega) \mapsto X_t(\omega)$ is jointly measurable on $[0,1] \times \Omega$ relative to the $dt\, dP$-measure. Thus (5) is a well-defined r.v. by Fubini's theorem. To find the d.f. of $Q(X)$, an obvious method is to use ch.f.s. Thus one calculates the latter function, and if it can be inverted, then the d.f. is obtained. The ch.f. is also useful to get at least an approximation to its d.f., using certain series expansions and related techniques. The analysis shows how several abstract results are needed here.

We first simplify $Q(X)$ by means of the expansion (3). Thus

$$\begin{aligned}
Q(X) &= \int_0^1 X_t^2 \, dt - 2\alpha \int_0^1 X_t q(t) \, dt + \alpha^2 \int_0^1 q^2(t) \, dt \\
&= \sum_{j,k=1}^{\infty} \xi_j \xi_k \int_0^1 \frac{u_j(t) u_k(t)}{\sqrt{(\lambda_j \lambda_k)}} \, dt - 2\alpha \sum_{j=1}^{\infty} \xi_j \int_0^1 \frac{q(t) u_j(t)}{\sqrt{\lambda_j}} \, dt \\
&\quad + \alpha^2 \int_0^1 q^2(t) \, dt \\
&= \sum_{j=1}^{\infty} [\xi_j \lambda_j^{-1/2} - \alpha \int_0^1 q(t) u_j(t) \, dt]^2 - \alpha^2 \sum_{j=1}^{\infty} \left(\int_0^1 q(t) u_j(t) \, dt \right)^2 \\
&\quad + \alpha^2 \int_0^1 q^2(t) \, dt \qquad \text{(by the orthonormality of the } u_j) \\
&= \sum_{j=1}^{\infty} \left(\xi_j \lambda_j^{-1/2} - \alpha \int_0^1 q(t) u_j(t) \, dt \right)^2,
\end{aligned}$$

since $\{u_n, n \geq 1\}$ is complete in $L^2(0,1)$ and $q \in L^2(0,1)$, so that by Parseval's equation we may cancel the last two terms. Consider the moment-generating function (m.g.f.) which exists and is more convenient here than the ch.f.,

$$\begin{aligned}
\psi(\tau) &= E(e^{-(\tau^2/2) Q(X)}) \\
&= E\left[\prod_{j=1}^{\infty} \exp\left\{ -\frac{\tau^2}{2} (\xi_j \lambda_j^{-1/2} - \alpha \int_0^1 q(t) u_j(t) \, dt)^2 \right\} \right] \\
&= \prod_{j=1}^{\infty} E\left(\exp\left\{ -\frac{\tau^2}{2} (\xi_j \lambda_j^{-1/2} - \alpha \int_0^1 q(t) u_j(t) \, dt)^2 \right\} \right)
\end{aligned}$$

(by independence and bounded convergence)

$$= \prod_{j=1}^{\infty}(1+\tau^2\lambda_j^{-1})^{-1/2}$$

$$\times \left(\exp\left\{ -\frac{\tau^2\alpha^2}{2\lambda_j} \left(\int_0^1 q(t)u_j(t)\,dt \right)^2 \Big/ (1+\tau^2\lambda_j^{-1}\ ) \right\} \right)$$

[since ξ_j is $N(0,1)$ so that $(\xi_j-\mu)^2$ has a "noncentral chi-square" d.f.,]

$$= [D_r(-\tau^2)]^{-1/2} \exp\left\{ -\frac{\tau^2\alpha^2}{2} \int_0^1 \int_0^1 \sum_{j=1}^{\infty} \frac{q(t)q(s)u_j(t)u_j(s)}{\lambda_j+\tau^2}\,dt\,ds \right\}, \tag{6}$$

where $D_r(\lambda) = \prod_{j=1}^{\infty}(1-\lambda\lambda_j^{-1})$, the Fredholm determinant of r, is an analytic function of λ in a small disc around the origin. The second term is still complicated. However, in the classical theory, one sets $r_1 = r$, and for $k > 1$, lets $r_k(s,t) = \int_0^1 r(s,x)r_{k-1}(x,t)\,dx$, to use induction. Then set

$$-R(s,t;\lambda) = \sum_{k=1}^{\infty} \lambda^{k-1} r_k(s,t), \tag{7}$$

which converges absolutely and uniformly if $|\lambda| \max_{s,t} |r(s,t)| < 1$ on the unit square. R is called the *reciprocal kernel* of r, which clearly satisfies

$$r(s,t) + R(s,t;\lambda) = \lambda \int_0^1 r(s,x)R(x,t;\lambda)\,dx.$$

Now using (2), it can be verified that

$$R(s,t;\lambda) = \sum_{j=1}^{\infty} u_j(s)u_j(t)(\lambda-\lambda_j)^{-1}; \tag{8}$$

the series converging absolutely and uniformly if $|\lambda|$ is as before. Thus (6) becomes

$$\psi(\tau) = [D_r(-\tau^2)]^{-1/2} \exp\left\{ \frac{\tau^2\alpha^2}{2} \int_0^1 \int_0^1 R(s,t;-\tau^2)q(s)q(t)\,ds\,dt \right\}. \tag{9}$$

If $\alpha = 0$ in (5), then the m.g.f. of $\int_0^1 X_t^2\,dt$ is $[D_r(-\tau^2)]^{-1/2}$. We thus have established the following result. (Some others are given as problems.)

Proposition 2 *If $\{X_t, 0 \le t \le 1\}$ is a Gaussian process with mean zero and a continuous covariance function r, then the m.g.f. of the distribution of* *the quadratic functional $Q(X)$ defined by* (5) *is given by the expression* (9).

Regarding the second direction, one considers processes with independent increments which generalize the second major property of the Brownian motion. To gain an insight into this class, let $\{X_t, t \in [0,1]\}$ be a process with independent increments. If $0 < s < t < 1$, we consider the ch.f. $\phi_{s,t}$ of $X_t - X_s$. For $0 < t_1 < t_2 < t_3 < 1$, one has, by the independence of $X_{t_3} - X_{t_2}$ and $X_{t_2} - X_{t_1}$, the following:

$$\phi_{t_1,t_3}(u) = \phi_{t_1,t_2}(u)\phi_{t_2,t_3}(u), \quad u \in \mathbb{R} \tag{10}$$

Suppose that the process is stochastically continuous; i.e., for each $\varepsilon > 0$,

$$\lim_{t \to s} P[X_t - X_s| \geq \varepsilon] = 0, \quad s \in (0,1).$$

Then it is not hard to verify (cf. Problem 4) that $\lim_{t\to s} \phi_{s,t}(u) = 1$ uniformly in u and s,t in compact intervals. Hence if $0 \leq s < t_0 < t_1 < \cdots < t_n < t \leq 1$, with $t_k = s + (k(t-s)/n)$, we get

$$\phi_{s,t}(u) = \prod_{i=0}^{n-1} \phi_{t_i,t_{i+1}}(u), \quad u \in \mathbb{R}, \tag{11}$$

and the factors can be made close to 1 uniformly if n is large enough. Thus $\phi_{s,t}$ is infinitely divisible (in the generalized sense, and hence in the ordinary sense; cf. Problem 17 in Chapter 5). Consequently, by the Lévy-Khintchine representation (cf. Theorem 5.2.5) with $s = 0 < t < 1, u \in \mathbb{R}$,

$$\phi_{0,t}(u) = \exp\left\{i\gamma_t u + \int_{\mathbb{R}} \left(e^{iuv} - 1 - \frac{iuv}{1+v^2}\right) \frac{1+v^2}{v^2}\, dG_t(v)\right\} \tag{12}$$

for a unique pair $\{\gamma_t, G_t\}$. The dependence of γ_t and G_t on t is continuous because of the assumed (stochastic) continuity of the process. Considering the subinterval $[s,t] \subset [0,1]$, and using (12), we get a new pair $\gamma_{s,t}$ and $G_{s,t}$ in (12) for $\phi_{s,t}$. However, using (10) applied to $0 < s < t < 1$, so that $\phi_{0,t} = \phi_{0,s} \cdot \phi_{s,t}$ we get (note that Log ϕ exists by Proposition 4.2.9)

$$\operatorname{Log} \phi_{s,t}(u) = \operatorname{Log} \phi_{0,t}(u) - \operatorname{Log} \phi_{0,s}(u). \tag{13}$$

Substituting (12) in (13), with the uniqueness of the formula, one can deduce that $\gamma_{s,t} = \gamma_t - \gamma_s$ and $G_{s,t} = G_t - G_s$. Thus

$$\phi_{s,t}(u) = \exp\left\{i(\gamma_t - \gamma_s)u + \int_{\mathbb{R}} \left(e^{iuv} - 1 - \frac{iuv}{1+v^2}\right) \frac{1+v^2}{v^2}\, d(G_t - G_s)(u)\right\}. \tag{14}$$

This formula can be used to analyze the processes with independent increments which are stochastically continuous. By choosing different such pairs, and hence different classes of these processes, one can study various properties. Note (by Example 3.3.3) that a process with independent increments is Markovian, and if it has one moment finite and then means zero (in particular, a Brownian motion), then it is also a martingale. These relations indicate the directions in which the above types of processes can be studied. (See Problems 5 and 6.) We now illustrate with some simple but important examples.

An interesting and concrete example of an additive process is obtained by considering a nonnegative integer-valued process $\{N_t, t \geq 0\}$ with independent increments. We now detail this view to explain the growth in new directions.

For instance, N_t can represent the total number of events that have occurred up to time t, so that such a process is often termed a *counting process.*

Now, unlike Brownian motion where it is assumed that increments of the process are Gaussian, we shall assume only that the independent increments are strictly stationary, with no distribution implied.

For the sake of simplicity, let $N_0 = 0$, and assume that the probability of an event occurring during an interval of length t depends upon t. One simple way to develop this dependence is to assume, as originally done by Poisson (c. 1800) that

$$P[N_{\Delta t} = 1] = \lambda \Delta t + o(\Delta t), \tag{15}$$

where λ is a nonnegative constant and Δt is the length of the interval $[0, \Delta t]$. For a small value of Δt, equation (15) implies

$$P[N_{\Delta t} \geq 2] = o(\Delta t) \tag{16}$$

and that events in nonoverlapping time intervals are independent.

[The assumption in (15) is the simplest condition one can place upon the process N_t. Other assumptions on this probability will lead to a development of other processes. Some of these are considered later as problems.]

The probabilities given by both equations (15) and (16) imply that during a small interval of time, the process $N_t, t \geq 0$ can have at most one event to occur.

The goal now is to determine the probability distribution for N_t. Letting

$$P_n(t) = P[N_t = n | N_0 = 0]$$

be the conditional probability of n events at time t given that there were none initially. (Note that, in this illustration, the conditioning events always have positive probability and so the difficulties considered in Section 3.2 do not arise.) It follows that at time $t + \Delta t$,

$$\begin{aligned}
P_0(t + \Delta t) &= P[N_{t+\Delta t} = 0 | N_0 = 0] \\
&= P[N_t = 0, N_{t+\Delta t} - N_t = 0 | N_0 = 0] \\
&= P[N_t = 0 | N_0 = 0] P[N_{t+\Delta t} - N_t = 0 | N_0 = 0] \\
&\quad \text{(by independent increments assumption)} \\
&= P_0(t)(1 - \lambda \Delta t + o(\Delta t)), \qquad \text{(by a consequence of (15))}.
\end{aligned}$$

Thus

$$P_0(t+\Delta t) - P_0(t) = (-\lambda \Delta t + o(\Delta t))P_0(t)$$

which is

$$\frac{P_0(t+\Delta t) - P_0(t)}{\Delta t} = -\lambda P_0(t) + \frac{o(\Delta t)}{\Delta t}$$

so upon letting $\Delta t \to 0$, one has $P_0'(t) = -\lambda P_0(t)$ since the last term tends to zero by definition of $o(\Delta t)$. This simple differential equation, with the assumption $N_0 = 0$, so that

$$P_0(0) = P[N_0 = 0 | N_0 = 0] = 1$$

gives

$$P_0(t) = e^{-\lambda t}. \tag{17}$$

Similarly, for $n \geq 1$

$$\begin{aligned} P_n(t+\Delta t) &= P[N_{t+\Delta t} = n | N_0 = 0] \\ &= P[N_t = n, N_{t+\Delta t} - N_t = 0 | N_0 = 0] \\ &\quad + P[N_t = n-1, N_{t+\Delta t} - N_t = 1 | N_0 = 0] \\ &\quad + P[N_{t+\Delta t} = n, N_{t+\Delta t} - N_t \geq 2 | N_0 = 0] \\ &= P[N_t = n | N_0 = 0] P[N_{t+\Delta t} - N_t = 0 | N_0 = 0] \\ &\quad + P[N_t = n-1 | N_0 = 0] P[N_{t+\Delta t} - N_t = 1 | N_0 = 0] \\ &\quad + P[N_{t+\Delta t} = n | N_0 = 0] P[N_{t+\Delta t} - N_t \geq 2 | N_0 = 0] \end{aligned}$$

where the last expression follows from the assumption of independent increments of N_t. Now, by the (strict) stationarity of N_t's increments, as well as the conditions (15) and (16), it follows that

$$\begin{aligned} P_n(t+\Delta t) &= P_n(t)P_0(\Delta t) + P_{n-1}(t)P_1(\Delta t) + o(\Delta t) \\ &= P_n(t)(1 - \lambda \Delta t) + P_{n-1}(t)\,\lambda \Delta t + o(\Delta t). \end{aligned}$$

Rearrangement of this expression and letting $\Delta t \to 0$ gives the system of differential equations

$$P_n'(t) = -\lambda P_n(t) + \lambda P_{n-1}(t) \quad \text{for } n \geq 1. \tag{18}$$

The assumption that $N_0 = 0$ gives $P[N_0 = n] = 0$ for $n \geq 1$ so that using (17) and recursively solving (18) it follows that

$$P_n(t) = e^{-\lambda t} \frac{(\lambda t)^n}{n!} \qquad \text{for } n \geq 0. \tag{19}$$

For each $t \geq 0$, $P_n(t)$ is the Poisson probability distribution with parameter $\lambda > 0$. Thus the nonnegative integer-valued process N_t that has independent and strictly stationary increments which satisfies (15), is called the *Poisson process* with rate parameter $\lambda > 0$.

The Poisson process is a member of a wide class of integer-valued continuous time stochastic processes collectively known as *birth-death processes*. An integer-valued process X_t with independent increments is a birth-death process if the conditional probabilities given as:

$$P[X_{t+\Delta t} = n+1 | X_t = n] = \lambda_n \Delta t + o(\Delta t) \quad \text{for } n \geq 0,$$
$$P[X_{t+\Delta t} = n-1 | X_t = n] = \mu_n \Delta t + o(\Delta t) \quad \text{for } n \geq 1,$$

and

$$P[X_{t+\Delta t} > n+1 | X_t = n] = o(\Delta t),$$
$$P[X_{t+\Delta t} < n-1 | X_t = n] = o(\Delta t)$$

so that during a time Δt, the process can only increase (a "birth") by one unit or decrease (a "death") by one unit. Birth-death processes have a wide variety of applications in the biological and physical sciences. A few examples of these processes are included in the exercises by considering various generalizations of equations (15) and (16).

An important application of the Poisson process occurs in queueing theory where the process N_t represents the number of arrivals to the queue and equation (15) gives the probability of an arrival to the queue during a small time interval. This is a specific example of the queueing model considered in Section 2.4. We now reconsider the process N_t from a slightly more advanced point of view.

Thus alternately the N_t process can be obtained as follows: Let X be an exponentially distributed random variable so that $P[X < x] = 1 - e^{-\lambda x}, x \geq 0, \lambda > 0$. If $X_1, \ldots, X_n$ are independent with the same distribution as X, let $S_n = \sum_{k=1}^n X_k$, be the partial sum and for $t \geq 0$, set $N_t = \sup\{n \geq 1 : S_n \leq t\}$ so that N_t is the last time before the sequence $\{S_n, n \geq 1\}$ crosses the level $t \geq 0$, where as usual $\sup(\emptyset) = 0$. Then N_t is an integer valued random variable, and its distribution is easily obtained. In fact, since S_n has a gamma distribution (c.f. Section 4.2) whose density is given by

$$f_{S_n}(x) = \frac{\lambda^n x^{n-1}}{\Gamma(n)} e^{-\lambda x}, \quad x \geq 0, n \geq 1, \lambda > 0,$$

we have for $n = 0, 1, 2, \ldots$ (set $S_0 = 0$), since $[N_t \geq n] = [S_n \leq t]$,

$$\begin{aligned} P[N_t = n] &= P[S_n \leq t, S_{n+1} > t] \\ &= \int\int_{[S_n \leq t, X_{n+1} + S_n > t]} f_{S_n}(x) f_{X_{n+1}}(y)\, dx\, dy \\ &\quad \text{since } S_n, X_{n+1} \text{ are independent} \\ &= \int_0^t f_{S_n}(x) dx \int_{[X_{n+1} > t-x]} f_{X_{n+1}}(y)\, dy \\ &= \int_0^t f_{S_n}(x)\, dx\, P[X_{n+1} > t-x] = e^{-\lambda t} \cdot \frac{(\lambda t)^n}{n!}. \end{aligned} \tag{20}$$

Alternatively (following Feller (1966), p.11),

$$\begin{aligned}P[N_t = n] &= P[(S_n \le t) \cap (S_{n+1} \le t)^c] \\ &= P[S_n \le t] - P[S_{n+1} \le t] \\ &= \int_0^t f_{S_n}(x)dx - \int_0^t f_{S_{n+1}}(x)dx \\ &= e^{-\lambda t}\frac{(\lambda t)^n}{n!},\end{aligned}$$

which is obtained by integrating the second term by parts and cancelling the resulting integrals. Thus $\{N_t, t \ge 0\}$ is a Poisson process. Moreover, it has the properties for $\omega \in \Omega$: (a) $N_0(\omega) = 0$, $\lim_{t\to\infty} N_t(\omega) = \infty$, (b) integer valued, nondecreasing, right continuous (i.e. $\lim_{s\downarrow t} N_s(\omega) = N_t(\omega)$), and at discontinuity points, say t_0, $N_{t_0}(\omega) - N_{t_0-}(\omega) = 1$. We may ask whether these properties characterize the Poisson process in the sense that such a process has independent stationary increments as well as the distribution given by equation (20). For a positive answer, we need to strengthen the hypothesis which can be done in different, but equivalent ways. This is presented as follows.

Theorem 3 *Let $\{N_t, t \ge 0\}$ be a nonnegative integer valued nondecreasing right continuous process with jumps of size 1 and support $\mathbb{Z}^+ = \{0, 1, 2, \ldots\}$. Then the following are equivalent conditions:*

1. *the process is given by $N_t = \max\{n : S_n \le t\}$, where $S_n = \sum_{k=1}^n X_k$, with the X_k as i.i.d and exponentially distributed, i.e.,, $P[X > x] = e^{-\lambda x}, x \ge 0, \lambda > 0$,*

2. *the process has independent stationary increments, each of which is Poisson distributed, so that (20) holds for $0 < s < t$ in the form*

$$P[N_t - N_s = n] = e^{-\lambda(t-s)}\frac{[\lambda(t-s)]^n}{n!}, \quad \lambda \ge 0, n = 0, 1, 2, \ldots (\lambda^0 = 1),$$

3. *the process has independent and stationary increments,*

4. *the process has no fixed discontinuities, and satisfies the Poisson (conditional) postulates: for each $0 < t_1 < \ldots < t_k$; and $n_k \in \mathbb{Z}^+$ one has for a $\lambda \ge 0$ as $h \searrow 0$*
   (i) $P[N_{t_k+h} - N_{t_k} = 1 | N_{t_j} = n_j, j = 0, 1, \ldots k] = \lambda h + o(h)$
   (ii) $P[N_{t_k+h} - N_{t_k} \ge 2 | N_{t_j} = n_j, j = 0, 1, \ldots k] = o(h)$.

Here we sketch a proof of the difficult part 1. ⇔ 2. following Billingsley (1995, p.300). It is of interest to observe that, as A. Prékopà proved, the stationarity of (independent) increments of the N_t-process is not implied by an integer valued independent increment process without fixed discontinuities although each increment has a Poisson distribution. The method is to show 1. ⇔ 2. ⇔ 3. and 2. ⇔ 4. The complete proof uses Prékopà's theorem also.

Proof of 1. $\Rightarrow$ 2. From definition of N_t, we note that the sets $[N_t \geq n] = [S_n \leq t]$, and as seen in (20) that $[N_t \geq n] = [S_n \leq t < S_{n+1}]$. For given $t \geq 0$ and $y \geq 0$, we have, on using the independence of S_n and X_{n+1} with exponential distribution for X_n and with $S_{n+1} = S_n + X_{n+1}$,

$$\begin{aligned} P[S_n \leq t < S_{n+1}, S_{n+1} - t > y] &= P[S_n \leq t, X_{n+1} > t + y - S_n] \\ &= \int_0^t P[X_{n+1} > t + y - x] f_{S_n}(x) dx \\ &= \int_0^t e^{-\lambda(t+y-x)} f_{S_n}(x) dx \\ &= e^{-\lambda y} \int_0^t P[X_{n+1} > t - x] f_{S_n}(x) dx \\ &= e^{-\lambda y} P[S_n \leq t, X_{n+1} > t - S_n]. \end{aligned} \tag{21}$$

On the other hand the properties of the independent X_n of hypothesis 1. imply for $y_j \geq 0$,

$$\begin{aligned} &P[S_{n+1} - t > y_1, X_{n+2} > y_2, \ldots, X_{n+k} > y_k, S_n < t < S_{n+1}] \\ &\quad = P[S_{n+1} - t > y_1, S_n \leq t < S_{n+1}] e^{-\lambda y_2} e^{-\lambda y_3} \ldots e^{-\lambda y_k} \\ &\quad = P[S_n \leq t < S_{n+1}] e^{-\lambda(y_1 + y_2 + \ldots + y_k)}, \end{aligned}$$

(since $S_{n+1} > t + y$ on $[S_n \leq t < S_{n+1}]$ implies $X_{n+1} > y_1$, and use (21))

$$= P[N_t = n] P[X_1 > y_1, \ldots, X_k > y_k]. \tag{22}$$

To simplify the left side, we define new random variables depending on the fixed (but arbitrary) t as follows. Let $X_1^{(t)} = S_{N_t+1} - t$, $X_2^{(t)} = X_{N_t+2}, X_3^{(t)} = X_{N_t+3}, \ldots$, and observe that for $0 < s < t$, $[N_{t+s} - N_t \geq m] = [S_{N_t+m} \leq t + s] = [\sum_{j=1}^m X_j^{(t)} \leq s]$. This implies

$$N_{t+s} - N_t = \sup\{m : \sum_{j=1}^m X_j^{(t)} \leq s\}$$

which brings in the increments of the N_t-process with the "new" partial sums $S_n^{(t)}$ in terms of $X_j^{(t)}$ random variables, and the increment process $\{N_{t+s} - N_t, s \geq 0\}$ is similar to the $\{N_t, t \geq 0\}$ process for the X_j random variables. Thus we have $[N_{t+s} - N_t = m] = [S_m^{(t)} \leq s < S_{m+1}^{(t)}]$. With this new definition, if $A = \times_{j=1}^k [y_j, \infty) \subset \mathbb{R}^k$ is a rectangle, then (22) becomes

$$P[N_t = n, (X_1^{(t)}, \ldots, X_k^{(t)}) \in A] = P[N_t = n] P[(X_1^{(t)}, \ldots, X_k^{(t)}) \in A]. \tag{23}$$

But such rectangles as A are generators of the Borel σ-algebra of $\mathbb{R}^k$, so that by a standard result (cf. Proposition 1.2.8), (23) holds for all Borel sets A.

The next step is to express the joint event $[N_{S_i} = m_i, i = 1, \ldots, \ell]$ as one of the events in (23) using $m_\ell + 1$ variables $X_j^{(t)}$'s. Thus if we consider $\tilde{A}$ as the rectangle $\times_{j=1}^{k}[x_1 + \ldots + x_{m_i} \leq s_i < x_1 + \ldots + x_{m_i+1})$, then we find

$$[N_{t+s_i} - N_t = m_i, i = 1, \ldots, \ell] = [(X_1^{(t)}, \ldots, X_{m_\ell+1}^{(t)} \in \tilde{A})],$$

and this implies (23) because

$$P[N_t = n, N_{t+s_i} - N_t = m, i = 1, \ldots, \ell] = P[N_t = n]P[N_{s_i} = m_i, i = 1, \ldots, \ell].$$

This is the key step in using induction with $\ell = 1, 2, \ldots,$ and $0 = t_0 < t_1 < \ldots < t_\ell$, since $N_0 = 0$, so that the following is trivial for $\ell = 1$, and then assume for $\ell = k$, to complete the induction:

$$P[N_t - N_{t_{i-1}} = n_i, i = 1, \ldots, \ell] = \prod_{i=1}^{\ell} P[N_{t_i - t_{i-1}} = n_i] \tag{24}$$

and obtain the result for $\ell = k + 1$. The equations (24) and (20) imply our assertion and gives 2.

Proof of 2. $\Rightarrow$ 1. Observe that $[N_0 = 0] = [X_1 > t]$ and so the distribution of N_0 in 2. gives $P[X_1 \geq t] = e^{-\lambda t}$. To get the distribution of X_1, X_2 ($S_1 = X_1, X_2 = S_2 - S_1$), let $0 = s_1 < t_1 < s_2 < t_2$ and observe that

$$\begin{aligned}
P\ [\ s_1 &< S_1 \leq t_1, s_2 < S_2 < t_2] \\
&= P[N_{s_1} = 0, N_{t_1} - N_{s_1} = 1, N_{s_2} - N_{t_1} = 0, \text{ and } N_{t_2} - N_{s_2} = 1] \\
&= P[N_{s_1} = 0]P[N_{t_1} - N_{s_1} = 1]P[N_{s_2} - N_{t_1} = 0]P[N_{t_2} - N_{s_2} = 1] \\
&\quad \text{(by independence of increments of the } N_t\text{-process)} \\
&= \lambda(t_1 - s_1)(e^{-\lambda s_2} - e^{-\lambda t_2}), \\
&\quad \text{(using the Poisson distribution of increments in the hypothesis 2.)} \\
&= \int\!\!\int_{[s_1 < y_1 \leq t_1] \times [s_2 < y_2 \leq t_2]} \lambda^2 e^{-\lambda y_2} dy_1 dy_2.
\end{aligned}$$

From this we get for any Borel set $A \subset [(y_1, y_2) : 0 < y_1 < y_2] \subset \mathbb{R}^2$ that

$$P[(S_1, S_2) \in A] = \int\!\!\int_A \lambda^2 e^{-\lambda y_2} dy_1 dy_2.$$

The result can now be extended to the sector $(0 < y_1 < \ldots < y_k)$ in $\mathbb{R}^k$ using the mapping $x_i = y_i - y_{i-1}$, computing the Jacobian, as in the proof of Theorem 3.3.9. One deduces that the X_i are independent and each has an exponential distribution with parameter $\lambda \geq 0$. This gives 1. and 1. $\Rightarrow$ 2. is established.

We omit the proof of the other equivalences, and refer the reader to Billingsley (1995). We shall however use all parts of the above theorem in the following discussion.

As Feller remarked in his book on Probability Theory (Vol. 1), "the three distributions, Bernoulli, Poisson and Normal and their ramifications in the subject as a whole are astounding." In fact the Poisson process has the same properties as the Brownian motion, except that it is integer valued and is continuous in probability with moving discontinuities of unit jumps (with at most countably many in number). We resume the general theme, strongly motivated by the Poisson case treated above.

Now formula (12) takes a particular form when G_t has no jump (so $\sigma^2 = 0$) at the origin (i.e., the Gaussian component is absent) so that the appropriate formula is that given by the Lévy measures (c.f. Section 5.2, equation (20)). In this case it takes the following simple form:

$$\varphi(t) = \exp\{i\gamma t + \int_0^\infty (e^{itx} - 1)\, dN(x)\}, \quad t \in \mathbb{R},$$

where $N(\{0\}) = 0, \gamma$ is a constant and $N(\cdot)$ is nondecreasing with

$$\int_{0+}^2 u^2\, dN(u) < \infty.$$

[One starts with $dN(x) = \frac{1+x}{x^2}\, dG(x)$ in (12).] To proceed further, let us rewrite (20) as:

$$\pi_\lambda(\cdot) = e^{-\lambda} \sum_{n=0}^\infty \frac{\lambda^n}{n!} \delta_n(\cdot) \tag{25}$$

where $t = 1$ and $\delta_n(\cdot)$ is the Dirac point measure (and $\pi_0 = \delta_0$, supp $(\pi_\lambda) = \{0, 1, 2, \ldots\} = \boldsymbol{Z}^+$). Since $\pi_\lambda(\cdot)$ is clearly a measure on the power set $\mathcal{P}(\boldsymbol{Z}^+)$, if $\lambda_1, \lambda_2 \geq 0$ one has the convolution

$$(\pi_{\lambda_1} * \pi_{\lambda_2})(A) = \int_{\boldsymbol{Z}^+} \pi_{\lambda_1}(A - x)\pi_{\lambda_2}(dx)$$

and its ch.f. gives, with $\hat{\pi}_\lambda(t) = \int_{\boldsymbol{Z}^+} e^{itx}\pi_\lambda(dx) = e^{-\lambda}\sum_{n=0}^\infty e^{itn}\frac{\lambda^n}{n!} = e^{\lambda(e^{it}-1)}$,

$$\widehat{(\pi_{\lambda_1} * \pi_{\lambda_2})}(t) = \hat{\pi}_{\lambda_1}(t)\hat{\pi}_{\lambda_2}(t) = \hat{\pi}_{\lambda_1+\lambda_2}(t)$$

so that $\{\pi_\lambda, \lambda \geq 0\}$ is a semi-group of probability measures, under convolution. But (25) motivates the following immediate extension.

Let $(S, \mathcal{B}, \nu)$ be a finite measure space and $0 < c = \nu(S) < \infty$. If $\tilde{\nu}(\cdot) = \frac{1}{c}\nu(\cdot)$, then $(S, \mathcal{B}, \tilde{\nu})$ is a probability space (usually) different from (Ω, Σ, P). Let $X_j : S \to \mathbb{R}$ be independent identically distributed random variables relative to $\tilde{\nu}$. Then $\delta_{X_j} : \mathcal{R} \to \mathbb{R}^+$ is a (simple) random measure on $(\mathbb{R}, \mathcal{R})$, the Borelian line, in the sense that for each $s \in S$, $\delta_{X_j(s)}(\cdot)$ is the Dirac point measure. We may use the adjunction procedure of Chapter 2, discussed following Corollary 2.1.8 and replace Ω by $\tilde{\Omega} = \Omega \times S$, $\tilde{\Sigma} = \Sigma \bigotimes \mathcal{B}$ and $\tilde{P} = P \bigotimes \tilde{\nu}$. We leave this explicit formulation, and assume that our basic space is rich enough to carry all these variables. If N is a Poisson random variable with intensity $c(= \nu(s))$ so that $P[N = n] = e^{-c}\frac{c^n}{n!}$, consider the measure $\pi_c(\cdot)$ in (25) as a generalized (or compound) variable as:

$$\tilde{\pi}(B) = \sum_{j=1}^{N} \delta_{X_j}(B), \quad B \in \mathcal{B}, \tag{26}$$

where N is the Poisson random variable with $\nu(B)$ as intensity noted above. Here N and X_j are independent. As a composition of N and X_j, all at most countable, $\tilde{\pi}(\cdot)$ is a random variable. In fact $[\tilde{\pi}(B) = n] = \bigcup_{m \geq n}[\sum_{j=1}^{m} \delta_{X_j}(B) = n] \cap [N = n]$, for each integer $n \geq 1$ so that $\tilde{\pi}(B)$ is measurable for Σ, and thus is a random element for all $B \in \mathcal{B}$. To find the distribution of $\tilde{\pi}(B)$ we proceed through its characteristic function, and establish the following statement

Lemma 4: *For each $B \in \mathcal{B}$, $\tilde{\pi}(B)$ is Poisson distributed with intensity $c \cdot \tilde{\nu}(B) = \nu(B)$, implying that $\tilde{\pi}(\cdot)$ is pointwise a.e. σ-additive. Moreover, the result extends even if the intensity measure $\nu(\cdot)$ is σ-finite.*

Proof In establishing this result, through ch.f.s, we employ the fact that $E(Y) = E(E(Y|Z))$ for any *integrable* (or *nonnegative*) random variable Y and any r.v. Z (cf. Proposition 3.1.2). In view of Corollary 4.2.2 (uniqueness), this is seen to establishe the result. First we assume that $0 < \nu(S) < \infty$, as in the statement so that $\tilde{\nu}(S) = 1$.

Thus denoting again by $\hat{\tilde{\pi}}_B(t) = E(e^{it\tilde{\pi}_B})$, and using the hypothesis that X_j are i.i.d. on $(S, \mathcal{B}, \tilde{\nu})$ which are independent of N, one has

$$\begin{aligned}
\hat{\tilde{\pi}}_B(t) &= E(e^{it\sum_{j=1}^{N}\delta_{X_j}(B)}) \\
&= E(E(e^{it\sum_{j=1}^{N}\delta_{X_j}(B)}|N)), \text{ by the above identity,} \\
&= \sum_{n \geq 0} E(e^{it\sum_{j=1}^{n}\delta_{X_j}(B)})P(N = n), \text{ since } N \text{ is discrete} \\
&\quad \text{so that the difficulties with multiple solutions noted} \\
&\quad \text{in Section 3.2 do not arise} \\
&= \sum_{n=0}^{\infty} e^{-c\tilde{\nu}(B)}\frac{(c\tilde{\nu}(B))^n}{n!}(\prod_{j=1}^{n} E(e^{it\delta_{X_j}(B)})), \\
&\quad \text{by the independence of the } X_j, \\
&= \sum_{n=0}^{\infty} e^{-c\tilde{\nu}(B)}\frac{(c\tilde{\nu}(B))^n}{n!}e^{itn}, \text{ since } X_1^{-1}(B) \neq \emptyset \text{ for } B = \emptyset
\end{aligned}$$

$$\text{so that} E(e^{it\delta_{X_1}(B)}) = e^{it\cdot 1},$$
$$= e^{-\nu(B)}e^{\nu(B)e^{it}} = e^{\nu(B)(e^{it}-1)}.$$

Comparing this with (25) and the following discussion, we conclude that $\tilde{\pi}_B(\cdot)$ is Poisson distributed with intensity $\nu(B)$.

Now, if $B = \cup_{k=1}^{\infty} B_k$, $B_k \in \mathcal{B}$, $0 < \nu(B_k) < \infty$, B_k disjoint, then $\nu(B) = \sum_{k=1}^{\infty} \nu(B_k) < \infty$, implying that $\nu(B_n) \to 0$ as $n \to \infty$ so that the ch.f. of $\tilde{\pi}_{B_n}(\cdot)$ tends to unity and hence $\tilde{\pi}_{B_n} \to 0$ in probability. Hence $\tilde{\pi}_{(\cdot)}$ is σ-additive in P-measure. It is also seen that $\tilde{\pi}_{B_n}$ are independent on disjoint sets (and ≥ 0 a.e.), it follows that $\tilde{\pi}_B = \sum_{n=1}^{\infty} \tilde{\pi}_{B_n}$, holds pointwise a.e.

Finally, let $\nu(\cdot)$ be σ-finite, and so writing $S = \cup_{n=1}^{\infty} S_n$, $\nu(S_n) < \infty$, S_n disjoint, let $\tilde{\pi}_n = \tilde{\pi}_{S_n}$ which are independent Poisson measures on $(S_n, \mathcal{B}(S_n), \nu(S_n \cap \cdot)), n \geq 1$, by the preceding paragraph. If $\tilde{\pi} = \sum_{n=1}^{\infty} \tilde{\pi}_{S_n}$, a sum of independent Poisson random measures, it is Poisson with intensity $0 < \nu(S_n) < \infty$ on S_n, and this depends only on ν. Thus the results holds for σ-finite $\nu(\cdot)$ also, completing the proof.

Hereafter we write $\pi(\cdot)$ for $\tilde{\pi}(\cdot)$ to simplify notation. Also the relation between the intensity parameters of N and $\pi(\cdot)$ should be noted.

This result implies several generalizations (the versatility of the Poisson measure!) of the subject, originally investigated by P. Lévy in 1937. We indicate a few consequences here. The above property of $\pi(\cdot)$ motivates the following generalization.

Definition 5 Let $L^0(P)$ be the space of all real (or complex) random variables on a probability space (Ω, Σ, P) and $(S, \mathcal{B})$ be a measurable space. A mapping $\mu : \mathcal{B} \to L^0(P)$ is called a **random measure**, if the following (abstraction of Poisson measure given in Lemma 4) holds:

(i) $A_n \in \mathcal{B}, n = 1, 2, \ldots$, disjoint, implies $\{\mu(A_n), n \geq 1\}$ is a mutually independent family of infinitely divisible random variables,

(ii) for A_n as above, $\mu(\bigcup_{n=1}^{\infty} A_n) = \sum_{n=1}^{\infty} \mu(A_n)$, the series converges in P-measure.

Since a Poisson random variable is infinitely divisible, and, by Proposition 5.2.2, a general infinitely divisible nonconstant random variable has an unbounded range, the above definition includes all these cases. An important subclass of the infinitely divisible d.f.'s is the *stable family*, also analyzed in Section 5.3 (cf. Theorem 5.3.16 and the structural analysis that follows there). In the present context, these are called *stable random measures*, and they include the Poisson case. Recall that a stable random variable $X : \mathbb{R} \to L^0(P)$ has its characteristic function $\varphi(t) = E(e^{itX}) = \int_\Omega e^{itX} dP$ to be given (by the Lévy formula) as:

$$\varphi(t) = \exp\{i\gamma t - c|t|^\alpha(1 - i\beta \operatorname{sgn} t \cdot \varpi(t, \alpha))\}, \tag{27}$$

where $\gamma \in \mathbb{R}, |\beta| \leq 1, c \geq 0, 0 < \alpha \leq 2$ and

$$\varpi(t,\alpha) = \begin{cases} \tan \frac{\pi\alpha}{2}, & \text{if } \alpha \neq 1 \\ -\frac{2}{\pi} \log |t|, & \text{if } \alpha = 1. \end{cases}$$

Here α is the characteristic exponent of φ (or X), and $\alpha > 2$ implies $c = 0$, to signify that X is a constant. The ch.f. φ of a stable random measure $\mu : \mathcal{B} \to L^0(P)$, (27), takes the following form:

$$\begin{aligned} \varphi_A(t) = E(e^{it\mu(A)}) &= \int_\Omega e^{it\mu(A)}\, dP, \\ &= \exp\{i\gamma(A)t - c(A)|t|^\alpha(1 - i\beta(A) \operatorname{sgn} t\, \varpi(t,\alpha))\}, 0 < \alpha \leq 2, \quad (28) \\ &= \exp\{-\psi(A,t)\}, \quad \text{(say)}, \quad (29) \end{aligned}$$

for all $A \in \mathcal{B}, \nu(A) < \infty$ where $\nu : \mathcal{B} \to \overline{\mathbb{R}}^+$ is a σ-finite measure. By the two conditions in the above definition, one finds that $\gamma(\cdot)$ and $c(\cdot)$ are σ-additive, and for $\beta(\cdot)$ if $A_i \in \mathcal{B}_0(= \{B \in \mathcal{B} : \nu(B) < \infty\})$, disjoint, such that $\bigcup_{i=1}^\infty A_i \in \mathcal{B}_0$, ($\mathcal{B}_0$ being a δ-ring) $\sum_{i=1}^\infty c(A_i)\beta(A_i) = c(\bigcup_{i=1}^\infty A_i)\beta(\bigcup_{i\geq 1}^\infty A_i)$, (and, of course, $|\beta(A_0)| \leq 1$). We leave this to the reader to verify. It is not necessary that $\nu(\cdot)$ and $c(\cdot), \beta(\cdot)$ have any relation except that $\mu(A)$ is defined as a real random variable for $A \in \mathcal{B}_0$. The function $\psi(\cdot,\cdot)$ is the exponent in (29). It is often called the *characteristic exponent* which is uniquely determined by the parameters (γ, c, α, and β) and conversely determines them to make (29) the Lévy formula.

Observe that the Poisson random measure $\pi : \mathcal{B} \times \Omega \to \mathbb{R}^+$, is a function of a set and a point, so that $\pi(A,\omega)(= \pi(A)(\omega))$ is a nonnegative number which is σ-additive in the first variable and a measurable (point) function in the second. It is sometimes termed a kernel. Its important use is motivated by the following considerations. In the classical literature (e.g., in *Trigonometric Series* of Zygmund (1959), Vol. I, p. 96), the Poisson kernel is utilized to define a Poisson integral which is used to study the continuity, differentiation and related properties of functions representable as Poisson integrals. Analogous study should (and could) be given for our case here. To make this point explicit, we recall the classical case briefly and then present the corresponding random integrals, indicating a substantial new development of great value, opening up for investigation. Thus the remainder of this Section is a survey of the evolving and interesting work for researchers given without detailed proofs.

For a Lebesgue integrable $f : [-\pi,\pi] \to \mathbb{R}$, using the orthonormal system $\frac{1}{2}, \cos nx, \sin nx, n = 1,2,\ldots$, consider the Fourier coefficients a_k, b_k given by

$$a_k = \frac{1}{\pi}\int_{-\pi}^{\pi} f(x)\cos kx\, dx, \quad b_k = \frac{1}{\pi}\int_{-\pi}^{\pi} f(x)\sin kx\, dx$$

and for $0 \leq r < 1$, set

$$f_r(x) = \frac{1}{2}a_0 + \sum_{k=1}^{\infty}(a_k \cos kx + b_k \sin kx)r^k.$$

Then the Poisson kernel $P(\cdot,\cdot)$ is given by

$$P(r,x) = \frac{1}{2}\sum_{k=1}^{\infty} r^k \cos kx \left(= \frac{1}{2}\frac{1-r^2}{1-2r\cos x + r^2}\right) \geq 0,$$

$$\frac{1}{\pi}\int_{-\pi}^{\pi} P(r,x)\,dx = 1,$$

and $f_r(\cdot)$ is representable as the convolution:

$$(Tf)(r,x) = f_r(x) = \frac{1}{\pi}\int_{-\pi}^{\pi} f(x)P(r,u-x)\,du, \quad 0 \leq r < 1. \tag{30}$$

The classical results assert that $f_r(x) \to f(x)$, for all continuous periodic functions f, uniformly as $r \to 1$. Thus, T is a continuous linear mapping on $L^1(-\pi,\pi]$. The study leads to a profound analysis in harmonic function theory and elsewhere, (cf. e.g., Zygmund (1959), p. 96).

Replacing $P(r,x)\,dx$ by $\pi(\omega, ds)$ or more inclusively $\mu(ds)(\omega)$ of Definition 5 above one could consider the corresponding analysis for random functions or process (or sequences) that admit integral representation, modeling that of (30) and then study the sample path behavior of them. Here the Lebesgue interval $[-\pi,\pi]$ is replaced by $(S,\mathcal{B},\nu)$ and ω (in lieu of r) varies in (Ω,Σ,P). Such a general study has been undertaken by P. Lévy when μ is a stable random measure, (cf. (28), (29)). The resulting class of processes is now called *Lévy processes.* Almost all the classical results have important nontrivial extensions and the ensuing theory has an enormous growth potential in many directions. We include here a glimpse of this aspect.

The basic step in the analysis is to define an integral of a scaler (nonrandom) function relative to a stable random measure $\mu : \mathcal{B} \to L^0(P)$. In the case of a Poisson random measure, the intensity measure $\nu : \mathcal{B} \to \overline{\mathbb{R}}^+$ (but σ-finite) is a natural one defining the triple $(S,\mathcal{B},\nu)$. In the general case (of a stable random measure) (28) or (29) we have $\gamma(\cdot), c(\cdot)$ and $\beta(\cdot)$ as set functions, with σ-additivity properties but are not generally related to ν of the triple. So the first simplification made is to *assume* that $\gamma(\cdot)$ and $c(\cdot)$ are related (or proportional) to ν and β is a constant. Thus, let $\gamma(A) = a\nu(A), (a \in \mathbb{R})\, c(A) = c\nu(A), (c \geq 0)$, and $|\beta| \leq 1$ is a constant, so that the characteristic exponent $\psi(\cdot,\cdot)$ of (29) becomes for $a \in \mathbb{R}, 0 < \alpha \leq 2$,

$$\psi(A,t) = ia\nu(A)t - c\nu(A)|t|^{\alpha}\{1 - i\beta \text{sgn}\, t \cdot \varpi(t,\alpha)\}, A \in \mathcal{B}_0, t \in \mathbb{R}. \tag{31}$$

It is now necessary (and nontrivial) to show that $\exp\{-\psi(A,\cdot)\}$ is a characteristic function. This is true and then one can establish the existence of an α-stable random measure into $L^0(P)$ on a probability space (Ω,Σ,P), using a form of Theorem 3.4.10. This will show that the random measure $\mu : \mathcal{B}_0 \to L^0(P)$ is "controlled" by ν in the sense that $\mu(A) = 0$, a.e. $[P]$ holds whenever $\nu(A) = 0$, and μ is governed by the quadruple (a, c, β, ν). With this a stochastic integral corresponding to the classical one defined by (29) can be introduced. It will be specialized to show a close relation to strictly stationary processes, represented as integrals, which also connects measure preserving mappings of the last chapter at the same time.

Thus if $f_n = \sum_{i=1}^{n} a_i \chi_{A_i}$, $A_i \in \mathcal{B}_0$, disjoint, so that f_n is a simple function and $f_n \in L^\alpha(S, \mathcal{B}, \mu)$, define as usual

$$\int_S f_n \, d\mu = \sum_{i=1}^{n} a_i \mu(A_i) \in L^0(P) \tag{32}$$

and if $f_n \to f$ pointwise and $\{\int_S f_n \, d\mu, n \geq 1\} \subset L^0(P)$ is Cauchy in probability (or equivalently in the metric discussed in Exercise 3 of Chapter 1), then we set

$$\int_S f \, d\mu = \lim_n \int_S f_n \, d\mu, \tag{33}$$

since a Cauchy sequence has a unique limit in a metric space. It may now be shown that the limit does not depend on the sequence $\{f_n, n \geq 1\}$. The method is standard but not trivial (cf., Dunford-Schwartz (1958), Sec. IV.10) and the uniqueness proof for (33) *depends on the availability of a controlling measure* ν. Using the existence of such a ν, it is possible to consider two measures μ_1, μ_2 and obtain a Lebesgue type decomposition as well as the Radon-Nikodým theory for them. This analysis has deep interest in applications (See Section 5.4 of Rao (2000) for an aspect of this work where the Lévy-Itô representation and related integral formulas are given.)

Thus $T : f \mapsto \int_S f \, d\mu$ is well-defined, and it may be verified that the integral (33) or the mapping T is linear. The next important concern is to characterize the class of μ-integrable functions $f \in L^0(S, \mathcal{B}, \nu)$. We state a result in this direction for an understanding of this new area and to relate it with the strictly stationary process introduced at the end of the preceding chapter. The following result is a substitute for (30) in the present context.

Theorem 6 *Let $(S, \mathcal{B}, \nu)$ be a σ-finite space and $\mu : \mathcal{B}_0 \to L^0(P)$, on a probability space (Ω, Σ, P), be an α-stable random measure with parameters $(a, c, \beta, \nu), 0 < \alpha \leq 2$. Let $\mathcal{F}_1^\alpha \subset L^0(\nu)$ be the set of (real) functions such that $f \in \mathcal{F}^\alpha$ provided $||f||_\alpha < \infty$ where*

$$||f||_\alpha = \begin{cases} ||f||_1 + ||f||_\alpha, & \text{if } 1 < \alpha \leq 2, \\ ||f||_1 + ||f||_\alpha^\alpha, & \text{if } 0 < \alpha < 1 \\ ||f||_1 + ||f||_{1+\frac{1}{e}}, & \text{if } \alpha = 1, \quad \text{(e=base of the natural log).} \end{cases} \tag{34}$$

Then $\{\mathcal{F}^\alpha, ||\cdot||_\alpha\}$ is a complete metric space and the mapping $T : \mathcal{F}^\alpha \to L^0(P)$ by (33) is well-defined, one-to-one and the range $\mathcal{R}_T = T(\mathcal{F}^\alpha) \subset L^0(P)$

consists of α-stable random variables. Moreover, for each $0 \leq \alpha_1 < \alpha_2 \leq 2$, the mapping $T : \mathcal{F}^{\alpha_2} \to L^{\alpha_1}(\nu)$ is continuous between these metric spaces.

In particular, if $a = 0 = \beta$, and $0 \leq \alpha_1 < \alpha_2 \leq 2$, then $T : L^{\alpha_2}(\nu) \to L^{\alpha_1}(P)$ is an isomorphism into, whose range is the set of α_2-stable symmetric random variables with ch.f.'s given by

$$E(e^{itT(f)}) = \exp\{-c|t|^\alpha \int_S |f|^{\alpha_2}\, d\nu\}, f \in L^{\alpha_2}(\nu). \tag{35}$$

A detailed proof of this result with extensions if the spaces are vector valued $L^{\alpha_1}(\nu; \mathcal{X})$ and $L^{\alpha_2}(P, \mathcal{Y})$ where $\mathcal{X}, \mathcal{Y}$ are certain Banach (or even Fréchet) spaces is given by Y. Okizaki (1979). We omit the details here. It uses various properties of ch.f.'s and the work of Chapter 5. For a special class of random measures, namely those defined by symmetric stable independent increment process, M. Schilder (1970) has given a simpler description of the stochastic integral (for a brief sketch of this procedure, see Problem 14). It is of interest to characterize the range $\mathcal{R}$ of the stochastic integral T as a subspace of $L^0(P)$, but this is not a familiar object. For instance, it may be shown that with ($\wedge = \min$, $\vee = \max$) the metric given by

$$||f|| = \int_S (1 \wedge |f(s)|)\, d\nu(s) + \int_\Omega 1 \wedge \left| \left(\int_S f(s)\, d\mu(s) \right) \right|(\omega)\, dP,\ f \in \mathcal{F}^\alpha, \tag{36}$$

$\{\mathcal{R}_T, ||\cdot||\}$ becomes a complete linear metric (or a Fréchet) space, where $a \wedge b = \min(a, b)$ for pairs of real numbers a, b and where μ is a symmetric α-stable random measure. (See also Exercise 5.25 (c).) We are now ready to introduce the classification of processes admitting integral representations.

Recall that a (real) process $\{X_t, t \in I\}$ is *strictly stationary* (as defined at the end of the last chapter) if for each finite set of indices $t_1, \ldots, t_n \in I$ with $t_1 + s, \ldots, t_n + s \in I$ for any $s \in I$ (and any index set I with such an algebraic structure), all the (joint) distributions of $(X_{t_1}, \ldots, X_{t_n})$ and $(X_{t_1+s}, \ldots, X_{t_n+s})$ are identical. Equivalently, their (joint) ch.f.'s satisfy

$$E(\exp[i\sum_{j=1}^n u_j X_{t_j}]) = E(\exp[i\sum_{j=1}^n u_j X_{t_j+s}]), \quad u_j \in \mathbb{R}. \tag{37}$$

Now consider this property for a class of α-stable processes in a stronger form leading to an interesting area of probability with many applications. For simplicity we treat here only the *symmetric* α-stable class. Thus, a process $\{X_t, t \in I\}$ is termed α-stable if each finite linear combination $\sum_{j=1}^n a_j X_{t_j}$ is α-stable, as noted in Exercise 5.25 (c). Consequently, for each $n \geq 1$, the finite dimensional ch.f. of $X_{t_1}, \ldots, X_{t_n}$ is representable as:

$$\varphi_{t_1,\ldots,t_n}(u_1, \ldots, u_n) = \exp\{-\int_{\mathbb{R}^n} \left| \sum_{j=1}^n u_j e^{i<t_j,\lambda>} \right|^\alpha dG_n(\lambda), \tag{38}$$

where we replaced the support of G_n from the unit sphere S by $\mathbb{R}^n$ so that $|e^{i<t_j\lambda>}| = 1$ and the G_n measure is now defined on the Borelian space $(\mathbb{R}^n, \mathcal{B}_n)$. But as n varies, the system of measure spaces $\{(\mathbb{R}^n, \mathcal{B}_n, G_n), n \geq 1\}$ changes, and the consistency of the finite dimensional distributions of the process implies (by the uniqueness theorem) that these G_n satisfy the conditions of Theorem 3.4.10 and, hence, there is a unique measure G on the cylinder σ-algebra $\mathcal{B}$ of $\mathbb{R}^I$ whose projection, or n-dimensional marginal, satisfies $G_n = G \circ \pi_n^{-1}$ where $\pi_n : \mathbb{R}^I \to \mathbb{R}^n$ is the coordinate projection. (This is *not* a random measure!) If such a G exists, it is called the *spectral measure* of the α-stable process. An α-stable symmetric process for which (38) holds with $G_n = G \circ \pi_n^{-1}$ is called a *strongly stationary α-stable* process. To use the word stationary again, we observe that it is automatically strictly stationary as defined before. This may be seen as follows. Let $t_i, s, t_i + s \in I \subset \mathbb{R}^n$. Then

$$\begin{aligned}
\varphi_{t_1+s,\ldots,t_n+s}(u_1, \ldots, u_n) &= \exp\left\{-\int_{\mathbb{R}^n} \left|\sum_{j=1}^{n} u_j e^{i<t_j+s,\lambda>}\right|^\alpha dG_n(\lambda)\right\} \\
&= \exp\left\{-\int_{\mathbb{R}^n} \left|e^{i<s,\lambda>} \sum_{j=1}^{n} u_j e^{i<t_j,\lambda>}\right|^\alpha dG_n(\lambda)\right\} \\
&= \exp\left\{-\int_{\mathbb{R}^n} \left|\sum_{j=1}^{n} u_j e^{i<t_j,\lambda>}\right|^\alpha dG_n(\lambda)\right\} \\
&= \varphi_{t_1,\ldots,t_n}(u_1, \ldots, u_n), \quad u_j \in \mathbb{R}. \qquad (39)
\end{aligned}$$

Thus a strongly stationary α-stable class is always strictly stationary. Now (38) can be expressed more conveniently as follows. Consider the subspace $\mathbb{R}^{(I)}$ of $\mathbb{R}^I$ consisting of all functions [or sequences if I is countable] that vanish outside a finite set, the latter varying with functions. Thus $\mathbb{R}^{(I)} \in \mathcal{B}$, $\mathbb{R}^n = \pi_n(\mathbb{R}^{(I)})[= \pi_n(\mathbb{R}^I)]$, and we can express (38) for each finite subset J of I as:

$$E(e^{i\sum_{t\in J} a_t X_t}) = \exp\left\{-\int_{\mathbb{R}^{(I)}} \left|\sum_{t\in J} a_t e^{i<t,\lambda>}\right|^\alpha dG(\lambda)\right\}, \quad a_t \in \mathbb{R}. \qquad (40)$$

If the process is complex valued, then it has the corresponding form (Re = real part) as:

$$E(e^{i\mathrm{Re}(\sum_{t\in J} \bar{a}_t X_t)}) = \exp\left\{-\int_{\mathbb{C}^{(I)}} \left|\sum_{t\in J} \bar{a}_t e^{i<t,\lambda>}\right|^\alpha dG(\Lambda)\right\}, \quad a_t \in \mathbb{C}. \quad (41)$$

Since the measure G on $(\mathbb{R}^{(I)}, \mathcal{B})$ is obtained through an application of the Kolmogorov-Bochner theorem, one may hope that all symmetric strictly stationary α-stable processes are also strongly stationary. However, it is shown by Marcus and Pisier (1984), who introduced and analyzed this class, that the inclusion is proper unless $\alpha = 2$ which corresponds to the Gaussian case in which they both coincide. The following (canonical) example of a strongly stationary α-stable process shows the motivation for this class and its close affinity with certain problems in (random) trigonometric series.

Example 7: Let $a_\lambda \in \mathbb{R}^n$ be such that $\sum_{\lambda \in \mathbb{R}^n} |a_\lambda|^\alpha < \infty$, and $\{\theta_\lambda, \lambda \in \mathbb{R}^n\}$ be a set of independent α-stable symmetric variables. Consider the process

$$X_t = \sum_{\lambda \in \mathbb{R}^n} a_\lambda \theta_\lambda e^{i<t,\lambda>}, \quad t \in \mathbb{R}^n. \tag{42}$$

It may be verified that $\{X_t, t \in I = \mathbb{R}^n\}$ is a strongly stationary α-stable process (or field) $0 < \alpha \leq 2$ with spectral measure $G(\cdot)$ given by

$$G(\cdot) = \sum_{\lambda \in \mathbb{R}^n} |a_\lambda| \delta_\lambda(\cdot) \tag{43}$$

where $\delta_\lambda(\cdot)$ is the Dirac measure at $\lambda \in \mathbb{R}^n$. We omit the details which are not difficult, although not entirely simple, and refer to the above noted paper.

An interesting outcome of this example is that if $\alpha = 2$ then θ_λ must be Gaussian by Theorem 5.3.2, and if $0 < \alpha < 2$ it is a stable process. However, if θ_λ are the Rademacher functions (i.e., independent sequence taking values +1 and -1 with equal probability), then (42) reduces to a series considered by Paley and Zygmund in 1932 about its convergence with probability one for every λ. If $\theta_\lambda = e^{2i\theta_n\lambda}$ where for $\lambda = n$, the θ_n are independent uniformly distributed random variables on [0,1], then the series becomes a Steinhaus series. Here the hypothesis of α-stability of θ_λ was not (and could not be) assumed. But the convergent series represents an α-stable process. These are not simple and both were considered in Paley-Zygmund papers and later by Salem and Zygmund in 1954. The subject was further detailed and generalized by Kahane (1968 and 1985). The point of this discussion is that a strongly stationary α-stable class contains these interesting results and a detailed analysis and characterizations are obtained by Marcus and Pisier (1984) noted above. This study moves in a different direction if $\alpha = 2$ where we can admit a much wider class of processes to be discussed in the following section, and it is called weak stationarity which will be of equal (and perhaps more) importance in applications. A surprising fact is that for ε_λ as Rademacher, Stienhaus or Gaussian i.i.d. variables the series represents an a.e. continuous function for t or a.e. unbounded function (all t). One has to employ different methods. We refer to Kahane (1985) for details.

We have noted in (33) above that integrals of the form $\int_S f \, d\mu$ can be defined for random measures μ (with independent values on disjoint sets) on $(S, \mathcal{S}, \nu)$ for $f : S \to \mathbb{R}$ (or $\mathbb{C}$) of bounded measurable class from $L^0(\nu)$. In particular, if $S = \mathbb{R}$ and $f_\lambda(s) = e^{is\lambda}, \lambda \in \mathbb{R}$, then one has

$$X_t = \int_{\mathbb{R}} e^{it\lambda} d\mu(\lambda), \quad t \in \mathbb{R}. \tag{44}$$

Processes admitting such a representation were introduced and studied by Y. Hosoya (1982), K. Urbanik (1968) and others. This class should be termed *strictly harmonizable.* Although this is similar to strict stationarity, neither includes the other. An immediate problem is to characterize processes $\{X_t, t \in \mathbb{R}\}$ that admit the representation (44) for a (unique) random measure μ, perhaps of symmetric α-stable class. Here a brief indication of this problem will be given and a more satisfactory second order case will be considered in the next section, for which the above discussion is also a strong motivation.

The first step is to obtain a vector measure $Z(\cdot)$ such that X_t admits a (generalized) Fourier transform of Z, and then seek conditions in order that it has independent values to render X_t a strictly harmonizable process, and this will demand that the index set have a group structure. We have just defined in (32) and (33), the integral of a (bounded) scalar measurable function relative to a random (or vector) measure. The same definition holds if the vector measure takes values in a Banach space when the convergence there is understood in terms of the metric of the Banach space instead of that of "in probability" (or the corresponding equivalent Fréchet metric). In this connection, the following concept and theorem originally due to S. Bochner in the scaler case and modified for the Banach space case by R.S. Phillips, are of special interest.

Definition 8 Let $\mathcal{X}$ be a Banach space and $f : G \to \mathcal{X}$ be a mapping, where G is a locally compact abelian group, so that $G = \mathbb{R}^n, n \geq 1$ is possible. Then f is said to be V*-bounded* (V for variation) provided the following three conditions hold:

(i) $f(G)$ is bounded, or equivalently contained in a ball of $\mathcal{X}$,
(ii) f is measurable relative to the Borel σ-algebras of $\mathcal{X}$ and G, and that the range of f is separable (also termed "strongly measurable"),.
(iii) the set

$$W = \left\{ \int_G f(t)g(t)\, dt : \|\hat{g}\|_\infty \leq 1, \ g \in L^1(G) \right\} \subset \mathcal{X}, \tag{45}$$

is such that its closure $\overline{W}$ in the weak topology of $\mathcal{X}$, is compact, where 'dt' is the invariant or *Haar* measure of G [the Lebesgue measure if $G = \mathbb{R}^n$], and $\hat{g}$ is the Fourier transform of g, [i.e., $\hat{g}(s) = \int_{\hat{G}} < g, \gamma > d\gamma$ or in the case that $G = \mathbb{R}^n, \hat{G} = \mathbb{R}^n$ and $\hat{g}(s) = \int_{\mathbb{R}} e^{is\lambda} g(\lambda)\, d\lambda$].

The point of this definition is that f is not required to be positive definite. Our aim is to get a corresponding representation of f as in Theorem 4.4.2 (or 4.5.8), for a process $\{X_t, t \in G\}$. Here is the solution of the first step noted when G is as above.

Theorem 9 *Let $X : G \to L^\alpha(P), \alpha \geq 1$ be a process (also called a random field). Then $X_t = \int_{\hat{G}} < t, s > dZ(s), t \in G$, (so it is strictly harmonizable) iff X is V-bounded and weakly continuous.* [Here $Z(\cdot)$ is simply a vector measure.]

Details of the above result can be found, for instance, in the first author's book, (Rao (2004), p. 550) and will be omitted. A consequence of this representation is that $\{X_t, t \in G\}$ under the stated conditions is an integral of a vector (particularly stochastic) measure $Z : \mathcal{B}(\hat{G}) \to L^\alpha(P), \alpha \geq 1$. But now one has to find the (probabilistic) properties of the measure Z related to the random field $\{X_t, t \in G\}$ under consideration. If $\{X_t, t \in G\}$ is strictly stationary, then some special properties of Z should be obtained. The following result for strictly stationary α-stable processes answers the above question and shows the basic role of the classical theory of probability here. It is essentially a restatement of similar analyses of K. Urbanik (1968) and of Y. Hosoya (1982) which is given using the new concept of strong stationarity introduced above.

Theorem 10 *Suppose $\{X_t, t \in \mathbb{R}\}$ is a strictly harmonizable α-stable process with its representing measure $Z : \mathcal{B}(\mathbb{R}) \to L^\alpha(P)$, (guaranteed by Theorem 9), which is also isotropic, meaning that for each $x, \lambda \in \mathbb{R}, Z(A_x)$ and $e^{i\lambda}Z(A_x)$ are identically distributed where $A_x = (-\infty, x)$. Then $\{X_t, t \in \mathbb{R}\}$ is strongly stationary α-stable. On the other hand, if the process is strongly stationary α-stable, $1 < \alpha < 2$, then it is V-bounded and hence is strictly harmonizable with the representing random measure isotropic in addition.*

A key ingredient in obtaining the desired properties of $Z(\cdot)$, is to employ a form of Fejer's theorem on summability of Fourier series and integrals. This was used in the Urbanik-Hosoya treatment if $\mathbb{R}$ is replaced by a compact group. If the X_t is not required to be the Fourier transform of a random measure, but is merely a representation of such a measure relative to some element $f_t \in L^\alpha(d\lambda)$, then there is a corresponding integral (of "Karhunen type") as

$$X_t = \int_G f_t(\lambda)\, dZ(\lambda), \quad t \in G \tag{46}$$

where $G \subset \mathbb{R}$ is a compact group, and $\{f_{t_j}(\cdot), j \geq 1\}$ is dense in $L^\alpha(d\lambda)$. This is due to M. Schilder and J. Kuelbs, but as yet there is no method of construction available for these f_t's. The structure and representation of these strict versions are somewhat intricate. This is further exemplified by the fact that if the process $\{X_n, n \in Z\}$ is of independent and strictly stationary elements, (hence i.i.d.), then it is α-stable for $\alpha = 2$ as a Gaussian sequence but not for $0 < \alpha < 2$. Then X_n cannot be a (finite) linear combination of α-stable random variables of the same type. These specializations will not be discussed further.

Although α-stability, $0 < \alpha \leq 2$, plays a key role in this work, it can be generalized further using the fact that it is a special class of the infinitely divisible family to get another strongly stationary process. This enlarges the previous case and yet not exhausting the strict stationarity. We indicate this notion, introduced by Marcus (1987), to round up these ideas known at this time.

Comparing formulas (35) and (40) it is clear that the characteristic exponent, $t \mapsto \psi(t)$ in $\exp(-\psi(t))$ is a nonnegative, nondecreasing function, and a similar statement holds for real symmetric infinitely divisible random variables in Lévy's form as seen in expression (20) of Section 5.2. In detail, let ξ be a symmetric infinitely divisible real random variable whose ch.f. can, therefore, be expressed as

$$\phi_\xi(t) = E(e^{it\xi}) = \exp\{-\psi(|t|)\}, \tag{47}$$

where we have set

$$\begin{aligned} \psi(|t|) &= \int_0^\infty \mathrm{Re}\left(e^{ixt} - 1 - \frac{ixt}{1+x^2}\right) dN(x) \\ &= \int_0^\infty (\cos xt - 1)\, d\tilde{N}(x) \end{aligned} \tag{48}$$

with $\tilde{N}(\{0\}) = 0$ and $\int_0^\infty (x^2 \wedge 1)\, d\tilde{N}(x) < \infty$, for the Lévy measure $\tilde{N}$. Based on this, (40) is reformulated as follows. If $\psi(\cdot)$ is the exponent of ξ, then a process $\{X_t, t \in I\}$ is termed *ξ-radial* if there is a (cylindrical) finite measure G on the cylindrical σ-algebra of the space $\mathbb{R}^{(I)}$, such that

$$\begin{aligned} \phi_{X_{t_1},\dots,X_{t_n}}(u_1,\dots,u_n) &= E(e^{i\sum_{j=1}^n a_{t_j} X_{t_j}}) \\ &= \exp\left\{-\int_{\mathbb{R}^{(I)}} \psi\left(\left|\sum_{t_j=1}^n a_{t_j} e^{i<t_j,\lambda>}\right|\right) dG(\lambda)\right\}, \end{aligned} \tag{49}$$

for all finite subsets $t_1,\dots,t_n \in I$. If $t_j + s \in I$ for each $t_j, s \in I$, then as before this definition implies that the *ξ-radial process $\{X_t, t \in I\}$ is also strongly stationary.* The earlier work of Marcus and Pisier (1984) applies here and shows that this enlarged class is still a proper subset (i.e. does not exhaust) of the strictly stationary family. This enlargement substantially expands the study of random Fourier series, and their sample path analysis has been advanced by Marcus (1987). One should also note an interesting extension of this work by Cuzick and Lai (1980) in this connection. Some interesting analysis (but no characterizations) of strictly harmonizable α-stable processes ($1 < \alpha \le 2$) are also in Weron (1985).

The restrictions can be substantially relaxed if we consider processes with two moments. Many nonGuassian classes can be included. Here V-boundedness of Definition 7, with $\mathcal{X}$ as a Hilbert space, plays a central role. This aspect of the work will now be outlined in the next and final section of the present chapter.

8.5 Second-Order Processes

The class of processes $\{X_t, t \in I\} \subset L^2(P)$ to be considered here uses the Hilbert space geometry fully. This in conjuction with probabilistic ideas gives an advantage for this class in applications in such subjects as prediction, filtering and signal detection.

Consider the natural parameters of a second-order process, namely, its mean and covariance functions m and r :

$$m(t) = E(X_t), \qquad r(s,t) = \mathrm{Cov}(X_s, X_t).$$

To make effective use of the Hilbert space geometry, we take the r.v.s as complex valued, (otherwise we complexify the process for convenience) so that

$$m(t) = E(X_t), \quad r(s,t) = E((X_s - m(s))\overline{(X_t - m(t))}), \tag{1}$$

where, as usual, the overbar denotes the complex conjugate. The process is called *weakly stationary* (or also termed K-stationary, K for Khintchine,– or *in the wide sense*) if $m(t)$= constant and $r(s,t) = \tilde{r}(s-t)$ with $\tilde{r}$ assumed as a Borel function in (1). Recall that we defined strictly stationary processes (or sequences) in Section 7.3 and studied in the above section, as those whose finite-dimensional distributions are invariant under a shift of the time axis. Thus a strict sense stationarity implies the weak sense version if the d.f.s have two moments finite. Since r defined by (1) is positive definite, if $I = \mathbb{R}$, then by the Bochner-Riesz theorem (cf. Theorem 4.4.5), for a weakly stationary process $\{X_t, t \in \mathbb{R}\}$ we have

$$r(s-t) = \int_{\mathbb{R}} e^{i(s-t)\lambda} F(d\lambda) \tag{2}$$

for almost all $s-t \in \mathbb{R}$ (Lebesgue), and if r is also continuous, then (2) holds for all $s-t \in \mathbb{R}$. Here F is a bounded nondecreasing nonnegative function, called the *spectral function* of the process, uniquely determined by r. Because of this connection again with the Fourier transform, one can consider the harmonic analysis of these processes. A very simple example of such a family is any complete orthonormal sequence $\{f_n, -\infty < n < \infty\}$ in (a separable) $L^2(P)$. Then $r(s-t) = \delta_{0,s-t}$, the Kronecker delta function, and clearly $F'(\lambda)\,d\lambda = d\lambda/2\pi$, F having a constant *spectral density* F' (relative to Lebesgue measure.) For a detailed analysis of stationary processes, we refer the reader to the books by Rozanov (1967), or Yaglom (1987). Hereafter a second order process is assumed measurable, i.e. $X : (t,\omega) \mapsto X_t(\omega)$ is a measurable mapping of $\mathbb{C} \times \Omega \to \mathbb{C}$, so that the mean and covariance functions are measurable.

For some problems, it is desirable to relax the condition that r be a function of the difference. The advantages of the Fourier analysis can still be retained in generalizing (2) to a class of second-order nonstationary processes. A process

$\{X_t, t \in \mathbb{R}\} \subset L^2(P)$ with means zero and covariance r is termed *Loève* (or *strongly*) *harmonizable* [introduced by Loève in an appendix to Lévy's book (1948)] if

$$r(s,t) = \int_{\mathbb{R}} \int_{\mathbb{R}} e^{is\lambda - it\lambda'} F(d\lambda, d\lambda'), \quad s,t \in \mathbb{R}, \tag{3}$$

(the two dimensional Lebesgue integral) where $F : \mathbb{R}^2 \to \mathbb{C}$ is a covariance function of bounded *Vitali variation* in the plane, i.e.,

$$|F|(\mathbb{R}^2) = v(F) = \sup\{\sum_{i=1}^{m}\sum_{j=1}^{m} |F(A_i, B_j)| : \{A_i\}_1^n, \{B_i\}_1^n \text{ are disjoint intervals of } \mathbb{R}\} < \infty. \tag{4}$$

Here $F(A,B) = \int\int_{\mathbb{R}^2} \chi_A(\lambda)\chi_B(\lambda')F(d\lambda, d\lambda')$ and F is again called the *spectral function* (*bimeasure*) of the process. Clearly every weakly stationary process is strongly harmonizable as can be seen when $F(\cdot,\cdot)$ concentrates on the diagonal $\lambda = \lambda'$. A very simple harmonizable process which is not weakly stationary is the following.

Let $f \in L^1(\mathbb{R})$ and $\hat{f}$ be its Fourier transform: $\hat{f}(t) = \int_{\mathbb{R}} e^{it\lambda} f(\lambda)\, d\lambda$. If ξ is an r.v. with mean zero and unit variance, and $X_t = \xi \hat{f}(t)$, then $\{X_t, t \in \mathbb{R}\}$ is such a process. Harmonizable processes are finding interesting applications in the areas mentioned above, and some basic theory of the strongly harmonizable class is given by Loève (1963).

Consider the simple stationary example given above. If $\mathcal{H}_1 = \overline{\text{sp}}\{f_n, n \geq 0\}$ is the closed linear span in $L^2(P)$, let $Q : L^2(P) \to \mathcal{H}_1$ be the orthogonal projection with range $\mathcal{H}_1$. If $g_n = Qf_n$, then $g_n = f_n$ for $n \geq 0$, and $=0$ for $n < 0$. Even though $\{f_n, -\infty < n < \infty\}$ is a very simple weakly stationary process, the g_n-sequence is not strongly harmonizable. A proof is not difficult, but is nontrivial. In fact, if $\{X_t, t \in \mathbb{R}\} \subset L^2(P)$ is weakly stationary, and T is a continuous linear mapping of $L^2(P)$ into itself, letting $Y_t = TX_t$, then $\{Y_t, t \in \mathbb{R}\}$ is generally not strongly harmonizable. However, every such process can be shown to be *weakly harmonizable* in the following sense.

A process $\{X_t, t \in \mathbb{R}\} \subset L^2(P)$ is *weakly harmonizable* if $E(X_t) = 0$, and its covariance can be represented as (3) in which the spectral function F is a covariance function of bounded variation in *Fréchet's sense*:

$$||F|| = \sup\{\sum_{i=1}^{n}\sum_{j=1}^{n} a_i\bar{a}_j F(A_i, A_j) : a_i \in \mathbb{C}, |a_i| \leq 1, \{A_i\}_1^n \text{ are disjoint Borel sets } A_i \subset \mathbb{R}\}. \tag{5}$$

It is easily seen that $||F|| \leq v(F) \leq \infty$, usually with a strict inequality between the first terms. [$v(F)$ is the *Vitali variation* cf. (4).] With this relaxation,

$$r(s,t) = \int_{\mathbb{R}} \int_{\mathbb{R}} e^{is\lambda - it\lambda'} F(d\lambda, d\lambda'), \quad s,t \in \mathbb{R}, \tag{6}$$

but now the integral here has to be defined in the (weaker) sense of M. Morse and W. Transue and it is not an absolute integral, in contrast to Lebesgue's definition used in (3). It is clear that each strongly harmonizable process is weakly harmonizable, and the above examples show that the converse does not hold. Most of the Loève theory extends to this general class, although different methods and techniques of proof are now necessary. These processes are of considerable interest in applications noted above. The structure theory of such processes and other results can be found in the literature [in particular, see the first author's paper (1982)] and some further details will be given in the problems section.

We present a brief characterization of weakly harmonizable processes to give a feeling for those classes and classifications to contrast with the strict sense classes discussed in the preceding section. Thus, we have the following direct characterization of weakly harmonizable processes, specializing Theorem 4.9 above to the present (Hilbert space) context. For simplicity we consider $G = \mathbb{R}$, and give the essential details of proof.

Theorem 1 *A process $X : \mathbb{R} \to L_0^2(P)$ is weakly harmonizable if and only if it is V-bounded and weakly continuous in the sense that $t \mapsto \ell(X(t))$ is a continuous scalar function for each continuous linear functional ℓ on $L^2(P)$.*

Proof If X is weakly continuous and V-bounded then, $\hat{f}$ being the Fourier transform of f, one has: ($\|\cdot\|_\infty$ denoting the uniform norm)

$$\left\| \int_{\mathbb{R}} f(t)X(t)\, dt \right\|_2 \leq C\|\hat{f}\|_\infty, \quad f \in L^1(\mathbb{R}). \tag{7}$$

Also, by the Riemann-Lebesgue lemma $\mathcal{Y} = \{\hat{f} : f \in L^1(\mathbb{R})\} \subset C_0(\mathbb{R})$, the complete (under uniform norm) space of continuous complex functions vanishing at '∞', and $\mathcal{Y}$ is uniformly dense in the latter since it separates points of $\mathbb{R}$ and the Stone-Weierstrass theorem applies. Set $e_t : \lambda \mapsto e^{it\lambda}$, and consider $T_1 : f \mapsto \int_{\mathbb{R}} f(\lambda)\bar{e}_t(\lambda)\, d\lambda, t \in \mathbb{R}$, so that $T_1 : L^1(\mathbb{R}) \to C_0(\mathbb{R})$ is one-to-one and contractive. If we set $T_0(\hat{f}) = \int_{\mathbb{R}} f(t)X(t)\, dt (\in L_0^2(P))$, then the mapping $T_2 : L^1(\mathbb{R}) \to L_0^2(P)$ is given by the commuting diagram

$$\begin{array}{ccc} L^1(\mathbb{R}) & \xrightarrow{\quad T_1 \quad} & \mathcal{Y} \\ & T_2 \searrow \quad \swarrow T_0 & \\ & L_0^2(P) & \end{array}$$

$$T_2(f) = (T_0 T_1)(f) = \int_{\mathbb{R}} f(t)X(t)\, dt.$$

Now T_0 is bounded and has a bound preserving extensions to all of $C_0(\mathbb{R})$, by the density of $\mathcal{Y}$, and thus is a bounded linear mapping into the Hilbert space

$L_0^2(P)$. We use the same symbol for the extension also. Hence, by a classical Riesz representation theorem (cf. Dunford-Schwartz (1958), IV.7.3) which is seen to extend to the locally compact case (such as $\mathbb{R}$ here), there exists a unique measure $Z : \mathcal{B}(\mathbb{R}) \to L_0^2(P)$, such that

$$T_0(f) = \int_{\mathbb{R}} f(t)Z(dt), \quad f \in C_0(\mathbb{R}). \tag{8}$$

Then $T_0 \leftrightarrow Z$ correspond to each other and ($||Z||_{\mathbb{R}}$, semi-variation of Z)

$$||T_0|| = \sup\left\{\left|\left|\int_{\mathbb{R}} f(t)\, Z(dt)\right|\right|_2 : f \in C_0(\mathbb{R}),\ ||f||_\infty \le 1\right\} = ||Z||(\mathbb{R}).$$

To see that $X(\cdot)$ is the Fourier transform of $Z(\cdot)$ so that it is weakly harmonizable, consider for any $\ell \in (L_0^2(P))^*$, a continuous linear functional, from (8) and the diagram relation

$$\ell\left(\int_{\mathbb{R}} \hat{f}(t)Z(dt)\right) = \ell\left(\int_{\mathbb{R}} f(t)X(t)\, dt\right). \tag{9}$$

Applying a well-known theorem of E. Hille (cf. Dunford-Schwartz (1958), p. 324 and p. 153) which allows commuting the integral and the functional ℓ, it follows that

$$\int_{\mathbb{R}} \hat{f}(t)(\ell \circ Z)(dt) = \int_{\mathbb{R}} f(t)(\ell \circ X)(t)\, dt. \tag{10}$$

Substituting for $\hat{f}$ and using Fubini's theorem on the left side (clearly valid for signed measures), one has

$$\int_{\mathbb{R}} f(t)\left(\int_{\mathbb{R}} e_t(\lambda)(\ell \circ Z)(d\lambda)\right) dt = \int_{\mathbb{R}} f(t)(\ell \circ X)(t)\, dt,$$

so that

$$\int_{\mathbb{R}} f(t)\left[\ell\left(\int_{\mathbb{R}} e_t(\lambda)Z(dt) - X(t)\right)\right] dt = 0, \quad f \in L^1(\mathbb{R}). \tag{11}$$

Since f is arbitrary, its coefficient function must vanish a.e., and it is actually everywhere by the (weak) continuity of that element. This establishes that $X(\cdot)$ is the Fourier transform of the (stochastic) measure $Z(\cdot)$, hence weakly harmonizable.

Conversely, let $X(\cdot)$ be weakly harmonizable so that it is the Fourier transform of a (stochastic or vector) measure $Z(\cdot)$. We claim it is V-bounded. Using the representation of X_t, it is seen that

$$||X_t||_2 = \left|\left|\int_{\mathbb{R}} e^{it\lambda} Z(d\lambda)\right|\right|_2 \le ||Z||(\mathbb{R}) = M_0 < \infty, t \in \mathbb{R}. \tag{12}$$

If $\ell \in (L_0^2(P))^*$, then one has $\ell(X)(\cdot)$ to be the Fourier transform of $(\ell \circ Z)(\cdot)$, and hence, for $f \in L^1(\mathbb{R})$, it follows that, as before, for the Bochner (or vector Lebesgue) integrable $(fX)(\cdot), e_t(\cdot) = e^{i\lambda t}$,

$$\begin{aligned}\ell\left(\int_{\mathbb{R}} f(t)X(t)\,dt\right) &= \int_{\mathbb{R}} f(t)\ell(X(t))\,dt \\ &= \int_{\mathbb{R}} f(t)\left[\int_{\mathbb{R}} \ell(e_t \circ Z)(\lambda)(d\lambda)\right]\,dt \\ &= \int_{\mathbb{R}}\int_{\mathbb{R}} f(t)e_t(\lambda)(\ell \circ Z)(d\lambda)\,dt, \quad \text{by Fubini's theorem,} \\ &= \int_{\mathbb{R}} \hat{f}(\lambda)(\ell \circ Z)(d\lambda) \\ &= \ell\left(\int_{\mathbb{R}} \hat{f}(\lambda)Z(d\lambda)\right), \quad \text{using the same argument} \\ &\quad \text{of the direct part above.} \end{aligned} \tag{13}$$

Since $\ell \in (L_0^2(P))^*$ is arbitrary, we conclude from (12) and (13) that

$$\int_{\mathbb{R}} f(t)X_t\,dt = \int_{\mathbb{R}} \hat{f}(\lambda)Z(d\lambda) \in L_0^2(P)$$

and with $M_0 = ||Z||(\mathbb{R})$,

$$\left|\left|\int_{\mathbb{R}} f(t)X_t dt\right|\right|_2 \leq ||\hat{f}||_\infty ||Z||(\mathbb{R}) = M_0||\hat{f}||_\infty, \quad f \in L^1(\mathbb{R}). \tag{14}$$

It follows that (14) implies (cf. Definition 4.7 above) that X is V-bounded, since $L_0^2(P)$ is reflexive and as a Fourier transform of $(\ell \circ Z)(\cdot) \mapsto (\ell \circ X_t)$, is continuous, i.e., $X(\cdot)$ is weakly continuous, completing the proof.

The same argument extends easily for any locally compact abelian group. It is not too difficult to show that the covariance function $r(\cdot, \cdot)$ of $X(\cdot)$ is representable as

$$r(s,t) = E(X_s\overline{X}_t) = \int\int_{\mathbb{R}^2} e_s(\lambda)\overline{e_t(\lambda')}F(d\lambda, d\lambda') \tag{15}$$

where F is a "bimeasure", and the integral now has to be defined *not* as a standard Lebesgue integral, but in a weaker (nonabsolute) sense using the work of M. Morse and W. Transue (1956). These will not be discussed here. Several other extensions of second-order processes are also possible. For some of these we refer to Cramér's lectures (1971), Rao (1982), Chang and Rao (1986), and, for the state of the art during the next decade, Swift (1997).

In this chapter, our aim has been to show some of the work on stochastic processes growing out of the standard probability theory that we have presented in the preceding analysis. The reader should get from this an idea of the vigorous growth of stochastic theory into essentially all branches of analysis, and will now be in a position to study specialized works on these and related subjects.

Exercises

1. Find an explicit form of an n-dimensional density of the Brownian motion process $\{X_t, 0 \leq t \leq 1\}$.

2. Let $\{X_t, t \geq 0\}$ be a Gaussian process with mean function zero and covariance r given by $r(s,t) = e^{-|s-t|}$, $s,t \geq 0$. If $0 < t_1 < \cdots < t_n$, show that $(X_{t_1}, \ldots, X_{t_n})$ has a density $f_{t_1,\ldots,t_n}$ given by

$$f_{t_1,\ldots,t_n}(\lambda_1,\ldots,\lambda_n) = \left[(2\pi)^n \prod_{i=2}^{n}(1 - e^{2(t_{i-1}-t_i)})\right]^{-1/2} \times \exp\left\{-\frac{\lambda_1^2}{2} - \frac{1}{2}\sum_{i=2}^{n}(1 - e^{2(t_{i-1}-t_i)})^{-1}(\lambda_i - e^{(t_{i-1}-t_i)}\lambda_{i-1})^2\right\}.$$

[Such a family is called an *Ornstein-Uhlenbeck* process.]

3. If $\{X_t, t \geq 0\}$ is the Brownian motion of Section 1 and $A_\alpha = \{(t,\omega) : X_t(\omega) > \alpha\}$, then verify that A_α is $\lambda \times P$-measurable, where λ is the Lebesgue measure on $\mathbb{R}^+$ and P is the given probability. Thus $X = \{X_t, t \geq 0\}$ is a (jointly) measurable random function on $\mathbb{R}^+ \times \Omega \to \mathbb{R}$.

4. Prove the remark following Eq. (4.10), namely: If $\{X_t, a \leq t \leq b\}$ is a process such that for each $\varepsilon > 0, a \leq t, u \leq b, \lim_{t\to u} P[|X_t - X_u| \geq \varepsilon] = 0$, then the ch.f. $\phi_{t,u} : v \mapsto E(e^{iv(X_t - X_u)})$ satisfies $\lim_{t\to u}\phi_{t,u}(v) = 1$ uniformly in v, and conversely the latter property implies the stochastic continuity, as given there.

5. Let $\phi : \mathbb{R} \to \mathbb{C}$ be an infinitely divisible ch.f. Show that there is a probability space (Ω, Σ, P) and a stochastic process $\{X_t, t \geq 0\}$ on it such that $X_0 = 0$ having (strictly) stationary independent increments satisfying $(\phi(u))^t = E(e^{iuX_t}), t \geq 0$. [*Hint*: For each $0 = t_0 < t_1 < \cdots < t_n$ define $\phi_{t_1,\ldots,t_n}$, an n-dimensional ch.f. of independent r.v.s $Y_1, \ldots, Y_n$ such that $E(e^{iuY_j}) = (\phi(u))^j$, and set $X_{t_1} = Y_1, X_{t_2} = Y_1 + Y_2, \ldots, X_{t_n} = \sum_{j=1}^n Y_j$. Verify that $\{\phi_{t_1,\ldots,t_n}, n \geq 1\}$ is a consistent family of ch.f.s.]

6. (Converse to 5) Let $\{X_t, t \geq 0\}$ be a stochastically continuous process, $X_0 = 0$, with strictly stationary independent increments, on a probability

space (Ω, Σ, P). Show that $\phi_t : u \mapsto E(e^{iuX_t}) = \exp\{t(\psi(u) + u\alpha)\}$ for a constant α, and $\psi(0) = 0, \psi : \mathbb{R} \to \mathbb{C}$ is continuous. Is stochastic continuity automatic when the rest of the hypothesis holds?

7. Let $\{f_n, n \geq 1\} \subset L^2(0,1)$ be any complete orthonormal set and $\{\xi_n, n \geq 1\}$ be independent $N(0,1)$ r.v.s. Show that the process $\{X_t, 0 \leq t \leq 1\}$ given by

$$X_t = \sum_{n \geq 1} \xi_n \int_0^t f_n(u)\, du$$

is a Brownian motion. (This observation is due to L.A. Shepp, and the proof is analogous to that of Theorem 1.2, but the uniform convergence is less simple.)

8. (A Characterization of Brownian Motion, due to P. Lévy) Let $\{X_t, 0 \leq t \leq 1\}$ be a process on (Ω, Σ, P) with strictly stationary independent increments and $X_0 = 0$. If the sample functions $t \mapsto X_t(\omega)$ are continuous for almost all $\omega \in \Omega$, then the process is Brownian motion. [*Hint*: Observe that the continuity hypothesis implies $P[\max_{1 \leq k \leq n} |X_{k/n} - X_{(k-1)/n}| \geq \varepsilon] \to 0$ for each $\varepsilon > 0$, and this gives $P[|X_{1/n}| \geq \varepsilon] = o(n^{-1})$. Verify that $\phi_t : u \mapsto E(e^{iuX_t})$ is an infinitely divisible ch.f. for which the Lévy-Khintchine pair (γ_t, G_t) satisfies $\gamma_t = t\gamma, G_t = tG$, where G is a constant except for a jump at the origin.]

9. Solve the system of differential equations

$$\begin{aligned} P_0'(t) &= -\lambda P_0(t) \\ P_n'(t) &= -\lambda P_n(t) + \lambda P_{n-1}(t) \quad \text{for } n \geq 1 \end{aligned} \tag{1}$$

with the conditions $P_0(0) = 1$ and $P_n(0) = 0$ for $n \geq 1$, recursively. The solution of this system (of differential equations) is the Poisson process of Equation (5) of Section 4 above.

An alternate method of solution is obtained by letting

$$\psi(s,t) = \sum_{n=0}^{\infty} P_n(t) s^n$$

which is the *probability generating function* (p.g.f.) for the system (1). Rewriting (1) on multiplying by an appropriate power of s show that the partial derivatives yield the p.g.f. ψ which satisfies

$$\frac{\partial \psi}{\partial t} = \lambda(s-1)\psi \tag{2}$$

with the boundary conditions that $\psi(s,0) = 1$ and $\psi(1,t) = 1$.

Solve the partial differential equation (2) to obtain

$$\psi(s,t) = e^{\lambda(s-1)t}.$$

Expanding $\psi(s,t)$ as a power series in s, obtain the Poisson process probabilities as coefficients of this power series. Alternately, one could recognize $\psi(s,t)$ as the p.g.f. of a Poisson probability distribution with parameter λt, assuming (or establishing) a uniqueness theorem for p.g.f.s.

10. Some properties of the Poisson process N_t are detailed in this problem.

(a) (Memoryless property of the exponential distribution) A random variable T is said to have the *memoryless property* if for each $s, t \in \mathbb{R}$

$$P(T > s + t | T > t) = P(T > s).$$

Show that T with a continuous distribution has the memoryless property if and only if T is an exponential r.v. [T exponentially distributed implying T memoryless is straightforward, and for the converse show that if $g : \mathbb{R} \to \mathbb{R}^+$ is a nondecreasing continuous function which satisfies

$$g(s+t) = g(s)g(t)$$

for all $s, t \in \mathbb{R}$, then (being a probability)

$$g(t) = e^{-\alpha t}$$

for some constant $\alpha \in \mathbb{R}^+$. Using this, deduce that T has an exponential distribution by considering

$$g(t) = P[T > t].]$$

(b) (Time of occurence in a Poisson process) Consider a Poisson process $\{X_t, t \geq 0\}$ [X_t integer valued!] with rate parameter $\lambda > 0$. Let T_1 be the time of the first event's occurence in this process. Show that the conditional probability $P[T_1 < s | X_t = 1]$ is the distribution function of a uniform $[0, t]$ random variable. Generalize this result to show for $0 < s < t$ and $k = 0, 1, \ldots, n$

$$P[X_s = k | X_t = n] = \binom{n}{k} \left(\frac{s}{t}\right)^k \left(1 - \frac{s}{t}\right)^{n-k}.$$

11. (The pure birth (linear birth) process) Consider a nonnegative integer-valued process $\{X_t, t \geq 0\}$ with independent increments. A pure birth process is obtained by assuming

$$P[X_{t+\Delta t} = n | X_t = n-1] = (n-1)\lambda \Delta t + o(\Delta t) \tag{3}$$

for $n = n_0, n_0 + 1, \ldots$ and $\lambda > 0$. [Compare this with equation (1) above.]

Letting

$$P_n(t) = P[X_t = n | X_0 = n_0]$$

and in a manner analogous to the derivation of the system of differential equations for the Poisson process, show that a pure birth process $\{X_t, t \geq 0\}$ has probabilities $P_n(t)$ which satisfy

$$P'_{n_0}(t) = -\lambda n_0 P_{n_0}(t)$$
$$P'_n(t) = (n+1)\lambda P_{n+1}(t) - n\lambda P_n(t), \qquad n \geq n_0 + 1.$$

Either solving this system recursively, or using the p.g.f. method detailed in Problem 9, show

$$P_n(t) = \binom{n-1}{n_0-1} e^{-n_0\lambda t}(1-e^{-\lambda t})^{n-n_0}, \quad \text{for } n \geq n_0.$$

Using $P'_n(t)$, the derivative, show that the expected value $m(t) = E(X_t)$ satisfies the differential equation

$$m'(t) = \lambda m(t), \quad \text{with } m(0) = n_0. \tag{4}$$

[Remark: The pure birth process can be motivated as a stochastic version of the simple deterministic population model described by (4). Under fairly nonrestrictive conditions, it is possible to formulate a stochastic version of a large variety of first order population models. For a further discussion of these ideas, see the second author's text (1999) and the papers by Swift (2001), and Switkes, Wirkus, Swift and Mihaila (2003).]

12. (The general birth-death process) Consider a nonnegative integer valued process $\{X_t, t \geq 0\}$ with independent increments. The general birth-death process is obtained by assuming

$$P[X_{t+\Delta t} = n+1 | X_t = n] = \lambda_n \Delta t + o(\Delta t) \quad \text{for } n = 0, 1, 2, \ldots$$

and

$$P[X_{t+\Delta t} = n-1 | X_t = n] = \mu_n \Delta t + o(\Delta t) \quad \text{for } n = 1, 2, \ldots$$

where $\lambda_n > 0$ for $n = 0, 1, \ldots$, and $\mu_n > 0$ for $n = 1, 2, \ldots$.
Letting

$$P_n(t) = P[X_t = n | X_0 = 0]$$

show that the general birth-death process $\{X_t, t \geq 0\}$ satisfies

$$P'_0(t) = -\lambda_0 P_0(t) + \mu_1 P_1(t) \tag{5}$$

$$P'_n(t) = \lambda_{n-1}P_{n-1}(t) - (\mu_n + \lambda_n)P_n(t) + \mu_{n+1}P_{n+1}(t) \quad \text{for } n \geq 1.$$

Obtaining a solution to this general system of differential equations is not easy without some conditions upon the λ_n's and μ_n's. However, it is possible to obtain the *steady-state distribution* as follows. Suppose

$$\lim_{t\to\infty} P_n(t) = P_n \quad \text{(say)} \quad \text{for } n = 0, 1, \ldots$$

exists. Then provided that the limit and the derivative can be interchanged, (formulate a simple [uniform] condition for this) show that the system (5) can be written as

$$-\lambda_0 P_0 + \mu_1 P_1 = 0 \tag{6}$$
$$\lambda_{n-1} P_{n-1} - (\mu_n + \lambda_n) P_n + \mu_{n+1} P_{n+1} = 0 \quad \text{for } n \geq 1.$$

Show further that the system (6) of difference equations can be solved recursively for P_n to obtain

$$P_n = \prod_{i=1}^{n} \frac{\lambda_{i-1}}{\mu_i} P_0 \quad \text{for } n = 1, 2, \ldots \tag{7}$$

with

$$P_0 = \frac{1}{1 + \sum_{n=1}^{\infty} \prod_{i=1}^{n} \frac{\lambda_{i-1}}{\mu_i}},$$

provided the series $\sum_{n=1}^{\infty} \prod_{i=1}^{n} \frac{\lambda_{i-1}}{\mu_i}$ converges. This steady-state distribution for the general birth-death process can be specialized to find the steady-state distribution for a wide class of birth-death processes. For instance, if $\lambda_n = \lambda$ for $n = 0, 1, 2, \ldots$, and $\mu_n = \mu (> \lambda \geq 0)$ for $n = 1, 2, \ldots$, then show that (7) can be used to obtain the steady-state distribution for the single-server queue:

$$P_n = \left(\frac{\lambda}{\mu}\right)^n \left(1 - \frac{\lambda}{\mu}\right) \quad \text{for } 0, 1, 2, \ldots.$$

Recently, S.K. Ly (2004) considered the class of birth-death processes with polynomial transition rates. Specifically, if $\lambda_j = \prod_{k=1}^{p} (\alpha_k j + \beta_k)$ for $j = 0, 1, \ldots$ and $\mu_j = \prod_{k=1}^{q} (\gamma_k j + \delta_k)$ for $j = 1, 2, \ldots$ with p, q positive integers $\alpha_k, \beta_k, \gamma_k, \delta_k$ real numbers, show that

$$P_0 = \left[1 + {}_pF_q\left(\left\{1, \frac{\alpha_1}{\beta_1}, \frac{\alpha_2}{\beta_2}, \ldots, \frac{\alpha_p}{\beta_p}\right\},\right.\right.$$
$$\left.\left.\left\{1 + \frac{\gamma_1}{\delta_1}, 1 + \frac{\gamma_2}{\delta_2}, \ldots, 1 + \frac{\gamma_q}{\delta_q}\right\}, \frac{\beta_1 \beta_2 \ldots \beta_p}{\delta_1 \delta_2 \ldots \delta_q}\right)\right]^{-1}$$

where ${}_pF_q(\cdot, \cdot, \cdot)$ is the *generalized hypergeometric function* defined as

$${}_pF_q(\{a_1, a_2, \ldots, a_p\}, \{b_1, b_2, \ldots, b_q\}, z) = \sum_{n=0}^{\infty} \frac{(a_1)_n (a_2)_n \ldots (a_p)_n}{(b_1)_n (b_2)_n \ldots (b_q)_n} \frac{z^n}{n!}$$

which converges for $|z| < \infty$ if $p \leq q$ and if $p + q = 1$, then the series converges if $|z| < 1$. The symbol $(a)_n$ is known as the *Pochhammer* notation and it is defined by

$$(a)_n = \frac{\Gamma(a+n)}{\Gamma(a)} \quad \text{with } (a)_0 = 1$$

and

$$\Gamma(a) = \int_0^{\infty} x^{a-1} e^{-x}\, dx, \; a > 0,$$

is the standard gamma function. Verify that this gives

$$P_n = \frac{\left(\frac{\alpha_1}{\beta_1}\right)_n \left(\frac{\alpha_2}{\beta_2}\right)_n \cdots \left(\frac{\alpha_p}{\beta_p}\right)_n}{\left(1+\frac{\gamma_1}{\delta_1}\right)_n \left(1+\frac{\gamma_2}{\gamma_2}\right)_n \cdots \left(1+\frac{\gamma_q}{\delta_q}\right)_n} \left(\frac{\beta_1\beta_2\ldots\beta_p}{\delta_1\delta_2\ldots\delta_q}\right)^n$$

$$\times \left[1 +_p F_q\left(\left\{1, \frac{\alpha_1}{\beta_1}, \frac{\alpha_2}{\beta_2}, \ldots, \frac{\alpha_p}{\beta_p}\right\}, \left\{1+\frac{\gamma_1}{\delta_1}, 1+\frac{\gamma_2}{\delta_2}, \ldots, 1+\frac{\gamma_q}{\delta_q}\right\}\right.\right.$$

$$\left.\left.\times \quad \frac{\beta_1\beta_2\ldots\beta_p}{\delta_1\delta_2\ldots\delta_q}\right)\right]^{-1} \quad \text{for } n \geq 1.$$

13. (A catastrophe process) An example of a non-birth-death process can be obtained by considering a nonnegative integer valued process $\{X_t, t \geq 0\}$ with independent increments as before, with its transitions given by $j \to j+1$ with probability $\alpha\Delta t + o(\Delta t)$ for $j \geq 0$ and $j \to 0$ and probability $\gamma\Delta t + o(\Delta t)$ for $j \geq 1$. Following the method of exercises 11 and 12, show that $\{X_t, t \geq 0\}$ satisfies

$$P_n'(t) = \alpha P_{n-1}(t) - (\alpha+\gamma)P_n(t)$$

where $P_n(t) = P[X(t) = n|X(0) = 0]$. Note that, since a catastrophe can occur for any value of $X(\cdot)$, we have that $P_0'(t) = \gamma\sum_{j=1}^{\infty} P_j(t) - \alpha P_0(t)$. This expression simplifies using

$$\sum_{j=1}^{\infty} P_j(t) = 1 - P_0(t),$$

to give $P_0'(t) = \gamma - (\alpha+\gamma)P_0(t)$. For simplicity, assume $X(0) = 0$. This implies that $P_0(0) = 1$ and $P_n(0) = 0$ for all $n \geq 1$. Show that this system of differential equations has a solution

$$P_0(t) = \frac{\gamma}{\alpha+\gamma} + \frac{\alpha}{\alpha+\gamma} e^{-(\alpha+\gamma)t}.$$

and

$$P_n(t) = e^{-(\alpha+\gamma)t} \int_0^{\infty} \alpha e^{(\alpha+\gamma)t} P_{n-1}(t)\, dt.$$

Show that in general by an induction argument that

$$P_n(t) = \frac{\alpha^n\gamma}{(\alpha+\gamma)^{n+1}} + \left(\frac{\alpha^{n+1}(\alpha+\gamma)^n t^n - n!\alpha^n\gamma\sum_{i=0}^{n-1}(\alpha+\gamma)^i\frac{t^i}{i!}}{n!(\alpha+\gamma)^{n+1}}\right) e^{-(\alpha+\gamma)t}.$$

Using the identity,

$$\Gamma(n, at) = \int_{at}^{\infty} x^{n-1}e^{-x}\, dx = (n-1)!e^{-ax}\sum_{i=0}^{n-1} a^i \frac{t^i}{i!}$$

(which can be obtained by repeated integration by parts, or found in a standard table of integrals) where $\Gamma(\cdot,\cdot)$ is the *incomplete gamma function* and $a > 0$ is a constant, that P_n is

$$P_n(t) = \frac{\alpha^n \gamma}{(\alpha+\gamma)^{n+1}} + \frac{\alpha^{n+1} t^n}{(\alpha+\gamma) n!} e^{-(\alpha+\gamma)t} - \frac{\alpha^n \gamma}{(\alpha+\gamma)^{n+1}(n-1)!} \int_{(\alpha+\gamma)t}^{\infty} x^{n-1} e^{-x}\, dx,$$

or more succinctly

$$P_n(t) = \frac{\alpha^n \gamma}{(\alpha+\gamma)^{n+1}} + \frac{\alpha^{n+1} t^n}{(\alpha+\gamma) n!} e^{-(\alpha+\gamma)t} - \frac{\alpha^n \gamma}{(\alpha+\gamma)^{n+1}(n-1)!} \Gamma(n, (\alpha+\gamma)t).$$

Deduce that the expected value of $X(t)$ is

$$E(X(t)) = \frac{\alpha}{\gamma} - \frac{\alpha}{\gamma} e^{-\gamma t}$$

and the variance is

$$V(X(t)) = \frac{\alpha\,(\alpha - e^{-2\,t\,\gamma}\,\alpha + \gamma - e^{-t\,\gamma}\,(1 + 2\,t\,\alpha)\,\gamma)}{\gamma^2}.$$

[This simple process was first considered by Swift (2000a), but has since found a wide range of additional applications. Indeed, this idea has been applied to queueing models by B.K. Kumar, and D. Arivudainambi, (2000) and A. Di Crescenzo, V. Giorno, A.G. Nobile, (2003) as well as population models by R.J. Swift (2001), M.L. Green (2004). This simple catastrophe process has also been extended to a multiple catastrophe process by I. Chang, A.C. Krinik, and R.J. Swift (2006), although a different method of solution is required.]

14. Let $(S, \mathcal{B})$ be a measurable space, (Ω, Σ, P) a probability space and $X : \mathcal{B} \to L^\alpha(P)$ a symmetric random measure in the sense of Section 4 so that $X(\cdot)$ is σ-additive in probability and takes independent values on disjoint sets of $\mathcal{B}$. Suppose $X(A)$ is a symmetric α-stable random variable, $0 < \alpha \le 2$, $A \in \mathcal{B}$ so that its ch.f. $\phi : (A, t) \mapsto E(e^{itX(A)})$ is given by: $\phi(A,t) = e^{-C(A)|t|^\alpha}$ where $C : \mathcal{B} \to \overline{\mathbb{R}}^+$ is a "rate measure", a σ-finite positive measure function on $\mathcal{B}$. Let $L^\alpha(C)$ be the Lebesgue space on $(S, \mathcal{B}, C)$. Then $C(A) = 0 \Rightarrow P[(X(A) = 0] = 1$ and $C(\cdot)$ is again called a *control measure* of $X(\cdot)$. If $f \in L^\alpha(C)$ is simple so $f = \sum_{i=1}^n a_i \chi_{A_i}$, $A_i \in \mathcal{B}$, disjoint, $a_i \in \mathbb{R}$, define

$$\int_S f_n dX = \sum_{i=1}^n a_i X(A_i) \ (\in L^\alpha(P)).$$

Verify that the integral is well-defined (show that it does not depend upon the representation of f) and is α-stable. [Here the availability of the control measure is needed.] Show for each $0 < p < \alpha \leq 2$, one has for each simple f_n in $L^\alpha(C)$,

$$E\left(|\int_S f_n dX|^p\right)^{\frac{1}{p}} = k(p,\alpha)\left(\int_S |f_n|^\alpha dC\right)^{\frac{1}{\alpha}}$$

for some constant $k(p,\alpha) > 0$ depending only on p and α. [This nontrivial fact was established independently by M. Schilder (1970) and in an equivalent form slightly earlier by J. Bretagnalle, D. Dacunha-Castelle and J.L. Krivine (1966).] If $f \in L^\alpha(C)$ is arbitrary, verify that there exist simple $f_n \in L^\alpha(C)$ such that $f_n \to f$ in α-norm, $1 \leq \alpha \leq 2$, and $\{\int_S f_n dX, n \geq 1\}$ is Cauchy in $L^p(P)$. If we set

$$Y_f = \int_S f dX = \lim_n \int_S f_n dX,$$

verify that Y is an α-stable symmetric random variable such that $||Y_f||_{p,P}$ and $||f||_{\alpha,C}$ are equivalent. [This needs a careful computation. The work is carried out using the "control measure C" as in Dunford - Schwartz (1958), Section IV.10 or see Schilder (1970). The integral is analogous to Wiener's original definition with $S = \mathbb{R}, \mathcal{B} =$ Borel σ-algebra of S and $C(\cdot)$ as Lebesgue measure, with $p = \alpha = 2$ (Brownian motion). But the fact that $C(\cdot)$ [and $X(\cdot)$] has atoms and $X(\cdot)$ need not be symmetric in general, introduces some significant difficulties. The latter aspect was considered by Okazaki (1978) where he also considered the case that f takes values in a topological vector space (such as a Fréchet or Banach space). Thus the integration with stable random measures (not Brownian motion) extends the work on Stochastic calculus significantly and nontrivially.]

15. Let $\{X_t, t \in \mathbb{R}\} \subset L^2(P)$ be a second-order process with means zero and continuous or measurable covariance r. It is said to be of *class* (KF) (after the authors J. Kampé de Fériet and F. N. Frankiel who introduced it) if

$$\tilde{r}(h) = \lim_{T\to\infty} \frac{1}{T}\int_0^{T-|h|} r(s, s+|h|)\, ds, \quad h \in \mathbb{R}$$

exists. This class was also independently given by Yu. A. Rozanov and E. Parzen, the latter under the name "asymptotic stationarity".) Show that r is positive definite, and hence by the Bochner-Riesz theorem, coincides a.e. (Lebesgue) with the Fourier transform of a positive bounded nondecreasing F, called the *associated spectral function* of the process. Verify that if X_t is real valued and stationary then X_t belongs to class (KF), so that the stationary processes are contained in class (KF). [It can be shown, that if $X_t = Y_t + Z_t$, where Z_t is a zero mean stationary process and Y_t a process with zero mean and periodic covariance i.e. $r(s+k, t+k) = r(s,t)$ for some k, $Y \perp Z$, these processes are known as *periodically correlated*, then X is in class (KF), but is not stationary, so class (KF) contains nonstationary processes.] Show that any strongly harmonizable process belongs to the class (KF). [It is also true that *some* weakly harmonizable processes are contained in the class (KF), but this proof is somewhat involved. In a slightly more general direction, a process X_t which has covariance of the form

$$r(t_1, t_2) = \int_{\mathbb{R}} \int_{\mathbb{R}} g(t_1, \lambda)\overline{g(t_2, \lambda')}\, dF(\lambda, \lambda') \tag{8}$$

where $g(\cdot, \lambda)$ is a uniformly almost periodic function relative to $\mathbb{R}$ and with $F(\cdot, \cdot)$ having finite Vitali variation then the process is called *almost strongly harmonizable.* The class of almost strongly harmonizable processes contains the class of strongly harmonizable processes. This can be immediately seen by setting $g(t, \lambda) = e^{i\lambda t}$. Further, if the spectral bimeasure $F(\cdot, \cdot)$ concentrates on the diagonal $\lambda = \lambda'$ the process will be termed *almost stationary.* The first author showed (Rao (1978)), that every almost strongly harmonizable process belongs to class (KF). These ideas were extended by the second author (Swift, (1997)) with the introduction of classes of nonstationary processes with covariance representation (8) and $g(\cdot, \lambda)$ satisfying the following Cesàro summability condition:

$$\lim_{T\to\infty} a_T^{(p)}(|h|, \lambda, \lambda')$$

exists uniformly in h and is bounded for all h, $p \geq 1$, where

$$a_T^{(p)}(|h|, \lambda, \lambda') = \begin{cases} \frac{1}{T}\int_0^T a_\alpha^{(p-1)}(|h|, \lambda, \lambda')d\alpha & \text{for } p > 1, \\ \frac{1}{T}\int_0^{T-|h|} g(s, \lambda)\overline{g(s+|h|, \lambda')}ds & \text{for } p = 1. \end{cases}$$

These families are termed (c, p)*-summable Cramér*, with $p \geq 1$. It can be shown that these nonstationary processes are contained in the class (KF, p) processes, $p \geq 1$ where a process is (KF, p) if its covariance r satisfies for $p \geq 1$ and for each $h \in \mathbb{R}$ the following limit condition:

$$\tilde{r}(h) = \lim_{T\to\infty} r_T^{(p)}(h)$$

where

$$r_T^{(p)}(h) = \begin{cases} \frac{1}{T}\int_0^T r_\alpha^{(p-1)}(h)d\alpha & \text{for } p > 1, \\ \frac{1}{T}\int_0^{T-|h|} r(s, s+|h|)ds & \text{for } p = 1.] \end{cases}$$

16. Finally we describe briefly, another direction of second order random measures arising in some applications. Thus let $(\mathbb{R}, \mathcal{B})$ be the Borelian line and $Z : \mathcal{B} \to L^2(P)$ on a probability space (Ω, Σ, P), be σ-additive such that $Z(\cdot)$ is additionally translation invariant in the sense that $Z(A) = Z(\tau_x A), x \in \mathbb{R}$, where $\tau_x A = \{x + y : x \in A\} \in \mathcal{B}$ for all $A \in \mathcal{B}$. Then $m(A) = E(Z(A)) = m(\tau_x A)$, and $b(A, B) = E(Z(A)\overline{Z(B)}) = b(\tau_x A, \tau_x B)$. Since the only translation invariant measure on $\mathbb{R}$ [or $\mathbb{R}^n$] is Lebesgue measure except for a constant of proportionality factor, say α, we have $m(A) = \alpha\mu(A)$, μ being Lebesgue measure. Suppose that the (clearly) scalar measures $b(\cdot, B)$, $b(A, \cdot)$ for each

$A, B \in \mathcal{B}$ are such that $b(A, B) = \tilde{b}(A \times B)$ where $\tilde{b}(\cdot) : \mathcal{B} \to \mathbb{C}$ is a scalar measure (σ-additive), so this gives $b(\cdot, B) = \tilde{b}(\cdot \times B) = \overline{\tilde{b}(B \times \cdot)} = \overline{b(B, \cdot)}$, being positive definite. [The general $b(\cdot, \cdot)$ is a bimeasure and $\tilde{b}$ results from the additional condition that b is translation invariant, is also termed weak stationarity of the random measure $Z(\cdot)$.]

Let $\mathcal{K}$ be the space of infinitely differentiable scalar functions on $\mathbb{R}$ vanishing outside of compact sets. Thus for every compact set A of $\mathcal{B}$, there exists $f_n \in \mathcal{K}$ such that $f_n \downarrow \chi_A$. This well-known result is from real analysis (c.f. e.g. Rao (1987, 2004), Proposition 1, p. 632). Then the preceding can be restated as $F(f) = \int_{\mathbb{R}} f(t) Z(dt) \in L^2(P)$ for each $f \in \mathcal{K}$ (the integral as in Equation (8)), and show that $m(A) = \lim_{n\to\infty} \int_{\mathbb{R}} f_n(t) dm(t), f_n \downarrow \chi_A$ for compact $A \subset \mathbb{R}$. Here $F : \mathcal{K} \to L^2(P)$ is called a *generalized random process* (and *field* if $\mathbb{R}$ is replaced by $\mathbb{R}^k, k \geq 1$). It is a classical result of K. Itô, in the mid-1950's, that the positive definite functional $\beta(f, g) = E(F(f)\overline{F(g)})$ admits an integral representation, for a unique positive measure $\nu : \mathcal{B} \to \mathbb{R}^+$ satisfying $\int_{\mathbb{R}} \frac{d\nu(x)}{(1+|x|^2)^k} < \infty$ for some integer $k \geq 0$ [such a measure ν is usually termed "tempered"], given by $\beta(f, g) = \int_{\mathbb{R}} \hat{f}(t)\overline{\hat{g}(t)} d\nu(t), \hat{f}, \hat{g}$ being Fourier transforms of $f, g \in \mathcal{K}$. Specializing this, show that for **bounded** Borel sets A, B one has

$$b(A \times B) = \beta(\chi_A, \chi_B) = \int_{\mathbb{R}} \hat{\chi}_A(t) \hat{\chi}_B(t) d\nu(t).$$

This representation was essentially given in Thornett (1978) with applications, and an n-dimensional version was discussed already in Vere-Jones (1971.) Extensions from the point of view of generalized random fields was obtained by Rao (1969), and that of Thornett's work, for some general harmonizable and Karhunen/Cramér random fields was presented by Swift (2000b). It is noted that the generalized random fields is the proper setting for this work and application. We merely draw the attention of the reader to the ideas described here and omit further analysis and extensions, with the remark that Thornett's work used some results of Argabright and Gil de Lamadrid (1974) on Fourier tansforms of *unbounded* measures which, as the latter authors indicated, is a form of Schwartz distribution theory.

The broad classes of nonstationary process briefly described in Problems 15 and 16, and in fact the last two sections above, give just a glimpse of the breadth of the emerging study of processes related to the harmonizable class and the broader Stochastic Analysis proper.

References

Argabright, L. and Gil de Lamadrid, J. (1974). Fourier analysis of unbounded measures on locally compact Abelian groups. *Memoirs of Am. Math. Soc.* **145**.

Barndorff-Nielsen, O. (1964). On the limit distributions of the maximum of a random number of independent random variables. *Acta Math. (Hung.)* **15**, 399-403.

Bhattacharya, R. N., and Rao, R. R. (1976). "Normal Approximation and Asymptotic Axpansions." Wiley, New York.

Billingsley, P. (1968). "Convergence of Probability Measures." Wiley, New York.

Billingsley, P. (1995). "Probability and Measure.", 3rd. ed., Wiley, New York.

Blackwell, D.,and Dubins,.L. E. (1963). A converse to the dominated convergence theorem. *Illinois J. Math.* **7**, 508-514.

Bochner, S. (1955). "Harmonic Analysis and the Theory of Probability." Univ. of California Press, Berkeley, Calif.

Bochner, S. (1975). A formal approach to stochastic stability. *Z. Wahr.* **31**, 187-198.

Breiman, L. (1968). "Probability." Addison-Wesley, Reading, Mass.

Bruckner, A. M. (1971). Differentiation of integrals. *Amer. Math. Monthly* **78**, 1-51 (Special issue, Nov., Part II.)

Cartan, H. (1963). "Elementary Theory of Analytic Functions of One or Several Complex Variables." Addison-Wesley, Reading, Mass.

Chang, D. K. and Rao, M.M. (1986). Bimeasures and nonstationary processes. *in* Real and Stochastic Analysis, 7-118, Wiley, New York.

Chen, Z., Rubin, H. and Vitale, A. (1997). Independence and Determination of Probabilities. *Proc. Amer. Math. Soc.*, **125**, 3721-3723.

Chung, K.L. and Ornstein, D.S. (1962). On the recurrence of sums of random variables. *Bull. Amer. Math. Soc.*, **68**, 30-32.

Chung, K. L. (1974). "A Course in Probability Theory," 2nd ed. Academic Press, New York.

Ciesielski, Z. (1961). Holder conditions for realizations of Gaussian processes. *Tran. Amer. Math. Soc.* **99**, 403-413.

Cramér, H. (1946). "Mathematical Methods of Statistics." Princeton Univ. Press, Princeton, New Jersey.

Cramér, H. (1970). "Random Variables and Probability Distributions," 3rd ed. Cambridge Univ. Press, New York.

Cramér, H. (1971). "Structural and Statistical Problems for a Class of Stochastic Processes." Princeton Univ. Press, Princeton, N.J.

Csörgö, M., and Revesz, P. (1981). "Strong Approximations in Probability and Statistics." Academic Press, New York.

Cuzick, J., and Lai, T.L. (1980). On random Fourier series. *Trans. Am. Math. Soc.* **261**, 53-80.

DeGroot, M. H., and Rao, M. M. (1963). Stochastic give-and- take. *J. Math. Anal. Appl.* **7**, 489-498.

Dharmadhikari, S. W., and Sreehari, M. (1976). A note on stable characteristic functions. *Sankhyā Ser. A.*, **38**, 179-185.

Di Crescenzo, A., Giorno, V., Nobile, A. G. (2003). On the M/M/1 Queue with Catastrophes and its Continuous Approximation, Queueing Systems, **43**, 329-347.

Donsker, M.D. (1951). An invariance principle for certain probability limit theorems. *Mem. Amer. Math. Soc.* **6**, 1-12.

Doob, J. L. (1953). "Stochastic Processes." Wiley, New York.

Dubins, L. E., and Savage, L. J. (1965). "How to gamble if you must." McGraw-Hill, New York.

Dunford, N., and Schwartz, J. T. (1958). "Linear Operators. Part I: General theory." Wiley-Interscience, New York.

Dvovetzky, A., Erdös, P., and Kakutani, S. (1961). Nonincreasing everywhere of the Brownian motion process. *Proc. Fourth Berkeley Symp. in Math. Statist. Prob.* **2**, 103-116.

Dynkin, E. B. (1961). "Foundations of the Theory of Markov Processes" (English translation). Prentice-Hall, Englewood Cliffs, N.J.

Edgar, G. A., and Sucheston, L. (1976). Amarts: A class of asymptotic martingales. *J. Multivar. Anal.* **6**, 193-221.

Edwards, W.F. (2004). "Dependent Probability Spaces." M.S. Thesis, California State Polytechnic University, Pomona.

Eisenberg, B. and Ghosh, B.K. (1987). Independent events in a discrete uniform probability space. *Amer. Stat.* **41**, 52-56.

Feller, W. (1943). The general form of the so-called law of the iterated logarithm. *Trans. Amer. Math. Soc.* **54**, 373-402.

Feller, W. (1957). "An Introduction to Probability Theory and its Applications," Vol. I (2nd Ed.). Wiley, New York.

Feller, W. (1966). "An Introduction to Probability Theory and its Applications," Vol. II. Wiley, New York.

Gikhman, I. I., and Skorokhod, A. V. (1969). "Introduction to the Theory of Random Processes" (English translation). Saunders. Philadelphia.

Gnedenko. B. V.. and Kolmogorov. A. N. (1954). "Limit Distributions for Sums of Independent Random Variables" (English translation). Addison-Wesley. Reading, Mass.

Green, M. L. (2004). The Immigration-Emigration with Catastrophe Model, *n* Stochastic Processes and Functional Analysis, A Volume of Recent Advances in Honor of M.M. Rao, A. C. Krinik, R. J. Swift, Eds., Vol. 238 in the Lecture Notes in Pure and Applied Mathematical Series, Marcel Dekker, New York, 149-159.

Gundy, R. F. (1966). Martingale theory and pointwise convergence of certain orthogonal series. *Trans. Amer. Math. Soc.* **124**, 228-248.

Haldane, J. B. S. (1957). The syādvādā system of prediction. *Sankhyā* **18**, 195-200.

Hall, P., and Heyde, C. C. (1980). "Martingale Limit Theory and its Application." Academic Press, New York.

Halmos, P.R. (1950). "Measure Theory." Van Nostrand, Princeton, N.J.

Hardy, G. H., Littlewood, J. E., and Polya, G. (1934). "Inequalities." Cambridge Univ. Press, London.

Hayes, C. A., and Pauc, C. Y. (1970). "Derivation and Martingales." Springer-Verlag, Berlin.

Hewitt. E., and Savage, L. J. (1955). Symmetric measures on cartesian products. *Trans. Amer. Math. Soc.* **80**, 470-501.

Hida, T. (1980). "Brownian Motion." Springer-Verlag, Berlin.

Hosoya, Y. (1982). Harmonizable stable processes. *Z. Wahrs.* **80**, 517-533.

Hsu, P. L., and Robbins, H. (1947). Complete convergence and the law of large numbers. *Proc. Nat. Acad. Sci.* **33**, 25-31.

Ibragimov, I.A., and Linnik, Ju. V. (1971) "Independent and Stationary Random Variables." Norodhoff Publishers, The Netherlands.

Ionescu Tulcea, C. (1949). Mesures dans les espaces produits. *Atti Accad. Naz. Lincei Rend. C1. Sci. Fis. Mat. Natur.* **7**, 208-211.

Ionescu Tulcea, A., and Ionescu Tulcea, C. (1969). "Topics in the Theory of Lifting." Springer- Verlag, Berlin.

Jessen, B., and Wintner, A. (1935). Distribution functions and the Riemann zeta function. *Trans. Amer. Math. Soc.* **38**, 48-88.

Kac, M., and Slepian, D. (1959). Large excursions of Gaussian processes. *Ann. Math. Statist.* **30**, 1215-1228.

Kahane, J. P. (1968). (2nd ed. 1985). "Some Random Series of Functions." Cambridge University Press, Cambridge, U.K.

Kakutani, S. (1951). Random ergodic theorems and Markoff processes with a stable distribution. *Proc. Second Berkely Symp. in Math. Statist, and Prob.*, 247-261.

Kendall, D. G. (1959). Unitary dilations of Markov transition operators, and the corresponding integral representations for transition-probability matrices, *in* "Probability and statistics" (The Harald Cramér volume). Wiley, New York, 139-161.

Kolmogorov, A. N. (1933). "Grundbegriffe der Wahrscheinlichkeitsrechnung." Springer-Verlag, Berlin.

Kuelbs, J. (1973). A representation theorem for symmetric stable processes and stable measures on compact groups. *Z. Wahrs.* **26**, 259-271.

Kumar, B. K, Arivudainambi, D. (2000). Transient Solution of an M/M/1 Queue with Catastrophes. *Computer and Mathematics with Applications*, **40**, 1233-1240.

Lai, T.L. (1974). Summability methods for i.i.d. Random Variables. *Proc. Amer. Math. Soc.* **45**, 253-261.

Lamb, C. W. (1974). Representation of functions as limits of martingales. *Trans. Amer. Math. Soc.* **188**, 395-405.

Lévy, P. (1937). (2nd ed. 1954). "Théorie de l'addition des variables" Gauthier-Villars, Paris.

Lévy, P. (1948). "Processus stochastiques et mouvement brownien." Gauthier-Villars, Paris.

Linde, W. (1983). "Probability in Banach Spaces—Stable and Infinitely Divisible Distributions." Wiley-Interscience, New York.

Loève. M. (1963). "Probability Theory" (3d ed.). Van Nostrand, Princeton. N.J.

Ly, S.K. (2004). "Birth-Death Processes with Polynomial Transition Rates." M.S. Thesis, California State Polytechnic University, Pomona.

McKean, H. P., Jr. (1969). "Stochastic Integrals." Academic Press, New York.

McLeish, D. L. (1974). Dependent central limit theorems and invariance principles. *Ann. Prob.* **2**, 620-628.

Mahalanobis, P. C. (1954). The foundations of statistics. *Dialecta* **8**, 95-111. (Reprinted in *Sankhyā* **18** (1957). 183-194.)

Maistrov, L. E. (1974). "Probability Theory: A Historical Sketch" (Translation). Academic Press. New York.

Marcus, M. B. and Pisier, G. (1984). Characterizations of almost surely continuous α-stable random Fourier series and strongly stationary processes. *Acta Math* **152**, 245-301.

Marcus, M. B. (1987). ψ-radial processes and random Fourier series. *Memoirs of the American Mathematical Society,* **Vol. 68**, 1-181.

Meyer, P. A. (1966). "Probability and Potentials." Blaisdell, Waltham, Mass.

Mooney, D.D. , and Swift, R.J. (1999). "A Course in Mathematical Modeling." The Math. Assoc. of America, Washington D.C.

Morse, M. and Trausue, W. (1956) Bimeasures and their integral extensions. *Ann. Math.* **64**, 480-504.

Neal, D.K. and Swift, R.J. (1999). Designing Payoffs for Some Probabilistic Gambling Games. *Miss. J. Math. Sci.*, **11**, 93-102.

Neuts, M. F. (1973). "Probability." Allyn & Bacon, Boston, Mass.

Neveu, J. (1965). "Mathematical Foundations of the Calculus of Probability." Holden-Day, Inc. San Francisco, CA.

Okazaki, Y. (1979). Wiener integral by stable random measure. *Fae. Sci., Kyushu Univ., A, Math.* **33**, 1-70.

Paley, R.E.A.C. and Wiener, N. (1934). "Fourier Transforms in the Complex Domain." Amer. Math. Soc., Providence, R.I.

Parthasarathy, K. R. (1967). "Probability Measures on Metric spaces." Academic Press, New York.

Parzen, E. (1962). On the estimation of a probability density and mode. *Ann. Math. Statist.* **33**. 1065-1076.

Pierre, P. A. (1971). Infinitely divisible distributions, conditions for independence, and central limit theorems. *J. Math. Anal. Appl.* **33**, 341-354.

Prokhorov, Yu. V. (1956). Convergence of random processes and limit theorems in probability theory. *Theor. Veroyat. Premenin.* 1, 157-214.

Ramachandran, B., and Rao, C. R. (1968). Some results on characteristic functions and characterizations of the normal and generalized stable laws. *Sankhyā, Ser. A.*, **30**, 125-140.

Ramaswamy, V., Balasubramanian. K., and Ramachandran, B. (1976). The stable laws re- visited. *Sankhyā, Ser. A.*, **38**, 300-303.

Rao, M. M. (1961). Consistency and limit distributions of estimators of parameters in explosive stochastic difference equations. *Ann. Math. Statist.* **32**, 195-218.

Rao, M. M. (1962). Theory of order statistics. *Math. Ann.* **147**, 298-312.

Rao, M.M. (1969). Representation theory of multidimensional generalized random fields. *Proc. Symp. Multivariate Analysis*, **2**, Acad. Press, New York, 411-435.

Rao, M. M. (1982). Harmonizable processes: Structure theory. *L'Enseian. Math.* (2nd Series) **28**, 295-351.

Rao, M. M. (1979). "Stochastic Processes and Integration." Sijthoff and Noordhoff, Alphen aan den Rijn, The Netherlands.

Rao, M. M. (1981). "Foundations of Stochastic Analysis." Academic Press, New York.

Rao, M. M. (1987). "Measure Theory and Integration." Wiley Interscience, New York. Second Edition, Enlarged, Marcel Dekker, Inc., New York. (2004).

Rao, M. M. (2000). "Stochastic Processes: Inference Theory." Kluwer Academic Publishers, Dordrecht, The Netherlands.

Rao, M.M. (2004a). Convolutions of vector fields - III: Amenability and spectral properties, in "Real and Stochastic Analysis: New Perspectives , Birkhauser, Boston, MI, 375-401.

Rényi, A. (1953). On the theory of order statistics. *Acta Math.* (*Hung.*) **4**, 191-232.

Rényi, A. (1960). On the central limit theorem for the sum of a random number of independent random variables. *Acta Math.* (*Hung.*) **11**, 97-102.

Royden, H. L. (1968). "Real Analysis" (2nd. ed.). Macmillan, New York.

Rozanov, Yu. A. (1967). "Stationary Random Processes" (English translation). Holden-Day, San Francisco.

Samorodnitsky, G. and Taqqu, M.S. (1994). "Stable Non-Gaussian Random Processes: Stochastic Models with Infinite Variance." Chapman & Hall, London, UK.

Schilder, M. (1970). Some structure theorems for the symmetric stable laws. Ann. Statist. **41**, 412-421.

Shiflett, R.C., and Schultz, H.S. (1979). An Approach to Independent Sets. Math. Spec. **12**, 11-16.

Shohat, J.,and Tamarkin, J. D. (1950). "The Problem of Moments " (2nd ed.). Amer. Math. Soc., Providence, R.I.

Sion, M. (1968). "Introduction to the Methods of Real Analysis." Holt, Rinehart, and Winston, New York.

Skorokhod, A. V. (1965). "Studies in the Theory of Random Processes" (English translation). Addison-Wesley, Reading, Massachusetts.

Spitzer, F. (1956). A combinatorial lemma and its application to probability theory. Trans. Amer. Math. Soc. **82**, 323-339.

Stomberg, K. (1994). "Probability for Analysts." Chapman & Hall, CRC Press.

Sudderth, W. D. (1971). A "Fatou equation" for randomly stopped variables. Ann. Math. Statist. **42**, 2143-2146.

Swift, R.J. (1997). Some Aspects of Harmonizable Processes and Fields, *in* Real and Stochastic Analysis: Recent Advances, Ed. M.M. Rao, 303-365, CRC Press.

Swift, R.J. (2000a). A Simple Immigration-Catastrophe Process, The Math. Sci., **25**, 32-36.

Swift, R.J. (2000b). Nonstationary Random Measures, Far East Journal of Th. Stat., **4**, 193-206.

Swift, R. J.(2001). Transient Probabilities for a Simple Birth-Death-Immigration Process Under the Influence of Total Catastrophe, Int. J. Math. Math. Sci., **25**. 689-692.

Switkes, J., Wirkus, S., Swift, R.J., and Mihaila, I. (2003). On the Means of Deterministic and Stochastic Populations, The Math. Sci., **28**, 91-98.

Thornett, M.L. (1978). A Class of Second-Order Stationary Random Measures, Stochastic Processes Appl. **8** 323-334.

Tjur, T. (1974). "Conditional Probability Distributions." Inst. Math. Statist., Univ. of Copenhagen, Denmark.

Tucker, H. G. (1967). "A Graduate Course in Probability." Academic Press, New York.

Urbanik, K. (1968). Random measures and harmonizable sequences. *Studia Math.* **31**, 61-88.

Vere-Jones D. (1974). An Elementary Approach to Spectral Theory of Stationary Random Measures. *in* Stochastic Geometry, E.A. Harding, D.G. Kendall Eds. Wiley, New York, 307-321.

Wald, A. (1947). "Sequential Analysis." Wiley, New York.

Weron, A. (1985). Harmonizable stable processes on groups: Spectral, Ergodic, and Interpolation Properties. *Z. Wahrs.* **68**, 473-491.

Wiener, N. (1923). Differential space, J. Math. Phys. (M.I.T.), **2**, 131-174.

Zaanen, A. C. (1967). "Integration" (2nd. ed.). North-Holland, Amsterdam, The Netherlands.

Zolotarev, V.M. (1986). "One-dimensional Stable Distributions.", Amer. Math. Soc. Trans. Monograph, Vol. 65, Providence, R.I.

Zygmund, A. (1959). "Trigonometric Series." Cambridge University Press, Cambridge, UK.

Author Index

Subject Index